Fast Spectrum Reactors

Alan E. Waltar · Donald R. Todd · Pavel V. Tsvetkov
Editors

Fast Spectrum Reactors

Foreword by Jacques Bouchard

Springer

Editors
Alan E. Waltar
Pacific Northwest National Laboratory
(PNNL), Retired
Richland, WA 98847, USA
alan.waltar@wildblue.net

Donald R. Todd
NuScale Power, Inc.
Corvallis, OR 97330, USA
donald.r.todd@gmail.com

Pavel V. Tsvetkov
Zachry Engineering Center MS3133
Department of Nuclear Engineering
Texas A&M University
College Station, TX 77843-3133, USA
tsvetkov@tamu.edu

ISBN 978-1-4419-9571-1 e-ISBN 978-1-4419-9572-8
DOI 10.1007/978-1-4419-9572-8
Springer New York Dordrecht Heidelberg London

Library of Congress Control Number: 2011935931

Printed on acid-free paper

Springer is part of Springer Science+Business Media (www.springer.com)

Foreword

It is very timely that Alan Waltar, Donald Todd and Pavel Tsvetkov have undertaken the effort to update the classic Fast Breeder Reactor text, written in 1981 by Alan Waltar and Albert Reynolds. Indeed, there is presently a strong and growing interest in building new nuclear plants as part of the carbon-free solution to the world energy needs and this brings the question of long term supplies, leading to the development of fast breeder reactors.

Energy, in abundant and affordable supplies, is universally recognized as basic to the advancement, and indeed survivability, of humanity. The progress of civilization can clearly be traced to the discovery and adoption of new energy sources that have relieved human drudgery and allowed more time for enhancing the quality of life. The energy needs are still growing, in relation with the increase of the world population and also to reduce the important fraction of humanity, near 25% today, which yet has access to any organized energy network.

Unfortunately, most of the energy we use today is coming from burning fossil fuels, resources that are in finite supply and not evenly distributed throughout the globe. Furthermore, the continual burning of carbon based fuels is widely considered to be polluting our atmosphere, in particular by enhancing the greenhouse effect, thus creating serious concerns about the risks associated with climate change.

The electrical component is a growing part of our total energy consumption, currently based nearly 70% upon the burning of fossil fuels. Therefore, many nations are now giving serious attention to developing non-carbon, renewable sources of providing electricity. As such, a plethora of windmill farms and solar plants are being constructed, largely financed with substantial subsidies from supportive government agencies. Such a movement can generally be embraced as progressive, for we will need to seriously address any form of producing electricity that can lessen our dependency on fossil fuels.

But the use of such renewable resources is limited because of their own characteristics: they rely on a very diffuse source of energy, rays of the sun, which means large collection areas to produce meaningful supplies of energy, and they are intermittent, thus requiring either massive and expensive storage units or the availability of other means to generate a continuous supply of electricity. Hence, it is essential to have another reliable means to generate electricity at any time, following the demand of consumers. Here is where nuclear power becomes so crucial.

Today, nuclear power contributes 16% of the electricity generated around the world. Much more will be required in the future if we really want to have a reliable, abundant and carbon-free supply of electricity to meet the worldwide growing needs. This fact is now recognized by most developed and developing nations. The International Atomic Energy Agency has received over 60 inquires from new nations desiring to acquire nuclear power, beyond the 30 nations that currently generate at least a portion of their electricity from nuclear power plants.

Such a strong and growing interest in building new nuclear power plants poses the question of long-term supplies. The world's current nuclear power plants use less than 1% of the natural uranium processed to fabricate their fuel. This lack of efficiency poses a legitimate question on the sustainability of nuclear power if there is no way to make a better use of a natural resource—which is abundant in the globe, but only in limited amounts that can be recovered at a reasonable cost.

Fortunately, it is well known within the scientific community that the non-fissile uranium, more than 99% of the mined resource, can be converted to fissile material in a fast spectrum reactor system, a system in which the neutrons released from the fission process are not slowed down as they are in current power reactors. With the successful deployment of such reactors, often referred to as breeders since they can be designed to generate more fissile material than they consume, nuclear power has the potential to provide the global electricity needed for untold millennia.

The technology required to build such fast spectrum reactors, and their associated fuel cycle facilities, is more complex and less experienced than the light water technology used for most of the existing power plants. It is the obvious reason why they have not been selected for the first industrial deployment of nuclear power plants. Nevertheless, several fast spectrum reactors have been built and successfully operated over the past few decades, so the reality of building such systems is well-proven. Further developments are still necessary to meet the high level requirements on economy, safety and reliability that are targeted for an optimized use of this energy source.

Unfortunately, the research and development required to build truly efficient fast spectrum systems has waned since the end of the 1980s, partly because of a slowdown in the construction of nuclear plants and partly because of concerns of potential risks in closing the overall fuel cycle, a necessary step to reap the advantages of the breeder reactor. But with the dawn of the new century, the long-term fuel supply issue has again become a concern, providing a renewed global interest in fast spectrum reactors. Furthermore, citizens in many countries remain concerned about dealing with nuclear waste. They fear that providing assured containment of the radioactive waste coming from the used fuels of nuclear power plants, particularly those elements that are very long lived, is too risky to allow a massive commitment to nuclear power. It is now becoming more widely recognized that most of the elements with the long half-lives can be very efficiently fissioned in a fast spectrum reactor, with the resulting fission products having much shorter half-lives. The deployment of fast reactors will also progressively reduce and eventually eliminate the need for enrichment of uranium, a step necessary for the fuel of light water reactors but which is widely recognized as the most sensitive point in the fuel cycle for proliferation risk.

A better use of natural resource, a reduction of the long-lived wastes, and the elimination of some proliferation risks are the main factors contributing to the renewed interest in fast breeder reactors. For several years now, they have been extensively discussed and are the goal of joint research actions within the framework of the Generation IV International Forum, a well-recognized global forum which I had the privilege to chair.

The present update of the Fast Breeder Reactor book includes contributions from a carefully selected set of fast reactor experts from many parts of the globe. It provides a solid framework for both young professionals now entering this vastly important field and for more seasoned professionals currently coming back to a technology that has been put on hold. The text is still centered around sodium cooling, given the extensive base of technology for such systems, but it includes new chapters on both gas-cooling and heavy metal-cooling. The book features major updates to introduce transmutation physics, advanced fuels and structural materials, and new approaches to safety, an area where special efforts are required to develop a harmonized global approach. Since the fundamentals have not changed, much of the original, well organized book is still intact. But the entire text has been reviewed to include updates and references to key technical developments over the past three decades, allowing the book to become a refreshing, up-to-date, and indispensible source for any professional becoming engaged in the exciting and rewarding field of fast spectrum reactors.

Director of Nuclear Energy, French CEA, Retired
Paris, France

Jacques Bouchard

Preface

During the decade of the 1970s, major commitments were made by many countries to acquire nuclear power for the production of electricity. The amount of global electricity generated by nuclear power plants increased from about 20 GWe in 1970 to about 150 GWe by the end of the decade, and there were wide-spread expectations that the amount of electricity that might be generated globally by nuclear power plants might rise to some 1,000 GWe by the end of the century. But it was not clear that sufficient uranium supplies could be found to service such a growing demand. Accordingly, several nations made significant investments in developing fast breeder reactors.

In an attempt to collect the experience gained in fast reactor technology, Professor Albert Reynolds and I published a text in 1981 entitled FAST BREEDER REACTORS, which, along with the Russian translation, became a major reference book used by new professionals coming into the field.

But on March 28, 1979, an accident occurred at the Three Mile Island Unit #2 near Harrisburg, PA in the United States and new orders for nuclear power plants were put on hold in many areas around the world. Specifically, no new orders for new plants were made in the United States and numerous plants already ordered were deferred or cancelled. This slowdown in the commitment to nuclear power allowed uranium mining activities to catch up with the needs for current systems and it became clear that sufficient supplies of pitchblende could be obtained to service commercial thermal spectrum reactors for several more decades. The April 26, 1986 accident at Chernobyl (Ukraine) further elongated the buildup of nuclear power plants. In addition, with the end of the Cold War, significant stocks of enriched uranium became surplus from military needs and were thus made available for use in commercial reactors.

However, in the 1990s and on into the new century, two new issues relevant to nuclear power began to dominate the energy landscape. First, global climate change concerns began to claim center state and public citizens around the world began to seek ways to reduce the dependence on burning fossil fuels for energy. Nuclear power was back into central view. However, this renewed global interest in nuclear energy was somewhat offset by a second issue; namely, growing public concerns over dealing with nuclear waste—particularly the long-lived high level waste generated from the used fuel discharged from nuclear power plants.

These two seemingly opposing factors—the recognized need for more non-polluting nuclear energy but the fears of dealing with nuclear waste—have sparked renewed interest in fast spectrum reactors. This is because of two unique characteristics of a fast spectrum system: (1) the ability to breed fuel, allowing nuclear fission reactors to provide a very long-term energy supply without polluting our atmosphere, and (2) the ability to fission or transmute objectionable constituents of "high level waste" into elements with much shorter half-lives—thereby reducing the nuclear waste storage process from potentially tens or hundreds of thousands of years down to considerably less than a thousand years. In recognition of these desirable features, interest in developing and constructing fast spectrum systems is now gaining strong international attention.

It is because of this renewed interest that I was approached by several young professionals requesting an update of the original FAST BREEDER REACTORS book, published back in 1981. Such an update should include international progress on fast spectrum systems, with particular updates for fuels, materials, and transmutation physics. Whereas most of the interest is still focused on sodium-cooled fast reactors, due to the relative maturity of this technology, there is also a growing interest in gas-cooling and heavy metal-cooling (lead and lead/bismuth systems). Unfortunately, Professor Reynolds felt he has been retired too long to make meaningful contributions to the update. However, I was fortunate to persuade Professor Pavel Tsvetkov of Texas A&M University and Dr. Donald R. Todd (NuScale Power, Inc., formerly AREVA NP Inc.) to join me in this arduous task. We were then able to recruit several other experts in fast spectrum technology to conduct updates of many of the specialized chapters, including Dr. Chaim Braun (Stanford University) to update the economics chapter, Dr. Massimo Salvatores (CEA, France) for transmutation, Dr. Baldev Raj (Indira Gandhi Center for Atomic Research, India) for materials, Dr. John Sackett (Argonne National Laboratory) for the entire safety section, Dr. Kevan Weaver (TerraPower, formerly Idaho National Laboratory) for gas-cooled reactors, and Dr. James Seinicki (Argonne National Laboratory) for lead-cooled reactors. We were also fortunate to enlist Dr. Russell Stachowski (GE-Hitachi Nuclear Energy) to provide a summary of the sodium-cooled PRISM pool reactor design and Professor Toshikazu Takeda (University of Fukui, Japan) with Dr. Yoshitaka Chikazawa (Japan Atomic Energy Agency) for the Japanese Loop-Type Sodium-Cooled Fast Reactor.

It is our sincere hope that this updated version appropriately reflects the progress made in fast spectrum reactor technology over the past three decades and will be of benefit to the new generation of professionals entering this fascinating and vastly important field.

Richland, WA — Alan E. Waltar

Richland, WA — Donald R. Todd
College Station, TX — Pavel V. Tsvetkov

Acknowledgments

In our efforts to appropriately update the Fast Breeder Reactors text (Pergamon Press), published in 1981 by Alan Waltar and Albert Reynolds, we first wish to express our deep gratitute to Professor Reynolds for allowing us to use major portions of the original text where the fundamentals have not significantly changed. Whereas Professor Reynolds felt he had been retired too long to contribute to the new, updated version, he gave his full permission to use much of the material that he originally created—and he strongly encouraged our efforts in preparing this new text.

Given the major international interest that has been revitalized in this important topic over the past three decades, we feel fortunate to have attracted an exceptionally strong cadre of world-class fast reactor specialists in providing key updates. Those making especially important contributions include the following:

- Dr. Chaim Braun, currently Consulting Professor of the Center for International Security and Cooperation (CISAC) at Stanford University and recognized internationally for his work in conducting economic analyses of nuclear reactor systems;
- Dr. Massimo Salvatores, Scientific Advisor to the Director of Nuclear Energy of the CEA in France and the acknowledged global leader in the development of transmutation analyses;
- Dr. Baldev Raj, Director of the Indira Gandhi Center for Atomic Research in India, and internationally known as a top scholar in the field of advanced nuclear materials;
- Dr. John Sackett, retired Director of Argonne National Laboratory West, and head of the EBR-II reactor where the ground-breaking passive safety tests were conducted;
- Dr. Kevan Weaver, currently Director of Research & Development at TerraPower, formerly the Department of Energy leader for advanced gas-cooled reactor development at the Idaho National Laboratory;
- Dr. James Sienicki, Manager of the Innovative Systems Development Section at Argonne National Laboratory and chief investigator for heavy liquid metal cooled fast reactors;
- Dr. Russell Stachowski, Chief Consulting Engineer of Nuclear and Reactor Physics at GE-Hitachi Nuclear Energy, designers of the advanced PRISM fast reactor;
- Professor Toshikazu Takeda, Director of the Institute of Nuclear Engineering, University of Fukui and 2003 recipient of the Science Achievement Award from the Atomic Energy Society of Japan; and Dr. Yoshitaka Chikazawa at the Japan Atomic Energy Agency; and
- Dr. Won Sik Yang, Section Head for Methods Development at Argonne National Laboratory, long recognized for major contributions in fast spectrum physics analyses.

In addition to those mentioned as making major contributions to various chapters, others deserving recognition for special reviews include Dr. Frank Garner, recently retired from the Pacific Northwest

National Laboratory, for Chapter 7 (materials); Dr. Roald Wigeland, Idaho National Laboratory (safety chapters); Dr. Jim Cahalan (safety chapters), Dr. Bob Hill and Dr. Temitope Taiwo (advanced simulations)—all of the Argonne National Laboratory. We are also deeply indebted to Joshua E. Hansel (Texas A&M University) for conducting the extensive data search needed in the preparation of Appendix A.

We wish to express our sincere gratitude to both Lauren Danaly and Elaine Tham who gave us the initial encouragement needed to proceed with Springer, our assistant editors at Springer; Andrew Leigh, who demonstrated incredible patience with three over-committed authors in preparing the manuscript for submittal and Allison Michael, who guided us down the homestretch as we finalized the manuscript, and finally Lydia Shinoj and her staff for directing us through the final proofing process.

Contents

Contributors

Chaim Braun Center for International Security and Cooperation (CISAC), Stanford University, Stanford, CA, USA, chbraun@att.net

Yoshitaka Chikazawa Japan Atomic Energy Agency, Tokaimura, Japan, chikazawa.yoshitaka@jaea.go.jp

Baldev Raj Indira Gandhi Center for Atomic Research, Kalpakkam, Tamil Nadu, India, dmg@igcar.gov.in

John Sackett Argonne National Laboratory (Retired), Argonne, IL, USA, johnflyer@3riversdbs.net

Massimo Salvatores CEA, Paris, France, massimo.salvatores@cea.fr

James Sienicki Innovative Systems Development Section, Argonne National Laboratory, Argonne, IL, USA, sienicki@anl.gov

Russell Stachowski GE-Hitachi Nuclear Energy, Wilmington, NC, russell.stachowski@gnf.com

Toshikazu Takeda University of Fukui, Fukui, Japan, t_takeda@u-fukui.ac.jp

Donald Todd NuScale Power, Corvallis, OR, USA, donald.r.todd@gmail.com

Pavel Tsvetkov Zachry Engineering Center, Department of Nuclear Engineering, Texas A&M University, College Station, TX, USA, tsvetkov@tamu.edu

Alan Waltar Pacific Northwest National Laboratory (PNNL), Retired, Richland, WA, USA, alan.waltar@wildblue.net

Kevan Weaver TerraPower, LLC, Bellevue, WA, USA, kweaver@terrapower.com

Won Sik Yang Argonne National Laboratory, Argonne, IL, USA, wyang@anl.gov

About the Editors

Dr. Alan E. Waltar retired as Senior Advisor and Director of Nuclear Energy for the Pacific Northwest National Laboratory (PNNL) in Richland, WA. and Professor and Head, Nuclear Engineering, Texas A&M University (1998–2002). He currently consults for the IAEA, the U.S. Department of Energy, and several private nuclear firms. He led People to People Ambassadors nuclear delegations to China (2007) and India (2009).

Dr. Waltar is a Fellow of the American Nuclear Society and served as ANS President (1994–1995). He holds a B.S. in electrical engineering (University of Washington, 1961), an M.S. in nuclear engineering (MIT, 1962), and a Ph.D. in engineering science (University of California, Berkeley, 1966).

His long career with Westinghouse Hanford Company included leadership positions in several areas of fast reactor technology. He served on the faculty at the University of Virginia where, with Professor Albert Reynolds, he co-authored the "Fast Breeder Reactors" textbook. He also authored "America the Powerless: Facing Our Nuclear Energy Dilemma" in 1995 and "Radiation and Modern Life: Fulfilling Marie Curie's Dream" in 2004.

Dr. Waltar was instrumental in the formation of the World Nuclear University Summer Institute (SI) and has served as a mentor and a member of the faculty for all Summer Institutes since their inception in 2005.

Dr. Donald R. Todd holds a B.S. and M.S. in nuclear engineering from Penn State University (1997 and 1999) and a Ph.D. in nuclear engineering from Texas A&M University (2002). During his time as a student, Dr. Todd was extensively involved in promoting the nuclear sciences; he led student delegations that promoted nuclear science and engineering to policy makers in Washington, DC, he chaired the 2001 American Nuclear Society Student Conference held at Texas A&M University, and he organized numerous technical sessions for students at professional conferences. More recently, Dr. Todd served on the Executive Committee of the American Nuclear Society's Thermal Hydraulics Division (2006–2009).

Dr. Todd currently is a group leader at NuScale Power, leading the team responsible for developing and assessing safety analysis computer programs used in licensing of nuclear reactors. Previously, he held leadership and technical positions at AREVA NP Inc., principally involved in code and methods development for light water reactor safety analysis. In 2010 AREVA honored Dr. Todd as "Engineer of the Year" in his division to recognize his outstanding technical contributions in thermal-hydraulics methods development.

Dr. Todd has written or contributed to numerous proprietary topical and technical reports related to thermal-hydraulic analysis of nuclear power plants.

Dr. Pavel V. Tsvetkov is an Associate Professor at the Department of Nuclear Engineering, Texas A&M University. He holds a nuclear engineer degree (M.S. equivalent) from Moscow State

Engineering Physics Institute (Russia, 1995) and a Ph.D. in nuclear engineering from Texas A&M University (2002).

Dr. Tsvetkov with his students develops high-fidelity integrated system methods for predictive simulations of complex engineered systems. More than 40 students at Ph.D., M.S., and B.S. research levels have been working under the advisorship of Dr. Tsvetkov since 2005. Dr. Tsvetkov offers an undergraduate research program to allow for an early engagement of undergraduate students in state-of-the-art projects including collaborative efforts and internships with national laboratories and industry.

Dr. Tsvetkov leads and actively participates in numerous collaborative research projects of recognized scientific value and national significance as well as in international educational programs marketing his development efforts in high-fidelity integrated systematics. He has authored and co-authored 80 peer reviewed publications including 11 refereed journal papers, 17 book titles and book chapters, 52 refereed conference papers and summaries as well as 76 technical reports, non-refereed conference papers and white-papers.

Dr. Tsvetkov serves on numerous national committees of the American Nuclear Society (ANS), as a reviewer for peer reviewed journals and federal funding agencies. His service at Texas A&M University ranges from University Library Council Chairmanship to Nuclear Engineering Ph.D. Curriculum Committee Chairmanship.

Part I
Overview

The first part of this book is intended as an overview of the Fast Spectrum Reactor field, from which the more detailed chapters to follow can be placed into perspective. The emphasis in this part is on the basic design considerations for breeders (the fast breeder reactor) because of the key role that fast spectrum reactors are expected to play in the long-term quest for extended fuel utilization. However, these basic design considerations equally apply to essentially all other applications of the fast spectrum reactor. One particular application, nuclear transmutation of transuranic elements, is gaining interest, and some discussions of basic design considerations for this application are provided.

Chapter 1 addresses the concept of breeding and the role that the fast breeder has grown to occupy in the energy structure of the industrialized nations. It also outlines the role of fast spectrum reactors for the partitioning and transmutation of potential nuclear waste. Chapter 2 provides a concise introduction to the basic design features that characterize a fast spectrum reactor. Chapter 3 is included to provide the principal economic concepts that must be employed in assessing ultimate commercialization aspects.

Chapter 1
Sustainable Development of Nuclear Energy and the Role of Fast Spectrum Reactors

Pavel Tsvetkov, Alan Waltar, and Donald Todd

1.1 Introduction

Energy, abundantly produced and wisely used, has always been needed for the advancement of civilization. Until the last few centuries, productivity was severely limited because only human and animal power were available as prime movers. By the early nineteenth century, wood burning, along with wind and water power, had considerably advanced the human capability to do work. Coal and then oil and natural gas sequentially replaced wood, water, and wind as the world's primary energy sources.

The discovery of nuclear fission in the late 1930s provided hope that nuclear energy could relieve the strain on rapidly depleting fossil fuels and provide an abundant, clean, and relatively inexpensive new form of energy. Although the basic sources for nuclear energy—uranium and potentially thorium—were plentiful, these raw resources could not be used directly in the fission process. Early experiments indicated that specific isotopes of uranium and plutonium must be used to exploit this new energy source.

By the early 1940s it was known that the isotopes ^{233}U, ^{235}U, and ^{239}Pu would fission when bombarded by neutrons with energies in the low and intermediate to fast ranges, i.e., at thermal energies (<1 eV) and at energies around 1 MeV. These isotopes are called *fissile* isotopes. It was also found that the abundant isotopes ^{238}U and ^{232}Th would fission only when struck by higher energy neutrons, of the order of 1 MeV or more. Fissile isotopes were found to be required to sustain a neutron chain reaction. But only a very small percentage (0.7%) of naturally occurring uranium is fissile; essentially all of the rest is ^{238}U. In order to derive full benefit from nuclear fission, it was soon recognized that a way must be found to utilize the remaining 99% of uranium's potential.

It was known that ^{238}U and ^{232}Th could *capture* neutrons at energies below the 1 MeV range and thereby convert ^{238}U into ^{239}Pu and ^{232}Th into ^{233}U. Therefore, ^{238}U and ^{232}Th were called *fertile* isotopes. If more fissile isotopes could be produced from fertile isotopes than were destroyed in the chain reaction, it would be possible to utilize the abundant fertile isotopes to make more fissile fuel. This process was shown to be possible. The name given to the process was *breeding.*

It was soon learned that the η value (where η for an isotope is the number of neutrons emitted in the fission process per neutron absorbed) for the fissile isotope ^{239}Pu is higher in a fast neutron spectrum than in a thermal spectrum. This higher number of emitted neutrons meant that more neutrons would be available for conversion of ^{238}U to ^{239}Pu. Hence, the idea developed that a breeder reactor operating on fast spectrum neutrons would utilize ^{238}U more efficiently than one operating on thermal spectrum neutrons. Although the ^{232}Th-^{233}U cycle was found to provide a possible path for breeding in either a

P. Tsvetkov (✉)
Zachry Engineering Center, Department of Nuclear Engineering, Texas A&M University, College Station, TX, USA
e-mail: tsvetkov@tamu.edu

A.E. Waltar et al. (eds.), *Fast Spectrum Reactors*, DOI 10.1007/978-1-4419-9572-8_1,

fast or thermal spectrum, the ^{238}U-^{239}Pu cycle in a fast spectrum provided considerably more efficient breeding. Hence, this cycle was generally adopted for the optimal breeder reactor path. Another reason for giving preference to the ^{238}U-^{239}Pu cycle for the fast breeder was that it employed the same fuel cycle as the present commercial thermal reactors.

Fast spectrum reactors, also commonly called *fast reactors*, are undoubtedly the most efficient system for the effective utilization of uranium resources, due to the possibility of using the uranium stored in used nuclear fuel[1] to be recycled while producing energy. It is also possible to adopt the strategy of burning plutonium along with the uranium stored as tailings from enrichment plants. With these unique features, the energy potential of uranium increases significantly (by a factor of approximately 60) compared to light water reactors (LWRs). In addition, the radioactive wastes containing long lived *minor actinides*[2] become practically insignificant. It is estimated that the ultimate accumulated long lived radioactive isotopes that need to be disposed contribute less than 0.1% of all the fission products. Hence, a fast spectrum reactor can minimize the mining effort and also reduce storage space and the time required in a used fuel repository (by a factor of ~1,000) to reduce the waste radiotoxicity to the level of the uranium that was originally mined from the ground.

Thus, the fast spectrum reactor has the flexibility to operate either as breeder to achieve the net creation of fissile fuel or operate as a *transmuter* to convert the long lived minor actinides and other radioisotopes to short lived ones.[3] An appropriately designed fast spectrum reactor has the flexibility to shift between these operating modes. The desired actinide management strategy will depend on a balance of waste management and resource extension considerations. By developing and introducing adequate advanced technologies for an optimal management of nuclear fission products and minor actinides, the fast spectrum reactor offers a realistic path forward to achieve energy sustainability consistent with environmental stewardship.

The first fast reactor was Clementine, built at Los Alamos in 1946. The next step was an experimental fast breeder reactor, EBR-I, designed by the Argonne National Laboratory. This reactor was built largely because of the enthusiasm of Enrico Fermi and Walter Zinn; and on December 20, 1951, EBR-I became the world's first nuclear plant of any type to generate electricity. The concept of the fast breeder reactor (FBR) was demonstrated and gave impetus to the prospect of a long-term reliance on nuclear fuel as a new energy source.

Approximately 30 years after EBR-I, four *liquid metal* (sodium) cooled *fast breeder reactors* (LMFBR's) in the 250–600 MWe power range began producing electricity (and desalinizing water) in three European countries, and construction on a 1,200 MWe LMFBR was nearing completion. At that time it was expected that LMFBRs in the 1,200–1,600 MWe range would soon be under construction in five countries. Hopes were high that the dream originating in the middle of the twentieth century of an economical, inexhaustible and practically independent energy source appeared likely to become a reality early in the twenty-first century for a many countries having few indigenous energy resources.

However, a major slowdown in new nuclear power construction began to emerge about two decades prior to the turn of the century. This was due partly to an excess of electrical production capacity in many of the developed countries and, more specific to the nuclear profession, the accident at

[1]The term "spent" fuel often appears in the literature to depict irradiated fuel that is discharged from a nuclear reactor. We prefer to use the term "used" fuel because a very large amount of nuclear fuel (plus other valuable radioisotopes for other uses) still remains in such fuel.

[2]"minor" actinides refers to all actinide isotopes other than uranium and plutonium.

[3]Though the term "nuclear transmutation" has a broader meaning, in the context of this book we use the term "transmutation" to indicate the process of converting minor actinides found in used nuclear fuel to more stable elements that are more readily handled and disposed.

Chernobyl on April 26, 1986. The latter event cast a significant anti-nuclear pall over most of Europe and America, and this brought nuclear programs to a near standstill in those regions. Progress was also slowed in some parts of Asia, although significant advancements continued to be made in China, India, Japan, and South Korea. As a result of this slowdown, interest in fast breeder reactors somewhat stagnated.

But around the turn of the century, concerns over global climate change and the rising price of petroleum spurred a renewed interest in alternate energy supplies. With conventional nuclear power plants then operating at near pristine conditions, worldwide, nuclear power was again elevated to prime consideration. One complication, however, was associated more with the issue of nuclear waste disposal[4] than on safety considerations. In particular, most of the public concerns seemed to focus on the longevity of the nuclear waste because of the very long half lives of some of the actinides and mobile fission products residing in used nuclear fuel. Having to design geologic repositories to safely protect high-level nuclear waste for periods up to a million years caused a good deal of public anxiety in many parts of the world.

Nuclear scientists and engineers have known since the early discovery of the fission process that many of the long-term radioisotopes remaining in used nuclear fuel could be very effectively transmuted to benign products in a fast spectrum reactor. Accordingly, a good deal of the interest in fast spectrum reactors over recent decades became focused more on transmutation for the near term, with the breeding capacity of such systems reserved for the longer term. The relative interest in these two key capabilities of the fast spectrum reactor systems varies somewhat from country to country, but there is now widespread global interest in developing the fast spectrum system because of its uniqueness to meet a variety of both short and very long term energy supplies.

1.2 International Focus on Fast Reactors

As noted above, most of the early interest in fast spectrum reactors was for developing breeding capability. Figure 1.1 provides an overview of the major fast reactor projects that have been built or are in the construction or detailed planning stages.

The projects are tabulated chronologically (from the date of first criticality) and by national origin. Dates associated with future projects are based on published estimates of first criticality. The illustration also shows the basic purpose of each project as follows:

Experimental and Test Reactors. Those with power levels typically up to 100 MWth, built to demonstrate the technology (experimental purposes or for fuels and materials testing) but often with steam plant and turbine-generators to allow operation as a small power station;

Demonstration or Prototype Reactors. Those generally in the 250–350 MWe range in which much of the scaling up required for a commercial station has been incorporated (in terms of both overall size and individual components);

Commercial-Sized Reactors. Those developed as lead stations to demonstrate the system's capability to operate in a utility environment.

[4] In the absence of a program to recycle used nuclear fuel, the discarded assemblies are "waste". With a recycling program in place, the uranium and plutonium are extracted and reused, but the remaining unseparated products are waste.

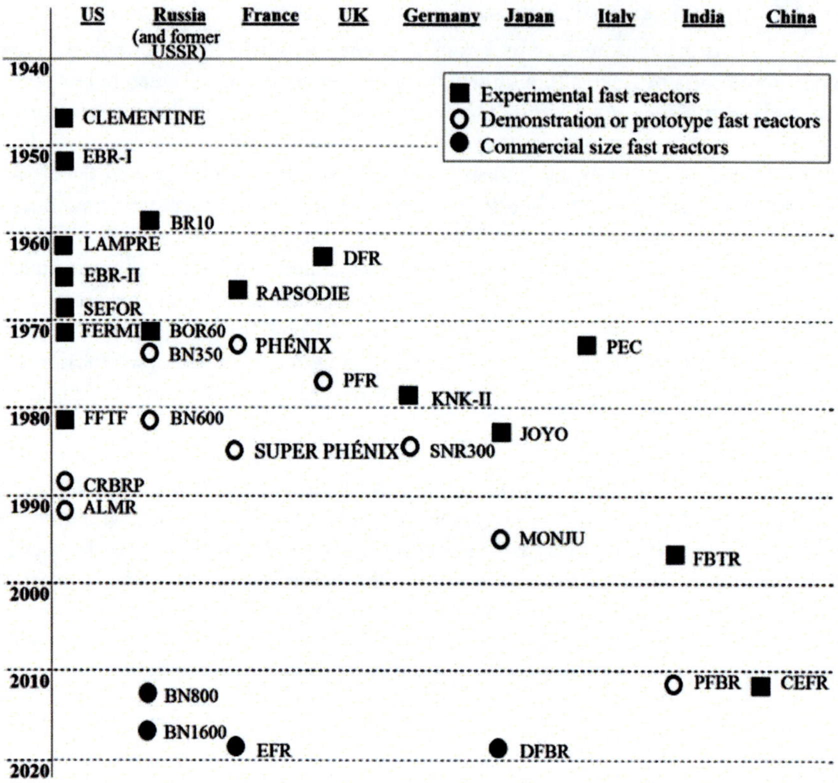

Fig. 1.1 International fast reactor development through the demonstration plant phase

The following three tables summarize the key features of the reactor systems shown in Fig. 1.1. Table 1.1 lists experimental and test reactors with information on electrical generating capability, if included. Table 1.2 lists the fast reactor prototype and demonstration reactors. Table 1.3 lists the full size commercial plants. Super Phénix-2, SNR-2, and CDFR are listed in Table 1.3 because they were well into detailed design prior to cancellation. The efforts expended on these designs were subsequently subsumed into the European Fast Reactor (EFR) design.

Two interesting observations are worth emphasizing from these tables. First, all of the fast reactors built and planned are based on the U-Pu fuel cycle. This is indicative of the advanced state of development of U-Pu fuel relative to Th-U fuel. The second observation is that all of the fast reactor plants built or planned use a liquid metal coolant; none uses a gas coolant, even though gas cooling for fast breeder reactors is still being considered (see Chapter 17). New interest in adapting lead cooling to advanced fast spectrum reactors is covered in Chapter 18.

Although Tables 1.1 through 1.3 contain some basic data for the major fast reactor projects, the interested reader may want more detailed information, particularly as design considerations are introduced in later chapters. Appendix A contains more detailed information on these reactors. Appendix B and Appendix C provide a description of two ongoing fast spectrum reactor developments projects in two organizations that have continued to engage in fast reactor technology.

Table 1.1 Experimental fast reactors

Reactor	Country	First critical	Power MWth	Power MWe	First full power	Type	Fuel	Coolant, at IHX inlet	Status
Clementine	US	1946	0.025	–	–	–	Pu	Hg	Shutdown
EBR-I	USA	1951	1.2	0.2	1951	Loop	U	Na, 230°C	Shutdown
BR-10	Russia	1958	8	0	1959	Loop	UN PuO_2, UC	Na, 470°C	Shutdown 2002
DFR	UK	1959	60	15	1963	Loop	U-7%Mo	NaK, 350°C	Shutdown 1977
LAMPRE	USA	1961	1.0	0	1961	Loop	Molten Pu Alloy	Na, 450°C	Shutdown
EBR-II	USA	1961	62.5	20	1965	Pool	U-Zr	Na, 473°C	Shutdown 1994
Fermi-1	USA	1963	200	61	1970	Loop	U-10%Mo	Na, 427°C	Shutdown 1972
Rapsodie	France	1967	40	0	1967	Loop	UO_2-PuO_2	Na, 510°C	Shutdown 1983
BOR-60	Russia	1968	55	12	1970	Loop	UO_2-PuO_2	Na, 545°C	In operation shutdown by 2015
SEFOR	USA	1969	20	0	1971	Loop	UO_2-PuO_2	Na, 370°C	Shutdown
KNK-II	Germany	1972	58	20	1978	Loop	UO_2-PuO_2	Na, 525°C	Shutdown 1991
FFTF	USA	1980	400	0	1980	Loop	UO_2-PuO_2	Na, 565°C	Shutdown 1992
JOYO	Japan	1977	140	0	1977	Loop	UO_2-PuO_2	Na, 500°C	In operation
FBTR	India	1985	40	13	–	Loop	PuC-UC	Na, 544°C	In operation
PEC	Italy	–	120	0	–	Loop	UO_2-PuO_2	Na, 550°C	Construction started in 1974, project cancelled
CEFR	China	2010	65	23.4	–	Pool	UO_2-PuO_2	Na, 516°C	In operation

Table 1.2 Demonstration or prototype fast reactors

Reactor	Country	First critical	Power MWth	Power MWe	First full power	Type	Fuel	Coolant, at IHX inlet	Status
BN-350	Kazakhstan	1972	750	130	1973	Loop	UO_2	Na, 430°C	Shutdown 1999
Phénix	France	1973	563	255	1974	Pool	UO_2-PuO_2	Na, 560°C	Shutdown 2009
PFR	UK	1974	650	250	1977	Pool	UO_2-PuO_2	NA, 550°C	Shutdown 1994
SNR-300	Germany	–	762	327	–	Loop	UO_2-PuO_2	Na, 546°C	Project cancelled
BN-600	Russia	1980	1,470	600	1981	Pool	UO_2 UO_2-PuO_2	Na, 550°C	In operation
PFBR	India		1,250	500	2009	Pool	UO_2-PuO_2	Na, 544°C	Under construction
Super-Phénix 1[a]	France	1985	2,990	1,242	1986	Pool	UO_2-PuO_2	Na, 542°C	Shutdown
MONJU	Japan	1994	714	280	1996	Loop	UO_2-PuO_2	Na, 529°C	Restarted 2010
CRBRP	USA	–	975	380	–	Loop	UO_2-PuO_2	Na, 535°C	Project cancelled
ALMR[a]	USA	–	840	303	–	Pool	U-Pu-Zr UO_2-PuO_2	Na, 498°C	Project cancelled

[a]Following IAEA, Super-Phénix 1 and ALMR are classified as "commercial size fast reactors" in Table A.3.

Table 1.3 Commercial size fast reactors[a]

Reactor	Country	First critical	Power MWth	Power MWe	First full power	Type	Fuel	Coolant, at IHX inlet	Status
Super-Phénix 2	France	–	3,600	1,440	–	Pool	UO_2-PuO_2	Na, 544°C	Subsumed into EFR
SNR 2	Germany	–	3,420	1,497	–	Pool	UO_2-PuO_2	Na, 540°C	Subsumed into EFR
DFBR	Japan	–	1,600	660	–	Loop	UO_2-PuO_2	Na, 550°C	To be determined
CDFR	UK	–	3,800	1,500	–	Pool	UO_2-PuO_2	Na, 540°C	Subsumed into EFR
BN-1600	Russia	–	4,200	1,600	–	Pool	UO_2-PuO_2	Na, 550°C	To be determined
BN-800	Russia	–	2,100	870	–	Pool	UO_2-PuO_2	Na, 544°C	Under construction
EFR	Europe	–	3,600	1,580	–	Pool	UO_2-PuO_2	Na, 545°C	To be determined

[a]In Table A.3, data for these and other representative designs are provided.

Generation IV International Forum

The Generation IV International Forum (GIF) is a cooperative international endeavor that is currently carrying out work to define and perform the research and development needed to establish the feasibility and performance capabilities of the next generation nuclear energy systems [1]. Three fast reactor systems are being pursued through international collaborations under the auspices of this program; the sodium-cooled fast reactor (SFR), the gas-cooled fast reactor (GFR), and the lead-cooled fast reactor (LFR) systems [2]. As indicated earlier in this section, the SFR system is by far the most tested of these three, and is extensively described in this book. The general characteristics of the GFR and LFR systems are addressed in Chapters 17 and 18, respectively.

Whereas earlier fast reactor development programs were generally driven by individual nations, members of the GIF are developing technology for a new generation of fast reactors through "systems arrangements" within GIF, which assures international collaboration. The GIF members select specific systems to work on that achieve desired goals for their situation (i.e. hydrogen generation for transportation applications, reduced nuclear waste volume and/or radiotoxicity for disposal, etc.). The goal of these systems arrangements is to integrate research and development projects contributing to the development of the system.

The membership in GIF, the "systems arrangements" participants, and technological information can be found in Ref. [1].

1.3 Basic Physics of Fast Reactors

1.3.1 Conversion Chains

Whereas the interest in fast spectrum reactors in the early part of the twenty-first century has turned primarily to minor actinide transmutation in many parts of the world, the ultimate goal of fast spectrum reactors is to breed fuel to enable nuclear reactors to power our planet for several millennia. Hence, we start with the basic physics for breeding.

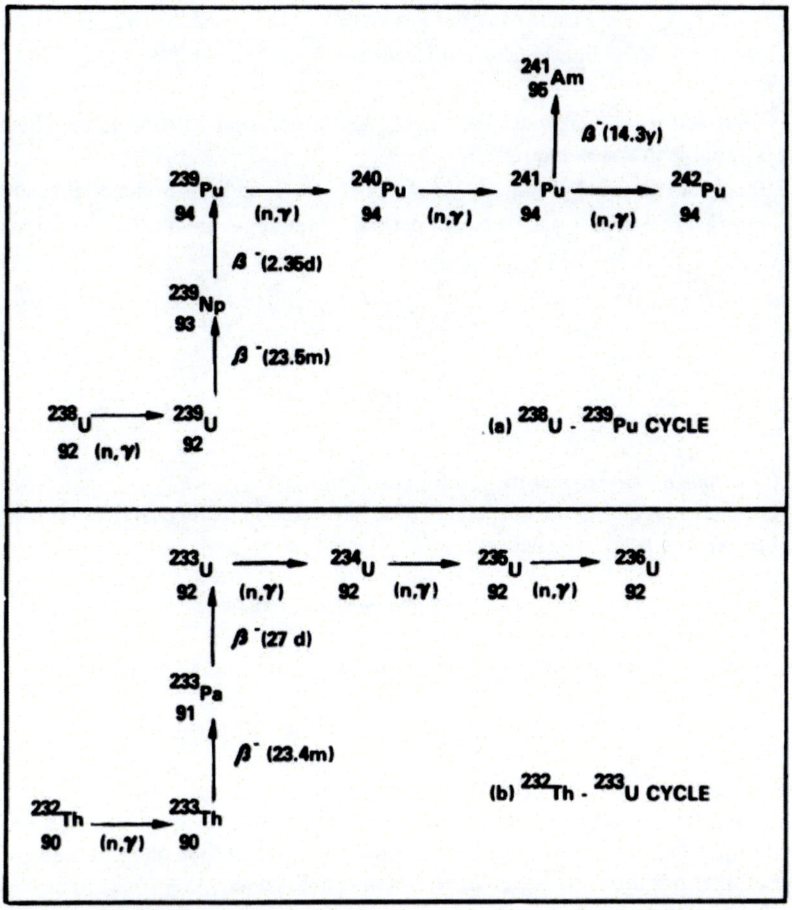

Fig. 1.2 ^{238}U-^{239}Pu and ^{232}Th-^{233}U conversion chains

In order to achieve breeding, a fertile isotope (^{238}U, ^{240}Pu, ^{232}Th, ^{234}U) must be converted via neutron capture (n, γ) into a fissile isotope (^{239}Pu, ^{241}Pu, ^{233}U, ^{235}U). The two important conversion chains are shown in Fig. 1.2. A more complete presentation of these chains is given in Chapter 7. All of the fertile and fissile isotopes appearing in Fig. 1.2 are long lived alpha emitters ($T_{1/2} \geq 10^4$ y) so that, with the exception of ^{241}Pu, they can be considered stable with regard to mass balances. Plutonium-241 is also a beta emitter with a half-life short enough to require consideration in fuel cycle calculations. Ignored in Fig. 1.2 are fission reactions (n, f) by all isotopes, a well as neutron capture reactions by the short lived beta emitters.

1.3.2 Conversion Ratio, Breeding Requirements and Fast Reactor Design Objectives

The degree of conversion that occurs in a reactor is denoted by the general term *conversion ratio,* CR, which is defined as

$$CR(\vec{r},t) = \frac{\text{Fissile Material Produced}}{\text{Fissile Material Destroyed}} = \frac{\text{FP}}{\text{FD}} = \frac{RR_c^{(FP)}(\vec{r},t)}{RR_a^{(FD)}(\vec{r},t)}, \tag{1.1}$$

where $RR_c^{(FP)}(\vec{r},t)$ and $RR_a^{(FD)}(\vec{r},t)$ are the capture and absorption reaction rates. The $CR(\vec{r},t)$ is a property that is variable in space and time.

Generally, the fissile material produced, $\text{FP} = RR_c^{(FP)}(\vec{r},t)$, and fissile material destroyed, $\text{FD} = RR_a^{(FD)}(\vec{r},t)$, during a fuel cycle—i.e., between periodic refueling—are used in the calculation:

$$\overline{CR} = \frac{\int_0^{T_f} dt \int_{V_c} RR_c^{(FP)}(\vec{r},t)\, dv}{\int_0^{T_f} dt \int_{V_c} RR_a^{(FD)}(\vec{r},t)\, dv}.$$

In this case, $\overline{CR}$ is also averaged over the reactor core volume.

A reactor is called a *breeder* if the conversion ratio is greater than unity. In this case, the conversion ratio is called a *breeding ratio,* BR; hence

$$\text{BREEDER: } \overline{BR} = \overline{CR} > 1.0. \tag{1.2}$$

A reactor for which the conversion ratio is less than unity is called a *converter,*

$$\text{CONVERTER: } \overline{BR} = \overline{CR} < 1.0. \tag{1.3}$$

Present commercial thermal reactors (LWRs) are converters.

In a breeder reactor it is possible for the in-core conversion ratio to be less than one, while the breeding ratio for the entire reactor—core plus *blanket*[5]—is greater than one.[6]

Another useful term is the *breeding gain*, $\overline{G}$, which is defined as

$$\overline{G} = \overline{BR} - 1, \tag{1.4}$$

Using the notation FBOC for the fissile inventory in the reactor at the beginning of a cycle (i.e., directly after refueling) and FEOC at the end of the cycle (i.e., when the reactor is shut down for refueling), the breeding gain can be written as

$$\overline{G} = \frac{\text{FEOC-FBOC}}{\text{FD}} = \frac{\text{FG}}{\text{FD}}. \tag{1.5}$$

where FG is the fissile material gained per cycle. Clearly the breeding gain must be greater than zero in order for the reactor to be a breeder.

It is possible for a nuclear reactor to breed over a broad neutron energy spectrum, but adequate breeding ratios can be achieved for a given energy spectrum only by carefully selecting the appropriate fissile isotopes for that spectrum. The reason for this becomes evident from a review of a few key properties of basic reactor physics. This review also shows that a high breeding gain can be obtained

[5]The blanket is the region of the reactor containing fertile fuel.

[6]The in-core conversion ratio of a breeder reactor has often been called the internal breeding ratio, despite the fact that it is generally less than one.

only with a fast neutron spectrum, although a low breeding gain characteristic of some thermal breeder designs is not necessarily an overriding disadvantage (as we shall see later).

Recall from the fission process that[7]

$\overline{\nu}_f$ = Number of neutrons per fission,
$\overline{\eta}$ = Number of neutrons produced per neutron absorbed,
$\overline{\alpha}$ = Capture-to-fission ratio $(\overline{\sigma}_c/\overline{\sigma}_f)$.

If $\overline{\eta}$ is the average number of new neutrons for each neutron absorbed in the fuel averaged over the neutron flux spectrum, these parameters are related by:

$$\overline{\eta} = \frac{\overline{\nu}_f \overline{\sigma}_f}{\overline{\sigma}_a} = \frac{\overline{\nu}_f}{1 + \frac{\overline{\sigma}_c}{\overline{\sigma}_f}} = \frac{\overline{\nu}_f}{1 + \overline{\alpha}}, \tag{1.6}$$

where, as defined above, $\overline{\alpha} = \frac{\overline{\sigma}_c}{\overline{\sigma}_f}$, the maximum possible value of $\overline{BR}$ is:

$$\overline{BR}_{\max} = \overline{\eta} - 1. \tag{1.7}$$

The parameters $\overline{\nu}_f$ and $\overline{\alpha}$ are measured quantities, while $\overline{\eta}$ is a derived quantity. The quantity $\overline{BR}_{\max}$ is called *breeding potential* of the fuel.

For each of the primary fissile isotopes, $\overline{\nu}_f$ is fairly constant for neutron energies up to about 1 MeV (about 2.9 for ^{239}Pu and about 2.5 for ^{233}U and ^{235}U) and slowly rises at higher energy. On the other hand, $\overline{\alpha}$ varies considerably with energy and between isotopes. For ^{239}Pu and ^{235}U, $\overline{\alpha}$ rises sharply in the intermediate energy range between 1 eV and 10 keV and then drops again at high energy; for ^{233}U, $\overline{\alpha}$ never rises appreciably. This behavior of ν and α leads to the variations of $\overline{\eta}$ with energy shown in Fig. 1.3.

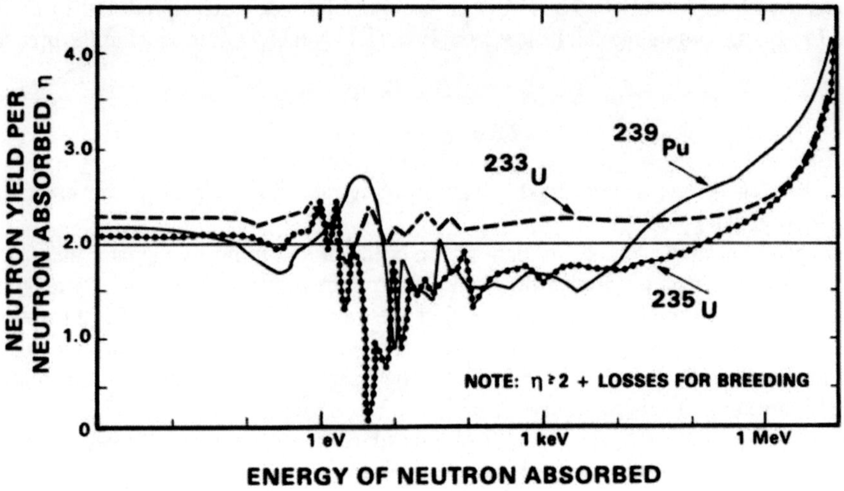

Fig. 1.3 Neutrons produced per absorption vs. energy for fissile isotopes

[7] Bars above symbols are used to signify proper cross section averaging over the flux spectrum.

In order to appreciate the significance of Fig. 1.3, it is useful to establish the minimum criterion for breeding. Consider a simple neutron balance for which the basis is one neutron absorbed by a fissile nucleus (which is equivalent to the destruction of 1 fissile nucleus). In order to breed, the next generation of neutrons must, at a minimum, replace that 1 destroyed fissile nucleus.

The number of neutrons produced by this absorbed neutron is η. Let us examine the fate of these η neutrons:

- One neutron must be absorbed in a fissile isotope in order to continue the chain reaction;
- L neutrons are lost unproductively, by parasitic absorption or by leakage from the reactor. For our present purpose, absorption by any material other than fissile or fertile materials is parasitic.

Hence, from the neutron balance, $[\eta - (1 + L)]$ is the number of neutrons captured by the fertile material.

Since this value represents the number of new fissile nuclei produced, and since the basis of our neutron balance is the destruction of one fissile nucleus,

$$\eta - (1 + L) \text{ must be } \geq 1 \tag{1.8a}$$

in order to replace the one destroyed fissile nucleus, i.e., in order for breeding to occur. This relation defines a minimum value for η for breeding.[8]

Rewriting this relation gives

$$\eta \geq 2 + L. \tag{1.8b}$$

Since the loss term is always greater than zero,

$$\eta > 2 \tag{1.9}$$

becomes a further simplified minimum criterion for breeding.

The quantity $[\eta - (1 + L)]$ in Eq. (1.8a) is the ratio of fissile nuclei produced to fissile nuclei destroyed for this simplified model; hence, from Eqs. (1.1) and (1.2), it is equal to the breeding ratio, i.e.,

$$\overline{BR} = \overline{\eta} - (1 + L)\,. \tag{1.10}$$

Therefore, a high value of η results in a high breeding ratio. This expression for breeding ratio is useful as a conceptual guide.

Now we can re-examine Fig. 1.3 in light of our criterion that η must be greater than 2.0 in order to breed. In a thermal reactor most of the fissile absorptions occur in the 0.01–1 eV range. In a fast spectrum reactor using mixed oxide (UO_2-PuO_2) fuel, about 90% of the fissile absorptions occur above 10 keV. Examination of Fig. 1.3 reveals that ^{239}Pu is the best choice for breeding in a fast reactor, with ^{233}U a possibility and ^{235}U doubtful. The figure also shows that ^{233}U is the only realistic candidate for breeding in a thermal reactor.

These observations are reinforced by Table 1.4, which compares η for the three fissile isotopes averaged over typical LWR and fast spectrum reactor spectra. If relative loss events characterized by L in Eq. (1.10) can be made about the same for a fast fast spectrum reactor and a thermal breeder

[8] It should be noted that the expressions are purposely simplified in order to elucidate the basic concept. In reality, $[\eta - (1 + L)]$ can be slightly smaller than unity for the breeding condition because of the fast fission effect in ^{238}U.

Table 1.4 Value for $\overline{\eta}$ averaged over fast and thermal spectra [3]

$\overline{\eta}$ -averaging spectrum type	^{239}Pu	^{235}U	^{233}U
Average over an LWR spectrum (0.025 eV)	2.04	2.06	2.26
Average over a typical oxide-fueled fast reactor spectrum	2.45	2.10	2.31

reactor, the results of Fig. 1.3 and Table 1.4 show that larger breeding ratios can be achieved with fast reactors. Moreover, even higher breeding ratios can be obtained from fast reactors that use metal (U-Pu), carbide (UC-PuC), or nitride (UN-PuN) fuel, partly because the average neutron energy for these fuels is higher and partly because these fuels have greater fissile and fertile material densities.

1.3.3 Doubling Time

An obvious economic incentive for obtaining a high breeding ratio is to maximize the production of excess fissile material for its sale value. Another perhaps more important incentive can be illustrated by introducing a parameter called the *doubling time*. Although there are several ways to define the doubling time (see Chapter 7 for more detail), the simplest concept is the *reactor doubling time*, RDT. This is the time required for a particular breeder reactor to produce enough fissile material in excess of its own fissile inventory to fuel an identical reactor. Hence, it is the time necessary to double the initial load of fissile fuel.

A constantly debated aspect of fast breeder reactors involves the rate at which they will be needed in a nation's or region's power system and when commercialization should begin. At issue is the rate the breeder economy should grow to meet projected electrical growth and to replace older fossil and LWR plants. The introduction date for the commercial FBR varies between countries, depending on many factors, including indigenous energy resources, energy growth rates, the supply of low-cost natural uranium, and the desire to become energy-independent. Indeed, this is a classical problem in systems engineering. Doubling times in the range of 10–15 years were generally considered as reasonable goals for breeder reactors during the 1970s, and meeting such a requirement appears to be possible for fast reactors. More recent electrical growth projections suggest that doubling times in the range of 15–20 years (or more) are probably adequate.

The reactor doubling time can be expressed quite simply in terms of the initial fissile inventory in a reactor, M_0 (kg) , and the fissile material gained during 1 year, $\dot{M}_g$, (kg/y), where $\dot{M}_g$, is a time-averaged difference between the fissile inventory at the beginning of a year and the fissile inventory at the end of the year, i.e.,

$$RDT = \frac{M_0}{\dot{M}_g}. \tag{1.11}$$

For example, if $\dot{M}_g$, were $0.1M_0$, and every year this $0.1M_0$ were set aside (since only M_0, is needed in the reactor), then after 10 years there would be $2M_0$ of fissile material M_0 still in the reactor and M_0 set aside.

Although Eq. (1.11) is simple, an accurate calculation of $\dot{M}_g$, is not. Computer techniques for this type of calculation are the subject of Chapter 7. Nevertheless, it is instructive to consider an approximation for $\dot{M}_g$, because this will correctly illustrate what parameters influence its value, and hence influence doubling time. The quantity $\dot{M}_g$ can be expressed in terms of the breeding gain, $\overline{G}$, rated power in megawatts, P, fraction of time at rated power, f, and $\overline{\alpha}$ as

$$\dot{M}_g = \overline{G} \cdot (\text{fissile mass destroyed/y}) \cong \overline{G} \cdot (1+\overline{\alpha}) \cdot (\text{fissile mass fissioned/y})$$

$$\dot{M}_g \cong \overline{G}(1+\overline{\alpha}) \left[\frac{\left(P \times 10^6\right)\left(2.93 \times 10^{10}\ \text{fissions/W} \cdot \text{s}\right) \times \left(3.15 \times 10^7\ \text{s/y}\right)(f)\,(239\ \text{kg/kg-mol})}{\left(6.02 \times 10^{26}\ \text{atoms/kg-mol}\right)} \right] \cong \frac{\overline{G}Pf\,(1+\overline{\alpha})}{2.7}. \quad (1.12)$$

Hence,

$$RDT \cong \frac{2.7M_0}{\overline{G}Pf\,(1+\overline{\alpha})}. \quad (1.13)$$

The doubling time in Eq. (1.13) is in years. It is proportional to the *fissile specific inventory*,[9] M_0/P, and inversely proportional to the breeding gain, $\overline{G}$. Allowances for time spent by the fuel in the fuel cycle outside the reactor, fuel cycle losses, fissions in fertile material, and variations during the burnup cycle need to be added to make Eq. (1.13) more precise, but the sensitivity of the doubling time to the breeding gain and fissile specific inventory is readily apparent. For example, an increase in breeding ratio from 1.2 to 1.4 results in a factor-of-two reduction in doubling time. The fissile specific inventory for a fast spectrum reactor with oxide fuel is in the range of 1–2 kg/MWth.

Equation (1.13) can also be used to estimate doubling times for thermal breeders. Two thermal breeder designs have received considerable development—the *light water seed-and-blanket breeder* and the *molten salt breeder*. Both utilize the ^{232}Th-^{233}U cycle. The light water breeder (Shippingport, Pennsylvania) has a calculated breeding ratio barely in excess of unity; hence, its doubling time is very long. The molten salt breeder has a breeding ratio higher than the light water breeder but still low relative to fast breeder reactors. However, since the fissile specific inventory of thermal reactors is generally significantly lower than for fast breeder reactors, the ratio of M_0/P to $\overline{G}$, and hence the doubling time of the molten salt breeder, may be comparable to that of a fast breeder.

Applying the above-introduced concepts of breeding and conversion ratios, the design objectives for fast reactors can be formulated as:

Breeders:

1. High breeding ratio, $\overline{BR} = \overline{CR} > 1.0$,
2. Low fissile inventory,
3. High burnup fuel.

Converters (transmuters):

1. Low conversion ratio, $\overline{CR} < 1.0$,
2. High fissile inventory,
3. High burnup fuel.

[9] The inverse of this term, P/M_0 is also often quoted. This ratio is called fissile specific power.

1.4 Breeder Strategy Analysis and Uranium Resources

1.4.1 The Breeder as an Inexhaustible Energy Source

Fuel cycle studies indicate that approximately 60–80 times more energy can be obtained from a given quantity of uranium in a FBR than in a light water reactor. While the ratio of ^{238}U to ^{235}U is 140, LWRs convert some ^{238}U to plutonium; this changes the ratio of uranium utilization in an FBR to that in a LWR to a value considerably below 140. Moreover, detailed fuel cycle analyses must account for losses of heavy metals throughout the fuel cycle, and the net result is about 60–80 times more efficient fuel utilization for the FBR. Improved LWR designs and fuel cycles are being considered that may reduce this ratio to closer to 60.

Since only a small fraction of the uranium is utilized in an LWR, the power cost is sensitive to the cost of the natural uranium, usually quoted as cost per kilogram or pound of the uranium oxide compound U_3O_8. Hence, low grade ores (i.e., of low uranium concentration) become economically unattractive for LWR use. In contrast, due to the FBR's factor-of-60 advantage in fuel utilization, the power cost for a FBR is insensitive to U_3O_8 costs. Therefore, low grade ores unacceptable for LWR use and even uranium from seawater could be used in FBR's.

Increased utilization of uranium, together with insensitivity to uranium cost, means that uranium used in FBR's represents an inexhaustible energy source, at least to the extent that many thousands of years satisfies the term inexhaustible. In the same way, thorium is another inexhaustible fuel for use in FBR's, though, as we observed in Section 1.3, it is less efficient for breeding than uranium. In summary, then, fission energy through the breeder reactor joins solar energy (together with renewable sources derived from the sun), fusion energy, and geothermal energy as the four inexhaustible energy sources available for development to meet the world's long-term future energy needs.

Even though low grade uranium ores may be economical for FBR's (as explained above), in practice further mining of uranium beyond that needed for LWR's may never be needed. For example, there is enough uranium-238 currently stored as tails[10] from the U.S. gaseous diffusion separation plants to satisfy all U.S. electrical energy requirements for several hundred years. Moreover, this stockpile will grow as long as ^{235}U is needed for LWR's.

1.4.2 Uranium Resources and Fast Reactors

While these long-term considerations are rather apparent, the near-term considerations that govern the timing for the introduction of the commercial FBR and the strategies for the transition to the FBR or to an FBR-LWR mix are more complex. Some of the issues involved were raised briefly in Section 1.3 in the discussion of doubling time.

Operating nuclear power reactors use uranium fuel, mostly low enriched uranium containing some 3–5% ^{235}U. Most of the reactors currently under construction will also be using low enriched uranium fuels. According to IAEA data [4], the identified uranium resources are currently estimated at 4.7 million metric tons of uranium at prices around or below \$130/kg. Additional uranium resources are expected to add up to over 10 million metric tons of uranium.

Table 1.5 illustrates current estimates of how long uranium resources are expected to last, assuming current burnup rates. The first row values are based on the assumption that new reactors will use the same technology as those in operation today. This technology utilizes only less than 2% of the natural uranium energy value. The second row values are based on the assumption that recycling technologies

[10]Tails represents the depleted ^{238}U that remains after completion of the enrichment process.

Table 1.5 Uranium availability for nuclear power [4]

Fuel cycle scenario	Uranium availability (years) Conventional resources	Total resources
Once-through fuel cycle with LWRs	80	270
Closed fuel cycle based on pure recycling in fast spectrum reactors	4,800–5,600	16,000–19,000

will eventually be introduced on an increasing scale as uranium prices go up. This expansion will lead to more energy being extracted. All numbers are expected to decrease as a function of expanding nuclear power capacities worldwide.

It is expected that as economical uranium resources are depleted, LWR construction will be replaced by FBRs such that energy supply will be sustained via recycling and breeding.

There is another technology pathway for fast reactors, one that does not require reprocessing facilities and offers an order of magnitude higher fuel efficiency than LWRs. This pathway leads to a so called breed-and-burn fast reactor operating with an open (once-through) fuel cycle. It thus realizes most of the benefits of a closed fuel cycle without any of the associated costs [5].

The first known proposal of a fast reactor design that could use an open fuel cycle was made in 1958 by Feinberg [6] who suggested that a breed-and-burn fast reactor could use only natural uranium or depleted uranium as fuel. Other similar concepts were proposed by Driscoll [7], Feoktistov [8], Teller et al. [9], and van Dam [10]. More recently, Fomin et al. [11] has completed work on the mathematical treatment of the space-dependent criticality in nuclear-burning waves. Takaki and Sekimoto [12] have also made great progress in demonstrating the strengths of this type of reactor. In 2006, TerraPower launched an effort to develop the first practical engineering design of a breed-and-burn fast reactor, producing a design concept now known as a traveling-wave reactor or TWR [13]. TWR designs are being developed for both low- to medium-power (~300 MWe) and large power (~1,000 MWe) applications.

1.5 Fast Reactor Transmutation Rationale[11]

1.5.1 The High Level Nuclear Waste Dilemma

As noted in the introduction to this chapter, public concerns regarding the nuclear waste issue appeared to overtake reactor safety as the primary concern for nuclear power in many parts of the world around the turn of the century. Political challenges have been encountered in almost every nation regarding the siting and development of a geologic repository for the long-term storage of high level nuclear waste.

Final repositories for low level radioactive waste from nuclear power plants and from medical, research, and other applications have been licensed and are in operation in many countries. However, there is no commercially operating repository for the final disposal of high level waste (HLW) from civilian nuclear power plants—despite the fact that the scientific and technical communities generally agree that such waste can be disposed of safely in stable geological formations. There is one operating

[11] Partitioning and transmutation overview is contributed by M. Salvatores. The extended discussion is given in Chapter 7.

geological repository for the disposal of long lived *transuranic* (TRU)[12] *waste* generated by the production of nuclear weapons; namely, the Waste Isolation Pilot Plant in New Mexico, USA. But this repository is not licensed for HLW from commercial nuclear power reactors.

Currently, used fuel generated by operating nuclear power plants is either recycled[13] or stored. The recycling process extracts usable uranium and plutonium from the used fuel for reuse in power plants. What remains is HLW that is currently stored pending final disposal. Several countries (notably France and the U.K.) actively maintain civilian programs to chemically process and recycle most of their used fuel, and that of other countries on a commercial basis. Other countries, such as Japan, are actively pursuing civilian recycling programs in their countries. Countries, such as Canada, Finland, Sweden and the USA have opted for the alternative of direct disposal of used fuel as HLW, although a third approach has been proposed in which used fuel would be recycled without extracting usable uranium and plutonium. The plutonium would immediately be "burned" to reduce the volume and toxicity of the waste requiring permanent disposal. Countries that have not yet chosen a strategy are currently storing used fuel and keeping abreast of developments associated with all alternatives.

There is now over half a century of experience with used fuel storage technology. Further, it is relatively easy to add incremental storage capacity and the used fuel from LWRs is very suitable for long-term storage.[14] Hence, there is no strong technical reason to expedite creation and operation of a deep geological repository. There may be good political and symbolic reasons to do so, but the relative ease of storage means that politicians and the public have time to exhaustively debate, explore and determine each country's preferred solution. Where politically acceptable, multinational disposal approaches can be considered as a potentially more cost effective option, especially for small countries with small nuclear programs and limited sites on which a repository could be located.

1.5.2 Partitioning and Transmutation

Fuel being irradiated in a reactor core produces fission products such as various isotopes of barium, strontium, cesium, iodine krypton, and xenon. Many of these isotopes, formed as fission products within the fuel, are highly radioactive—having short half lives. In addition to these smaller and lighter atoms formed during fission, various TRU isotopes are formed by neutron capture. These include ^{239}Pu, ^{240}Pu, ^{241}Pu, and ^{241}Am, as well as other less common isotopes. All are radioactive (normally with very long half lives) and apart from the fissile plutonium that is "burned". The transuranic isotopes and other actinides form most of the long-lived portion of the HLW.

The partitioning and transmutation (P&T) approach has been developed within radioactive waste management strategy studies to substantially reduce the time required for managing the HLW. By reducing the time when radiotoxicity is an issue (e.g. from tens of thousands or even a million years down to less than one thousand years), the consequences of accidental scenarios (e.g. human intrusion) in the repository are greatly reduced—as well as the heat load in the repository. Within this common generic interest for P&T, different objectives can be pursued via three general categories [14]: (1) sustainable development of nuclear energy and waste minimization by separating TRU from LWR fuel and repeatedly reusing it in fast spectrum reactors by also recycling fast spectrum reactor fuel;

[12]Transuranic isotopes all have atomic numbers greater than 92 (the atomic number of uranium).

[13]Recycling and reprocessing are words often interchanged in the literature.

[14]This perspective—which in reality represents a major attribute of nuclear energy—is woefully misunderstood by a large majority of the general public.

(2) reduction of *minor actinide*[15] (MA) inventory and reusing the separated Pu as fuel in LWRs; and (3) reduction of TRU inventory as separated from LWR used fuel.

The P&T strategies are very powerful, unique tools to reduce drastically the radiotoxicity level of the wastes, thus reducing the time needed to reach the reference level where there is widespread confidence of appropriately providing radiological protection (say from ~ 100,000 years to few hundred years). Furthermore, P&T provides a reduction of the residual heat that must be addressed in geological repository design and operation, with a correspondingly significant reduction in the repository size and improvement in characteristics.

In Chapter 7 the P&T approach is discussed in detail to emphasize (1) the physics arguments that reveal fast reactors to be the most adaptable system for minimizing radioactive wastes; (2) the different options to implement appropriate fast reactor types in order to achieve specific transmutation objectives; (3) the P&T implementation strategy; (4) the problem of nuclear data uncertainties and their impact on the nominal performances of the different transmutation systems; and (5) more detail on the role of the different types of fast reactors described in the previous paragraphs, according to the different P&T objectives and implementation scenarios.

Studies have shown that advanced fuel cycles with P&T offer significant potential benefits to a deep geological storage of used fuel. Specifically, they have shown a reduction of the potential source of radiotoxicity in a deep geological storage (of relevance in the so-called "intrusion" scenarios) and they have shown a reduction of the heat load so that a larger amount of waste material can be stored in the same repository (depending on the host formations).

As for the first point (i.e. radiotoxicity reduction), while P&T will not replace the need for appropriate geological disposal of HLW, several studies have confirmed that different transmutation strategies could significantly reduce, i.e. a hundred-fold, the long-term radiotoxicity [15]. The radiotoxicity reduction is comparable (i.e. higher than a factor 100) in fuel cycle scenarios discussed previously, depending upon the level of losses during reprocessing.[16] As illustrated in Fig. 1.4, the goal to reduce

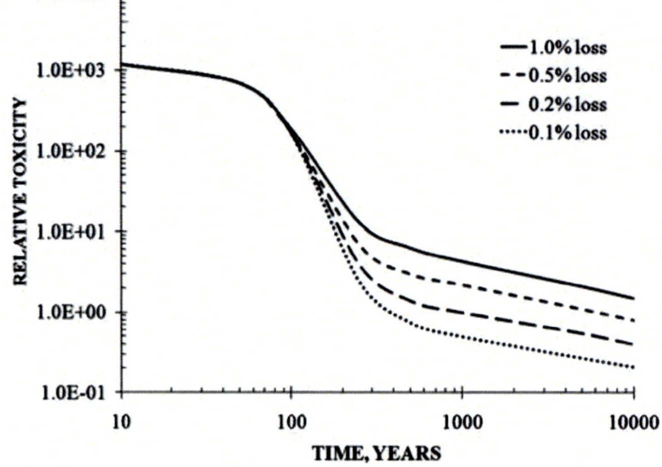

Fig. 1.4 Reduction of radiotoxicity as a function of reprocessing losses [15]

[15]Minor actinides (described earlier) are essentially the same isotopes as found in TRU but without plutonium, which is considered a "major actinide" along with uranium.

[16]*Reprocessing losses* are associated with inefficiencies in the reprocessing streams. Some reprocessing losses are inevitable, since 100% separation chemistry is never fully attainable. However, achieving losses down to the 0.1–0.2% range appear quite feasible.

the radiotoxicity to the level of natural uranium ore after ~300 years cannot be reached if the losses at TRU reprocessing are higher than 0.2%

As for the second point (i.e. heat load reduction), the HLW arising from the advanced fuel cycle scenarios associated with P&T generate appreciably less decay heat than the LWR used fuel. This is important in the case of disposal in hard rock, clay and tuff formations in which the maximum allowable disposal density is determined by thermal limitations.

Several studies have underlined the role of heat load, and its potential reduction, in repository design. This reduction provides the possibility, for scenarios with full Pu and MA recycling, of large gains in the reduction of the thermal load to the repository and on its associated storage density by delaying the time from used fuel processing until insertion to the repository to 100–200 years or more. Similar reduction on the HLW thermal power can be gained at shorter times by separating the Sr and Cs from the HLW [16].

If it is our desire to shorten the time for the radiotoxicity to be significantly reduced, this can be accomplished by first removing the remaining uranium and plutonium from the used fuel. That is seen to reduce the time in which the "nuclear waste" becomes less radiotoxic than natural uranium to about 10,000 years. This is still a very long time, but the time scale starts to become compatible with historical recordings of human life on earth.

If we then further remove the minor actinides (principally neptunium, americium, californium, and curium), the time at which the material inserted into the repository becomes less radiotoxic than natural uranium drops to well under 1,000 years. The most radioactive substances (mainly the fission products) all decay away in about 300 years. If we recycle used nuclear fuel to remove essentially all of the above radioactive substances and reinsert the "clean" fuel into a thermal reactor, the minor actinides start building up again because the transuranics preferentially absorb thermal neutrons and transmute to *higher actinides*.[17] However, if we reinsert this "clean" fuel into a fast spectrum reactor, the actinides preferentially fission and the objectionable long-lived radioisotopes disappear. These observations provide the primary motive for recycling used nuclear fuel and inserting the resulting recycled material in a fast spectrum reactor.

Studies performed in Japan [17, 18], have confirmed the possible benefit of P&T of MA and fission products (FP) in the context of the waste management by varying several conditions of fuel cycle parametrically, i.e. the reactor type, the cooling time before reprocessing and the scheme of reprocessing: PUREX,[18] MA-recycling, FP partitioning, and full P&T for both MA and FP. It was found that MA transmutation in fast reactors and FP partitioning reduces the repository size by a factor ~4–5 and that MA transmutation plus FP partitioning and long-term storage of Cs and Sr reduces the repository size by more than a factor 100.

As will be illustrated in Chapter 7, fast spectrum systems are more "efficient" in destroying actinides because fewer neutrons are lost to capture reactions before eventual fission. Furthermore, the generation rate of higher actinides is significantly reduced at each transmutation step. For example, the ^{241}Pu content is seven times lower in an operating fast reactor, compared to a typical LWR, and the higher actinides (americium and curium) continue to build-up with LWR recycle. These higher actinides tend to be more radioactive and can be problematic for fuel handling and fabrication in a closed thermal reactor fuel cycle. However, use of the separated plutonium and MA as a feedstock for "accelerated incineration" of TRU's in a fast reactor requires special reactor core designs,

[17] The term "higher actinides" refers to actinides with progressively higher atomic numbers that are successively created by neutron capture.

[18] The PUREX acronym stands for Plutonium—URanium Extraction. This is the reference aqueous nuclear reprocessing method for the recovery of uranium and plutonium from used nuclear fuel.

advanced fuel types (oxides, metals, nitrides) and possibly new or advanced recycling techniques (e.g. pyrochemical reprocessing).

It is because of the ability of fast neutrons to fission higher actinides, rather than to allow neutron capture to generate even more long-life elements, that fast spectrum reactors have regained prominence in the scientific community. The detailed physics describing this capability are covered in Chapter 7.

1.5.3 Fast Burner Reactor Scenarios

The wide range of conversion characteristics achievable in a fast spectrum system provides for the flexibility of two basic strategies:

Fast breeder reactor configuration. $\left(\overline{BR} = \overline{CR} > 1.0\right)$ The neutron balance in a FBR allows a net production of fissile material from ^{238}U. The traditional fuel cycle approach exploits this behavior by employing excess fertile material (blankets) to breed additional fissile material. This FBR configuration is envisioned for a rapidly expanding nuclear power economy where fissile material is scarce and expensive.

Fast burner reactor configuration. $\left(\overline{CR} < 1.0\right)$ The fast reactor can readily be configured to achieve a net destruction (burning) of TRUs.

Given the current status of the nuclear fuel cycle with stockpiles of excess weapons material, separated civil plutonium, and TRU-containing spent fuel, fast burner reactor configurations have been targeted in recent studies to reduce the global TRU inventory.

As shown in multiple strata studies, a limited capacity of fast burner systems can be utilized to stabilize the total TRU inventory. This approach may be advantageous as it allows exploitation of current enriched uranium resources (i.e. the current LWR system) without continued buildup of once-through fuel cycle wastes. For either the stockpile reduction or waste transmutation mission, high net TRU consumption rates are advantageous; the number of systems required to complete the mission will be directly dictated by the consumption rate.

Fast burner reactor designs have been developed for a variety of missions. For example, burner configurations have been developed for weapons plutonium disposition. In particular, a moderate burner design ($\overline{CR} \cong 0.5$ using conventional fuel forms) and a pure burner design ($\overline{CR} \cong 0.0$ using non-uranium fuel form) have been developed. In the French CAPRA study, moderate burners were considered with enrichments as high as 45% plutonium to heavy metal (Pu/HM) [15]. To summarize these original studies:

- The conversion ratio was constrained to $\overline{CR} \cong 0.5$ by conventional fuel composition limits;
- Some evaluations of non-uranium fuels were conducted, and significant perturbations in the reactivity coefficients were observed.

In more recent fast burner studies, the entire range of TRU fuel compositions (not just conventional 0–30% TRU/HM) was investigated. Eliminating the TRU content limit provides more flexibility in the design of burner fast reactor systems. Furthermore, this allows the development of low conversion ratio systems ($\overline{CR} < 0.5$) without resorting to the extreme limit of non-uranium fuels (with associated safety implications). As the conversion ratio is reduced, the fast reactor reactivity feedbacks will vary.

A primary mission identified for the SFRs under the GIF, described in Section 1.2, is the management of high-level wastes and, in particular, management of plutonium and other actinides (termed actinide management). SFR systems, configured for both high conversion ratio (greater than 1.0) and low conversion ratio (<0.5) have been extensively developed. However, the GIF systems are targeting toward a high conversion ratio, consistent with the sustainability (uranium resource) goals. For actinide transmutation missions, low conversion ratio systems have been analyzed primarily in the U.S. Advanced Fuel Cycle program.

The best achievable SFR fuel cycle employs complete recycle of the TRU and uranium. This significantly extends fuel resources as only depleted uranium is required as a make-up feed. Furthermore, TRUs are nearly eliminated from waste (~0.1% recycle loss each pass). The actinide-free waste has significantly reduced TRU mass, long-term heat, and radiotoxicity, compared to once through fuel cycles and placement of the used fuel in a permanent repository. Thus, these features lead to exceptional performance with regard to the GIF sustainability criteria related to resource utilization and waste management.

The SFR burner designs pursued recently by U.S. Department of Energy have continued to be modular (840 MWth) ternary metal (U/Pu-10Zr) fueled SFR systems. Two options were considered to achieve core designs with a low conversion ratio:

- Geometric spoiling (increasing the effective radius of the core and/or decreasing core height) to increase neutron leakage.
- Directly reducing the fertile material content of the fuel to reduce the production of TRU elements.

Parametric studies have revealed that in reducing the conversion ratio by modifying the core geometry, the effective core radius can become unacceptably large; a practical $\overline{CR}$ limit for this approach is about 0.5. However, very low conversion ratios can be achieved by modest reductions in the fuel pin diameter while maintaining the assembly size; that is, reducing the fuel volume fraction. Core designs with conversion ratios varying from 0.5 to 0.0 have been considered, using transuranics extracted from spent LWR fuel.

References

1. "Generation IV International Forum," http://www.gen-4.org
2. Generation IV International Forum, *GIF R&D Outlook for Generation IV Nuclear Energy Systems* (August 21, 2009).
3. G. R. Keepin, *Physics of Nuclear Kinetics*, p. 4, Addison-Wesley Publishing Co., Inc., Reading, MA (1965).
4. "Nuclear Power and Sustainable Development", Ed. M. ElBaradei, April 2006, IAEA, http://www.iaea.org/books (2006).
5. T. Ellis, R. Petroski, P. Hejzlar, et al., "Traveling-Wave Reactors: A Truly Sustainable and Full-Scale Resource for Global Energy Needs", *Proceedings of the ICAPP10*, June 13–17, 2010, Paper 10189, San Diego, CA (2010).
6. S. M. Feinberg, "Discussion Comment," *Rec. of Proc. Session B-10*, ICPUAE, United Nations, Geneva, Switzerland (1958).
7. M. J. Driscoll, B. Atefi, and D. D. Lanning, "An Evaluation of the Breed/Burn Fast Reactor Concept," MITNE-229 (December 1979).
8. L. P. Feoktistov, "An Analysis of a Concept of a Physically Safe Reactor," Preprint IAE-4605/4, in Russian (1988).
9. E. Teller, M. Ishikawa, and L. Wood, "Completely Automated Nuclear Power Reactors for Long-Term Operation," *Proceedings of the Frontiers in Physics Symposium*, American Physical Society and the American Association of Physics Teachers Texas Meeting, Lubbock, TX (1995).
10. H. van Dam, "The Self-stabilizing Criticality Wave Reactor," *Proceedings of the Tenth International Conference on Emerging Nuclear Energy Systems* (ICENES 2000), p. 188, NRG, Petten, Netherlands (2000).
11. S. P. Fomin, A. S. Fomin, Y. P. Mel'nik, V. V. Pilipenko, and N. F. Shul'ga, "Safe Fast Reactor Based on the Self-Sustained Regime of Nuclear Burning Wave," *Proceedings of Global 2009*, Paper 9456, Paris, France (September 2009).

12. N. Takaki and H. Sekimoto, "Potential of CANDLE Reactor on Sustainable Development and Strengthened Proliferation Resistance," *Progress in Nuclear Energy, 50* (2008) 114.
13. J. Gilleland, C. Ahlfeld, D. Dadiomov, R. Hyde, Y. Ishikawa, D. McAlees, J. McWhirter, N. Myhrvold, J. Nuckolls, A. Odedra, K. Weaver, C. Whitmer, L. Wood, and G. Zimmerman, "Novel Reactor Designs to Burn Non-Fissile Fuel," *Proceedings of the 2008 International Congress on Advances in Nuclear Power Plants* (ICAPP 2008), ANS, Anaheim, CA, Paper 8319 (2008).
14. M. Salvatores and J. Knebel, "Overview of Advanced Fuel Cycles for the 21st Century," Jahrestagung Kerntechnik 2008, Hamburg (May 27–29, 2008).
15. *Accelerator-Driven Systems (ADS) and Fast Reactors (FR) in Advanced Nuclear Fuel Cycles*, A Comparative Study, OECD-NEA (2002).
16. R. A. Wigeland, et al., "Separations and Transmutation Criteria to Improve Utilization of a Geological Repository", *Nuclear. Technology, 154* (2006) 95.
17. H. Oigawa, et al., "Partitioning and Transmutation Technology in Japan and Its Benefit on High-Level Waste Management", *Proceedings of the International Conference on GLOBAL '07*, Boise (September 2007).
18. K. Nishihara, et al., "Impact of Partitioning and Transmutation on LWR High-Level Waste Disposal", *Journal of Nuclear Science and Technology, 45*(1) (2008) 84–97.

Chapter 2
Introductory Design Considerations

Pavel Tsvetkov, Alan Waltar, and Donald Todd

2.1 Introduction

Before discussing the neutronics, systems, and safety considerations involved in designing fast spectrum reactors, it is appropriate to follow the lead of Wirtz [1] in sketching the bases for fast spectrum reactor designs. In this orientation, care will be taken to indicate the principal differences relative to thermal reactor systems with which the reader is more likely familiar.

This introduction to design begins with a brief discussion of major design objectives, followed by an overview of the mechanical and thermal systems designs of fast spectrum reactors (with an emphasis in this chapter on fast breeder reactors, since most other applications of fast spectrum systems—such as waste transmutation—will be optimized if the reactor has a high internal conversion ratio and/or a high breeding ratio).

Because of the position occupied by the sodium-cooled fast reactor (SFR) in the international fast spectrum reactor community, that system will be used for purposes of illustration. Chapter 17 focuses on another leading design, the gas-cooled fast reactor (GFR), and Chapter 18 is focused on another liquid metal cooling system, the lead-cooled fast reactor (LFR); utilizing lead or lead-bismuth for coolant. This introductory description will be followed by considerations that have led to particular choices of fuels and major core parameters, coolant, and structural materials.

2.2 Major Design Objectives

Table 2.1 lists the principal goals of fast spectrum reactor design and the specific reactor design objectives that accompany the goals. Whereas breeding is an important long-term goal, there is substantial interest in the nearer term for transmuting higher actinides in fast spectrum systems, thereby removing the need to address these long-lasting radioisotopes in the design and treatment of ultimate waste forms. However, we refer mainly to breeding considerations in several of these chapters because high breeding performance often provides a framework for long-term performance required for any application.

Safe operation is mandatory for any nuclear reactor system. In terms of design objectives, safe operation translates into conscious awareness of safety at every stage of design, with the result of providing reliable components, passive safety features, and adequate safety margins for potential accident situations. Part IV of this book (Safety) addresses these considerations in detail.

P. Tsvetkov (✉)
Zachry Engineering Center, Department of Nuclear Engineering, Texas A&M University, College Station, TX, USA
e-mail: tsvetkov@tamu.edu

A.E. Waltar et al. (eds.), *Fast Spectrum Reactors*, DOI 10.1007/978-1-4419-9572-8_2,

Table 2.1 Goals and design objectives for the breeder reactor

Goals	Accompanying design objectives
Safe operation	– Reliable components – Passive safety features – Adequate containment margin
High breeding ratio	– High neutron energy
Low doubling time	– High breeding gain – Low fissile specific inventory
Low cost	– High burnup – Compatibility with LWR cycle – Minimum capital cost

A high breeding ratio (BR) is needed to reduce fuel doubling time and thereby allow a sufficiently rapid rate of introduction of breeder reactors after uranium resources become scarce and U_3O_8 costs become excessive. As noted from Fig. 1.3 and Table 1.4, the best BR can be obtained with a fast neutron spectrum, although a reasonable value could be obtained in a thermal reactor if ^{233}U were used for fissile fuel.

A low doubling time is required in order for the breeder reactor to provide a significant fraction of the world's capacity for electricity generation during the present century. It is not sufficient simply to double the number of breeder reactors at the doubling rates of electrical demand. In order to continually increase the fraction of electrical capacity provided by breeder reactors, the doubling time must be shorter than the doubling time of electrical demand. The importance of high breeding gains and low fissile specific inventories in order to achieve low doubling time was discussed in the previous chapter.

Fast reactor systems are quite adaptable to the transmutation missions. These systems can be configured as transuranic (TRU) sustainers or as TRU burners. The former mission is important in fuel cycles where fuel material utilization is a major consideration. In this mode, the fuel could be continuously recycled with minimal need for uranium enrichment, since either depleted or natural uranium would be used as make-up feed to the fuel cycle. TRU burners on the other hand are required in fuel cycles with a need to burn down waste materials; this could be either to complement once-through systems that are producing TRU or for the purpose of drawing down the accumulated legacy nuclear waste. The base fuel cycle for the fast reactor concepts (SFR, GFR, and LFR) is a closed fuel cycle using recycled TRU and depleted uranium fuels. Waste management gains from complete recycle are substantial, with the final disposition heat load determined by processing losses. The base Generation IV concepts discussed in Chapter 1 allow consumption of ^{238}U, significantly extending uranium resources (up to 100 times) [2, 3].

A very important objective of any energy generating system, of course, is low cost. From the standpoint of operational costs, overall fuel cycle considerations are of paramount importance. Hence, high burnup becomes an important objective, both for optimum utilization of high-fissile-content fuel and to minimize downtime for refueling. An additional cost objective is to mesh the breeder fuel cycle with that of the existing light water reactor (LWR) cycle, which yields a considerable advantage for the ^{238}U-^{239}Pu system compared to the ^{232}Th-^{233}U system. Capital costs are directly related to hardware requirements and, therefore, put a premium on systems that require the least expensive capital equipment and still meet reliability criteria.

2.3 Overview of Mechanical and Thermal Systems Designs

This section provides an overview of some of the key features in a typical fast spectrum system, as a prelude to the more detailed discussions in later chapters. Included are brief descriptions of core and blanket arrangements, the fuel configuration, vessel internals, and heat transfer systems. Discussions related to several of the design options under consideration are included as appropriate.

2.3.1 Core and Blanket Arrangements

Since the ability to breed fuel is the principal feature that distinguishes fast reactors from thermal converter reactors, it is appropriate to ask how fertile and fissile fuels should be arranged to optimize breeding potential.

Two basic choices exist regarding where the breeding takes place. In the *external* breeding concept, all the fertile material is contained in the blanket surrounding the core; hence, all breeding takes place external to the core. In the *internal*, or *in-core* breeding concept, some fertile fuel is mixed with fissile fuel within the core fuel assemblies. Figure 2.1 illustrates the two basic breeding configuration concepts.

Some design consideration was given to the external breeding configuration for the small, early fast breeder reactors. Such an arrangement results in a very hard spectrum, a low fissile inventory, and a good breeding ratio. However, the lack of in-core breeding results in a rapid reactivity loss with burnup (requiring frequent fuel changes), low burnup, a small Doppler effect, few fast fissions in fertile material, and requires a high fissile enrichment and a thick blanket. Consequently, all current designs allow for considerable internal or in-core breeding.

Whereas present reactors all allow internal breeding, they normally have an in-core breeding ratio less than unity. However, substantial breeding also takes place in the surrounding blankets, and this additional breeding raises the overall breeding ratio to well in excess of unity.

The configuration shown in Part (b) of Fig. 2.1 is known as a *homogeneous* core design because all assemblies containing pure fertile fuel are located in the radial and axial blanket regions. This leaves a relatively uniform or homogeneous mixture of fertile and fissile fuel spread throughout the core. An interesting variation of this internal breeding configuration is the *heterogeneous* core, in which blanket assemblies (containing pure fertile material) are distributed through the core region. These designs yield higher breeding ratios and reduced sodium void coefficients, but require higher fissile fuel inventories. Appendix A lists the fast reactors incorporating the heterogeneous core design.

A schematic top and side view of a typical homogeneous SFR core is given in Fig. 2.2a. The central region indicates the core itself (i.e., the region containing the initial load of fissile/fertile fuel), and the outer region represents the radial blanket and shielding arrangement. The core shown in this figure contains two enrichment zones. The outer zone incorporates higher fissile fuel content than the inner

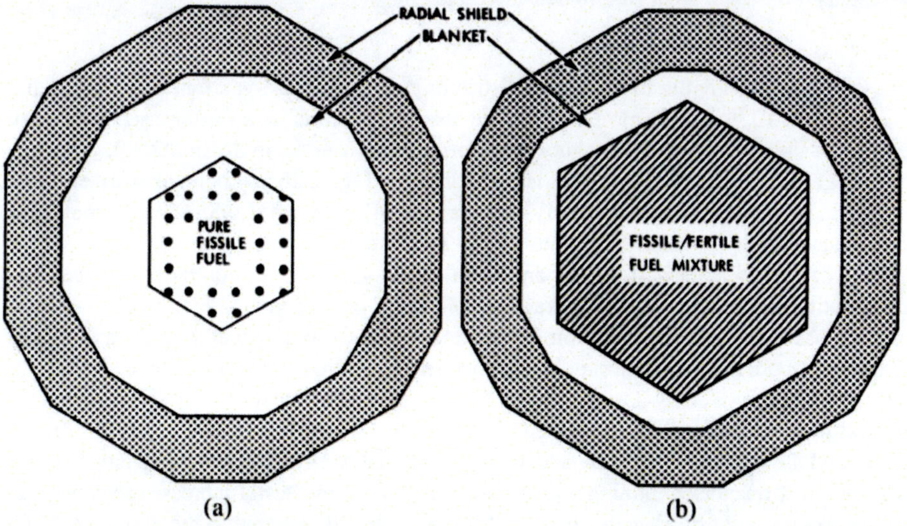

Fig. 2.1 External vs. internal breeding core/blanket configurations. **(a)** *External* breeding concept, **(b)** *Internal* or *in-core* breeding concept

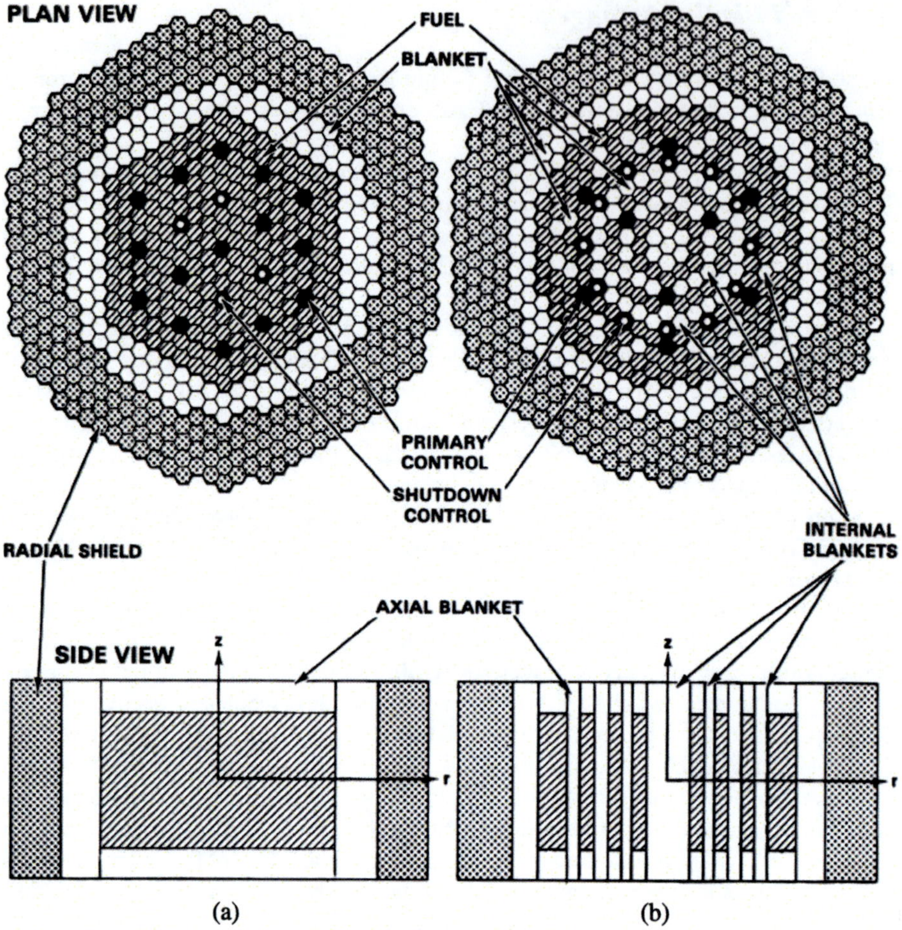

Fig. 2.2 Typical homogeneous and heterogeneous SFR core/blanket arrangements wherein both schemes incorporate internal breeding. (**a**) Homogeneous core, (**b**) Heterogeneous core

zone in order to effect a flatter radial power distribution. Regions reserved for control are indicated on the figure. Boron-10, in the form of B_4C, is the absorber material most often used for this purpose. The reason for the triangular, or hexagon, shape of the core is given in Section 2.3.2.

A heterogeneous core configuration is illustrated in Fig. 2.2b, with blanket assemblies at the center and in concentric rings. An early comparison [4] between 1,000-MWe homogeneous and heterogeneous designs provided the results shown in Table 2.2.

The core of a fast reactor is smaller than that of a thermal reactor of comparable power. A thermal reactor is optimized at some particular fuel-to-moderator ratio, and any core smaller than optimum represents a less economic configuration. For a fast reactor, there are great incentives to minimize the size of the core, both for neutronics and cost considerations. No moderator is needed; in fact, a moderator is normally excluded by design in order to maintain a hard spectrum. By squeezing out as much coolant and structural material as possible to increase the fuel volume fraction, the neutron leakage is decreased, and the fissile fraction can be decreased. Viewed from a different perspective, reducing the core size of a fast reactor containing a given amount of fissile material increases reactivity. From an optimal neutronics point of view, the height of the core should be equal to the diameter, $H_c/D_c \approx 1$.

Table 2.2 Comparison of homogeneous and heterogeneous cores (1,000-MWe designs) [4]

	Homogeneous	Heterogeneous
Number of fuel assemblies[a]	276	252
Number of in-core blanket assemblies[b]	–	97
Number of ex-core blanket assemblies[b]	168	144
Number of control assemblies	19	18
Number of enrichment zones	3	1
Fissile mass (kg)	3,682	4,524
Pu fraction by zone (%)	13.5/14.9/20.3	20.2

[a]271 pins/fuel assembly, 1.22 m core height, 7.9 mm pin diameter, 0.36 m axial blankets.
[b]127 pins/blanket assembly, 13 mm pin diameter.

However, to minimize the effect of sodium voiding on reactivity, there is incentive to reduce this ratio. In practical fast reactor designs, the height is often about 1.0–1.2 m, whereas the diameter is about 2.4 m or more. The arguments for the core height, aside from sodium void arguments, come from considerations regarding the temperature rise of the coolant as well as its pressure drop, depending on available pumps, etc.

In a typical burner-core design, the net consumption of TRU is achieved by minimizing the uranium inventory—e.g., by eliminating blanket assemblies and by reducing the core height (pancaking the core shape) to further reduce fertile material and to reduce the probability of neutron capture by increasing the leakage [2].

2.3.2 Fuel Lattice

Because of the incentive to minimize the specific fissile inventory in the breeder reactor, a "tight" fuel lattice that maximizes fuel volume fraction is desirable. A triangular lattice arrangement intrinsically allows a higher fuel volume fraction than a square lattice. The higher fuel volume fraction minimizes fissile loading mainly by reducing reactor leakage. Hence, the triangular lattice or hexagonal structure has been universally selected for SFR design (except for the early Fermi reactor, located near Detroit, Michigan). However, as noted in Chapter 18, design considerations specific to the chosen coolant of LFRs has lead to selecting an open square lattice in some designs.

A LWR typically employs a square lattice because relatively larger spaces between fuel rods are needed to optimize the water-to-fuel ratio. The square lattice provides such space and also provides for easier mechanical assembly.

2.3.3 Fuel Assembly

Figure 2.3 is a simplified sketch of a typical SFR fuel assembly (also called a subassembly). For this standard triangular pitch lattice, the pins are separated by a spiral wire wrap and assembled as a cluster of 217 pins within the assembly duct. Alternatively, grid spacers can be used to separate the fuel pins. Fuel pellets make up the active core region, and blanket pellets provide the axial boundaries. For the design shown, the fission gas plenum is located above the upper blanket region, although there are substantial arguments in favor of a lower gas plenum region (below the lower blanket) as discussed in Section 8.2.1.

For neutronics calculations, it is often convenient to define a unit cell, shown in Fig. 2.3, as a 1/12 segment of an assembly. The figure shows the geometric arrangement of the fuel, structure, and

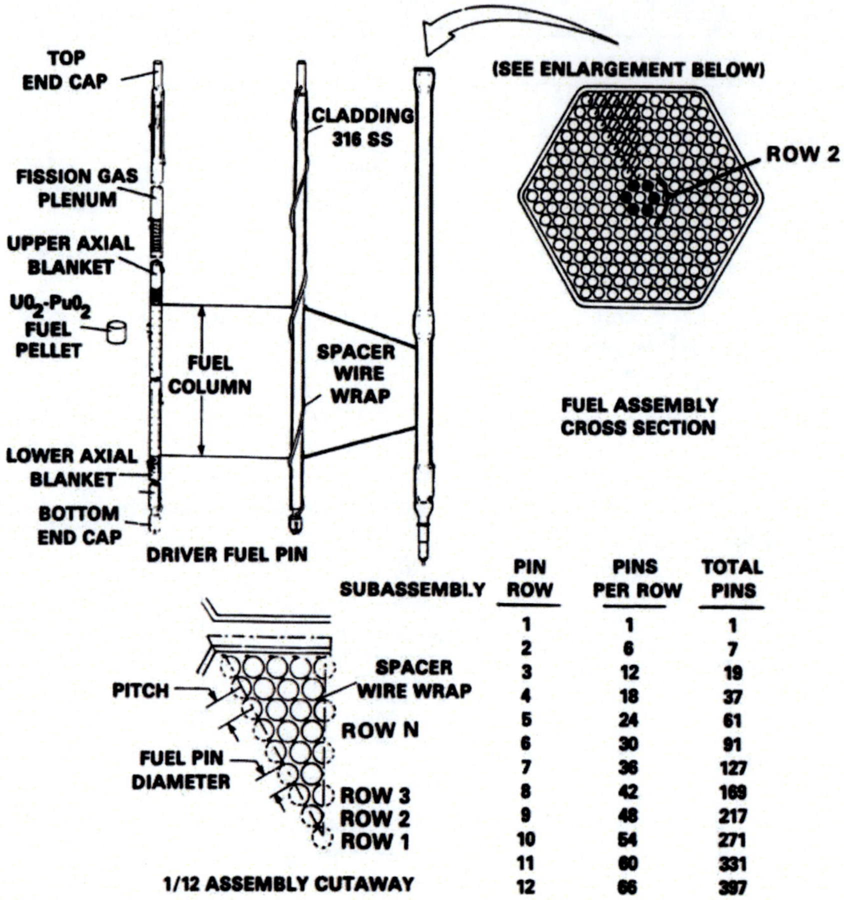

PIN ROW	PINS PER ROW	TOTAL PINS
1	1	1
2	6	7
3	12	19
4	18	37
5	24	61
6	30	91
7	36	127
8	42	169
9	48	217
10	54	271
11	60	331
12	66	397

Fig. 2.3 Typical SFR fuel assembly system

coolant. A table is also included to provide a ready reference for the relationship between concentric (hexagonal) rows of pins and the number of pins in an assembly. This same table can be used to determine the number of assembly positions in a core, given the number of assembly rows.

2.3.4 Vessel Internals

The core of a SFR is located on top of a core support structure, which is normally hung from the reactor vessel, or reactor tank, as shown in Fig. 2.4.

Because the SFR does not require pressurization to keep the coolant in a liquid state, outlet pressures are near atmospheric; consequently, the reactor vessel need only be thick enough to satisfy standard weight-bearing structural and safety requirements. A typical thickness may be the order of 30 mm, as compared to 300 mm for LWR systems.

Assemblies are slotted into positioning holes in the core support structure, and radial core restraint is normally provided at two axial locations (generally just above the active core and again near the top of the assemblies). Control rods enter from the top of the core, with the drive mechanisms located

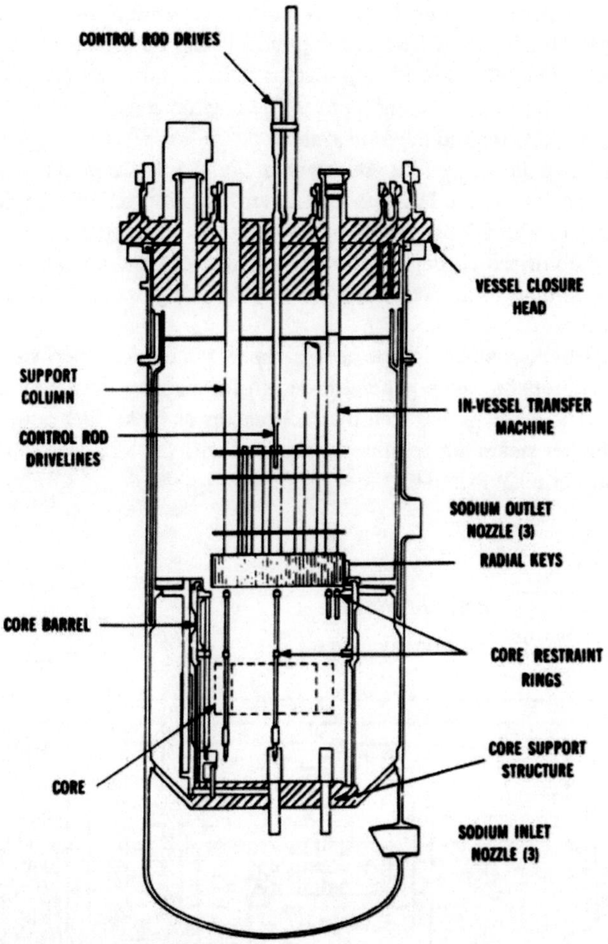

Fig. 2.4 Typical loop-type SFR vessel internals

atop the vessel closure head. The vessel normally hangs from a support ledge, and the head is bolted to this structure. However, several variations in vessel support design have been used.

Sodium enters the lower inlet plenum, traverses upward through the core and blanket regions, and collects in a large reservoir above the core before exiting to the intermediate heat exchanger (IHX). An inert cover gas, usually argon, separates the sodium pool from the closure head.

2.3.5 System Heat Transfer

Neutron activation of the sodium coolant in the *primary* heat transport system of a SFR (discussed in Section 2.4.2) normally requires that a *secondary*, or intermediate, heat transport system also be employed in a SFR. Hence, sodium coolant in a primary loop is pumped through the core and circulated through an intermediate heat exchanger and a secondary sodium loop transports heat from the IHX to the steam generator. The temperature of the sodium leaving the reactor (of the order of 550°C) is substantially higher than the coolant in a LWR (~300°C); hence, steam temperatures in the SFR are higher.

In contrast to the secondary sodium loop, the radioactive sodium in the primary loop must be shielded from plant personnel. Two different arrangements can be used to accomplish this; the IHX and the primary pump can be placed inside the reactor tank, or these two components can be located in adjacent hot cells with pipes connecting them to the reactor vessel. The first option is called the *pool system,* and the second is termed the *loop system.*

Simplified pool and loop design options are shown in Fig. 2.5. In the pool system, hot sodium flows directly into the IHX and then into a large pool that surrounds the vessel internals. A primary pump then recirculates this cool sodium into the core inlet plenum. In the loop system, two important design questions are the location of the primary pump (hot or cold leg) and the vessel penetration location for coolant inlet piping. The pros and cons of pool vs. loop designs and the two pump locations will be covered in Chapter 12.

Figure 2.6 is a simplified schematic of the heat transport system for a pool design. With the exception of special considerations in steam generator design for a sodium/water system, the steam cycle side of the SFR system is quite similar to a thermal reactor or fossil-fuel power plant. It should be noted, however, that higher steam temperatures associated with the SFR lead to a higher overall plant thermal efficiency than for a LWR plant.

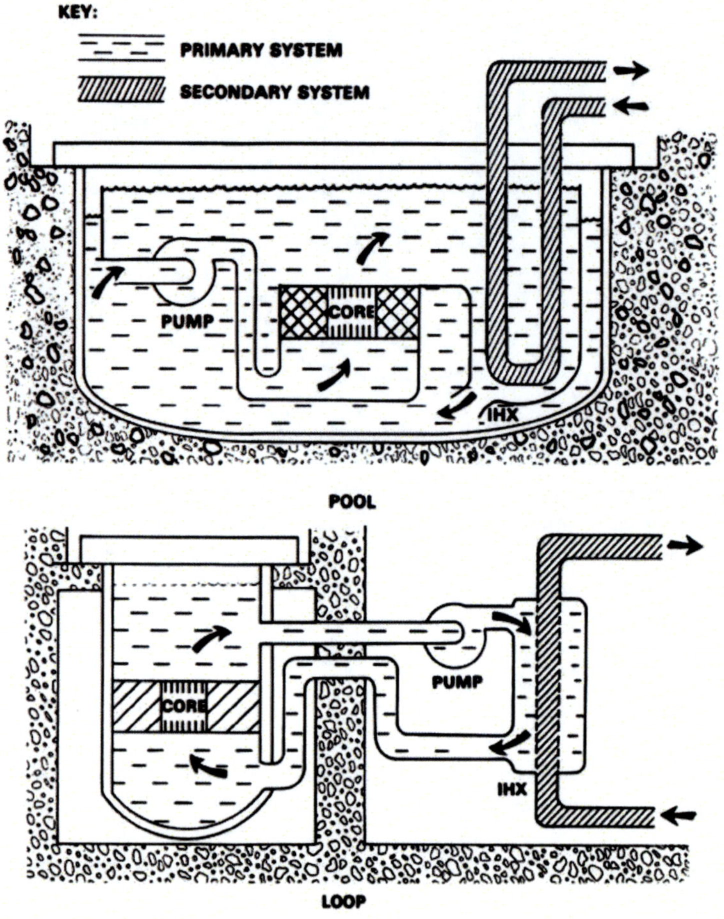

Fig. 2.5 Pool vs. loop design systems

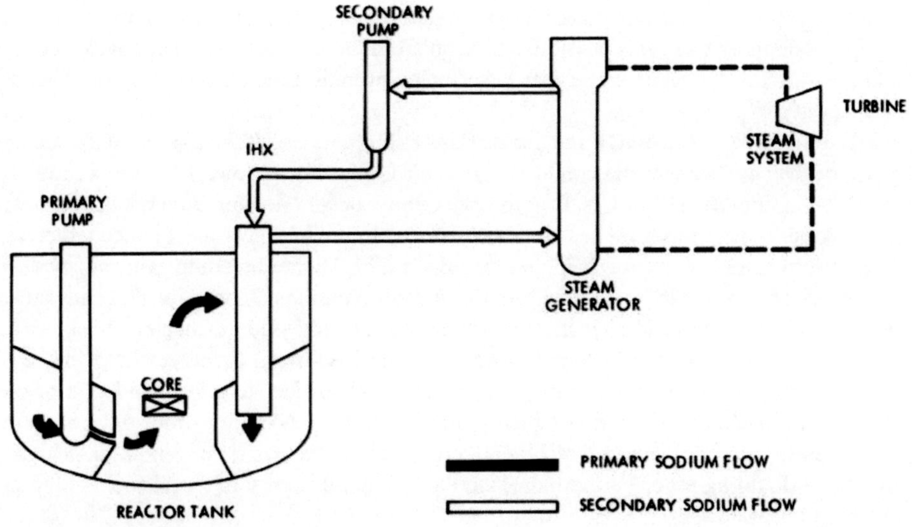

Fig. 2.6 Pool-type reactor heat transport system

2.4 Selection of In-Core Materials and Parameters

Having introduced the basic mechanical and thermal systems designs of a SFR, we can next address several obvious questions. What fuel is preferable for fast reactors? Why has sodium been so widely proposed as the coolant and are other alternatives attractive? What materials are used for cladding and other structures? Principal considerations that influence the answers to these questions are discussed briefly in this section. Chapter 11 deals with materials properties and selections in more detail.

In addition, we shall discuss basic core parameters related to the fuel—in particular the fuel fissile fraction, the fuel-pin size, and fuel burnup.

2.4.1 Fuel

2.4.1.1 Fuel Candidates and Properties

The prime candidates for SFR fuel are uranium-plutonium oxide, carbide, metal, and possibly nitride. Thorium-uranium-233 fuel can also be used in SFR's, but higher breeding ratios and more favorable economics can be achieved with uranium-plutonium.

Generally desirable properties for fuels are the capability for irradiation to high exposures in terms of energy generated per unit mass (i.e., high burnup) and high values of linear power (i.e., kilowatts per unit length of fuel pin). As shown later in this section, high linear power implies either high thermal conductivity or high melting point.

Oxide fuel (i.e., a mixture of UO_2 and PuO_2) became the favored reference fuel in the early 1960s because of the demonstrated high-burnup capability (perhaps in excess of 100 MWd/kg). The large body of oxide fuel experience developed within the LWR program also provided considerable impetus to selecting this fuel for the base fast reactor program. Oxide fuel has a high melting temperature (~2,750°C), which largely compensates for its poor thermal conductivity. The neutron spectrum of an oxide core is softer than that of a metal-fueled breeder due to moderation by the oxygen; this softer spectrum allows a strong negative Doppler coefficient. Oxide fuel leads to lower power density compared to metal fuels, and this places less stringent requirements on the cooling system (indeed, allowing gas cooling to become a possibility). Apart from its low thermal conductivity, the principal disadvantage of oxide fuel is its somewhat lower breeding ratio (due to the softer neutron spectrum and lower density).

Metal (U-Pu), carbide (UC-PuC), and nitride (UN-PuN) fuels are alternatives to oxide fuel, primarily because of their higher thermal conductivity. Their higher linear powers lead to a lower specific fissile inventory. For either UC or UN, the presence of only one moderator atom per heavy metal atom instead of two for oxide, plus a higher metal density due to the more compact lattice structure, leads to a harder spectrum and, therefore, to higher breeding ratios. Higher breeding gain and lower specific fissile inventory both result in lower doubling times. However, the database for these advanced fuels (particularly in the case of nitrides) is in development and is addressed in Chapter 11.

The early U.S. and U.K. experimental fast reactors utilized metal fuel. Metal fuel has a distinct advantage over oxide and a slight advantage over carbide or nitride from the standpoint of breeding ratio. The neutron spectrum with metal fuel is hard since no moderation from oxygen, carbon, or nitrogen is present, and this leads to high breeding ratios. Also, the fissile concentration is high, which leads to a good doubling time. Furthermore, the thermal conductivity of metal fuel is high and this leads to a high linear power.

Although the melting temperature of metal fuel is low, its high thermal conductivity offsets this to provide sufficiently high linear power. A disadvantage of metal fuel may be the difficulty in achieving high burnup. Metal fuel swells even at low burnups, and the only way to accommodate such swelling is to build in considerable porosity. Although this tends to negate some of its principal advantages, recent favorable burnup experience with metal fuel allows it to remain a strong contender for SFR fuel. Another disadvantage is that the neutron spectrum is so hard that the Doppler effect (important for safety reasons) is small. A strong negative Doppler coefficient is not crucial for metal fuels, however, since metal fuel is generally believed to exhibit a more reliable thermal axial expansion coefficient than ceramic fuels. Also, it should be noted that a strong negative Doppler coefficient becomes a liability for loss of coolant accidents.

2.4.1.2 Fuel Fissile Fraction

The fissile fraction, or "enrichment," required to attain criticality in a fast reactor is considerably higher than that required for thermal reactors because of the low fission cross sections at high neutron energies. Plutonium fractions typical of commercial size SFR's are in the 12–30% range, depending on reactor size and configuration, with about 75% of the plutonium being fissile material (i.e., ^{239}Pu or ^{241}Pu). Hence, fissile fractions are in the range of 9–23%.

Figure 2.7 shows the fission cross sections of the principal fissile fuels as a function of energy. Note that σ_f for the fast reactor is approximately two barns, whereas the comparable cross section in a thermal reactor is of the order of 500 barns. The effect of this decrease in cross section at high neutron energies can be partially offset by fast fission in fertile fuel. (See Fig. 2.8.) It must also be recognized that all parasitic absorption cross sections are lower for a high energy spectrum. Nonetheless, the low fission cross section results in a fast reactor requirement of four to five times the fissile fraction normally found in a thermal reactor.

2.4.1.3 Fuel-Pin Diameter

In LWRs the fuel-pin diameter is controlled by heat-transfer considerations. The same is not the case for SFRs, and it is useful to understand this important difference. One of two heat transfer

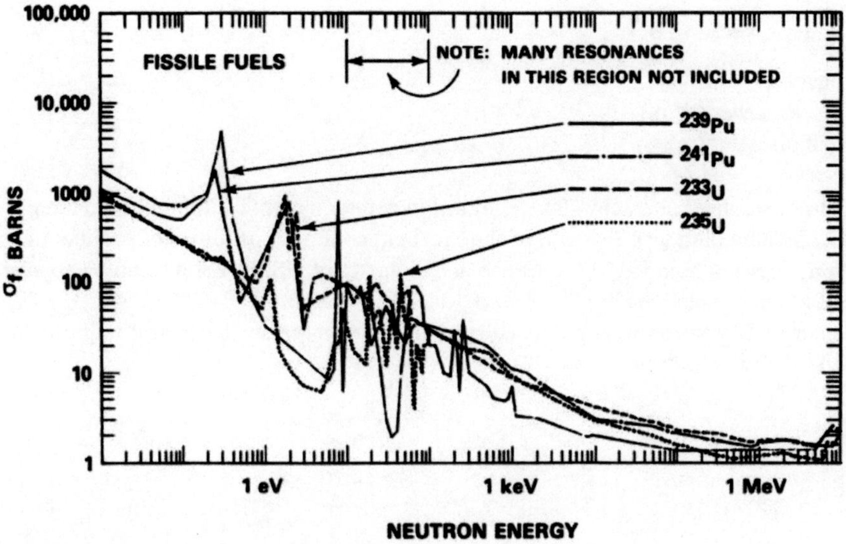

Fig. 2.7 Fission cross sections vs. energy for common fissile fuels (adapted from Ref. [5])

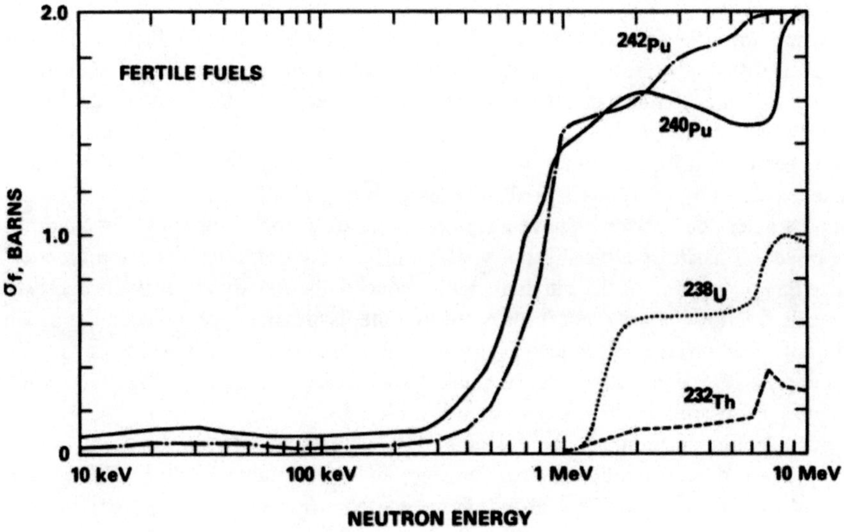

Fig. 2.8 Fission cross sections vs. energy for common fertile fuels (adapted from Ref. [5])

considerations controls fuel-pin design—either maximum allowable heat flux or maximum center fuel temperature, but only the first of these affects pin diameter.

The surface heat flux can be expressed as

$$q = \frac{\chi}{\pi D}, \tag{2.1}$$

where

q = heat flux (W/m^2)
χ = linear power (W/m)
D = pin diameter (m).

For a water-cooled thermal reactor, this relationship defines a practical minimum to the pin diameter (to prevent cladding burnout). Because of the excellent cooling capabilities of sodium, however, this burnout limit does not hold for SFRs. Hence, the diameter of a SFR fuel pin can be appreciably less than that of a water-cooled reactor fuel pin.

The center fuel temperature is governed by the *linear power*, or linear heat rating, which can be determined from the expression[1]:

$$\chi = 4\pi \int_{T_s}^{T_0} k\,(T)\; dT, \tag{2.2}$$

where

$k\,(T)$ = thermal conductivity of fuel (watts/m·°C)
T_s = fuel surface temperature (°C)
T_0 = fuel center temperature (°C).

This expression limits the linear power in a SFR since T_0 must be below the fuel melting temperature. Note, however, that this expression is independent of pin diameter; heat extracted from the pin is a function only of fuel center and surface temperatures and thermal conductivity of the fuel.

It is for these reasons that heat transfer in SFRs does not control fuel-pin diameter. It is also of interest to observe from Eq. (2.2) that for a given power rating, the total length of pins in the core is to a great extent fixed by the fuel selected and the maximum T_0 allowed.

The pin diameter does, however, have a significant influence on economics. For example, it has an important effect on fissile specific inventory, and a primary design objective for a breeder reactor is to minimize this parameter. As the pin diameter is reduced, the fuel mass per pin is decreased and the required fissile fraction is increased. It turns out that the fissile mass per pin (i.e., the product of fuel mass and fissile fraction) is also reduced as the pin diameter is reduced. Hence, from the point of view of minimizing fissile specific inventory, a small pin diameter is desirable. Breeding ratio, however, decreases with decreasing pin diameter (e.g., see Table 7.6).

Other considerations are involved in placing a lower limit on fuel pin diameter, such as fabrication costs, pitch-to-diameter ratio, neutron leakage, and fuel volume fraction. These considerations typically lead to SFR mixed-oxide fuel-pin diameters from 6 to 9 mm, or approximately two-thirds those of water cooled reactor fuel pins. (See Section 8.2.2 for a more detailed discussion of fuel-pin diameter.)

2.4.1.4 Fuel Burnup

Burnup requirements for an economical breeder reactor are considerably more demanding than for a LWR. This arises mainly from the difference in fissile enrichment involved. Whereas a burnup of 45–55 MWd/kg for a typical LWR assembly is acceptable, and burnups as low as 7 MWd/kg can

---

[1] This expression will be derived in Chapter 9.

be accommodated for a natural uranium reactor, burnups of the order of 100 MWd/kg or higher are required for economic operation of the SFR, resulting directly from the higher fissile fraction.

This high burnup requirement for an economical breeder is the principal reason for the world-wide shift in emphasis during the early 1960s from metal-fueled to oxide-fueled reactors. Relative to the LWR, this requirement translates into a need to get more energy out of each kilogram of heavy metal fabricated and loaded into a SFR before returning the fuel for reprocessing and refabrication.

2.4.2 Coolant

The primary requirements for a fast spectrum reactor coolant are as follows: (1) it must minimize neutron moderation, (2) it must remove heat adequately from a high-power-density system (approximately a factor of four higher than for a LWR), and (3) it must minimize parasitic neutron absorption. The first requirement automatically eliminates water and any organic coolant, but does allow gases or liquid metals of intermediate atomic number to be considered. Principal fast spectrum coolant candidates, including a summary of their primary advantages and disadvantages, are discussed below. More detailed discussions of coolants appear in Chapters 9 and 11.

2.4.2.1 Principal Coolant Candidates

Liquid metals were the first candidates investigated for cooling a fast spectrum reactor because they easily met the cooling requirements. However, helium and steam were also given considerable thought, primarily in connection with the switch from metal to ceramic fuels, in which accompanying lower power densities allow the gas coolant to fulfill the basic heat transfer requirements.

2.4.2.2 Property Considerations for Coolants

With regard to liquid metals, sodium, sodium/potassium eutectic (NaK), mercury, bismuth and lead have all been considered. The latter three were once generally discarded because their high densities (and, consequently, high mass flow rates) require excessive pumping power. However, lead and lead-bismuth eutectics have made a comeback with the recognition that more open lattice designs could largely alleviate the high pumping requirements (see Chapter 18). The sodium/potassium eutectic was used in early designs because of its low melting point. Sodium melts at 98°C, requiring heaters somewhere in the system, whereas NaK is a liquid at room temperature. Potassium, however, is a fairly strong neutron absorber. Given these considerations, sodium is the liquid metal most generally considered for cooling fast spectrum reactor plants.

Sodium has excellent thermal properties, has a high boiling point (880°C at one atmosphere pressure), and is quite compatible with the standard cladding material being employed. The high boiling point of sodium allows low pressure operation—in marked contrast to the pressurized systems required for LWR operation. The prime disadvantages of sodium are its neutronic activation properties (since ^{23}Na absorbs neutrons to form ^{24}Na, which decays with a 15-h half life, and ^{22}Na, with a 2.6 year-half life), its opaqueness, which somewhat complicates certain operating procedures such as refueling, and its chemical incompatibility with water and air. The 1.4- and 2.8-MeV gamma rays from the ^{24}Na decay process are sufficiently penetrating that an intermediate cooling loop is normally required to eliminate the possibility of radioactive contamination in the steam generator. This loop further protects the core from pressure surges or reactivity effects of hydrogen moderation should a steam generator leak result in a sodium-water reaction.

Helium gas can be used as a coolant for ceramic-fueled breeder reactors, with even less net degradation on the neutron spectrum than sodium. Although the atomic weight of He is lower than for Na, its atom density in gas form is much lower. Cladding compatibility with He is excellent since He is inert, and there is no neutron activation problem. Hence, the intermediate loop normally employed in a sodium cooled system would not be necessary. On the other hand, a He coolant system must be highly pressurized (~8 MPa, or 1,200 psi) in order to meet cooling requirements, and considerably more pumping power is required than for a sodium system.

Steam received substantial attention in early designs, but was ruled out because of excess cladding corrosion and problems related to high pressure operation. (Fuel pins required internal pressurization.) Steam cooling also causes appreciable neutron spectrum degradation, which results in low breeding ratio and fuel cycle cost penalties.[2]

At present, essentially all countries actively pursuing the fast spectrum reactor option have selected sodium as the reference coolant. Steam has been discarded, but He is still actively being pursued, primarily in the U.S. and in the U.K. (where considerable experience in gas cooling technology exists). Russia has developed a fast breeder design utilizing N_2O_4, as a gas coolant.

2.4.3 Structure

The prime requirements for the structural material used in the core of a fast spectrum reactor are (1) maintenance of integrity at high temperatures, (2) capability to withstand high fluence,[3] (3) compatibility with fuel and coolant, and (4) low parasitic neutron absorption. As in the case of coolant candidates, a summary of the major considerations is given below, with details left for coverage in Chapter 11.

2.4.3.1 Principal Structure Candidates

Stainless steel is one of the materials considered for fast reactor cladding and assembly duct material, although considerable work has focused on specific variations in composition and heat treating to best match the particular structural function with material requirements. Zircaloy, the standard cladding material for LWR application, is not suitable because its strength is not compatible with the high-temperature fields in a fast spectrum reactor.

2.4.3.2 Property Considerations for Structural Materials

Stainless steel has good strength and corrosion-resistant properties at high temperature and fluence. It also absorbs few neutrons. Most stainless steels swell significantly under the high fluences encountered in SFR operation, however, and a substantial developmental program has been directed toward solving this problem. Stainless steel 316, 20% cold worked, is the material most commonly used for cladding and assembly ducts because of its good high temperature properties and its resistance to swelling, but ferritic materials such as HT9 are becoming more widely used—along with other advanced materials as discussed in Chapter 11.

---

[2]Despite these early concerns, substantial progress has been made in Japan is assessing fast reactors cooled with supercritical-pressure water. Confer the book Super Light Water Reactors and Super Fast Reactors, by Yoshiaki Oka, Seiichi Koshizuka, Yuki Ishiwatari, and Akifumi Yamaji, Springer Books, 2010.

[3]Fluence is time integrated flux (πt or nvt), which represents a measure of the neutron bombardment that must be withstood by the structural material.

2.5 Selection of Structural Materials for Out-of-Core Components

Major factors to be considered for the selection of structural material for fast reactor components (other than core) are (a) operating conditions and plant life, (b) structural integrity, (c) ease of fabrication, (d) availability & cost and (e) international experience. The mechanical properties such as tensile strength, creep behavior, low/high cycle fatigue strength, creep–fatigue interaction, and elastic stability are important considerations in the choice of structural materials.

2.5.1 Principal Candidates for Structural Materials

Austenitic stainless steel is the major structural material used for the main vessel, inner vessel, safety vessel, core support structure, grid plate and other reactor internals. In the case of the Top Shield/Roof Slab, carbon steel with controlled impurities and specified properties is generally used as structural material.

2.5.2 Property Considerations for Structural Materials

Austenitic stainless steel is used as the major structural material in view of adequate high temperature mechanical properties, compatibility with liquid sodium coolant, and good irradiation resistance. SS 316 LN & 304 LN are the most commonly used austenitic stainless steel for this purpose. Carbon steel used for the Roof Slab/Top Shield should have good mechanical strength in the operating temperature range, good weldability and high ductility in both longitudinal and short transverse thickness directions.

2.6 Generation IV Sodium-Cooled Fast Reactors for Waste Management Scenarios

Several Generation IV sodium-cooled fast reactors are targeted for nuclear waste management applications.[4] The goal is the effective management of high-level wastes—specifically, plutonium and minor actinides (neptunium, americium, and curium). With capital cost reductions achieved from design innovations, the reactors can also be used for electricity production.

Several design configurations are being considered for the Generation IV sodium-cooled fast reactor. They range from modular systems of a few hundred MWe to large power units exceeding 1,500 MWe. Sodium core-outlet temperatures are expected to be around 550°C. Table 2.3 provides the system reference values for SFR concepts [3].

[4]The Generation IV International Forum (GIF) and the Generation IV fast spectrum reactor systems are introduced in Section 1.2.

Table 2.3 Reference ranges of the Generation IV SFR design parameters

System parameters	Reference values
Inlet temperature (°C)	366–395
Outlet temperature (°C)	510–550
Pressure (atmospheres)	~1
Power rating (MWe)	150–1,500
Fuel	Oxide or metal alloy: U-TRU-10%Zr Alloy MOX(TRU bearing)
Cladding	Ferritic (HT9) or ODS ferritic
Average burnup (MWd/KgHM)	~70–200
Conversion ratio	0.5–1.3
Average power density (MWth/m^3)	~350
Plant efficiency (%)	38–42
Fuel reload batch, batches	3–4
Cycle length, months	18

References

1. K. Wirtz, *Lectures on Fast Reactors,* Kernforschungszentrum, Karlsruhe (1973).
2. *Use of Fast Reactors for Actinide Transmutation*, IAEA-TECDOC-693, IAEA, Vienna (1993).
3. T. A. Taiwo and R. N. Hill, *Summary of Generation-IV Transmutation Impacts*, ANL-AFCI-150, Argonne National Laboratory, 9700 South Cass Avenue, Argonne, IL, USA (2005).
4. E. Paxon, Editor, *Radial Parfait Core Design Study,* WARD-353, Westinghouse Advanced Reactors Division, Waltz Mill, PA (June 1977).
5. D. I. Garber and R. R. Kinsey, *Neutron Cross Sections, Volume II, Curves*, BNL 325, Third Edition, Brookhaven National Laboratory, New York (January 1976).

Chapter 3
Economic Analysis of Fast Spectrum Reactors

Chaim Braun

3.1 Introduction

Economics always plays a pivotal role in the decision process associated with the construction of any project. Accordingly, we shall endeavor in this chapter to outline the key factors that influence the cost of building a nuclear power plant—with special emphasis regarding fast spectrum reactors.

It is very difficult at the present time, however, to provide firm cost estimates for fast spectrum systems because very few reactors of this type have been built in recent years and none of these could be considered fully commercial. Hence, we shall focus our discussion on comparative evaluations, i.e. citing the differences between fast spectrum systems and the current generation of light water reactors (LWRs). Specific mathematical techniques for determining actual costs, once key input numbers are known, are contained in Appendix D. Specific techniques for defining such items as the time value of money, levelized costs, etc. are detailed in this Appendix.

The basic expectation of fast spectrum reactor economics is that the lifetime–levelized fast spectrum reactor total generation costs should be similar to the lifetime–levelized total generation costs of the competing LWRs. Total generation costs of a power plant are the sum of: (1) annual levelized fixed charges on the initial investment in the plant, (2) annual operating and maintenance (O&M) charges, and (3) levelized fuel costs. For the comparative evaluations undertaken in this chapter, we shall refer to an LWR operated on a closed fuel cycle rather than the current once-through cycle. If used (spent) LWR fuel is not being reprocessed and the plutonium contained therein is not extracted, than the plutonium fuel inventory for new fast spectrum reactor introduction cannot be obtained, and the fast spectrum reactor fuel cycle cannot be established. Fast spectrum reactors can be started with enriched uranium fuel, as was done in Russia; however, used fast spectrum reactor fuel must still be reprocessed and recycled to maintain the integrity of the closed fast spectrum reactor fuel cycle. Long-term operation of fast spectrum reactors on a once-through enriched uranium cycle is not sustainable.

In theory, should the total generation costs of both the LWR (with fuel recycling) and the fast spectrum reactor prove to be equal, then electric utilities would be indifferent to the choice of an LWR or a fast spectrum reactor as their next base load plant. Under such conditions, the long-term benefits of fast spectrum reactor introduction in terms of reduced reliance on natural uranium supply availability and concerns with price increases or price volatility, as well as further concerns with the availability and cost of uranium enrichment services, might swing utility commitment decisions in favor of the fast spectrum reactor. The socioeconomic value associated with transmutation of transuranic isotopes and reducing high level waste volumes that can be achieved with fast spectrum reactor systems can add additional incentive to the fast spectrum reactor introduction.

C. Braun (✉)
Center for International Security and Cooperation (CISAC), Stanford University, Stanford, CA, USA
e-mail: chbraun@att.net

A.E. Waltar et al. (eds.), *Fast Spectrum Reactors*, DOI 10.1007/978-1-4419-9572-8_3,

While this basic principle represents the theoretical approach to fast spectrum reactor economics analysis, it is currently difficult to produce reliable cost figures for either LWRs (with fuel recycle) or fast spectrum reactors that will make such computations feasible. First, with regards to LWRs: LWR initial capital investment costs have risen considerably, particularly in western countries over the last decade or so. This is accounted for by the rising commodities prices, by the lack of sufficient cadres of trained engineers or craft-workers such as nuclear pipe welders, by incomplete plant designs and under-developed equipment supply chains, and by uncertainties in the plant licensing process, which might make nuclear plant construction a high-risk venture requiring risk premium on capital expenditures. The lack of sufficient heavy components manufacturing infrastructure has increased production costs in the few facilities capable of producing such components worldwide. In fact, due to the above factors, LWR plant cost estimates have more than doubled during the first decade of the current century. The lack of a supporting infrastructure of second and third tier suppliers is not something to be easily dismissed and, in fact, may be the major reason that in the resurgence of nuclear power, the first few plants will see higher costs until this infrastructure is reestablished. Such factors will also play out during the eventual transition to a fast spectrum reactor based future nuclear energy system.

Offsetting these effects is the recent interest in Small Modular Reactors (SMRs). At present there are prospective Nuclear Steam Supply System (NSSS) suppliers for integrated Pressurized Water Reactor (iPWR) versions, High Temperature Gas Reactor (HTGR) versions, and liquid metal-cooled reactor versions. These smaller-sized reactors, with their lower capital costs, offer an opportunity to break the cost impasse. But calculating the production cost of electricity from these reactors is a complex computational process and requires the use of the equations shown in Appendix D. The reason for the complexity is that while the initial investment or capital cost and interest during construction should be lower, the output or electricity produced by the SMR plant is considerably less. Unless carefully accomplished with a serious concern for unit production costs, the time-tested paradigm of economies of scale will be encountered in an adverse manner. The economy of scale law states that a plant twice as large does not cost twice as much and so there is an incentive to design and construct a reactor plant as large as possible. Nonetheless, in those market areas where investment costs are a major concern and market prices for electricity are high, the SMR may have an attractive role if costs are carefully assessed using the principles outlined in this chapter. If current trends of relatively low natural gas prices of about 4.0 \$/MMBTU[1] persist in the U.S., it is not clear whether a market for SMRs or even the larger and more capital-intensive LWRs will emerge in the U.S. Better prospects emerge in high gas prices regions.

We should note that for SMRs the investment (capital cost) should be lower because the basic construction commodities such as steel and concrete are less for a smaller reactor. The total interest paid during construction should be lower because the construction time should be shorter. In addition, it may be possible to rely upon factory fabrication for major systems in contrast to field construction. The field work in this case would be largely a major system assembly process, rather than a field construction process, and so the reliance on expensive field construction labor would be less.

In general, fast spectrum reactors are more amenable to larger plant sizes than the SMR plants size range. The need for a closed fuel cycle including fuel reprocessing and recycling facilities will favor larger sized fast spectrum reactors so that the fuel recycling expenses are distributed over the larger amount of MWh generated, thus reducing the unit fuel cycle costs. This trend is evident in India, where a 500 MWe fast spectrum reactor was developed first to be followed by a commercial 1,000 MWe sized standard fast spectrum reactor. A similar tend is evident in China, where there is interest in commercializing the Russian 800 MWe BN800 reactor design.

[1] One MMBTU is the equal to one million British thermal units. The unit of \$/MMBTU is commonly used in the energy industry, and is the trading unit for New York Mercantile Exchange (NYMEX) for natural gas futures.

Nuclear fuel costs based on the closed fuel cycle are difficult to obtain on a reliable commercial basis worldwide. This is because there are relatively few commercial reprocessing and plutonium fabrication plants and those that are operating are not necessarily fully commercial (rather, they are joint government-commercial enterprises). Thus, a fully developed competitive commercial fuel recycling industry does not yet exist. This is also true for LWR fuel recycling facilities. Fast spectrum reactor-specific commercial-sized fuel recycling plants have not yet been designed or built, due to the uncertainties in the evolution of the fast spectrum reactor systems, and the still evolving fast spectrum reactor fuel recycling technology.

The current costs for reprocessing and plutonium mixed oxide fabrication most likely reflect this unique monopolistic situation of the moment rather than a mature, competitive, market situation. Thus, it is difficult to project fuel cycle costs for the closed (recycling fuel) cycle. And the situation is not much better in regards to reliably estimating the total generation costs of the fast spectrum reactor as there are currently no new fully commercial fast spectrum reactor construction projects that could provide mature capital cost estimates.

The situation is about the same with regards to fast spectrum reactor fuel cycle costs. Reprocessing plants for fast reactor fuel will most likely rely on wet chemistry processes such as the PUREX[2] process. But pyrochemical processes, which are thought to be simpler, could be more economically attractive when fully commercialized if the balance-of-plant design requirements will not substantially increase the total plant costs.

In terms of fast spectrum reactor annual operating and maintenance costs, we have even less reliable operational data due to the lack of commercial experience. On the one hand, the higher number of components in the fast spectrum reactor plant design (as compared with the LWR) might require more maintenance and equipment replacement costs. On the other hand, the fast spectrum reactor operates at much reduced pressures as compared with the LWR, thus allowing ease of operation and avoiding the need for high pressure resistant components. Furthermore, the fast spectrum reactor operates at a higher temperature range and a higher thermal efficiency than then the LWR, thus allowing for commercial turbines use, avoiding the need for special, lower efficiency nuclear turbines. The combined effects of both the positive and negative factors might eventually result in a similar range of O&M costs for the LWR and the fast spectrum reactor, though this assertion will need to be tested in the future. The higher thermal efficiency of the fast spectrum reactor will, however, provide a salutary effect for the entire fast spectrum reactor levelized generation costs, since all costs expressed on a per KWh basis will be reduced as the thermal efficiency increases and more useful electrical KWh are extracted from a similar thermal output of the reactor's core.

Consequently, it is now difficult to conduct a reliable cost estimates comparison of LWR and fast spectrum reactor total (levelized) generation costs. This situation should change within the next decade or two as more LWRs are built worldwide, and as some fast spectrum reactors come on line. In the meantime we must confine our discussion here to qualitative issues only. With this in mind, we can now proceed to discuss the basic principles of fast spectrum reactor economics.

3.2 Basic Assumption of Fast Spectrum Reactor Economics

The basic assumption related to the comparative economics of the fast spectrum reactor and the LWR is that fast spectrum reactor capital costs will be somewhat higher than LWR capital costs, due to factors listed below. This assertion will certainly be true in the early days of worldwide fast spectrum

[2]The PUREX acronym stands for Plutonium—URanium Extraction. This is the reference aqueous nuclear reprocessing method for the recovery of uranium and plutonium from used nuclear fuel.

reactor introduction. As more fast spectrum reactors are constructed and construction experience accumulates, the capital cost differential will be reduced and ultimately, possibly eliminated. Before we reach that state, however, we must contend with higher fast spectrum reactor capital costs that should be more than compensated for by the expectation that fast spectrum reactor fuel cycle costs will ultimately be lower than LWR once-through and also closed fuel cycle costs. This will eventually be the case since the fast spectrum reactor does not depend on fresh uranium resources (whose prices are volatile and will increase in time) or on uranium enrichment services. Thus, the lower fast spectrum reactor fuel cycle costs would compensate for the higher fast spectrum reactor capital cost (relative to the LWR), resulting in cost parity with the LWR and eventually with a fast spectrum reactor generation cost advantage.

One can, as has been done in the past, compute the equivalent break even fast spectrum reactor capital cost increment over the LWR, measured in $/KWe units,[3] that will exactly compensate the increase in uranium fuel price, measured in units of $/kg (or $/lb) U_3O_8. The reasons for and the validity of the expectations of higher fast spectrum reactor capital costs, and lower fast spectrum reactor fuel cycle costs, in comparison with the LWR, are discussed next.

3.3 Fast Spectrum Reactor Capital Costs

It has always been assumed that fast spectrum reactor plant capital costs should be somewhat higher than LWR capital costs. The cost increment over the LWR has generally been assumed to range between 10 and 25% for mature plants. In some cases it has been assumed that with a transition to a mature and optimized fast spectrum reactor design, the expected capital cost increment over the LWR might eventually disappear. Why then do we start with an assumption of higher fast spectrum reactor capital costs?

The basic reason for this is the fact that the fast spectrum reactor, due to its very high core power density, has to be cooled with a high thermal conductivity fluid such as sodium.[4] In order to provide a separation between the highly radio-activated sodium coming out of the sodium-cooled fast reactor (SFR) core and the water-based steam cycle, the SFR operates on three heat transfer cycles as described elsewhere. The first cycle delivers sodium from the reactor core to the intermediate heat exchanger (IHX) and back to the core. The non-radioactive sodium coming out of the IHX is transported through the intermediate heat transfer loop to the steam generator, where it exchanges heat with the water in the tertiary (steam cycle) loop.[5] Thus, unlike the LWR, which operates on one (in boiling water reactors) or two heat transport loops (primary water from the core to the steam generator and back and secondary water from the steam generator to the turbine and back), the SFR requires three heat transport loops. The extra heat transport loop, including the IHX and its associated piping and associated increase in vessel sizes and building volume, add to the capital cost of the SFR, in comparison with the LWR.

The SFR reactor vessel system is more complicated than the LWR pressure vessel. In the LWR design a single pressure vessel, though very thick to withstand the high pressures of the Pressurized Water Reactor (PWR) cycle and prevent boiling of the primary cycle water, is required. We should

[3] Currency in this chapter is always based on US dollars ($) and cents (¢).

[4] As we note in Chapter 17, it is possible to cool a fast spectrum reactor with a gas such as helium, but the power density must necessarily be lower, resulting in a significantly larger core.

[5] As mentioned in Chapter 13, some designers are proposing to use super-critical CO_2 in a Brayton Cycle to replace the intermediate loop and directly drive the turbines to produce electricity. Such a concept would reduce the cost and complexity of the SFR design, but is yet unproven.

mention here that for LWRs designed for passive safety enhancements, a large body of water in the reactor vessel is required to provide the first level of core cooling in case of a major pipe break. For these situations, a very long pressure vessel is required, with the core located at the bottom of the vessel and covered by a large volume of water. In the SFR design, on the other hand, even though the system works at atmospheric (or near-atmospheric) pressures, a two or three concentric reactor vessels system might be needed, in order to provide emergency cooling, safety barrier, and primary system barriers, as described elsewhere in this book. These three vessels (even though not as thick as the single LWR pressure vessel) add to the cost of the plant and the complexity of its construction. Additionally the top plate covering all three concentric SFR core vessels represents a very complicated equipment item with multiple openings for IHXs (assuming this is a pool-type system), electromagnetic pumps (another equipment item unique to the SFR), rotating plugs for fuel charge and discharge, and control rod penetrations. Such a unique design heavy equipment item is most likely fabricated on-site, thus further adding to its highly specialized cost.

The basic conclusion here is that the SFR requires more and unique equipment items than does the LWR, due to the requirements of operating with two different heat transfer fluids—sodium and water—contained within three separate heat transport loops. There also exist other specialized systems for pumping, storing, cleaning, heating and looking through the sodium coolant. Due to the special design features, the large equipment items are manufactured or at least assembled in an on-site factory whose cost adds to the total cost of the SFR plant. On the other hand, the LWRs that operate at very high pressures and require thick and expensive pressure vessels and primary piping, which are manufactured only in a small number of special facilities worldwide. Furthermore, the SFR operates at higher temperatures than the LWR and, thus, could utilize standard fossil fired steam turbines, without special requirements for steam re-heating in between the high- and low-pressure turbine stages, or for moisture separators and reheaters. The beneficial effect of the higher thermal efficiency in computing capital costs in units of $/KWe has been mentioned above. The LWRs, which operate at lower temperatures than the SFRs, also require special design (larger sized) nuclear-type steam turbines. Thus, while the SFR has inherently more unique equipment items in the nuclear island, resulting in higher equipment costs in comparison with the LWR, it still enjoys some relative advantages that have to do with lower primary system pressure, higher steam cycle temperatures, and simplified steam cycle design.

In time, as more SFRs are installed and the various equipment items become more standardized and mass-produced, the capital cost disadvantage of the greater equipment items content will shrink and be compensated by the lower pressure higher steam temperature advantages. Until that time arrives, we must assume that the high component prices effect will predominate, and that SFR capital cost will remain higher than the LWR capital cost.

3.4 SFR Fuel Cycle Costs

As has been mentioned before, neither fuel cycle costs of the LWR operating on a recycling mode nor the fast spectrum reactor fuel cycle costs are known with any degree of certainty. Accordingly, we are again restricted to limiting our discussion to qualitative assertions only.

In the LWR fuel cycle we know quite well the quantitative cost components of the front end of the fuel cycle, including uranium supply, conversion, enrichment and fabrication. We do not reliably know the cost component for fuel reprocessing and recycling. This is because only two large-sized commercial reprocessing plant now operate worldwide—the UP2 plant (dedicated to French reactors) and the UP3 plant (dedicated to other reactors), both located in the La Hague site in Northwestern France. If we assume the range of reprocessing costs at UP3 as 500–1,000 $/kg, than at an average

Table 3.1 Cost of enriched UO_2 LWR fuel

Service	Amount	Unit cost	Fuel cost component $/kg
Uranium	9 kg U_3O_8	110 $/kg (50 $/lb)	990
Conversion	7.6 kg U	13 $/kg	100
Enrichment	7 SWU	150 $/SWU	1,050
Fabrication	1 kg	300 $/kg	300
Total			2,440

fuel burnup of 50,000 MWthd/MTU, this translates into a fuel cycle component cost range of 1.2–2.5 Mills/KWh (0.12–0.25 ¢/KWh). Other reprocessing charges will scale the reprocessing contribution to fuel cycle cost accordingly. Used fuel management charges are covered in the U.S. by the mandatory 1.0 Mills/KWh collected by the government and by used fuel storage charges for an intermediate storage period 20–50 years, which are estimated at 1.0–2.0 Mills/KWh (0.1–0.2 ¢/KWh). A sample components cost distribution for one kilogram of enriched UO_2 fuel destined for use in a 1,000 MWe class LWR (assuming 4% enrichment and a 0.25% tails assay at the enrichment plant), is shown in Table 3.1.

Considering the total fuel cost component from Table 3.1, a core average discharge burn-up of 45,000 MWth·d/MTU, and a thermal efficiency of 33.3%, this translates to a fuel cost (FC) of about 0.68 ¢/KWh generated, or 6.8 $/MWe·h. If we assume a closed fuel cycle we have to add 2.2–3.5 Mills/KWh (0.22–0.35 ¢/KWh) for the back end of the fuel cycle costs, not considering credits for uranium or plutonium recycling. Such credits related to reduced natural uranium supply and reduced enrichment requirements could reduce the uranium and enrichment components of the closed lWR fuel cycle by about one third, compared with the costs of such fuel components in the LWR once-through fuel cycle. Thus in the recycle case the front end of the cycle costs ate lower, but the back end of the fuel cycle costs are higher, all compared to the once-through fuel cycle. The fuel cycle price is derived as follows:

$$FC = \frac{FCC \times 1{,}000 \text{ kg/MT}}{BU \times \varepsilon \times 24 \text{ hr/d}}, \tag{3.1}$$

where:

FC = Fuel cycle cost ($/MWe·h)
FCC = Fuel cost component ($/kg),
BU = fuel burnup (MWth·d/MTU),
ε = thermal efficiency (MWe/MWth).

In the interests of simplicity, the calculation above does not consider time sequences of expenditures or of electricity generation and is used only to define fuel costs on a per unit generation basis. More complete fuel cycle costs computational techniques, including the use of discounting and the time value of money, are shown in Appendix D.

As an illustrative example, the impact of varying the uranium price, keeping all other cost components constant, is shown in Fig. 3.1, assuming no alteration to the tails assay at the enrichment plant.

Doubling the uranium price from 55 to 110 $/kg (25–50 $/lb) U_3O_8 increases the fuel cost from 0.54 up to 0.68 ¢/MWh, an increase of 26%, and the cost of generation of the best nuclear plants from 1.3 to 1.44 ¢/KWh (an increase of 10%). In other words, an increase of about 55 $/kg (25 $/lb) U_3O_8 in the price of natural uranium will result in a 0.15 ¢/KWh increase in fuel cycle costs. Substantial and sustained uranium price rises to above 220 $/kg (100 $/lb) U_3O_8 will be required to support higher fast

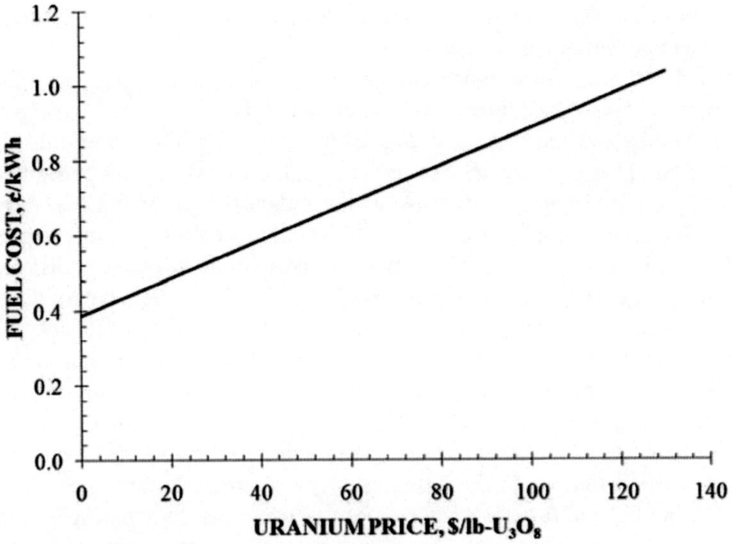

Fig. 3.1 Sensitivity of front end fuel cost to the uranium price

spectrum reactor back-end of the fuel cycle costs (minus the important credit for savings in uranium and enrichment charges due to fuel recycling) as well as the fast spectrum reactor/LWR capital cost differential.

As has been mentioned above, we are not sure yet of the price of fast spectrum reactor fuel reprocessing and recycling since mature commercial industries providing such services do not yet exist. We can only note that the fast spectrum reactor does not require fresh uranium supplies (even for the blankets) and neither does it require uranium enrichment services. These two cost components, which comprise more than 75% of the LWR fuel cycle cost, as seen in Table 3.1, are completely missing in the fast spectrum reactor fuel cycle. This is the basis for our above assertion that the fast spectrum reactor fuel cycle costs should be lower than the LWR fuel cycle costs despite the expected higher reprocessing and re-fabrication costs in the fast spectrum reactor fuel cycle as compared with the LWR recycling fuel cycle.

Qualitatively, assuming reprocessing of fast spectrum reactor fuel to be accomplished using a wet chemistry processes such as variants of the PUREX or UREX[6] families of processes, one would expect somewhat higher costs compared with LWR fuel reprocessing. This is due to the fact that fast spectrum reactor fuel will contain higher plutonium concentrations, thus requiring smaller process vessel dimensions to reduce the chance of criticality accidents, and also requiring more detailed, stringent and costly materials protection control and accounting (MPC&A) practices. It is quite possible that the transition to electro-refining and pyro-processing processes will reduce reprocessing and waste management charges, as claimed by the technology developers. We do not, however, have detailed commercial-sized plant flow-sheets to make accurate cost estimates, nor do we have real-world large-scale experience in the operation of this promising technology.

Fast spectrum reactor fuel re-fabrication will pose its own unique set of technical challenges, since the more radioactive recycle plutonium (as compared with natural uranium) will not permit direct-contact fabrication work, and will require carrying out the re-fabrication steps in special large

---

[6]The URanium Extraction (UREX) process is a modified version of PUREX in which a clean stream of uranium is extracted from used fuel, leaving a stream of plutonium mixed with other transuranic isotopes available for use as fuel in breeder or burner reactors.

glove boxes or hot cells. Thus, remote re-fabrication of fast spectrum reactor fuels will result in somewhat higher component fuel cycle cost.

To these we will eventually have to add the cost of the fissile plutonium supply. Currently, when the market for plutonium is rather limited, plutonium could be considered a free good provided at no cost so it could be burned and destroyed. One of the features of a fast spectrum system is that it can be designed to be either a net supplier of fissile fuel (a breeder) or a net consumer of plutonium (a burner). Once a breeder economy develops, and a market for plutonium fuel develops between LWRs providing plutonium converted from ^{238}U and new fast spectrum reactors requiring plutonium stockpile for their initial inventories, market prices for plutonium will be established. (It should be noted, parenthetically, that different isotopic content plutonium fuels might have a different monetary value; however, this is beyond the scope of this discussion). In general an LWR operating on a closed fuel cycle, recycling uranium (thus saving some uranium and enrichment requirements), will obtain a credit from the sale of plutonium. The higher the plutonium sale price, the lower the LWR fuel cycle cost. On the other hand, a fast spectrum reactor requiring external plutonium source for initial inventory and for first few annual fuel reloads (when averaged or levelized over lifetime) will see its fuel prices increase in relation to the changing value of the plutonium fuel. By equating LWR fuel cycle cost (selling Pu) and fast spectrum reactor fuel cycle cost (purchasing Pu) an equivalent or market-clearing price for plutonium could be established, as has been proposed by Eugene Escbach more than 50 years ago. Plutonium prices then were considered in the range of 10–25 $/g. It would be more difficult to estimate this value today, but it will eventually impact the commercial fast spectrum reactor fuel cost at some future-established equivalence value.

Summarizing this discussion, we can say that the combination of a lower LWR capital cost and higher LWR front-end of the fuel cycle cost (minus credit for uranium recycling) will likely provide the LWR with a cost advantage relative to the higher fast spectrum reactor fuel recycling costs and higher capital costs, at least in the early days of fast spectrum reactor market penetration. As noted above, the fast spectrum reactor recycling costs might be higher than the back-end cost of the LWR fuel cycle. However, the total fast spectrum reactor fuel cost should be lower than the total LWR fuel cost, due to the lack of uranium supply and enrichment services cost components. The exact numerical values of the relevant parameters cannot be well estimated today since all the related technologies have not yet been developed or commercialized on an industrial scale. The basic tenet of fast spectrum reactor economics—the equivalence of the total generation costs of the fast spectrum reactor and the LWR—is thus justified by the different values of the cost components involved (Capital charges, O&M costs, Fuel expenses) of the two different reactor systems. The variation in the values of the different components expected over time, e.g. fast spectrum reactor capital costs decline, LWR fuel costs increase, etc., will ensure maintaining such a dynamic total costs equivalence.

We should also note that there has been a recent renewed interest in the possibility of uranium extraction from sea water. Should such technical development prove real and be commercialized (yet a great unknown), a maximum cost for uranium could be established. This is because while extraction of uranium from seawater will most likely be the most expensive process, the supply will be essentially infinite. Currently the cost as demonstrated by small scale engineering demonstrations is on the order of 1,000 $/kg (450 $/lb) of uranium, a maximum value that is largely meaningless.

As indicated in the text above, while the economic assessment principles are known, their precise quantitative application is a matter for the future. Nonetheless, it is imperative to both the supplier and the purchaser of a new nuclear power plant that these economic assessments be both accurate and reliable. As nuclear power makes it resurgence and the unit costs become better known, the mathematical methods contained in Appendix D will allow an accurate and reliable assessment of the economics of a reactor plant upon which both the buyer and the seller can agree.

Part II
Neutronics

The neutron physics aspects of the fast spectrum reactor deserve special attention because it is the hard neutron spectrum required for efficient breeding and transmutation that provides a principal distinguishing feature from the more common thermal spectrum used with most power reactors. Chapter 4 addresses nuclear design techniques, where multigroup diffusion theory is emphasized. The methodology for providing the multigroup cross sections is then included in Chapter 5. Kinetics features unique to the fast reactor are included in Chapter 6, and fuel management concepts—which are needed for the determination of breeding ratios, doubling times, and transmutation—are covered in Chapter 7.

Whereas many computational tools exist for performing stand alone analyses of each of the areas addressed by Chapters 4 through 7, our primary goal in these chapters is to provide sufficient information to enable the reader to gain a firm basic understanding of the key neutron physics concepts associated with fast spectrum reactors. However, the challenge for designers is to address each of these analyses in an integrated, self-consistent framework. This requires the use of complex tools with a very high degree of sophistication that is beyond the interest of the general student. Considering this, and the fact that such tools are constantly being adapted and improved, a summary of modern neutron physics tools for fast reactor simulations is provided in Appendix E. The purpose of this appendix is to allow advanced readers direct access to integrated computation systems currently available for performing integrated neutronics calculations.

Chapter 4
Nuclear Design

Pavel Tsvetkov, Alan Waltar, and Donald Todd

4.1 Introduction

The nuclear design of a reactor plant involves defining the nuclear environment that will exist inside the reactor core. This phase of design work has a great bearing on thermal-hydraulic and mechanical analyses; hence, close coordination with these activities must be maintained. Safety and control requirements are closely tied to the nuclear design effort and must be considered for all phases of the fuel cycle. Power distributions are needed in order to determine peak-to-average power factors for thermal-hydraulics analysis. Other neutronics calculations are needed to obtain the following kinds of information: required fissile fraction and inventories, fuel cycle data, shielding data, and transient response.

The topics considered in this chapter include multigroup diffusion theory, geometrical considerations for obtaining input to the diffusion equation, and spatial power distributions and neutron flux spectra encountered in typical fast reactor designs. Two sets of multigroup cross sections are supplied in Appendix F as reasonably good representations of effective cross sections for the typical SFR designs specified in the problems.

4.2 Multigroup Diffusion Theory

The principal nuclear design tools available to the fast reactor designer include diffusion theory, transport theory, and Monte Carlo techniques. Diffusion theory has been found to be adequate for most fast reactor design calculations. The primary reason for the validity of this relatively simple technique is that the mean free path of fast neutrons is generally long relative to the dimensions of the fuel pins and coolant channels. Hence, there is little localized spatial flux depression or flux peaking over most of the energy range. On the other hand, the core size is large relative to the mean free path.[1] Transport corrections are often made for specialized problems, such as certain plate geometries in critical-facility analyses, but the general diffusion theory approach can still be successfully employed. The emphasis in this chapter will be concerned with diffusion theory.

A detailed treatment of the neutron energy spectrum is clearly very important in fast reactor neutronics analysis because of the wide range of neutron energies over which most capture and fission

[1]Transport theory may be required for the analyses of small metal fast assemblies where the core dimensions are not large compared to the neutron mean free path.

P. Tsvetkov (✉)
Zachry Engineering Center, Department of Nuclear Engineering, Texas A&M University, College Station, TX, USA
e-mail: tsvetkov@tamu.edu

A.E. Waltar et al. (eds.), *Fast Spectrum Reactors*, DOI 10.1007/978-1-4419-9572-8_4,

reactions occur in fast reactors as compared to thermal spectrum reactors. Multigroup analysis must begin, therefore, with many groups in the resonance and high energy regions. The starting point for the many groups is a complete data file of cross sections for fast reactor materials, which is almost continuous in energy. One such file is ENDF/B-VII [1], the Evaluated Nuclear Data File, which is commonly used in the U.S.

Multigroup cross sections are obtained from discrete group cross-section data by the use of conventional theories of resonance absorption and scattering. These theories are similar whether one is analyzing a fast reactor or a thermal reactor, although in some cases emphasis is placed on different methods due to differences in energy spectra of importance. Procedures peculiar to fast reactor design are now in common use, however, and these will be described in Chapter 5.

Spatial distributions of neutron flux and power density are as important in fast reactor analysis as in thermal reactor analysis. Fortunately, spatial analysis is easier for fast reactors as a result of the long neutron mean free path. A further consequence of the long mean free path is that a fast power reactor is more tightly coupled neutronically than a large thermal reactor. Hence, the fundamental mode of the neutron flux solution (e.g. the first eigenvalue solution) is more dominant, the neutron leakage from the core is higher, and spatial convergence of the diffusion equations is easier to achieve. Despite the general need for detailed energy analysis for fast reactors, many special space-dependent problems requiring two- and three-dimensional analyses can be solved adequately with the use of relatively few energy groups—even as few as four in some cases.

4.2.1 The Multigroup Equation

For the reasons noted above, multigroup diffusion theory occupies a central role among the computational techniques employed for fast spectrum reactor neutronics analyses. Because of this widespread applicability, it is important that the physical basis for each of the terms appearing in the multigroup equation be clearly understood.

The neutron diffusion equation is simply a neutron balance equation, with losses equal to sources. The equation is discussed here for steady state; a transient term will be added in the discussion of reactor kinetics in Chapter 6. Any imbalance between losses and sources is accounted for by the effective multiplication factor k_{eff}, an eigenvalue for this problem.

Since the multigroup equation is a neutron balance equation, the fluxes calculated from it are relative rather than absolute values; the fluxes are known only to within an arbitrary constant. The absolute value of the neutron flux can be calculated from the reactor power level. (See Section 4.7.2.)

Suppose the energy spectrum is divided into G energy groups. Let us consider the gth group. The *loss terms* for group g, in terms of neutrons per unit volume per second, include leakage ($-D_g\nabla^2\phi_g$), absorption ($\sum_{ag}\phi_g$ – which includes capture and fission since $\sum_{ag} = \sum_{cg} + \sum_{fg}$), removal by elastic scattering ($\sum_{erg}\phi_g$), and removal by inelastic scattering ($\sum_{irg}\phi_g$). Elastic or inelastic scattering within the group does not appear in the version of the multigroup equation presented in this book; hence, the total inelastic cross section is different from the value of $\sum_{irg}$, used here. (Some versions of multigroup theory do include in-group scattering; the only requirement is that internal consistency be maintained between the scattering cross section and the scattering matrix.)

The *source terms* include neutrons produced by fission and neutrons scattered into group g by elastic and inelastic scattering at energies above group g. The rate of production of neutrons by fission (per cm^3 per second) is $\sum_{g'=1}^{G}(\nu\sum_f)_{g'}\phi_{g'}$. The summation is taken over all G energy groups since fission occurs at all energies; the subscript g' is used since groups other than group g are included in this summation. The fraction of those neutrons produced in fission that appear in group g (or are "born into" group g) is χ_g. Hence, the fission source for group g is $\chi_g\sum_{g'=1}^{G}(\nu\sum_f)_{g'}\phi_{g'}$. In the steady-state multigroup balance equation, the balance between production and loss is achieved by multiplying this source by $1/k_{eff}$ so that the effective production rate for group g from fission is $(1/k_{eff})\chi_g \sum_{g'=1}^{G}(\nu\sum_f)_{g'}\phi_{g'}$. Hence, if the reactor is supercritical $(k_{eff} > 1)$, so that the fission source is greater than the sum of the loss terms, multiplying the source by $1/k_{eff}$ reduces the production rate so that a neutron balance is achieved.

The elastic and inelastic scattering source terms are $\sum_{g'=1}^{g-1}\sum_{eg'\to g}\phi_{g'}$ and $\sum_{g'=1}^{g-1}\sum_{ig'\to g}\phi_{g'}$ where $\sum_{g'\to g}$ represents scattering from g' into g. (In this book group 1 will denote the fastest energy group and group g the slowest; hence, the scattering source terms imply that only down-scattering from group g' into group g is allowed. Up-scattering is used in thermal reactor analysis in the thermal energy range, but in fast reactors no neutrons survive the slowing-down process to thermal energies.)

The multigroup equation can now be expressed as the following balance:

$$\begin{aligned}&\left[-D_g\nabla^2\phi_g\right]+\left[\left(\Sigma_{ag}+\Sigma_{erg}+\Sigma_{irg}\right)\phi_g\right]=\\&\text{LEAKAGE}\qquad\text{REMOVAL}\\&\frac{1}{k_{eff}}\chi_g\sum_{g'=1}^{G}\left(\nu\Sigma_f\right)_{g'}\phi_{g'}+\sum_{g'=1}^{g-1}\Sigma_{eg'\to g}\phi_g+\sum_{g'=1}^{g-1}\Sigma_{ig'\to g}\phi'_g.\\&\text{FISSION SOURCE}\qquad\text{SCATTERING IN}\end{aligned} \tag{4.1}$$

4.3 Spatial Solutions to the Multigroup Equation

One-dimensional, two-dimensional, and three-dimensional solutions to the multigroup diffusion equation are all regularly used in fast reactor design, always with appropriate compromises between the numbers of groups and dimensions and the economics of computer solutions for the problem of interest. Zero-dimensional (or fundamental mode) solutions were even historically used for special purposes, as for example in cross section and fuel cycle calculations.

In this book, our purpose is to develop an understanding of the most important features of the energy spectrum and spatial distributions in fast reactors and not to devote excessive attention to multi-spatial solution techniques. Hence, a one-dimensional computer formulation of the diffusion equation will be sufficient for our purposes. Even more useful as a tool for understanding the effects of energy spectrum in a fast reactor is a zero-dimensional solution, which the student can program on a computer in a reasonably short time. One two-dimensional topic will be briefly discussed, a two-dimensional triangular mesh, because this mesh is widely used for analysis of the hexagonal fast reactor geometry and is not used for light-water-reactor analysis.

First, we will discuss the approximation to be used for leakage in the directions perpendicular to the direction of the solution. This provides the basis for introducing the zero-dimensional solution. Next, a one-dimensional solution will be presented, and the section is then concluded with a discussion of the two-dimensional triangular mesh.

4.3.1 Transverse Leakage Approximation

In one-dimensional diffusion theory, a technique for treating neutron leakage in the transverse directions is needed since spatial effects in these directions are not treated rigorously. In zero dimensions this technique must be applied for all directions. The method is not different from techniques used for thermal reactors.

To illustrate the technique, consider a single energy diffusion equation for a critical reactor:

$$-D\nabla^2\phi + \left(\Sigma_a - \nu\Sigma_f\right)\phi = 0. \tag{4.2}$$

In terms of geometric buckling B^2, the equation is

$$\nabla^2\phi + B^2\phi = 0. \tag{4.3}$$

We will consider cylindrical geometry since fast spectrum reactors (like thermal power reactors) are normally calculated as right circular cylinders. Hence, Eq. (4.3) becomes

$$\frac{\partial^2\phi}{\partial r^2} + \frac{1}{r}\frac{\partial\phi}{\partial r} + \frac{\partial^2\phi}{\partial z^2} + B^2\phi = 0. \tag{4.4}$$

By separation of variables,

$$\phi = R(r)\,Z(z)\,, \tag{4.5}$$

where R is a function of r only and Z is a function of z only. Dividing Eq. (4.4) by Eq. (4.5) gives

$$\frac{1}{R}\left(\frac{d^2R}{dr^2} + \frac{1}{r}\frac{dR}{dr}\right) + \frac{1}{Z}\frac{d^2Z}{dz^2} = -B^2. \tag{4.6}$$

Since the sum of the two terms on the left-hand side is a constant, and the first term is a function of r only and the second term is a function of z only, and each term can be varied independently of the other, it follows that each term on the left-hand side is a constant. Hence,

$$\frac{1}{R}\left(\frac{d^2R}{dr^2} + \frac{1}{r}\frac{dR}{dr}\right) = -B_r^2 \tag{4.7}$$

$$\frac{1}{Z}\frac{d^2Z}{dz^2} = -B_z^2, \tag{4.8}$$

where

$$B_r^2 + B_z^2 = B^2; \tag{4.9}$$

and it follows that

$$\frac{\partial^2\phi}{\partial r^2} + \frac{1}{r}\frac{\partial\phi}{\partial r} = -B_r^2\phi, \tag{4.10}$$

and

$$\frac{\partial^2\phi}{\partial z^2} = -B_z^2\phi. \tag{4.11}$$

The simplest boundary conditions are (1) symmetry, and (2) disappearance of the flux at some extrapolated distance, δ, beyond the boundary; or

$$(1)\quad \frac{dR}{dr} = 0 \text{ at } r = 0, \text{ and } + \frac{dZ}{dz} = 0 \text{ at z} = 0. \tag{4.12}$$

$$(2)\quad R = 0 \text{ at } r = R_0 + \delta_r, \text{ and Z} = 0 \text{ at z} = \frac{H}{2} + \delta_z. \tag{4.13}$$

The solutions to Eqs. (4.7) and (4.9) for these boundary conditions are a Bessel function and a cosine, respectively. Hence the solution to Eq. (4.2), to within an unspecified constant A, is

$$\phi = AJ_0\left(B_r r\right)\cos\left(B_z z\right), \tag{4.14}$$

where

$$B_r = \frac{2.405}{R_0 + \delta_r} \text{ and } B_z = \frac{\pi}{H + 2\delta_z}. \tag{4.15}$$

Now we can introduce the required approximations for transverse leakage. Equation (4.2) can be written

$$-D\left(\frac{\partial^2\phi}{\partial r^2} + \frac{1}{r}\frac{\partial\phi}{\partial r} + \frac{\partial^2\phi}{\partial z^2}\right) + \left(\Sigma_a - \nu\Sigma_f\right) = 0. \tag{4.16}$$

The terms with the r derivatives represent radial leakage; the $-D\partial^2\varphi/\partial z^2$ term represents axial leakage.

For a one-dimensional solution in the r direction, the axial leakage is replaced by Eq. (4.11), or

$$-D\left(\frac{d^2\phi}{dr^2} + \frac{1}{r}\frac{d\phi}{dr}\right) + DB_z^2\phi + \left(\Sigma_a - \nu\Sigma_f\right)\phi = 0. \tag{4.17}$$

In the z direction the radial leakage is replaced by Eq. (4.10) so that

$$-D\frac{d^2\phi}{dz^2} + DB_r^2\phi + \left(\Sigma_a - \nu\Sigma_f\right)\phi = 0. \tag{4.18}$$

For a zero-dimensional solution, both r and z-derivatives are replaced by Eqs. (4.10) and (4.11), so that

$$DB_r^2\phi + DB_z^2\phi + \left(\Sigma_a - \nu\Sigma_f\right)\phi = 0. \tag{4.19}$$

In order to use these approximations for leakage, it is necessary to evaluate δ_r, and δ_z , which appear in the second boundary condition, Eq. (4.13). These parameters are similar to a reflector savings which, for a large reflector, is given approximately by $D_c M_r/D_r$, where D and M are the diffusion coefficient and migration length, respectively and subscripts c and r refer to the core and reflector. In a fast spectrum reactor the core is surrounded by a blanket for which the diffusion coefficient is about the same as that of the core. Hence the δ's would be about equal to M. The migration area, M^2, is approximately D/Σ_a. For a neutron energy of 100 keV, which is the order of magnitude for the median fission energy in a fast reactor, the diffusion coefficient in the blanket is about 1.3 cm and the absorption cross section is of the order of 0.003 cm^{-1}. This gives a value for M and for δ of the order of 20 cm.

The effective value of δ will vary with energy and with the buildup of fissile material in the blanket. Hence, appropriate values for δ for fast reactors must be determined by experience that can only be obtained by comparing actual multigroup spatial leakage calculations with the approximations represented by $DB^2\phi$ discussed here. This does, in fact, result in values for δ of about 15–20 cm over a relatively wide energy range, from several 10's of kilo electron-volts to 1 MeV. Values of δ can be estimated from the flux shapes shown later in this chapter, in Section 4.7.4.

4.3.2 Zero-Dimensional Solution

The zero-dimensional, or fundamental mode, solution is easier than the one-dimensional solution because of the simple treatment of leakage. This simplifying approximation results in a non-iterative solution for the effective criticality factor, k_{eff}, a situation not possible for the one-dimensional solution.

Replacing the leakage term, $-D\nabla^2\phi_{g'}$ in Eq. (4.1) with $D_g B_g^2 \phi_g$ as discussed in the previous part of this section, the multigroup equations can be written as

$$D_g B_g^2 \phi_g + \Sigma_{rg}\phi_g = \frac{1}{k_{eff}}\chi_g \sum_{g'=1}^{G} \left(\nu\Sigma_f\right)_{g'} \phi_{g'} + \sum_{g'=1}^{g-1} \Sigma_{eg'\to g}\phi_{g'} + \sum_{g'=1}^{g-1} \Sigma_{ig'\to g}\phi_{g'}, \tag{4.20}$$

where

$$B_g^2 = B_{rg}^2 + B_{zg}^2, \tag{4.21}$$

and

$$\Sigma_{rg} = \Sigma_{ag} + \Sigma_{erg} + \Sigma_{irg}, \tag{4.22}$$

To solve for the fluxes, the first step is to specify that

$$\frac{1}{k_{eff}} \sum_{g'=1}^{G} \left(\nu\Sigma_f\right)_{g'} \phi_{g'} = 1. \tag{4.23}$$

[Recall from an earlier discussion that the (ϕ_g)'s are relative values, a fact that is quite apparent from Eq. (4.23).]

As a result of Eq. (4.23), the equation for group 1 is

$$D_1 B_1^2 \phi_1 + \Sigma_{r1}\phi_1 = \frac{1}{k_{eff}}\chi_1 \sum_{g'=1}^{G} \left(\nu\Sigma_f\right)_{g'} \phi_{g'} = \chi_1. \tag{4.24}$$

The only unknown in this equation is ϕ_1; hence it can be evaluated immediately.

The equation for group 2 is

$$D_2 B_2^2 \phi_2 + \Sigma_{r2}\phi_2 = \chi_2 + \Sigma_{e1\to 2}\phi_1 + \Sigma_{i1\to 2}\phi_1. \tag{4.25}$$

Since ϕ_1, has by now been evaluated, we can solve for ϕ_2. We proceed in a similar manner until all (ϕ_g)'s are known.

Knowing all of the group fluxes, we can use Eq. (4.23) to solve for k_{eff}, or

$$k_{eff} = \sum_{g'=1}^{G} (\nu\Sigma_f)_{g'}\phi_{g'}. \tag{4.26}$$

Note that the solution is not iterative because this first value of k_{eff} is consistent with our original fission source defined in Eq. (4.23).

4.3.3 One-Dimensional Solution

4.3.3.1 Cylindrical Geometry (Radial Direction)

We present here the simple one-dimensional solution of the multigroup diffusion equation to illustrate some of the characteristic features. An early example of a computer code solving these equations is 1DX [2]. The complex three-dimensional solutions have been implemented in codes for whole-core reactor calculations. As a notable example, DIF3D [3] should be mentioned. In DIF3D, The neutron diffusion equation is solved using a nodal scheme with one mesh cell (node) per hexagonal assembly. Cartesian geometry node sizes are specified by the user. The nodal equations are derived using higher order polynomial approximations to the spatial dependence of the flux within the hexagonal or Cartesian node. The final equations, which are cast in response matrix form, involve spatial moments of the node-interior flux distribution plus surface-averaged partial currents across the faces of the node. These equations are solved using a fission source iteration with coarse-mesh rebalance acceleration. Equivalence Theory parameters are permitted with hexagonal nodal models [4]. Additional information regarding DIF3D is included in Appendix E.

In our simple one-dimensional example, the reactor is divided into N mesh intervals, as shown in Fig. 4.1. Note that for this formulation mesh points are not located on the boundaries. Equally valid methods are available with mesh points on the boundaries and at the center. The equation is solved numerically by approximating the axial leakage by the fundamental mode solution for $\nabla^2\phi\,(z)$ (i.e. $-B_z^2\phi$) and integrating over each mesh volume. Boundary conditions are needed at the center of the core and at the outside of the reactor.

In the following analysis, the superscript k is the mesh index, while the subscript g continues to be the energy group index.

Integration over a mesh volume around point k gives

$$\begin{aligned} &-\int_k D_g^k \nabla^2 \phi_g^k dV + \int_k D_g^k B_{zg}^2 \phi_g^k dV + \int_k \Sigma_{rg}^k \phi_g^k dV \\ &= \frac{1}{k_{eff}} \int_k \chi_g \sum_{g'=1}^{G} (\nu\Sigma_f)_{g'}^k \phi_{g'}^k dV + \int_k \sum_{g'=1}^{g-1} \Sigma_{g'\to g}^k \phi_{g'}^k dV, \end{aligned} \tag{4.27}$$

where Σ_{rg}^k denotes all of the removal terms listed on the left-hand side of Eq. (4.1) and $\Sigma_{g'\to g}^k$ denotes the sum of both elastic and inelastic scattering from group g' to group g.

Using the divergence theorem,

$$\int D\nabla^2\phi dV = \int D\nabla\phi \cdot \overrightarrow{dA},$$

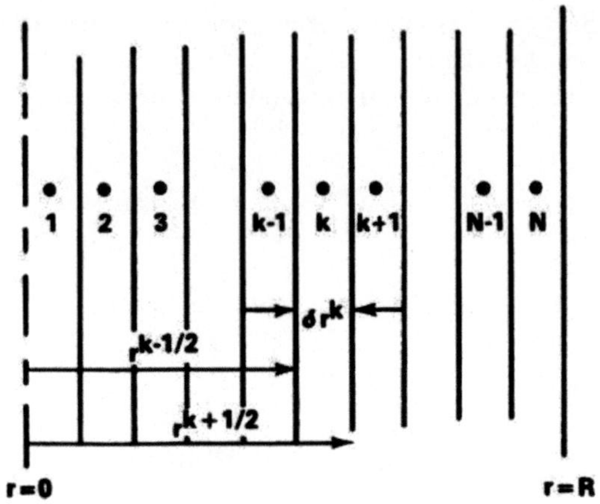

Fig. 4.1 Mesh geometry for one-dimensional radial solution

and evaluating the derivatives halfway between the mesh points gives

$$-\int_k D_g^k \nabla \phi_g^k \cdot \overrightarrow{dA} = D_g^{k-1,k} A^{k-1,k} \frac{\phi_g^k - \phi_g^{k-1}}{r^k - r^{k-1}} - D_g^{k,k+1} A^{k,k+1} \frac{\phi_g^{k+1} - \phi_g^k}{r^{k+1} - r^k}. \tag{4.28}$$

The minus sign disappears in front of the first term on the right-hand side because $\nabla\phi \cdot \overrightarrow{dA}$ is negative (whereas $\nabla\phi \cdot \overrightarrow{dA}$ is positive in the second term). The difference $\phi_g^k - \phi_g^{k-1}$ in the first term is generally negative, in which case the first term would represent a source, or leakage, into the mesh interval at *k*. The second term would then be positive and would represent a loss, or leakage, out of the interval.

The coefficient $D_g^{k-1,k}$ is an effective diffusion coefficient between mesh points *k*–1 and *k;* it is based on the volume-averaged macroscopic transport cross section Σ_{trg} between mesh points. This volume-averaged value for Σ_{trg} is

$$\Sigma_{trg}^{k-1,k} = \Sigma_{trg}^{k-1} \frac{\delta r^{k-1}}{\delta r^{k-1} + \delta r^k} + \Sigma_{trg}^k \frac{\delta r^k}{\delta r^{k-1} + \delta r^k}.$$

Since $D = 1/3\Sigma_{tr}$,

$$\frac{1}{D_g^{k-1,k}} = \frac{1}{D_g^{k-1}} \frac{\delta r^{k-1}}{\delta r^{k-1} + \delta r^k} + \frac{1}{D_g^k} \frac{\delta r^k}{\delta r^{k-1} + \delta r^k}$$

or

$$D_g^{k-1,k} = \frac{D_g^{k-1} D_g^k \left(\delta r^{k-1} + \delta r^k\right)}{D_g^{k-1} \delta r^k + D_g^k \delta r^{k-1}}. \tag{4.29}$$

Values for the area and mesh volume are

$$A^{k-1,k} = 2\pi r^{k-1/2} H, \tag{4.30}$$

$$V^k = \pi \left[\left(r^{k+1/2}\right)^2 - \left(r^{k-1/2}\right)^2 \right] H = \pi v^k H, \tag{4.31}$$

where H is the axial dimension, and v^k is defined by Eq. (4.31) and is therefore a volume per unit height, divided by π.

Next we define a source S_g^k as

$$S_g^k = \frac{1}{k_{eff}} \chi_g \sum_{g'=1}^{G} (\nu \Sigma_f)_{g'}^k \phi_{g'}^k + \sum_{g'=1}^{g-1} \Sigma_{g' \to g}^k \phi_{g'}^k. \tag{4.32}$$

Assuming that all cross sections and χ's are constant over the mesh volume, substituting Eqs. (4.28) through (4.32) into (4.27) gives

$$D_g^{k-1,k} 2r^{k-1/2} \frac{\phi_g^k - \phi_g^{k-1}}{r^k - r^{k-1}} - D_g^{k,k+1} 2r^{k+1/2} \frac{\phi_g^{k+1} - \phi_g^k}{r^{k+1} - r^k}$$
$$+ D_g^k B_{zg}^2 \phi_g^k v_k + \Sigma_{rg}^k \phi_g^k v^k = S_g^k v^k.$$

Rearranging gives

$$\begin{aligned} &-D_g^{k-1,k} \frac{2r^{k-1/2}}{r^k - r^{k-1}} \phi_g^{k-1} + \left(D_g^{k-1,k} \frac{2r^{k-1/2}}{r^k - r^{k-1}} + D_g^{k,k+1} \frac{2r^{k+1/2}}{r^{k+1} - r^k} \right. \\ &\left. + D_g^k B_{zg}^2 v_k + \Sigma_{rg}^k v^k \right) \phi_g^k - D_g^{k,k+1} \frac{2r^{k+1/2}}{r^{k+1} - r^k} \phi_g^{k+1} = S_g^k v^k. \end{aligned} \tag{4.33}$$

We next define α and β such that

$$\alpha_g^k = D_g^{k-1,k} \frac{2r^{k-1/2}}{r^k - r^{k-1}}, \tag{4.34}$$

$$\beta_g^k = \alpha_g^k + \alpha_g^{k+1} + D_g^k B_{zg}^2 v_k + \Sigma_{rg}^k v^k, \tag{4.35}$$

so that Eq. (4.33) becomes

$$-\alpha_g^k \phi_g^{k-1} + \beta_g^k \phi_g^k - \alpha_g^{k+1} \phi_g^{k+1} = S_g^k v^k. \tag{4.36}$$

We next consider the boundary conditions for the two most usual cases:

(1) $\frac{d\phi}{dr} = 0$ at $r = 0$
(2) $\phi = 0$ at $r = R + 0.71\lambda_{tr}$

where R is the outermost boundary of the reactor and λ_{tr} is the transport mean free path.

For boundary condition (1), one can imagine a mesh interval ($k = 0$) to the left of center and equal in width to mesh interval 1. The boundary condition is satisfied by setting $\phi^0 = \phi^1$. The first term and the first part of the second term of Eq. (4.33) cancel, so that Eq. (4.33) becomes

$$\left(D_g^{1,2} \frac{2r^{3/2}}{r^2 - r^1} + D_g^1 B_{zg}^2 v^1 + \Sigma_{rg}^1 v^1 \right) \phi_g^1 - D_g^{1,2} \frac{2r^{3/2}}{r^2 - r^1} \phi_g^2 = S_g^1 v^1,$$

or $\alpha_g^1 = 0$, and

$$\beta_g^1 \phi_g^1 - \alpha_g^2 \phi_g^2 = S_g^1 v^1. \tag{4.37}$$

For boundary condition (2), one can imagine a mesh point, N+1, a distance $0.71\lambda_{tr}$, to the right of the outermost boundary, R, where $\phi^{N+1} = 0$.[2] The distance between the mesh point N and N+1 is $\frac{1}{2}\delta r^N + 0.71\lambda_{tr}$.

The value of α_g^{N+1}(for use in β^N) is

$$\alpha_g^{N+1} = D_g^N \frac{2R}{\frac{1}{2}\delta r^N + 0.71\lambda_{trg}},$$

where $R = r^N + \frac{1}{2}\delta r^N$. Since $\phi_g^{N+1} = 0$, Eq. (4.33) becomes

$$-\alpha_g^N \phi_g^{N-1} + \beta_g^N \phi_g^N = S_g^N v_g^N. \tag{4.38}$$

This system of multigroup equations can be written in matrix form as

$$A\phi = C, \tag{4.39}$$

where A is a square matrix with $N \times G$ rows and columns:

$$A = \begin{bmatrix} \beta_1^1 & -\alpha_1^2 & & & & & & & & & & \\ -\alpha_1^2 & \beta_1^2 & -\alpha_1^3 & & & & & & & & & \\ & & \ddots & & & & & & & & & \\ & & -\alpha_1^k & \beta_1^k & -\alpha_1^{k+1} & & & & & & & \\ & & & & \ddots & & & & & & 0 & \\ & & & & -\alpha_1^N & \beta_1^N & & & & & & \\ & & & & & & \beta_2^1 & -\alpha_2^2 & & & & \\ & & & & & & -\alpha_2^2 & \beta_2^2 & -\alpha_2^3 & & & \\ & & & & & & & & \ddots & & & \\ & 0 & & & & & & & -\alpha_g^k & \beta_g^k & -\alpha_g^{k+1} & \\ & & & & & & & & & & \ddots & \\ & & & & & & & & & -\alpha_G^{N-1} & \beta_G^{N-1} & -\alpha_G^N \\ & & & & & & & & & & \alpha_G^N & \beta_G^N \end{bmatrix}$$

and ϕ and C are the following vector matrices, each with $N \cdot G$ elements:

$$\phi = \left\{ \phi_1^1, \phi_1^2 \cdots \phi_1^N, \phi_2^1, \phi_2^2 \cdots \phi_g^k \cdots \phi_G^{N-1}, \phi_G^N \right\},$$

$$C = \left\{ S_1^1 v^1, S_1^2 v^2 \cdots S_1^N v^N, S_2^1 v^1, S_2^2 v^2 \cdots S_g^k v^k \cdots S_G^{N-1} v^{N-1}, S_G^N v^N \right\}.$$

This equation can be solved by inversion from the relation

$$\phi = A^{-1} C. \tag{4.40}$$

---

[2]Note that the extrapolation distance $0.71\lambda_{tr}$ is different from the fundamental mode distance δ at which the flux extrapolates to zero, as described in the first part of this section.

The method of solution is iterative. To describe the method, we will first define another term, the fission source, denoted by F, and given by

$$F = \sum_{k=1}^{N} \sum_{g'=1}^{G} F_{g'}^{k} = \sum_{k=1}^{N} \sum_{g'=1}^{G} (\nu \Sigma_f)_{g'}^{k} \phi_{g'}^{k} V^{k}. \tag{4.41}$$

In order to converge on the criticality factor, k_{eff}, it is necessary first to guess a flux distribution (ϕ_g^k) to be used in the expression for S_g^k and for F, and then to iterate until convergence is reached to within a specified criterion. A typical starting flux distribution is a fission spectrum, uniform in space.

New fluxes are obtained from Eq. (4.40), and a new fission source, F, is calculated. Each new calculation of fluxes and the fission source is an iteration.

To see how convergence is obtained, let $F^{\nu+1}$ be the fission source obtained from the fluxes calculated in the νth iteration. A multiplication factor for the next iteration, $\lambda^{\nu+1}$, is defined as the ratio of the fission source of the $(\nu+1)^{\text{th}}$ iteration to that of the previous iteration, or

$$\lambda^{\nu+1} = \frac{F^{\nu+1}}{F^{\nu}}. \tag{4.42}$$

The effective multiplication factor $k_{eff}^{\nu+1}$ for the next iteration is then obtained by multiplying the previous iteration value by the new λ, or

$$k_{eff}^{\nu+1} = k_{eff}^{\nu} \lambda^{\nu+1}. \tag{4.43}$$

This value of k_{eff}, together with the new values of the neutron fluxes, is then used to obtain the new source terms, $\left(S_g^k\right)^{\nu+1}$, which are in turn used for the next iteration.

As the solution converges, the fluxes and, hence, the fission source F converge toward the same values on successive iterations. Thus, λ tends toward unity. A convergence criterion ε is specified such that, when

$$|1 - \lambda| < \varepsilon, \tag{4.44}$$

the solution is said to have converged.

The actual convergence technique used in computer codes such as 1DX is more sophisticated than that presented above in that fission-source *over-relaxation* is employed to accelerate convergence. The procedure is as follows. After the new fission source is calculated (call it $F_1^{\nu+1}$), a second "new" value, $F_2^{\nu+1}$, is computed by magnifying the difference between the new and the old fission sources by a factor of β, the over-relaxation factor, as follows:

$$F_2^{\nu+1} = F^{\nu} + \beta\left(F_1^{\nu+1} - F^{\nu}\right). \tag{4.45}$$

$F_2^{\nu+1}$ is then normalized to give the same total source as $F_1^{\nu+1}$. The original value for β used in 1DX was 1.4.

4.3.3.2 Slab Geometry

Slab geometry would be used for a one-dimensional calculation in the axial direction. The only differences between slab and cylindrical geometry arise in the equations for $A^{k-1,k}$ and V^k, Eqs. (4.30) and (4.31), and in replacing B_{zg}^2 with B_{rg}^2. Assuming the axial solution is desired for a cylindrical reactor, the equations for $A^{k-1,k}$ and V^k are

$$A^{k-1,k} = \pi R^2, \tag{4.30a}$$

$$V^k = \pi R^2 \delta z^k. \tag{4.31a}$$

Replacing δr^k by δz^k in the definition of $D^{k-1,k}$, Eq. (4.33) becomes

$$\begin{aligned} &-\frac{D_g^{k-1,k}}{z^k - z^{k-1}}\phi_g^{k-1} + \left(\frac{D_g^{k-1,k}}{z^k - z^{k-1}} + \frac{D_g^{k,k+1}}{z^{k+1} - z^k} + D_g B_{rg}^2 \delta z^k + \Sigma_{rg}^k \delta z^k\right)\phi_g^k \\ &-\frac{D_g^{k,k+1}}{z^{k+1} - z^k}\phi_g^{k+1} = S_g^k \delta z^k. \end{aligned} \tag{4.33a}$$

In slab geometry,

$$\alpha_g^k = \frac{D_g^{k-1,k}}{z^k - z^{k-1}}, \tag{4.34a}$$

$$\beta_g^k = \alpha_g^k + \alpha_g^{k+1} + D_g^k B_{rg}^2 \delta z^k + \Sigma_{rg}^k \delta z_k, \tag{4.35a}$$

$$-\alpha_g^k \phi_g^{k-1} + \beta_g^k \phi_g^k - \alpha_g^{k+1} \phi_g^{k+1} = S_g^k \delta z^k, \tag{4.36a}$$

$$\beta_g^1 \phi_g^1 - \alpha_g^2 \phi_g^2 = S_g^1 \delta z^1, \tag{4.37a}$$

$$\alpha_g^{N+1} = \frac{D_g^N}{\frac{1}{2}\delta z^M + 0.71\lambda_{trg}},$$

$$-\alpha_g^N \phi_g^{N-1} + \beta_g^N \phi_g^N = S_g^N \delta z^N. \tag{4.38a}$$

4.3.4 Two-Dimensional Triangular Mesh

A geometry often used for two-dimensional analysis in fast reactors but not encountered in light-water-reactor analysis is the triangular mesh. This mesh arrangement is useful for fast reactor hexagonal assemblies.[3] It is used in place of the conventional *x–y* or *r–θ* mesh.

In hexagonal geometry, a symmetric fast reactor can often be conveniently represented by one-sixth of the core and the blanket. Radial power distributions are frequently reported in this form, i.e., for each hexagonal assembly in a sextant of the core and blanket. Examples of this representation are given in Section 4.7.

An example of a 3×4 triangular mesh representing two hexagons is shown in Fig. 4.2. A mesh point of coordinates (*i, j*) is located at the center of each triangle. The mesh boundaries x_i, and y_j, are fixed by the flat-to-flat width, *W*, of the hexagons. The mesh spacings Δx and Δy are constant throughout the mesh.

[3]This mesh is also used for analysis of the hexagonal assemblies in High Temperature Gas Reactors (HTGR).

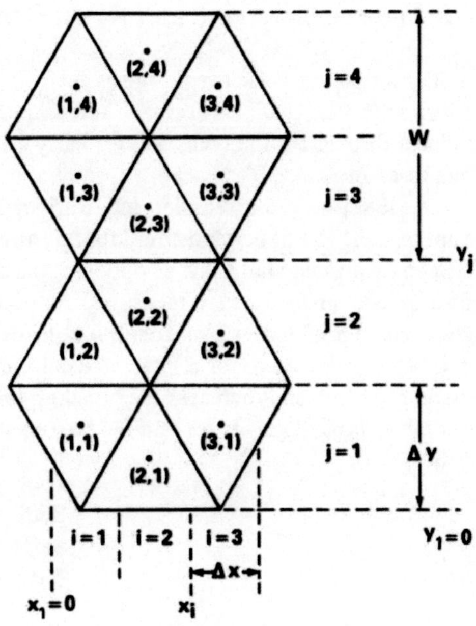

Fig. 4.2 Triangular mesh, showing mesh points (i, j)

The mesh height, Δy, is simply $W/2$, from Fig. 4.2. The mesh width, Δx must be defined to conserve the area of the triangular mesh,

$$(\Delta x)\,(\Delta y) = \frac{1}{2}\cdot\frac{W}{2}\cdot\frac{W}{\sqrt{3}}, \tag{4.46}$$

so that

$$\Delta x = \frac{W}{2\sqrt{3}}. \tag{4.47}$$

Setting x_1, and y_1, equal to zero, the mesh boundaries are

$$x_i = (i-1)\,\frac{W}{2\sqrt{3}}, \tag{4.48}$$

$$y_j = (j-1)\,\frac{W}{2\sqrt{3}}. \tag{4.49}$$

4.4 Treatment of Localized Absorbers

Usually absorber rods are made of enriched boron in the form of a single rod or cluster of rods. Estimating the spatial self shielding within the absorber rod is important for finding the power in the fuel assemblies surrounding the absorber rod as well as in the absorber rod itself. Collision probabilities methods are well suited to handle this problem.

4.5 Volume Fractions and Atom Densities

A starting point in nuclear design is the calculation of volume fractions, *F*, and atom densities, *N*. These must be calculated for each region in the core and blankets. Control assemblies will be distributed throughout the core, and for early scoping calculations these can be homogenized with the fuel assemblies.

The fuel pins are assembled into a triangular array in hexagonal fuel assemblies in a fast spectrum reactor. The basic motivation for this arrangement is neutronics performance. The "tight" lattice offered by a triangular array, as opposed to a square lattice, allows a larger value for the fuel volume fraction and smaller core dimensions. As pointed out in Chapter 2, decreasing the core size with a given amount of fertile plus fissile fuel leads to a lower fissile fraction and lower fissile inventory in a fast reactor. Also, for a fixed size core, increasing the fuel volume fraction lowers the required fissile inventory and increases the breeding ratio; hence, increasing fuel volume fraction is generally desirable. Typical ranges for volume fractions for materials in SFR cores are

Fuel	30–45%
Sodium	35–45%
Steel	15–20%
B_4C	1–2%

The basic repeating cell for calculating volume fractions is the assembly. Volume fractions are calculated for fuel assemblies, radial blanket assemblies, and control assemblies. Fuel in the axial blankets is contained in the same pins as the fuel in the core; hence, volume fractions in axial blankets are the same as for the core. Core-wide average volume fractions can be approximated by weighting the volume fractions for fuel and control assemblies by the numbers of each type of assembly present. More detailed calculations can account for the spatial distribution of control assemblies by homogenizing an appropriate number of fuel assemblies with the control assemblies at a particular radial position to form an annular core region. More expensive but more precise calculations involving all three dimensions could be employed to allow individual modeling of each assembly in the reactor. Such a model would not require homogenization of adjacent assemblies.

Volume fractions are calculated from the geometrical cross-sectional dimensions, since the axial dimension of all materials is the same. A cross section of a fuel assembly was illustrated in Fig. 2.3. The repeating lattice is illustrated in the one-twelfth section shown in Fig. 4.3. As noted from this figure, the outer boundary of the repeating lattice is in the sodium halfway between adjacent assemblies. The fuel is assumed to occupy all the space inside the cladding. Structural material is

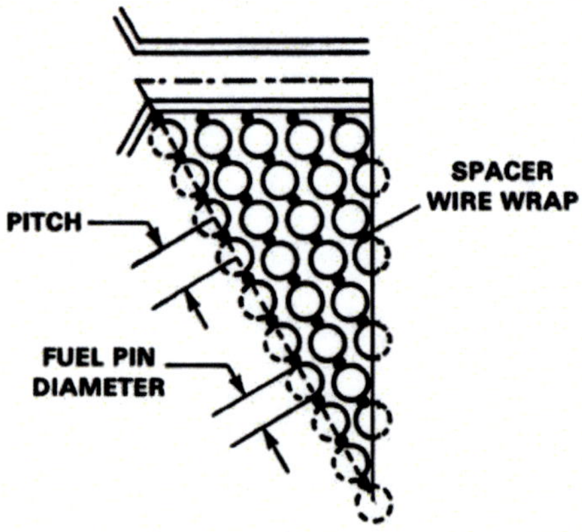

Fig. 4.3 Repeating lattice

present in the cladding and hexcan walls and in the spacers between fuel pins, either in the form of wire wrap or grid spacers. Often all the structural material is the same, generally some selected type of stainless steel. The sodium fills all of the remaining spaces inside the hexcan and is also present between the hexcans of adjacent assemblies. Useful geometric relationships for the flow areas in SFR fuel assemblies are given by Eq. (9.31).

When translating dimensions to volume fractions, a question arises regarding whether the dimensions are for the assemblies as-fabricated (hence, at room temperature) or at full power in the reactor. Here careful communication between the neutronics designer, the mechanical designer, and the thermal-hydraulics analyst is important. The neutronics designer wants to calculate neutronics parameters (hence, atom densities) for full-power conditions. Consider the fuel pin diameter, for example, for Type 316 stainless steel. The coefficient of thermal expansion $\left[\alpha = (1/L)\,(dL/dT)\right]$ between 20°C and 500°C (a typical cladding temperature at full power) is about 2×10^{-5}/K. The fractional change in the inside and outside diameter of the cladding between the pin as-fabricated and the pin at full power is about 1%. The cross sectional area of the structure (hence, the structure volume fraction) and the area inside the cladding (which controls the fuel volume fraction) would increase by $2\alpha\Delta T$, or 2% between 20°C and 500°C.[4]

A parameter called *smear density* is used in calculating atom densities for the fuel. The smear density is the density of the fuel if it were uniformly spread or smeared throughout the inside of the cladding. During fabrication of fuel pins, pellets of a given density and at room temperature are loaded into the cladding, with a gap between the pellet and the cladding. For calculation of atom densities, the neutronics designer does not need to know the details of the pellet design e.g., the exact diameter, whether the pellets are *dished* (i.e. have concave cavities ground into the faces between pellets), or how the pellets will crack and expand when the reactor is taken to power. The designer only needs to know the space available for the fuel and the mass of fuel in that space and must, therefore, know the inside diameter of the cladding and the smear density of the fuel at full power.

The fuel fabricator is concerned with another type of density called the *percent of theoretical density*. Mixed oxide fuel with 20–25% PuO_2 in UO_2, as in SFRs, has a theoretical (crystalline) density of about 11 g/cm^3 with some variation depending on Pu/U ratio and stoichiometry (i.e., the average number of oxygen atoms in each molecule, which can vary 1% or 2% around two oxygen atoms per heavy metal nuclei). Oxide fuel pellets are designed to achieve a specified percent of this theoretical density, usually around 95% or a little higher. The smear density generally lies in the range of 85–90% of theoretical density.

Atom densities for material *m* are calculated from the relation,

$$N_m = \frac{F_m \rho_m N_A}{M_m}, \tag{4.50}$$

where

N_m = atom density (atoms/cm^3)[5]
F_m = volume fraction
ρ_m = density (g/cm^3)
M_m = molecular weight (g/g-mol)
N_A = Avogadro's number (6.023×10^{23} atoms/g-mol).

[4]Consideration of as-fabricated (e.g. "cold dimension"), versus as-operated (e.g. "hot dimension") geometry is important in all aspects of fast reactor analysis. The actual size of the core and flow areas through the assemblies will change, affecting nuclear and thermal-hydraulic analyses if appropriate consideration is not made when preparing inputs to the codes based on manufacturing drawings and other sources of "cold dimension" data.

[5]Neutronics designers generally use cgs units for atom densities, cross sections, and neutron fluxes, and this procedure will be followed for these parameters in this book.

For each material in the metal alloys used for structures, such as iron, nickel and chromium in stainless steel, the ρ_m, is replaced by the product of the alloy density and the weight percent of the material. The atomic weight of the particular material is used for M_m.

Care must be taken to be consistent in methods of calculating fuel atom densities. Different organizations develop their own methods. The method suggested here will be used throughout the fuel cycle analysis of Chapter 7. We use the smear density for ρ_f, and a single molecular weight for the fuel, recognizing that the fuel is composed of isotopes from ^{235}U to ^{242}Pu plus even higher actinides for the U-Pu cycle. The single molecular weight suggested for the mixed oxide U-Pu cycle is 270, i.e., 238 for ^{238}U plus 32 for O_2. Though the average M may be slightly higher, additional accuracy is not generally warranted.

During irradiation of fuel in the reactor, some fuel atoms fission and others are transmuted to higher actinides by neutron capture. When fission occurs, a pair of fission products is formed. The number of fission-product pairs is exactly equal to the number of fissions. Hence, if a fission-product pair is considered to be one material (instead of two), the number density of heavy metal atoms ($\Sigma_i N_i$) plus fission-product pairs (N_{fp}) remains constant throughout the time the fuel remains in the reactor. The fact that the average molecular weight of this mixture is changing is irrelevant. Thus, at all times during fuel irradiation,

$$N_f = \frac{F_f \rho_f N_A}{270} = \sum_{\substack{heavy \\ metals,\, i}} N_i + N_{fp} = \text{constant}. \tag{4.51}$$

For a particular heavy metal isotope N_i the atom density is

$$N_i = I_i N_f, \tag{4.52}$$

where I_i is the fraction of fuel atoms that are isotope type i.

Let us next consider the units for the atom densities that are to be used in the multigroup equations. Macroscopic cross sections, $\sum$, have units of cm^{-1} and they are the product of an atom density and a microscopic cross section. Microscopic cross sections are given in barns, where 1 barn $=10^{-24}$ cm^2. The barn is such a convenient unit that it is an unnecessary nuisance to convert it to cm^2. Moreover, atom densities are in terms of atoms/cm^3, and there are generally 10^{23} or 10^{24} of them per cm^3. The logical solution is to calculate atom densities in "atoms per barn centimeter," so that the product $N\sigma$ is in cm^{-1} when σ is in barns. To do this, one uses for N_A the value 0.6023 instead of 0.6023×10^{24}. This way one never has to deal with 10^{24} s and 10^{-24} s again. "Well," as the Captain said in *H.M.S. Pinafore*, "hardly ever." We will have to carefully reintroduce a 10^{-24} in a few places in the fuel cycle calculations. But that's a few chapters away.

4.6 Neutron Balance

Questions in nuclear analysis frequently arise that can be answered by examining the neutron balance for the core and blankets. A neutron balance for a large (1,200 MWe) homogeneous SFR, with two core zones of equal volume, is provided in Table 4.1.[6] The basis is 100 neutrons produced. Values are

[6]Table 4.1 and Table 4.2 are compliments of C. L. Cowan, General Electric Company, Sunnyvale, California, 1979, 1981.

Table 4.1 Neutron balance for 1,200 MWe homogeneous SFR design (equilibrium cycle, mid-burnup)

Event	Material	Zone 1	Zone 2	Axial blanket	Radial blanket	Shielding	Total
Captures	^{238}U	19.15	11.79	7.65	7.13	0	45.72
	^{235}U	0.09	0.06	0.05	0.05	0	0.25
	^{239}Pu	3.46	2.64	0.25	0.23	0	6.58
	^{240}Pu	1.41	1.19	0.01	0.01	0	2.62
	^{241}Pu	0.43	0.40	0	0	0	0.83
	^{242}Pu	0.13	0.11	0	0	0	0.24
	F.P.	1.25	0.67	0.09	0.07	0	2.08
	O	0.10	0.07	0.02	0.02	0	0.21
	Na	0.14	0.10	0.07	0.06	0.07	0.44
	Fe	0.69	0.48	0.30	0.29	0.66	2.42
	Ni	0.29	0.20	0.09	0.09	0.24	0.91
	Cr	0.29	0.21	0.12	0.11	0.28	1.01
	^{10}B	0.34	0.21	1.10	0	0	1.65
	Total	**27.77**	**18.13**	**9.75**	**8.06**	**1.25**	**64.95**
Fissions	^{238}U	2.71	2.06	0.48	0.45	0	5.70
	^{235}U	0.29	0.20	0.14	0.14	0	0.77
	^{239}Pu	11.44	9.48	0.55	0.48	0	21.95
	^{240}Pu	0.91	0.90	0	0	0	1.81
	^{241}Pu	2.20	2.14	0	0	0	4.34
	^{242}Pu	0.09	0.09	0	0	0	0.18
	Total	**17.64**	**14.87**	**1.17**	**1.07**	**0.0**	**34.75**
Radial leakage							**0.03**
Axial leakage							**0.27**
Total							**100.00**

given for an equilibrium cycle, at mid-burnup. The captures in boron represent captures in the control rods. A neutron balance for a 1,000 MWe heterogeneous design at the end of an equilibrium cycle is provided in Table 4.2.

Note that for these reactor designs, the fraction of fissions in each isotope in the core fuel assemblies (not including the internal blankets) is apportioned as follows:

	Fraction fissions, %	
Fuel isotope	Homogeneous	Heterogeneous
^{235}U	1.5	0.6
^{238}U	14.7	10.6
^{238}Pu	–	0.6
^{239}Pu	64.3	70.2
^{240}Pu	5.6	5.4
^{241}Pu	13.3	12.1
^{242}Pu	0.6	0.5

These values are fairly typical for UO_2-PuO_2 SFR cores, except that the fraction of fissions in the ^{238}U in the homogeneous core may be slightly higher than typical. The core of this particular 1,200 MWe design was larger than normal so that the fission fraction was somewhat lower than typical. The

Table 4.2 Neutron balance for 1,000 MWe heterogeneous SFR design (end of equilibrium cycle)—axial blanket extension

Event	Material	Core fuel	Inner blankets	Driver fuel	Inner blanket	Radial blanket	Shielding	Total
Capture	O	0.24	0	0.01	0.01	0.02	0	0.28
	Na	0.12	0.04	0.03	0.01	0.03	0.05	0.28
	Cr	0.42	0.14	0.09	0.03	0.09	0.17	0.94
	Mn	0.18	0.08	0.08	0.03	0.08	0.34	0.79
	Fe	1.07	0.35	0.23	0.07	0.24	0.50	2.46
	Ni	0.77	0.23	0.12	0.04	0.13	0.23	1.52
	Mo	0.32	0.12	0.09	0.03	0.10	0.27	0.93
	^{235}U	0.05	0.06	0.01	0.02	0.02	0	0.16
	^{238}U	16.57	12.32	5.31	2.61	8.42	0	45.23
	^{238}Pu	0.10	0	0	0	0	0	0.10
	^{239}Pu	5.34	0.81	0.15	0.07	0.34	0	6.71
	^{240}Pu	1.95	0.04	0	0	0.01	0	2.00
	^{241}Pu	0.61	0	0	0	0	0	0.61
	^{242}Pu	0.20	0	0	0	0	0	0.20
	F.P.	1.70	0.19	0.01	0.01	0.05	0	1.96
	^{10}B	0	0	0.65	0	0	0	0.65
	Total	**29.64**	**14.38**	**6.78**	**2.93**	**9.53**	**1.56**	**64.83**
Fission	^{235}U	0.17	0.12	0.07	0.03	0.10	0	0.49
	^{238}U	2.93	1.38	0.31	0.14	0.59	0	5.35
	^{238}Pu	0.16	0	0	0	0	0	0.16
	^{239}Pu	19.42	2.42	0.31	0.15	0.84	0	23.14
	^{240}Pu	1.50	0.02	0	0	0.00	0	1.52
	^{241}Pu	3.33	0.01	0	0	0	0	3.34
	^{242}Pu	0.14	0	0	0	0	0	0.14
	Total	**27.65**	**3.95**	**0.69**	**0.32**	**1.53**	**0**	**34.14**
Radial leakage								**0.16**
Axial leakage								**0.87**
Total								**100.00**

trend is correct, however; lower fissile fractions in homogeneous designs generally lead to higher ^{238}U fission fractions than in heterogeneous designs.

For this homogeneous core design, the fraction of fissions occurring in the core is 93.6%, with only 6.4% in the blankets. Neutron leakage from the outer boundaries of the blankets/shielding is very small (0.3%). However, the radial neutron leakage from the outer core enrichment zone to the radial blanket is approximately 10%, and the leakage to the axial blankets is 11%. Hence, the total neutron leakage from the core into the surrounding blankets is 21%.

4.7 Power Density

In the multigroup solutions discussed in Section 4.3, the neutron flux values were relative, not absolute. In order to obtain absolute fluxes, it is necessary to relate neutron flux to power density. Factors required to do this, together with representative values of neutron flux and power distribution, are presented in this section.

Table 4.3 Approximate energy per fission in a fast reactor

	Contribution	Energy (MeV)
Prompt	Fission fragment kinetic energy	174
	Neutron kinetic energy	6
	Fission gammas	7
	Gammas from (n, γ) reactions	13
Delayed	Betas from fission product decay	6
	Gammas from fission product decay	6
	Betas from ^{239}U and ^{239}Np decay[a]	1
	Total	**213**

[a]Energy release due to this source is not always included in such tabulations.

4.7.1 Energy Per Fission

The energy generated per fission that is ultimately transferred as heat to the coolant in a UO_2-PuO_2 fueled fast reactor is about 213 MeV. The source of this energy is given in Table 4.3. The neutrino energy (~9 MeV) is not included in the table, since this energy is not absorbed in the reactor.

The values in Table 4.3 vary slightly with the isotope fissioned. While ^{241}Pu gives about the same fission fragment energy as ^{239}Pu (~175 MeV), ^{238}U gives only about 169 MeV. On the other hand, the beta and gamma energies from fission products for ^{238}U are higher than for ^{239}Pu. Considering the distribution of fissions between ^{239}Pu, ^{241}Pu, and ^{238}U in a fast reactor, the results in Table 4.3 represent reasonable average values for the reactor. Useful reviews of fission energy for various fissionable isotopes are given in Refs.[5] through [8].

The gamma energy from (n,γ) reactions is equivalent to the binding energy of the target nucleus; note from Table 4.1 that the dominant target nucleus for neutron capture is ^{238}U. There are about 1.9 (n,γ) reactions per fission in a fast reactor.

The higher value of 213 MeV for fast reactors relative to lower values generally used for thermal reactors results primarily from the difference in fission fragment kinetic energy for ^{239}Pu relative to ^{235}U. For ^{235}U, this relative energy is only ~169 MeV. Moreover, the large number of captures in hydrogen in a light water reactor leads to a lower value for the gamma energy from (n,γ) reactions.

Fission product kinetic energy and beta energy are absorbed in the fuel. Kinetic energy of the neutrons is transformed into gamma energy from inelastic scattering, and into kinetic energy of target nuclei from elastic scattering. Gamma rays are absorbed throughout the reactor, often distant from the source; the relative absorption of gammas by each material in the core is approximately proportional to the mass of material present.

4.7.2 Relation Between Power Density and Neutron Flux

There are 1.602×10^{-13} J/MeV. The relation between power density and absolute neutron flux is given by

$$\begin{aligned} p\,(\text{W/cm}^3) &= \left(213\frac{\text{MeV}}{\text{fission}}\right)\left(1.602\times10^{-13}\frac{\text{J}}{\text{MeV}}\right)\left(\sum_g \Sigma_{fg}\phi_g\frac{\text{fissions}}{\text{cm}^3\cdot\text{s}}\right) \\ &= \sum_g \Sigma_{fg}\phi_g\frac{\text{fissions}}{\text{cm}^3\cdot\text{s}}\Big/ 2.93\times10^{10}\frac{\text{fissions}}{\text{W}\cdot\text{s}}, \end{aligned} \tag{4.53}$$

where ϕ is in n/cm^2·s, and $\sum_f$ is in cm^{-1}.

Table 4.4 Power distribution for the early CRBRP homogeneous core design [9]

	Fraction of power in region	
Region	Beginning of equilibrium cycle	End of equilibrium cycle
Inner core	0.50	0.46
Outer core	0.38	0.38
Total core	**0.88**	**0.84**
Radial blanket	0.08	0.10
Axial blanket	0.03	0.05
Beyond blankets	0.01	0.01

In this relation the power density p is proportional to the fission distribution. From the previous discussion we know that the distribution of energy transferred to the coolant is slightly different from the fission distribution, due to diffusion and transport of neutrons and gamma rays.

4.7.3 Power Distributions

Power distributions can be shown in several ways. The fraction of total power transferred to the coolant in each region for the early CRBRP homogeneous core design is given in Table 4.4 for both the beginning and end of an equilibrium cycle [9]. The shift to the blankets during the cycle reflects the buildup of plutonium in the blankets. The small power deposited beyond the blankets results from neutron leakage and gamma transport.

The radial power distribution calculated for a homogeneous 1,200 MWe SFR design[7] at the core midplane at the middle of the equilibrium fuel cycle is plotted in Fig. 4.4. This sawtooth shape is typical of a two-zone homogeneous core design. The discontinuity between the inner and outer enrichment zones is a direct consequence of the higher fissile fuel content in the outer zone. The axial power distribution for this 1,200 MWe SFR design is shown in Fig. 4.5.

The radial power distribution for a heterogeneous 1,000 MWe SFR design [10] at the middle of the equilibrium cycle is plotted in Fig. 4.6.

Discontinuities in the curves at assembly row boundaries result from different residence times during the fuel cycle. The fraction of power generated in each region of this core is

Central blanket	0.016
Inner driver	0.230
Internal annular blanket	0.062
Outer driver	0.617
Radial blanket	0.056
Axial blanket	0.019

A useful way to display the radial power distribution is illustrated in Fig. 4.7 [11]. Here the CRBRP radial peaking factors for the heterogeneous core are listed in each hexagonal assembly, where the peaking factor is the ratio of the local power density to the average power density in the core or in the blanket. Values are listed for both the average pin and for the peak-power pin in each assembly. Values are shown in Fig. 4.7 for the beginning of Cycle 3 when peaking factors are quite high.

[7]Data for the 1,200 MWe homogeneous SFR design used for constructing Figs. 4.4, 4.5, 4.8, 4.9, and 4.11 were supplied by D. R. Haffner, R. W. Hardie, and R. P. Omberg of the Hanford Engineering Development Laboratory, Richland, WA, 1978.

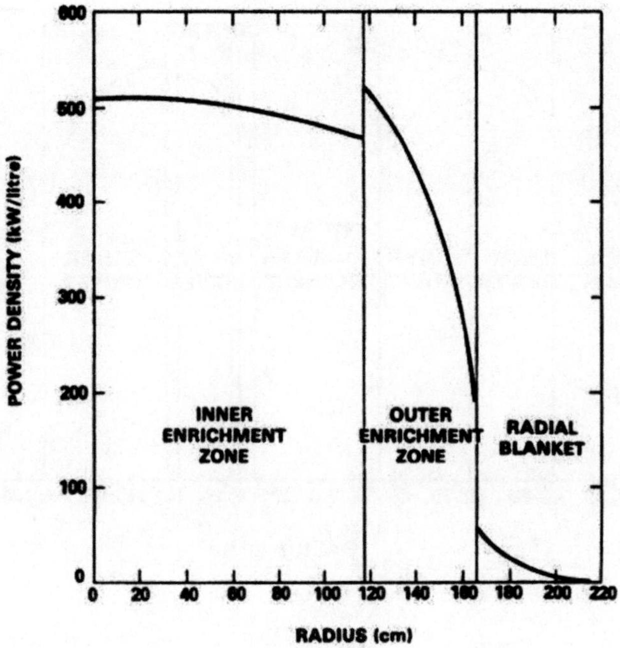

Fig. 4.4 Radial power density at core midplane (1,200 MWe homogeneous core design)

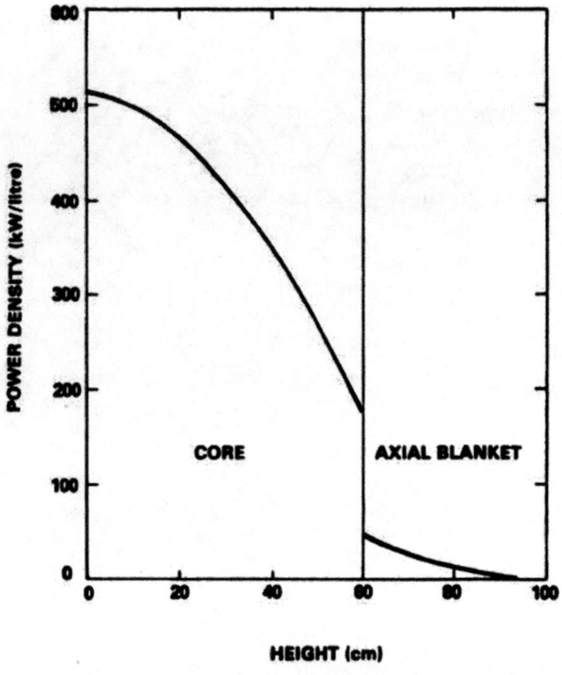

Fig. 4.5 Axial power density at core midplane (1,200 MWe homogeneous core design)

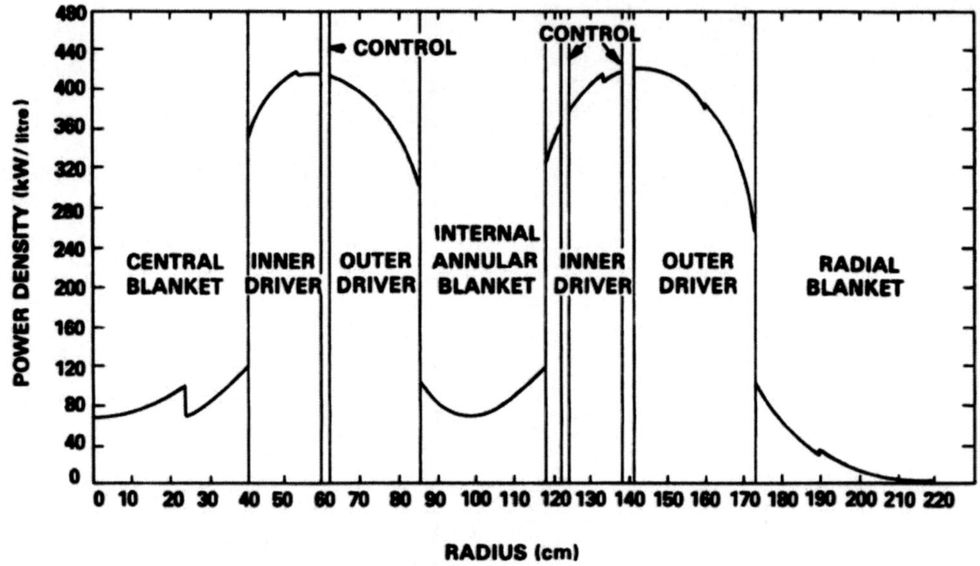

Fig. 4.6 Radial power density for 1,000 MWe heterogeneous core [10]

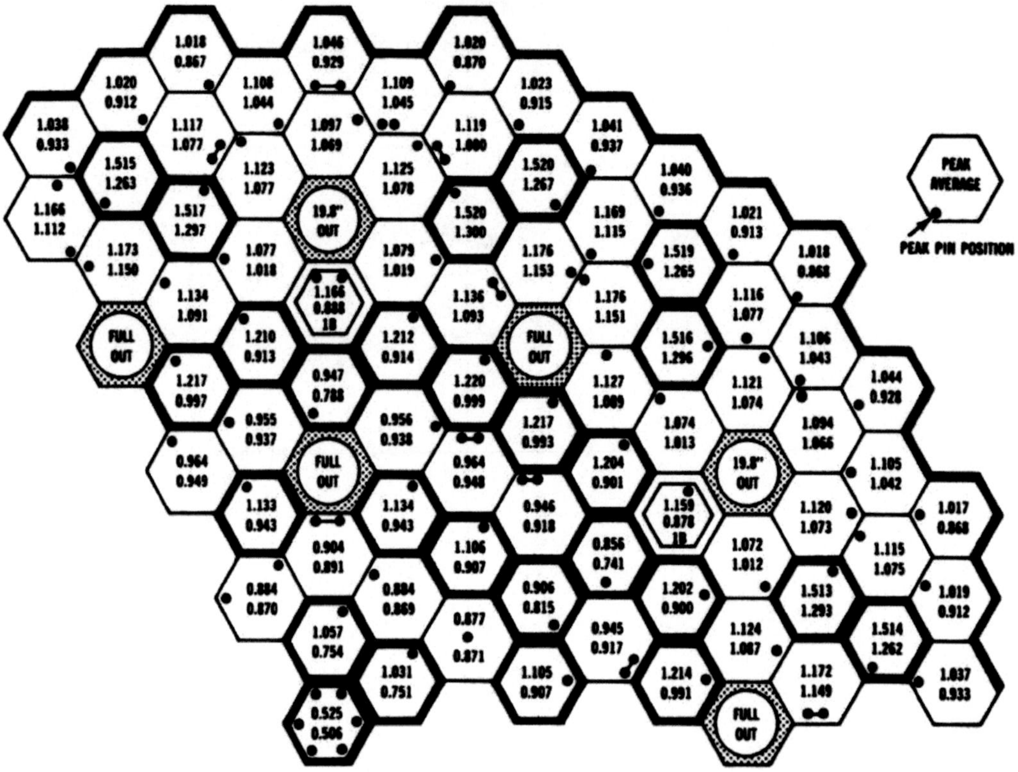

Fig. 4.7 Assembly radial power factor and peak rod radial power factor at the beginning of Cycle 3 [CRBRP heterogeneous design [11]]. Heavy hexagons denote blanket assemblies

4.7.4 Neutron Flux Distributions

Radial and axial flux distributions for a four-group calculation of a SFR are plotted in Figs. 4.8 and 4.9. The reactor is the same 1,200 MWe homogeneous core for which power densities were plotted in Figs. 4.4 and 4.5. The radial fluxes are plotted at the core axial midplane, and the axial fluxes correspond to the core radial centerline. For this case, the total flux at the center of the core is 7.0×10^{15} n/cm^{2}s. The energy group structures employed in these calculations are shown on the figures.

The axial fluxes follow a cosine shape relatively well, and group-dependent extrapolation distances for use in axial leakage approximations in radial diffusion theory calculations are relatively well defined. The use of two enrichment zones in the radial direction, however, causes considerable flattening of the radial flux in the inner zone and, consequently, a large deviation from the uniform cylindrical Bessel function flux shape. This design measure is often employed to maximize total power production capabilities without exceeding the peak linear power constraint in the hottest fuel pins.

4.8 Neutron Spectrum

The neutron flux of most fast reactors tends to peak around 200 keV. However, substantial variations in the neutron spectrum do exist, especially for low-energy flux, depending upon the coolant and fuel forms employed.

Neutron flux spectra for several fast reactor designs are compared in Fig. 4.10 as relative flux per unit lethargy [12]. The effect of oxygen and carbon in the fuel on softening the spectrum relative to metal-fueled SFR's is readily observed. The spectrum is softer for oxide than for carbide fuel because

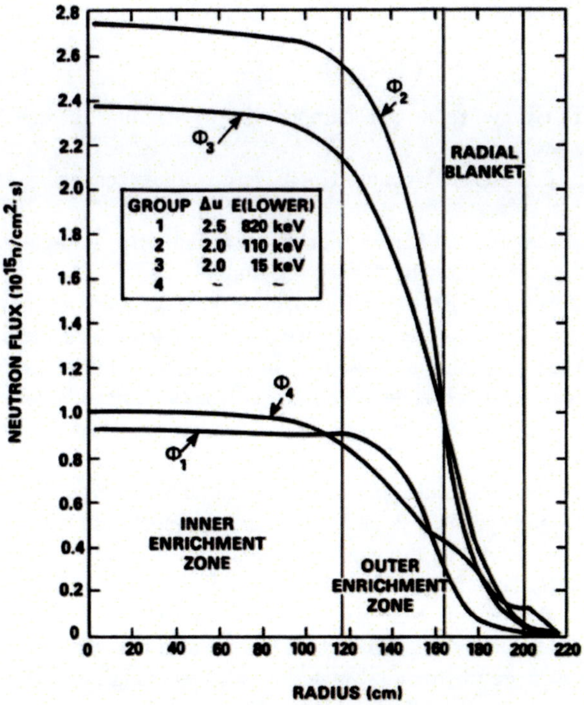

Fig. 4.8 Radial four-group flux distribution, core midplane, 1,200 MWe homogeneous core

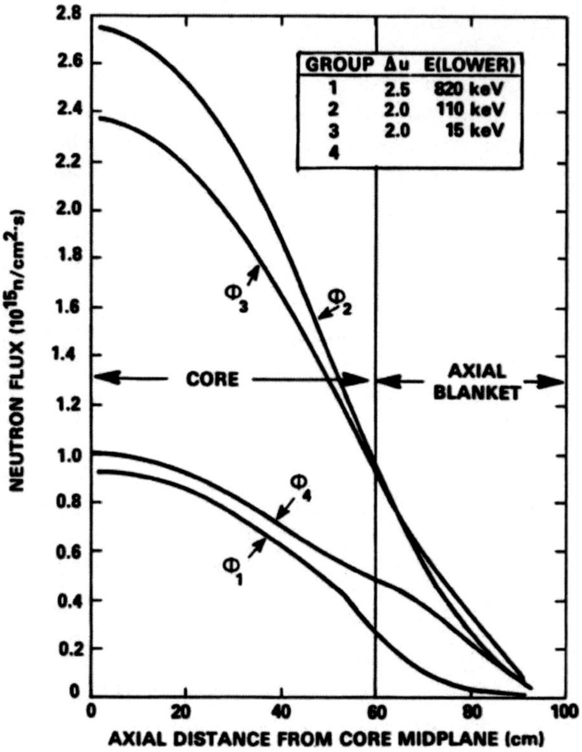

Fig. 4.9 Axial four-group flux distribution, core centerline. 1,200 MWe homogeneous core

of the higher density of oxygen relative to carbon; with M standing for metal, oxide fuel is MO_2, whereas carbide fuel is MC.

The depression in the flux at 3 keV is caused by the large sodium scattering resonance at this energy level.

The spectrum of a gas-cooled fast reactor (not shown in Fig. 4.10) is slightly harder than a sodium cooled reactor due to the lack of slowing down by sodium.

Twelve-group flux spectra are plotted in Fig. 4.11 at three locations of the 1,200 MWe homogeneous core-at the core center, at the outer radial edge of the core at the core midplane, and at the middle of the radial blanket at the core midplane. We may note that there is little difference in the spectrum across this homogeneous core, but that the spectrum becomes significantly softer in the blanket. For a heterogeneous core, the internal blankets cause a more pronounced spatial variation in neutron spectra than in a homogeneous core.

A $1/E$ slowing down spectrum would appear as a horizontal line in Figs. 4.10 and 4.11. In the oxide-fueled SFR of Fig. 4.11, the spectrum is fairly close to a $1/E$ spectrum between 10 keV and 1 MeV despite the large absorption rates in this energy range. Below 10 keV the flux drops rapidly. This represents a dramatic difference from a thermal reactor. Essentially no neutrons survive to thermal energies in the core of a fast reactor. Note also from Fig. 4.10 that the $1/E$ region is narrower for fast reactor designs with harder spectra, i.e., for carbide and metal-fueled reactors.

Although radiation damage in structural materials such as cladding is caused by neutrons of all energies in a fast reactor, the neutron flux above 0.1 MeV is widely used in estimates of the lifetime of structural components. At the center of the 1,200 MWe design for which the spectrum appears in

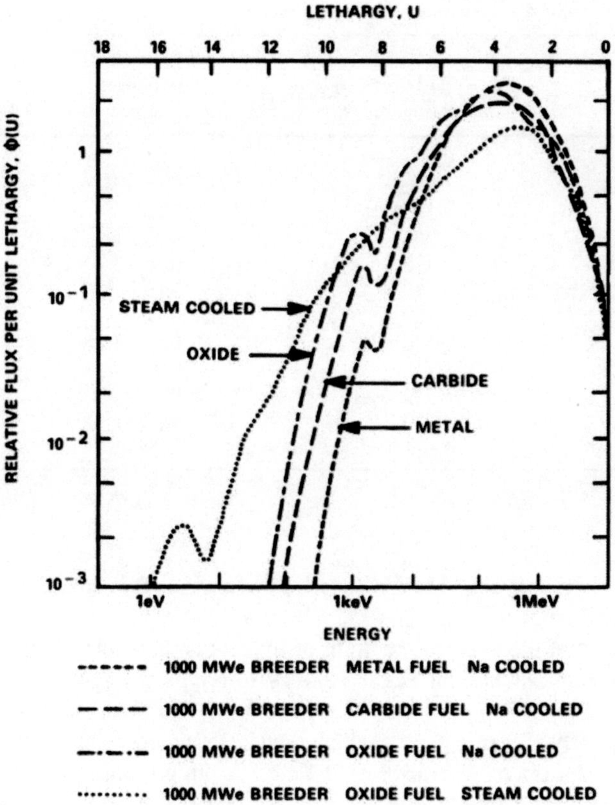

Fig. 4.10 Neutron flux spectra of fast reactors [12]

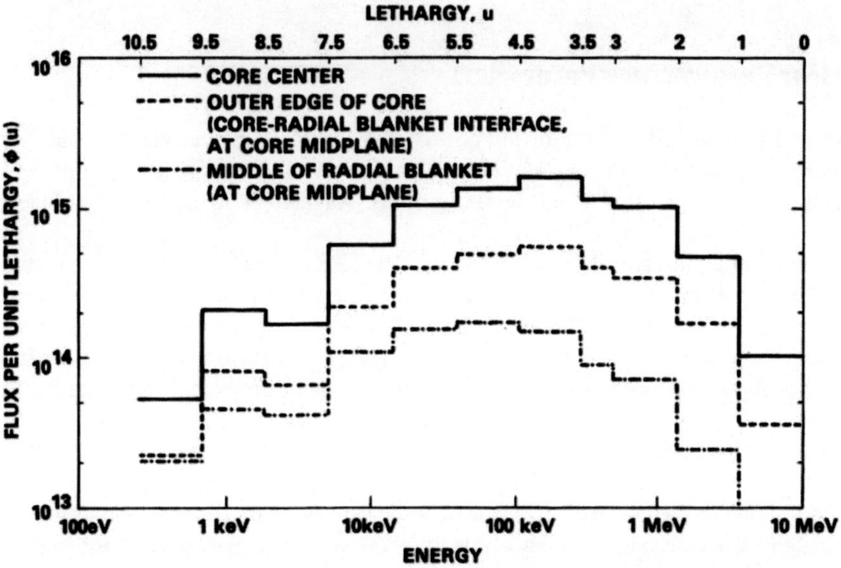

Fig. 4.11 Twelve-group flux spectra at three radial locations, 1,200 MWe homogeneous core (UO_2-PuO_2 fuel)

Table 4.5 Fraction of fissions at each energy, 1,200 MWe SFR

Group	Δu	Lower energy of group (E_i)	Fraction fissions in group	Fraction fissions above E_i
1	1.0	3.7 MeV	0.05	0.05
2	1.0	1.35	0.18	0.23
3	1.0	500 keV	0.12	0.35
4	0.5	300	0.05	0.40
5	1.0	110	0.14	0.54
6	1.0	41	0.13	0.67
7	1.0	15	0.11	0.78
8	1.0	5.5	0.08	0.86
9	1.0	2.0	0.03	0.89
10	1.0	740 eV	0.07	0.96
11	1.0	280	0.03	0.99
12	–	–	0.01	1.00

Fig. 4.11, the neutron flux above 0.1 MeV is $3.7 \times 10^{15} n/\text{cm}^2 \cdot \text{s}$. This compares with the total flux at the center of $7.0 \times 10^{15} n/\text{cm}^2 \cdot \text{s}$.

An additional aspect of the neutron spectrum of interest is the energy at which fissions occur in a fast reactor. The fraction of fissions at each energy in a typical large oxide-fueled SFR is given in Table 4.5. The median fission energy is about 150 keV. The median fission energy in a metal-fueled SFR is significantly higher, e.g., several hundred keV. Moreover the relatively large fraction of fissions below 10 keV for the oxide reactor of Table 4.5 is not present in a metal-fueled fast reactor, and, as will be discussed in Chapter 6, this large fraction of low energy fissions is an important factor in contributing to the much larger Doppler coefficient in an oxide-fueled fast reactor relative to a metal-fueled reactor.

4.9 Nuclear Performance Parameters

It is interesting to summarize some of the characteristic nuclear parameters that vary with different SFR core designs. Values from Ref. [13] are particularly illuminating in showing how nuclear design parameters vary as a function of fuel pin size for both homogeneous and heterogeneous cores.

The results reported in Table 4.6 are for 1, 200 MWe SFR designs, for which the thermal rating is about 3,300 MWth. Common to all designs listed are peak linear power for fresh fuel (44.3 kW/m) and residence time of the fuel in the core (2 years). Average burnup after 2 years is low for the large pin designs; hence residence time might be increased for them. However, results in Ref. [13] for these cases for 3-year residence times do not change much from the 2-year results, except for burnup.

The large variation of design parameters with fuel pin size is illustrated by these results. Also the large penalty in fissile mass for the heterogeneous design is apparent. Comparative results for sodium loss reactivity show the compensating advantage of the heterogeneous designs that makes it attractive despite the high fissile mass. Results show little difference in doubling time since the higher breeding ratio in the heterogeneous design tends to offset the higher fissile mass requirement.

Table 4.6 Variation of nuclear design parameters with fuel pin diameter for homogeneous and heterogeneous designs (as adapted from Ref. [13])

	Electrical rating	1,200 MWe	
	Thermal rating	~3,300 MWth	
	Linear power (fresh fuel)	44.3 kW/m	
	Core fuel residence time	2 years	
Common parameters	Fuel smear density	88% TD	
Pin diameter		Homogeneous	Heterogeneous
6.35 mm	Pin pitch-to-diameter ratio	1.28	1.28
(0.25 in.)	Core fissile inventory[a] (kg)	3,171	4,041
	Fissile fraction, average (%)	15.9	22.8
	Fissile specific inventory (kg/MWth)	1.16	1.51
	Volume fractions		
	Fuel	0.3164	0.3863[b]
	Sodium	0.4593	
	Structure	0.2243	
7.62 mm	Pin pitch-to-diameter ratio	1.21	1.21
(0.30 in.)	Core fissile inventory[a] (kg)	3,949	5,992
	Fissile fraction, average (%)	12.8	18.8
	Fissile specific inventory (kg/MWth)	1.38	1.94
	Volume fractions		
	Fuel	0.3845	0.4436[b]
	Sodium	0.3848	
	Structure	0.2157	
8.38 mm	Pin pitch-to-diameter ratio	1.16	1.16
(0.33 in.)	Core fissile inventory[a] (kg)	4,528	6,704
	Fissile fraction	11.3	16.4
	Fissile specific inventory (kg/MWth)	1.51	2.09
	Volume fractions		
	Fuel	0.4375	0.4891[b]
	Sodium	0.3566	
	Structure	0.2059	

[a]Average during an equilibrium cycle.
[b]Fuel volume fraction after homogenizing core with internal blankets. Core assemblies in the heterogeneous cores are identical to homogeneous core assemblies, but the fuel volume fraction in the blanket assemblies is higher than in the core assemblies.

References

1. M. B. Chadwick, et al., "ENDF/B-VII.0: Next Generation Evaluated Nuclear Data Library for Nuclear Science and Technology", *Nuclear Data Sheets*, *107* (12) (Special Issue) (2006) 2931–3060.
2. R. W. Hardie and W. W. Little, Jr., *1DX, A One Dimensional Diffusion Code for Generating Effective Nuclear Cross Sections*, BNWL-954, Battelle Northwest Laboratory, Washington, DC (March 1969).
3. K. L. Derstine, *DIF3D: A Code to Solve One-, Two-, and Three-Dimensional Finite-Difference Diffusion Theory Problems*, ANL-82-64, Argonne National Laboratory, Argonne, IL (1984).
4. P. J. Finck and K. L. Derstine, "The Application of Nodal Equivalence Theory to Hexagonal Geometry Lattices", *Proceedings of the International Topical Meeting on Advances in Mathematics, Computations and Reactor Physics*, Pittsburgh, PA, Vol. 4 (1991) 16.1 4-1.
5. M. F. James, "Energy Released in Fission," Journal of Nuclear Energy, *23* (1969) 516–536.
6. J. P. Unik and J. E. Grindler, *A Critical Review of the Energy Release in Nuclear Fission*, ANL-774R, Argonne National Laboratory, Argonne, IL (March 1971).
7. F. A. Schmittroth, *Decay Heat for the Fast Test Reactor (FTR)*. HEDL-TME 77-13, ENDF-771, UC-79d, IAEA, Vienna, Austria (June 1977).

8. R. Sher and C. Beck, *Fission Energy Release for 16 Fissioning Nuclides.* EPRI NP-1771, Research Project 1074-1, Final Report, EPRI, Palo Alto, California, USA, Accessible through DOE Information Bridge at http://www.osti.gov (March 1981).
9. *Preliminary Safety Analysis Report,* Clinch River Breeder Reactor Plant, Project Management Corporation (1974).
10. *Liquid Metal Fast Breeder Reactor Conceptual Plant Design.* 1000 MWe, TID-27701-2, Vol. 11, Rockwell International and Burns and Roe. Accessible through DOE Information Bridge at http://www.osti.gov (May 1977).
11. *Preliminary Safety Analysis Report,* Clinch River Breeder Reactor Plant, Amendment # 51 (September 1979) 4.3–15.4.
12. W. Häfele, D. Faude, E. A. Fischer, and H. J. Laue, "Fast Breeder Reactors," *Annual Review of Nuclear Science,* Annual Reviews, Inc., Palo Alto, CA (1970).
13. W. P. Barthold and J. C. Beitel. "Performance Characteristics of Homogeneous Versus Heterogeneous Liquid-Metal Fast Breeder Reactors," *Nuclear Technology, 44* (1979) 45.

Chapter 5
Nuclear Data and Cross Section Processing

Pavel Tsvetkov and Alan Waltar

5.1 Introduction

The common approach of representing energy dependence of neutron–nucleus interactions consists of discretizing the energy dependence in a number of energy groups. During the fission reactions neutrons are emitted with an average energy of approximately 2 MeV. Neutron numbers are negligibly small above 10–15 MeV, making the maximum energy of interest in most nuclear fission reactors to be of about 15 MeV. Neutrons are then slowed down to much lower energies by collisions with the reactor components. The energy-dependent neutron spectrum in a reactor takes into account probabilities of absorption events, slowing down by elastic and inelastic collisions,

$$
{}_0^1n + {}_Z^AX \rightarrow \begin{cases} \left({}^{A+1}_{\ \ Z}X\right)^* \rightarrow \begin{cases} \textit{Fission } (n,F) : {}_{Z_1}^{A_1}X_1 + {}_{Z_2}^{A_2}X_2 + \nu_{fp} \cdot {}_0^1n + \sim 200\,MeV, \\ \textit{Capture} : \begin{cases} \textit{Radiative Capture } (n,\gamma) : {}^{A+1}_{\ \ Z}X + \gamma, \\ \textit{Capture } (n,\alpha) : {}_{Z-2}^{A-3}Y + \alpha, \end{cases} \\ \textit{Inelastic Scattering } (n,n') : {}_0^1n' + {}_Z^AX + \gamma, \end{cases} \\ \textit{Elastic Scattering } (n,n) : {}_0^1n + {}_Z^AX, \end{cases}
$$

and of leakage out of the system. It is important to determine the rate at which various types of neutron-nuclear reactions occur within the reactor. Because of the wide range of neutron energies and the very strong energy dependence of some neutron-nucleus cross-sections, the number of energy groups should be sufficiently high to capture necessary spectral details. Since the energy dependence of the cross-sections within these groups cannot be neglected, group average cross-sections have to be produced. The actual numbers of energy groups to be used in multigroup calculations vary according to the complexity of the calculations being performed and the accuracy desired.

Multigroup microscopic cross sections of the materials in a fast reactor depend on the particular composition of each region in the reactor; hence, a single set does not suffice for all designs. For detailed design, a multigroup cross section set of *effective cross sections* is developed for the particular reactor composition and design being analyzed. For scoping studies, however, less care is required and a single set can apply to a range of designs, provided the compositions are not too different from the composition for which the set was originally developed.

P. Tsvetkov (✉)
Zachry Engineering Center, Department of Nuclear Engineering, Texas A&M University, College Station, TX, USA
e-mail: tsvetkov@tamu.edu

A.E. Waltar et al. (eds.), *Fast Spectrum Reactors*, DOI 10.1007/978-1-4419-9572-8_5,

The main approach used in the United States to obtain fast reactor cross sections for design purposes is the *Bondarenko shielding factor* approach [1, 2], named for the lead scientist who developed the system first in Russia. Shielding factors for a given material are factors that account for *energy self-shielding*, i.e., for the effects of other materials on the cross sections of the given material. The shielding factor approach as used for design among reactor vendors in the United States will be described in this chapter. An alternate approach, developed at Argonne National Laboratory, is represented by the MC^2 code [3]; and while this approach is expensive for ordinary design, it has been useful in providing a check on results obtained from the shielding factor method.

For both approaches, the starting point for calculating fast reactor cross sections is a reference set of cross sections and nuclear data, used and developed in the United States, called the Evaluated Nuclear Data Files (ENDF) [4]. These files went through various versions with names like ENDF/B-III, ENDF/B-IV, ENDF/B-VI, and ENDF/B-VII [5]. The formats were upgraded with each version to handle new features, for example, the extension from the original upper limit of 15–20 MeV, the addition of photon production information, the introduction of new resonance formats, or the addition of charged-particle data. The data used in the methods described in this chapter are contained in the ENDF/B set.

Two books related to neutron cross sections that cover some of the methods presented in this chapter are worth special reference. An early discussion of fast reactor cross sections appears in the first three chapters of *Reactivity Coefficients in Large Fast Power Reactors* by Hummel and Okrent [6]. Second, Dresner's classic *Resonance Absorption in Nuclear Reactors* [7] provides an excellent background on resonance absorption.

5.2 Evaluated Nuclear Data

All nuclear reactor calculations rely on the knowledge of the various relevant neutron-nuclear cross sections. The complicated dependence of such cross sections on neutron energy and angle of incidence combined with the large number of nuclides involved in nuclear reactor analysis implies that large databases are needed for storing nuclear data compilations. Such compilations have been accumulated over the past few decades from both experimental programs and theoretical calculations.

Nuclear cross section professionals have decided to consolidate and standardize all of the cross section information into the Evaluated Nuclear Data Files (ENDF). The ENDF system contains both neutron and photon cross-section data. These data are stored in four Internet-accessible nuclear reaction databases:

- EXFOR, alias CSISRS (Cross Section Information Storage and Retrieval System), is a database containing experimental data. This data set contains unevaluated raw data from experimental measurements. The data presently included in the EXFOR database include compilations of experimental neutron-induced reaction data, charged-particle induced reaction data, and photon-induced reaction data. It covers more than 18,600 experiments and spans nearly all of neutron induced reactions experimental works worldwide
- ENDF (Evaluated Nuclear (reaction) Data File) databases:
  - ENDF/A data set contains both complete and incomplete sets of nuclear data as soon as they become available. For each isotope there may be more than one data set for a particular reaction, or there may be none at all for certain reactions of interest.
  - ENDF/B is the main nuclear reaction database containing evaluated (recommended) data from the ENDF/B-VII library (also JEFF, JENDL, ENDF/B-VI, BROND, ROSFOND and CENDL). It provides data in the ENDF-6 format, covering all nuclides of practical relevance (393 in total)

for neutrons up to 20 MeV and partly up to 150 MeV. It serves as principal input for neutronics calculations, including nuclear reactor design, national security, accelerators, criticality safety, shielding, radiation protection and detector simulations. The ENDF/B data set contains only complete, evaluated sets of nuclear data presented in a form that can be efficiently accessed and processed.

- RIPL (Reference Input Parameter Library) database contains input parameters for theoretical calculations of nuclear reactions.
- SIGMA (ENDF retrieval and plotting) database contains evaluated (recommended) nuclear reaction and decay data from ENDF/B-VII.0, JEFF-3.1, JENDL-4.0, JENDL-3.3, CENDL-3.1, ROSFOND 2008 and ENDF/B-VI.8 libraries.

In addition to the nuclear data retrieval systems, there are two Internet-accessible bibliographical databases:

- CINDA (Computer Index of Nuclear (reaction) Data) database contains bibliographical neutron induced reaction information, including experimental, theoretical and evaluation works. It contains references to 275,000 reactions from 55,000 works worldwide.
- NSR (Nuclear Science References) database contains bibliographical nuclear physics information including more than 200,000 nuclear science articles, indexed according to content. It spans 100 years of research, and currently covers 80 journals with about 3,800 new articles added each year worldwide.

Appendix H summarizes the available internet resources including WWW access links to these databases.

When ENDF/B-VI was being prepared, it was noted that the ENDF formats were coming into wide use around the world, including the JEF files in Europe, the JENDL files in Japan, and the BROND files in Russia. It was decided to decouple the ENDF formats from the ENDF/B libraries in order to make their international use easier. Since then, it is referred as the "ENDF-6 format" to distinguish from the "ENDF/B-VI library". The current ENDF-6 format can represent cross sections for neutrons, photons, and charged particles, including particle yields and distributions in angle and energy, for energies up to several hundred MeV, the radioactive decay properties of reaction products, and estimated errors and covariances of the various nuclear parameters.

Today, the ENDF data representation format and the ENDF/B data set are the accepted data presentation standard and the reference data set that are used worldwide for nuclear reactor data representation and for reactor analysis applications. Control over the ENDF formats has been retained by the US Cross Section Evaluation Working Group (CSEWG). The format specifications are published through the National Nuclear Data Center at the Brookhaven National Laboratory.

The uses of the ENDF data sets have evolved over the years. The first few versions were largely intended for thermal-reactor applications. ENDF/B-IV and ENDF/B-V shifted the emphasis toward fast-reactor and fusion applications. For ENDF/B-VI, additional extensions have been made for charged-particle and accelerator applications.

In recent years, the ENDF system has also gained a role as a mode of publication and archiving of basic low-energy nuclear physics data. The five comprehensive data libraries for evaluated neutron reaction data are available worldwide as well as several other specific data libraries (ENDF/B-VI.8 300 K (2005), JENDL-3.3 300 K (2002), JEFF-3.1 300 K (2005), ENDF/HE-VI High Energy, and etc.). These are supplemented by two data libraries (FENDL/E and EFF) that had been developed for fusion applications but, of course, can be used for other applications as well:

- United States—ENDF/B is the library available as basic data with resonance-parameters, and/or as point data where resonance-parameters have been converted to cross-sections. This library is regarded as the standard source of nuclear data for use in nuclear reactor analysis in US. Current version—ENDF/B-VII.
- Russia—ROSFOND is the Russian evaluated neutron data library in ENDF format. It is continuously being updated since its release. Current version—ROSFOND-2010.
- Europe—JEFF is the evaluated nuclear data library of the OECD Nuclear Energy Agency (European Union) released in ENDF format. It is continuously being updated since its release. Current version—JEFF-3.1.
- Japan—JENDL is the Japanese evaluated nuclear data library released in ENDF format. It is continuously being updated since its release. Current version—JENDL-4.0.
- China—CENDL is the Chinese evaluated neutron data library issued in ENDF format. It is continuously being updated since its release. Current version—CENDL-3.1.

The NJOY Nuclear Data Processing System is typically used to convert evaluated nuclear data in the ENDF format into forms useful for applications [8]. Because the ENDF format is now widely used around the world, even though NJOY was originally designed to work with the US ENDF/B libraries, it now provides a universal capability to work with nuclear data libraries all over the world.

Sensitivity studies over a wide range of integral parameters demonstrate that existing uncertainties in nuclear cross-section data have significant impact on nuclear system calculations [9]. Reactor physics parameters affected by nuclear data variations include k_{eff}, Doppler reactivity, coolant void reactivity, burnup, transmutation rate, peak power, spent fuel decay heat, radiation source level, radiotoxicity, and others. A number of materials have been identified as needing improved cross-section evaluations with corresponding uncertainty (i.e., covariance) data: ^{232}Th, ^{233}U, ^{234}U, ^{235}U, ^{236}U, ^{238}U, ^{237}Np, ^{238}Pu, ^{239}Pu, ^{240}Pu, ^{241}Pu, ^{242}Pu, ^{241}Am, ^{242m}Am, ^{243}Am, ^{242}Cm, ^{243}Cm, ^{244}Cm, ^{245}Cm, Pb, Bi, ^{56}Fe, ^{57}Fe , ^{58}Ni, ^{52}Cr, Zr, Mo, ^{15}N, Si, C, O, Na, ^{10}B, Ti, Rb, etc. The data needs for these materials span the entire energy range. The specific nuclear data needs are [10, 11]:

- Cross-section covariance data.
- Cross-section evaluations for actinide fission and capture, prompt neutrons per fission, delayed neutrons, and photon production.
- Various measurements to include fission, capture, delayed neutron fractions, and other decay data for both major and minor actinides at accuracies of 2–4%.
- Inelastic cross-section data for structural and coolant materials. The main data needs are for Fe, Na, Pb, and Si cross sections.
- Development of sensitivity analysis tools for both Monte Carlo and deterministic methods.
- Various data for material detection and assay (accountability issues) and criticality safety.

Table 5.1 Experimental errors of common cross sections

Cross section type	Measurement error
Total cross sections	<1%
Fission cross sections	2–5%
Gamma-production cross sections	5–10%
$(n, 2n)$ etc., from gamma production	5–20%
(n, xp), $(n, x\alpha)$	10–15%
Neutron emission including fission neutrons	15% or better
(n, γ)	1–10%

The level of available accuracy of cross section measurements depends on the type of measurements. Table 5.1 summarizes typically quoted experimental errors associated with various cross section measurements [9–11]:

5.3 Shielding Factor Method

The Bondarenko shielding factor approach consists of two parts. In the first, *generalized* multigroup cross sections are obtained for a given material, as a function of a *background cross section* σ_o (to be defined quantitatively later), and for the abundant fuel nuclides, as a function of temperature T. These cross sections are generally obtained over a wide range of values of σ_o for materials, and at three temperatures for fuel nuclides. The generalized multigroup cross sections need not relate to any particular reactor composition. Methods for obtaining these cross sections are presented in Sections 5.4 through 5.6.

In the second part of the shielding factor method, a cross section set is obtained from the generalized cross section set for a *particular* reactor composition at a given fuel temperature. The actual values of σ_o are calculated for each material, and cross sections for the given composition are interpolated from the cross sections in the generalized set. For detailed design calculations, different cross sections are obtained for different regions of the core and the blanket. The second part of the shielding factor method is described in Section 5.7.

The legacy computer codes that incorporated the first capabilities to perform this two-part calculation are MINX [12] and SPHINX [13], the former was developed at Los Alamos Scientific Laboratory and the latter at Westinghouse (Advanced Reactors Division). The MINX-SPHINX technology is an extension of two earlier sets of codes developed for the shielding factor method during the late 1960s—the ETOX [14] and 1DX [15] codes developed at the Hanford Engineering Development Laboratory, and the ENDRUN [16] and TDOWN [17] codes developed at General Electric.

5.3.1 Shielding Factor, *f*

Self-shielding is an important effect in the resonance energy range where total cross sections become large enough to influence the local neutron flux and reaction rates. The explanation of the effect is given in Section 5.4 in the derivation of the slowing down flux, $\phi(E)$. The energy range where self-shielding begins to influence the results is between 100 keV and 1 MeV. Although effective cross sections must be obtained in both the high-energy range and the resonance region, the shielding factor method gets its name from the necessarily careful treatment of the resonance region.

The principal idea of the shielding factor method is the evaluation of effective cross sections in the resonance energy range by means of a *shielding factor*, *f*, for each material, *m*, and each cross section for which resonances make energy self-shielding important. The shielding factor is defined as the ratio of the *effective* cross section for reaction *x*, $\sigma_{xmg}(\sigma_o)$, to an infinitely dilute cross section, $\sigma_{xmg}(\infty)$. The effective cross section, and hence the shielding factor *f*, will be dependent on a background cross section, σ_o. Thus,

$$f_{xmg}(\sigma_o) = \frac{\sigma_{xmg}(\sigma_o)}{\sigma_{xmg}(\infty)}. \tag{5.1}$$

The background cross section and the methods for calculating $\sigma_{xmg}(\infty)$ and $\sigma_{xmg}(\sigma_o)$ are described in the next section.

5.4 Generalized Cross Sections

We start with the generalized multigroup cross sections calculated in the first part of the Bondarenko shielding factor method (e.g., in the legacy codes like MINX, ETOX, and ENDRUN codes, and in the general purpose code like NJOY). In this part, the following infinitely dilute cross sections are calculated for each material and for each group: σ_t, $\sigma_f, \sigma_c, \sigma_e$, and σ_i. Effective cross sections and shielding factors are calculated as functions of σ_o and material temperature for $\sigma_t, \sigma_f, \sigma_c$, and σ_e. Other quantities calculated by material and group include ν, $\overline{\mu}$, ξ, χ, and the inelastic scattering matrix.

5.4.1 Group Cross Sections and Group Flux

5.4.1.1 Group Flux

The group neutron flux, $\phi_g \left(n/\text{cm}^2 \cdot \text{s}\right)$, is defined as

$$\phi_g = \int_{E_g}^{E_{g-1}} \phi\left(E\right) dE = \int_g \phi\left(E\right) dE, \tag{5.2}$$

where $\phi\left(E\right)$ is the flux per unit energy, E_g is the lower energy of the group, and $\int_g$ refers to integration over the energy group.

5.4.1.2 Capture, Fission, and Scattering Cross Sections

For capture, fission and elastic and inelastic scattering reactions, the average cross section for a group is the cross section which, when multiplied by the group flux, will give the correct reaction rate. The reaction rate for material *m* and for group *g* [where *x* stands for capture (*c*), fission (*f*), elastic scattering (*e*), or inelastic scattering (*i*)] is

$$\int_g \sigma_{xm}\left(E\right) \phi\left(E\right) dE.$$

The reaction rate is said to be *conserved* when $\sigma_{xmg}\phi_g$ is equal to this rate. Hence, the cross section for material *m* for group *g* is

$$\sigma_{xmg} = \frac{\int_g \sigma_{xm}\left(E\right) \phi\left(E\right) dE}{\int_g \phi\left(E\right) dE}. \tag{5.3}$$

5.4.1.3 Transport and Total Cross Sections

The microscopic total cross section, σ_t, is used to calculate the transport cross section, σ_{tr}, which is in turn used in evaluating the diffusion coefficient *D*. The different nature of leakage relative to other nuclear reactions requires that σ_t be averaged in a manner different from the other cross sections. The following relationships between *D*, σ_{tr} and σ_t are recalled:

$$D = \frac{\lambda_{tr}}{3} = \frac{1}{3\Sigma_{tr}} = \frac{1}{3\sum_m N_m \sigma_{trm}}, \tag{5.4}$$

$$\begin{aligned} \sigma_t &= \sigma_a + \sigma_i + \sigma_e, \\ \sigma_{tr} &= \sigma_a + \sigma_i + \sigma_e\left(1 - \overline{\mu}\right), \end{aligned} \tag{5.5}$$

$$= \sigma_t - \sigma_e \overline{\mu}, \tag{5.6}$$

where $\overline{\mu}$ is the average cosine of the scattering angle for elastic scattering in the laboratory system ($\overline{\mu} = 2/3A$), where A is the atomic weight of the scattering nuclide). The transport and total cross sections are nearly equal since, $\overline{\mu} \ll 1$ for materials in fast reactors. Hence, the total and transport cross sections should be averaged the same way.

For a proper treatment of leakage, the energy dependent transport mean free path, $\lambda_{tr}(E)$, should be weighted with the flux in order to obtain an average transport mean free path for an energy group, or

$$\lambda_{trg} = \frac{\int_g \lambda_{tr}(E)\,\phi(E)\,dE}{\int_g \phi(E)\,dE}. \tag{5.7}$$

In order to obtain this result, the microscopic transport cross section of each material should be weighted with the current density, $D\nabla\phi$, in obtaining average microscopic transport cross sections, or

$$\sigma_{trmg} = \frac{\int_g \sigma_{trm}(E)\,D(E)\,\nabla\phi(E)\,dE}{\int_g D(E)\,\nabla\phi(E)\,dE}. \tag{5.8}$$

By assuming separability in space and energy, the *energy* dependence of $\nabla\phi(E)$ is the same as for $\phi(E)$. Hence, Eq. (5.8) can be written (replacing D with $\lambda_{tr}/3$)

$$\sigma_{trmg} = \frac{\int_g \sigma_{trm}(E)\,\lambda_{tr}(E)\,\phi(E)\,dE}{\int_g \lambda_{tr}(E)\,\phi(E)\,dE}. \tag{5.9}$$

Since, σ_t is close to σ_{tr} we calculate σ_t also with Eq. (5.9). By replacing λ_{tr} in both the numerator and denominator of Eq. (5.9) by$1/\Sigma_{tr}$, and then by $1/\Sigma_t$, since Σ_{tr} and Σ_t are approximately equal, we obtain

$$\sigma_{tmg} = \frac{\int_g \sigma_{tm}(E)\,\frac{\phi(E)}{\Sigma_t(E)}dE}{\int_g \frac{\phi(E)}{\Sigma_t(E)}dE}. \tag{5.10}$$

The crucial factor in calculating effective cross sections is the proper evaluation of the flux, $\phi(E)$. It is in the evaluation of the flux that the energy self-shielding appears.

5.4.2 Energy Dependency of the Neutron Flux, ϕ (E)

5.4.2.1 Fission Spectrum Flux

Two different flux energy spectra are used to evaluate cross sections. At high energy the flux follows a fission spectrum. At lower energy the flux spectrum is governed by slowing down, and it approaches the familiar $1/E$ spectrum. A cutoff energy is defined as the interface between the fission spectrum and the slowing down spectrum. A typical value of the cutoff energy is 2.5 MeV.

The fission spectrum for fissile isotope *m* has the form

$$\phi(E) = C\sqrt{\frac{E}{(kT_m)^3}}\, e^{-E/kT_m}, \tag{5.11}$$

where T_m is a characteristic nuclear temperature for the fissile nuclide *m*. For ^{239}Pu, kT_m, is equal to ~1.4 MeV. The constant *C* provides the proper units and must have a numerical value such that the transition to the $1/E$ spectrum at the cutoff energy is continuous. Since $\phi(E)$ appears in both the numerator and denominator of the equations for the cross sections, however, this constant will eventually cancel out.

5.4.2.2 Slowing Down Flux

The relation for $\phi(E)$ below the fission spectrum is obtained from slowing down theory, and it is necessary to derive $\phi(E)$ in order to understand the shielding factor method. For this derivation, reactions between neutrons and atomic nuclei have to be considered to account for phenomena at resonance energies.

All reactions between neutrons and atomic nuclei can be divided into two main categories: *resonance* reactions and *potential* scattering. In resonance reactions, incident neutron and target nucleus interact with neutron absorption to form a compound nucleus. This compound nucleus is in an excited state with an energy consisting of the binding energy between nucleus and neutron and of the neutron kinetic energy. In heavy nuclides there is a high density of possible excited states. Compound nuclei are easily formed when the sum of binding and collision energy corresponds to quantum states of the compound nucleus so that the cross-section presents sharp peaks around these energies (resonances). The compound nucleus decays in different ways ("channels"), including neutron emission (scattering), γ-emission (capture), and fission—each mode having a different probability of occurrence. If λ_i is the decay constant of the compound nucleus along the *i*-th decay channel, the probability of decay along this channel is

$$\frac{\lambda_i}{\sum\limits_i \lambda_i} = \frac{\Gamma_i}{\Gamma},$$

where:

$\Gamma_i = \hbar\lambda_i$ is the partial width,
$\Gamma = \sum\limits_i \Gamma_i$ is the total level width. Note that the total width Γ is equal to $\Gamma = \Gamma_n + \Gamma_\gamma + \Gamma_f$.

These partial widths correspond to the energy uncertainty of each compound state:

Neutron width, Γ_n : $\frac{\Gamma_n}{\Gamma}$ is the probability of neutron emission,
Radiation width, Γ_γ : $\frac{\Gamma_\gamma}{\Gamma}$ is the probability that the compound nucleus will decay by γ-radiation, and
Fission width, Γ_f : $\frac{\Gamma_f}{\Gamma}$ is the probability that the compound nucleus will decay by fission.

In potential scattering reactions, incident neutron and target nucleus interact without neutron absorption or the formation of a compound nucleus. The neutron is deflected by the potential field of the nucleus. If the neutron wavelength is much greater than the radius of the nucleus, the potential scattering cross-section is approximately energy independent. Although quantum mechanical (s-wave) in nature, this process may be visualized and treated as a classical hard-sphere scattering event away from resonance energies. Near resonance energies, there is quantum mechanical interference between the *potential* and *resonance scattering*. In the case of resonance scattering, the two possibilities include elastic scattering, in which after neutron emission the nucleus returns to normal unexcited state, and inelastic scattering, in which case part of the neutron energy is retained as excitation energy of the nucleus. In light nuclides, excited states have energy levels too high to be of interest in nuclear reactors, while in heavy nuclides these energies are much lower. The cross sections involved for a given nuclide with a resonance are illustrated schematically in Fig. 5.1.

As noted above, the resonance can be a capture, fission, or scattering resonance, and, furthermore, the *potential scattering cross section,* σ_p, varies slowly with energy and can be considered constant in the vicinity of the resonance. One would expect something dramatic to happen to the neutron flux around this resonance energy if the nuclide with the resonance is present in the reactor to an appreciable extent. This, we shall see, is indeed the case.

The derivation of $\phi(E)$ will be based on the *narrow resonance approximation* since the neutron energy in a fast reactor is confined to the energy range over which the narrow resonance approximation applies. *Wide resonances* exist below the energies of relevance to a fast reactor. In the narrow resonance approximation, the energy width of the resonance is considered to be so narrow that neutrons scattered into the resonance come from energies which are not influenced by the resonance.

The present derivation of the slowing down flux will include the influence only of elastic scattering since inelastic scattering becomes small at the low energies where the resonances are large enough to cause appreciable self-shielding.

Let us consider the energy picture of Fig. 5.2 where E describes the neutron energy. If a resonance occurs, we suppose that it occurs in the vicinity of energy E.

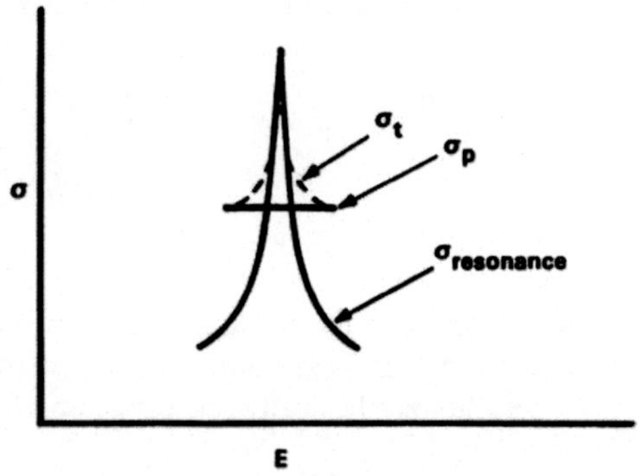

Fig. 5.1 Schematic illustration of cross sections around a resonance

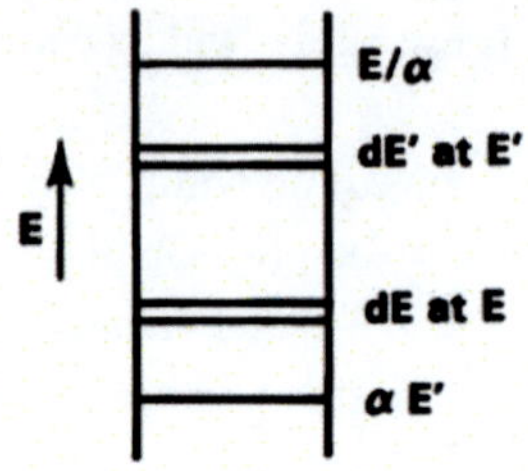

Fig. 5.2 Energy diagram for scattering analysis

The α in the figure is defined such that $(1-\alpha)$ is the maximum fraction of energy that can be lost in an elastic scattering collision. Thus, when a neutron of energy E suffers an elastic collision, the minimum energy of the neutron after the collision is αE. The quantity α is given by

$$\alpha = \left(\frac{A-1}{A+1}\right)^2, \tag{5.12}$$

where A is the atomic weight of the target isotope.

We define a *collision density,* $F(E)$, at E as the number of reactions per unit volume, or

$$F(E) = \Sigma_t(E)\,\phi(E)\,. \tag{5.13}$$

At energies higher than those affected by the resonance, the macroscopic total cross section, Σ_t, and the macroscopic scattering cross section, Σ_s, are equal, so that outside the resonance,

$$F(E') = \Sigma_t(E')\,\phi(E')\,.$$

We can write an expression for the number of scattering reactions occurring in dE' at E' on Fig. 5.2 as follows:

$$\text{Neutrons scattered out of } dE' = \Sigma_s(E')\,\phi(E')\,dE' = F(E')\,dE'.$$

For the purpose of deriving an expression for $\phi(E)$, we assume that the scattering is elastic and isotropic. Corrections can be made later for small deviations from this assumption. This assumption will be used to obtain the local neutron flux in the vicinity of the resonance in order to perform the integrations in Eqs. (5.3) and (5.10).

As a consequence of the assumption of isotropic elastic scattering, there is an equal probability that the energy of the neutron scattered at energy E' will lie in any given interval dE between E' and $\alpha E'$. Hence, the fraction of the neutrons scattered at E' that are scattered into dE at E is

$$\text{Fraction scattered into } dE = \frac{dE}{E' - \alpha E'}.$$

(The α is the value for the target scattering nucleus, and there may be several nuclides accounting for the scattering in a fast reactor. For the purpose of deriving the form of $\phi(E)$, it is not necessary to consider the details involved in obtaining an effective average value of α.)

Since neutrons elastically scattered into dE must have had energies ranging from E to E/α before the collision,

$$\text{Neutrons scattered into } dE = \int_{E}^{E/\alpha} \Sigma_s(E')\,\phi(E')\,dE' \frac{dE}{E' - \alpha E'}$$
$$= \int_{E}^{E/\alpha} \Sigma_s(E')\,\phi(E') \frac{dE'}{E'(1-\alpha)} dE.$$

A steady-state balance of neutrons in dE requires that the collision density in dE equals the rate that neutrons enter dE from scattering, or

$$\Sigma_t(E)\,\phi(E)\,dE = \int_{E}^{E/\alpha} \Sigma_s(E')\,\phi(E') \frac{dE'}{E'(1-\alpha)} dE. \tag{5.14}$$

The dE on each side cancels, leaving

$$F(E) = \Sigma_t(E)\,\phi(E) = \int_{E}^{E/\alpha} \Sigma_s(E')\,\phi(E') \frac{dE'}{E'(1-\alpha)}. \tag{5.15}$$

In the *absence* of resonances at E or anywhere in the energy range between E and E/α, $\Sigma_t(E) = \Sigma_s(E)$. We can also assume that $\Sigma_s(E)$ is constant in the energy range E to E/α since this is a small energy range for materials in fast reactors and Σ_s is slowly varying with energy. For these conditions the solution of the above integral equation is the familiar *asymptotic flux per unit energy*, $\phi_o(E)$:

$$\phi_o(E) = \frac{C_1}{E}. \tag{5.16}$$

This solution can be verified by inserting C_1/E' for $\phi(E')$ into Eq. (5.15) and performing the integration.

With a resonance at E, the solution for $F(E)$ is

$$F(E) = \frac{C_2}{E},$$

which again can be verified by direct substitution, remembering that $\Sigma_s(E')\,\phi(E') = F(E')$ in the integral and that $F(E) = \Sigma_t(E)\,\phi(E)$ at the energy E of the resonance. The fact that $\Sigma_s(E')$ is used throughout the energy range of the integral in Eq. (5.15) despite the resonance at E, the tail of which actually will spread into the energy range between E and E/α, is a consequence of the narrow resonance approximation.

The collision density at E in the absence of a resonance at E is

$$F(E) = \Sigma_s(E)\,\phi_o(E). \tag{5.17}$$

If there *is* a resonance at E, $F(E)$ will be given by $\Sigma_t(E)\,\phi(E)$. The value of $F(E)$ for this case will be identical to that in Eq. (5.17), however, because $F(E)$ is equal to the integral on the right-hand side of Eq. (5.15) *regardless* of whether there is a resonance at E or not (as a consequence of the narrow resonance approximation). Thus, the actual flux and the asymptotic flux are related by

$$\Sigma_t(E)\,\phi(E) = \Sigma_s(E)\,\phi_o(E)$$

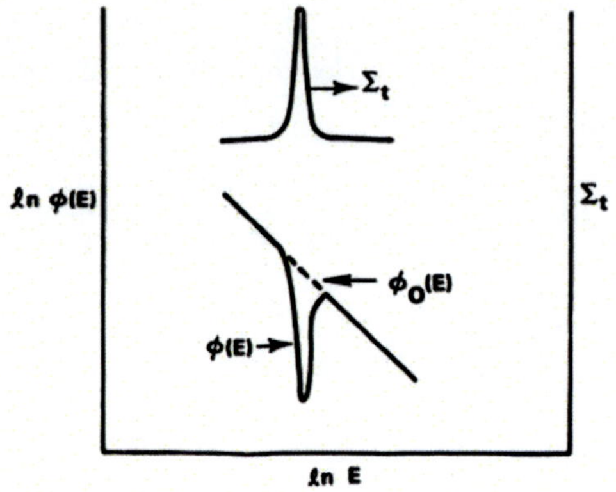

Fig. 5.3 Schematic illustration of flux depression caused by a resonance

or

$$\phi(E) = \frac{\Sigma_s(E)}{\Sigma_t(E)}\phi_o(E). \tag{5.18}$$

Thus, there is a sharp decrease in the actual flux at the energy of the resonance relative to the flux without a resonance. Schematically the flux will appear as shown in Fig. 5.3.

The effect of the resonance is to depress the flux in the vicinity of the resonance by the ratio $\frac{\Sigma_s(E)}{\Sigma_t(E)}$. The effect of this flux depression is referred to as *energy self-shielding*.

5.4.3 Self-Shielded and Background Cross Sections

We are now in a position to calculate this self-shielding effect. We can insert Eq. (5.18) for the flux into Eq. (5.3) to obtain the following equation for the effective microscopic cross section:

$$\sigma_{xmg} = \frac{\int_g \sigma_{xm}(E)\frac{\Sigma_s}{\Sigma_t(E)}\phi_o(E)\,dE}{\int_g \frac{\Sigma_s}{\Sigma_t(E)}\phi_o(E)\,dE} = \frac{\int_g \frac{\sigma_{xm}(E)}{\Sigma_t(E)}\frac{dE}{E}}{\int_g \frac{1}{\Sigma_t(E)}\frac{dE}{E}}, \tag{5.19}$$

where we have assumed that Σ_s, is independent of energy.

The total macroscopic cross section, Σ_t, is composed of the contribution from the resonance in material *m* plus the scattering cross section of all of the other materials that comprise the composition for which the effective cross section is being obtained, or

$$\Sigma_t(E) = N_m\sigma_{tm}(E) + \sum_{\substack{\text{other}\\ \text{materials}}} N_{\text{other}}\,\sigma_{t,\,\text{other}}(E), \tag{5.20}$$

Dividing by N_m gives

$$\frac{\Sigma_t(E)}{N_m} = \sigma_{tm}(E) + \sigma_{om}(E), \tag{5.21}$$

where

$$\sigma_{om}(E) = \frac{1}{N_m} \sum_{\substack{\text{other} \\ \text{materials}}} N_{\text{other}}\, \sigma_{t,\,\text{other}}(E). \tag{5.22}$$

Hence, $\sigma_{om}(E)$ is the macroscopic scattering cross section of other materials per atom of material m per cm^3, i.e., per N_m. This parameter, $\sigma_{om}(E)$, is called the *background cross section.*

Equation (5.19) becomes, since $\Sigma_t(E)$ can be divided by N_m in both the numerator and denominator,

$$\sigma_{xmg}(\sigma_{om}) = \frac{\int_g \frac{\sigma_{xm}(E)}{\sigma_{tm}(E) + \sigma_{om}(E)} \frac{dE}{E}}{\int_g \frac{1}{\sigma_{tm}(E) + \sigma_{om}(E)} \frac{dE}{E}}, \tag{5.23}$$

where the $\sigma_{xmg}(\sigma_{om})$ notation emphasizes the dependence of the effective cross section on the background cross section.

Similarly, using Eq. (5.18) for $\phi(E)$, Eq. (5.10) can be written as follows:

$$\sigma_{tmg}(\sigma_{om}) = \frac{\int_g \frac{\sigma_{tm}(E)}{[\sigma_{tm}(E) + \sigma_{om}(E)]^2} \frac{dE}{E}}{\int_g \frac{1}{[\sigma_{tm}(E) + \sigma_{om}(E)]^2} \frac{dE}{E}}. \tag{5.24}$$

Equations (5.23) and (5.24) are the expressions used in the three codes—MINX, ETOX, and ENDRUN—to calculate effective microscopic cross sections.

5.4.4 Infinitely Dilute Cross Sections

The infinitely dilute cross section is the cross section of a material that is present in such small quantities that its resonances do not affect either its own effective cross section or that of other materials present. This cross section is denoted by $\sigma_{xmg}(\infty)$. As a consequence of the narrow resonance approximation and the infinitely dilute amount of material present, the flux at the resonance of material m is equal to the flux without resonances, $\phi_o(E)$, which was defined by Eq. (5.16). Using this flux in Eq. (5.3) gives $\sigma_{xmg}(\infty)$

$$\sigma_{xmg}(\infty) = \frac{\int_g \sigma_{xm}(E) \frac{dE}{E}}{\int_g \frac{dE}{E}}. \tag{5.25}$$

This cross section applies for the total cross section as well as capture, fission, elastic scattering and inelastic scattering.

We now have all of the information needed to obtain the shielding factor *f*, as defined in Eq. (5.1).

5.4.5 Inelastic Scattering

In an inelastic scattering collision, the nucleus is excited to a higher energy state by an incident neutron of energy E' and then returns to a ground state by emitting a neutron of energy E and a gamma ray. Inelastic scattering generally occurs at high energies where resonances are small and self-shielding has little effect. The inelastic scattering cross section can therefore be calculated by Eq. (5.3) with the fission spectrum or the asymptotic flux. The inelastic scattering cross section $\sigma_i(E)$ is a part of the total cross section $\sigma_i(E)$, as given by Eq. (5.5).

The matrix for inelastic scattering from group g' to group g is obtained from the relation:

$$\sigma_{img' \to g} = \frac{\int\limits_{g}\int\limits_{g'} \sigma_{im}\left(E'\right) W\left(E' \to E\right) \phi\left(E'\right) dE' dE}{\int\limits_{g}\int\limits_{g'} W\left(E' \to E\right) \phi\left(E'\right) dE' dE}, \tag{5.26}$$

where the dE' integration is over group g' and the dE integration is over group g. The evaluation of $W\left(E' \to E\right)$ from the ENDF/B files is complex and will not be reproduced here; details are given in Refs. [12, 14, 16]. Another reaction that is generally included in the inelastic scattering matrix is the $(n, 2n)$ reaction.

5.4.6 Calculation of χ, ν, $\overline{\mu}$, ξ

Values for the fission spectrum fraction, χ, the number of neutrons emitted per fission, ν, the average cosine for elastic scattering, $\bar{\mu}$, and the average logarithmic energy decrement, ξ, are calculated for the generalized cross section set from the ENDF/B files.

An accurate expression for χ_{mg} is

$$\chi_{mg} = \frac{\int\limits_{g}\int\limits_{g'} \nu_m\left(E'\right) \sigma_{fm}\left(E'\right) \chi_m\left(E' \to E\right) \phi\left(E'\right) dE' dE}{\int\limits_{g}\int\limits_{g'} \nu_m\left(E'\right) \sigma_{fm}\left(E'\right) \phi\left(E'\right) dE' dE}, \tag{5.27}$$

where $\chi_m\left(E' \to E\right)$ is obtained from the ENDF/B files and $\phi\left(E'\right)$ is the fission spectrum or the asymptotic flux. The spectrum $\chi_m\left(E' \to E\right)$ is a slowly varying function of E' over the important energy range. Therefore, the MINX code allows the use of a simple approximation for χ_{mg},

$$\chi_{mg} = \int\limits_{g} \chi_m\left(E^* \to E\right) dE, \tag{5.28}$$

where $E^* = 1$ MeV.

The integrals for ν, $\overline{\mu}$, and ξ are

$$\nu_{mg} = \frac{\int\limits_{g} \nu_m(E)\, \sigma_{fm}(E)\, \phi(E)\, dE}{\int\limits_{g} \sigma_{fm}(E)\, \phi(E)\, dE}, \tag{5.29}$$

$$\bar{\mu}_{mg} = \frac{\int\limits_g \bar{\mu}_m(E)\,\sigma_{em}(E)\,\phi(E)\,dE}{\int\limits_g \sigma_{em}(E)\,\phi(E)\,dE}, \tag{5.30}$$

$$\xi_{mg} = \frac{\int\limits_g \xi_m(E)\,\sigma_{em}(E)\,\phi(E)\,dE}{\int\limits_g \sigma_{em}(E)\,\phi(E)\,dE}, \tag{5.31}$$

where $\phi(E)$ is the fission spectrum or the asymptotic flux.

5.5 Resonance Cross Sections

As discussed above, except for the lightest isotopes, elastic, capture, fission, and sometimes inelastic cross sections show sharp peaks and valleys due to resonances. For the lighter isotopes, it is practical to include all the pointwise detail of the cross sections directly. Good examples of this approach include carbon, ^{16}O, and ^{27}Al in the ENDF/B libraries.

For heavier isotopes, this approach would lead to very large files with tens of thousands of energy points. Furthermore, in the resonances, cross sections vary so rapidly with energy that it is unreasonable to include values at enough energy points in a data file such as ENDF/B to evaluate the integrals in Eqs. (5.23) and (5.24). Instead analytical expressions for resonance cross sections are used in these integrals.

Resonance cross sections of a material are dependent on the material temperature. The source of this dependence is the relative motion between target nuclei and neutrons, the phenomenon referred to as the *Doppler effect*. As the temperature of a material is raised, the thermal motion of the material nuclei increases, thus modifying the relative motion and changing the effective cross section.

The Doppler effect is very important in a fast reactor with ceramic fuel, as will be discussed in Chapter 6. The most important cross section with regard to the Doppler effect is the capture cross section of the fertile isotope, such as ^{238}U in the U-Pu fuel cycle. Of lesser importance are the capture and fission cross sections of ^{239}Pu and the capture cross section of ^{240}Pu The fuel isotopes are important because of the magnitude of the resonances and also because the fuel temperature changes over a wide range both between zero power and full power and in the event of an accidental power transient.

Temperature has only a small effect on the capture cross section of materials other than fuel and on resonance scattering cross sections. However, the temperature dependence of these cross sections can also be evaluated in fast reactor cross section codes such as MINX and the general purpose NJOY code.

The starting point for calculating absorption and scattering resonance cross sections is the Breit-Wigner single-level formula.[1] The resonance parameters in this relation are provided in the ENDF/B file. This relation is summarized in this section. (For an understanding of such parameters as wave numbers, statistical factors, angular momentum wave numbers, etc., the reader must go to more basic texts in reactor theory or nuclear physics.) The method used to modify the Breit-Wigner relation to account for the Doppler effect for the fuel isotopes is presented in Section 5.6.

Resonances can be subdivided into two groups—*resolved* and *unresolved*. For the resolved resonances the resonance parameters have been determined (or "resolved") experimentally. The unresolved resonances occur at higher energies, where the peak resonance cross sections are lower and

[1] Multilevel theory can also be used, and this method is available in addition to the single-level formulation in the legacy MINX code and in the general purpose NJOY code. Single-level theory is usually adequate, however.

the resonance parameters are not well established. For ^{238}U, capture resonances are resolved up to ~10 keV; for ^{239}Pu, fission and capture resonances are resolved only to a few tenths of a keV. Methods for obtaining effective cross sections in both the resolved and unresolved regions are described in this section.

The ENDF-format nuclear data evaluations use resonance cross section formulas to calculate the elastic, capture, and fission cross sections over a defined "resonance range." Only resonance parameters to use in these formulas for resolved and unresolved resonance ranges are given in ENDF sets. Several analytical models are supported:

- Single-Level Breit-Wigner,
- Multi-Level Breit-Wigner,
- Reich-Moore,
- Adler-Adler,
- Hybrid R-Function,
- Generalized R-Matrix.

The resonance data for each nuclide may be divided into several incident neutron energy ranges, given in order of increasing energy. The energy ranges for an isotope should not overlap; each may contain a different representation of the cross sections.

5.5.1 Resolved Resonances

The expressions for resonance cross sections are written here first for s-wave interaction only, for which the angular momentum quantum number, l, is zero, and then for the general case that includes higher angular momentum values. Below 100 keV, interactions with $l > 0$ contribute little.

If the resonances are well separated, the cross-section for a reaction x is given by the Breit-Wigner single level formula ($l = 0$):

$$\sigma_x(E) = \pi \lambda\!\!\!^{-2} g \frac{\Gamma_n \Gamma_x}{(E - E_0)^2 + \frac{1}{4}\Gamma^2}, \tag{5.32}$$

where:

$g = \frac{2J+1}{2(2I+1)}$ is the statistical weight,
J is the spin of compound nucleus (target + neutron),
I is the spin of target nucleus,
E_0 is the resonance energy,
$\lambda\!\!\!^{-}$ is the reduced de Broglie neutron wavelength,
E is the energy of incoming neutron or relative energy between neutron and target nucleus if it is not at rest,
$\Gamma = \Gamma_n + \Gamma_\gamma + \Gamma_f$ is the total width,
$\frac{\Gamma_n}{\Gamma}$ is the probability that the compound nucleus decays to the ground state by neutron emission,
$\frac{\Gamma_x}{\Gamma}$ is the probability that the compound nucleus decays to the ground state by x-channel.

For example, for absorption (capture plus fission), the Breit-Wigner single-level formula for the resonance cross section is:

$$\sigma_a(E) = \pi \lambda\!\!\!^{-2} g \frac{\Gamma_n(\Gamma_\gamma + \Gamma_f)}{(E - E_0)^2 + \frac{1}{4}\Gamma^2},$$

where $\Gamma_a = \Gamma_\gamma + \Gamma_f$.

Using $\frac{\Gamma_n}{\Gamma}$ explicitly, Eq. (5.32) can be re-written as:

$$\sigma_x(E) = \left(4\pi \lambda\!\!\!^{-2} g\right)\left[\frac{\Gamma_n}{\Gamma}\right]\frac{\Gamma_x}{\Gamma}\frac{1}{\left(\frac{2(E - E_0)}{\Gamma}\right)^2 + 1}. \tag{5.33}$$

Consequently, the Breit-Wigner single level formula for neutron absorption becomes:

$$\sigma_a(E) = \left(4\pi \lambda\!\!\!^{-2} g\right)\left[\frac{\Gamma_n}{\Gamma}\right]\frac{\Gamma_\gamma + \Gamma_f}{\Gamma}\frac{1}{\left(\frac{2(E - E_0)}{\Gamma}\right)^2 + 1},$$

where:

$\sigma_0 = \left(4\pi \lambda\!\!\!^{-2} g\right)\left[\frac{\Gamma_n}{\Gamma}\right]$ is the maximum value of the total cross section at $E = E_0$,

$\frac{\Gamma_\gamma + \Gamma_f}{\Gamma}$ is the probability that the compound nucleus will decay by γ-radiation emission or fission.

Using σ_0 and denoting $x = \frac{2(E - E_0)}{\Gamma}$,

$$\sigma_x(E) = \sigma_0 \frac{\Gamma_x}{\Gamma}\frac{1}{x^2 + 1}, \tag{5.34}$$

and the neutron absorption cross section:

$$\sigma_a(E) = \sigma_0 \frac{\Gamma_\gamma + \Gamma_f}{\Gamma}\frac{1}{x^2 + 1}.$$

As one can see from Eq. (5.33), the cross-section $\sigma_x(E)$ has a maximum at $E = E_0$ and reaches half of this maximum value at $E - E_0 = \Gamma/2$, so that Γ is the width at half height of the resonance. From Eq. (5.34) it follows that the cross section is symmetrical around the resonance energy $E = E_0$. The symmetrical function $1/\left(x^2 + 1\right)$ is called the *natural line shape*.

The Breit-Wigner formula cannot be extrapolated to $E \to 0$ because the parameter Γ_n can be considered independent of energy only in small intervals around the resonance. At very low energy:

$$\Gamma_n = \Gamma_{n0}\sqrt{E}.$$

This formula substituted into the Breit-Wigner expression gives the 1/v-behavior of the low-energy cross-section of most nuclides. This 1/v-part can often be calculated from the known resonance parameters, but for some nuclides (e.g. ^{232}Th) negative energy resonances (bound levels) have to be considered in order to reproduce the experimental 1/v-cross-sections.

In order to include contributions for higher-order angular-momentum numbers (i.e., $l > 0$), the cross section is

$$\sigma_x(E) = \sum_l \sigma_x^l(E), \tag{5.35}$$

where

$$\sigma_x^l(E) = \pi \lambda^2 \sum_J g_J \sum_{r=1}^{NR_J} \frac{\Gamma_{nr}\Gamma_{xr}}{(E - E_r')^2 + \frac{1}{4}\Gamma_r^2}. \tag{5.36}$$

The summation on J extends over all possible J-states for a particular l - state [18],

$$g_J = \frac{2J+1}{2(2I+1)(2l+1)}.$$

The r-summation limit, NR_J, is the number of resonances for a given pair of $\{l, J\}$ values. Details of the evaluation of the Γ_{nr}'s and Γ_r's and the difference between E_0 and E_r are given in Refs. [12, 14, 16, 19].

The scattering cross-section is composed of a fairly constant potential scattering ($\sigma_{s,pot.}$) and a strongly energy-dependent resonance scattering ($\sigma_{s,res.}$):

$$\sigma_{s,pot.} + \sigma_{s,res.}. \tag{5.37}$$

In the vicinity of resonances these two terms interfere so that the scattering cross-section can be smaller than the potential part alone. In general, we have

$$\sigma_s = \sigma_{s,pot.} + \sigma_{s,res.} + \sigma_{s,\text{int.}}, \tag{5.38}$$

where $\sigma_{s,\text{int.}}$ is the interference part of the scattering cross-section.

Considering as a reference frame a coordinate system centered on the center of mass of the two particles (nucleus and neutron) involved in the collision and defining $\cos\theta_{CM}$ as the cosine of the scattering angle, θ_{CM}, in this reference system, we can expand the scattering cross-section in Legendre polynomials:

$$\sigma_s(E, \cos\theta_{CM}) = \sum_{l=0}^{+\infty} \sigma_s^l(E) P_l(\cos\theta_{CM}). \tag{5.39}$$

The $l = 0$ gives the isotropic component (s-wave scattering) while the component with $l = 1$ is called p-wave scattering. The l-th partial wave of the scattering cross-section is given by:

$$\sigma_s^l(E) = \overbrace{\frac{\pi\lambda^2 g}{(E-E_0)^2 + \frac{1}{4}\Gamma^2}}^{\text{Resonance}} \left[\Gamma_n^2 - \overbrace{2\Gamma_n\Gamma_x \sin^2\delta_l + 2\Gamma(E-E_0)\sin 2\delta_l}^{\text{Interference}}\right] + \overbrace{4\pi\lambda^2 g(2l+1)\sin^2\delta_l}^{\text{Potential}},$$

where δ_l is the phase shift associated with the potential scattering.

For $l = 0$, the elastic scattering cross section is given by:

$$\sigma_s(E) = \sigma_0 \frac{\Gamma_n}{\Gamma} \frac{1}{1+x^2} + \left(\sigma_0 \sigma_{s,pot.} g \frac{\Gamma_n}{\Gamma}\right)^{1/2} \frac{2x}{1+x^2} + \sigma_{s,pot.},$$

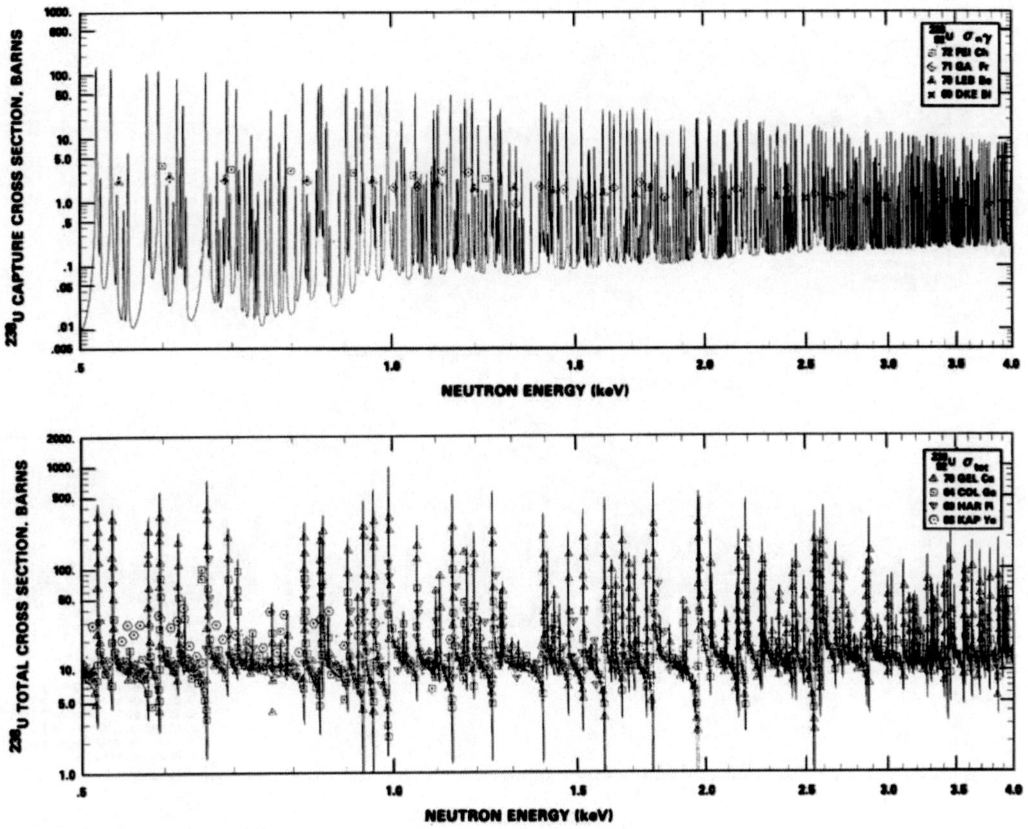

Fig. 5.4 Capture and total resonance cross sections for ^{238}U in the low keV energy range [20]

where $\sigma_{s,pot.}$ is the potential scattering cross section (which is energy independent).

Examples of resonance cross sections are given in Figs. 5.4 through 5.6 over resonance ranges of particular interest for fast reactors. Capture and total resonance cross sections are shown in Fig. 5.4 for ^{238}U. The scattering cross sections often contribute more to the total cross sections than the capture; hence, the large negative contribution from the interference line shape can be observed. Figure 5.5 shows results for iron. Figure 5.6 gives resonances for sodium.

5.5.2 Unresolved Resonances

At lower energies, resonances are well separated. But they become broader and closer together at higher energies so that closer together single resonance parameters can no longer be determined experimentally. The lower energy parameters are extrapolated for these higher energies and statistical distributions are used for the width and energy separation of these unresolved resonances.

The resolved resonances provide information on the number of resonances in a given energy range, or the *average spacing, D* (in units of energy), between resonances. They also provide estimates of the resonance parameters. Studying the statistical distribution of resolved resonance parameters, Porter and Thomas found that the parameters followed a chi-squared distribution in energy with varying degrees of freedom [21].

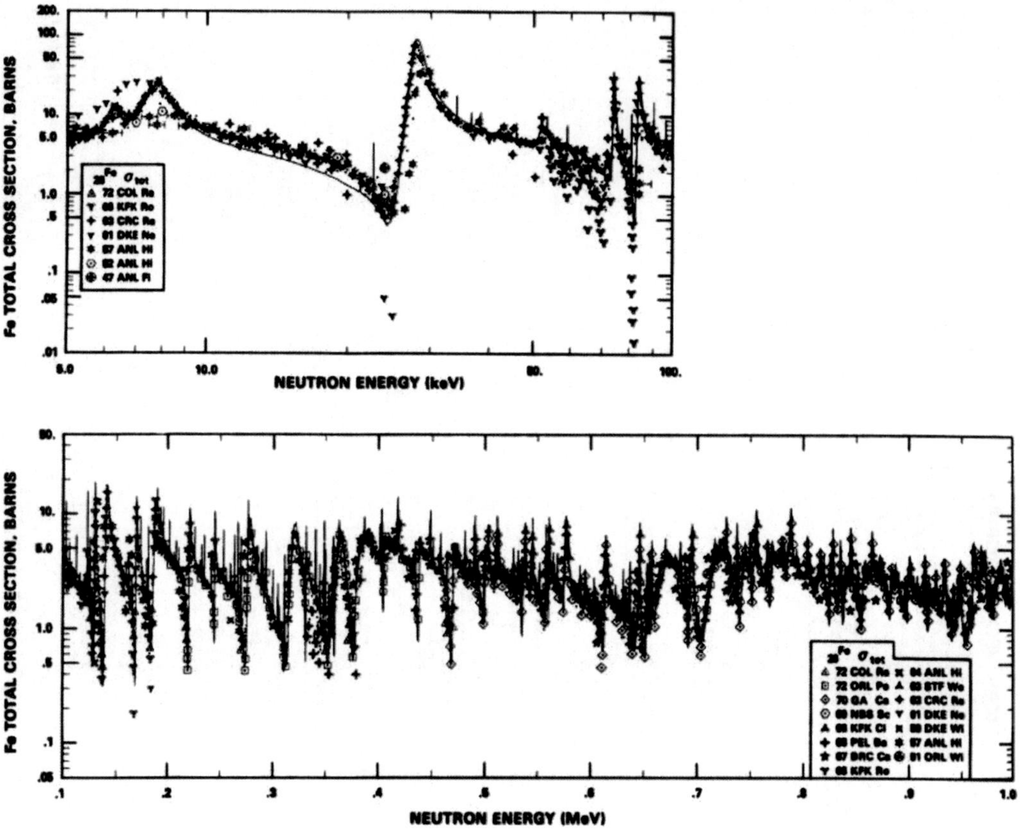

Fig. 5.5 Resonance cross sections for Fe [20]

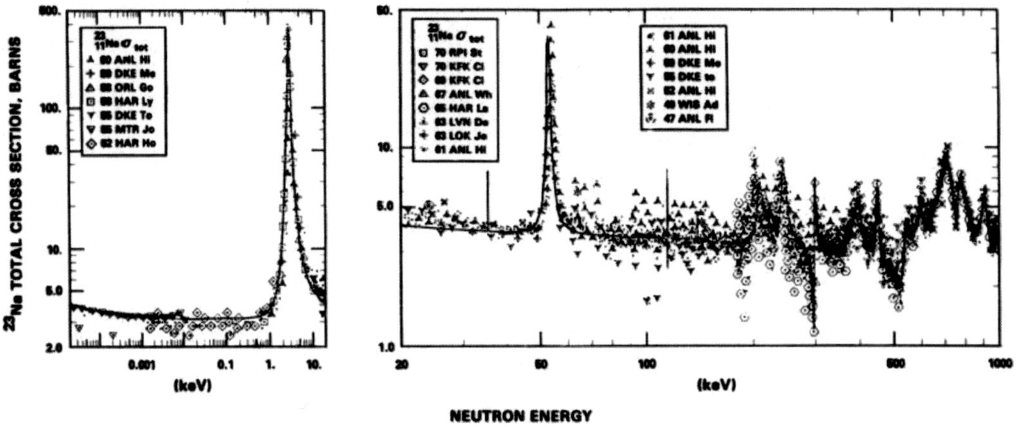

Fig. 5.6 Total cross section for sodium [20]

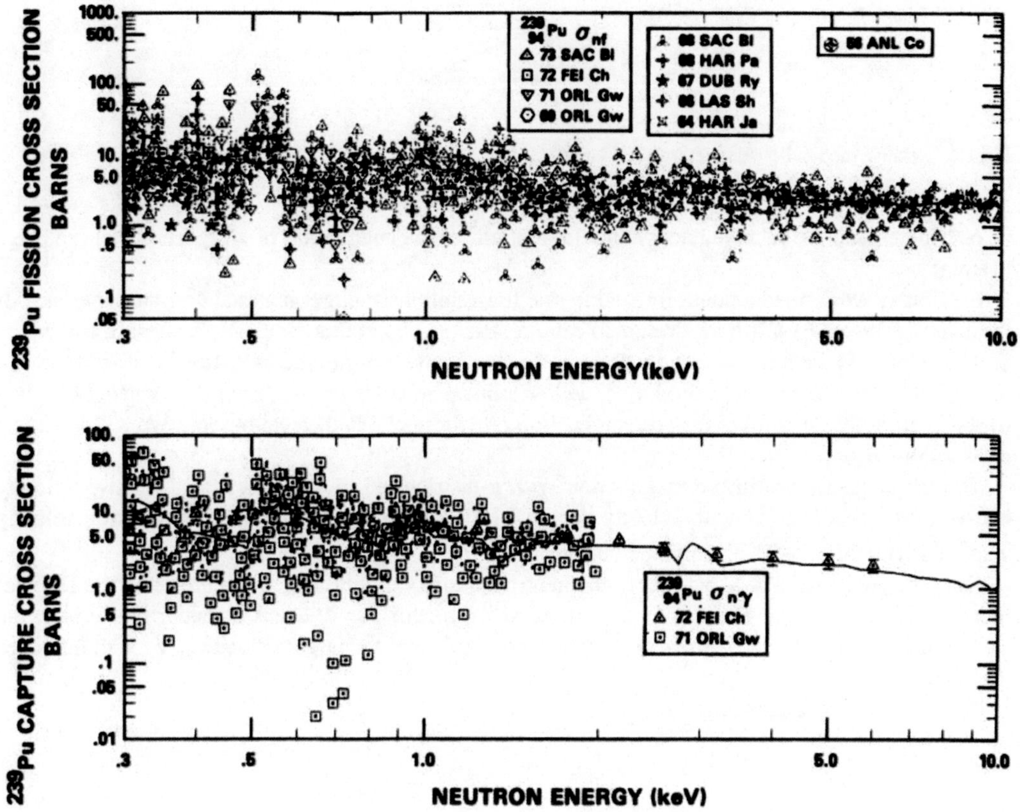

Fig. 5.7 Fission and capture cross sections in the unresolved resonance region for ^{239}Pu [20]

Instead of giving individual resonances with their energies and characteristic widths, ENDF-format evaluations give average values for the resonance spacing and the various characteristic widths, together with the probability distributions needed to describe the quantities [5].

Fission and capture cross sections for ^{239}Pu in the unresolved resonance range of principal interest are shown in Fig. 5.7. Below about 0.3 keV, the resonances are resolved. The difficulty of measuring α, the ratio of σ_c to σ_f, in this energy range led for a long time to a rather large uncertainty in the breeding ratio for a fast oxide-fueled reactor; however, this uncertainty has been substantially reduced as a result of extensive experimental investigations.

5.6 Doppler Broadened Cross Sections

5.6.1 General Theory

The energy E appearing in the Breit-Wigner formula is the relative energy between target nucleus and neutron. Therefore, the effective cross section of a material will be influenced by the thermal motion of the resonance nuclei. This means that the thermal motion of the nuclei of reactor material should

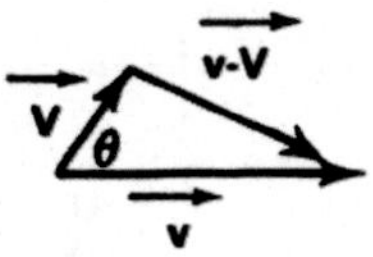

Fig. 5.8 Relative motion between neutron and target

be considered and the relative energy should be used for the calculation of all reactions taking place in a reactor.

Previously we defined a neutron reaction rate for a stationary target material (per atom per cm^3 of the material) as $\sigma_x(E)\,\phi$, where x refers to capture, fission, absorption, or elastic scattering and where $\phi = n\mathrm{v}$. Here n is the neutron density and v is the neutron velocity in the laboratory system. We must now redefine a cross section, called the *Doppler broadened* cross section, such that, when multiplied by the neutron flux ϕ, it gives the correct reaction rate for nuclei in thermal motion. We will label this cross section $\sigma_{x,Dop}$.

The target nuclei are in thermal motion with a distribution of velocities. The relative velocity between the target nucleus with velocity vector $\vec{\mathrm{V}}$ in the laboratory system and the neutron velocity vector $\vec{v}$ is equal to $\overrightarrow{\mathrm{v}-\mathrm{V}}$, as illustrated in Fig. 5.8.

Only the component of $\vec{\mathrm{V}}$ parallel to the neutron velocity $\vec{\mathrm{v}}$ will have an influence on the Doppler broadened cross section. Since we can assume any direction for $\vec{\mathrm{v}}$, let us consider that $\vec{\mathrm{v}}$ is in the z direction. Hence, only the distribution of the z-component of the target velocities, V_z, will influence the result.

The Doppler broadened cross section is

$$\sigma_{x,Dop}\phi = \sigma_{x,Dop}n\mathrm{v} = n\int_0^{\infty}\left|\overrightarrow{\mathrm{v}-\mathrm{V}}\right|\sigma_x(\mathrm{v}-\mathrm{V}_z)\,P(\mathrm{V}_z)\,d\mathrm{V}_z, \tag{5.40}$$

because σ_x, will vary with the z-component of the target velocity. Here $\left|\overrightarrow{\mathrm{v}-\mathrm{V}}\right|$ is the magnitude of the relative velocity. $P(\mathrm{V}_z)$ represents the Maxwellian distribution[2] of the z-component of the target velocities so that the fraction of target nuclei with velocities in $d\mathrm{V}_z$ is

$$P(\mathrm{V}_z)\,d\mathrm{V}_z = \sqrt{\frac{M}{2\pi kT}}\,e^{-\frac{M\mathrm{V}_z^2}{2kT}}\,d\mathrm{V}_z, \tag{5.41}$$

where M is the mass of the target nucleus, T is the temperature, and k is Boltzmann's constant.

Since $\mathrm{V} \ll \mathrm{v}$, $\left|\overrightarrow{\mathrm{v}-\mathrm{V}}\right|$ is sufficiently close to v that the v and $\left|\overrightarrow{\mathrm{v}-\mathrm{V}}\right|$ on the two sides of Eq. (5.40) can be cancelled:

$$\sigma_{x,Dop} = \int_0^{\infty}\sigma_x(\mathrm{v}-\mathrm{V}_z)\,P(\mathrm{V}_z)\,d\mathrm{V}_z; \tag{5.42}$$

[2]The Maxwellian distribution holds strictly only for a gas, whereas the resonance absorber is a solid. Above a temperature referred to as the Debye temperature (based on Debye's model for solids) this distribution is satisfactory for solids. The Debye temperature for most solids is of the order of 300 K: hence, the distribution holds for the entire range of fast reactor temperatures.

and, recognizing dependence on T in Eq. (5.41), we must have:

$$\sigma_{x,Dop}\left(\mathrm{v}_{rel.},T\right)=\int_{0}^{\infty}\sigma_{x}\left(\mathrm{v}_{rel.}\right)P\left(\mathrm{V}_{z},T\right)d\mathrm{V}_{z},\tag{5.43}$$

where:

$\mathrm{v}_{rel.}=\mathrm{v}-\mathrm{V}_{z}$ is the relative neutron-nucleus velocity, v is the neutron velocity, V_{z} is the velocity of the nucleus;
$P\left(\mathrm{V}_{z},T\right)$ is the probability of having the z-component of the target velocities, V_{z} ; and
T is the temperature.

In practice, one uses effective cross-sections in which only the neutron energy appears. We will next switch to an energy variable because we want $\sigma_{x,Dop}(E)$ for use in Eqs. (5.23) and (5.24), where E is the neutron energy in the laboratory system. Hence, we must consider the relationship between E, E_{CM} (the neutron energy in the center-of-mass system) and the velocity component V_{z}. The energy E_{CM}, is

$$E_{CM}=\frac{1}{2}\frac{Am_{n}}{A+1}\left|\overrightarrow{\mathrm{v}-\mathrm{V}}\right|^{2},\tag{5.44}$$

and

$$\begin{aligned}\left|\overrightarrow{\mathrm{v}-\mathrm{V}}\right|^{2}&=\mathrm{v}^{2}-2\left(\vec{\mathrm{v}}\cdot\vec{\mathrm{V}}\right)+\mathrm{V}^{2}\\&\cong\mathrm{v}^{2}-2\mathrm{vV}\cos\theta,\ \text{where } \theta \text{ is given in Fig. 5.8}\\&\cong\mathrm{v}^{2}-2\mathrm{vV}_{z},\ \text{since } \vec{\mathrm{v}} \text{ is postulated to be in the } z\text{ - direction.}\end{aligned}\tag{5.45}$$

The energy E is

$$E=\frac{1}{2}\frac{Am_{n}}{A+1}\mathrm{v}^{2}.\tag{5.46}$$

Substituting Eqs. (5.45) and (5.46) into Eq. (5.44) gives:

$$E_{CM}=E-\mathrm{V}_{z}\sqrt{2E\frac{Am_{n}}{A+1}},\tag{5.47}$$

and

$$d\mathrm{V}_{z}=\frac{dE_{CM}}{\sqrt{2E\dfrac{Am_{n}}{A+1}}}.\tag{5.48}$$

Substituting Eqs. (5.41), (5.47), and (5.48) into Eq. (5.42) gives:

$$\sigma_{x,Dop}(E)=\int_{0}^{\infty}\sigma_{x}\left(E_{CM}\right)\sqrt{\frac{M}{4\dfrac{Am_{n}}{A+1}\pi kTE}}\,e^{-\frac{M}{4\left(\frac{Am_{n}}{A+1}\right)kTE}\left(E-E_{CM}\right)^{2}}\,dE_{CM};\tag{5.49}$$

and recognizing dependence on T:

$$\sigma_{x,Dop}(E,T)=\int_0^\infty \frac{A\sigma_x(E_{CM})}{\sqrt{TE}}e^{-\frac{\pi A^2}{T}\frac{(E-E_{CM})^2}{E}}dE_{CM},\tag{5.50}$$

where $A=\sqrt{\frac{M}{4\frac{Am_n}{A+1}\pi k}}=$ const. Using Eq. (5.50) one can demonstrate that if $\sigma_x(E_{CM})$ follows the $1/\text{v}$-law, $\sigma_{x,Dop}(E,T)$ will also follow the $1/\text{v}$-law, independently of the temperature T. Of course, $\sigma_{x,Dop}\to\sigma_x$ if $T\to 0\text{K}$.

In the case of resonances, temperature has the effect of lowering the maximum of the resonance and broadening it at the same time (Doppler broadening). Substituting the Breit-Wigner expressions, one can obtain the Doppler broadened formulation of the resonance cross-section:

$$\sigma_{x,Dop}=\frac{\Gamma_x}{\Gamma}\sigma_0\psi(\zeta,\tau).$$

Therefore, the cross sections considered above become:

- absorption: $\sigma_{a,Dop}=\frac{\Gamma_\gamma+\Gamma_f}{\Gamma}\sigma_0\psi(\zeta,\tau)$;
- capture: $\sigma_{c,Dop}=\frac{\Gamma_\gamma}{\Gamma}\sigma_0\psi(\zeta,\tau)$;
- scattering: $\sigma_{s,Dop}=\frac{\Gamma_n}{\Gamma}\sigma_0\psi(\zeta,\tau)+\sigma_{s,pot.}+\sigma_{s,int.}$,

where:

$\sigma_0=\left(4\pi\lambda\!\!\!^{-2}g\right)\left[\frac{\Gamma_n}{\Gamma}\right]$ is the max. total cross section at $E=E_0$,
$\sigma_{s,\text{int.}}=\left(\frac{\Gamma_n}{\Gamma}\sigma_0\sigma_{s,pot}\right)^{1/2}\chi(\zeta,\tau)$is the interference scattering,
$\sigma_{s,pot.}$ is the potential scattering,
$\psi(\zeta,\tau)$ and $\chi(\zeta,\tau)$ are the so called shape functions:

$$\psi(\zeta,\tau)=\frac{\zeta}{4\pi}\int_{-\infty}^{+\infty}dy\frac{e^{-(\zeta^2/4)(\tau-y)^2}}{1+y^2},$$

$$\chi(\zeta,\tau)=\frac{\zeta}{4\pi}\int_{-\infty}^{+\infty}dy2y\frac{e^{-(\zeta^2/4)(\tau-y)^2}}{1+y^2}=2\tau\psi(\zeta,\tau)+\frac{4}{\zeta^2}\frac{\partial\psi(\zeta,\tau)}{\partial\tau},$$

$$\tau=\frac{2(E-E_0)}{\Gamma},$$

$$y=\frac{2(E_{CM}-E_0)}{\Gamma},$$

$$\zeta=\frac{\Gamma}{\left(\frac{4E_0kT}{A+1}\right)^{1/2}}=\frac{\Gamma}{\Gamma_D}\text{ is the natural to Doppler width,}$$

E_0 is the resonance energy.

In these relationships, the limits of integration of y correspond to the limits of integration of E_{CM}. The function $\psi(\zeta,\tau)$ is the Doppler broadened line shape replacing the natural line shape (see Eq. (5.34)). The function $\chi(\zeta,\tau)$ is the Doppler broadened counterpart of the interference line shape, $2x/\left(1+x^2\right)$. Like the interference line shape, $\chi(\zeta,\tau)$ is asymmetrical around E_0.

5.6.2 Absorption Cross Sections

Next, let us consider the Doppler broadened absorption cross section, $\sigma_a(E)$. For compactness, we drop the subscript "Dop" in the remainder of this section:

$$\sigma_a(E) = \frac{\Gamma_\gamma + \Gamma_f}{\Gamma}\sigma_0\psi(\zeta,\tau). \tag{5.51}$$

Here we will consider only s-wave interactions ($l = 0$). Like the natural line shape, the Doppler broadened line shape is symmetrical around the resonance energy E_0. Its value at E_0, however, is less than the value of the natural line shape at E_0. On the other hand its value far from E_0 is greater than the natural line shape. Thus, the effect of Doppler broadening on $\sigma_a(E)$ is to decrease the peak cross section at the resonance energy while increasing the cross section away from the resonance, i.e., "in the wings" of the resonance. This is illustrated schematically in Fig. 5.9.

A number of properties of Doppler broadened cross sections are described by Dresner [7] and others. For example, as the temperature, T, approaches zero, the Doppler broadened line shape, $\psi(\zeta,\tau)$, does indeed approach the natural line shape, as it should.

Another interesting property is that the integral over a resonance of a Doppler broadened absorption cross section weighted with an asymptotic energy spectrum, i.e., $\int \sigma_a(E)\, dE/E$—called the *resonance integral*—is the same as the resonance integral for the non-Doppler broadened cross section. The result is curious because, since dE/E is the weighting factor for the infinitely dilute cross section, this result says that the infinitely dilute group cross section is unaffected by Doppler broadening and, hence, by the temperature of the absorbing material. On the other hand, the important integral for effective cross sections, $\int \sigma_a(E)\,\phi(E)\,dE$, where $\phi(E)$ is the self-shielded flux given by Eq. (5.18), *does* depend on the temperature. The actual absorption rate is higher for the Doppler broadened cross section because the Doppler cross section is higher in the wings (where the flux, $\phi(E)$, is not greatly depressed), whereas *both* the Doppler and the non-Doppler cross sections are so high at the center of the resonance that the absorber is essentially black to neutrons regardless of Doppler broadening.

In other words, the lower peak Doppler cross section is just about compensated by a higher effective neutron flux at the center of the resonance, whereas the reverse is not the case out in the wings.

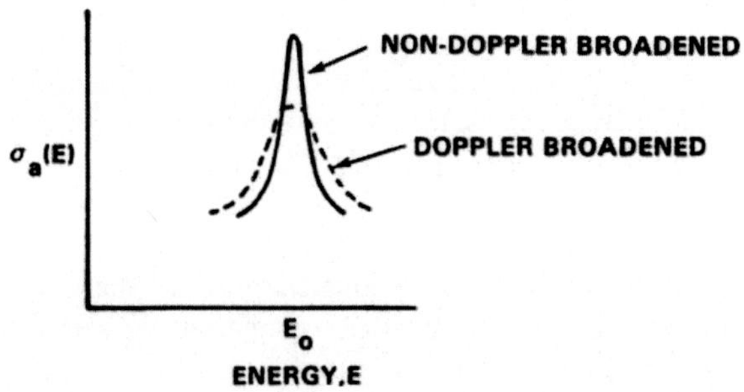

Fig. 5.9 Effect of Doppler broadening on $\sigma_u(E)$

The important result of this is that self-shielding of the neutron flux is responsible for the variation of the effective group cross section with temperature and for Doppler reactivity effects that will be discussed in Chapter 6. Without self-shielding, there would be no Doppler reactivity effect from temperature changes.

The influence of a temperature change on the effective cross section of a resonance absorber depends not only on the magnitude of the temperature change but also on the concentration of the absorber. This is because the self-shielding depends on the *macroscopic* cross section, Σ_t, and hence the atom density of the absorber. If there is very little of the resonance absorber present, the effective group cross section will hardly change with temperature. For this reason the change in the effective capture cross sections with temperature for the fertile material ^{240}Pu will be much lower than for the fertile material ^{238}U. Further subtleties in the behavior and effects of Doppler broadening appear when the effective cross sections are combined with atom densities and adjoint fluxes for the calculation of Doppler reactivities. That will be the subject of Chapter 6.

5.6.3 Resonance Overlap

A complication in the calculation of effective cross sections from Eqs. (5.23) and (5.24) not discussed in Section 5.4 is that there may be more than one resonance (i.e., from different materials) at about the same energy. Thus, parts of the two resonances may *overlap* and the assumption that σ_{om} is constant across a resonance in material *m* is, therefore, inaccurate.

The influence of resonance overlap on the absolute magnitude of an effective cross section is not great. However, the influence of overlap on the temperature dependence of cross sections, and hence on the Doppler effect, is quite important because here one is concerned with small changes in effective cross sections with temperature. This problem has been extensively analyzed by contributors to the theory of the Doppler effect in fast reactors, including Greebler and Hutchins, Nicholson, and Hwang and Hummel in the U.S. and Fischer, Rowlands, and Codd and Collins in Europe. These investigators showed, for example, that it was necessary to account for resonance overlapping in order to calculate the contribution of the fissile isotopes to the Doppler effect. A review of resonance overlapping is given by Hummel and Okrent [6]. The analysis is too complex to describe in this text, but it is accounted for in the generalized cross section codes—MINX, ETOX, ENDRUN, NJOY and all other modern cross section processing tools.

5.6.4 ENDF-Format Resonance Processing and Doppler Broadening

As noted above, the ENDF-format nuclear data evaluations contain only resonance parameters to use in analytical formulas. This poses challenges for cross section processing codes to consistently account for temperature changes.

To overcome the difficulties of Doppler broadening of R-Matrix cross sections, pole approximations equivalent to R-Matrix forms that facilitate the use of the familiar complex probability function have been developed [22, 23]. The real and imaginary parts of this function are the usual $\psi(\zeta,\tau)$ and $\chi(\zeta,\tau)$ functions as discussed above. This makes it easy to continue the use of many of the existing methods in reactor applications that depend on traditional Doppler broadening functions and the rigor of cross sections based on R-Matrix formalism can be preserved by the use of these rigorous pole representations. Apart from the convenience, the use of these familiar functions has intuitive appeal. In this connection, a method based on the application of Steffensen's inequality has been developed [24] that is capable of computing a complex probability function to any desired accuracy.

5.7 Cross Sections for a Specific Reactor Composition

5.7.1 General Method

The second part of the shielding factor method is the calculation of effective cross sections for each material from the generalized cross sections obtained in the previous part. This is done in the SPHINX [13] 1DX [15] and TDOWN [17] codes. To do this, it is necessary first to calculate the background cross section, σ_o for each material and then to interpolate for the appropriate shielding factor from the values of $f(\sigma_o)$ found in the generalized cross section calculations.

An evaluation of the effective cross section for each material is an iterative process because the correct σ_o for a given material depends on the effective total cross sections of all the other materials present. The iterative scheme proceeds as follows.

First iteration:

- Make a first guess for σ_{tmg} for each material in each energy group. The infinitely dilute value represents a reasonable first choice.
- Calculate σ_{omg} for all materials, using the first guesses for σ_{tmg}.
- Obtain the shielding factor for the total cross section for each material, f_{tmg} from the generalized cross section correlations of f_{tmg} *vs* σ_o.
- Calculate σ_{tmg} for each material and each group from Eq. (5.1).

Second and subsequent iterations:

- With the new values of σ_{tmg} recalculate all σ_{omg} 's.
- Obtain new values of f_{tmg}.
- Obtain new values of σ_{tmg}.
- Compare each σ_{tmg} with the σ_{tmg} from the previous iteration. Continue the iteration procedure until the fractional change in every value of σ_{tmg} between iterations is less than a selected convergence criterion.

After convergence is achieved, we have the final values of both σ_{tmg} and σ_{omg} for each m. At this time all other f_{xmg} factors (i.e., those for capture, fission, and elastic scattering) and σ_{xmg} values are calculated. The process usually converges in a few iterations.

The inelastic scattering removal cross section, together with the inelastic matrix, is carried over directly from the generalized cross section set; it is not composition dependent. (Recall that inelastic scattering occurs at high enough energies that self-shielding is not important.)

5.7.2 Elastic Removal Cross Section

For most cross sections, the flux distribution $\phi(E)$ used in the generalized calculations is adequate for obtaining the specific composition cross sections. The elastic removal cross section, however, may be sensitive to the energy variation in flux across an energy group. particularly if the energy width of the group is wide relative to the maximum energy loss in an elastic collision—a situation frequently encountered in fast reactor analysis.

5.7.3 Correction for Heterogeneous Geometry

The expressions derived thus far have assumed a homogeneous reactor composition. Little error results from this assumption for most fast reactor calculations because the mean free paths are large relative to the fuel-rod and coolant-channel dimensions. In the resonance region, however, mean free paths are shorter where the resonance material is located—for example, in the fuel pin when a ^{238}U or Pu resonance is being considered. Hence, heterogeneity will have some influence in the resonance region.

A simple correction to account for the effect of heterogeneity on self-shielding for the fuel isotopes is based on the *rational approximation* (sometimes called the Wigner or canonical approximation) [24]. This method is based on an approximation of the neutron leakage probability from a fuel rod. For repeating lattices of rods separated by moderator or, in the case of fast reactors, by cladding and sodium, a modification of the rational approximation provided by Bell [25] gives an appropriate correction.

A detailed treatment of the core lattice on a scale sufficiently fine to account for the spatial variation of the flux in the neighborhood of a given fuel element is clearly a computationally expensive procedure, since it would require an unmanageably large array of mesh points in a multigroup diffusion calculation. Indeed, due to their strongly absorbing nature, fuel and control elements in the core frequently require a more accurate treatment of neutron transport than that provided by diffusion theory. Hence, a computationally efficient, yet sufficiently accurate modeling technique, must be developed and implemented to account for the heterogeneity effects. Among the various factors under consideration, the choice of heterogeneity treatment mainly depends on the specific purpose of the calculation (as it determines the required accuracy), and on the specific reactor features (as it determines the potential complexity). The primary goal of a heterogeneity treatment is usually to calculate effective group constants that have been spatially averaged over the flux distribution in a lattice cell and, therefore, can be used to characterize the cell in subsequent multigroup diffusion calculations in which the cell structure is ignored.

5.8 Collapsing of Multigroup Cross Sections

5.8.1 Fine-Group Cross Section Libraries and Cross Section Collapsing

It is now routine for reactor calculations to be performed using detailed spectrum calculations in a high number of energy groups, $\vec{\Phi}_{\text{fine}}$, with only rough approximation of the spatial dependence, followed by a few-group space-dependent calculations for $\vec{\phi}_{\text{broad}}(\vec{r})$ with broad-group cross sections $\{\Sigma_{\text{broad}}\}$ calculated by averaging the fine group cross sections on the spectrum $\vec{\Phi}_{\text{fine}}$:

$$\boxed{\begin{matrix}\vec{\Phi}_{\text{fine}}\\ \{\Sigma_{\text{fine}}\}\end{matrix}} \to \boxed{\begin{matrix}\vec{\phi}_{\text{broad}}(\vec{r})\\ \{\Sigma_{\text{broad}}\}\end{matrix}}.$$

To optimize computer time expenses, in the initial stages of the design of a reactor, the computations are usually carried out with only a few groups. These are sufficient for providing information regarding the general properties of the system. More detailed multigroup calculations are performed in the later stages in the design process. Detailed multigroup calculations involving fine group libraries with many energy groups are also frequently performed for the sole purpose of providing values of the group constants for a few-group calculations with broad group libraries. It is sometimes useful to use few-group cross sections, or even, for fuel cycle analysis, one-group cross sections. For one dimensional design calculations for specific reactor compositions, it may be sufficient to use 10–30 groups, and for two- or three-dimensional analysis even fewer groups. For fuel cycle analysis, relative reaction rates are important and sufficient accuracy can be achieved with one-group capture and absorption cross sections—although a single effective one-group value may have to be changed for different stages in the burnup cycle. For these reasons it is necessary to have techniques for collapsing cross sections from many groups to fewer groups.

The ENDF/B file contains almost continuous data on cross sections as a function of energy. The generalized cross sections may involve from 100 to 500 energy groups, and in certain regions of rapidly varying cross sections, even more groups. It is necessary to have techniques for:

- generating the generalized multi-group cross sections, and
- collapsing cross sections from many groups to fewer groups.

Such calculations begin with a fine-group library $\{\Sigma_{\text{fine}}\}$ in a fine energy structure (hundreds or thousands groups) including cross-sections for all reactions of interest, scattering data, resonance parameters, fission spectra, and etc. The fine group libraries must be produced and regularly updated starting from the ENDF/B cross-section sets. Special processing codes are used in order to generate from ENDF/B a fine group cross-section library, $\{\Sigma_{\text{fine}}\}$, usable in computer codes for multi-group calculations [8]. The fine group cross sections must be obtained from the point values of ENDF/B, assuming a certain form of the flux within the fine energy groups. When resonance data are available, resonance contributions can be calculated using above-described methods. It is possible in some processing codes to reduce the data storage requirement by means of polynomial fittings. Using $\vec{\Phi}_{\text{fine}}$ as a weighting function, the fine-group library is then collapsed into the broad-group cross section library, $\{\Sigma_{\text{broad}}\}$.

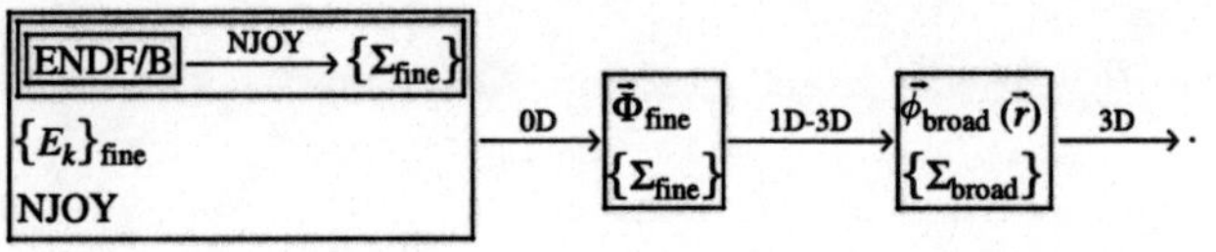

The broad group structure is assumed to be obtained from the fine group structure by collapsing several fine groups into one broad group. Two approaches to collapsing cross sections can be followed. The first conserves reaction rates. The second conserves reactivity worths. The first is more often used, and it will be described here.

The multigroup cross section collapsing is equivalent to the cross section averaging over the appropriate energy intervals employing fine group scalar flux components, ϕ_g, and fine group reaction rate densities, $\sigma_{xg}\phi_g$, as weighting functions. It has to be noted that fine and broad group cross section libraries contain group cross sections in the microscopic form. Since microscopic cross sections are composition independent, the same cross section libraries can be used for modeling of reactor configurations with different compositions if configurations are characterized by similar neutron spectra. Figure 5.10 illustrates the process of cross section collapsing.

Suppose several *fine groups,* with index g are to be collapsed into one *broad group* with index G as shown in Fig. 5.10. The lower energy of the new broad group is E_G. For capture, fission and elastic scattering for each material, the collapsed cross section is computed conserving the corresponding

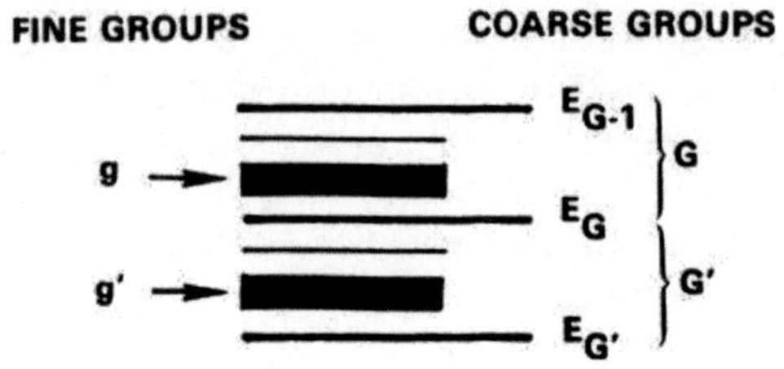

Fig. 5.10 Group structure for group collapsing

reaction rates (omitting the material index, *m*):

$$\sigma_{xG}\phi_G = \int_{E_G}^{E_{G-1}} \sigma_x(E)\,\phi(E)\,dE = \sum_{g\in G} \sigma_{xg}\phi_g, \tag{5.52}$$

where $g \in G$ means that the fine group g is contained in the broad group G. Since $\phi_G = \sum_{g\in G} \phi_g$,

$$\sigma_{xG} = \frac{\sum_{g\in G} \sigma_{xg}\phi_g}{\sum_{g\in G} \phi_g}. \tag{5.53}$$

For σ_{tr}, the leakage rate is to be conserved. The collapsed inelastic scattering cross section matrix is

$$\sigma_{iG\to G'} = \frac{\sum_{g\in G} \phi_g \sum_{g'\in G'} \sigma_{ig\to g'}}{\sum_{g\in G} \phi_g}, \tag{5.54}$$

where G' is any group other than G.

5.8.2 Energy Group Structures

Although some attempts of standardization of the fine group structure have been made, there are considerable differences between various $\{\Sigma_{\text{fine}}\}$—libraries. These differences have historical and physical reasons. Usually group structures are defined by trying to separate, if possible, different types of nuclear reactions or other significant phenomena. This separation, which is sometimes necessary for accuracy reasons, provides better information on the effect of different reactions. It is then possible to distinguish a range of the fission source spectrum, ranges of unresolved and of resolved resonances, and a thermal energy range.

In order to limit the computer time, the number of broad groups in $\{\Sigma_{\text{broad}}\}$ used in diffusion and transport calculations should be kept as low as possible, as long as sufficient accuracy can be guaranteed. This leads to different group structures for the treatment of different problems. While k_{eff} can be calculated with sufficient accuracy with relatively few energy groups, an accurate power distribution in regions with strong space-energy dependence of the neutron flux (for example, core-reflector interface regions) requires a high number of energy groups.

Particular energy partitions are used in the interpretation of experiments. A high number of fast groups may be required if the core calculation has to provide a neutron source to be used in subsequent shielding calculations. Considerable detail in the fast spectrum may also be required for the calculation of neutron damage in reactor materials. In the case of calculations with a very limited number of groups, care must be taken to avoid the artificial transfer of neutrons to energy ranges they cannot physically reach. The accuracy of multigroup calculations naturally increases as the number of groups is increased, and computations involving several hundreds of groups have been carried out routinely.

5.8.3 Homogenization of the Lattice Cell

In spectrum calculations, one has the choice between a multigroup space-dependent transport treatment, and a calculation on a homogenized unit cell. The homogenization of the lattice cell implies a previous few-group transport calculation on the unit cell. Once the fine structure of the flux within the unit cell is known, the homogenization can be performed multiplying the cross-sections by energy-dependent flux disadvantage (self-shielding) factors:

$$\sigma_{x,g}^{(Cell)} = \sum_{k=1}^{\text{Cell Regions}} \sigma_{x,g}^{(k)} \cdot \left(\frac{\phi_g^{(k)}}{\phi_g^{(Cell)}} \right) \cdot \left(\frac{V_k}{V_{Cell}} \right). \tag{5.55}$$

The region-wise averaged flux ratios $\frac{\phi_g^{(k)}}{\phi_g^{(Cell)}}$ perform the required cross section adjustment to account for the spatial variation of the neutron flux within the lattice. These self-shielding factors are defined for each region in such a way as to give the same reaction rate in the homogenized calculation as in the real heterogeneous unit cell:

$$RR_{x,g}^{(Cell)}\Big|_{\substack{\text{Equivalent}\\ \text{Homogeneous Cell}}} = RR_{x,g}^{(Cell)}\Big|_{\substack{\text{Heterogeneous}\\ \text{Fuel-Moderator Cell}}} . \tag{5.56}$$

The homogenized calculation deals only with a reference flux while in reality the flux depends on the position within unit cell.

5.8.4 Problem-Specific Multigroup Cross Section Libraries

The procedure to generate problem-specific multigroup cross sections typically consists of the following steps:

- ENDF file processing to extract the composition-specific nuclear data,
- Processing of the resolved resonance data,
- Processing of the unresolved resonance data,
- Preparation of the multigroup master library (fine-group library $\{\Sigma_{\text{fine}}\}$),
- Fine-group transport calculations or Monte Carlo calculations to collapse cross sections with the proper problem-specific energy spectrum and prepare the multigroup working library (broad-group library $\{\Sigma_{\text{broad}}\}$),

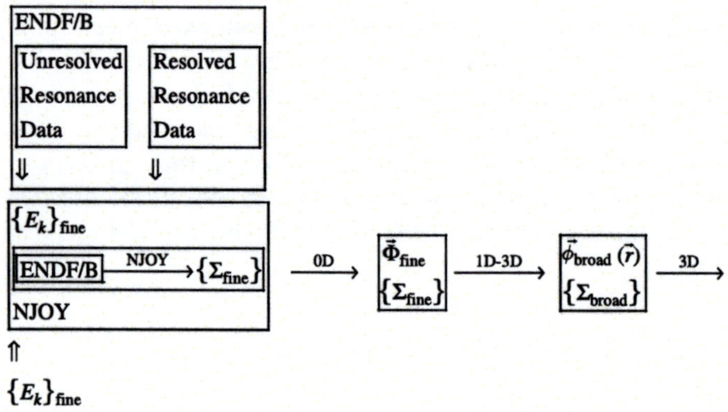

The details of the modern fast reactor calculations are discussed in Appendix E.

References

1. I. I. Bondarenko, et al., *Group Constants for Nuclear Reactor Calculations*. Translation—Consultants Bureau Enterprises, Inc., New York, NY (1964).
2. R. B. Kidman, R. E. Schenter, R. W. Hardie, and W. W. Little, "The Shielding Factor Method of Generating Multigroup Cross Sections for Fast Reactor Analysis", *Nuclear Science and Engineering*, **48** (1972) 193–201.
3. H. Henryson, B. J. Toppel, and C. G. Stenberg, "MC^2-2: A Code to Calculate Fast Neutron Spectra und Multigroup Cross Sections", ANL-8144, Argonne National Laboratory, Argonne, IL (1976).
4. M. K. Drake, "Data Formats und Procedures for the ENDF Neuron Cross Section Library", BNL-50274, Brookhaven National Laboratory, New York, NY (April 1974 Revision).
5. M. B. Chadwick, P. Oblozinsky, M. Herman, N. M. Greene, R. D. McKnight, D. L. Smith, P. G. Young, R. E. MacFarlane, G. M. Hale, et al., "ENDF/B-VII.0: Next Generation Evaluated Nuclear Data Library for Nuclear Science and Technology", *Nuclear Data Sheets*, **107** (2006) 2931–3060.
6. H. H. Hummel and D. Okrent, *Reactivity Coefficients in Large Fast Power Reactors*, American Nuclear Society, LaGrange Park, IL (1970).
7. L. Dresner, *Resonance Absorption in Nuclear Reactors*, Pergamon Press, New York, NY (1960).
8. R. E. MacFarlane and D. W. Muir, "The NJOY Nuclear Data Processing System Version 91", LA-12740-M, Los Alamos, NM (1994).
9. A. Fessler and D. L. Smith, "Parameter Sensitivities in Nnuclear Reaction Cross-Section Calculations", *Annals of Nuclear Energy*, **29** (4)(2002) 363–384.
10. P. Finck, D. Keyes, and R. Stevens, "Workshop on Simulation and Modeling for Advanced Nuclear Energy Systems", , DOE, Washington, DC (August 15–17, 2006).
11. L. Schroeder and E. Lusk, "Report of the Nuclear Physics and Related Computational Science R&D for Advanced Fuel Cycles Workshop", Bethesda, MD, (August 10–12, 2006).
12. C. R. Weisbin, P. D. Soran, R. E. MacFarlane, D. R. Morris, R. J. LaBauve, J. S. Hendricks, J. E. White, and R. B. Kidman, *MINX: A Multigroup Integration of Nuclear X-Sections from ENDF/B*. LA-6486-MS, Los Alamos Scientific Laboratory, Los Alamos, NM (1976).
13. W. J. Davis, M. B. Yarborough, and A. D. Bortz, "SPHINX : A One-Dimensional Diffusion and Transport Nuclear Cross Section Processing Code", WARD-XS-3045-17, Westinghouse, Advanced Reactors Division, Madison, PA (1977).
14. R. E. Schenter, J. L. Baker, and R. B. Kidman, "ETOX, A Code to Calculate Group Constants for Nuclear Reactor Calculations", BNWL-1002, Battelle Northwest Laboratory, Washington, DC (1969).
15. R. W. Hardie and W. W. Little Jr., "1DX, A One Dimensional Diffusion Code for Generating Effective Nuclear Cross Sections", BNWL-954, Battelle Northwest Laboratory, Washington, DC (1969).
16. B. A. Hutchins, C. L. Cowan, M. D. Kelly, and J. B. Turner, "ENDRUN-II, a Computer Code to Generate a Generalized Multigroup Data File from ENDF/B", GEAP-13704, General Electric Company, USA (March 1971).
17. C. L. Cowan, B. A. Hutchins, and J. E. Turner, "TDOWN—A Code to Generate Composition and Spatially Dependent Cross Sections", GEAP-13740, General Electric (1971). (See also R. Protsik, E. Kujawski, and C. L. Cowan, "TDOWN-IV", GEFR-00485, General Electric, 1979).

18. K. O. Ott and W. A. Bezella, *Introductory Nuclear Reactor Statics*, American Nuclear Society, La Grange Park, IL (1989).
19. K. Gregson, M. F. James, and D. S. Norton, "MLBW—A Multilevel Breit- Wigner Computer Programme", UKAEA Report AEEW-M-517, UKAEA, UK (1965).
20. D. I. Garber and R. R. Kinsey, *Neutron Cross Sections, Vol. II, Curves*, 3rd ed. BNL 325, Brookhaven National Laboratory, Upton, NY (January 1976).
21. C. E. Porter and R. G. Thomas, "Fluctuation of Nuclear Reaction Widths", *Physical Review*, **104** (1956) 483.
22. K. Devan and R. S. Keshavamurthy, "Extension of Rational Approximations to p-wave Collision Amplitudes in Reich-Moore Formalism", *Annals of Nuclear Energy*, **28** (2001) 1013.
23. R. N. Hwang, "An Overview of Resonance Theory in Reactor Physics Applications", *Transactions of the American Nuclear* Society, **91** (2004) 735.
24. W. M. Stacey, *Nuclear Reactor Physics*, 439–452, Wiley-VCH, Berlin (2007).
25. G. I. Bell, "A Simple Treatment of the Effective Resonance Absorption Cross Sections in Dense Lattices", *Nuclear Science and Engineering*, **5** (1958) 138.

Chapter 6
Kinetics, Reactivity Effects, and Control Requirements

Pavel Tsvetkov, Alan Waltar, and Donald Todd

6.1 Introduction

Values for reactivity effects are required both for transient safety analysis and for control requirements during normal operation. Reactivity effects of importance in fast reactor design and safety include (1) effects of dimensional changes in core geometry, (2) the Doppler effect, (3) effects of sodium density changes or loss of sodium, and (4) long-term reactivity loss from fuel burnup.

The reactor control system must compensate for these reactivities during normal operation and provide sufficient margin to handle off-normal situations.

We begin this chapter with a review of the reactor kinetics equations (Section 6.2). We then proceed to discuss adjoint flux and perturbation theory (Section 6.3) since these are needed for an understanding of reactivity effects. Kinetics parameters $\overline{\beta}$ and l, the effective delayed neutron fraction and the neutron lifetime respectively, are then discussed and the differences in these values between fast and thermal reactors are presented (Section 6.4). Sections 6.5 through 6.7 cover the first three categories of reactivity effects. Section 6.8 is addressed to reactivity worth distribution, and the final section discusses the control requirements for a fast reactor. A detailed discussion of the fourth category of reactivity, that associated with fuel burnup, will be delayed until Chapter 7, although sufficient information will be summarized in this chapter to define the control requirements.

6.2 Reactor Kinetics

Reactor kinetics equations for both fast and thermal reactors are identical. However, point kinetics approximations can be used more effectively for fast reactors than for thermal reactors because fast reactors are more tightly coupled neutronically. Tighter coupling implies that the neutron flux is more nearly separable in space and time, which is a necessary condition for point kinetics approximations to be valid. Fast reactor safety codes to date have therefore generally employed point kinetics.[1] Inevitably, as commercial fast reactors reach the 1,000–2,000 MWe range, and particularly if they employ heterogeneous cores, problems involving space-time kinetics will arise. Such considerations, however, are beyond the scope of this text.

[1] Though the point kinetics formulation is normally employed in fast reactor safety codes, the reactivity feedback terms due to reactor temperature variations and material motion generally contain spatial effects.

P. Tsvetkov (✉)
Zachry Engineering Center, Department of Nuclear Engineering, Texas A&M University, College Station, TX, USA
e-mail: tsvetkov@tamu.edu

A.E. Waltar et al. (eds.), *Fast Spectrum Reactors*, DOI 10.1007/978-1-4419-9572-8_6,

6.2.1 Point Kinetics Equations

The point kinetics equations can be written as

$$\frac{dn}{dt} = \frac{\rho - \bar{\beta}}{\Lambda} n + \sum_{i=1}^{6} \lambda_i C_i \tag{6.1}$$

$$\frac{dC_i}{dt} = \frac{\bar{\beta}_i}{\Lambda} n - \lambda_i C_i \quad (i = 1 \text{ to } 6) \tag{6.2}$$

where

n = neutron density (n/cm^3),
C_i = delayed neutron precursor concentration for the ith group (precursors/cm^3),
$\rho = (k-1)/k = \delta k/k$ = reactivity,
$\overline{\beta}$ = effective delayed neutron fraction,
$\overline{\beta_i}$ = effective delayed neutron fraction for the ith group,
Λ = neutron generation time (s),
λ_i = decay constant for the ith delayed neutron group (s^{-1}).

The neutron generation time, Λ (which measures the neutron birth-to-birth time) is related to the neutron lifetime, l (which measures neutron birth-to-death) as follows:

$$l = k\Lambda\ . \tag{6.3}$$

This relationship implies that l exceeds Λ when the power is rising ($k > 1$) and that Λ exceeds l when the power is dropping ($k < 1$). Because k is always very close to unity, even for large postulated accident conditions, both terms are loosely referred to as neutron lifetime.

As in the case of thermal reactors, reactivity for fast reactors is often reported in units of the *dollar*,[2] which is obtained by dividing the absolute value of ρ by $\overline{\beta}$. For reactivities less than 1 dollar (i.e., $\rho < \overline{\beta}$), the first term on the right hand side of Eq. (6.1) is small relative to the second term and the reactor kinetics equations are controlled by delayed neutrons. This is as valid for fast reactors as it is for thermal reactors. Hence, normal startup and shutdown and power level changes proceed in a fast reactor exactly as in a thermal reactor.

It is only when the net reactivity approaches or exceeds $\overline{\beta}$ that the behavior of a fast reactor differs significantly from that of a thermal reactor. This difference arises because of the very much smaller prompt neutron lifetime for the fast spectrum reactors ($\sim 4 \times 10^{-7}$ s) relative to that of the LWR ($\sim 2 \times 10^{-5}$ s). Even for prompt critical conditions, however, the transient behavior of an fast spectrum reactors is smooth and predictable. This can be shown by observing the solution to Eqs. (6.1) and (6.2) for neutron density (using a one-delayed group precursor approximation):

$$\frac{n}{n_0} = \frac{\bar{\beta}}{\bar{\beta} - \rho} \exp\left(\frac{\lambda \rho}{\bar{\beta} - \rho} t\right) - \frac{\rho}{\bar{\beta} - \rho} \exp\left(-\frac{\bar{\beta} - \rho}{\Lambda} t\right) \tag{6.4}$$

where λ represents a weighted decay constant for all six delayed neutron groups. For $\rho < \overline{\beta}$ the second term drops out rapidly, leaving the first term to express the long term transient response. Within this region, the reactor period, τ, is approximated by

[2]Reactivity is sometimes reported in units of cents. One cent is 0.01 dollars.

$$\tau \cong \frac{\bar{\beta} - \rho}{\lambda \rho} \quad \left(\text{for } \rho < \bar{\beta}\right).$$

For $\rho > \overline{\beta}$ the second term in Eq. (6.4) becomes positive in sign and large in magnitude (due to the very small value for Λ). The reactor period is then approximated by

$$\tau \cong \frac{\Lambda}{\rho - \bar{\beta}} \quad \left(\text{for } \rho > \bar{\beta}\right).$$

A plot of reactor period vs. reactivity is shown in Fig. 6.1. Note that for a reactivity of less than about \$0.90 the period is unaffected by neutron lifetime (i.e., fast and thermal reactors behave identically). Even for prompt critical conditions, behavior is smooth and reactor response can be accurately predicted—as graphically demonstrated in the SEFOR program [1, 2], in which the reactor was purposely made super prompt critical to demonstrate the capability of the Doppler effect to arrest a transient.

The absolute value for the reactivity corresponding to one dollar is lower for a fast reactor (using plutonium as the principal fissile material) relative to a thermal reactor (using principally enriched uranium) since β for ^{239}Pu is only 0.00215 compared to 0.0068 for ^{235}U. Delayed neutron data for fast fission in these and other relevant isotopes are listed in Table 6.1. Methods for calculating β and $\overline{\beta}$ are presented later in Section 6.4.

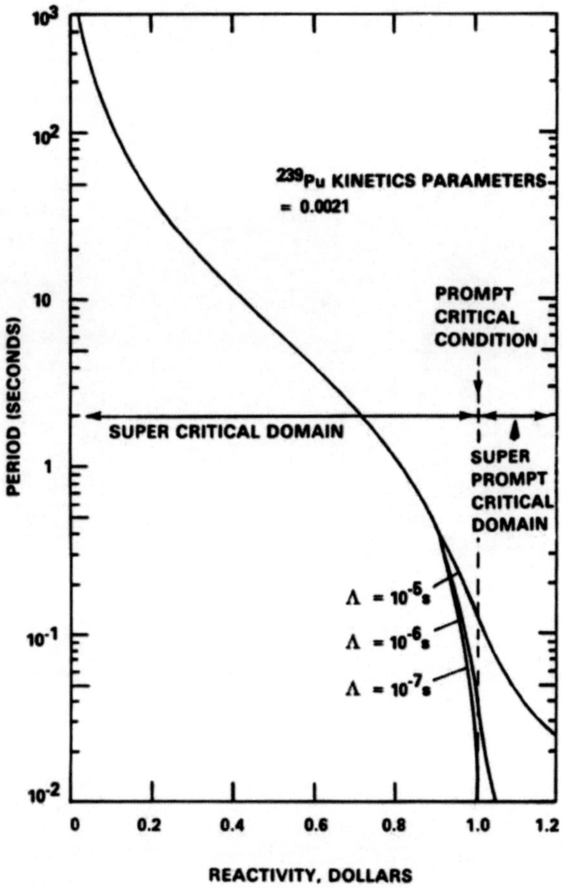

Fig. 6.1 Plot of reactor period vs. reactivity for ^{239}Pu fueled fast spectrum reactors

Table 6.1 Delayed neutron parameters

Isotope	Delayed[a] neutrons per fission	Total[b] neutrons per fission, ν	β	Group	$\lambda_i{}^c(s^{-1})$	β_i/β^c
^{232}Th	0.0531 ±0.0023	2.34	0.0227	1	0.0124	0.034
				2	0.0334	0.150
				3	0.121	0.155
				4	0.321	0.446
				5	1.12	0.172
				6	3.29	0.043
^{233}U	0.00731 ±0.00036±	2.52	0.0029	1	0.0126	0.086
				2	0.0334	0.274
				3	0.131	0.227
				4	0.302	0.317
				5	1.27	0.073
				6	3.13	0.023
^{235}U	0.01673 ±0.00036	2.45	0.0068	1	0.0127	0.038
				2	0.0317	0.213
				3	0.115	0.188
				4	0.311	0.407
				5	1.40	0.128
				6	3.87	0.026
^{238}U	0.0439 ±0.0010	2.77	0.0158	1	0.0132	0.013
				2	0.0321	0.137
				3	0.139	0.162
				4	0.358	0.388
				5	1.41	0.225
				6	4.02	0.075
^{239}Pu	0.00630 ±0.00016	2.93	0.00215	1	0.0129	0.038
				2	0.0311	0.280
				3	0.134	0.216
				4	0.331	0.328
				5	1.26	0.103
				6	3.21	0.035
^{240}Pu	0.0095 ±0.0008	3.07	0.0031	1	0.0129	0.028
				2	0.0313	0.273
				3	0.135	0.192
				4	0.333	0.350
				5	1.36	0.128
				6	4.04	0.029
^{241}Pu	0.0152 ±0.0011	2.95	0.00515	1	0.0128	0.010
				2	0.0299	0.229
				3	0.124	0.173
				4	0.352	0.390
				5	1.61	0.182
				6	3.47	0.016

Table 6.1 (continued)

Isotope	Delayed[a] neutrons per fission	Total[b] neutrons per fission, ν	β	Group	$\lambda_i{}^c(s^{-1})$	β_i/β^c
^{242}Pu	0.0221 ±0.0026	3.05	0.0072	1	0.0128	0.004
				2	0.0314	0.195
				3	0.128	0.161
				4	0.325	0.412
				5	1.35	0.218
				6	3.70	0.010

[a]Ref. [3]. Delayed neutrons per fission rather than β is the parameter measured. These values are for fast fission, meaning fissions in the few keV to 4 MeV range. This parameter is not very sensitive to the neutron energy causing the fission.
[b]The average value of ν (to be used with delayed neutrons per fission to obtain β) varies slightly with the fast spectrum reactors neutron spectrum. The values here are consistent with the values in Appendix F and the neutron spectrum for which the Appendix F cross sections were obtained. Since β is the ratio of the first two columns, β will differ slightly for a neutron spectrum different from that assumed here.
[c]Ref. [4]. Uncertainties in λ_i, and β_i/β are reported in that reference.

For many safety calculations, it is convenient to change the principal variable in the kinetics equations from neutron density, *n*, to reactor power density, p(W/cm^3). This can be done by introducing the following definitions and approximations:

$$p = \frac{\Sigma_f n v}{2.93 \times 10^{10}},$$

$$y_i = \frac{\Sigma_f C_i v}{2.93 \times 10^{10}}.$$

With these variables, the kinetic equations become[3]

$$\frac{dp}{dt} = \frac{\rho - \bar{\beta}}{\Lambda} p + \sum_{i=1}^{6} \lambda_i y_i \tag{6.5}$$

$$\frac{dy_i}{dt} = \frac{\bar{\beta}_i}{\Lambda} p + \lambda_i y_i \quad (i = 1 \text{ to } 6). \tag{6.6}$$

6.2.2 Prompt Jump Approximation

A very convenient approximation that can be used to estimate the change in power density resulting from a change in reactivity (for reactivities less than about \$0.9) is called the prompt jump approximation. For small values of reactivity (ρ < \$0.9), the rate of change of power is sufficiently slow that the derivative *dp/dt* is negligible compared to the two terms on the right hand side of Eq. (6.5). Hence, the power density can be approximated from Eq. (6.5) as

[3]This formulation ignores the difference in timing between prompt and delayed energy generation.

$$p = \frac{\Lambda \sum_{i=1}^{6} \lambda_i y_i}{\bar{\beta} - \rho}. \quad (6.7)$$

Prior to the reactivity change, the derivative dy_i/dt in Eq. (6.6) is zero and the power density is the initial or steady-state value, p_0. Hence, $\lambda_i y_i$ at the time of the reactivity change is $\left(\bar{\beta}_i/\Lambda\right)p_0$ from Eq. (6.6). Inserting this expression into Eq. (6.7) gives the value of p immediately after the reactivity change, or

$$p \simeq \frac{p_0 \sum_{i=1}^{6} \beta_i}{\bar{\beta} - \rho} = \frac{p_0 \bar{\beta}}{\bar{\beta} - \rho} = \frac{p_0}{1 - \left(\frac{\rho}{\beta}\right)}.$$

Hence,

$$p \simeq \frac{p_0}{1 - \rho\,(\$)}, \quad (\text{for } \rho < \$0.9) \quad (6.8)$$

where the reactivity is in dollars. Equation (6.8) represents the almost instantaneous change (or "jump") in power density due to prompt neutrons when a reactivity change less than \$0.9 is made, yet the result does not contain the neutron generation time Λ. It is further noted that this approximate value for p/p_0, is simply the coefficient of the first term in Eq. (6.4) i.e., the term that remains after the rapidly decaying second term of Eq. (6.4) dies away. This is yet another way of recognizing that the transient behavior of an fast spectrum reactors is essentially independent of the prompt neutron lifetime for reactivities less than the prompt critical condition.

6.3 Adjoint Flux and Perturbation Theory

The adjoint flux, ϕ^*, is introduced in this chapter because of its importance to the calculation of reactivity effects. We shall not take the time here to derive the adjoint flux or the perturbation expression for reactivity; the theory is identical for fast and thermal spectrum reactors and appears in standard reactor theory textbooks. We will, however, discuss the physical meaning of the adjoint flux and will use the adjoint flux in multigroup perturbation expressions for reactivity effects.

6.3.1 Adjoint Flux

The adjoint flux is often called the importance function, a name that describes its physical meaning. The adjoint flux is proportional to the importance of the neutron in continuing the neutron chain reaction. A neutron of high importance has a high chance of causing a fission, and the neutrons resulting from the fission have a high chance of causing other fissions, and so on. Spatially, a neutron near the center of the core has a higher importance than one near the edge because the center neutron has less chance to leak from the core—hence, more chance to cause fission. In fact, we may recall that for a bare core the spatial shapes of the flux and the adjoint flux are identical, being highest at the center.

In a fast reactor, the *spatial* shapes of the flux and the adjoint flux in the most important energy ranges are generally quite similar, even though the core is surrounded by blankets.

Of particular interest in a fast reactor is the *energy* dependence of the adjoint flux, and here we find a large difference in behavior between flux and adjoint flux. We also find significant differences in behavior between fast and thermal reactors in this regard. In a fast reactor, the neutron importance

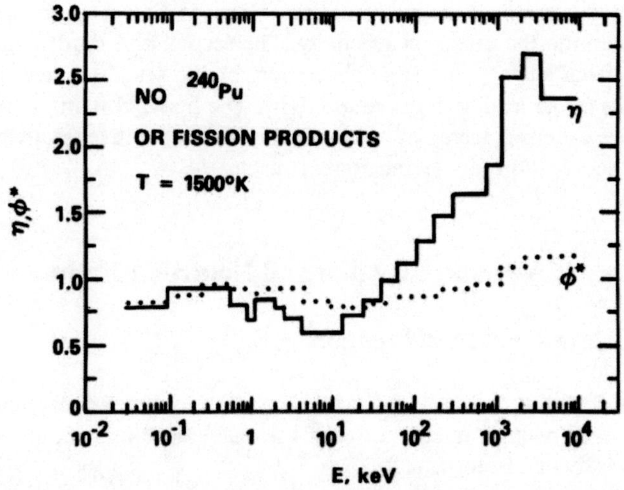

Fig. 6.2 Energy dependence of adjoint flux and effective η in a large fast oxide reactor (adapted from Ref. [5])

tends to increase with energy above some energy in the low keV range. This is primarily because of the increase in η of the fissile isotopes with increasing energy—the same property that led to high breeding ratios for fast reactors fueled with plutonium. Above the fission threshold energy of the fertile materials, the importance also rises because of fission by these materials.

At low energies, below the few-keV range, the importance function begins to rise again. This rise is caused by the increasing fission cross section of the fissile material relative to other competing cross sections; we are referring here to resonance rather than thermal cross sections because neutrons do not survive to thermal energies in a fast reactor. For a large fast reactor with oxide fuel, the neutron energy of minimum importance is just above 10 keV.

The energy dependence of the adjoint flux is illustrated in Fig. 6.2 for a large oxide-fueled fast reactor. Also plotted in the figure is η versus neutron energy, which shows the rise above 10 keV. The η plotted is an effective η for the fissile and fertile materials. The sharp rise above 1 MeV reflects the effect of fertile material fission, while the rise at low energy reflects the rise in fissile material fission cross section relative to fertile capture cross section.

We will soon see that the energy dependence of the adjoint flux plays an important role in all fast reactor safety parameters, including Doppler, sodium, and fuel compaction reactivities and the effective delayed neutron fraction.

6.3.2 Reactivity from Perturbation Theory

A reactivity effect based on multigroup first order perturbation theory is calculated as follows:

$$\rho = \frac{\left[\begin{array}{l}\int \sum\limits_{g} \phi_g^* \chi_g \sum\limits_{g'} \delta\left(\nu\Sigma_f\right)_{g'} \phi_{g'} dV + \int \sum\limits_{g} \nabla\phi_g^* \delta D_g \nabla\phi_g dV \\ - \int \sum\limits_{g} \phi_g^* \delta\Sigma_{rg} \phi_g dV + \int \sum\limits_{g} \phi_g^* \sum\limits_{g'<g} \delta\Sigma_{g'\to g} \phi_{g'} dV\end{array}\right]}{\int \sum\limits_{g} \phi_g^* \chi_g \sum\limits_{g'} \left(\nu\Sigma_f\right)_{g'} \phi_{g'} dV} \tag{6.9}$$

where the integrals are taken over the entire reactor. Here reaction rates are being weighted with adjoint fluxes to determine the effect on reactivity. The second and third terms in the numerator represent loss terms from leakage and from absorption, elastic scattering, and inelastic scattering; increases in these loss terms lead to negative reactivity. The first and fourth terms represent source terms from fission and scattering; increases in these terms provide positive reactivity. The denominator is a weighted fission source term for the unperturbed reactor.

6.4 Effective Delayed Neutron Fraction and Neutron Lifetime

6.4.1 Effective Delayed Neutron Fraction

The delayed neutron fraction, β, can be estimated from a zero-dimensional neutronics calculation by weighting the delayed neutron fraction of each fissionable isotope with the number of neutrons produced from fissions by that isotope, as follows:

$$\beta = \frac{\sum_m \left(\beta_m \sum_g (\nu\Sigma_f)_{mg} \phi_g \right)}{\sum_g (\nu\Sigma_f)_g \phi_g}. \tag{6.10}$$

A more accurate space-dependent calculation must include the adjoint function in the weighting procedure and the energy spectrum of the delayed neutrons, as described below [Eq. (6.11)]. Values of β for fast fission for the principal fissionable isotopes that may be in fast reactors were given in Table 6.1.

A typical value for β for a fast oxide reactor can be estimated from values of β_m and ν_m in Table 6.1 by assuming a typical distribution of fissions between the fissionable isotopes. For the fission fractions in the core of the 1,200 MWe homogeneous design with the neutron balance in Table 4.1, an approximate value of β is 0.0046. For the 1,000 MWe heterogeneous design of Table 4.2, the approximate β is 0.0040.

An accurate calculation of β for use in the kinetic equations must account for both the spatial distribution of fissions and, more important, for the difference in energy spectra of delayed neutrons and prompt neutrons. The resulting expression for β is called the *effective beta*, $\overline{\beta}$:

$$\overline{\beta} = \frac{\sum_m \sum_{i=1}^{6} \beta_{im} \int \sum_g \phi_g^* \chi_{dg} \sum_{g'} (\nu\Sigma_f)_{mg'} \phi_{g'} dV}{\int \sum_g \phi_g^* \chi_g \sum_{g'} (\nu\Sigma_f)_{g'} \phi_{g'} dV}. \tag{6.11}$$

The summation over m includes all fissionable materials m, for which β and $\nu\Sigma_f$ will differ. We assume that the spectrum of delayed neutrons, χ_d, from each material is identical. The summation over i refers to the six delayed-neutron groups. Note too that the weighted source term in the denominator is identical to the denominator of Eq. (6.9).

In a fast reactor, $\overline{\beta}$ is less than β by about 10%, a condition that differs from a thermal reactor. Delayed neutrons are born at energies lower than the energy of prompt neutrons. Since the adjoint flux in a fast reactor rises with energy (as in Fig. 6.2), a delayed neutron is worth less in a fast reactor than a prompt neutron, so that $\overline{\beta}/\beta < 1$. In a thermal reactor a delayed neutron is worth more than a prompt neutron because the delayed neutron has less chance to leak from the reactor, while there is little chance (relative to a fast reactor) that either a delayed or a prompt neutron will cause a fission at high energy; hence, $\overline{\beta}/\beta > 1$ for a thermal reactor.

For the estimates of β given above for the homogeneous and heterogeneous designs of Tables 4.1 and 4.2, the values of $\overline{\beta}$ are: $\overline{\beta} \simeq 0.0042$ for the homogeneous design and $\overline{\beta} \simeq 0.0036$ for the heterogeneous design. These values would represent the reactivity dollar for these designs. The values are fairly typical for fast spectrum reactors although the homogeneous design value is slightly higher than those usually reported. The reason for the difference between the two designs is the larger fraction of ^{238}U fissions in the homogeneous design, which results from the considerably lower fissile fraction in that design.

6.4.2 Neutron Lifetime

The neutron lifetime is given by the following relation for multigroup theory:

$$l = \frac{\int \sum_g \phi_g^* \frac{1}{v_g} \phi_g dV}{\int \sum_g \phi_g^* \chi_g \sum_{g'} (\nu \Sigma_f)_{g'} \phi_{g'} dV}. \tag{6.12}$$

A reasonable value for v_g for group g can be obtained by setting

$$\frac{1}{v_g} \int_{E_{lg}}^{E_{ug}} \phi(E) dE = \int_{E_{lg}}^{E_{ug}} \frac{1}{v(E)} \phi(E) dE, \tag{6.13}$$

where subscripts u and l refer to the upper and lower energy boundaries of the group. Letting $\phi(E) = C/E$, $v(E) = \sqrt{2E/m}$, and solving for v_g, gives

$$v_g = \frac{v_{ug} v_{lg} \ln(E_{ug}/E_{lg})}{2(v_{ug} - v_{lg})}, \tag{6.14}$$

although at low energy, $\phi(E)$ decreases with energy faster than $1/E$.

A typical value for the neutron lifetime of a SFR is 4×10^{-7} s.

6.5 Reactivity from Dimensional Changes

Changes in reactor core dimensions occur during normal operation of a SFR and might occur during off-normal transients. Fuel will expand axially when its temperature is increased, as in normal startup from zero to full power. The structure that supports the core will expand radially if the inlet sodium temperature increases, thus causing a radial expansion of the core.

A major role of axial expansion of fuel in the normal solid fuel pin geometry is to provide a prompt negative reactivity feedback at the start of a power transient. This mechanism is the principal prompt negative feedback available in a metal-fueled fast reactor. (Doppler feedback is also available for the metal-fueled system, but it is relatively small due to the hard neutron spectrum.) For a ceramic-fueled reactor, lack of structural integrity (due to cracking) in high-burnup fuel renders the axial expansion mechanism somewhat unreliable so that more dependence is placed on the Doppler effect for a prompt negative feedback.

If fuel melts in a severe accident, fuel might slump downward tending to compact the core, or it might be swept upward or forced outward causing an expansion of the core. For these reasons, it is important to understand the reactivity effects caused by changes in core dimensions.

One of the most important differences between a fast and a thermal power reactor is that fuel compaction in a fast reactor leads to a large positive reactivity effect, while in a thermal reactor it does not. A fast reactor is not designed to operate in an optimum configuration from a reactivity point of view; the fissile mass could always be reduced, for example, if a way were found to reduce the coolant volume fraction. A light water reactor, on the other hand, is designed to operate at an optimum moderator-to-fuel ratio.

6.6 Doppler Effect

6.6.1 Computational Methods

It is important to have a prompt negative reactivity feedback that reverses a power transient if the reactor becomes prompt critical. Mechanical action by control rods is too slow after prompt criticality is reached, whether one is talking about thermal or fast reactors. A prompt negative feedback is particularly important for fast reactors because two mechanisms, fuel compaction and sodium loss, have the potential to make the reactor super prompt critical.

The Doppler effect provides this prompt negative reactivity feedback for a fast reactor fueled with ceramic fuel. In a power excursion, excess fission energy quickly raises the fuel to a high temperature. As we observed in Chapter 5, the temperature rise in the principal fertile isotope (^{238}U or ^{232}Th) results in a relatively large increase in the effective parasitic capture cross section of this isotope. This provides a large negative reactivity effect. The existence of the Doppler effect in fast oxide-fueled reactors and our ability to predict its magnitude and its feedback effect on transients with reactivities above prompt critical was fully demonstrated by the SEFOR[4] reactor experiment in the U.S. in the late 1960 [1, 2].

The feedback response is prompt only if the fissile and fertile materials are mixed intimately together. Most of the fission energy comes from slowing down of the fission products; hence, the fertile isotopes must be close enough to the fissile isotopes that the loss in kinetic energy by the fission products will promptly cause an increase in temperature and thermal motion of the fertile nuclei. This requires that the distance between the fertile and fissile nuclei be no greater than the order of the slowing-down distance, or range, of the fission products. This range is of the order of 10 μm. This intimate mixture of fertile and fissile materials is achieved by the fabrication processes generally used for making mixed oxide, carbide, and nitride fuels.

As discussed in Chapter 5, the Doppler reactivity comes predominantly from the capture of low-energy neutrons. Energy self-shielding is required for an increase in the temperature of the fertile material to increase the effective cross section of that material. At high energy there is little self-shielding. Ceramic-fueled reactors, due to the presence of oxygen, carbon, or nitrogen in the fuel, have a soft enough neutron spectrum to have a large Doppler effect. Reactors with hard spectra, particularly the early small metal test reactors, have small Doppler effects. Even large metal-fueled fast reactor

[4] An interesting historical application of the interaction of Doppler feedback and core axial expansion and radial neutron leakage was provided by the SEFOR reactor. The reactor was constructed for the specific purpose of measuring the Doppler effect. Consequently, it was desirable to eliminate the axial and radial expansion effects entirely. This was successfully done by inserting a gap in each fuel pin at just the right axial position (about 2/3 of the way up the core) so that the natural increase in radial leakage with fuel expansion was exactly offset by closing the gap.

designs have significantly smaller Doppler effects than ceramic-fueled reactors. This represents one of the important advantages of ceramic over metal fuel to arrest overpower transients. As we shall see in the Safety Section, a large negative Doppler coefficient can be a liability for loss of cooling accidents.

The Doppler reactivity effect can be calculated from perturbation theory, or it can be obtained from two successive criticality calculations with effective cross sections at different temperatures. A multigroup perturbation expression for the Doppler reactivity is

$$\rho = \frac{-\int \sum_{g} \phi_g^* \delta \Sigma_{ag} \phi_g dV + \int \sum_{g} \phi_g^* \chi_g \left[\sum_{g'} \delta \left(\nu \Sigma_f\right)_{g'} \phi_{g'} \right] dV}{\sum_{g} \phi_g^* \chi_g \left[\sum_{g'} \left(\nu \Sigma_f\right)_{g'} \phi_{g'} \right] dV}, \tag{6.15}$$

where the $\delta\Sigma_a$ and $\delta\left(\nu\Sigma_f\right)$ terms reflect changes in the absorption and fission cross sections due to heating. The integrals are over the entire reactor volume. The first integral in the numerator accounts for the increase in the effective absorption cross sections of the fertile and fissile isotopes (including both capture and fission); this is a negative reactivity effect for an increase in fuel temperature (i.e., $\delta\Sigma_a$ is positive for this case). The second term accounts for the positive contribution of the increase in the effective fission cross section of the fissile isotopes with an increase in fuel temperature. Note that in the first term the increased absorption rate, $\delta\Sigma_{ag}\phi_g$, is weighted with the importance, ϕ_g^*, of the same energy group in which the absorption occurs because the neutron is lost from the reactor at that energy. On the other hand, the neutrons produced in the second integral are weighted with the importance of the energy group into which they are born.

The importance of the shape of the adjoint flux curve in Fig. 6.2 can be seen from Eq. (6.15). In a ceramic-fueled fast reactor, the main contribution to $\delta\Sigma_{cg}\phi_g$ for the principal fertile material occurs at low energy, i.e., in the 0.1–10 keV range. From Fig. 6.2, we can see that the adjoint flux levels out at about 10 keV and rises again below that energy.

Most of the Doppler effect comes from the fertile isotope. For the fissile isotopes, the effects of $\delta\Sigma_c$, and $\delta\Sigma_f$ almost cancel, an effect first observed experimentally and later explained by the application of resonance-overlap theory. Also, the minor fertile isotopes, such as ^{240}Pu, contribute little to the Doppler effect because of their small quantities (hence, small atom densities, N). Furthermore, as a result of their small quantities, their energy self-shielding is small and the effective $\delta\sigma_{cg}$ values are consequently small.

The criticality factor for a ceramic-fueled reactor as a function of absolute fuel temperature has the shape illustrated in Fig. 6.3. Whereas the absolute value of the variation in k about 1.0 is small, this difference can be large in terms of reactivity increments. The slope of the curve at any temperature is

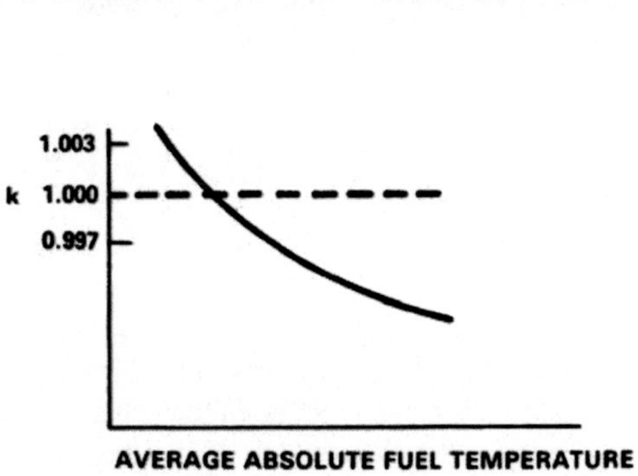

Fig. 6.3 Decrease of k with increasing fuel temperature

the *Doppler coefficient*, dk/dT, at that temperature. The coefficient is always negative, but dk/dT is not constant; it is a function of temperature, decreasing in magnitude with increasing T.

It has been found that, for oxide-fueled fast reactors, the temperature dependence of dk/dT is almost exactly

$$\frac{dk}{dT} = \frac{K_D}{T} \tag{6.16}$$

where K_D is a constant. For this reason, the constant K_D has taken on the title *Doppler constant,* where

$$K_D = T\frac{dk}{dT}. \tag{6.17}$$

This constant is the value that characterizes the Doppler effect for fast oxide reactors. (Frequently this constant is referred to in the literature as the Doppler coefficient, but we prefer to reserve the term "coefficient" for the derivative *dk/dT*.)

For a uniform change in fuel temperature from T_1 to T_2, the Doppler reactivity effect can be calculated from Eq. (6.16) as follows:

$$\rho = \int_{T_1}^{T_2} \frac{dk}{dT} dT = \int_{T_1}^{T_2} \left(\frac{K_D}{T}\right) dT = K_D \ln\left(T_2/T_1\right). \tag{6.18}$$

In the reactor there is a fuel temperature distribution that must be taken into account in calculating the reactivity effect. This can enter Eq. (6.15) through the changes in cross sections $\delta\Sigma_{ag}$ and $\delta\Sigma_{fg}$, that are space-dependent since the temperature changes are space-dependent. Moreover, one must be careful to consider the type of power change one is calculating. One type of change is from one steady-state power level to another. This type of change must be analyzed to determine steady-state power coefficients or the required reactivity compensation by the control rods for a given change in power level. Another type of change would be a postulated accidental power excursion. Here the spatial change in fuel temperature for a given change in power level would differ from the spatial temperature variation between the same power levels at steady-state.

These temperature variations are often accounted for by the expression

$$\rho = \sum_i \frac{K_{Di}}{V_i} \int_{V_i} W(r,z) \ln \frac{T(r,z)}{T_0(r,z)} dV, \tag{6.19}$$

where the summation is over regions i of the reactor (including the blankets), K_D is the contribution to the Doppler constant for that region, and $T_0(r,z)$ is the initial temperature distribution. The weighting factor $W(r,z)$ is a normalized spatial variation of the Doppler effect, such that

$$\frac{1}{V_i} \int_{V_i} W(r,z)\, dV = 1. \tag{6.20}$$

Studies of the effect of temperature variation within an actual fuel rod indicate that the use of an effective temperature equal to or slightly less than the mean gives results of adequate accuracy[5].

While accounting for subtleties is a requirement for final design and final safety analysis, these differences are generally second-order effects. For steady-state power changes, calculational experience indicates that the magnitude of the actual reactivity change from a power level with a fuel temperature distribution $T_1(r)$ for which the average fuel temperature is $\overline{T_1}$ to a power with fuel at $T_2(r)$ with average fuel temperature $\overline{T_2}$ is about 10% higher than the calculated value for a uniform temperature change from $\overline{T_1}$ to $\overline{T_2}$ i.e.,

$$\rho\,[T_1(r) \rightarrow T_2(r)] \approx 1.1 K_D \ln\left(\overline{T}_2/\overline{T}_1\right). \tag{6.21}$$

The reason that dk/dT varies so closely with $1/T$ for fast oxide reactors is due more to convenient good luck than to some guiding law of nature. For a reactor with a much harder neutron spectrum, such as a metal fueled fast reactor, the theory for the effective Doppler cross section indicates that dk/dT should vary as $1/T^{3/2}$. For a reactor with a much softer spectrum, such as a light water reactor, resonance theory indicates that dk/dT should vary as $1/T^{1/2}$. The spectrum of the fast oxide reactor lies in between, and by chance, $1/T$ appears to describe the variation remarkably well.

6.6.2 Effect of Coolant Loss on Doppler Feedback

We have learned that the magnitude of the Doppler coefficient, or Doppler constant, is relatively large for a ceramic-fueled fast reactor because of the large neutron flux below 10 keV. This soft spectrum results from the combination of the moderating effect of sodium and the low-mass material (oxygen, carbon, or nitrogen) in the fuel. In an unprotected accident of the type described in Chapter 15, however, the liquid sodium may have boiled and been expelled from all or part of the core, leaving behind only low-density sodium vapor. In this event the neutron spectrum will be hardened and the magnitude of the Doppler coefficient will be reduced. Figure 6.4 illustrates the cause of the reduced Doppler coefficient due to the loss of neutrons in the Doppler resonance region. This reduced Doppler coefficient must be used in safety analyses when sodium is lost from the core. Values of the Doppler constant for the CRBRP design will be given later in this section in order to indicate the influence of the loss of sodium on K_D.

One might also argue that the change in spectrum should alter the form of the temperature variation of dk/dT, since the exponent m in $1/T^m$ is spectrum-dependent. The spectrum change due to sodium loss is insufficient to change the variation significantly, however, and values of K_D (with the implied $1/T$ variation in dk/dT) are quoted for oxide fueled designs both with and without sodium.

In calculating the Doppler reactivity with sodium removed, it is important to account for the change in all effective cross sections due to the change in self-shielding caused by the absence of sodium scattering. The most important change, of course, is the modification of the Doppler-broadened cross sections for the fuel as a function of temperature.

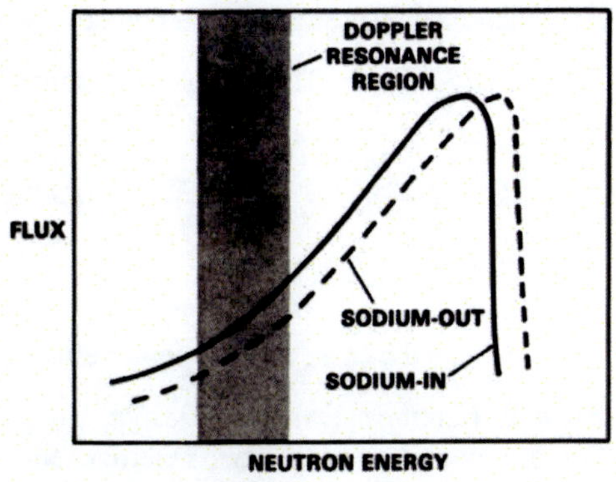

Fig. 6.4 Flux reduction in the Doppler resonance region when sodium is lost from the core

6.6.3 Changes in the Doppler Effect During the Fuel Cycle

The Doppler coefficient changes during the fuel cycle. The magnitude and even the direction of the change, however, are difficult to estimate a priory because a number of subtle changes are occurring simultaneously. First, boron absorption is being removed with the withdrawal of the shim control rods. Second, fission products are being produced. Each of these effects modifies both ϕ and ϕ^* at low energies, where the Doppler is important, relative to ϕ and ϕ^* at high energies. Both types of absorbers reduce the low-energy ϕ's and ϕ^*'s more than the high-energy ones, but there is competition between the two as to which has the dominant influence. The fuel plays two roles. The small decrease in ^{238}U concentration decreases the Doppler slightly through a decrease in $\delta\Sigma_c$, but more importantly the internal conversion ratio influences the extent to which boron absorber must be removed during the fuel cycle, and hence, changes the relative influence between boron and fission products on ϕ and ϕ^*. The internal conversion ratio is strongly influenced by reactor size (or rated power); because of this, the influence of burnup on the Doppler coefficient may vary between small and large fast reactors.

6.6.4 Sample Results

Typical values of the Doppler constant for the early CRBRP homogeneous design [6] are listed below, together with the influences of reactor size (i.e., power level), sodium loss, fuel cycle, and heterogeneous core design. All results in this section are for UO_2-PuO_2 fueled SFR's. Sample values of K_D appear in Table 6.3 in the section on sodium loss reactivity and in Appendix A.

Effect of Reactor Size (or Power Level). Values can be compared for the early CRBRP homogeneous design [6] and a 1,200 MWe homogeneous design [7]

$$\begin{aligned} K_D\,(\text{CRBRP}, 350\text{ MWe}) &= -\,0.0062 \\ K_D\,(1{,}200\text{ MWe}) &= -\,0.0086. \end{aligned}$$

An increase in K_D with size is expected predominantly because the neutron spectrum for a large reactor is softer (due to decreased leakage).

Effect of Sodium Loss. At the beginning of the equilibrium fuel cycle, the Doppler constants with and without sodium present are

$$\begin{aligned} K_D\,(\text{with Na}) &= -\,0.0062 \\ K_D\,(\text{without Na}) &= -\,0.0037. \end{aligned}$$

Note that the reduction in Doppler due to sodium loss is quite large.

Effect of Fuel Cycle. The Doppler constants with sodium present at the beginning and the end of a cycle are

$$\begin{aligned} K_D\,(\text{beginning of cycle}) &= -\,0.0062 \\ K_D\,(\text{end of cycle}) &= -\,0.0070. \end{aligned}$$

We observe that, for the CRBRP-size design, the Doppler effect increases slightly during the cycle. This was also the case for the Doppler effect without sodium present.

Effect of Heterogeneous Core Design. Doppler constants are reported in Ref. [7] for a 1,200 MWe homogeneous core and for two heterogeneous core designs. The results are

$$K_D \text{ (homogeneous, 7.26 mm pin)} = -0.0086$$
$$K_D \text{ (heterogeneous, 7.26 mm pin)} = -0.0088$$
$$K_D \text{ (heterogeneous, 5.84 mm pin)} = \sim -0.008.$$

Thus, the heterogeneous design appears to have little influence on the Doppler constant. However, it should be noted that blankets respond more slowly than driver fuel; thus, the effective Doppler feedback is less for the heterogeneous design-especially for rapid transients.

Doppler Contribution by Region. The integrations in Eq. (6.15) are over the entire reactor. By separating the integrations in the numerator into regions, one can find the contributions to the Doppler constant from individual regions. A complete summary of these contributions for the CRBRP is given in Table 6.2.

Table 6.2 Doppler constants by region for the early CRBRP homogeneous design [6]

	Contribution to—K_D			
	Beginning of cycle		End of cycle	
Region	With Na	Without Na	With Na	Without Na
Inner core	0.0034	0.0016	0.0037	0.0019
Outer core	0.0011	0.0006	0.0013	0.0008
Radial blanket	0.0011	0.0010	0.0012	0.0012
Axial blanket	0.0006	0.0005	0.0008	0.0005
Total	0.0062	0.0037	0.0070	0.0044

6.7 Sodium Loss Reactivity

The loss of sodium from a large fast reactor can result in a large positive reactivity effect. Sodium might be expelled from the core in the unlikely event of an unprotected transient, as described in Chapter 15, in which sodium boiling results from under cooling the reactor. This condition presents an important safety problem for a SFR, a problem not present in thermal reactors. This problem is also not present in gas-cooled fast reactors, which is one of the major advantages of gas cooling. It is also not a concern for lead-cooled reactors, because of the boiling point of lead (or lead-bismuth) is exceptionally high.

The sodium loss reactivity effect is exceedingly space-dependent. Sodium loss from the center of the core yields a highly positive reactivity effect whereas sodium loss from near the edge gives a negative effect. An understanding of such behavior is revealed by observing individually the four phenomena that contribute to the overall sodium void effect:

(1) spectral hardening
(2) increased leakage
(3) elimination of sodium capture
(4) change in self-shielding.

The first two effects are large and of opposite sign. The last two are small. Hence, most of the sodium loss reactivity results from the difference of two large numbers, as illustrated in Fig. 6.5, and this situation makes an accurate calculation of the net effect difficult. Each of these contributions to the sodium loss reactivity will be discussed later. Before doing so, however, we will discuss briefly the methods used to evaluate the sodium loss effect.

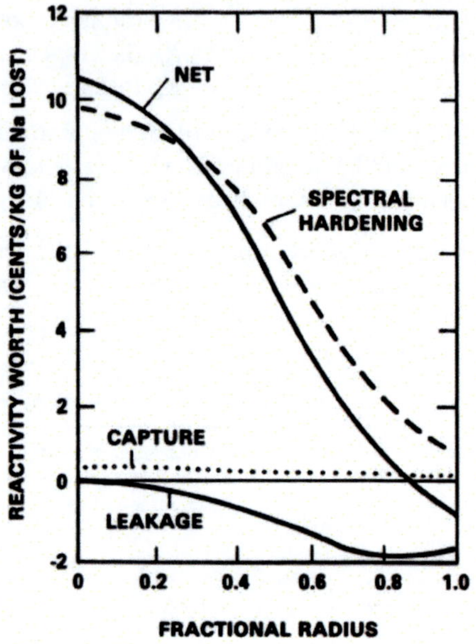

Fig. 6.5 Components of sodium void coefficient for a small cermet-fueled SFR (adapted from Ref. [8])

6.7.1 Method for Calculating Sodium Loss Reactivity

The method generally used to calculate sodium loss reactivity is to perform successive multi-dimensional multigroup calculations (one with sodium present and the second with sodium removed from the zone of interest) and to compare the criticality factors. Care must be taken to use the proper self-shielded cross sections in the zone from which sodium is removed; all effective cross sections of the fuel and structure will change when the scattering effect of the sodium is removed. This technique does not yield the four individual contributions listed above; consequently, it is easy to lose sight of the fact that the result is still, in effect, the difference between two large reactivity effects. Large errors could accrue unless extraordinary care is taken.[5]

An alternate method for calculating the sodium loss effect is to use perturbation theory and calculate the four contributions separately. The calculation is more difficult, but it provides useful insight into the physical processes. It also provides a convenient data base for use in conducting transient calculations where only small localized perturbations occur during the time scale of interest. The appropriate perturbation expression for each contribution is given in the following discussion.

Spectral Hardening. The spectral hardening effect is positive. The loss of sodium from the core results in decreased moderation of the neutrons so the average neutron energy increases. This produces a positive reactivity effect because of the increase in neutron importance with increasing energy, as illustrated in Fig. 6.2.

The denominator of the perturbation theory expression for each of the four contributions is the same weighted source term that has appeared previously. Hence, it will be replaced in each expression by the integral S, where

[5]While the zero-dimensional calculation introduced in Chapter 4 can provide reasonable results for many fast reactor problems and is useful as an educational tool, it gives misleading results for a process as spatially dependent as sodium loss reactivity.

$$S = \int \sum_{g} \phi_g^* \chi_g \sum_{g'} (\nu \Sigma_f)_{g'} \phi_{g'} dV \tag{6.22}$$

The perturbation theory expression for the spectral hardening contribution can be written as

$$\rho = \frac{1}{S} \int \sum_{g} \phi_g^* \left[\left(\sum_{g'<g} \delta \Sigma_{g' \to g} \phi_{g'} \right) - \left(\delta \Sigma_{erg} + \delta \Sigma_{irg} \right) \phi_g \right] dV. \tag{6.23}$$

An alternate, but equivalent, expression also appears in the literature:

$$\rho = \frac{1}{S} \int \sum_{g} \sum_{g'>g} \left(\phi_{g'}^* - \phi_g^* \right) \delta \Sigma_{g' \to g} \phi_g dV. \tag{6.24}$$

Note that this contribution to the sodium loss effect is proportional to the product of ϕ and ϕ^*, both of which peak at the center of the core. Hence, this contribution would be highest per unit volume of sodium lost near the center of the core and lowest near the edge of the core.

Increased Leakage. Sodium loss results in increased leakage, which in turn yields a negative contribution to sodium loss reactivity. The strong spatial dependence of the sodium loss effect is due primarily to the different spatial behavior of the leakage component relative to that of the spectral hardening component. Sodium loss from a unit volume of core near the center of the core adds little to the leakage contribution because the flux gradient is low, and hence, leakage per cubic centimeter is low. The effect of sodium loss near the center of the core is highly positive, since the large spectral hardening effect is not counterbalanced by a large leakage contribution. The same sodium loss near the edge of the core, however, increases the leakage considerably and adds much to the negative leakage contribution to the sodium loss effect and little to the positive spectral hardening contribution.

The source of the spatial dependence of the leakage contribution can be deduced from the perturbation expression for this contribution:

$$\rho = \frac{1}{S} \int \sum_{g} \nabla \phi_g^* \delta D_g \nabla \phi_g dV. \tag{6.25}$$

Elimination of Sodium Capture. The elimination of sodium capture results in a small positive reactivity effect, given by

$$\rho = -\frac{1}{S} \int \sum_{g} \phi_g^* \delta \Sigma_{cg,Na} \phi_g dV. \tag{6.26}$$

Change in Self-Shielding. The elimination of sodium scattering results in a small reactivity effect, given by the following expression, where the changes in the cross sections are those due only to the change in self-shielding:

$$\rho = \frac{1}{S} \int \sum_{g} \phi_g^* \left[\chi_g \sum_{g'} \delta \left(\nu \Sigma_f \right)_{g'} \phi_{g'} - \delta \Sigma_{ag} \phi_g \right] dV. \tag{6.27}$$

6.7.2 Sodium Loss Reactivity Values and Methods to Reduce This Reactivity

For a commercial-size SFR with UO_2-PuO_2 fuel and a homogeneous core, the positive reactivity from loss of sodium from the core is large. Values for a number of designs are presented in Table 6.3, which shows reactivities in the \$5 to \$7 range for commercial-size homogeneous cores. Comparison of the early CRBRP homogeneous design with the commercial-size plants shows the increase in sodium loss reactivity with size. The reason for this trend is that, as the core becomes larger, the (negative) leakage effect becomes less pronounced, while the (positive) spectral hardening effect is less affected.

Numerous attempts have been made to reduce the sodium loss reactivity by design. Most design modifications have focused on increasing the leakage component of the sodium loss reactivity. The most promising appears to be a concept first proposed by the French [11], generally referred to as the heterogeneous core, and described earlier in Chapters 2 and 4 . In this method, annular rings of blanket assemblies are placed in the core (as in Fig. 2.2) to achieve a high neutron leakage rate from the core to the blankets.

The effect of the heterogeneous core design on sodium loss reactivity can be seen in Table 6.3. For sodium loss from the core assemblies only, the reactivity drops to the \$2 range. In an accident involving loss of sodium, sodium expulsion from the blanket assemblies would likely lag the core assemblies if it occurred at all, so that the core reactivity feedback would control the accident sequence. This is a significant safety advantage for the heterogeneous design. On the other hand, we saw earlier the disadvantage of a higher fissile inventory for this design (Table 4.5).

Two other methods to decrease the sodium loss reactivity by increasing leakage were explored early in SFR development; these were the "pancake core" and the "modular core." As seen by the

Table 6.3 Selected reported values of sodium loss reactivity

Reactor	Geometry	Core fuel	Core sodium loss reactivity, \$	Doppler constant $\left(T\frac{dk}{dT}\right)$
CRBRP (350 MWe)	Homogeneous [6]	UO_2-PuO_2	3.3	–0.0062
	Heterogeneous	UO_2-PuO_2	2.3	–0.008
1,000 MWe [9]	Heterogeneous	UO_2-PuO_2	2.9 (core only) 4.1 (core plus internal blankets)	
1,000 MWe [10]	Homogeneous	UO_2-PuO_2	6.9	
		ThO_2-PuO_2	2.3	
		$^{238}UO_2$-$^{233}UO_2$	1.5	
1,200 MWe [7]	Homogeneous	UO_2-PuO_2	5.0	–0.0086
	Heterogeneous, tightly coupled	UO_2-PuO_2	1.5	–0.0088
	Heterogeneous, tightly coupled, small pin[a]	UO_2-PuO_2	1.7	~–0.008
	Heterogeneous, less tightly coupled	UO_2-PuO_2	1.5	–
	Heterogeneous, modular island	UO_2-PuO_2	0.25	–0.0060
	Homogeneous, pancake height/diameter = 0.1	UO_2-PuO_2	1.75	–0.004
	Homogeneous, BeO moderated	UO_2-PuO_2	2.2	–0.0116

[a]Small pin diameter is 5.84 mm. Pin diameter for other cases in Ref. [7] is 7.26 mm.

example results for these concepts in Table 6.3, the sodium loss reactivity is indeed reduced, but the economic penalties for both concepts are greater than for the heterogeneous core.

Given a particular fuel type, the spectral component of the sodium loss reactivity is difficult to modify. One early suggestion to accomplish this was to deliberately soften the neutron spectrum by adding diluents such as BeO. The introduction of such a light element moderator reduces the dependency of the spectrum on the presence of sodium; when sodium is lost, the BeO keeps the spectrum soft. This softened neutron spectrum also enhances the magnitude of the Doppler coefficient, but economic penalties and a large drop in breeding ratio made this design undesirable.

The sodium loss reactivities for a core utilizing thorium for the fertile material (even with plutonium as the fissile material) and for a core utilizing uranium-233 as the fissile material (even with uranium-238 as the fertile material) are both lower than for a $^{238}UO_2$-PuO_2 reactor. The lower value for the thorium fuel results from the lower fast fission fraction in thorium relative to uranium-238; hence, the spectral hardening component in the thorium reactor is smaller. The lower value for the uranium-233 fuel results from the lower rate of increase of η with energy for ^{233}U than for ^{239}Pu below 1 MeV.

Finally, it is of interest to observe how the different designs used to modify the sodium loss reactivity affect the Doppler coefficient. Note from Table 6.3 that little difference exists in the Doppler constant, $T\,dk/dT$, between the homogeneous and the heterogeneous cores.

6.8 Reactivity Worth Distributions

In performing safety calculations, it is often of interest to know the reactivity implications of moving a particular reactor material such as fuel, steel, or sodium from its original location to another. A direct way to compute such an effect is to perform two eigenvalue calculations, one for the original core configuration and one for the new configuration. However, this computation is expensive and impractical to perform for many transient domains of interest.

As long as the amount of material that moves is relatively small, perturbation calculations can be performed to provide a good approximation of the expected reactivity change. Given the flux and adjoint flux distributions, the perturbation worth curves for any material of interest can be determined using Eq. (6.9). The unit of the perturbation worth calculations is usually given as ρ per kg of material (although ρ per unit volume of material is sometimes used). Figures 6.6 through 6.11 are included to illustrate typical results (shown in this case for the FFTF [12]).

Figures 6.6 and 6.7 show the flux and power distributions, respectively, for the central core assembly. They are included to provide a frame of reference. Figures 6.8 through 6.11 show respectively the fuel, sodium void, steel and Doppler weighting perturbation curves for the same central assembly. It should be noted that the FFTF is relatively small core (1,000 liters) and has axial reflectors, rather than blankets. The slight axial skew in the flux and power is due to the beginning-of-life core configuration in which the control rods are banked about halfway into the core. Of particular interest is the sodium void worth, which shows the typical positive region in the central regions of the core, and the negative region near the core periphery where the leakage component dominates. The steel worth curve has a shape similar to that of sodium (note that Fig. 6.9 is for the sodium *void* worth), due to the same processes. The Doppler weighting distribution (Fig. 6.11) is determined by altering the cross section of ^{238}U in Eq. (6.6) to reflect a temperature increase. This curve indicates the increased effectiveness of Doppler feedback due to heating of the fuel near the axial reflectors, where the neutron spectrum is softer and more neutrons are in the Doppler resonance region. A curve of this type can be used in transient computer codes to assess the Doppler feedback due to non-uniform heating of the core.

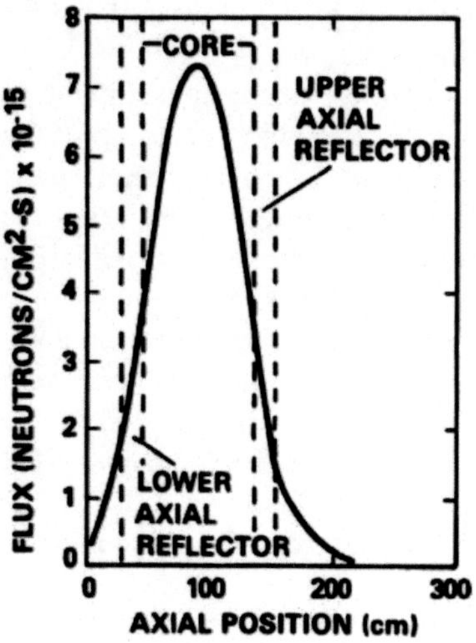

Fig. 6.6 FFTF total flux, central assembly

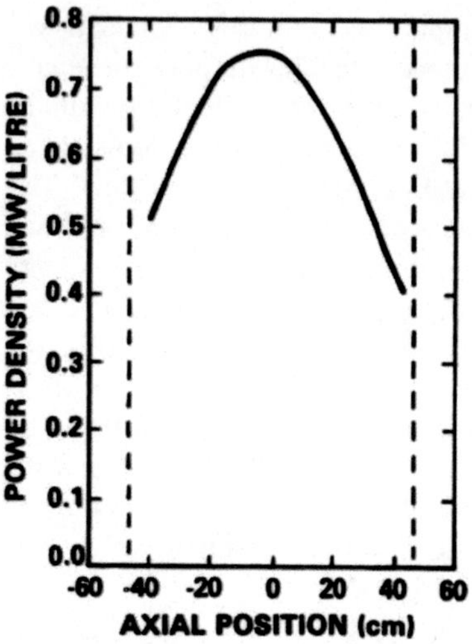

Fig. 6.7 FFTF power profiles, central assembly

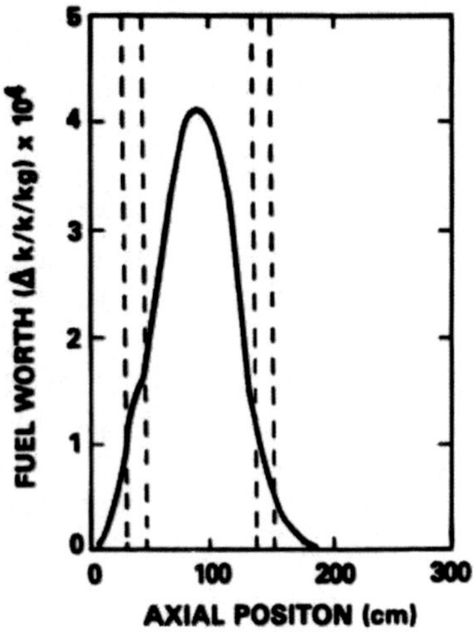

Fig. 6.8 FFTF fuel worth, central assembly

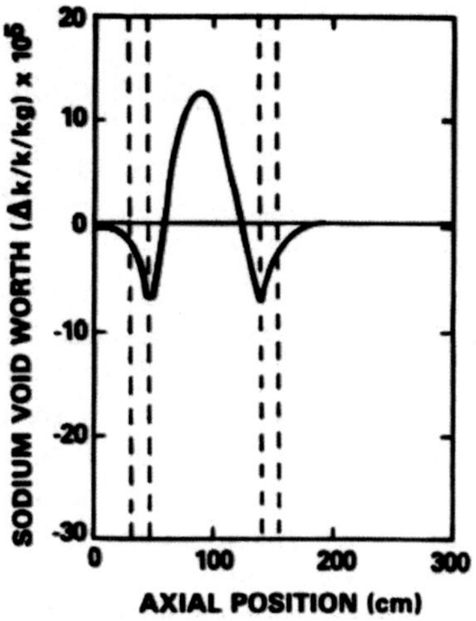

Fig. 6.9 FFTF sodium void worth, central assembly

6.9 Reactivity Control Requirements

Reactivity is controlled by control assemblies, usually containing rods of enriched B_4C, as described in Chapters 2 and 8. The reactivity control systems are part of the Plant Protective System, which is described in Chapter 14. For our purposes here, it is sufficient to know that two independent reactivity control systems—a primary and a secondary system—are used in fast reactors in order to assure that fuel design limits are not exceeded. Independent operation means that each system must operate in a manner such that the capability of either system to fulfill its control function is not dependent on the operation of the other system. The design of the two systems must be diverse as well as independent

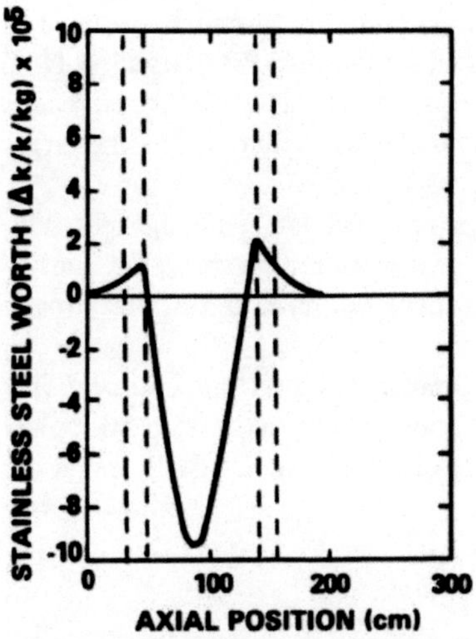

Fig. 6.10 FFTF stainless steel worth, central assembly

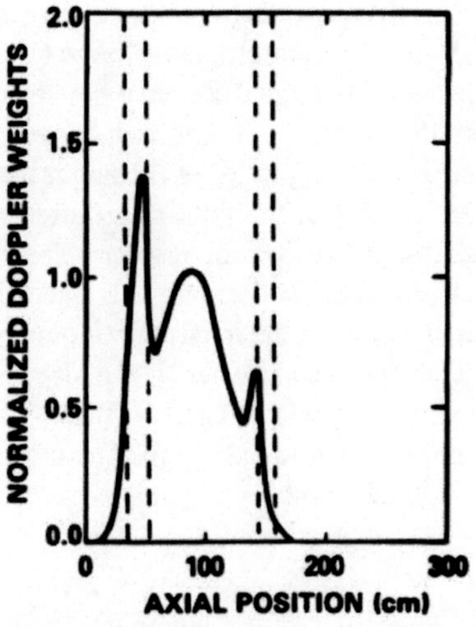

Fig. 6.11 FFTF Doppler weight, central assembly

in order to reduce the potential for common mode failure. These concepts are discussed further in Chapter 14.

Although detailed design criteria for the two control systems continue to evolve, the criteria are well enough established to indicate the function of each system. The primary system must be able to shut down the reactor from any operating condition to the refueling temperature, with the most reactive control assembly inoperative (i.e., failed, or stuck out of the core); this requirement is commonly called the "stuck rod criterion." Any operating condition, in this context, means an *overpower condition*[6] together with a *reactivity fault.* Such a fault is a positive reactivity introduced by some accident situation. The basis often used to quantify this reactivity fault is the maximum worth of a control assembly because the uncontrollable withdrawal of that assembly could be the fault that must be overridden. The primary system also serves to compensate for the reactivity effect of the fuel cycle, or fuel burnup. Moreover, reactivities associated with uncertainties in criticality and fissile inventory are accommodated by the primary control system.

The secondary system must be able to shut down the reactor from any operating condition to the *hot standby* condition, also with the most reactive assembly inoperative. In the hot standby condition, the reactor power is zero but the coolant is at the inlet temperature for full power operation. The fuel, consequently, is at the coolant temperature. The secondary system does not have to duplicate the primary system's ability to hold down the excess reactivity for the fuel cycle because this excess reactivity does not represent additional reactivity to be overridden in an accident. The reasoning here is that, although the secondary system must shut down the reactor without insertion of the primary control assemblies, it is not necessary to assume that the primary assemblies are removed from the core during an accident situation. The secondary system should, however, be able to override the uncontrollable withdrawal of one primary control assembly that is being used for burnup control. Hence, the reactivity fault is included in the secondary requirements. Since uncertainties in fuel loading are accommodated by the primary system, these reactivity uncertainties are not a part of the secondary system requirements.

---

[6]SFRs are generally designed for ~115% overpower operation (i.e., 15% above nominal full power)—to be discussed in Section 10.4.

Reactivity worth requirements for both the primary and secondary control systems are listed in Table 6.4 for a representative large (1,000 MWe) SFR plant [10] and for the CRBRP design [6]. Note especially the large difference between the fuel cycle reactivity requirements. This difference results from the difference in size and, hence, core leakage between the two plants—with the resulting larger internal breeding ratio for the larger plant.

In Table 6.5, the control requirements are compared for these two reactors with the reactivity available in the control systems. The shutdown margins shown for the primary systems are considered adequate. Little margin is needed for the secondary system because uncertainties are already accounted for in the maximum requirements.

The *temperature defect* is the reactivity effect due to changes in coolant and fuel temperatures. This temperature defect appears in Table 6.4 as the reactivity gained in cooling the reactor from full power to the refueling temperature. The contributions to the temperature defect for the CRBRP design are listed in Table 6.6.

Table 6.4 Reactivity worth requirements ($) (equilibrium cycle)

	Representative 1,000 MWe SFR [10]		Early CRBRP homogeneous design [6]	
	Primary	Secondary	Primary	Secondary
1. Temperature defect:				
a. Full power to hot standby	1.6	1.6	2.4 ± 0.8	2.4 ± 0.8
b. Hot standby to refueling	0.8	–	0.8 ± 0.3	–
2. Overpower	0.3	0.3	0.2 ± 0.1	0.2 ± 0.1
3. Reactivity fault	3	3	2.8 ± 0.4	0.2 ± 0.4
4. Fuel cycle excess reactivity	5	–	18.2	–
5. Allowance for refueling uncertainties	–	–	0.8 ± 0.4	–
6. Uncertainties				
a. Criticality	+1	–	–	–
b. Fuel tolerances	+0.8	–	0.4	–
Subtotal	11	5	25.2	5.4
Propagated uncertainties	1.2	–	1.3	0.9
Maximum requirement	12	5	26.5	6.3

Table 6.5 Comparison of control requirements and available reactivity worths (representative 1,000 MWe SFR and early CRBRP homogeneous design)

	Representative 1,000 MWe SFR [10]		CRBRP [6]	
	Primary	Secondary	Primary	Secondary
Number of control assemblies	13	6	15	4
Reactivity worth of control system ($)	20	8	31.8	8.4
Worth of 1 stuck rod ($)	3	2.5	2.8	2.0
Reactivity worth available ($)	**17**	**5.5**	**29.0**	**6.4**
Maximum requirements ($) (from Table 6.4; includes uncertainties)	12	5	26.5	6.3
Shutdown margin ($)	5	0.5	2.5	0.1

Table 6.6 Temperature defect, early CRBRP homogeneous design [6]

Contribution	Full power to hot standby	Hot standby (316°C) to refueling (205°C)
Doppler effect ($)	1.9	0.3
Axial fuel expansion ($)	0.15	0.05
Radial core expansion ($)	0.3	0.4
Coolant density ($)	0.01	0.01
Total ($)	2.4	0.8

References

1. W. Häfele, K. Ott, L. Caldarola, W. Schikarski, K. P. Cohen, B. Wolfe, P. Greebler, and A. B. Reynolds, "Static and Dynamic Measurements on the Doppler Effect in an Experimental Fast Reactor," *Proceedings of the Third International Conference on the Peaceful Uses of Atomic Energy*, Geneva, 1964.
2. L. D. Noble, G. Kassmaul, and S. L. Derby, *SEFOR Core I Transients*, GEAP-13837, General Electric Co., August 1972, and *Experimental Program Results in SEFOR Core 11*, GEAP-13833, Sunnyvale, CA, June 1972.
3. R. J. Tuttle, "Delayed Neutron Yields in Nuclear Fission," *Proceedings of the Consultants' Meeting on Delayed Neutron Properties,* IAEA, Vienna, March 1979.
4. R. J. Tuttle, "Delayed-Neutron Data for Reactor-Physics Analysis," *Nucl. Sci. Eng., 56* (1975) 70.
5. H. H. Hummel and D. Okrent, *Reactivity Coefficients in Large Fast Power Reactors,* American Nuclear Society, LaGrange Park, IL, 1970 87–88, 148.
6. *Preliminary Safety Analysis Report,* Clinch River Breeder Reactor Plant, Project Management Corp., Oak Ridge, TN, 1974, 43–61.
7. H. S. Bailey and Y. S. Lu, "Nuclear Performance of Liquid-Metal Fast Breeder Reactors Designed to Preclude Energetic Hypothetical Core Disruptive Accidents," *Nucl. Tech., 44* (1979) 81.
8. D. Okrent, "Neutron Physics Considerations in Large Fast Breeder Reactors," *Power Reactor Tech., 7* (1964) 107.
9. E. Paxson, ed., *Radial Parfait Core Design Study.* WARD-353, Westinghouse Electric Corp., Madison, PA, June 1977, 79.
10. B. Talwar, *Preconceptual Design Study of Proliferation Resistant Homogeneous Oxide LMFBR Cores,* GEFR-00392 (DR)-Rev. 1. General Electric Co., Sunnyvale, CA, September 1978, 5.14.
11. J. C. Mougniot. J. Y. Barre, P. Clauzon, C. Giacomette, G. Neviere, J. Ravier, and B. Sicard, "Gains de Regeneration des Reacteurs Rapides á Combustible Oxyde et á Réfrigerant Sodium," *Proc. European Nuclear Conf., 4* (April 1975) 133.
12. J. V. Nelson, R. W. Hardie, and L. D. O'Dell, *Three Dimensional Neutronics Calculations of FTR Safety Parameters*, HEDL-TME 74-52, Hanford Engineering Development Laboratory, Richland, WA, August 1974.

Chapter 7
Fuel Management

Pavel Tsvetkov, Alan Waltar, and Massimo Salvatores

7.1 Introduction

Fuel management deals with the irradiation and processing of fuel. An analysis of the fuel cycle is necessary to estimate fuel costs and to define operational requirements such as initial fuel compositions, how often to refuel, changes in power densities during operation, and reactivity control. The great flexibility of fast spectrum systems allows them to either "breed" desired fuel or "burn" undesired wastes—particularly the minor actinides (MA),[1] which constitute the greatest long-term contribution to radiotoxicity and heat load in geologic waste repositories.

Fuel costs represent one contribution to the total power costs, as discussed in Chapter 3. Unlike the light water reactor (LWR), fuel costs for a fast spectrum reactor are insensitive to U_3O_8, price. Hence, this contribution to total power cost is predicted to be lower for a fast breeder reactor than for a thermal reactor as the price of U_3O_8, rises. Since more fissile material is produced in a breeder reactor configuration than is consumed, the basic (feed) fuel for the fast breeder reactor is depleted uranium, which is available for centuries without further mining of uranium ore.

In this chapter we describe methods to calculate isotopic compositions in fuel being loaded into the reactor and fuel discharged from the reactor. These figures are needed in order to calculate fuel costs and to determine the amount of fissile material that must be loaded in order to achieve the desired cycle length and power level. The physics arguments will be presented that make fast spectrum reactors the most adaptable system to minimize radioactive wastes. Methods will also be presented for calculating breeding ratios and doubling times. They provide convenient figures of merit for comparison of various fast breeder concepts and designs.

7.2 Fuel Management Concepts

There are a number of choices regarding the loading and removal of fuel from a reactor. Consider first the concept of the *equilibrium cycle* versus the *first-core cycle.* Initially, the entire reactor is loaded with unirradiated fuel, also called *fresh,* or *new fuel,* in preparation for the first-core cycle. After an initial period of operation, some irradiated fuel is removed and replaced with more fresh fuel.

Section 7.8 Physics of Transmutation contributed by Massimo Salvatores

[1]"Minor actinides" refers to the actinide elements heavier than uranium typically found in used nuclear fuel, but does not include uranium or plutonium, which are "major actinides". Minor actinides are typically Np, Am, and Cm.

P. Tsvetkov (✉)
Zachry Engineering Center, Department of Nuclear Engineering, Texas A&M University, College Station, TX, USA
e-mail: tsvetkov@tamu.edu

A.E. Waltar et al. (eds.), *Fast Spectrum Reactors*, DOI 10.1007/978-1-4419-9572-8_7,

After several years, a transition is made to an equilibrium fuel cycle. The equilibrium cycle is said to be reached when, after each refueling, the composition of materials remaining in the core after used fuel has been discharged and fresh fuel reloaded is the same as the composition of materials at the start of the previous irradiation cycle. The details of transition to the equilibrium cycle are complex and will not be considered here. Instead we will only be concerned with analysis of the equilibrium cycle itself.

When fuel is irradiated, a net decrease in heavy atoms occurs due to fission and the fuel is said to be *burned.* The term *burnup* is used as the measure of either the energy obtained from the burned fuel or the fraction of fuel that has fissioned (i.e., burned). Units of burnup are described in Section 7.3.

We will use the term *fuel cycle* in a specific way in the remainder of this chapter; the term will refer to the irradiation process that begins at the time of reactor startup after refueling and ends when the reactor is shut down for subsequent refueling. This use of fuel cycle is in contrast to the broader use of the term to describe all the processes involving fuel, from mining to fabrication, irradiation, reprocessing, and storage. The duration of the fuel cycle, i.e. the time between refuelings, is called the *refueling interval.*

Fuel can be loaded into nuclear reactors either *continuously* or *batchwise.* In batch loading, some fraction of the irradiated fuel is replaced by a fresh batch of fuel at periodic shutdowns for refueling. Fast breeder reactors, like light water reactors, always utilize batch loading. If one-third of the fuel is replaced, there would be three batches of fuel in the reactor, each with its separate irradiation history. Generally the refueling interval for proposed commercial fast breeder reactors is 1 year, primarily because that is convenient for utilities, although it could be much longer (several years or even decades) if specific means are taken to keep the reactor sealed up for purposes such as enhanced nonproliferation attributes. Shorter refueling intervals (less than a year) have generally been used in the prototype plants. A typical residence time for fuel in the regions of highest power in a sodium-cooled fast reactor (SFR) is 2 years. The combination of a 1-year refueling interval and a 2-year residence time corresponds to a two-batch loading.

Another fuel management choice involves *scatter loading* versus *zone loading.* In scatter loading, fuel assemblies in a batch are loaded at various positions throughout the core. In zone loading, irradiated fuel is removed from one zone of the core only, while all fresh fuel is loaded in a different zone. Before loading the fresh fuel, it is necessary to *shuffle* partially irradiated fuel from the space into which the fresh fuel is to be placed to the positions from which the fully irradiated fuel was removed. Light water reactors use a combination of these two loading schemes. For fast reactors it is generally expected to use scatter loading. This has the disadvantage that highly irradiated fuel with lower fissile inventories and power densities will be located adjacent to fresh fuel with higher power densities, but the time-consuming step of shuffling fuel will not be required.

Non-fissile "diluent" assemblies made of either fertile material or steel can be used for (1) suppressing initial excess reactivity of fresh fuel and (2) power flattening to achieve higher initial power. In a fresh core, the reactivity may be too high to be controlled by the absorber rods, which are generally designed for the equilibrium core conditions. Strategically located diluent assemblies serve the purpose of reducing the reactivity without reducing the absorber rod worths. For example, steel assemblies were used in the Super Phénix first cycle [1]. In the Indian FBTR, depleted uranium bearing assemblies are being used for flattening the centrally peaked power distribution—thus making it easier to reach the design power within linear heat rating limits.

The fresh fuel loaded into the core can come either from thermal reactors or from fast breeder reactors. In the early stages of a mixed fast/thermal reactor economy, the fissile plutonium (or ^{233}U) produced in a thermal reactor can be more economically utilized as fuel for a fast reactor than as recycled fuel for a thermal reactor. As we will determine later from Section 7.9, it would take about 12–15 light water reactors operating for 1 year to generate enough plutonium to start up a fast spectrum reactor of comparable electrical rating. When a breeder reactor economy is established, the breeders develop enough fissile materials both to feed themselves and to start up new reactors, at a rate depending on doubling time (or, of course, demand for new electrical generating capacity). Eventually these breeders may produce more fuel than is needed for new fast reactor power plants, in which case the excess fissile material bred can be used in burner or thermal reactors. Thermal reactors will presumably always be less expensive to build; hence, it may be economical to continue to build and operate thermal converter reactors to run on fuel from breeders long after the mining of uranium ore becomes too expensive.

Whereas the main interest in fast spectrum systems continues to be tied to the ability of such systems to greatly extend fuel supplies (by converting fertile fuel such as ^{238}U or ^{232}Th into fissile fuel, thereby creating fuels to last for several millennia), much of the interest over the past decade or so has been to deploy such systems to substantially alleviate issues associated with nuclear waste disposition. Numerous studies have been performed in order to identify appropriate "Partitioning and Transmutation" (P&T) strategies to address the prospects of reducing the burden on a geological storage of high-level nuclear waste. The fission process is always in competition with other processes, and in particular with neutron capture. Neutron capture does eliminate nuclide A, but it transforms it into nuclide (A+1), which can still be radioactive. Nuclide (A+1) can in turn be fissioned or transmuted into nuclide (A+2), and so on. The advantage of a fast spectrum is that the fission to capture cross section for the minor actinides of interest is quite high—much larger than that existing for a thermal spectrum. This is illustrated later in the chapter in Fig. 7.5. A key requirement for designing a dedicated reactor for such a "burning" process is to be able to load fuels with potentially very different mixtures of Pu and MA, according to the chosen approach and the objective of the P&T strategy, and to accomplish this without diluting its safety and operability characteristics.

One can imagine an equilibrium fuel cycle in which the feed from a breeder reactor comes from another breeder just like it, and each cycle is exactly like the previous one. In this equilibrium condition, all depletion stages of the fuel are present at the same time. Periodic power variations due to load following do not disturb the equilibrium condition if their period is short compared with the decay constants of the nuclides considered. Computer codes are available to calculate such a cycle: iteration is required to find a feed nuclide composition that is consistent with the discharge composition. The problem is complicated by having the choice of either recycling fuel from the core only and using fuel from the blanket to start up new reactors, or mixing all the fuel discharged from both core and blankets before using it for recycle and new startups. As one considers these and other options, the possibilities multiply and the computer codes and logic correspondingly become very complex.

Equilibrium cycle calculations are usually the first calculations performed when designing and optimizing a reactor. In this case, one has to iterate in order to obtain a critical configuration. Usually a calculation is started with a guess on the initial composition or on the final discharge composition and the average composition is obtained. This can be done for each core region separately (e.g. inner and outer core regions) or for the whole core (point model calculations). Once the core equilibrium composition is known a criticality calculation is performed. First survey studies are performed with simpler methods and later more detailed calculations are performed in which various effects due to neighboring fuel elements, control rods, reflector, etc. can be properly treated. For more detailed equilibrium depletion modeling, space dependence has to be considered. In space-dependent equilibrium depletion models, it is possible to treat various core regions iterating on the initial or discharge fuel composition of these regions in order to flatten power distribution.

Examples of historical fast reactor fuel cycle codes used widely in the United States are FUMBLE [2], REBUS [3], 2DB [4], and 3DB [5]. To support the development of fast reactors in US, an extensive suite of computer codes has been developed and validated at Argonne National Laboratory. Criticality calculations with DIF3D [6, 7] and whole-core burnup calculations with REBUS-3 provide the performance characteristics. The obtained burnup-dependent data can be used to calculate reactivity coefficients and kinetics parameters. The DIF3D is a versatile code system for carrying out multigroup, three-dimensional, whole core neutronics calculations. It is used to compute the reactor multiplication factor, flux and power distributions, and other functionals of the neutron flux. The system models Cartesian, curvilinear, and hexagonal core geometries either by means of the finite difference approach or by a highly efficient nodal option. The REBUS-3 is a reactor burnup and fuel cycle analysis code that makes use of DIF3D as a flux calculation module. This code models nuclide transmutations on a three-dimensional, region-dependent basis and provides considerable flexibility for specifying operational constraints and fuel management strategies for both in-core and out-core portions of the fuel cycle. Calculations can be performed in an equilibrium fuel cycle mode as well as in a cycle-by-cycle core follow mode. More discussion of DIF3D and neutronics calculations methodologies is provided in Appendix E.

In this book, as a practical example, we will restrict ourselves to a single type of a relatively simple fuel cycle calculation. We will suppose that we know the isotopic composition of the plutonium feed. We will then develop a technique for calculating the plutonium fraction required to allow the reactor to operate for a refueling interval and will calculate the composition of fuel discharged from the reactor. We will *not* calculate the buildup of plutonium in the blanket, although the technique to do this, if we were using a space-dependent calculation, should be apparent. The proposed calculation will demonstrate most of the basic logic in fuel cycle analysis and will illustrate the approximate discharge compositions from the core. Yet it will be simple enough for the student to program in a reasonable length of time. Our objective will then have been accomplished—to illustrate basic fuel management logic and concepts. To consider further details of any of the numerous actual codes in use in the industry would be too tedious for our purposes here.

Other aspects of the fuel cycle that must be considered in more detailed analyses involve disposition of fuel outside the reactor. The first requirement of P&T strategies is the deployment of used fuel reprocessing techniques (aqueous or dry). This could be based on either the current commercial technology, which is now being implemented at La Hague in France, Sellefield in the UK, or Rokkasha in Japan, (where Pu is separated up to 99.9%) or by adopting innovative approaches such as pyrochemistry. The requirement for advanced separation techniques is to extend the 99.9% separation performance of Pu to Np, Am, and Cm, with these elements either kept together or separated. In any case, they need to be decontaminated from the lanthanide elements as much as possible.[2] The separated transuranics (TRU)[3] can then be "transmuted" in a neutron field. The essential mechanism is to fission them, transforming them into much shorter lived or stable fission products.

After fuel is removed from the core and blankets, it is generally allowed to remain in the reactor vessel to allow the short-lived fission products to decay. It may then be placed in an ex-vessel storage pool for further decay before being sent to a reprocessing plant. The time required for storage, shipping, reprocessing, fabrication, and return to the reactor all affect fuel cost analysis and doubling time.

[2]Lanthanide elements such as gadolinium are used as neutron absorbers in the manufactured fuel for reactivity control, and so they are likely present in the used fuel. Separation of these elements from the others is important to assure that neutron absorbers do not affect subsequent usage of the partitioned materials.

[3]Transuranic isotopes all have atomic numbers greater than 92 (the atomic number of uranium).

Moreover, there will inevitably be a small wastage rate of fissile material in reprocessing and fabrication, which also influences doubling time and makes the accountability of fissile inventory more difficult.

7.3 Fuel Burnup

The two units most widely used for reporting burnup of fuel, MWd/kg (or MWd/tonne) and atom percent, are measures of fundamentally different quantities. MWd/kg measures the energy obtained from irradiating fuel; it is most frequently used, at least in the United States, by design engineers performing fuel cycle and fuel cost analyses. Atom percent calculations measure the fraction of heavy atoms in the fuel that undergo fission. This unit is more useful to a research engineer investigating fuel damage or behavior as a function of irradiation, although it is widely used in Europe for fuel cycle design calculations. We will define both units quantitatively and will show how the two units can be related to one another. We will, however, use the specific energy unit (MWd/kg) in the rest of this chapter.

When burnup is reported in MWd/kg, the fuel mass in the denominator includes the heavy atoms only. It does not include the oxygen in oxide fuel (nor the carbon, nitrogen, etc. found in other fuel compounds or alloys); hence, the oxide fuel mass must be multiplied by the ratio of the fuel atomic weight to the oxide molecular weight, i.e., 238/270 for UO_2-PuO_2, in order to obtain the heavy atom mass for use in calculating burnup.

In calculating burnup, one must be careful to distinguish between *chronological time* and *time at rated power.* Time at rated power is the product of a load factor (or capacity factor), f, and chronological time. The refueling interval is given in chronological time. Suppose the thermal power of a fast reactor core containing 30,000 kg of oxide fuel is 3,000 MWth, the load factor is 0.7, and the refueling interval t_c is 1 year. The average burnup, B, during a fuel cycle is

$$\begin{aligned} B\left(\frac{\text{MWd}}{\text{kg}}\right) &= \frac{P\,(\text{MW})\,f\,t_c\,(\text{d})}{m\,(\text{kg oxide})\,\dfrac{\textit{mass}\text{ metal}}{\textit{mass}\text{ oxide}}} \\ &= \frac{(3{,}000\text{ MW})\,(0.7)\,(1\text{yr})\,(365\text{ d/y})}{(30{,}000\text{ kg oxide})(238\text{ kg metal/270 kg oxide})} \\ &= 29\text{ MWd/kg}. \end{aligned} \tag{7.1}$$

Burnup in MWd/kg can be related to neutron flux, fission rate, and chronological time, t, as follows:

$$B\,(t) = \frac{\Sigma_f\phi\left(\dfrac{\text{fissions}}{\text{cm}^3\text{core}\cdot\text{s}}\right) f\,t\,(\text{d})\cdot 10^3\,(\text{g/kg})}{2.9\times 10^{16}\left(\dfrac{\text{fissions}}{\text{MW}\cdot\text{s}}\right) F_f\left(\dfrac{\text{cm}^3\text{ oxide}}{\text{cm}^3\text{ core}}\right)\rho_{oxide}\left(\dfrac{\text{g oxide}}{\text{cm}^3\text{ oxide}}\right)\dfrac{238}{270}\left(\dfrac{g\text{ metal}}{g\text{ oxide}}\right)}, \tag{7.2}$$

where F_f is the fuel volume fraction and ρ_{oxide} is the fuel smear density. This relation assumes a constant $\Sigma_f\phi$ with time. In reality, $\Sigma_f\phi$ is both strongly space-dependent and slightly time-dependent. Even the average core value of $\Sigma_f\phi$ [and P in Eq. (7.1)] is slightly time-dependent due to the increase in power in the blankets during a fuel cycle.

Burnup in atom percent as a function of chronological time is

$$B\,(\text{at.}\%) = 100\frac{\sum_m N_m \sigma_{fm}\phi ft}{\sum_m N_{m,0}}, \tag{7.3}$$

where sums are over all heavy atoms and $N_{m,0}$ are the atom densities at $t = 0$.

The two burnup units are related to one another; the ratio of the specific energy value to the percent fission value is

$$\begin{aligned}\frac{B\left(\frac{\text{MWd}}{\text{kg}}\right)}{B\,(\text{at.}\%)} &= \frac{B\left(\frac{\text{MWd}}{\text{kg heavy metal}}\right)}{B\left(\frac{\%\text{ fissions}}{\text{atoms heavy metal}}\right)} \\ &= \frac{6.023\times 10^{26}\left(\frac{\text{atoms heavy metal}}{\text{kg-mol heavy metal}}\right)}{\left[100\left(\frac{\%}{\text{absolute}}\right)\cdot 2.9\times 10^{16}\left(\frac{\text{fissions}}{\text{MW}\cdot\text{s}}\right) \cdot 0.864\times 10^{5}\left(\frac{\text{s}}{\text{d}}\right)\cdot 238\left(\frac{\text{kg heavy metal}}{\text{kg-mol heavy metal}}\right)\right]} \\ &= 10.\end{aligned} \tag{7.4}$$

Fuel exposure in the reactor is usually limited by fuel damage or by cladding or structural damage or swelling, not by reactivity control. Design exposure goals for SFR's with oxide fuel are of the order of 100 MWd/kg or higher, although these may be reduced in early SFRs due to limitations on the cladding or hexagonal ducts.

The time that fuel must remain in the core in order to reach the design burnup depends on the specific power (MWth/kg metal), which in turn depends on the spatial distribution of the power density. In order to consider the chronological time that fuel can remain in the core, consider the radial power density distribution illustrated in Fig. 7.1 and the following numerical example.

Suppose that the radial peak-to-average power ratios for assemblies A, B, and C are 1.3, 0.8, and 0.65. Suppose that the axial peak-to-average ratio is 1.25. Assume typical values for specific inventories in the inner and outer cores of 1.2 and 1.6 kg fissile/MWth, respectively, and for fissile atom fractions of 0.11 and 0.15 for the inner and outer cores. Finally, assume a load factor of 0.7. The peak burnup for assembly A after a chronological time in the core of 2 years is

$$\left(\frac{1}{1.2}\frac{\text{MWth}}{\text{kg fissile}}\right)\left(0.11\frac{\text{kg fissile}}{\text{kg metal}}\right)(2\text{yr})\,(365\text{ d/yr})\,(0.7)\,(1.25) = 76\text{ MWd/kg}.$$

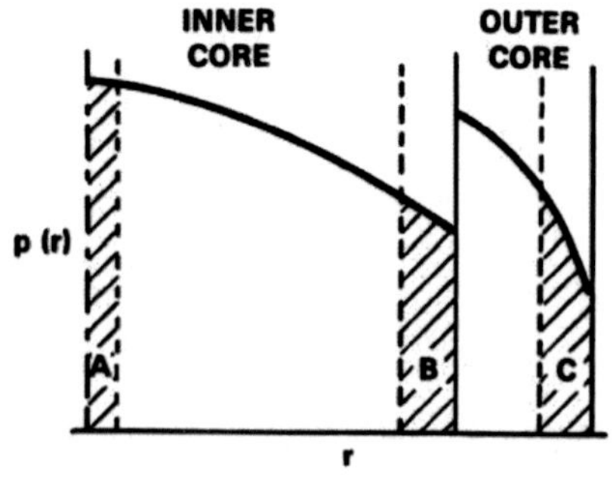

Fig. 7.1 Radial power distribution and location of assemblies A, B, and C

Table 7.1 Typical exposures for various residence times at core positions A, B, and C in Fig. 7.1

	Burnup (MWd/kg)					
	Average			Peak		
Residence time	A	B	C	A	B	C
2 years	61			76		
3 years	91	56	47	114	70	58
4 years		75	62		94	78

Average and peak exposures for assemblies A, B, and C as a function of residence time are shown in Table 7.1.

For an allowable peak burnup of 80 MWd/kg, most of the fuel would stay in the core only 2 or 3 years, with the outermost assemblies remaining for 4 years. Residence time for radial blanket assemblies would be longer, varying perhaps from 4 to 6 years.

7.4 Burnup Equations

The central calculation in a fuel management code is the depletion and production of nuclides as a function of time. It is extremely important to monitor (and be able to accurately model in computational studies) the nuclide composition in the reactor core during reactor operation, since changes in the core composition affect the core multiplication as well as flux and power distributions. The prediction of fuel depletion and conversion in the core is essential for the determination of fuel loading requirements.

Fortunately, changes in core composition occur relatively slowly (on time scale of hours, days, months) so it is possible to consider them as intermediate and longtime problems such that the reactor core can always be kept in a critical state by control system adjustments. This means that although a time-dependent analysis of the reaction rate equations describing changes in nuclide atomic density is necessary, the neutronics behavior can be studied by performing a sequence of instantaneous static criticality calculations for each core composition interspersed with time-dependent depletion computations to determine the atomic densities throughout the reactor of all nuclides affected by the depletion process.

The differential equations governing the burnup of fuel (fuel composition changes) are called the *Bateman equations* or the *burnup equations*. Sometime, these equations are also called the production-destruction (or depletion) equations. If all cross sections and fluxes are appropriately averaged one-group values, the Bateman equations can be written as:

$$\frac{dN_k}{dt} = \overbrace{\phi\sum_{i=l}^{m} N_i\sigma_{f,i}y_{i\to k}}^{\text{PRODUCTION DUE TO FISSION}} + \overbrace{\phi\sum_{z=r}^{q} N_z\sigma_{c,z}\gamma_{z\to k}}^{\text{PRODUCTION DUE TO CAPTURE}} + \overbrace{\sum_{j=n}^{p} N_j\lambda_j\alpha_{j\to k}}^{\text{PRODUCTION DUE TO DECAY}}$$

$$- \underbrace{\lambda_k N_k}_{\text{DESTRUCTION DUE TO DECAY}} - \underbrace{\phi N_k\sigma_{a,k}}_{\text{DESTRUCTION DUE TO ABSORPTION}},$$

where:

N_k = atomic density of nuclide k during the depletion time step,
ϕ = flux for the depletion time step,
$\sigma_{f,i}$ = microscopic fission cross section of nuclide i,
$\sigma_{c,z}$ = microscopic capture cross section of nuclide z including (n,xn) reactions
$\sigma_{a,k}$ = microscopic absorption cross section of nuclide k,
λ_j, λ_k = radioactive decay constants of nuclides j and k,
$y_{i\to k}$ = yield of nuclide k due to fission of nuclide i,
$y_{z\to k}$ = probability that a neutron capture in nuclide z produces nuclide k,
$\alpha_{j\to k}$ = probability that the radioactive decay of nuclide j produces nuclide k,
k = nuclide index for which an equation is written,
i = summation index over all precursor nuclides m yielding nuclide k through neutron-induced fission events $i = l, \ldots, m$,
z = summation index over all precursor nuclides q yielding nuclide k through neutron capture events, $z = r, \ldots, q$,
j = summation index over all precursor nuclides p yielding nuclide k through radioactive decay events, $j = n, \ldots, p$,

l, r, n are specific first nuclides undergoing fission, capture and radioactive decay events, respectively.

These equations and their solutions are described in this section. Either an analytical solution or a numerical integration procedure can be used to calculate atom densities from the burnup equations. An analytical solution assumes constant neutron fluxes and cross sections during the period of the calculation. The numerical integration technique assumes the same thing during a time step.

The constant-flux cross section approximation applies for longer periods of time in a fast reactor than in a thermal reactor because the cross sections are not as sensitive to energy changes and fission product buildup. In the iterative calculation illustrated later in Fig. 7.4, the flux and cross sections are constant throughout the fuel cycle.

In this section the burnup equations will first be presented. The two methods of solution (numerical and analytical) will then be described.

The two potential fertile-to-fissile conversion chains for the U-Pu cycle and the Th-U cycle were given in Chapter 1. We will assign material numbers to each of the nuclides in the chain according to Table 7.2. In commercial codes all nuclides are accounted for, such as ^{239}U, ^{239}Np, ^{237}Np, ^{238}Pu, ^{235}U, ^{236}U, and the isotopes of Am and Cm. The more complete conversion chains are shown in Figs. 7.2 and 7.3. Inclusion of ^{239}U and ^{239}Np makes little difference since their half-lives are so short (23.5 min and 2.35 days, respectively). Inclusion of ^{233}Pa in the Th-U cycle is more important because of its relatively long half-life (27.4 days).

Table 7.2 Material numbers for conversion chains

Material number	U-Pu cycle	Th-U cycle
1	^{238}U	^{232}Th
2	^{239}Pu	^{233}U
3	^{240}Pu	^{234}U
4	^{241}Pu	^{235}U
5	^{242}Pu	^{236}U
6	^{243}Am	^{237}Np
7	^{241}Am	–
8	f.p. pairs	f.p. pairs

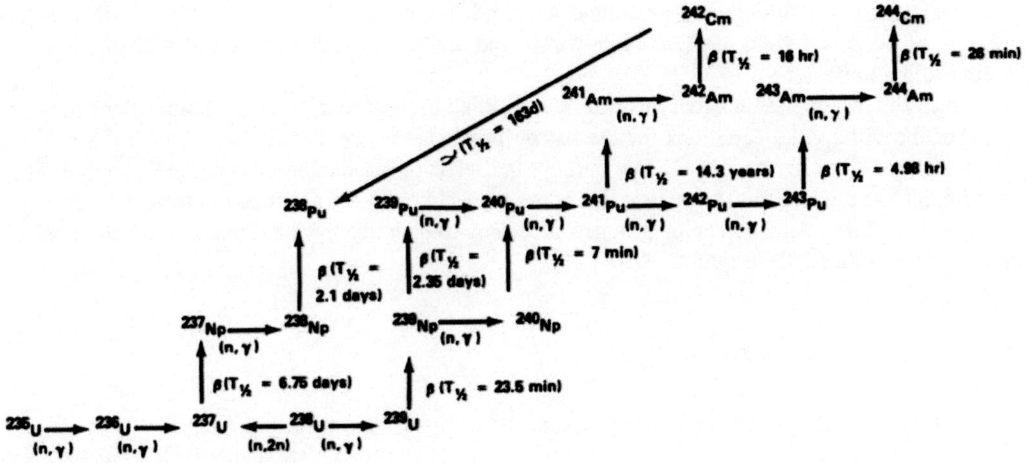

Fig. 7.2 ^{238}U-^{239}Pu conversion chain

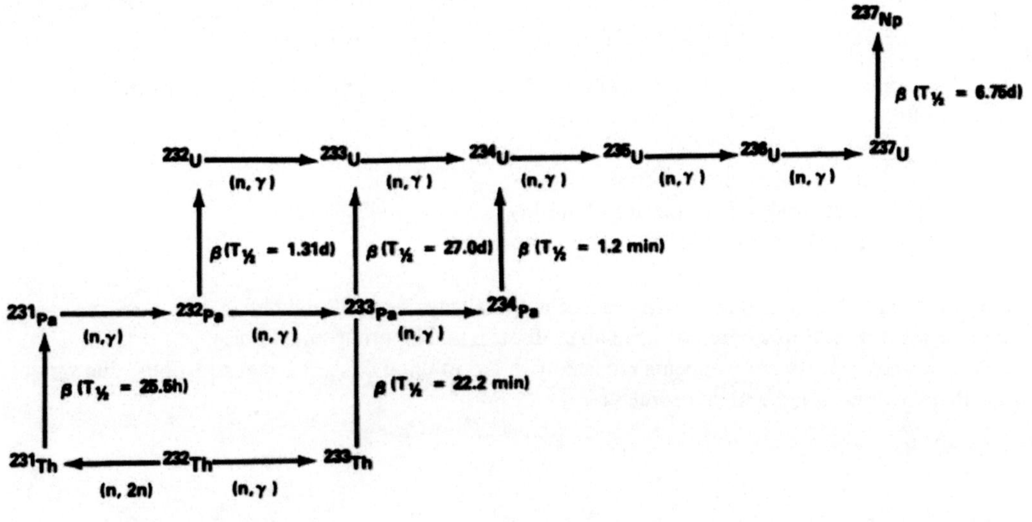

Fig. 7.3 ^{232}Th ^{233}U conversion chain

Nuclear systems intended for long term operation and for high burnup levels require detailed modeling of nuclear transformations of higher actinides up to isotopes of californium. In the derivations that follow, we will use the U-Pu cycle as typical because of its more widespread interest.

Plutonium-238 is frequently included in fuel cycle results (e.g., see Table 7.5 near the end of this chapter). Two sources of ^{238}Pu are (1) neutron capture by ^{237}Np followed by beta decay of ^{238}Np, and (2) alpha decay of ^{242}Cm that comes from beta decay of ^{242}Am, which in turn results from neutron capture by ^{241}Am. There are also two principal sources of ^{237}Np: (1) $(n, 2n)$ reaction by ^{238}U with subsequent beta decay of ^{237}U, and (2) (n, γ) reactions in ^{235}U and ^{236}U, followed by beta decay of ^{237}U.

Not all fission products need to be treated in a depletion calculation, but only those having a non-negligible absorption cross section, or decaying into nuclides with non-negligible absorption cross section. For this second case, the parent nuclide needs only to be treated if its decay time is sufficiently long. Otherwise, it is sufficient to treat the decay product attributing to its cumulative yield (which is the sum of the direct yield of the nuclide considered and of the yields of the nuclides of the decay chain leading to it).

In many cases, depletion equations can be simplified by neglecting the interconnections between some of the differential equations for the fission products, i.e. neglecting some links of the decay chains. This can be reasonable for low burnup, but leads to intolerable errors if the burnup is very high. When the nuclear transformation chains are neglected it is not necessary to treat independently all fission products, but they can be grouped according to their absorption cross section into a certain number of pseudo fission products.

7.4.1 The Differential Equations

The ^{238}U (and ^{232}Th) concentration (or mass) is obtained from the equation:

$$\frac{dN_1}{dt} = -N_1\sigma_{a1}\phi f, \tag{7.5}$$

where:

N_1 = concentration or mass of ^{238}U in any convenient unit such as atoms per barn cm or per cm^3, or kg;
t = chronological time (s),
σ_{a1} = ^{238}U absorption cross section (cm^2),
f = load factor (fraction of time at full power),
ϕ = one-group absolute value of neutron flux (n/cm^2·s).

It is noted that σ must be in cm^2 here because it will always be multiplied by φ; hence, cross sections in barns must be multiplied by 10^{-24} cm^2/barn for use in the burnup equations.

The variable $f\phi t$ (or $fnvt$) appears repeatedly in the solution of the burnup equations; this variable is called the *fluence* and will be denoted by z:

$$z = f\phi t. \tag{7.6}$$

Equation (7.5) can be written in terms of fluence as

$$\frac{dN_1}{dz} = -N_1\sigma_{a1}. \tag{7.7}$$

The ^{239}Pu (and ^{233}U) concentration is obtained from the equation:

$$\frac{dN_2}{dz} = -N_2\sigma_{a2} + N_1\sigma_{c1}. \tag{7.8}$$

Note that only the capture rate of fertile material 1 is included, whereas the absorption rate, $N_1\sigma_{a1}\phi$ was used in Eq. (7.5).

The equations for ^{240}Pu, and ^{242}Pu, and ^{243}Am are similar to the equation for ^{239}Pu, so that

$$\frac{dN_m}{dz} = -N_m\sigma_{am} + N_{m-1}\sigma_{c,m-1} \quad (m = 2,3,5,6). \tag{7.9}$$

Depletion of ^{243}Am will be ignored in the model described here, so that $\sigma_{a6} = 0$.

The equation for ^{241}Pu must include the β-decay of ^{241}Pu, for which the half-life is 14.7 years. The equation is written first in terms of chronological time, *t*, and then in terms of fluence, *z*.

$$\frac{dN_4}{dt} = -N_4\sigma_{a4}\phi f - \lambda_4 N_4 + N_3\sigma_{c3}\phi f, \tag{7.10}$$

$$\frac{dN_4}{dz} = -N_4\left(\sigma_{a4} + \frac{\lambda_4}{\phi f}\right) + N_3\sigma_{c3}. \tag{7.10a}$$

For the model described here, we will ignore all depletion of ^{241}Am. Hence, the equation for ^{241}Am is

$$\frac{dN_7}{dz} = \frac{\lambda_4}{\phi f}N_4. \tag{7.11}$$

A fission-product pair is produced every time a fission occurs. Therefore, the fission product concentration can be written easily if destruction of fission products by absorption is ignored:

$$\frac{dN_8}{dz} = \sum_{m=1}^{5} N_m\sigma_{fm} \tag{7.12}$$

where only materials 1 through 5 are considered fissionable in the present model.

7.4.2 Numerical Solution of the Burnup Equations

Several numerical integration procedures are available for solving the burnup equations. The simplest is the Euler method. This is an explicit method, in which the value of each atom density, *N*, at time step *j*+1, N_{j+1}, is expressed explicitly in terms of its value, N_j at time *j* and its time derivative, $\left(\Delta N/\Delta t\right)_j$ at *j* and the time step δt:

$$N_{j+1} = N_j + \left(\frac{\Delta N}{\Delta t}\right)_j \delta t. \tag{7.13}$$

A more accurate integration method is an implicit one, such as the following used in the 2DB and 3DB codes:

$$N_{j+1} = N_j + \frac{\delta t}{2}\left[\left(\frac{\Delta N}{\Delta t}\right)_j + \left(\frac{\Delta N}{\Delta t}\right)_{j+1}\right]. \tag{7.14}$$

Since $\left(\Delta N/\Delta t\right)_{j+1}$ is a function of the unknown N_{j+1}, $\left(\Delta N/\Delta t\right)_{j+1}$ is also an unknown. The solution for N_{j+1} must therefore be iterative.

7.4.3 Analytical Solution of the Burnup Equations

We will present analytical solutions of the burnup equations used in some fuel cycle codes such as FUMBLE [2]. One-group collapsed cross sections for capture and fission are used to obtain reaction rates. For fast reactor analysis, the effective one-group cross section does not change very rapidly with irradiation during a fuel cycle, so reasonably accurate results can be obtained by using a single set of one-group cross sections for the entire cycle. For more detailed analysis (as in FUMBLE), new one-group cross sections, together with new fluxes, would be obtained at various times during the cycle.

The solutions are given in terms of the initial values of material m, $N_{m,0}$, loaded into the fresh fuel, i.e., at $z = 0$. The solution of Eq. (7.5) for ^{238}U is

$$N_1 = N_{1,0}e^{-\sigma_{a1}z}. \tag{7.15}$$

Equation (7.8) can be solved by using an integrating factor, $p = \exp\left(\int \sigma_{a2}dz\right) = \exp(\sigma_{a2}z)$ to give

$$N_2 = \frac{1}{p}\int pN_1\sigma_{c1}dz = e^{-\sigma_{a2}z}N_{1,0}\sigma_{c1}\left[\frac{e^{(\sigma_{a2}-\sigma_{a1})z}}{\sigma_{a2}-\sigma_{a1}} + C\right]. \tag{7.16}$$

To solve for the constant C, we use the initial condition that at $z = 0$, $N_2 = N_{2,0}$. Thus,

$$C = \frac{N_{2,0}}{N_{1,0}\sigma_{c1}} - \frac{1}{\sigma_{a2}-\sigma_{a1}},$$

and

$$N_2 = N_{2,0}e^{-\sigma_{a2}z} + N_{1,0}\frac{\sigma_{c1}}{\sigma_{a2}-\sigma_{a1}}\left(e^{-\sigma_{a1}z} - e^{-\sigma_{a2}z}\right). \tag{7.17}$$

In general, the solutions to the equations for materials 1 through 6 as defined in Table 7.2 have the following form, which can be written as *recursion relationships*:

$$\begin{aligned}
N_1 &= A_{11}e^{-\sigma_{a1}z},\\
N_2 &= A_{21}e^{-\sigma_{a1}z} + A_{22}e^{-\sigma_{a2}z},\\
N_3 &= A_{31}e^{-\sigma_{a1}z} + A_{32}e^{-\sigma_{a2}z} + A_{33}e^{-\sigma_{a3}z},\\
&\vdots\\
N_m &= \sum_{n=1}^{m} A_{mn}e^{-\sigma_{an}z}, (m \leq 6),
\end{aligned} \tag{7.18}$$

where

$$A_{mn} = \frac{\sigma_{c,m-1}}{\sigma_{am}-\sigma_{am}}A_{m-1,n}, (1 \leq n \leq m-1), \tag{7.19}$$

and A_{mn} is determined from the relation:

$$N_{m,0} = \sum_{n=1}^{m} A_{mn}. \tag{7.20}$$

In these equations [and in all subsequent equations in this chapter except Eqs. (7.31) and (7.32)], the "σ_{a4}" for ^{241}Pu must be the sum of the actual σ_{a4} and $\lambda_4/\phi f$. The solution of Eq. (7.11) for ^{241}Am does not follow the recursion formula of Eq. (7.18). Using Eq. (7.18) for N_4, the solution for N_7 becomes:

$$N_7 = \frac{\lambda_4}{\phi f} \sum_{n=1}^{4} A_{4n} \int_0^z e^{-\sigma_{an} z'} dz' = \frac{\lambda_4}{\phi f} \sum_{n=1}^{4} \frac{A_{4n}}{\sigma_{an}} \left(1 - e^{-\sigma_{an} z}\right). \tag{7.21}$$

The solution of Eq. (7.12) for fission-product pairs, N_8 is:

$$N_8 = \int_0^z \sum_{m=1}^{5} N_m \left(z'\right) \sigma_{fm} dz'. \tag{7.22}$$

Substituting Eq. (7.18) for N_m gives:

$$N_8 = \sum_{m=1}^{5} \left[\sigma_{fm} \sum_{n=1}^{m} A_{mn} \int_0^z e^{-\sigma_{an} z'} dz' \right] = \sum_{m=1}^{5} \left[\sigma_{fm} \sum_{n=1}^{m} \frac{A_{mn}}{\sigma_{an}} \left(1 - e^{-\sigma_{an} z}\right) \right]. \tag{7.23}$$

Equation (7.23) has the same form as Eq. (7.21), except that (7.23) is summed over all five fissionable nuclides whereas the source in (7.21) is only one nuclide, $m = 4$ (see Eqs. (7.11) and (7.12)).

Accounting for all heavy nuclides provides a useful internal check on the correct operation of a fuel cycle computer program. Since it is assumed that there is no destruction of ^{241}Am, ^{243}Am, or fission-product pairs, the total number of heavy nuclides is $\sum_{m=1}^{8} N_m$, where we have included fission-product pairs in this total. According to Table 7.2 for the considered simplified chain model, this total must remain constant throughout the irradiation; and, in fact, the total is the initial fuel atom density, N_f, as calculated by the methods of Chapter 4. For commercial fuel cycle codes, the same type of check is available; the difference is that higher-order actinides [and fission-product (n, γ) progenies] become the last nuclides in the chain.

7.4.4 Average Atom Densities During the Fuel Cycle

For various nuclear design calculations it is useful to know average atom densities at various stages during a fuel cycle—e.g., at the beginning or end of an equilibrium cycle or at the middle or some average point in the cycle. Furthermore, for more accurate fuel cycle calculations, it may be important to recalculate multigroup neutron fluxes and one-group collapsed cross sections at various stages in the cycle.

For these reasons there is a need to calculate average atom densities during a fuel cycle. The average atom density for material m during an irradiation to fluence z is:

$$\overline{N}_m (z) = \frac{1}{z} \int_0^z N_m \left(z'\right) dz'. \tag{7.24}$$

For materials 1 through 6, the result is:

$$\overline{N}_m (z) = \frac{1}{z} \sum_{n=1}^{m} \frac{A_{mn}}{\sigma_{an}} \left(1 - e^{-\sigma_{an} z}\right), \ (1 \le m \le 6). \tag{7.25}$$

To be certain that the reader correctly understands the relationship between all of these N_m's, the following exercise in logic may be useful at this point. One can reason that the atom density for the fission-product pairs, N_8, at fluence z should be the sum of the products of the $\overline{N}_m, \sigma_{fm}$, and z over m from 1 to 5, or:

$$N_8(z) = \sum_{m=1}^{5} \overline{N}_m \sigma_{fm} z. \tag{7.26}$$

Substituting Eq. (7.25) for N_m into Eq. (7.26) shows that Eqs. (7.26) and (7.23) are indeed equivalent.

For materials 7 and 8:

$$\overline{N}_7(z) = \frac{1}{z}\frac{\lambda_4}{\phi f}\sum_{n=1}^{4}\frac{A_{4n}}{\sigma_{an}}\left[z - \frac{1}{\sigma_{an}}\left(1 - e^{-\sigma_{an} z}\right)\right], \tag{7.27}$$

$$\overline{N}_8(z) = \frac{1}{z}\sum_{m=1}^{5}\sigma_{fm}\sum_{n=1}^{m}\frac{A_{mn}}{\sigma_{an}}\left[z - \frac{1}{\sigma_{an}}\left(1 - e^{-\sigma_{an} z}\right)\right]. \tag{7.28}$$

7.5 Initial and Discharge Compositions and Fuel Cycle Reactivity

In this section an approximate method will be presented to calculate the required plutonium fraction in the fresh fuel, the composition of the discharge fuel, the reactivity loss during a fuel cycle, and a basis for estimating the breeding ratio. This simplified model is based on a zero-dimensional core calculation; hence, only the core will be treated in detail. Space-dependent calculations would obviously be required for analysis of the blankets. The logic for fuel cycle calculations will be effectively presented, however, by this simplified model.

Our primary task is to calculate an initial plutonium fraction that will lead to a criticality factor of unity at the end of the fuel cycle with all control rods withdrawn. Let us begin with some definitions:

E = atom fraction of Pu in the fuel,
Q = number of batches; also equal to the number of cycles for the discharge fuel,
q = cycle index, with $q = 0$ for fresh fuel, $q = 1$ for fuel after 1 cycle, etc., $q = Q$ for discharge fuel,
$N_m(q)$ = atom density of material in a batch after q cycles if all of the core were composed of this batch,
$N_{m,b}$ = average atom density (i.e., averaged over all batches) at the beginning of the cycle,
$N_{m,e}$ = average atom density (i.e., averaged over all batches) at the end of the cycle,
$\overline{N}_m$ = average atom density during the cycle (i.e., averaged over all batches *and* over the duration of the cycle),
k_b = criticality factor at the beginning of the cycle, with control rods withdrawn,
k_e = criticality factor at the end of the cycle, with control rods withdrawn,
$E^{(v)}, N^{(v)}, k^{(v)}$ = variables for the v-th iteration,
$\Sigma_{c,nf}$ = macroscopic capture cross section for the non-fuel material,
DB^2 = product of one-group diffusion coefficient and buckling,
t_c = refueling interval, or duration of one cycle, in chronological time in seconds,
B = burnup (MWd/kg metal).

The average atom densities at the beginning and end of the cycle are

$$N_{m,b} = \frac{1}{Q}\sum_{q=0}^{Q-1} N_m^{(q)}, \tag{7.29}$$

$$N_{m,e} = \frac{1}{Q}\sum_{q=1}^{Q} N_m^{(q)}. \tag{7.30}$$

The criticality calculation is approximated by a one-group neutron balance, whereas in an actual design fuel cycle code the calculation would be returned to a spatial multigroup code for each criticality calculation;

$$k_b = \frac{\sum_{m=1}^{5} \nu_{fm}\sigma_{fm}N_{m,b}}{\sum_{m=1}^{8} \sigma_{a,m}N_{m,b} + \Sigma_{c,nf} + DB^2}, \tag{7.31}$$

$$k_e = \frac{\sum_{m=1}^{5} \nu_{fm}\sigma_{fm}N_{m,e}}{\sum_{m=1}^{8} \sigma_{a,m}N_{m,e} + \Sigma_{c,nf} + DB^2}, \tag{7.32}$$

where the cross sections and D are all one-group collapsed values. Also, it should be noted that σ_{a4} (for ^{241}Pu) in these two equations are the actual σ_{a4} values, not $(\sigma_{a4} + \lambda_4/\phi f)$.

We assume that the isotopic composition of the feed plutonium fuel is known, so that all initial fuel atom densities are known once the plutonium atom fraction is specified. We assume that the refueling interval, t_c, and the load factor, f, are given; t_c will generally be 1 year.

The flow sheet for the iterative calculation of the required plutonium fraction in the fresh fuel is given in Fig. 7.4, where the equations in the BURNER module are those given in Section 7.4. We begin with an initial and a second guess for the plutonium fraction, $E^{(1)}$ and $E^{(2)}$, for use in the first two iterations. For the third and subsequent iterations, $E^{(\nu+1)}$ will be interpolated (or extrapolated) from comparison of $E^{(\nu)}$ and $E^{(\nu-1)}$ with $k_e^{(\nu)}$ and $k_e^{(\nu-1)}$ according to a user-selected convergence scheme. When $k_e^{(\nu)}$ is sufficiently close to unity (i.e., to within a convergence criterion, ε), the iterative procedure is terminated.

The neutron flux should be recalculated for each iteration, even though its change between iterations will be small:

$$\phi^{(\nu+1)} = \frac{2.9 \times 10^{16} P_{core}\,(\text{MW})}{V_{core}\sum_{m=1}^{5} \overline{N}_m^{(\nu)}\sigma_{fm}}. \tag{7.33}$$

Note that the average atom density, $\overline{N}_m$, is used in the denominator; hence, the flux is an average flux during the cycle. (Little error would be introduced by using the arithmetic average of $N_{m,b}$ and $N_{m,e}$ instead of $\overline{N}_m$.) For the-first iteration, $\phi^{(1)}$ is calculated with the guessed values of $N_{m,0}$ instead of $\overline{N}_m$.

The fluence for the cycle, needed for the BURNER calculations, is

$$z^{(\nu)} = \phi^{(\nu)} f t_c. \tag{7.34}$$

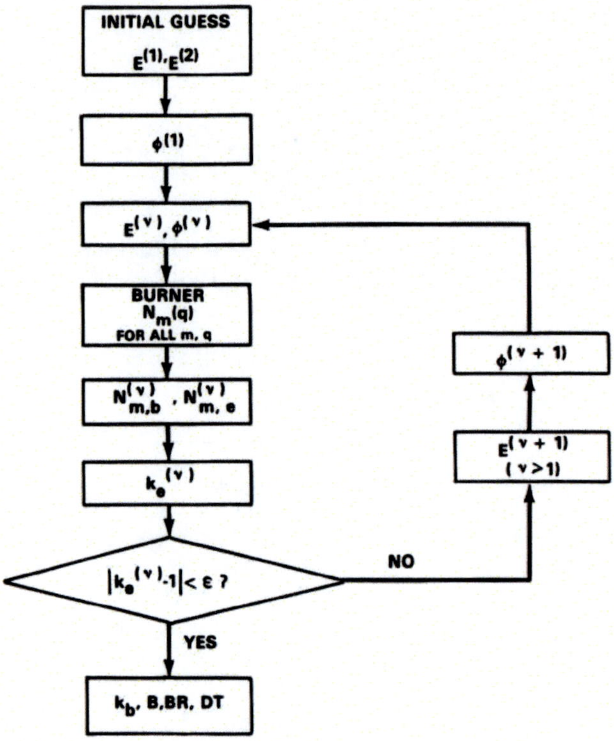

Fig. 7.4 Iterative calculation of the initial plutonium fraction, *E*

At the conclusion of the iterative calculation, the initial criticality factor, burnup, and breeding ratio (discussed in Section 7.6) can be calculated. The discharge compositions, $N_{m,d}$, are already available from the BURNER calculation as $N_m^{(Q)}$.

The reactivity for the cycle is:

$$\Delta k_{cycle} = k_b - 1. \tag{7.35}$$

We are interested in the burnup of the discharge fuel, in MWd/kg metal. The number of days at full power for the discharge fuel is

$$Qft_c \times \frac{1}{86400 \text{ s/d}}.$$

The mass of heavy atoms (U+Pu) in the core is

$$\rho_f F_f V_{core} \times 238/270,$$

where ρ_f in kg/m^3, V_{core} in m^3, and F_f is the fuel volume fraction. Thus, the burnup of the discharge fuel is given by

$$B\left(\frac{\text{MWd}}{\text{kg}}\right) = \frac{P_{core}\,(\text{MW})\,Qft_c/86\,400}{\rho_f F_f V_{core} 238/270}. \tag{7.36}$$

7.6 Breeding Ratio

A number of special quantities will be needed to calculate both breeding ratio and doubling time. These quantities are defined at the beginning of this section:

FP = fissile material produced per cycle,
FD = fissile material destroyed per cycle,
FG = fissile material gained per cycle,
FBOC = fissile material in the core and blankets at the beginning of the fuel cycle,
FEOC = fissile material in the core and blankets at the cycle end,
FLOAD = fissile material loaded into the core as fresh fuel at the beginning of the fuel cycle,
FDIS = fissile material in the fuel discharged from the core and blankets at the end of the cycle,
FE = fissile material external to the reactor per cycle,
FL = fissile material lost external to the reactor per cycle,
FPL = fissile material lost in processing per cycle,
EF = out-of-reactor (or ex-reactor) factor,
PLF = processing loss fraction,
RF = refueling fraction,
^{241}PuED = out-of-reactor decay of ^{241}Pu per cycle,
^{241}PuBOC = ^{241}Pu in the core and blankets at the beginning of the cycle,
^{241}PuEOC = ^{241}Pu in the core and blankets at the end of the cycle
C = cycles per year ($= 1/t_c$).

The units for the fissile material are the same as for N_m in the previous sections; the units can be, for example, kg, or atoms/barn-cm. Expressions for breeding ratio and doubling time always contain ratios of fissile material so that these units eventually cancel themselves.

The breeding ratio was defined in Chapter 1 as

$$\mathrm{BR} = \frac{\mathrm{FP}}{\mathrm{FD}}. \tag{7.37}$$

Despite the apparent simplicity of the definition, the breeding ratio has been calculated by several different methods, as reviewed for example, by Wyckoff and Greebler [8]. We will adopt their proposed definition and method for calculating breeding ratio. The definitions here apply to an equilibrium fuel cycle.

The breeding ratio is to be averaged over a fuel cycle. The numerator includes all fissile material produced during the cycle, not the net material produced. Therefore, the fissile material that is produced but later destroyed during the cycle is included in the numerator. Hence, the numerator can be written as

$$\mathrm{FP} = \mathrm{FD} + \mathrm{FEOC} - \mathrm{FBOC}. \tag{7.38}$$

We digress for a moment at this point to discuss another commonly used basis for the breeding ratio. A different expression for fissile material produced, FP′, based on the fuel entering and leaving the reactor is:

$$\mathrm{FP}' = \mathrm{FD} + \mathrm{FDIS} - \mathrm{FLOAD}.$$

Some would argue that this is a more logical basis for defining breeding ratio; Wyckoff and Greebler, however, prefer the definition based on all of the fuel in the reactor at the beginning and end of each cycle. The reason for the slight difference in breeding ratio calculated by the two methods is that ^{239}Pu and ^{241}Pu production rates are not linear with fluence.

Before returning to Eq. (7.38), we can add one further complicating factor by noting yet another attempt to improve the consistency of the definition of breeding ratio. This concept introduces a provision for the difference in the value of the various fissile nuclides. For example, ^{241}Pu is a better fuel than ^{239}Pu; it has a higher reactivity value and, hence, less is required for criticality. Moreover, the ^{240}Pu and the ^{242}Pu that are bred have some value, since these are fissionable. Hence, Ott suggested the concept of a breeding ratio based on *equivalent* ^{239}Pu [9]. In this way greater credit could be given for ^{241}Pu, instead of simply lumping all the fissile nuclides together, as in Eq. (7.38). In Ref. [8], it is suggested that ^{241}Pu is worth 1.5 times as much as ^{239}Pu with regard to reactivity value, primarily because of its higher fission cross section.

With Eq. (7.38), the breeding ratio is

$$\mathrm{BR} = 1 + \frac{\mathrm{FEOC} - \mathrm{FBOC}}{\mathrm{FD}}. \tag{7.39}$$

We further observe that, since breeding gain (G) is defined as (BR − 1), the breeding gain is given by the second term on the right-hand side of Eq. (7.39). The difference between FEOC and FBOC is the net increase in fissile material (i.e., the fissile material *gained*) during a fuel cycle, so that

$$\mathrm{G} = \frac{\mathrm{FEOC} - \mathrm{FBOC}}{\mathrm{FD}} = \frac{\mathrm{FG}}{\mathrm{FD}}. \tag{7.40}$$

To calculate the terms in Eq. (7.39), we introduce the subscript k as a region index because the quantities in Eq. (7.39) must be summed over all parts of the core and the radial and axial blankets. With the region volume expressed as V_k and with the quantities in Eqs. (7.29) and (7.30) in atoms or mass per unit volume, the beginning and end-of-cycle fissile inventories are

$$\mathrm{FBOC} = \sum_k \left(N_{2,b} + N_{4,b}\right)_k V_k, \tag{7.41}$$

$$\mathrm{FEOC} = \sum_k \left(N_{2,e} + N_{4,e}\right)_k V_k, \tag{7.42}$$

where subscripts 2 and 4 refer to ^{239}Pu and ^{241}Pu (or ^{233}U and ^{235}U).

With $z_k^{(Q)}$ signifying the fluence for the discharge fuel in region k, the total fissile destroyed in one cycle is

$$\begin{aligned}\mathrm{FD} &= \sum_k \sum_{m=2,4} \left(\int_0^{z_k^{(Q)}} N_m \sigma_{am} dz \right)_k V_k/Q_k \\ &= \sum_k \left[\sigma_{a2} \sum_{n=1}^{2} \frac{A_{2nk}}{\sigma_{an}} \left(1 - e^{-\sigma_{an} z_k^{(Q)}}\right) \right. \\ &\quad \left. + \sigma_{a4} \sum_{n=1}^{4} \frac{A_{4nk}}{\sigma_{an}} \left(1 - e^{-\sigma_{an} z_k^{(Q)}}\right) \right] V_k/Q_k,\end{aligned} \tag{7.43}$$

where N_{mk} and A_{mnk} are in atoms or mass per unit volume in region k as if all of the region were composed of this composition.[4] The breeding radio is then computed directly by substituting Eqs. (7.41), (7.42), and (7.43) into Eq. (7.39).

7.7 Doubling Time

The concept of doubling time was introduced in Chapter 1. Doubling time is one of the figures of merit used to compare breeder reactor designs, different fast reactor fuels, and fuel cycle systems involving many breeder reactors. Considerable confusion has existed regarding this parameter, however, because it can be defined in many different ways. Following a long history of proliferation of increasingly complex definitions. some of which were quite ingenious, Wyckoff and Greebler attempted to standardize the doubling-time concept by providing three useful definitions [8]. The first (reactor doubling time, or RDT) is the one discussed in Chapter 1. In the remainder of this section, all three of these definitions will be applied to an equilibrium fuel cycle. Despite the definitions here and in Ref. [9], other expressions are still widely used, e.g., Ref. [10]. An expression for a compound inventory doubling time that includes the approach to equilibrium has also been developed [11], but treatment of non-equilibrium cycles is beyond the scope of this text.

Wyckoff and Greebler's three proposed definitions of doubling time are given below, followed by a mathematical description of each:

RDT = reactor doubling time,
= time required for a specific reactor to produce enough fissile material in excess of its own fissile inventory to fuel a new, identical reactor;

SDT = system doubling time,[5]
= time required for a specific reactor to produce enough fissile material in excess of its own fissile inventory to supply its requirements of fissile material *external to the reactor* and to fuel a new, identical reactor;

CSDT = compound system doubling time,
= time required for a system of identical breeder reactors to double the fissile material in the system, assuming that the number of reactors is increasing at a rate such that all of the fissile material is being utilized.

7.7.1 Reactor Doubling Time

The reactor doubling time is given by

$$\mathrm{RDT} = \frac{\mathrm{FBOC}}{(\mathrm{FG})\,(C)}, \tag{7.44}$$

where FBOC was defined in Section 7.6 and FG is the net increase in fissile material during one fuel cycle as defined in Eq. (7.40). The fuel cycles per year, C, is equivalent to $1/t_c$, where t_c, was defined

[4]One can integrate over the fluence from zero to discharge here (i.e., for Q cycles) for a single batch as long as N_m is calculated as if all the region were composed of this batch. This is equivalent to the sum of the fissile material destroyed in each of the Q batches in the region for one cycle.

[5]This doubling time is called fuel-cycle-inventory doubling time (IDT) in Ref. [8].

in Section 7.5 as the refueling interval. This result for RDT is equivalent to M_0/M_g in Eq. (1.11). This definition for doubling time is sometimes referred to as a *simple* or *linear* reactor doubling time, to distinguish it from definitions that consider the fuel cycle external to the reactor and compounding growth rates of fissile material.

As shown in Chapter 1 [by Eq. (1.13)) the RDT can be approximated by

$$\mathrm{RDT} \approx \frac{2.7\mathrm{FBOC}}{\mathrm{G}Pf\,(1+\alpha)}, \tag{7.45}$$

where FBOC here is mass, in kg, and P is power, in MW. This relation shows the strong dependence of RDT on breeding gain, G, and on fissile specific inventory, FBOC/P.

7.7.2 System Doubling Time

The Reactor Doubling Time does not account for the processes in the fuel cycle external to the reactor. These processes are part of the breeder reactor system. Fissile inventory is tied up in these processes and this inventory must be produced in the reactor. The time required for external processing also affects doubling time. In addition, there are two types of losses of fissile material external to the reactor that affect doubling time. There is loss of material in fabrication and reprocessing, and (for the U-Pu cycle) there is loss from the β decay of ^{241}Pu.

It is estimated in Ref. [8] that the out-of-reactor fissile inventory is in the range of 60% of the in-reactor inventory. This includes fuel that is (a) being fabricated, (b) being shipped to the plant, (c) in storage at the plant awaiting loading or in a cool-down period, (d) in shipment for reprocessing, and (e) being reprocessed. By denoting the fissile inventory in these processes as *fissile external*, FE, the effect of this inventory is accounted for through a factor (the *ex-reactor factor*) defined as

$$\mathrm{EF} = \frac{\mathrm{FBOC} + \mathrm{FE}}{\mathrm{FBOC}}. \tag{7.46}$$

Its value is approximately 1.6.

The *processing losses*, FPL, are usually based on a *processing loss fraction*, PLF, which is the fraction of material lost in fabrication and reprocessing operations; this fraction may range from 1% to a few percent. With *refueling fraction*, RF, being defined as the fraction of core fuel assemblies replaced at each refueling (and, therefore, equal to 1/Q), the processing losses per cycle are

$$\mathrm{FPL} = \frac{1}{2}\,(\mathrm{FBOC} + \mathrm{FEOC})\,(\mathrm{RF})\,(\mathrm{PLF})\,. \tag{7.47}$$

The ^{241}Pu decay external to the reactor during one fuel cycle is

$$^{241}\mathrm{PuED} = \frac{1}{2}\left(^{241}\mathrm{PuBOC} + {}^{241}\mathrm{PuEOC}\right)(\mathrm{EF}-1)\left(1-e^{-\lambda_4 t_c}\right), \tag{7.48}$$

where $\left(1-e^{-\lambda_4 t_c}\right)$ is the fraction of the ^{241}Pu in the fissile material external to the reactor that decays during one fuel cycle. The expression for ^{241}PuED assumes that the average level of ^{241}Pu in an out-of-reactor fuel assembly can be represented by the average level of ^{241}Pu in an in-reactor fuel assembly.

With the results of Eqs. (7.47) and (7.48), we can define all of the losses external to the reactor per cycle as

$$\text{FL} = \text{FPL} + {}^{241}\text{PuED}. \tag{7.49}$$

Finally, with Eqs. (7.46) and (7.49), we are in a position to define the System Doubling Time, SDT:

$$\text{STD} = \frac{(\text{FBOC})\,(\text{EF})}{(\text{FG} - \text{FL})\,(C)}. \tag{7.50}$$

7.7.3 Compound System Doubling Time

If the excess fissile material produced in a system of breeder reactors were used to start up a new breeder as soon as the fuel from each cycle was reprocessed and fabricated into new fuel assemblies, the fissile inventory in the system of reactors would grow exponentially, as is the case with compound interest. The governing equation for this growth rate is

$$\frac{dM}{dt} = \lambda M, \tag{7.51}$$

where

M = fissile mass in the system as a function of time (kg), and
λ = fractional increase in fissile mass in the system per unit time (y^{-1}).

The constant λ can be expressed in terms previously defined as

$$\lambda = \frac{(\text{FG} - \text{FL})\,(C)}{(\text{FBOC})\,(\text{EF})}. \tag{7.52}$$

It is recognized from Eq. (7.50) that λ is the inverse of the System Doubling Time, or

$$\lambda = 1/\text{SDT}. \tag{7.53}$$

Eq. (7.51) can be integrated to give

$$\frac{M}{M_0} = e^{\lambda t}, \tag{7.54}$$

or, from Eq. (7.53),

$$\frac{M}{M_0} = e^{t/\text{SDT}}, \tag{7.55}$$

where M_0 = initial fissile mass in the system of reactors. [M_0 could also be thought of as the number of reactors in the system at time zero×(FBOC)×(EF).]

The compound system doubling time is the time required for M/M_0 to equal 2. Substituting 2 for M/M_0 and CSDT for t in Eq. (7.55) gives

$$\text{CSDT} = \text{SDT} \ln 2 = 0.693\text{SDT}. \tag{7.56}$$

Even though the concept of CSDT represents an idealization that will never be realized in practice, it is a useful figure of merit for a potential breeder system. Occasionally the 0.693 factor has been applied to the simple RDT, but a compound doubling time for a single reactor is misleading since it ignores the time and other requirements of the fuel cycle external to the reactor. The three doubling times defined here—RDT, SDT, and CSDT—all have merit in comparisons of breeder reactors, but only when it is clearly understood which doubling time is being considered.

7.8 Physics of Transmutation

Nuclear transmutation is continually taking place inside a nuclear reactor. The transmutations that we are most interested in for this section are those identified by the Bateman equations introduced earlier in this chapter that describe the isotopic chains shown in Figs. 7.2 and 7.3 (see Section 7.4). Any type of transmutation event is dictated by the spectral dependence of the neutron cross-section. If our objective is to transmute and eliminate nuclear wastes, we place a high premium on fission. The competition between the capture and fission processes is then of high relevance. The variation of transmutation behavior with energy spectrum is illustrated in Fig. 7.5 where the fission/absorption ratio is compared for dominant actinides in the pressurized water reactor (PWR) and the SFR spectra.

Table 7.3 provides an insightful tabulation of the ratios $\alpha = \bar{\sigma}_c/\bar{\sigma}_f$ of the average capture and fission cross-sections of several relevant nuclides. Here we see that the fission/capture ratios are consistently higher for the fast spectrum SFR, relative to thermal spectrum reactors. For fissile nuclides (^{235}U, ^{239}Pu, and ^{241}Pu) over 80% of fast neutron absorptions result in fission, as compared to 60–80% in the PWR spectrum. In addition, the fast spectrum fission fraction can rise to 50% for fertile nuclides as observed for ^{240}Pu, while remaining low (<5%) in a thermal spectrum. This is illustrated in Fig. 7.5, where the much higher probability of fission in actinides is seen to occur with a hard spectrum. Thus, in a fast spectrum, actinides are preferentially fissioned, rather than being transmuted into higher

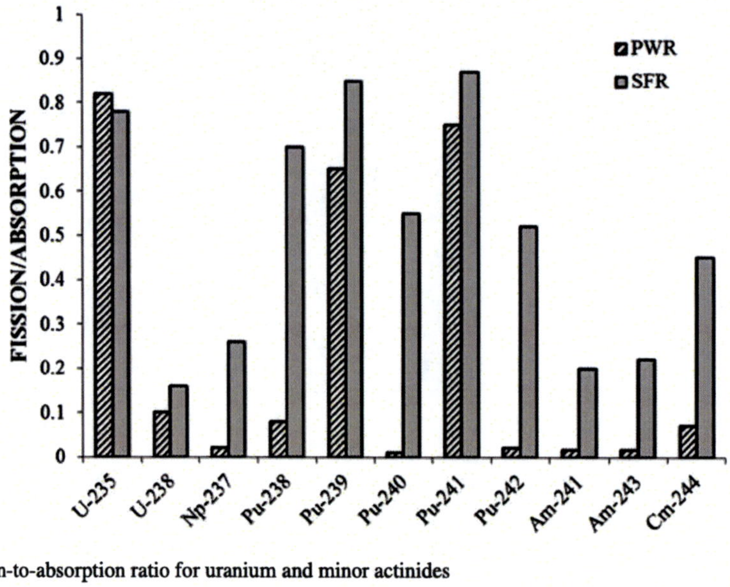

Fig. 7.5 Fission-to-absorption ratio for uranium and minor actinides

Table 7.3 Spectrum-averaged cross sections and capture-to-fission ratio for uranium and minor actinides[a]

Nuclide	PWR spectrum			Fast neutron spectrum		
	$\bar{\sigma}_f$	$\bar{\sigma}_c$	$\alpha = \frac{\bar{\sigma}_c}{\bar{\sigma}_f}$	$\bar{\sigma}_f$	$\bar{\sigma}_c$	$\alpha = \frac{\bar{\sigma}_c}{\bar{\sigma}_f}$
^{237}Np	0.520	33.000	63.000	0.320	1.700	5.300
^{238}Np	134.000	13.600	0.100	3.600	0.200	0.050
^{238}Pu	2.400	27.700	12.000	1.100	0.580	0.530
^{239}Pu	102.000	58.700	0.580	1.860	0.560	0.300
^{240}Pu	0.530	210.200	396.600	0.360	0.570	1.600
^{241}Pu	102.200	40.900	0.400	2.490	0.470	0.190
^{242}Pu	0.440	28.800	65.500	0.240	0.440	1.800
^{241}Am	1.100	110.000	100.000	0.270	2.000	7.400
^{242}Am	159.000	301.000	1.900	3.200	0.600	0.190
^{242m}Am	595.000	137.000	0.230	3.300	0.600	0.180
^{243}Am	0.440	49.000	111.000	0.210	1.800	8.600
^{242}Cm	1.140	4.500	3.900	0.580	1.000	1.700
^{243}Cm	88.000	14.000	0.160	7.200	1.000	0.140
^{244}Cm	1.000	16.000	16.000	0.420	0.600	1.400
^{245}Cm	116.000	17.000	0.150	5.100	0.900	0.180
^{235}U	38.800	8.700	0.220	1.980	0.570	0.290
^{238}U	0.103	0.860	8.300	0.040	0.300	7.500

[a]Average cross section (barn): $\bar{\sigma} = \frac{\int \sigma(E)\,\phi(E)\,dE}{\int \phi(E)\,dE}$

actinides. This implies that fast systems are more "efficient" (from the point of view of neutron economy) in destroying actinides because fewer neutrons are lost to capture reactions before eventual fission.

Furthermore, higher actinides (americium, curium, etc) continue to build-up with LWR recycle. These higher actinides tend to be more radioactive and can be problematic for fuel handling and fabrication in a closed fuel cycle, as will be discussed later. The hardest spectra are the most suitable, if, as we have indicated, fission is to be desired. An inspection of Fig. 7.6 makes this point very clear. This figure shows the fission cross-sections of the Am and Cm isotopes, respectively, most of them being of the threshold type.

7.8.1 Neutron Consumption Per Fission

For a more complete understanding of the transmutation potential of different neutron fields, a new parameter has been defined [12]; namely, the neutron consumption/fission (D_J) of a nuclide J.

Let us consider an infinitely large homogeneous core (with a given neutron flux φ) that is fed by several actinides at a given rate S (nuclides/s) and that there is a continuous discharge of part of its fuel inventory for reprocessing and/or storage in a repository. The processes of transmutation of the incoming nuclides (the "fathers") by the neutron flux according to their nuclear chains gives rise to "families" that are independent. Hence, one can consider the "transmutation behaviour" of each "family" separately. During its life in a core, the father and his family are producing neutrons up to their "death" (complete disappearance). The reasons for the nuclide destruction of the family can be: nuclide fission, nuclide capture, natural decay, or fuel cycle procedures (either discharge to storage or fuel losses during reprocessing).

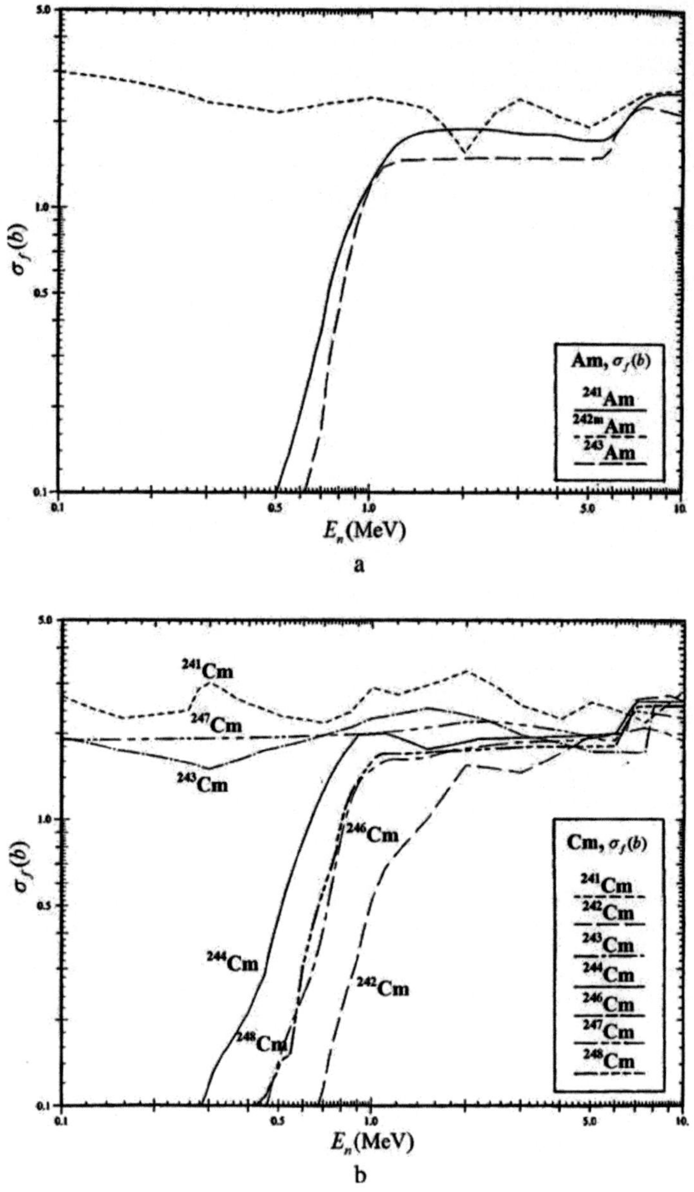

Fig. 7.6 The fission cross sections in the MeV region. (**a**) Americium, (**b**) Curium

One can calculate how many neutrons each father (together with his family) is able to consume/produce during its irradiation in a core with a chosen fuel cycle. Multiple nuclear reactions lead to different branches of family members: some of these reactions consume neutrons through neutron capture [e.g., (n, γ) reactions], other paths produce more neutrons [e.g., fission, (n, $2n$) reactions)] than consumed, and some [e.g., (n, n) reactions, decay)] do not influence the total neutron balance.

The total number of neutrons D_J consumed by the given J-family can serve as an indicator of the capability of a core to achieve destruction of a given J-feed if there is neither "parasitic" neutron consumption nor neutron leakage. In the case of a negative D (i.e., when the J-family produces more neutrons than it consumes), the core fed by J-nuclides produces enough neutrons to destroy the source material at equilibrium if the neutron excess compensates for parasitic captures (e.g. by structures, fission products etc) and for neutron leakage. In the case of a positive D, the neutron consumption in the fuel dominates over neutron production and the core requires a supplementary neutron source to support transmutation.

The procedure of the D_J calculation [13] can be illustrated by the following scheme:

$$D_J = \sum_{\mathrm{J1_i}} P_{\mathrm{J} \to \mathrm{J1_i}} \left\{ R_{\mathrm{J} \to \mathrm{J1_i}} + \sum_{\mathrm{J2_k}} P_{\mathrm{J1} \to \mathrm{J2_k}} \times \left[R_{\mathrm{J1} \to \mathrm{J12_k}} + \sum_{\mathrm{J3_n}} P_{\mathrm{J2_k} \to \mathrm{J3_n}} (\ldots) \right] \right\} \tag{7.57}$$

where $P_{JNr \to J(N+1)s}$ is the probability of transmutation of the nuclide J_{JNr} (which belongs to the number N of the chain J) into nuclide $J\,(N+1)$s (which belongs to the number N+1 of the chain J). All these nuclides are the members of the J-family. The neutron consumption factor, $R_{A \to B}$, shows the number of neutrons consumed during transition A→B and depends on the nuclear reaction type. In practice, starting with the father, one can evaluate the probability of each nuclear reaction, calculating at the same time the number of neutrons consumed as a result of each reaction. This procedure is systematically repeated up to (almost) complete disappearance of the J-family. The following simplified example presents the ^{235}U chain (family structure) to illustrate these calculations:

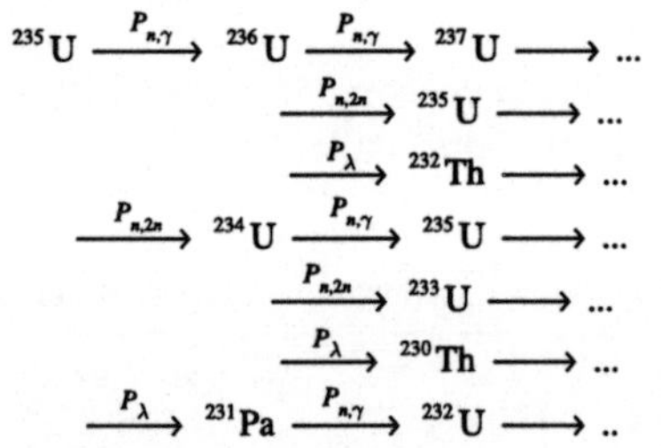

The neutron consumption factor values for each reaction type are defined as follows:

- neutron capture (n, γ) with 1 neutron being captured (consumed),
- neutron capture and subsequent multiplication (n, mn) with $(1\text{–}m)$ neutrons being produced,
- fission with $(1\text{–}\nu)$ neutrons being produced,
- natural decay with (0) neutrons being captured (consumed),
- discharge and nuclide loss with zero neutrons being captured (consumed).

Positive D means "consumption" and negative D means "production".

The linearity of the nuclide concentration equation with respect to the fuel feed allows a very simple algorithm of the D-evaluation for cores loaded with different "father" nuclides simultaneously. Taking into account that all feed components (families) are irradiated independently, one gets the total neutron consumption of the core at equilibrium (D_{fuel}) as:

$$D_{fuel} = \sum_J \varepsilon_J D_J, \tag{7.58}$$

where ε_J is the fraction of J-family in the feed stream.

The D-factor accounts only for the neutron balance of the actinide "families" as modelled in Eq. (7.57). However, the global neutron balance of a core must consider the total neutron production (consumption) of the fuel families, the parasitic captures of other core components (C_{par}) and of the accumulated fission products (C_{FP}) and neutron leakage (L_{core}). The general equation for the Neutron Surplus (NS_{core}) then becomes:

$$NS_{core} = -D_{fuel} - C_{par} - C_{FP} - L_{core}, \qquad (7.59)$$

with all quantities of Eq. (7.59) normalized to one disappeared actinide nuclide of the fuel feed. For a critical system, the neutron surplus must equal zero; and the neutron production of the fuel feed (as calculated in Eq. 7.58) must be sufficient to overcome the other loss terms. Thus, D_{fuel} is a useful parameter for quantifying the transmutation potential of a given isotopic fuel mix.

7.8.2 Neutron Balance Intercomparison (D-Factor)

The above physics approach has been used to compare the transmutation potential of a large variety of reactors, both with thermal and fast spectra [13, 14]. In Table 7.4 the following systems are compared: (1) a standard PWR with MOX fuel containing variations in the LWR moderator-to-fuel ratio, r, ranging from 1.4 to 4; (2) a VHTR; and (3) fast reactors with differing fuels and coolants. In the case of LWRs, there is a significant dependence of the D_J on the level of the flux. This is tabulated in Table 7.4, where the standard PWR D_J values are shown for two different flux levels (1×10^{14} and 2.5×10^{14} n/cm$^2\cdot$s). This effect is due to the competition between absorption rates, which depend on the flux level, and decay (independent of the flux level).

Table 7.4 D (neutron consumption/fission) for actinide-fueled cores

Nuclide	MOX-LWR[a] $r^b = 1.4$	MOX-LWR[a] $r^b = 2$	MOX-LWR[c] $r^b = 2$	MOX-LWR[a] $r^b = 4$	VHTR	He-cooled carbide fuel FR[d]	Super-Phénix[d]	Lead-cooled nitride fuel FR[d]	Na-cooled oxide fuel FR[d]	Na-cooled metal fuel FR[d]
^{235}U	−0.31	−0.38	−0.43	−0.55	−0.53	−0.84	−0.86	−0.92	−0.95	−1.04
^{238}U	0.104	0.068	−0.06	−0.007	0.23	−0.63	−0.62	−0.71	−0.79	−0.90
^{237}Np	0.91	0.93	0.75	0.96	1.11	−0.51	−0.56	−0.65	−0.73	−0.88
^{238}Pu	0.014	0.024	−0.16	0.038	0.12	−1.25	−1.33	−1.36	−1.41	−1.50
^{239}Pu	−0.60	−0.64	−0.79	−0.73	−0.72	−1.44	−1.46	−1.58	−1.61	−1.71
^{240}Pu	0.65	0.56	0.14	0.38	0.12	−0.93	−0.91	−1.02	−1.13	−1.27
^{241}Pu	−0.26	−0.37	−0.80	−0.58	−0.88	−1.25	−1.21	−1.26	−1.33	−1.39
^{242}Pu	1.27	1.22	0.73	1.13	0.79	−0.65	−0.48	−0.73	−0.92	−1.13
^{241}Am	0.92	0.93	0.71	0.95	0.90	−0.56	−0.54	−0.65	−0.77	−0.91
^{242m}Am	−1.55	−1.56	−1.66	−1.56	–	−2.03	−1.87	−2.08	−2.10	−2.16
^{243}Am	0.44	0.36	−0.15	0.25	−0.20	−0.84	−0.65	−0.85	−1.01	−1.15
^{242}Cm	0.004	0.014	−0.18	0.026	–	−1.26	−1.34	−1.37	−1.41	−1.51
^{244}Cm	−0.51	−0.60	−1.12	−0.71	−1.19	−1.54	−1.44	−1.53	−1.64	−1.71
^{245}Cm	−2.46	−2.46	−2.44	−2.44	–	−2.70	−2.69	−2.71	−2.74	−2.77

[a] $\varphi = 1\times10^{14}$ n/cm$^2\cdot$s.
[b] r = moderator-to-fuel ratio.
[c] $\varphi = 2.5\times10^{14}$ n/cm$^2\cdot$s.
[d] $\varphi = 1\times10^{15}$ n/cm$^2\cdot$s—Sodium-cooled burner configuration, GFR and LFR conventional concepts.

The results allow a comparison of the feasibility of transmutation of the different nuclides in each reactor concept. As an example, in the case of the Am isotopes, the ^{241}Am transmutation is a neutron-consuming process in any LWR concept, relatively independent from the moderator-to-fuel ratio (r). For a flux value of 1×10^{14} $n/\text{cm}^2\cdot\text{s}$, ^{243}Am is also a neutron-consuming process in a LWR, but a higher r value is preferable because the corresponding D value is positive but smaller with lower values of r. However, at $\phi = 2.5\times10^{14}$ $n/\text{cm}^2\cdot\text{s}$, the transmutation of ^{243}Am becomes a neutron-production process. ^{242}Am transmutation is a neutron-production process, whatever the spectrum, with a slight advantage for fast neutron spectra.

Of particular interest is the neutron balance of ^{238}U, which comprises the bulk of natural uranium resources. The VHTR exhibits a significant neutron defect for ^{238}U (0.23 net consumption); this is because the fission peak is much less pronounced in the graphite-moderated systems, effectively suppressing the direct threshold fission of ^{238}U. There is a slight neutron excess in the LWR, although the –0.02 will not overcome parasitic losses that will require roughly 0.38 neutrons [12]. This implies that the use of TRU fuel in LWRs requires the addition of fissile nuclides (an economic penalty) and that the amount of MA should be kept low, reducing the transmutation performance.

Conversely, the fast reactors have a neutron excess for ^{238}U ranging from 0.62 to 0.90. Thus, the fast systems have the potential to efficiently transmute the base ^{238}U resources, while the thermal systems will require, as indicated above, an additional source of neutrons (fissile feed) to drive the transmutation of the ^{238}U.

Significant variations in the D-factor are observed between the fast reactor concepts. In general, a harder neutron spectrum leads to a more favorable neutron balance; thus, the metal-fuelled SFR provides the most excess neutrons for every actinide isotope. However, all of the fast reactor systems exhibit a significantly more favorable neutron balance compared to the MOX-LWR results. In summary, these results confirm the trends that fast neutron spectrum systems have a marked advantage in terms of neutron balance and transmutation destruction efficiency (i.e., better fission/absorption ratios and consequently lower build-up of higher mass nuclei) over thermal systems [12].

7.8.3 Impact of Nuclear Data Uncertainties on Transmutation

Most of the reported transmutation performances or impacts on the fuel cycle parameters are significantly affected by nuclear data uncertainties and it is important to quantify the resulting uncertainties on the integral parameters of the fast systems foreseen for transmutation, in particular since their fuel can be heavily loaded with MA, whose data are certainly less known [15, 16].

For the present purpose, two systems—a SFR with MA content in the fuel of ~10% and an accelerator driven system (ADS) (more on this system later) with MA content ~70%—are of particular interest, since their fuel is loaded with large amounts of MA [17]. For these systems, the uncertainty (due to nuclear data uncertainties) of the most important integral parameters has been assessed using recent covariance data compilations [17]. The resulting uncertainties are fairly significant but still acceptable in preliminary design studies. However, in final design phases, when uncertainties have a very significant impact on margins and, consequently, on economy and overall design optimization, tight accuracy requirements should be met. A preliminary compilation of design target accuracies for fast reactor design parameters has been attempted within a OECD-NEA international expert group [17]. In order to assess quantitative nuclear data needs, the following design target accuracies have been specified at 1-sigma accuracy:

- Multiplication factors at the beginning of life are desirable to be known with accuracy of 0.3% $\Delta k/k$ or better,

- Power peaking factors at the beginning of life are desirable to be known with accuracy of 2% or better,
- Burnup reactivity swings are desirable to be known with accuracy of 0.3% Δk/k or better,
- Reactivity coefficients (coolant void and Doppler) at the beginning of life are desirable to be known with accuracy of 7% or better,
- Major nuclide densities at the end of an irradiation cycle are desirable to be known with accuracy of 2% or better,
- Other nuclide densities at the end of an irradiation cycle are desirable to be known with accuracy of 10% or better.

To meet these target accuracies, uncertainty reduction for the different type of reactions of the different nuclides have been quantified. Table 7.5 summarizes the results for the SFR system [17].

The previous discussion was focused on "neutron cross section" data needs. Some fuel cycle related data needs are implicitly accounted for, since nuclide densities at the end of irradiation were included in the analysis. However, fuel cycle data needs should be specified for the very specific fuel cycle(s) to be associated to a dedicated minor actinide burner. In this respect, a specific analysis should be made on the decay heat data needs (both in the reactor and at fuel discharge/storage) when a fuel highly loaded (e.g. ~50% MA) is considered [15, 16].

Table 7.5 Uncertainty reduction requirements needed to meet integral parameter target accuracies for the SFR system

Nuclide	σ	Energy range	Initial uncertainty (%)	Required uncertainty (%)
^{241}Pu	σ_{fiss}	1.35–0.498 MeV	16.6	3.4
		498–183 keV	13.5	2.6
		183–67.4 keV	19.9	2.6
		24.8–9.12 keV	11.3	3.5
		2.04–0.454 keV	12.7	4.4
^{56}Fe	σ_{inel}	2.23–1.35 MeV	25.4	3.3
		1.35–0.498 MeV	16.1	3.2
	σ_{capt}	2.04–0.454 keV	11.2	5.3
^{23}Na	σ_{inel}	1.35–0.498 MeV	28.0	4.0
^{244}Cm	σ_{fiss}	1.35–0.498 MeV	50.0	5.1
^{242m}Am	σ_{fiss}	1.35–0.498 MeV	16.5	4.2
		498–183 keV	16.6	3.1
		183–67.4 keV	16.6	3.1
		67.4–24.8 keV	14.4	4.0
		24.8–9.12 keV	11.8	4.2
		2.04–0.454 keV	12.2	5.1
^{240}Pu	σ_{fiss}	1.35–0.498 MeV	5.8	1.8
	ν	1.35–0.498 MeV	3.7	1.5
^{238}Pu	σ_{fiss}	2.23–1.35 MeV	33.8	5.6
		1.35–0.498 MeV	17.1	3.3
		498–183 keV	17.1	3.6
	ν	1.35–0.498 MeV	7.0	2.7
^{242}Pu	σ_{fiss}	2.23–1.35 MeV	21.4	4.9
		1.35–0.498 MeV	19.0	3.5
^{245}Cm	σ_{fiss}	183–67.4 keV	47.5	6.7
^{242}Pu	σ_{capt}	24.8–9.12 keV	38.6	8.4
^{238}U	σ_{capt}	24.8–9.12 keV	9.4	4.3

7.9 Fuel Cycle Results

7.9.1 U-PU Cycle and Comparison with Light Water Reactors

Selected results from SFR and PWR fuel cycles reported in the literature are provided in this section, both to gain an idea of the numbers for fast breeder reactors and to allow some comparison with water reactors.

Results for the designs described in Section 4.9 are given in Tables 7.6 and 7.7. These results show the extent to which breeding ratio and doubling time can be varied by core design and fuel pin diameter. The breeding ratio is considerably larger for the heterogeneous designs. This translates into lower doubling times for the small pin diameter cases but this advantage erodes for the larger pin diameters. Due to the breeding ratio advantage of the heterogeneous design, fuel management optimization studies have been carried out (by Dickson and Doncals of Westinghouse) that go beyond the general comparative studies reported in Table 7.6. Such studies show that doubling times for oxide fuel heterogeneous designs can be reduced to the 15–20 year range.

Plutonium inventories and fuel management data for a uranium-fueled PWR and a plutonium-fueled PWR and for two SFR designs are given in Table 7.7 [19]. The SFR results are based on early 1,000-MWe designs reported by Atomics International and General Electric (GE). The isotopic composition of plutonium in fuel discharged from SFR's and LWR's (Pressurized and Boiling Water Reactors) is given in Table 7.8 [19].

Several observations from Tables 7.7 and 7.8 are of interest. The plutonium discharged annually from a SFR is five to ten times the amount discharged from a uranium-fueled LWR, yet the net plutonium produced (i.e., Pu discharged minus Pu charged) is not very different for the two. The plutonium discharged from a uranium-fueled LWR each year will supply about 10% of the initial plutonium needed to start up a SFR and from 15 to 25% of the annual needs thereafter. The excess plutonium produced in two SFR's of the GE design would provide approximately enough fuel for a plutonium-fueled PWR.

The fissile fraction of the plutonium discharged from the core and axial blanket of a SFR is 0.7, which is similar to the value for fuel discharged from an LWR. The split between ^{239}Pu and ^{241}Pu is different, however, between a SFR and an LWR, with more ^{241}Pu in the LWR discharge. The ^{240}Pu content is similar in both SFR's and LWR's.

The isotopic composition of plutonium in the radial blanket of the SFR is considerably different from that of LWR discharge fuel; namely, the ^{239}Pu content is high and the ^{240}Pu content is low. The principal reasons for this result are the low capture-to-fission ratio for ^{239}Pu in a fast neutron spectrum and the relatively low reaction rate in the radial blanket of a fast spectrum reactor. For the GE and AI designs shown in Table 7.8, the radial blanket was to be removed before a large amount of ^{240}Pu was produced. The axial blanket plutonium in the outer part of the axial blanket is also low in ^{240}Pu, but this fuel is present in the same fuel pins as the core fuel. Hence, it would likely be processed and mixed directly with the core fuel during reprocessing.

Breeding ratios and doubling times for some of the operating and proposed SFR's are listed in Appendix A, where available, even though it is difficult to be sure of the basis for the results. The breeding ratio value of 1.16 for Phénix is probably the most reliable because it is the only one reported as being measured (from irradiated fuel). All of the demonstration and prototype reactors listed in Appendix A use or plan to use UO_2-PuO_2, fuel.

Predicted breeding ratios of carbide, nitride, and metal-fueled SFR's are significantly higher than for oxide-fueled SFR's, and doubling times are correspondingly shorter. Results of a study of fuel cycles for an oxide, a carbide, and a metal-fueled 1,000-MWe SFR are given in Table 7.9. While

Table 7.6 Fuel cycle results as a function of fuel pin diameter for homogeneous and heterogeneous designs (adapted from Ref. [18])

Common parameters:		
	Fuel	UO_2-PuO_2
	Electrical rating (MWe)	1,200
	Thermal rating (MWth)	3,300
	Residence times (y)	
	core and internal blankets	2.0
	Radial blanket	5.0
	Fuel cycle interval (y)	1.0
	Capacity factor	0.7

Configuration	Homogeneous						Heterogeneous					
Fuel pin diameter, mm	6.35		7.62		8.38		6.35		7.62		8.38	
	BOEC	EOEC	BOEC	EOEC	BOEC	EOEC	BOEC	EOEC	BOEC	EOEC	BOEC	EOEC
Fissile inventory (kg)												
Core	3,319	3,023	4,019	3,878	4,551	4,504	4,292	3,791	6,233	5,751	6,908	6,500
Internal blanket							255	710	278	797	258	743
Radial blanket	436	637	391	574	350	516	410	596	456	670	453	667
Axial blanket	114	325	120	349	121	353	65	190	77	226	81	238
Total	**3,869**	**3,985**	**4,530**	**4,801**	**5,022**	**5,373**	**5,022**	**5,287**	**7,044**	**7,444**	**7,700**	**8,148**
Discharge burnup (MWd/kg)												
Core average	82		56		46		77		52		43	
Core peak	116		80		66		109		78		70	
Fraction power (%)												
Core	93	89	94	91	94	93	83	74	87	81	88	84
Internal blanket							10	18	7	13	6	10
Radial blanket	4	6	3	4	3	3	5	6	4	4	4	4
Axial blanket	3	5	3	5	3	4	2	2	2	2	2	2
Fuel cycle reactivity (absolute)	–0.058		–0.023		–0.009		–0.019		–0.023		–0.020	
Breeding ratio	1.11		1.26		1.33		1.24		1.35		1.39	
Doubling time, CSDT (y)	43.5		19.3		16.5		22.3		20.7		20.2	

Table 7.7 1,000-MWe reactor charges, discharges, and inventories of plutonium (adapted from Ref. [19])

Reactor	Fuel fraction replacedper charge	Average residence time, days	Plutonium Discharged,[a] kg	Plutonium charged,[a] kg	Maximum plutonium inventory, kg	Average amount of Plutonium, kg		Average burnup of core fuel, MWd/kg
						Discharged per year	Charged per year	
PWR (U-fueled)	1/3	1,100	256		512	256		33
PWR (fueled with Pu from U-fueled PWR)	1/3	1,200	442	800	2,042	403	730	33
Atomics international[b] (AI) reference oxide SFR (fueled with Pu from U-fueled PWR)								
Core and axial blanket	1/2	540	1,270	1,380	2,740	1,716	1,865	80
Radial blanket	0.28	970	223		560	302		
Total			**1,493**	**1,380**	**3,300**	**2,018**	**1,865**	
General electric (GE)[b]	0.46	796	1,304	1,094	2,713	1,304	1,094	100
SFR (fueled with Pu from U-fueled BWR) Core and axial blanket								
Radial blanket	0.29	1,260	157		356	157		
Total			**1,461**	**1,094**	**3,069**	**1,461**	**1,094**	

[a] Refers to plutonium charged or discharged at an actual refueling. Refueling occurred annually for the PWR (U-fueled) and GE SFR reactors, so that these numbers agree with the average annual amounts for these reactors. Refueling was not annual for the PWR (Pu-fueled) and AI SFR reactors.

[b] The AI and GE designs are the 1,000-MWe *follow-on* designs developed for the USAEC in 1968.

Table 7.8 Isotopic composition of plutonium in discharge fuels (wt.%): Plutonium-fueled reactors[a] (adapted from Ref. [19])

	Burnup, MWd/kg	Weight percent				
		^{238}Pu	^{239}Pu	^{240}Pu	^{241}Pu	^{242}Pu
PWR (U-fueled)	33.0	1.8	58.7	24.2	11.4	3.9
BWR (U-fueled)	27.5	1.0	57.2	25.7	11.6	4.5
PWR (Pu-fueled)	33.0	2.7	39.3	25.6	17.3	15.1
SFR	80.0[c]					
(AI reference oxide)[b]						
Core and axial blanket		0.9	61.5	26.0	7.2	4.5
Radial blanket		0.02	97.6	2.33	0.04	
Core plus blankets (average)		0.8	66.8	22.5	6.2	3.8
SFR	100.0[c]					
(GE design)[b]						
Core and axial blanket			67.5	24.5	5.2	2.8
Radial blanket			94.9	4.9	0.2	
Core plus blankets (average)			70.5	22.4	4.6	2.5

[a]Plutonium from U-fueled PWR.
[b]The AI and GE designs are the 1,000-MWe *follow-on* designs developed for the USAEC in 1968.
[c]Core.

Table 7.9 Summary of breeding performance for oxide, carbide and Metal fueled SFR's under various fuel cycles [20]

Cycle		Pu–U	Pu fueled-Th	^{233}U/^{238}U-Th	^{233}U-Th
Core fissile/fertile		LWR Pu/Dep U	LWR Pu/Th	^{233}U/^{238}U	^{233}U-Th
Blanket fertile		Dep U	Th	Th	Th
Oxide	Doubling time[a]	16	29	23	112
	Breeding ratio	1.28	1.20	1.16	1.041
Carbide	Doubling time[a]	9	20	15	91
	Breeding ratio	1.42	1.23	1.23	1.044
Metal	Doubling time[a]	6	12	12	43
	Breeding ratio	1.63	1.38	1.30	1.11

[a]Reactor doubling time, years (75% capacity factor).

calculated breeding ratios for the oxide reactor were in the 1.2–1.3 range, breeding ratios for the carbide reactor were ~1.4 and for the metal reactor ~1.6. Doubling times (simple reactor doubling times—RDT) are reported as being reduced from 15 years for oxide fuel to 9 years for carbide fuel and 6 years for metal fuel.

7.9.2 Homogeneous Transmutation in Fast Reactors

The recycling of plutonium and MA in fast spectrum critical systems can be carried out both in a homogeneous and in a heterogeneous mode [21]. In the homogeneous recycling mode, unseparated TRU are recycled together and distributed homogeneously throughout the core fuel system. In the

Table 7.10 Comparison of fast reactor homogeneous transmutation performance [22] (CR = 0.25)

System	SFR	GFR	LFR
Net TRU destruction, g/MWt-day	0.74	0.76	0.75
System power, MWt	840	600	840
Outlet temperature, °C	510	850	560
Thermal efficiency, %	38	45	43
Power density, W/cc	300	103	77
TRU inventory, kg	2,250	3,420	4,078
Fuel volume fraction, %	22	10	12
TRU enrichment, % TRU/HM	44–56	57	46–59
Fuel burnup, GWd/t	177	221	180

heterogeneous mode the minor actinides are separated from plutonium and placed as targets in specific assemblies and managed independently from the standard Pu-bearing fuel.

Fast reactor systems are designed to operate on TRU-based fuels with continuous recycle. Recycle technologies avoid direct disposal of the used fuel. Rather, the TRUs are removed from the used fuel (reducing the long-term heat, dose, and radiotoxicity) and recycled in advanced fast reactors for consumption.

Differences in TRU destruction rates among various fast reactor concepts based on different coolants (SFR: Na cooled; GFR: gas cooled; LFR: lead-cooled) are tabulated in Table 7.10, where a very low conversion ratio (CR) of 0.25 is employed in the design. A few parameters relevant to different particular concepts are included. For comparison, the net TRU destruction in an SFR with a CR = 0.5 would drop to 0.47 g/MWt-day [22].

It is interesting to note from Table 7.10 that, since the transmutation physics behavior is similar for the fast burner concepts, the TRU destruction rate and compositions are very similar for a given conversion ratio. However, variations in other fuel cycle performance parameters are observed because of design differences. In particular, the mean power density in the core of the SFR is much higher; while low power density is required for the GFR to enable safe decay heat removal and by the LFR due to slow coolant flow rates. This fact impacts on the TRU inventory: lower power density means a higher TRU inventory. However, despite these inventory differences the burners can have a similar discharge burn up.

The compact SFR approach has some economic benefits, although higher thermal efficiency and design simplifications (e.g., removal of secondary loop) are being pursued for the GFR and LFR designs.

Each fast spectrum reactor system type has its specific requirement regarding the fuel type (oxide, metal, nitride, carbide). However, the choice of the chemical form of fuel (oxide, nitride, carbide) has only limited effects on the transmutation performances, as was shown in Table 7.4, where the *D*-factors were compared for different fast spectrum reactor concepts. Each design differs regarding fuel type and spectrum hardness—the last characteristic being related to the conversion ratio value. Dense fuels (like metal and nitride fuels) as well as low conversion ratios allow potentially better transmutation performance.

A comparison of MA transmutation effectiveness in different fuels and coolant systems has been reported to assess such points [23, 24]. In that work, three types of fast spectrum reactor fuel, oxide, nitride and metal, were considered. In order to provide a consistent and fair comparison, the referenced work assessed MA transmutation in fast reactor power reactors operating at the same power level. Table 7.11 shows the evaluated MA transmutation effectiveness of each fuel-type core where minor actinides were loaded at 5 wt.% of total fuel. The total transmutation effectiveness of nitride and metal fuelled cores is 9.9 and 9.7% per year, respectively, both a little better than that of an oxide-fuelled

Table 7.11 MA homogeneous transmutation in different fuel-type cores (MA loading: 5 wt.% of total fuel) [23, 24]

	Oxide core			Nitride core			Metal core		
	Transmutation			Transmutation			Transmutation		
Element	Initial amount at BOEC (kg)	Amount (kg)	Effectiveness (%)	Initial amount at BOEC (kg)	Amount (kg)	Effectiveness (%)	Initial amount at BOEC (kg)	Amount (kg)	Effectiveness (%)
Np	764	112	14.7	756	114	15.1	728	105	14.4
Am	746	72	9.7	728	83	11.4	695	78	11.3
Cm	150	−38	−25.4	148	−36	−24.4	139	−32	−23.1
MA total	1,659	147	8.8	1,631	161	9.9	1,562	151	9.7

core. The difference can be attributed to the harder neutron spectrum of the newer fuel-type cores. Note that the curium levels continue to increase for all homogeneous transmutation cases.

In another aspect of the same study [23], three types of coolant (sodium, lead, and gas) were compared with respect to their impact on MA transmutation effectiveness. Although there are many differences in design parameters besides coolant type (e.g. thermal power, operation cycle length, fuel-type, etc.), a rough evaluation of MA transmutation characteristics is possible by normalizing the results with respect to thermal power and operating period. The MA composition loaded in this survey is the same as in the fuel-type survey above.

Table 7.12 summarizes the MA transmutation effectiveness of each coolant-type core [23, 24]. After normalization, the MA transmutation effectiveness of these cores is almost identical with a value of 7.5–7.7% per year. The ratio of transmuted MA to loading is a little worse in the case of lead-coolant, but the difference is rather small compared with the dependency to other core parameters like core-fuel inventory. One can conclude that the effect of coolant choice in fast spectrum reactor designs will likely be negligible from the viewpoint of the MA transmutation.

The impact of MA on reactivity coefficients varies according to the reactor size and the type of coolant. In fact, the introduction of Am and Np in the core has, in general, a beneficial effect on the reactivity variation with burn-up (which becomes less negative), but can have the negative effect on some reactivity coefficients. In general, the Doppler reactivity coefficient becomes less negative and, in the case of a SFR, the coolant void reactivity coefficient becomes more positive. Moreover, the delayed neutron fraction will become smaller [23, 24].

Two sodium-cooled fast spectrum reactors with different core size and different MA content have been analyzed, namely SFR, i.e. a burner fast reactor, with a MA/Pu ratio equal to approximately 0.1 and CR~0.25 and a standard large size fast reactor (EFR: European Fast Reactor) with a limited content of MA in the fuel, equivalent to the equilibrium value [24]. Table 7.13 provides the key distinctive features of both configurations in their reference core states.

Figure 7.7 illustrates the impact of a variable MA content on reactivity coefficients and delayed neutron fractions evaluated in direct calculations of the relative composition perturbations [24]. The sodium void reactivity coefficients (Na-void), Doppler reactivity coefficients and delayed neutron fractions are shown for the SFR and EFR systems as a function of MA content.

These results show that doubling the amount of MA in EFR (i.e. from 1.2 to 2.4%), the void reactivity will only increase by roughly 0.1% $\Delta k/k$. In the much more extreme case of SFR, doubling the MA content from 10.8 to 21.6% corresponds to an increase of the positive Na-void effect of ~0.3% $\Delta k/k$. This value is significant, but still relatively modest. In any case, the impact of MA increase on reactivity coefficients is very much related to the approach taken in core and fuel design,

Table 7.12 MA homogeneous transmutation in different coolant-type cores (MA loading: 5 wt.% of total fuel) [23, 24]

Element	Sodium-cooled core (commercial type)			Lead-cooled core (BREST-300 type)			CO_2 gas-cooled core (ETGBR type)		
	Transmutation			Transmutation			Transmutation		
	Initial amount at BOEC (kg)	Amount (kg)	Effectiveness (%)	Initial amount at BOEC (kg)	Amount (kg)	Effectiveness (%)	Initial amount at BOEC (kg)	Amount (kg)	Effectiveness (%)
Np	1,058	201	19.0	326	28	8.5	961	118	12.3
Am	1,262	148	11.7	364	25	6.8	1,176	81	6.9
Cm	321	–52	–16.2	64	–8	–12.1	228	–34	–15.0
MA total	2,641	297	11.3	754	45	5.9	2,366	165	7.0
Normalized MA	695 kg per GWth	53 kg per GWth per year	7.6% per year	1,077 kg per GWth	83 kg per GWth per year	7.7% per year	657 kg per GWth	49 kg per GWth per year	7.5% per year

Table 7.13 Sodium-cooled fast spectrum reactor configurations for studies of MA effects on reactivity coefficients [24])

System	SFR	EFR
Power (MWth)	840	3,600
Core configuration	U-TRU-Zr metal fuel	U-TRU oxide fuel
	SS reflector	U blanket
TRU content (%)	56	22
MA content (%)	10.8	1.22
Doppler reactivity coefficient range (K)	850-to-300	1,520-to-300
Void reactivity coefficient definition	Na-void of the core	Na-void of the core and blankets

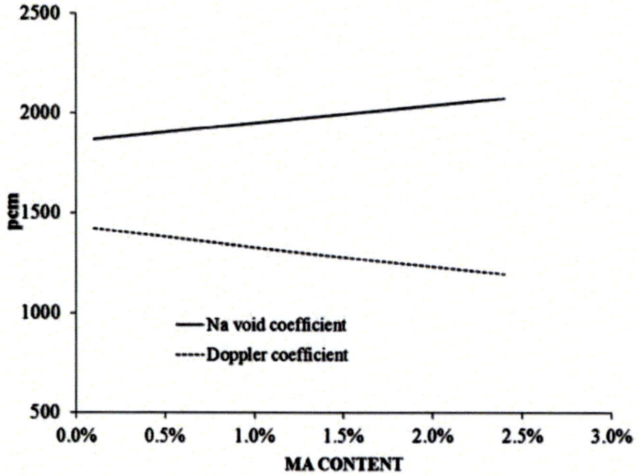

Fig. 7.7 Impact of MAs on reactivity coefficients

and it can be optimized according to pre-defined objectives and constraints. As shown in Table 7.14, the corresponding sensitivity analysis provides some insight into the observed trends for the sodium void reactivity variations.

The values shown are the sensitivity coefficients, integrated over energy, relative respectively to the Na-void reactivity coefficient in the SFR and EFR systems. A detailed sensitivity analysis shows that the global effect of each isotope (indicated under "SUM" in Table 7.14) can be interpreted as the combination of competing effects:

- The increase of capture at high (e.g. E > 100 keV) energy results in a flattening of the high energy slope of the adjoint flux (responsible for the positive reactivity Na-void effect). On the contrary, the capture increase at low (e.g. E < 100 keV), results in a hardening of the spectrum and, consequently in an increase of the positive component of the same reactivity effect. The sum of the two effects results in an increase of the Na reactivity coefficient.

As for fission, the increase at high energy of the fission cross section will result in an increase of the energy slope of the adjoint flux at high energy (and then of the associated positive reactivity effect), that is counterbalanced, in the case of fissile isotopes, by a corresponding effect of opposite sign at low energy. This means that "fertile" isotopes with threshold fission cross sections will induce

Table 7.14 Relative sensitivity coefficients of the sodium void reactivity [24]

(a) SFR system

Nuclide	Capture	Scattering Elastic	Scattering Inelastic	Neutrons per fission	Fission	Sum
^{238}U	0.356	0.038	−0.101	0.083	0.037	0.413
^{237}Np	0.113	0.002	−0.007	0.072	0.040	0.220
^{239}Pu	0.339	0.013	−0.022	−1.064	−0.857	−1.591
^{240}Pu	0.326	0.018	−0.031	0.676	0.391	1.380
^{241}Pu	0.052	0.004	−0.006	−0.662	−0.495	−1.107
^{242}Pu	0.099	0.006	−0.009	0.203	0.121	0.420
^{241}Am	0.142	0.002	−0.011	0.072	0.045	0.250
^{242m}Am	0.003	0.000	−0.001	−0.041	−0.032	−0.071
^{243}Am	0.118	0.002	−0.011	0.045	0.028	0.182
^{244}Cm	0.042	0.001	−0.003	0.083	0.053	0.176
^{245}Cm	0.007	0.000	−0.001	−0.098	−0.078	−0.170
^{56}Fe	0.112	1.020	−0.218	0.000	0.000	0.913
^{23}Na	0.078	0.154	0.515	0.000	0.000	0.747

(b) EFR system

Nuclide	Capture	Scattering Elastic	Scattering Inelastic	Neutrons per fission	Fission	Sum
^{238}U	0.421	0.096	−0.184	0.124	0.027	0.485
^{237}Np	0.006	0.000	0.000	0.003	0.002	0.010
^{239}Pu	0.313	0.005	−0.012	−1.089	−0.923	−1.706
^{240}Pu	0.149	0.003	−0.009	0.121	0.047	0.311
^{241}Pu	0.012	0.000	−0.001	−0.154	−0.120	−0.263
^{242}Pu	0.014	0.000	−0.001	0.013	0.006	0.031
^{241}Am	0.034	0.000	−0.003	0.012	0.006	0.049
^{242m}Am	0.000	0.000	0.000	−0.007	−0.006	−0.012
^{243}Am	0.009	0.000	−0.001	0.002	0.001	0.011
^{244}Cm	0.003	0.000	0.000	0.006	0.003	0.012
^{245}Cm	0.000	0.000	0.000	−0.007	−0.006	−0.013
^{56}Fe	0.066	0.022	−0.033	0.000	0.000	0.055
^{23}Na	0.055	0.362	0.272	0.000	0.000	0.690

only the increase of the positive reactivity effect at high energy, while the predominance of low energy effects for fissile isotopes give a global decrease of the positive reactivity.

Regarding a cross comparison of the sensitivity coefficients in SFR and in EFR, the qualitative indications given above are confirmed in both cases. However, the quantitative effects differ due to the different amounts in the two cores of the individual nuclides (e.g. of ^{242m}Am or ^{241}Am) and to the spectrum hardness (e.g. in the case of ^{240}Pu).

In summary, the sensitivity analysis shows that any increase in the amount of fertile nuclides results in an increase of the positive Na-void reactivity coefficient. Inversely, any increase in the amount of fissile nuclides results in a decrease of the Na-void reactivity. This feature clearly explains the effects of MA content increase, i.e. positive for ^{241}Am, ^{243}Am, ^{237}Np, and ^{244}Cm and negative for ^{242m}Am and ^{245}Cm.

7.9.3 Heterogeneous Transmutation in Fast Reactors

The homogeneous multi-recycle of TRU fuels discussed above (i.e. with no separated TRUs) in a fast reactor will end up with the production and the slow accumulation of Cm isotopes before reaching equilibrium. This build-up will be present regardless of the Pu/MA ratio in the fuel, but will be certainly somewhat mitigated by a high Pu/MA ratio. One significant consequence can be the strong increase of the neutron sources during fuel fabrication with respect to the case of a Pu-only multi-recycle. In fact, if no specific measure is taken (cooling times optimization, blending of fresh and irradiated fuels etc), the neutron source can increase by a factor 100 or more, essentially due to the build up of ^{244}Cm, a relatively strong neutron emitter from spontaneous fission. A more detailed description of this point will be given below. The obvious concern to the reactor and fuel cycle operators is the potential penalties regarding a fuel fabrication plant cost increase, higher doses to the workers, etc.

An alternative to the homogeneous recycle in fast reactors could be to separate out the less radioactive components of the LWR used fuel transuranics (e.g., Pu and Np) in order to make driver fuels and configure the remaining minor actinides (primarily Am and Cm) into target fuels/assemblies [25]. or stored as waste. Consequently, the driver and target fuels could then be managed separately in the fuel cycle. The heterogeneous recycling affords the possibility of disconnecting the minor actinides cycle from the conventional fuel cycle. The fabrication of the assemblies with MA could be done in a separated plant, and the specific constraints due to the minor actinides handling (heat, neutron sources) would not be applied to the conventional fuel fabrication plant. This separate management and recycle of the Pu-Np driver and MA target fuels constitutes the so-called heterogeneous recycle.

Some potential advantages of the heterogeneous recycle concept include:

- use of standard technology for Pu and Np co-extraction processes and for driver fuel fabrication. This can help an early deployment of advanced fuel cycle technology (permits time for additional R&D to find solutions to the handling of the high dose MA);
- the utilization of the heterogeneous recycle approach employing driver and target fuel assemblies might allow reducing the number of minor actinide-containing fuel assemblies to be fabricated and handled prior to core loading;
- potential to confine the remote fabrication of MA-containing fuels with lower throughput to a dedicated facility for target fabrication; and
- separate management of MA loading in the core.

Potential difficulties include:

- recycling, handling, and fabricating target assemblies;
- immature state of target technology (fabrication, irradiation performance, etc.);
- fuel behavior under irradiation (swelling and helium production);
- high specific heat for assembly manufacturing;
- high decay heat level for in-core and out-core assembly handling; and
- high neutron source level at the fuel treatment step.

As for the target material choice, it has been suggested to use an inert matrix material as a diluent for non-fertile TRU oxide targets in order to maintain specific heat generation levels within the proper range [26]. Desired properties of the matrix material are: (1) chemical compatibility with the reactor coolant, (2) chemical and physical compatibility with the fuel, (3) resistance to radiation damage, and (4) a low neutron absorption cross-section. Both metal and ceramic matrix materials have been considered for this application. Ceramic materials include zirconia (ZrO_2 stabilized in the fluorite cubic structure by additions of CaO, Y_2O_3 or MgO), magnesia, and spinels such as $MgAl_2O_4$. A dispersion of TRU oxides in zirconium or in a refractory metal such as molybdenum offers the advantage of a low fuel operating temperature, but provisions must be made to prevent a chemical interaction between the fuel and matrix (e.g. $PuO_2 + Zr = ZrO_2 + Pu$) that could lead to undesirable redistribution of the fissile constituents. The ceramic matrix must be stable under irradiation and capable of accommodating the very large quantities of helium produced in minor actinide-bearing fuels. A concern of MgO is the identified potential for significant vaporization or destabilization at ~2,100 K. This could lead to a disintegration of the pin under a severe transient or to an in-pin rearrangement of the TRUs with a subsequent compaction. It should be noted that under irradiation conditions the thermal conductivity of MgO based fuel deteriorates.

For the Mo CERMET based fuel, the melting temperature of Mo is given as 2,896 K. With AmO_2 having a melting point of 2,448 K, the matrix would be stable beyond the fuel phase "melting" point. However, detailed information on radiation induced degradation of Mo properties is still scarce. Regarding the matrix behavior under irradiation, the CEA has obtained valuable results from the past SUPERFACT experiment [27]. The SURERFACT 1 experiment consisted of irradiating solid solutions of uranium and Minor Actinide compounds with the following contents: $(U_{0.55}, Np_{0.45})O_2$ and $(U_{0,6}, Am_{0,2}, Np_{0,2})O_2$. Strong swelling accompanied by the formation of very high porosity was observed on the irradiated fuel from a burnup of 4.5 at.%. This important swelling is linked to the formation of numerous helium bubbles (helium being formed through the alpha decay of ^{242}Cm, coming from neutron capture of ^{241}Am), and led to an interaction between the fuel and the cladding, and finally to the mechanical deformation of the cladding. As for the MgO matrix, irradiation results obtained to date show that the MgO-UO_2 pellets display large swelling after irradiation, which is partly related to the presence of macroscopic defects in the pellets before irradiation (cracks and porous zones) [26, 27]. Improvements of the pellet microstructure will likely be required.

In order to cope with these difficulties, a support matrix of UO_2 appears to be a viable matrix for the following reasons:

- Targets can be reprocessed together with standard fuel assemblies;
- The risks of proliferation are reduced, due to the important production of ^{238}Pu

Recent results report the quantification of some specific features and issues of the heterogeneous recycle [28]. The following hypotheses have been made:

1. UO_2 matrix
2. The MA mix in the targets appropriately accounts for Np, Am and Cm.

Two values for the MA content have been studied: 10 and 40%. The major conclusions are as follows:

Global transmutation effectiveness. In the case of 40% MA, the transmutation rate is ~40% and the fission rate is ~19%. This means a consumption of ~11 Kg/TW$_e$h, and the scenario studies show that one needs to put these targets in 33% of the critical fast reactors (SFR) in the power fleet in order to stabilize the MA inventory. In the case of 10% MA, one obtains a transmutation rate of ~40%, a fission rate of ~11%, and a consumption ~3.5 kg/TW$_e$h. In this case, the scenario study indicates that 100% of the SFRs should be equipped with targets.

Thermo-mechanics and thermo-hydraulics. In the case of targets with 40% MA, the study indicates that only an ODS (oxide dispersion strengthened F-M steels, see Chapter 11) cladding can be envisaged to comply with cladding design rules and to keep the irradiation time long enough. The value used in the study was 4,100 days. Moreover, due to the strong He production, significant fuel swelling can be expected, unless specific solutions are envisaged (e.g. optimized microstructure to release the He produced).

The impact on the fuel cycle will be discussed later in this section.

7.9.4 Safety Challenges of Critical Transmuters and Subcritical Systems

Potential safety related problems were indicated in early transmutation studies in the case of a core loaded with only TRU (i.e. conversion ratio equal zero) and with a high content of MAs. In fact, in these types of fast reactor cores, the absence or very low content of Uranium produces as a consequence both a very low fraction of delayed neutrons (as illustrated in Table 7.15), and a very low Doppler reactivity coefficient (in general, mostly attributed to neutron capture rates in ^{238}U). Moreover, it has been observed previously that a high content of MAs, like isotopes of Am and Np, leads to a deterioration of the void reactivity coefficient (in case of liquid metal coolants).

Sub-critical systems (Accelerator Driven Systems = ADS and fusion-fission hybrid systems), were "rediscovered" in 1980s since they provide a technical way out from these potential challenges. These systems are being investigated worldwide to further refine their designs, transmutation efficiencies and deployment feasibility and scenarios [29].

In terms of transmutation performance, it was found that ADSs and fast spectrum reactors are comparable. As for the choice of the coolant and of the fuel type, the indications given above for critical fast spectrum reactors will hold also for ADSs [23]. Despite the fact that one could in principle reach the theoretical maximum TRU consumption rate in an ADS (which is about 1 g/MWt-d), in practice lower TRU consumption rates of the order of ~0.75 g/MWt-d have been found, very close to the performance of a critical burner fast reactor with conversion ratio (CR) equal to 0.25. Similar values are obtained for various fusion-fission hybrid concepts [30, 31].

Fertile-free fuels can be envisioned for subcritical transmutation systems in order to maximise the subcritical system support ratio in the power producing energy complex. As indicated previously, the neutronics properties of fertile-free TRU or MA fuels (and in particular the associated low delayed neutron fraction), make it much more difficult to design a critical fast reactor with this type of fuel. Therefore, the subcritical transmuter concept has a better capability to accept U-free fuels and to reach high transmutation performances. In particular, this means transmutation scenarios with subcritical systems will require a smaller number of transmuters to handle the TRUs arising from the electricity producing reactor fleet. This approach will also allow one to conceive of a separate stratum of the fuel cycle, leaving the fuel cycle stratum devoted to electricity production "non-contaminated" by the presence of MAs.

Subcritical transmuters can also provide an option if the critical fast reactor deployment would be delayed in time. This type of system can provide a tool to decrease TRU or MA stocks within a wide range of fuel types and compositions.

However, feasibility issues related to the development of U-free fuels remains a significant concern. No definitive candidate has been found to date, even though some promising candidates have been suggested (see above).

Table 7.15 Effective delayed neutron fraction values for selected actinides

Nuclide	$\bar{\beta}$
^{238}U	0.01720
^{237}Np	0.00388
^{238}Pu	0.00137
^{239}Pu	0.00214
^{240}Pu	0.00304
^{241}Pu	0.00535
^{242}Pu	0.00664
^{241}Am	0.00127
^{243}Am	0.00233
^{242}Cm	0.000377

7.9.5 Low Conversion Ratio Fast Reactors

Regarding the specific issue of U-free fuels, which have been advocated in order to maximize TRU consumption, studies devoted to the Pu consumption in fast reactors have revealed a gradual saturation of the Pu consumption with the reduction of the amount of uranium in the fuel [32]. Using a 1,000 MWth Advanced Burner Reactor (ABR) core design, the relation between TRU consumption rate and TRU conversion ratio has been investigated [33]. Both metal and oxide core designs were investigated for TRU conversion ratios of 1.0, 0.75, 0.50, 0.25, and 0.0 at the equilibrium cycle in which the TRU recovered from ABR spent fuel was used as the primary feed and the TRU from an external supply was used for makeup, both with Pu/MA~9 and makeup TRU feed with Pu/MA~1).

Figure 7.8 shows the TRU consumption rates relative to the maximum theoretical value of uranium-free fuel as a function of TRU fraction in the charged fuel for both cases. Similar trends are shown for both cases (i.e. makeup TRU feed from LWRs with Pu/MA~9 and makeup TRU feed with Pu/MA~1), and for both types of fuel. The slope of TRU consumption rate with respect to TRU fraction in the fuel decreases with increasing TRU fraction. The TRU consumption rate reaches ~80% of the maximum theoretical value when the TRU faction is ~60%, which corresponds to TRU conversion ratio in the range ~0.25–0.35.

The effective delayed neutron fraction, Doppler constant, radial expansion coefficient, axial expansion coefficient, and sodium void worth at end of cycle (EOC) have been studied as a function of the TRU fraction in heavy metal (HM) [33]. As an example, Fig. 7.9 shows that the effective delayed neutron fraction decreases monotonically because of reduced fission of ^{238}U as the TRU fraction in HM increases.

In summary, from the results reported in other studies it appears that cores with conversion ratios in the range 0.25–0.40 are likely to be feasible, as parameters relevant for the safety performance of these cores become comparable to those of cores that have already been proven to be feasible [34]. Since these cores allow TRU consumption close to 80% of the maximum theoretical consumption, whatever the Pu/MA ratio and fuel type, it appears that U-free fuels can be avoided for any realistic scenario and specific P&T strategy.

Moreover, even in the case of a heterogeneous recycle of MA in a fast spectrum reactor, the MA support can be U (as indicated previously), thus confirming that the inert matrix support can be avoided

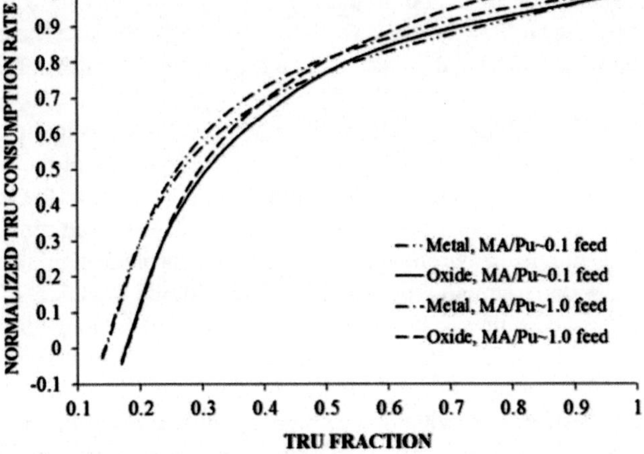

Fig. 7.8 TRU consumption vs. TRU fraction in the fuel

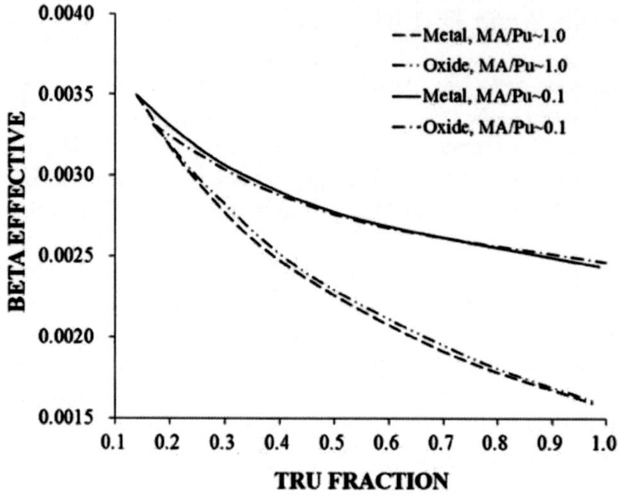

Fig. 7.9 Delayed neutron fraction vs. TRU fraction in the fuel

in practically all P&T implementation options. A final potential consequence is that the option of using ADS and fusion-fission hybrids should be carefully compared to the use of critical fast reactors with low conversion ratio.

7.9.6 Transmutation of Long Lived Fission Products (LLFP)

A detailed analysis of FP radiotoxicity reveals that there are only a few long-lived radionuclides (e.g. ^{129}I, ^{99}Tc) that contribute to the very long-term radiotoxicity. However, they tend to be fairly mobile in that they could potentially move to considerable distances from a geologic repository should they penetrate the waste can barrier. Fortunately, their absolute radiotoxicity magnitude remains below the TRU levels and even below the radiotoxicity of the natural ore materials removed to fabricate enriched uranium. The total radiotoxicity of fission products is about 1.4×10^7 Sv/ton U (enriched) 100 years after discharge, but decreases to 875 Sv/ton U (enriched) after 1,000 years. Thereafter, it is stabilized at that level for a long time (~100,000 years) i.e. at a level much lower than the reference level for natural ore [23, 25].

Regarding the possibility and limits to transmute fission products in order to reduce radiotoxicity, transmuting fission products is generally of little interest. The majority of the fission products have decayed after about 250 years, and their contribution to the radiotoxicity of the spent fuel, which was very high during the first 100 years of storage, has become small. However, some fission products are very mobile in certain geological environments and can thus contribute significantly to the radiological effects of disposal in underground repositories. In addition, the treatment of used fuel results in releases through gaseous and liquid effluents that also contributes to the long-term radiological effects of nuclear power generation. The fission products that deserve most attention in this respect are ^{129}I, ^{135}Cs, ^{79}Se, ^{99}Tc and ^{126}Sn. It may be possible to effectively transmute ^{129}I and ^{135}Cs in properly configured fast reactors if such transmutation could help in enhancing geologic repository effectiveness [35]. Unlike transuranics, fission products in a transmutation process produce no supplementary neutrons. They are purely consumers.

7.9.7 Thorium-Uranium Cycle

The results in Table 7.9 also provide predicted breeding ratios and doubling times for SFRs that utilize the thorium-uranium cycle and mixed Th-U-Pu cycles. We note, however, that the breeding ratio is considerably reduced for thorium systems, down to 1.04, for the ^{233}U-Th oxide cycle, and the doubling time is therefore increased.

7.9.8 Fuel Cycle Consequences and Transmutation

As it was indicated above, the consequences of the TRU multi-recycle can significantly affect the fuel cycle. Table 7.16 [36] summarizes the consequences at the fuel fabrication step by comparing the inputs of some of the fast reactor concepts considered previously with the impacts of TRU recycle in PWRs.

In the case of homogeneous recycling in PWRs and fast spectrum reactors, the large difference in the neutron source at fabrication is essentially due the impact of the ^{252}Cf spontaneous neutron fission ($\sim 10^{12}$ n/gs) contribution, due to the different mechanisms of its build-up in the two different types of spectrum.

The ^{252}Cf, which is a very powerful neutron emitter as shown in Table 7.17, results in an unacceptably high neutron source at fuel fabrication. This is one of the important points that have suggested avoiding full TRU transmutation in thermal reactors [36].

The case of heterogeneous recycle with 40% MA in the targets implies a very high thermal power of the fresh assemblies (~20 kW) and high decay heat (~35 kW after 300 days of cooling). The case with 10% MA in the targets allows decreasing these values to 5–6 kW. In Table 7.18, the values related to the fresh fuel are shown to illustrate major nuclide contributors.

A specific (and expected) feature is the very high value of neutron sources in cases, 40% MAs and 10% MAs: ~ 2–8.0×10^{10} n/s with respect to $\sim 4.0\times 10^{7}$ n/s for a standard fresh MOX assembly. Table 7.19 illustrates major contributors to the yields. These very high values will have to be studied in relation to the levels of neutron sources at current installations to define maximum admissible levels, consequences for the operation, and associated costs [36].

Finally, fuel handling is an area where a major impact is foreseen. In fact the calculated thermal and decay power are substantially higher than the maximum allowable values allowed for fuel handling in Phénix and Super Phénix (with or wt.% external Na container for cooling). All steps of the handling chain are involved: at fresh fuel transport, due to the need of an increased cooling; at storage inside the power plant, due to increased criticality, cooling and radioprotection problems, but also due to need of defining a strategy in case of clad failure; at fuel unloading (need of a Na container) etc. In general, new rules and new solutions should be envisioned and tested. These considerations have to

Table 7.16 Consequences at the fuel fabrication step in the case of recycling strategies in fast spectrum reactors and PWRs [36]

Reactor type	PWR		Fast spectrum reactor			ADS
Parameter \ Fuel type	MOX (Pu only, reference)	Full TRU recycle	Pu only	Full TRU (homogeneous recycle)	MA targets (heterogeneous recycle)	MA-dominated fuel
Decay heat (Wg^{-1} HM)	1	×3	×0.5	×2.5	×20–80	×90
Neutron source (ns^{-1}g^{-1} HM)	1	×8,000	×~1	×150	×1,000–4,000	×20,000

Table 7.17 Properties of Cm and Cf

		n/g·s				
Nuclide	Half life	Spontaneous fissions (SF)	(α,n)	Eγ (keV)	γ	nSv/Bq (ICFR-72)
^{242}Cm	163 d	1.72×10^7	4.18×10^6	1.4		12
^{243}Cm	30.0 y		6.09×10^4	133.2	104 keV (24%) 228 keV (11%) 278 keV (14%)	150
^{244}Cm	18.1 y	1.01×10^7	8.84×10^4	1.3		120
^{245}Cm	8.50×10^3 y			93.8	104 keV (30%) 100 keV (18%)	210
^{246}Cm	4.73×10^3 y	$\approx 7 \times 10^6$	<<SF	3.0		210
^{247}Cm	1.60×10^7 y			302.8	403 keV (69%)	190
^{248}Cm	3.40×10^5 y	$\approx 3 \times 10^7$	<<SF	579.1	579 keV (100%)	770
^{249}Cf	351 y		n.a.	329.2		350
^{250}Cf	13.1 y	$\approx 8 \times 10^9$	<<SF	6.3		160
^{251}Cf	898 y		n.a.	120.3		360
^{252}Cf	2.64 y	$\approx 10^{12}$	<<SF	217.4		90

Table 7.18 Thermal power of the assemblies containing MA targets ("fresh" fuel) [36]

	Fresh fuel assembly		
	SFR homogeneous (UPu+MAs)	Radial blanket 40% MAs	Radial blanket 10% MAs
Thermal power (kW)	0.7	21.6	5.4
Nuclide	Contributors (%)		
^{238}Pu	26		
^{239}Pu	3		
^{240}Pu	6		
^{241}Am	7	14	15
^{242}Cm	4	6	8
^{244}Cm	51	79	76

be put in perspective by comparing heat loads associated with different fuels and targets with the ones considered in the CEA study [36]. Table 7.20 shows decay heat values for different target compositions of interest.

All steps of the fuel handling system are affected, as discussed above. In particular, the solution with 40% of MA in the target, which strongly reduces the number of assemblies to be handled, is very penalizing. In fact, transport and maintenance operations under gas are impossible. A solution with Na cooling is needed. Moreover, after irradiation, the target assemblies should cool down for approximately 12 months to comply with current regulations. A long decay delay should also be planned in a separate drum, to comply with the present criteria for assembly cleaning (<7.5 kW). In the case of 10% MA targets (which implies a much larger number of target assemblies to be handled), the transport operations under gas could be possible. After irradiation, a relatively quick evacuation could be envisioned, if the solution of handling under Na is chosen. If handling under gas is chosen, it is necessary to allow ~12 month's decay in the reactor. In both cases, it will be necessary to cool down the assemblies before cleaning.

Table 7.19 Neutron sources in the assemblies containing MA targets ("fresh" fuel) [36]

	Fresh fuel assembly		
	SFR homogeneous (UPu+MAs)	Targets in radial blanket 40% MAs	Targets in radial blanket 10% MAs
Neutron source (*n*/s)	1.7×10^9	8.0×10^{10}	1.9×10^{10}
Nuclide	Contributors (%)		
^{244}Cm	80	80	83
^{246}Cm	10	8	12
^{248}Cm		2	1
^{250}Cf	7	6	2
^{252}Cf	3	4	1

Table 7.20 Heat loads of 40 and 10% MA targets after irradiation [36]

		Decay heat of irradiated target assembly (kW)		
Target	Fresh fuel power (kW)	3 month cooling	12 month cooling	36 month cooling
40% MAs	21	52	31	21
10% MAs	5.4	15	9	6

To avoid costly redundancies of the storage facilities, a research and development effort is needed, both for the transport operations and the management of assemblies with a high decay heat and for the cleaning (washing) installation. The, objective is to wash at higher temperature and quicker, which could imply necessary improvements in the assembly design to avoid Na retention.

In the case of homogeneous recycle of TRU (with a high MA/TRU ratio) in low conversion ratio critical cores, the results for a full set of configurations (i.e., metal or oxide fuel, MA/TRU ratio of 0.1 and 0.5, and all values of CR) considered are shown in Table 7.21—both for the "fresh" fuel (decay heat and neutron source) and for the unloaded fuel (decay heat, neutron source and radiotoxicity at ingestion). All values are normalized to one gram of fuel [34, 37].

The results show several significant features:

Decay heat at fuel loading/unloading and decay heat in the repository and its evolution with time. The decay heat at equilibrium increases by a factor of ~5–7 when the CR decreases from CR = 1 to CR = 0.5 and it increases by another factor of ~4 when the CR decreases from CR = 0.5 to CR = 0. This trend is similar both for metal and oxide fuels, even if oxide fuels have slightly higher decay heat values at short cooling times. Moreover, there is a further increase of a factor ~3 when the MA/TRU ratio increases. In summary, the decay heat increases by a factor of ~15–25 (depending on the type of fuel), going from a core with CR = 1 and MA/TRU = 0.1 to a core with CR = 0.5 and MA/TRU = 0.5. There is a further increase by a factor of 3, if the CR is equal to zero. These increases are significant and can have a significant impact on different features of the fuel cycle (spent fuel unloading procedures, decay heat in a repository etc).

Neutron sources at fuel fabrication. This parameter also rises significantly with decreasing CR and increasing MA/TRU ratio. In fact, the neutron source increases by a factor~30–50 (depending on the type of fuel), going from a core with CR = 1 and MA/TRU = 0.1 to a core with CR = 0.5 and MA/TRU = 0.5. There is a further increase by a factor of ~3–4, if the

Table 7.21 Fuel cycle consequences of low conversion ratio cores [34, 37]

Decay heat (MeV/s·g) for the fresh fuel of equilibrium cycle										
Metal (MA/TRU = 0.1)						Oxide (MA/TRU = 0.1)				
	CR = 0.00	CR = 0.25	CR = 0.50	CR = 0.75	CR = 1.0	CR = 0.0	CR = 0.25	CR = 0.50	CR = 0.75	CR = 1.0
$t = 0$	1.00E+12	5.14E+11	2.37E+11	1.05E+11	3.24E+10	1.02E+12	5.33E+11	2.83E+11	1.40E+11	5.80E+10
Metal (MA/TRU = 0.5)						Oxide (MA/TRU = 0.5)				
	CR = 0.00	CR = 0.25	CR = 0.50	CR = 0.75	CR = 1.0	CR = 0.0	CR = 0.25	CR = 0.50	CR = 0.75	CR = 1.0
$t = 0$	2.68E+12	1.85E+12	8.20E+11	3.18E+11	3.24E+10	2.72E+12	1.83E+12	8.88E+11	3.63E+11	5.80E+10
Neutron source (*n*/sg) for the fresh fuel of equilibrium cycle										
Metal (MA/TRU = 0.1)						Oxide (MA/TRU = 0.1)				
	CR = 0.00	CR = 0.25	CR = 0.50	CR = 0.75	CR = 1.0	CR = 0.0	CR = 0.25	CR = 0.50	CR = 0.75	CR = 1.0
$t = 0$	3.82E+05	1.88E+05	8.32E+04	3.36E+04	6.96E+03	4.14E+05	2.11E+05	1.08E+05	4.96E+04	1.67E+04
Metal (MA/TRU = 0.5)						Oxide (MA/TRU = 0.5)				
	CR = 0.00	CR = 0.25	CR = 0.50	CR = 0.75	CR = 1.0	CR = 0.0	CR = 0.25	CR = 0.50	CR = 0.75	CR = 1.0
$t = 0$	1.28E+06	8.68E+05	3.85E+05	1.45E+05	6.96E+03	1.30E+06	9.12E+05	4.41E+05	1.73E+05	1.67E+04

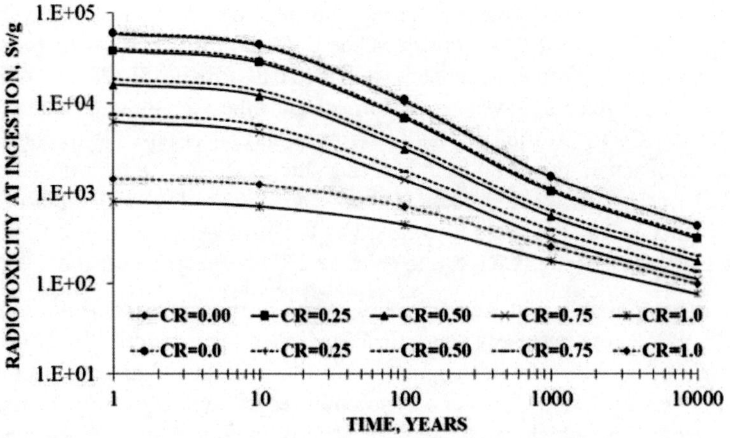

Fig. 7.10 Unloaded fuel radiotoxicity for different CRs (MA/TRU = 0.50, *dashed lines* = oxide fuel, *continuous lines* = metal fuel)

CR is equal to zero. These increases are clearly substantial. The increase in the neutron source is essentially due to the ^{244}Cm build up (for the short cooling times) and mostly to ^{246}Cm at longer cooling times (e.g., 100 years).

Thus, the high increase of the decay heat and of the neutron source in the fuel cycles associated with low conversion ratios should be a warning of potentially significant difficulties in handling the corresponding fuels, both in the case of oxide and metal fuels and whatever the strategy used for fuel makeup. From the analysis performed it seems that cores with CR below ~0.8, and in particular for high MA/TRU ratios in the feed fuel, could induce significant difficulties in the associated fuel cycles.

Figure 7.10 shows that for the radiotoxicity in the deep geological storage (i.e. the radiotoxicity associated to the losses at reprocessing), there is also an increase of the values with the decrease of the CR. For example, this can rise up to a factor ~6–8, according to the fuel type, at $t_{cool} = 10^4$ y. This increase can have an impact on the time needed for the radiotoxicity in the repository to go back to the value associated with the initial ore (e.g., from few hundred years to ~1,000). Once more this effect is due to the higher concentration of TRU in the fuel and in fact the highest values are found once more in the case with MA/TRU = 0.5 and CR = 0. This means, as indicated previously, that a compromise should be found between the transmutation performance and the impact on practically all parameters of interest in the fuel cycle.

7.9.9 Implementation of Transmutation Scenarios

According to the transmutation objectives, fast spectrum reactors offer flexible options for implementation scenarios, since a fast spectrum reactor core can be designed with the objective of breeding or burning fissile material—within a wide range of conversion ratio values. In meeting the above objectives, the three following generic scenarios can be defined. All three scenarios go beyond the strategy of the "once-through", or "open", fuel cycle that involves final storage of once irradiated fuel without fuel reprocessing. Their specific characteristics are summarized below.

Scenario A: Sustainable development of nuclear energy and waste minimization. In this case (e.g. multi-recycle of TRU in fast spectrum reactors) the ratio Pu/MA is ~8.5/1.5 for most used fuels as unloaded from UOX or MOX PWRs or BWRs. Two options can be foreseen:

Option 1: Homogenous TRU recycling in a critical fast reactor. The fuels could be rather standard mixed oxide or metal fuels, with MA content of the order of a few percent (e.g. definitely <5%). For this type of fuel a few experimental demonstrations (the SUPERFACT [28] and METAPHIX [38] experiments in the Phénix reactor) have been performed, and others are foreseen within the framework of Generation-IV (GACID [39] project). As for reprocessing, it is possible to consider a grouped TRU recovery without separation of Pu from MA (e.g., the GANEX [40] hydrometallurgical process developed at CEA or the UREX+1a process [41] developed in the US) with enhanced proliferation resistance. Within this option, the flexibility offered by the fast reactor neutronics allows the designer to tune the core Conversion Ratio (CR), e.g. to enhance TRU burning as required.

Option 2: MA targets, possibly with an inert matrix (with or without a moderator in the assembly), to be loaded (e.g. at the core periphery) in critical Pu-fuelled fast reactors. The MA content should be defined according to reactor core design, mass reduction criteria and fuel cycle requirements. The use of a U-matrix would provide an easier way to multi-recycle the targets. As for reprocessing, the separation of Pu from MA (which can be kept together, or implementing a process of separation of Cm from Am, with Cm storage in a specific installation) is required for this option, with potential drawbacks in terms of proliferation resistance.

For both options, the objective is a stabilization of the TRU inventory in the reactors and in the fuel cycle, together with the minimization of the masses sent to the repository (in practice limited to the losses at reprocessing). Potential advantages and disadvantages of both options are currently investigated within major R&D programs on TRU recycle. Figure 7.11 summarizes main features of this scenario.

Scenario B: Reduction of MA inventory. This is the case of a strategy driven by the decision to reduce drastically the MA inventories, while Pu is considered a resource. With respect to scenario A, the hypothesis is that the implementation of fast reactors is somewhat delayed in time and a transition scenario has to be envisaged, in order to avoid a buildup of MA that could jeopardize the successive implementation of fast spectrum reactors. With respect to option 1 of scenario A, the chemical separation process needs to be changed: in fact the separation of Pu from MA (which can be kept together, or implementing a process of separation of Cm from Am, with Cm storage in a specific installation) is required. To implement this scenario, the so-called "double strata" strategy can be envisioned:

- MA fuels in ADS with Pu/MA$\sim$1 (optimum value, according to numerous previous studies [23]) and CR=0. However, a U-matrix can be considered as alternative to U-free fuels [34], opening the way to a possible use of a "critical" burner fast spectrum reactor, as described above.
- Pu from LWRs will be recycled in MOX-LWRs.

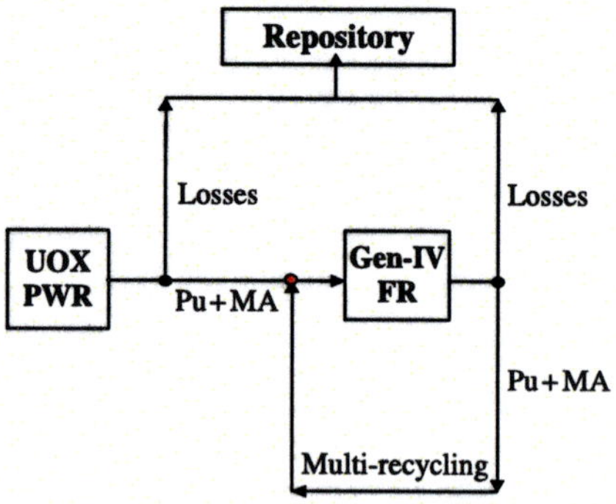

Fig. 7.11 Scenario A: sustainable development of nuclear energy and waste minimization

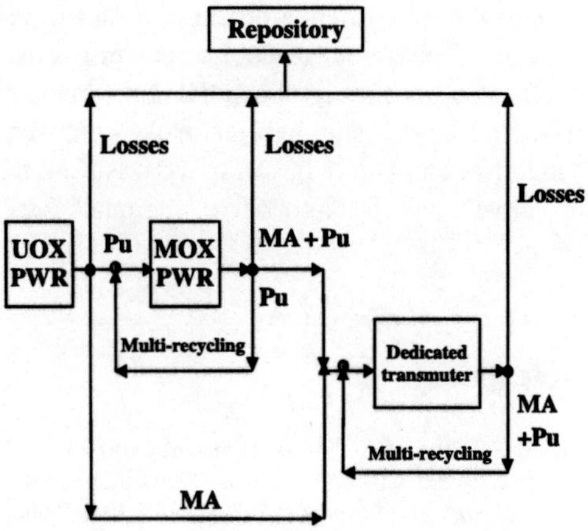

Fig. 7.12 Scenario B: reduction of MAs

The main interest of this scenario, as summarized in Fig. 7.12, is to keep the management of MA independent from the commercial fuel cycle, where Pu is multi-recycled. The expected reduction of radiotoxicity is similar to that in scenario A above, if the chemical separation performance (e.g. losses during reprocessing, or TRU recovery rate) is approximately the same in the two scenarios.

Scenario C: Reduction of TRU inventory as unloaded from LWRs. This is the case for reducing TRU stockpiles as a legacy of previous operation of LWRs. In this case, the ratio Pu/MA is also ~8.5/1.5. As for reprocessing, a grouped TRU recovery without separation of Pu from MA can be envisioned. To maximise consumption, a U-free fuel (inert matrix) in a fast neutron spectrum device (essentially an ADS or a fusion-fission-hybrid) with conversion ratio $CR = 0$, can be envisioned. However, a conversion ratio $CR \sim 0.5$ (which corresponds to a ratio $U/TRU \sim 1$) or less (if fuel cycle impacts can be limited), that allows ~75% (or more) of the maximum theoretical TRU consumption, can also be envisioned as discussed previously [34]. This also means that, in this case, an alternative to U-free fuels is the use of a U-matrix, i.e. a mixed oxide or metal fuel, opening the way once more to the possible use of a "critical" burner fast spectrum reactor. Figure 7.13 summarizes main features of this scenario.

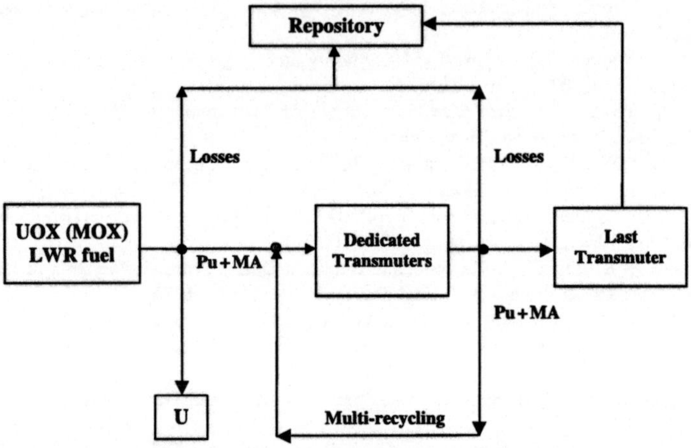

Fig. 7.13 Scenario C: reduction of TRUs as unloaded from LWRs

This scenario offers a potential means of reducing stockpiles of Pu and MA in spent fuel, e.g. in the case of phase-out of nuclear power plants. However, this scenario, if implemented by a country in isolation, implies a substantial deployment of new installations (fuel reprocessing and fabrication, ADS or fusion-fission hybrids, etc). Moreover, after a ~100 years of operation, ~20% of the initial TRU inventory would be left in the wastes. It has been shown that a better approach to reach the stated objective would be to conceive "regional" P&T scenarios [42].

References

1. J. Gourdon, et al., "An Overview of SUPERPHENIX Commissioning Tests," *Nucl. Sci. Eng.*, **106**, p. 1 (1990).
2. P. Greebler and C. L. Cowan, *FUMBLE: An Approach to Fast Power Reactor Fuel Management und Burnup Calculations*, November 1970, GEAP-13599. General Electric Co., USA (1970).
3. J. Hoover, G. K. Leaf, D. A. Meneley, and P. M. Walker, "The Fuel Cycle Analysis System, REBUS," *Nucl. Sci. Eng.*, **45**, pp. 52–65 (1971).
4. W. W. Little, Jr. and R. W. Hardie, *2DB User's Manual*, BNWL-831, Pacific Northwest Laboratory, Richland, WA (1968).
5. R. W. Hardie and W. W. Little, Jr., *3DB, Three-Dimensional Diffusion Theory Burnup Code,* BNWL-1264, March 1970, Pacific Northwest Laboratory, Richland, WA (1970).
6. K. L. Derstine, *DIF3D: A Code to Solve One-, Two-, and Three-Dimensional Finite-Difference Diffusion Theory Problems*, ANL-82-64, Argonne National Laboratory, Argonne, IL (1984).
7. G. Palmiotti, E. E. Lewis, and C. B. Carrico, *"VARIANT" VARIational Anisotropic Nodal Transport for Multidimensional Cartesian and Hexagonal Geometry Calculation*, ANL-95140, Argonne National Laboratory, Argonne, IL (1995).
8. H. L. Wyckoff and P. Greebler, "Definitions of Breeding Ratio and Doubling Time," *Nucl. Technol.*, **21**, pp. 158–164 (1974).
9. K. Ott, "An Improved Definition of the Breeding Ratio for Fast Reactors," *Trans. Am. Nucl. Soc.*, **12**, p. 719 (1969).
10. W. P. Barthold and Y. I. Chang, "Breeding Ratio and Doubling Time Definitions Used for Advanced Fuels Performance Characterization," *Trans. Am. Nucl. Soc.*, **26**, p. 588 (1977).
11. D. R. Marr. R. W. Hardie, and R. P. Omberg, "An Expression for the Compound System Doubling Time Which Explicitly Includes the Approach to Equilibrium," *Trans. Am. Nucl. Soc.*, **26**, p. 587 (1977).
12. M. Salvatores, I. Slessarev, and M. Uematsu, "A Global Physics Approach to Transmutation of Radioactive Nuclei," *Nucl. Sci. Eng.*, **116**, p. 1 (1994).
13. M. Salvatores, R. Hill, I. Slessarev, and G. Youinou, "The Physics of TRU Transmutation – A Systematic Approach to the Intercomparison of Systems," *Proc. PHYSOR 2004 – The Physics of Fuel Cycles and Advanced Nuclear Systems: Global Developments*, Chicago, IL, April 25–29 (2004).
14. R. N. Hill and T. A. Taiwo, "Transmutation Impacts of Generation-IV Nuclear Energy Systems," *Proc.Int. Conf. PHYSOR,* Vancouver (2006).
15. G. Aliberti, et al., "Nuclear Data Sensitivity, Uncertainty and Target Accuracy Assessment for Future Nuclear Systems," *Ann. Nucl. Ener.*, **33**, pp. 700–733 (2006).
16. G. Aliberti, et al., "Impact of Nuclear Data Uncertainties on Transmutation of Actinides in Accelerator-Driven Assemblies," *Nucl. Sci. Eng.*, **46**, pp. 13–50 (2004).
17. M. Salvatores, et al., "OECD/NEA WPEC Subgroup 26 Final Report: Uncertainty and Target Accuracy Assessment for Innovative Systems Using Recent Covariance Data Evaluations," OECD-NEA Report, Paris (2008).
18. W. P. Barthold and J. C. Beitel, "Performance Characteristics of Homogeneous Versus Heterogeneous Liquid-Metal Fast Breeder Reactors," *Nucl. Tech.*, **44**, pp. 45, 50–52 (1979).
19. C. A. Erdman and A. B. Reynolds, "Radionuclide Behavior During Normal Operation of Liquid-Metal-Cooled Fast Breeder Reactors. Part I: Production," *Nucl. Safety*, **16**, pp. 44–46 (1975).
20. Y. I. Chang, C. E. Till, R. R. Rudolph, J. R. Deen, and M. J. King, *Alternative Fuel Cycle Options: Performance Characteristics und Impact on Nuclear Power Growth Potential*, ANL-77 70. Argonne National Laboratory, Argonne, IL (1977).
21. M. Salvatores, "Nuclear Fuel Cycle Strategies Including Partitioning and Transmutation," *Nucl. Eng. Design*, **235**, p. 805 (2005).
22. M. A. Smith, et al., *Physics and Safety Studies of Low Conversion Ratio Sodium Cooled Fast Reactors*, PHYSOR 2004, Chicago, IL (2004).

23. *Accelerator-Driven Systems (ADS) and Fast Reactors (FR) in Advanced Fuel Cycles. A Comparative Study*. OECD-NEA Report, Paris (2002).
24. G. Palmiotti, M. Salvatores, and M. Assawaroongruengchot, "Innovative Fast Reactors: Impact of Fuel Composition on Reactivity Coefficients," *Proc. Int. Conf. on Fast Reactors*, FR09, December 2009, Kyoto, Japan (2009).
25. T. Wakabayashi, "Transmutation Characteristics of MA and LLFP in a Fast Reactor," *Progr. Nucl. Energy*, **40**, No. 3–4 (2002).
26. J. M. Bonnerot, et al., "Progress on Inert Matrix Fuels for Minor Actinid Transmutation in Fast Reactor," *Proceedings Global 2007*, Boise, ID, USA (2007).
27. J. F. Babelot and N. Chauvin, *Joint CEA/ITU Synthesis Report of the Experiment SUPERFACT 1*, Report, JRC-ITU, Karlsruhe, TN-99/03 (1999).
28. L. Buiron, et al., "Minor Actinide Transmutation in SFR Depleted Uranium Radial Blanket. Neutronics and Thermal-Hydraulics Evaluation," *Proc. Int. Conf. GLOBAL 2007,* Boise, ID, USA (2007).
29. W. Maschek, "Report on Intermediate Results of the IAEA CRP on Studies of Advanced Reactor Technology Options for Effective Incineration of Radioactive Waste," *Energy Conversion Manag*, **49**, pp. 1810–1819, Elsevier (2008).
30. K. Noack, A. Rogov, et al., "The GDT-Based Fusion Neutron Source as Driver of a Minor Actinides Burner," *Ann. Nucl. Energy*, **35**, pp. 1216–1222, Elsevier (2008).
31. T. A. Mehlhorn, et al., "Fusion–Fission Hybrids for Nuclear Waste Transmutation: A Synergistic Step Between Gen-IV Fission and Fusion Reactors," *Fusion Eng. Design*, **83**, pp. 948–953, Elsevier (2008).
32. J. Rouault and M. Salvatores, "The CAPRA Project: Status and Perspectives," *Nuclear Europe Worldscan* (1995).
33. E. A. Hoffman, W. S. Yang, and R. N. Hill, "A Study on Variable Conversion Ratio for Fast Burner Reactor," *Trans. Am. Nucl. Soc.*, **96** (2007).
34. C. Fazio, M. Salvatores, and W. S. Yang, "Down-Selection of Partitioning Routes and of Transmutation Fuels for P&T Strategies Implementation," *Proc.Int. Conf. GLOBAL 2007*, September 2007, Boise, ID (2007).
35. R. E. Schenter and M. E. Korenko, "Transmuting Very Long Lived Nuclear Waste into Valuable Materials," *Trans. Am. Soc.*, **99**, pp. 229–230 (2008).
36. M. Delpech, et al., "Scenarios Analysis of Transition from Gen II/III to Gen IV Systems. Case of the French Fleet," *GLOBAL 2005*, Tsukuba, Japan (2005).
37. C. Chabert, "An Improved Method for Fuel Cycle Analysis at Equilibrium and Its Application to the Study of Fast Burner Reactors with Variable Conversion Ratios," *Proc. Int. Conf. PHYSOR '08*, September 14–19, 2008 Interlaken, Switzerland (2008).
38. M. Kurata, et al., "Fabrication of U-Pu-Zr Metallic Fuel Containing Minor Actinides" *GLOBAL 97 Conference*, October 5–10 1997, Yokohama, Japan. See also L. Breton, et al., "METAPHIX-1 Non Destructive Post-Irradiation Examinations in the Irradiated Element Cells at PHENIX," *Proc. Int. Conf. Global 2007*, September 9–13,Boise, USA, and H. Otha, et al. "Irradiation Experiment on Fast Reactor Metal Fuels Containing Minor Actinides up to 7% At. Burn-up", ibidem.
39. F. Nakashima, et al., "Current Status of the Global Actinide Cycle International Demonstration Project," *Proc. Int. GIF Symposium*, September 9–10, 2009, OECD Report, Paris (2009).
40. J. M. Adnet, et al., "Development of New Hydrometallurgical Processes for Actinide Recovery: GANEX Concept," *Proc. Int. Conf. GLOBAL '05*, October 9–13, 2005, Tsukuba, Japan (2005).
41. C. Pereira, et al., "Preliminary Results of the Lab-Scale Demonstration of the UREX+1a Process Using Spent Nuclear Fuel," *2005 AIChE National Meeting*, Cincinnati (2005).
42. M. Salvatores, et al., "Fuel Cycle Synergies and Regional Scenarios," *Proc. Int. Conf. IEMPT-10*, October 2008, Mito, Japan (2008).

Part III
Systems

A fast spectrum reactor differs substantially from a thermal reactor with regard to core design and the heat transport system. This difference arises mainly from the much higher flux and power densities encountered in the fast spectrum reactor, the high energy neutron spectrum, and the desire to take advantage of higher possible coolant temperatures to improve the overall thermal efficiency.

Five chapters are included to provide coverage for the principal systems design questions of interest. Chapters 8 through 10 address the mechanical and thermal aspects of fast reactor core design. Although actual core design procedures involve a close coupling of these topics, an artificial separation is helpful for instructional purposes. Hence, Chapter 8 is focused on the mechanical aspects of fuel pin and assembly design, Chapter 9 is addressed to the thermal aspects of a single fuel pin, and Chapter 10 is then broadened to discuss full assembly and overall thermal hydraulics aspects. Whereas certain core materials properties must be introduced in the earlier chapters, the more detailed materials questions are concentrated in Chapter 11. Overall fast reactor plant systems, including the heat transport system, are included in Chapter 12.

Chapter 8
Fuel Pin and Assembly Design

Alan Waltar and Donald Todd

8.1 Introduction

This chapter deals with the mechanical designs of fuel pins and assemblies. These core components must be designed to withstand the high temperature, high flux environment of a fast spectrum reactor for a long irradiation exposure time. In this chapter we will describe many of the factors that influence this design, and we will examine in some detail the stress analysis of the fuel pin.

We begin in Section 8.2 with the basic geometric and heat transfer relationships for the fuel pin, and then discuss some topics related to fuel and fission gas that must be considered in analysis of steady-state fuel-pin performance. The discussion of fuel-pin design is continued in Section 8.3, in which failure criteria and stress analysis are presented. Discussion is then shifted in Section 8.4 to grouping the pins into a fuel assembly. This will include discussion of mechanical design problems such as fuel-pin spacing and duct swelling. Limited attention is then directed to the design of other assemblies, including blanket, control and shielding assemblies. The final section of the chapter will provide a description of methods to restrain the assemblies and to control duct bowing.

8.2 Fuel Pin Design Considerations

Fuel pin design is a complex process that involves an integration of a wide range of phenomena. The design procedure must integrate the thermal analysis of the pin with an assessment of the characteristics of the fuel and cladding as a function of temperature and irradiation history and with the stress analysis of the fuel-cladding system. Many of the phenomena that affect fuel pin performance are illustrated schematically in Fig. 8.1. Each of these will be treated in Chapters 8 and 9, with additional discussions in Chapter 11. It should be readily apparent that there is no single way to order the examination of so many interacting phenomena. Each sequence selected to cover these tightly coupled processes will inherently suffer from the need for input from processes not yet discussed. In actual design practice, all of the governing processes are integrated in large time-dependent pin analysis codes such as LIFE [1].

Our presentation will begin with the geometrical arrangement of the various sections of a fuel pin, including the function of each section. We will then extend the discussion of pin geometry to the selection of pin diameter. This will involve the concept of linear power. Fuel restructuring is then introduced, followed by fission gas release and the associated length of the gas plenum to establish

A. Waltar (✉)
Pacific Northwest National Laboratory (PNNL), Retired, Richland, WA, USA
e-mail: alan.waltar@wildblue.net

A.E. Waltar et al. (eds.), *Fast Spectrum Reactors*, DOI 10.1007/978-1-4419-9572-8_8,

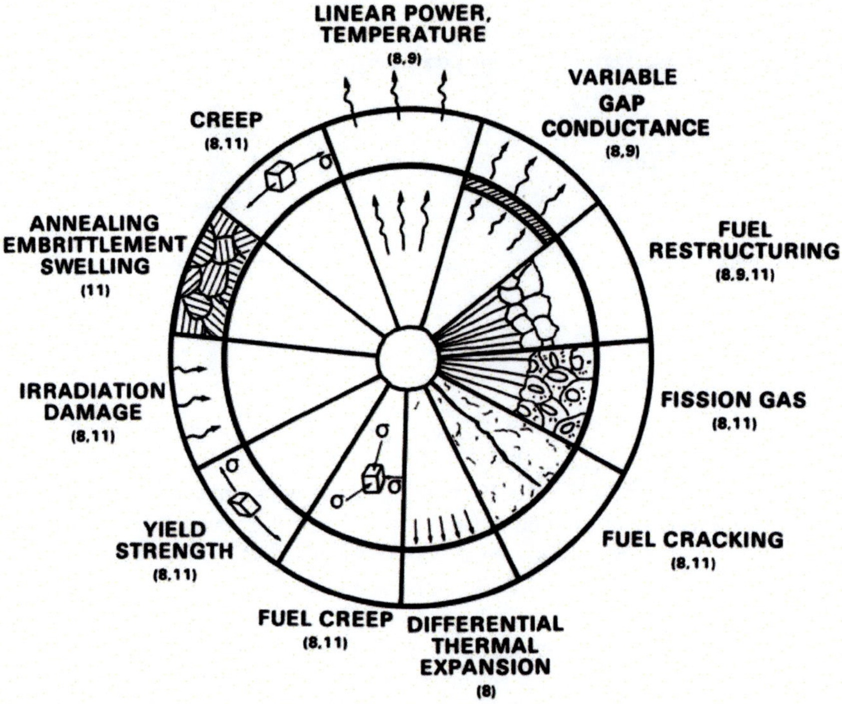

Fig. 8.1 Phenomena affecting fuel pin performance (numbers in parentheses denote chapters where the indicated phenomena are discussed)

the basis for the cladding loading function. Such discussions lead directly to the stress analysis of the fuel-cladding system, the subject of Section 8.3. These structural considerations are required to establish the requirements for cladding thickness, fuel porosity, and initial gap size (wherein the latter two items define the fuel smear density, given the theoretical density of the fuel material, at steady-state operating conditions). In addition, it is necessary to discuss fuel pin lifetime or design criteria in order to understand how the results of the stress analysis are used in the pin design.

The cladding is stressed by fission gas pressure when the fuel-cladding gap is open, and it may be stressed directly by the fuel if the gap is closed. A quantitative determination of this fuel-loading effect requires a thermal analysis of the fuel and cladding in order to calculate thermal expansions. (This thermal analysis is the subject of Chapter 9.) It also requires knowledge of the creep strength of the fuel, since for very slow transients the fuel can deform sufficiently to prevent significant loading on the cladding. Once the fuel pin temperature field is known, the temperature-dependent stress-strain relations can be incorporated.[1] We will introduce the basic factors required for the fuel-pin stress analysis, including some aspects of material behavior, in this chapter, but a more detailed discussion of materials properties will be delayed for Chapter 11.

Emphasis in this chapter will be placed on the behavior of mixed-oxide fuel since this is the fuel used in most SFR systems. Certain aspects of other fuels are discussed in Chapter 11.

---

[1]It is of interest to note that the temperature distribution in the fuel may be dependent upon the stress-strain relationships between fuel and cladding. The reason for this is that gap conductance is a function of the fuel-cladding interface pressure. As we shall see later, this gap conductance directly affects the fuel surface temperature, which in turn influences the temperature profile within the fuel pin.

8.2.1 Materials and Geometrical Arrangement

The fuel pin, or element, for a ceramic fueled SFR consists of an axial stack of cylindrical fuel pellets encased in a metal tube called the cladding. The cladding provides structural integrity for the fuel pin and serves to separate the fuel from direct contact with the coolant, thereby preventing the fission products from entering the primary coolant. Typical values for the outside diameter of the cladding are 6–8 mm. Each fuel pellet is approximately 7 mm in length and 6 mm in diameter. The pellet diameter is fabricated to be slightly less than the inside cladding diameter, so that there is an initial gap between the fuel and the cladding.

The fuel for many fast reactors is mixed uranium-plutonium oxide (UO_2-PuO_2). A second ceramic fuel, mixed carbide (UC-PuC), has also been investigated. Designs also exist that use metallic fuel and mixed thorium-^{233}U ceramic fuel. The mechanical designs described in this chapter apply principally to the oxide fuel designs, although the same general configuration is used for carbide fuels. All the fuels, including the metallic form, which is becoming increasingly popular (and is one of the choices for the S-PRISM and JSFR designs described in Appendices B and C, respectively), are described in more detail in Chapter 11.

The axial configuration of a SFR fuel pin is quite different from a light water reactor (LWR) pin. In the SFR, the zone containing the fissile fuel (which will form the reactor core) is only about a third of the pin axial length, whereas the fuel region of an LWR typically represents 90% of the overall pin length. The SFR core is generally about one meter long, while the entire fuel pin is about three meters long. Because of the much higher fueled core height for a similarly powered LWR, overall pin lengths for the LWR and SFR systems are comparable.

Cutaway views of two SFR pins are shown in Fig. 8.2. Axial blanket pellets, which are fabricated from depleted uranium oxide, are located above and below the core pellets. Typical heights for the axial blankets are 0.3–0.4 m. The spring above the pellet stack is used to hold the pellets in place during shipment and prevent pellet separation during insertion of the fuel assembly into the core; this spring provides no structural function once the pin has been brought to power.

A fission gas plenum is located in the pin as a reservoir for gaseous fission products produced during irradiation. The fission gas plenum is normally long, approximately equal to the height of the core. As noted in Fig. 8.2, the plenum can be located either above or below the core. For example, in the early US designs (FFTF and CRBRP), the plenum is above the core; the French Phénix and Super Phénix designs contain a small upper and a large lower plenum. The advantage of the above-core location is that a cladding rupture in the plenum region would not allow fission gas to pass through the core (since the sodium flow is upward). The disadvantage is that the coolant is at its highest temperature above the core and the plenum length (or volume) required to accommodate the fission gas pressure (since the fission gas is at the coolant temperature) is larger than would be the case for a plenum in the cooler region below the core. The longer plenum length adds to the total length of the fuel pin, which in turn leads to a higher pressure drop across the fuel assembly, higher required pumping power, and potentially a larger reactor vessel and related vessel internal components.[2]

Figure 8.3 contains a schematic diagram of the lengths of fuel pins for several fast reactors. The pins are drawn to scale, with the core centerline used for vertical normalization. The various design

[2]It is necessary to conduct a total systems evaluation before deciding the best location for the fission gas plenum. It is possible, for example, that a below-core plenum location could actually lead to an increase in the overall length of the assembly. This possibility arises from the general desire to keep the outlets of all core assemblies at about the same elevation. Since control rods normally enter from the top of the core, the control ducts extend well above the core to accommodate withdrawn absorber positions. That vertical distance, plus the added fuel assembly distance below the core to accommodate a below-core fission gas plenum location, may result in a longer fuel assembly length than if the above-core ducting were utilized to house a somewhat larger fission gas plenum.

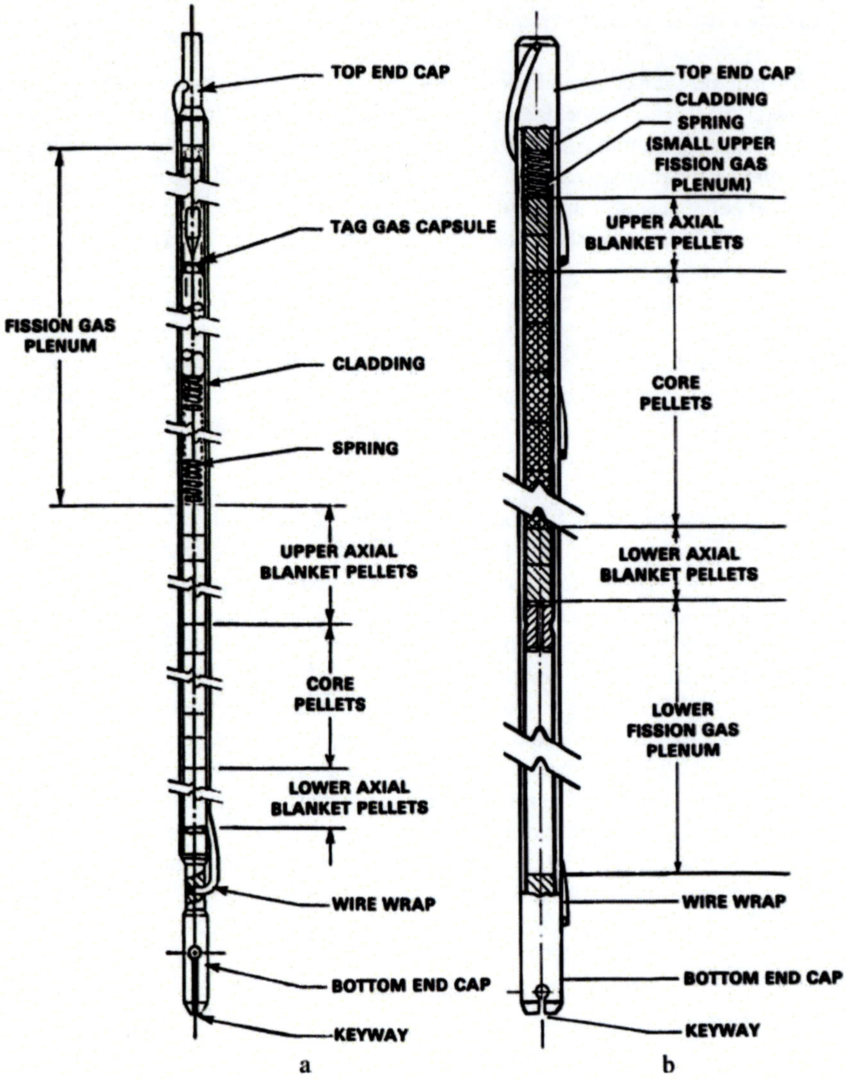

Fig. 8.2 Typical fast reactor fuel pins. (**a**) CRBRP [2], (**b**) Super Phénix [3]

options are readily apparent, particularly with regard to the placement of the fission gas plenum. It can be observed from this figure that the length of the plenum is approximately the same as that of the core.

Tag gas capsules are sometimes included in the plenum region to assist in the location of a failed fuel pin. These capsules in the fuel pins contain an isotopic blend of inert gases that is unique to that assembly. At the time of final fabrication, the capsule is punctured so that the gas mixture enters the fission gas plenum. Should a break occur in the cladding during irradiation, the unique blend of gases would escape with the fission product gases into the primary loop where they could be collected and analyzed to identify the assembly containing the failed pin. The effectiveness of this technique has been demonstrated in both EBR-II and FFTF. (See Section 12.6.2 and Fig. 12.31 for a more complete presentation of this technique.)

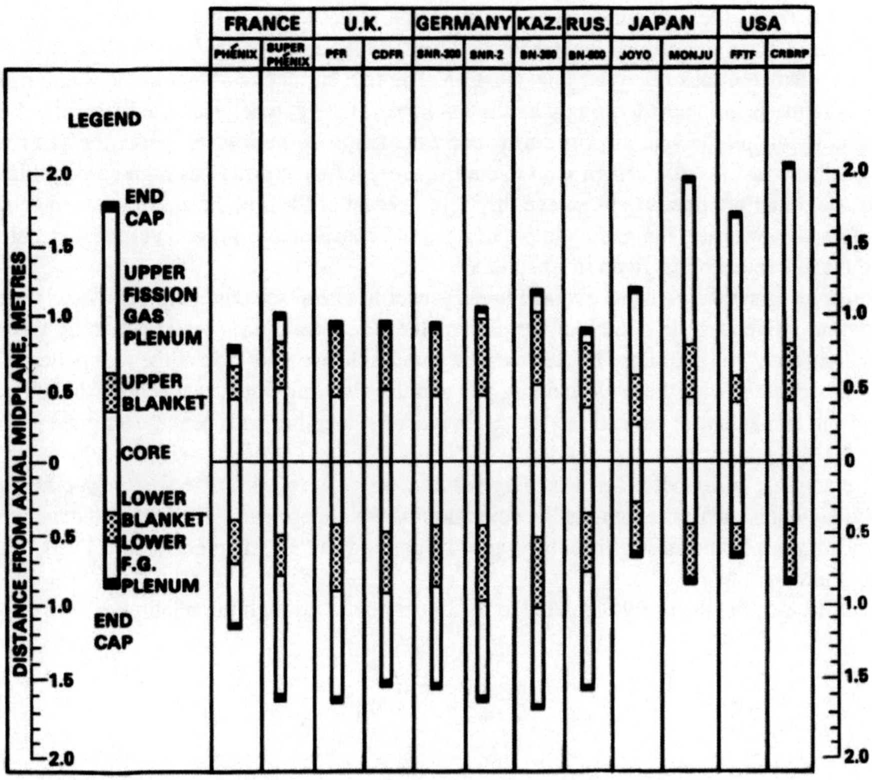

Fig. 8.3 Intercomparison of fuel pin lengths for selected fast reactor systems (updated from Ref. [4])

Above and below the plenum and blankets are solid end caps. In designs employing a wire wrap pin spacing technique, the wire is normally pulled through a hole in each end cap and then welded at that point. For some designs the wire is merely laid along the side of the end cap and then welded.

Austenitic 316 stainless steel, normally cold worked to approximately 20%, was selected as the early reference cladding material for many SFRs. This steel alloy contains 16–18 wt.% Cr, 10–14% Ni, 2–3% Mo, and the balance Fe (65%) except for small amounts of C, Nb, Si, P, and S. Characteristics of this and other candidate cladding and structural materials, definition of cold working, and reasons for the selection of 316 stainless steel as the reference cladding and structural material are discussed in Section 11.3.

8.2.2 Linear Power and Fuel-Pin Diameter

Linear power (also called linear power density and linear heat rate) governs the total length of all the fuel pins in the core—or, given a specified core height and a total core thermal power level, the total number of pins in the core. The linear power was introduced briefly in Chapter 2, through Eq. (2.2), as

$$\chi = 4\pi \int_{T_s}^{T_0} k(T)\, dT,$$

where T_s and T_0 are the fuel surface and center temperatures and k is the fuel thermal conductivity. This equation will be derived in Chapter 9 when the detailed thermal analysis is presented. Note that the fuel pin diameter does not enter this equation. The fuel surface temperature does enter, however. This temperature is influenced strongly by the thickness of the gap between the fuel and the cladding and, hence, by the pin design, at least during early operation of the reactor before the gap closes. An objective of any fast reactor design is to make the average linear power as high as possible without allowing the center temperature to exceed the melting point of the fuel. Selection of average and peak values of linear power is a major step in pin design, and considerable attention to the thermal analysis required for this selection is given in Chapter 9.

For now, let us suppose that an average linear power has been selected. Our next step is to see how this selection influences the choice of a pin diameter. The term *pin diameter* generally refers to the outside diameter of the cladding. In the next few paragraphs we will look at the fuel pellet radius, R_f. This radius, together with the fuel-cladding gap and the cladding thickness, defines the pin diameter. Values of pin diameters for several fast spectrum reactors, together with peak linear powers, are listed in Table 8.1.

In selecting a pin diameter for a breeder reactor, we seek to minimize *fissile specific inventory*, M_0/P, where M_0 is the fissile mass in the core and P is the core power. Minimizing the fissile specific inventory minimizes the fissile inventory and, as pointed out in Chapter 1, Eq. (1.13), minimizes doubling time.

The fissile specific inventory is related to the linear power through the relation

$$\chi = e\rho_f \pi R_f^2 \frac{P}{M_0}, \tag{8.1a}$$

where

χ = linear power,
e = fissile mass fraction in the fuel,
ρ_f = fuel density in the pellet,
R_f = fuel pellet radius.

Table 8.1 Pin diameter and linear power in SFRs using mixed oxide fuel

		Pin diameter (mm)		
Reactor	Country	Core	Radial blanket	Peak linear power[a] (kW/m)
BN-350	Kazakhstan	6.1	14.2	44
Phénix	France	6.6	13.4	45
PFR	UK	5.8	13.5	48
SNR-300	Germany	6.0	11.6	36
FFTF	US	5.8	–	42
BN-600	Russia	6.9	14.2	53
Super Phénix	France	8.5	15.8	47
MONJU	Japan	6.5	11.6	36
CDFR[b]	UK	6.7	13.5	42
SNR-2[b]	Germany	7.6	11.6	42
CRBRP[b]	US	5.8	12.8[c]	42

[a] Values for linear power change slightly during the burnup cycles. The CRBRP (heterogeneous core) and FFTF values quoted are for the beginning of the first cycle.
[b] These plants were cancelled before construction began.
[c] Radial and internal blankets.

Rearranging gives for the fissile specific inventory:

$$\frac{M_0}{P} = \frac{\pi \rho_f e R_f^2}{\chi}. \tag{8.1b}$$

Having selected χ, we can minimize M_0/P by modifying ρ_f, e, and R_f. Little flexibility is available for varying ρ_f since several considerations limit ρ_f to a narrow range. It is desirable to make ρ_f as near the theoretical density (TD) as possible to enhance the thermal conductivity (and hence, maximize χ) and to decrease the required fissile inventory [note in Eq. (8.1b) that increasing ρ_f results in a sufficient reduction in neutron leakage to cause a larger decrease in e than the relative increase in ρ_f]. On the other hand, ρ_f must be low enough to accommodate fuel swelling during steady-state operation and volumetric expansion during thermal upset conditions.[3] These considerations lead to a typical pellet density for mixed oxide SFR's in the range of 85–95% TD and a smear density in the range of 75–90% TD.

The parameters e and R_f are tied together. Reducing R_f will increase the required fissile fraction, e, but this will not offset the reduction in R_f^2 as long as the fuel volume fraction does not have to be decreased significantly. Therefore, reducing R_f reduces eR_f^2 and reduces M_0/P. These trends are illustrated in Table 4.5. Consequently, reducing R_f is a desirable objective and this is the reason that fast reactor fuel pins have such a small diameter.

One lower limit to an acceptable fuel pin diameter might be the ability of the coolant to handle the ever-increasing heat flux. The heat flux, q(W/m^2), diameter, D, and heat transfer coefficient, h, are related by

$$q = \frac{\chi}{\pi D} = h\left(T_{co} - T_b\right), \tag{8.2}$$

where T_{co} and T_b are the outside cladding and bulk coolant temperatures, respectively. With sodium as the coolant, h remains high even for low flow velocity without the problems of burnout—which are of concern in an LWR. Hence, sodium heat transfer is not the limiting factor in reducing the pin diameter.[4]

There must eventually be some limiting factor on how small the pin diameter can be. One important limitation is that beyond some point small pins become increasingly expensive to fabricate. A second limitation involves pitch-to-diameter ratio. As pin diameter is reduced, the point is reached eventually when further reduction requires an increase in P/D ratio, which leads to a significant reduction in fuel volume fraction and increased fissile inventory. The P/D ratio must eventually increase as D gets small for two reasons. First, there is a lower limit on spacing between pins due to hot spots with wire wrap spacers or fabrication with grid spacers; when this spacing is reached, further reduction in D necessarily increases P/D. Second, as D is lowered, P/D must eventually be increased in order to hold axial pressure drop (and hence, pumping power) down.

An important incentive for a large pin diameter is breeding ratio. The breeding ratio is reduced with reduced pin diameter (see Table 7.6). Other more subtle factors also come into play. For example, the increased fissile fraction, e, that accompanies reduced R_f would reduce the Doppler coefficient and

---

[3] UO_2 expands approximately 10% upon melting.

[4] The arguments relating to Eq. (8.1b) apply also to a gas-cooled fast reactor, so small pins are proposed for GCFR's; however, the pumping power and system pressure for the gas coolant must both be far greater than for a SFR to achieve the required heat transfer coefficient. (See Table 11.9 and Chapter 17 for further discussion of these points.)

require greater atom percent burnup in the fuel for a given burnup in MWd/kg of total fuel. Reducing the pin size increases the neutron flux, leading to higher fluences for a specific irradiation time.[5]

In summary, Eq. (8.1) identifies the principal factors which must be balanced in determining the fuel pin diameter, but a detailed optimization procedure involves several secondary effects which can become quite important.

8.2.3 Fuel Restructuring

An analysis of a fuel pin at high temperature and during its irradiation history is strongly influenced by the structure of the fuel. Like all ceramic and metallic materials, oxide fuel is composed of *grains*. A grain is a single crystal within which all the atoms are arranged in some particular pattern characterized by a *unit cell*. When the fuel is fabricated, the shapes of the grains are irregular, with the shape of each grain being controlled by the surrounding grains. The *grain boundary* is a zone of transition between crystalline alignments in adjacent grains. For curved boundaries, the atoms are more stable on concave than on convex surfaces; hence, thermal agitation at high temperatures will cause atoms to transfer from grains with convex surfaces to those with concave surfaces. As illustrated in Fig. 8.4, small grains have more convex surfaces, so atoms from small grains tend to move to larger grains. This process is called *grain growth*. Since an increase in temperature increases the thermal agitation of the atoms, which in turn facilitates grain growth, a large amount of grain growth might be expected at the high temperatures encountered by the fuel in a reactor.

Immediately upon reactor startup and attainment of full-power fuel temperature distributions, substantial alterations in the structure of the fuel do indeed take place. Above a certain temperature band, called the *equiaxed* temperature, the grains grow very rapidly and eventually reach many times their original size. These grains are called *equiaxed grains*. No radiation field is required to induce this change; it is an effect of temperature only. At an even higher temperature closer to the radial centerline of the fuel pin, the temperature gradient causes the internally fabricated *pores*[6] to collect and migrate up the temperature gradient (i.e., toward higher temperatures). Very long grains, called *columnar grains,* begin to emanate from the center like spokes of a wheel. The flow of pores to the radial

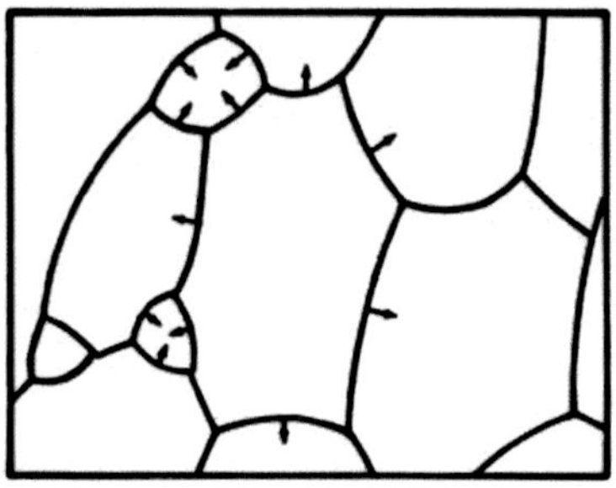

Fig. 8.4 Grain growth. The boundaries move toward the center of curvature (*arrows*). As a result the small grains eventually disappear [5]

[5]The flux increase results from the relation $\Sigma_f \phi V_{core} = 2.9 \times 10^{16} P_{core}\,(MW) =$ constant. As the pin diameter is reduced, V_{core}, decreases faster than $\sum_f$ increases; hence, ϕ increases. This is of benefit to the FFTF, which is a test reactor with the objective of maximizing ϕ in order to maximize fluence. It has smaller diameter pins (5.8 mm) than a typical power SFR.

[6]Pores are tiny void spaces purposely fabricated into the fuel to allow space for fuel swelling and expansion during steady-state and thermal upset conditions. *Porosity* is the fraction of the fuel volume that is void.

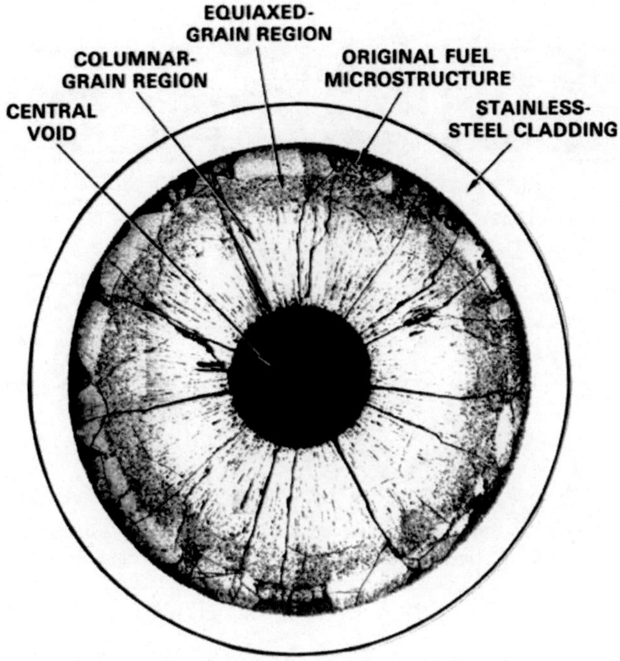

Fig. 8.5 Cross section of mixed-oxide fuel pin irradiated to 27 MWd/kg. No melting [6]

center of the pellet leaves a more fully dense fuel material behind and this can result in a large hole in the center of the pellet. Most of this *restructuring* occurs in a matter of hours at full power.

Figure 8.5 is a cross section of a mixed-oxide fuel pellet irradiated at a linear power of 56 kW/m to a burnup of ~27 MWd/kg. The void in the center of the element is clearly visible, as well as the very long grains associated with the columnar region and the large grains of the equiaxed region. The microstructure near the cooler edges of the pin remains unaltered from its fabricated condition; this fuel is called *unrestructured*. It should be noted that the large black traces extending from the central void all the way to the inner cladding wall are cracks that most likely occurred during the cooldown from operating temperature. These cracks were probably not present during most of the fuel lifetime. Small cracks in the unrestructured and equiaxed fuel do appear during irradiation, however, due to stress in the fuel.

Another feature illustrated by Fig. 8.5 is the absence of the gap at the fuel–cladding interface in an irradiated pin. Fuel swelling and cracking cause the gap to disappear after a short irradiation time.

Since restructuring patterns are directly dependent upon temperature and temperature gradients, it is expected that the differing axial temperature distributions in a SFR pin would result in a corresponding axial variation in fuel microstructure. This indeed is the case, as graphically illustrated by Fig. 8.6. The fuel pin shown was irradiated in EBR-II to a burnup of 50 MWd/kg at a peak linear power of 36 kW/m. Even though EBR-II has an active core length of only 34.3 cm, there is sufficient axial temperature variation to produce substantial differences in local fuel microstructure. We shall see in the later safety chapters that such differences play an important role in determining the transient response of the fuel pin to thermal upset conditions.

The microstructural differences described above affect fission gas retention patterns. Tracks left by pore migration within the columnar region allow fission gas to be easily vented to the central void region. Hence, it is of considerable interest to predict where such restructuring takes place, and

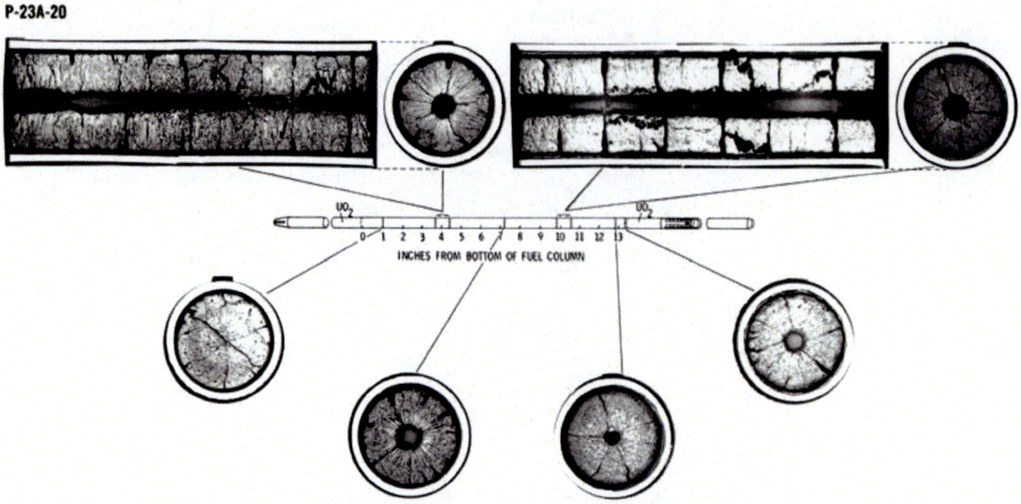

Fig. 8.6 Cutaway vide of microstructure patterns observed in mixed-oxide fuel irradiated in EBR-II. Compliments of: L. A. Lawrence, J. W. Weber, and J. L. Devary, Hanford Engineering Development Laboratory, Richland, WA, 1979

how to assess the position of the boundaries physically separating the various microstructures. This is done by estimating the temperature boundaries that separate (1) the equiaxed from the unrestructured fuel (T_{eq}), and (2) the columnar from the equiaxed fuel (T_{col}). Approximate expressions for these temperatures can be written as (Ref. [7])

$$T_{eq} = \text{Equiaxed Temperature} = \frac{62000}{2.3\log_{10} t + 26}, \tag{8.3}$$

$$T_{col} = \text{Columnar Temperature} = \frac{68400}{2.3\log_{10} t + 28}, \tag{8.4}$$

where

t = time measured in hours,
T = absolute temperature (K).

Figure 8.7 shows the drop in columnar temperature T_{col} as a function of time, as well as the fractional radius of columnar grain growth for three different centerline temperatures T_0. The equiaxed temperature, from Eq. (8.3), lies below the columnar temperature by 58 K at 1 h, and extends to 78 K less at 10,000 h.

Whereas the above formulation contains an explicit time dependence for the restructuring temperatures, other experience developed in the formulation of Ref. [8] suggests that in-pile mixed oxide fuel data can be fit quite adequately by using the following simplified forms, which are time-independent:

$$T_{eq} = 1500\ ^{\circ}\text{C}, \tag{8.3a}$$

$$T_{col} = 1900\ ^{\circ}\text{C} - 5\chi, \tag{8.4a}$$

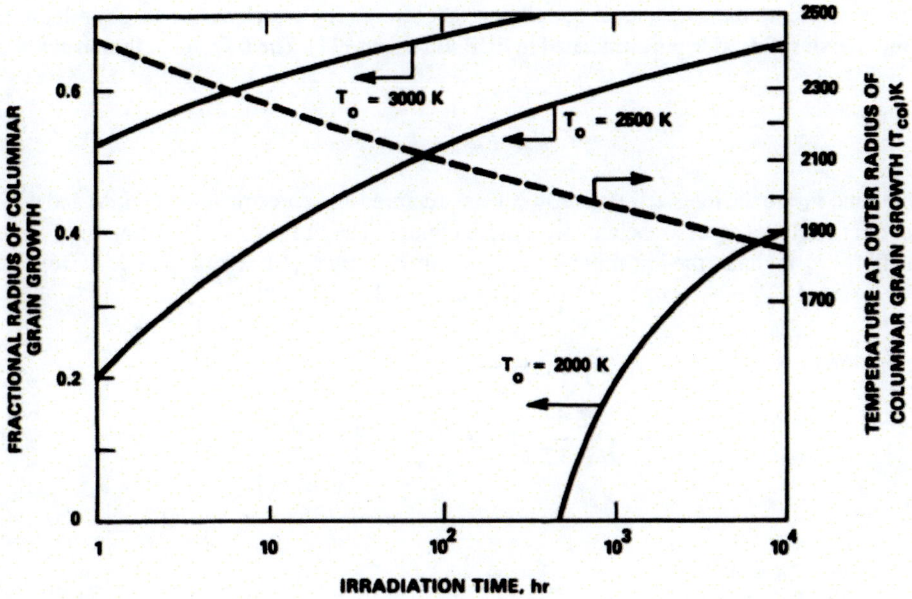

Fig. 8.7 Location and temperature of the outer boundary of the columnar-grain region in pure UO_2 as a function of irradiation time [7]

where χ is the linear power (kW/m). The reason these simplified forms apply is that the fuel-cladding gap closes very early in irradiation, causing a sufficient drop in the fuel pin temperature gradient to offset continued pore migration.

8.2.4 Fission Gas Release and Gas Plenum Length

Fission product gas is produced in the fission process, both directly from fission and from the decay of radionuclides to stable gases, as in the decay of iodine to xenon. Most of the fission gas is xenon; the second most abundant gas is krypton. The yield of stable fission product gas, i.e., the fraction of fission products that become stable gas atoms, for a mixed UO_2-PuO_2 fast reactor is about 0.27.

Much of the fission gas produced in the unrestructured fuel, where temperatures are lowest, is retained in the fuel grains during normal operation. In the equiaxed and columnar grain regions, most of the fission gas escapes to the central void region or through cracks to the fuel-cladding interface. (The gas locked in the unrestructured fuel could be released in the event of fuel melting, a phenomenon that will influence the safety analysis of unprotected transients, as discussed in Chapter 15.)

The processes by which fission gas bubbles nucleate, grow, diffuse, and eventually collect and condense at grain boundaries are extremely complex. A description of these processes is beyond the scope of this text. Suffice it to say that below a temperature of about 1,300 K, fission gas mobility is very low and there is essentially no gas escape. Between 1,300 and 1,900 K, atomic motion allows some diffusion to take place such that an appreciable amount of gas can reach escape surfaces over a long period of time. Above 1,900 K, thermal gradients can drive gas bubbles and pores over distances comparable to grain sizes in days or months. Gas release then occurs when cavities reach a crack or other surface connected directly with free volume. Early attempts to simulate the physics of these processes have been made with the FRAS2 [9] and POROUS [10] computer codes.

A relatively simple mathematical correlation that effectively describes fission gas release from a large number of solid oxide pins irradiated in EBR-II follows [11]. The fraction of fission gas released, F, is

$$F = F_r A_r + F_u A_u, \tag{8.5}$$

where F_r and F_u are the release fractions for the restructured and unrestructured regions and A_r and A_u are the fractional fuel areas associated with these regions. The fit for F_r is a function only of burnup, B. The fit for F_u includes the linear power, χ, in addition to burnup. Useful expressions from Ref. [8] are

$$F_r = 1 - \frac{4.7}{B}\left(1 - e^{-B/5.9}\right), \tag{8.6a}$$

$$F_u = 1 - \frac{25.6}{B - 3.5}\left[1 - e^{-(B/3.5-1)}\right] e^{-0.0125\chi} F'(B), \tag{8.6b}$$

where

$F'(B) = 1 \ (B < 49.2)$
$= e^{-0.3(B-49.2)} \ (B \geq 49.2)$
$B =$ local burnup (MWd/kg)
$\chi =$ local linear power (kW/m)

and F_u is set equal to zero whenever Eq. (8.6b) yields a negative number. A plot of fission gas release vs. burnup and linear power using the model is shown in Fig. 8.8. For the illustration, the 23 kW/m case was assumed to have zero restructuring (i.e., $A_u = 1.0$, $A_r = 0.0$), whereas for the 42.6 kW/m

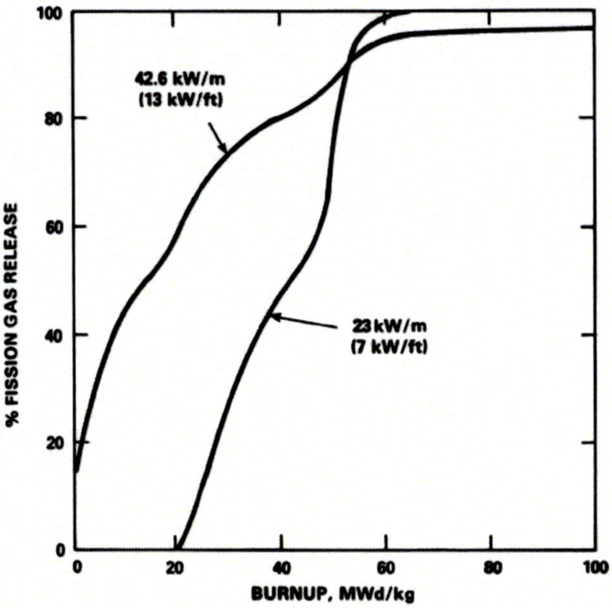

Fig. 8.8 Fission gas release vs. burnup

case the values $A_u = 0.3$ and $A_r = 0.7$ were used. It should be noted that, whereas this figure illustrates the character of the simplified fission gas release model just presented, a determination of the fission gas released from a full fuel pin during irradiation would show a smoother set of curves (wherein the actual restructuring areas associated with an axially varying power distribution are taken into account).

Fission gas, once released from the fuel matrix and vented to open collecting zones (such as the central void or the fuel-to-cladding gap), will not reenter the solid form. The gas is collected in the fission gas plenum above or below the fuel stack. The pressure in the fission gas plenum is normally in approximate equilibrium with pressures throughout the fuel pin. This pressure is manifest in the fuel-cladding gap or in the cracks in the fuel adjacent to the cladding if the fuel located there is well cracked and has little remaining structural integrity. Under these conditions the fission gas plenum pressure provides the loading pressure on the inside of the cladding.

For pin design, a plenum pressure at the end of pin life, P_p, is specified, consistent with the pin failure criteria (as described in Section 8.3). Typical fission gas plenum pressures at discharge are in the range of 6–10 MPa. The next problem to be treated here is to develop a relationship between plenum length, L_p and P_p. A common technique for calculating the plenum pressure is to calculate the volume, V_0 of fission gas at a standard temperature T_0 =273 K and pressure $P_0 = 1$ atm $= 1.013 \times 10^5$ Pa and to use the perfect gas relation

$$\frac{P_p V_p}{T_p} = \frac{P_0 V_0}{T_0}, \tag{8.7}$$

where subscript p refers to plenum conditions at rated power at the end of life.

A new parameter α_0 is defined as

α_0 = volume of fission gas released per cubic meter of fuel at standard temperature and pressure (273 K, 1 atmosphere).

The use of this parameter will, conveniently, introduce the active fuel length, L_f (i.e., the core height) into the analysis so that the plenum length will ultimately be expressed in terms of the core height. The volume V_0 is related to the active fuel volume, V_f as

$$V_0 = \alpha_0 V_f. \tag{8.8}$$

Hence,

$$\frac{P_p V_p}{T_p} = \frac{\alpha_0 V_f P_0}{273}. \tag{8.9}$$

Since the gas plenum is contained within the same cladding as the fuel,

$$\frac{V_p}{V_f} = \frac{L_p}{L_f}$$

so that

$$P_p L_p = \alpha_0 L_f \frac{T_p P_0}{273}. \tag{8.10}$$

The value of α_0 is related to the fission gas release fraction, F, and burnup, B, as follows:

$$\alpha_0 = \frac{Fn\mathbf{R}T_0}{P_0}, \tag{8.11}$$

where

n = kg-mol fission gas produced/m^3 fuel
R = universal gas constant (8,317 J / kg-mol·K).

For an oxide fuel theoretical density of 11 g/cm^3 (= 11,000 kg/mm^3) and a smear density of 85% TD, and for a fission gas yield of 0.27,

$$n = B\frac{\text{MWd}}{\text{kg heavy metal}} \times \frac{2.93\times 10^{16}\dfrac{\text{fissions}}{\text{MWs}}86400\dfrac{\text{s}}{\text{d}}}{6.023\times 10^{26}\dfrac{\text{molecules}}{\text{kg-mol}}} \times (11000)\,(0.85)\,\frac{\text{kg oxide}}{\text{m}^3\text{ fuel}}$$
$$\times\frac{238}{270}\frac{\text{kg heavy metal}}{\text{kg oxide}} \times 0.27\frac{\text{kg-mol fission gas}}{\text{kg-mol fissioned}} = 0.94\times 10^{-2}B.$$

This gives for α_0,

$$\begin{aligned}\alpha_0 &= \frac{\left(0.94\times 10^{-2}\right)(8317)\,(273)}{1.013\times 10^5}FB,\\ &= 0.21\,FB\left(\text{m}^3\text{ fission gas at STP / m}^3\text{ fuel}\right),\end{aligned} \tag{8.12}$$

where B is in MWd/kg, and F can be obtained as a function of the linear power at which the irradiation occurred from Eqs. (8.5) and (8.6).

8.3 Failure Criteria and Stress Analysis for Pin Design

The primary objectives of fuel pin design are to design a pin that will maintain its structural integrity during its lifetime in the reactor and will not expand or swell beyond the geometrical limits required for adequate cooling. A pin is considered to have failed if a breach occurs in the cladding that allows fission products to escape into the primary coolant.

Two approaches that have been used to predict cladding failure are described in this section. These predictions require the calculation of cladding strain and hoop stress. Hence, cladding stress analysis methods are presented after the introduction of the failure criteria.

8.3.1 Strain Limit Approach

A method commonly used to determine cladding failure for early fast reactor designs was the *strain limit* approach. This method was based on the observation that a combination of internal fuel pin gas pressure and/or mechanical interaction between the fuel and the cladding would at some point cause sufficient permanent cladding strain to cause rupture.

Though straightforward in concept, several difficulties exist in attempting to provide pin failure criteria in such simple terms. The principal difficulty is that the actual allowable cladding strain that can be accommodated prior to cladding rupture is a strong function of temperature, fluence, and strain rate. Hence, failure conditions are strongly dependent upon the location of the cladding in the core, as well as steady-state vs. accident conditions. Nonetheless, conservative bounds can be established to envelop worst-case conditions. The cladding strain limits established for the FFTF design provide such an example: 0.2% inelastic strain for steady state at goal exposure and 0.3% inelastic strain for thermal upset conditions. An inelastic strain limit of 0.7% was established as the integrity limit for cladding during emergency conditions.

8.3.2 Cumulative Damage Function

A somewhat more sophisticated guideline for pin lifetime is the *cumulative damage function* (*CDF*), which utilizes the linear *life fraction rule* and includes damage for both steady-state and transient operation. The steady-state portion of the cumulative damage function is based on *time-to-rupture*, t_r (also called time-to-failure) at the operating stress and temperature. If the cladding is subjected to a given stress at a constant temperature, the cladding will rupture at t_r. The life fraction rule assumes that if the cladding is subjected to that stress at that temperature for a short time, δt, the ratio of this duration to the time-to-rupture will produce a damage fraction, D, or the life fraction at that stress and temperature, i.e.,

$$D = \delta t / t_r. \tag{8.13}$$

The fractional damage is assumed to accumulate linearly. During the lifetime of a pin, the cladding may be subjected to a series of different values of stress and temperature for times, δ_i, for which the time-to-failure is t_{ri}. The cumulative damage function for irradiation at varying conditions is

$$CDF = \sum_i \left(\frac{\delta t}{t_r}\right)_i. \tag{8.14}$$

The life fraction rule states that failure will occur when the cumulative damage fraction reaches some value, generally unity, or

$$\sum_i \left(\frac{\delta t}{t_r}\right)_i \simeq 1. \tag{8.15}$$

This rule is entirely empirical, but high-temperature design experience indicates that satisfactory correlations for t_r are often obtained for fast reactors. Additional experience will be necessary before it can be determined whether the effect of irradiation hardening influences the linear addition process generally employed.

The designer must in addition consider two transient conditions: thermal cycling during normal anticipated startups and shutdowns, and off-normal, or accidental, transients. It is possible to treat off-normal transients by a life fraction rule based on fraction of time-to-failure at a given stress level and temperature during the transient. Hence, the $\sum_i \left(\delta t / t_r\right)_i$ term in Eq. (8.15) would include the off-normal transients. The appropriate time-to-failure relations for transients are determined from cladding temperature tests that duplicate the transient thermal conditions.

A proposed method for treating fatigue from thermal cycling is to determine the number of cycles to failure, N_r, for a particular type of cycle, say type j, and to define a cycle-fraction equal to the number of this type of cycle expected during the cladding lifetime, N, to the allowable cycles, or $\left(N / N_r\right)_j$. The cyclic damage fractions could then be accumulated linearly along with the steady-state damage fractions, thus defining a new value for the cumulative damage function and life fraction rule as follows:

$$CDF = \sum_i \left(\frac{\delta t}{t_r}\right)_i + \sum_j \left(\frac{N}{N_r}\right)_j = 1. \tag{8.16}$$

It might be noted that a strain-fraction rule, such as that mentioned in Section 8.3.1 above, could be added to the life fraction rule by adding a third item $\sum(\delta\varepsilon/\varepsilon_r)$, where $\delta\varepsilon$ is the cladding strain increment during a transient and ε_r is the allowable strain to rupture.

8.3.3 Correlations for Time-to-Failure

The time-to-failure must account for stress level, temperature, fluence, and any other mechanism degrading to cladding strength. Whereas it would be desirable to account for such mechanisms in a deterministic fashion, the present state-of-the-art requires t_r to be determined from correlations of stress rupture and transient tests performed on fuel cladding. These correlations may be based on one of several stress-rupture parameters or they may be simply a fit of the time-to-rupture data as a function of temperature, fluence and stress. One of the commonly used correlation parameters is the *Larson-Miller parameter* (LMP),

$$\text{LMP} = T\left(\log_{10} t_r + A\right), \tag{8.17}$$

where

t_r = failure time (hours)
T = temperature of the stressed cladding (K)
A = constant.

Another correlation commonly employed is the *Dorn parameter*, θ, given by

$$\theta = t_r e^{-Q/\mathbf{R}T}, \tag{8.18}$$

where

t_r = failure time (hours)
Q = activation energy (J/kg-mol)
$\mathbf{R}$ = universal gas constant, 8,317 J/kg-mol·K.

For steady-state unirradiated performance of 20% cold worked 316 stainless steel cladding without fuel, the time-to-rupture correlation utilizing the Dorn stress-rupture parameter is given by [12]

$$\ln\theta = A + \frac{1}{\lambda}\ln\,\ln\frac{\sigma^*}{\sigma}, \tag{8.19}$$

where λ^{-1} is the slope of an experimental correlation of $\ln\theta$ vs $\ln\,\ln\sigma^*/\sigma$. Combining Eqs. (8.18) and (8.19) gives

$$\ln t_r = A + \frac{B}{T} + C\ln\,\ln\frac{\sigma^*}{\sigma}, \tag{8.20}$$

where [12]

$A = -42.980$
$B = \frac{Q}{R} = 42{,}020$ (K)
$C = \frac{1}{\lambda} = 9.5325$
$\sigma^* = 930$ MPa
σ = cladding hoop stress (MPa) [See Eq. (8.29)].

Extensive radiation experience with mixed oxide fuel clad in 20% cold worked 316 stainless steel has shown that the fuel pin will fail sooner than predicted with the above correlation, especially at low irradiation temperatures [13]. This fuel pin failure at substantially lower cumulative damage fractions is illustrated in Fig. 8.9. In this figure each data point represents the calculated value of the CDF for an irradiated fuel pin, i.e., calculated from Eqs. (8.20) and (8.15) using hoop stress values from time-dependent pin analysis codes. Some of these pins failed; some did not. At 800°C the fuel pin failures correspond to a calculated damage fraction of nearly one, as should be the case. However, at a lower irradiation temperature, for example, at 600°C, the calculated damage fractions of the failed pins are only about 0.005 instead of 1. This means that there is a further reduction in time-to-failure at these low irradiation temperatures. The effect may be attributed to cladding damage caused by chemical reactions between fission products and the cladding. This phenomenon has been termed the *Fuel Adjacency Effect* (FAE), and has been shown to not occur in other cladding types such as HT9 [14] that was tested in FFTF and planned for use in newer SFR designs, such as S-PRISM as discussed in Appendix B. In order to provide a quantitative basis for relating this observed in-pile fuel failure

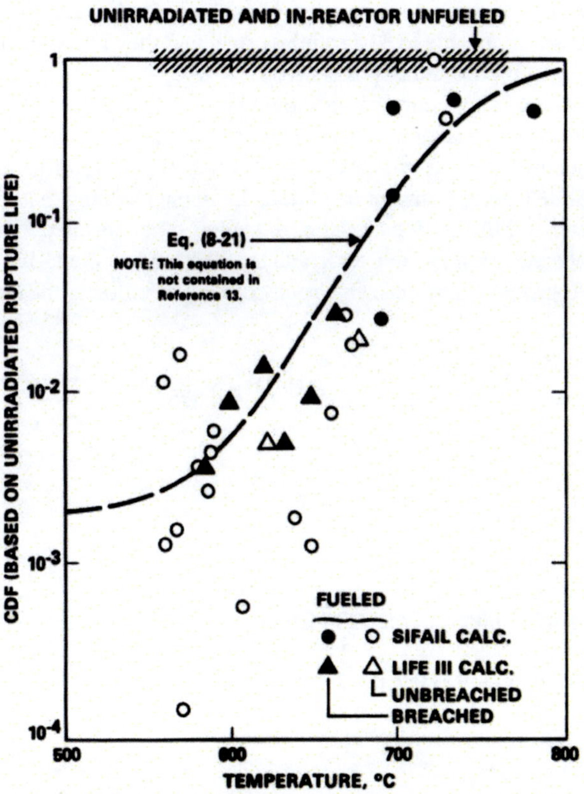

Fig. 8.9 Calculated stress rupture damage fraction vs. cladding temperature for cold worked 316 breached and selected unbreached EBR-II test fuel pins (adapted from Ref. [13])

experience to the CDF methodology, the following expression has been superimposed on the data for 316 stainless steel:

$$\mathrm{FAE} = M + \frac{1}{1 + 10^{(5-\alpha N)}}, \tag{8.21}$$

where

M = minimum CDF at failure $\simeq 2 \times 10^{-3}$,
$\alpha = 0.017$.
$N = T - 450$,
T = temperature (°C).

Therefore, the actual steady-state lifetime may be predicted by multiplying the time-to-failure from Eq. (8.20) and the FAE term from Eq. (8.21) for utilization in the CDF summation Eq. (8.16) as follows:

$$\mathrm{CDF} = \sum_i \left(\frac{\delta t}{\mathrm{FAE} \cdot t_r} \right)_i + \sum_j \left(\frac{N}{N_r} \right)_j = 1. \tag{8.22}$$

The reader should be cautioned that occurrence of the FAE phenomenon appears to occur in some cladding types, such as 316 stainless steel, and not others [14, 15]; hence, the details of this analysis will depend upon the material type. The appearance or lack of appearance of the FAE phenomenon demonstrates the critical need for detailed testing of any proposed cladding material. Nevertheless, the methods presented are applicable to 316 stainless steel and they introduce many of the important concepts in an extremely complex and important field.

For transient analysis the failure correlation should be based on test data acquired for the specific cladding material to be used in-reactor under temperature transient conditions that are similar to those of the reactor transients involved. As with the steady-state fuel pin performance, there is also a Fuel Adjacency Effect exhibited under transient conditions for some cladding materials [16, 17]. Both the Larson-Miller parameter and the Dorn parameter, as well as other correlation approaches, have been used to describe the dependence of fast reactor irradiated 20% cold worked 316 stainless steel cladding on the irradiation and transient conditions. One such correlation utilizing the Dorn parameter is given by [18].

$$\ln t_r = \frac{42800}{T} - 37.4201 \left(\dot{T}\right)^{0.00743} + 8.8754 \ln \ln \frac{\sigma^*}{\sigma} + F \cdot A, \tag{8.23}$$

where

T = temperature (K)
$\dot{T}$ = heating rate (K/s)
σ = cladding hoop stress (MPa)
σ^* = reference cladding hoop stress (930 MPa)

$$F = \left\{ \exp\left[-0.44 \left(\ln \ln \frac{\sigma^*}{\sigma} - 0.2 \right)^2 \right] \right\}^{0.8}$$

$$A = -\left(0.03677 + 1.53366X + 1.79437X^2 - 0.47756X^3 + 0.039902X^4 - 0.001099X^5\right) \Big/ \left(\ln \dot{T}\right)^{0.74}$$

X = neutron fluence, in units of 10^{22} $n/\mathrm{cm}^2 > 0.1$ MeV.

This equation incorporates the influence of the Fuel Adjacency Effect as well as the temperature and stress dependence of cladding failure under transient heating conditions.

8.3.4 Stress Analysis

In order to calculate the cumulative damage fraction, it is necessary to analyze the stress-strain history of the cladding. The analysis involves the time-dependent elastic-plastic response of both fuel and cladding to loading pressures, and must include creep deformations and swelling. The analysis is performed with computer codes in time steps during the irradiation and at various axial positions of the fuel pin. The damage fraction $(\delta t/t_r)$ is calculated at each time step and position, and is added to the previous value of the cumulative damage fraction.

Accurate stress-strain analyses are extremely complex. The most sophisticated US codes for this analysis are the LIFE [1] and PECT [19] codes and their derivatives such as LIFE-METAL [20, 21], that was developed for modeling metal fuel and blanket assemblies and used with the S-PRISM reactor. Simplified codes are also available, such as PECS [22] and SIEX-SIFAIL [8]. These codes perform thermal as well as stress analyses since the two always have to be coupled.

We will present here a simplified model for stress analysis that can be readily understood and used by the student to demonstrate the major principles of design analysis. While this model is somewhat similar to the simplified design codes, especially PECS, the reader must recognize that the model is only a beginning, and that fuel-pin design methods are continually being revised as data are obtained and evaluated.[7]

A flow chart for the simplified stress-strain analysis is shown in Fig. 8.10. It is assumed that the fission gas pressure, P_p is known at the beginning of the time step. At each time step the stresses in the cladding are calculated. These are used later to calculate elastic strains. In addition, inelastic strains from creep and swelling are calculated.

An evaluation of the stresses depends on the nature of the loading on the cladding. As discussed in Section 8.2, the cladding can be loaded either by fission gas or by the fuel cladding mechanical interaction (FCMI) pressure, P_{fc}, of the fuel pushing against the cladding, depending on whether the fuel-cladding gap is closed. Hence, the first step in the calculation is to determine whether the gap is closed.

It is possible for the fuel-cladding gap to be closed but for the fuel to be so cracked that it is incapable of exerting a large force on the cladding. In this case the fission gas will continue to provide the loading on the cladding. For such conditions, only during a rapid startup early in the history of the irradiation will the fuel maintain enough structural integrity for the fuel-cladding interface pressure to be of dominant importance.

8.3.4.1 Calculation of Gap Thickness

It is necessary to calculate the relative positions of the outer fuel surface and the inner cladding surface at rated power in order to determine the hot gap thickness. There are two steps to the calculation.

The first step is to determine the *residual* gap thickness, G_c. This is the cold gap that would exist if the irradiated fuel pin were removed from the reactor and cooled to room temperature. Initially, G_c is

[7] In the United States, most of the early experimental data came from EBR-II, which is a relatively low-flux fast reactor. The FFTF provided a step increase in the US ability to understand cladding behavior. The Europeans—with Phénix, BN-350, and PFR—have been obtaining these types of data since the early 1970s.

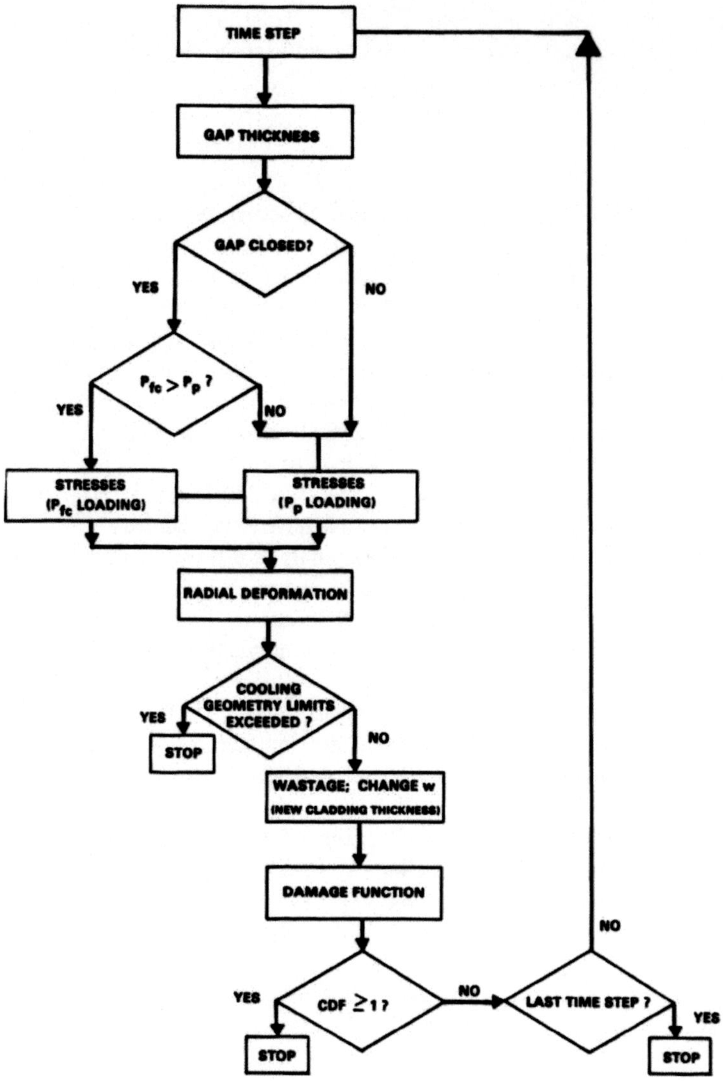

Fig. 8.10 Flow chart of simplified fuel pin stress analysis

simply G_0, the gap in the original fuel as fabricated. In fast reactor design the following correlation is widely used [23]

$$\begin{aligned} G_c = G_0 \{ & 1 - 4.851 \times 10^{-4} \chi\, (\chi - 19.1) \left(1 - e^{-C}\right) \\ & - 0.365 \left[1 - \exp\left(-2.786 \times 10^{-2} \chi B\right)\right] \\ & - 8.81 \times 10^{-5} \tfrac{B}{G_0} \}, \end{aligned} \tag{8.24}$$

where

G_c = residiual gap thickness (mm)
G_0 = fabricated gap thickness (mm)
χ = linear power (kW/m)
B = burnup (MWd/kg)
C = power cycles.[8]

The second step is to calculate the thermal expansion of the fuel and cladding from room temperature up to operating temperatures. The thermal expansion of uncracked UO_2-PuO_2 is used here. If the fuel is cracked, it may close the gap. Letting G refer to the hot gap,

$$G = G_c - \left(\frac{\delta R_f}{R_f} R_f - \frac{\delta R_{ci}}{R_{ci}} R_{ci}\right), \tag{8.25}$$

where R_f and R_{ci} are the cold pellet and inner cladding radii.

The fractional change in the inner cladding radius is obtained using the mean linear thermal expansion coefficient α_c as follows:

$$\frac{\delta R_{ci}}{R_{ci}} = \alpha_c \left(T_c - 298\right), \tag{8.26}$$

where for type 316 stainless steel [24],

$$\alpha_c \left(T\right) = 1.789 \times 10^{-5} + 2.398 \times 10^{-9} T_c + 3.269 \times 10^{-13} T_c^2,$$

and T_c average cladding temperature (K).

The thermal expansion of the fuel surface can be approximated by using a simplified model that neglects the internal forces due to differential thermal growth. Such a model, determined by summing up the thermal expansion of the radial fuel increments, can be mathematically expressed as

$$\frac{\delta R_f}{R_f} = \frac{\int_{R_0}^{R_f} \left[\alpha_f \left(T_f - 273\right)\right] r\, dr}{\int_{R_0}^{R_f} r\, dr}, \tag{8.27}$$

where for 75% UO_2—25% PuO_2 mixed oxide fuel [24], the mean linear thermal expansion coefficient is

$$\alpha_f = 6.8 \times 10^{-6} + 2.9 \times 10^{-9} \left(T_f - 273\right),$$

and T_f = fuel temperature (K), where T_f is a function of r as calculated by methods explained in Chapter 9.

If the calculated value of G is negative, the gap is closed. Since physically the gap cannot be negative, the fuel and cladding must deform to accommodate this situation, and this gives rise to fuel cladding mechanical interaction.

---

[8]One power cycle represents zero to full power and back to zero power. NOTE: If $\chi < 19.1$, the term in Eq. (8.24) containing the power cycle correction should be set to zero.

8.3.4.2 Plenum Gas Loading

When the plenum gas provides the loading on the cladding, the cladding may be conveniently approximated by a *thin-walled* tube *closed* at both ends. The geometry for the stress-strain analysis is shown in Fig. 8.11.

The stresses are calculated from thin-walled vessel theory, for which $w \ll R$ and the difference between the inside cladding radius and the mean cladding radius is ignored. The thin-walled vessel stresses can be obtained from Fig. 8.12.

The hoop stress, σ_θ, is obtained from the following force balance in the tangential (θ) direction

$$(P_i - P_o)\,2RL = \sigma_\theta \cdot 2wL, \tag{8.28}$$

where L is the length of the thin-walled vessel of thickness w. Solving for the hoop stress gives

$$\sigma_\theta = (P_i - P_o)\,\frac{R}{w}. \tag{8.29}$$

For a closed-end cylindrical vessel there is an axial stress σ_z. The geometry is shown in Fig. 8.12b. An axial force balance gives

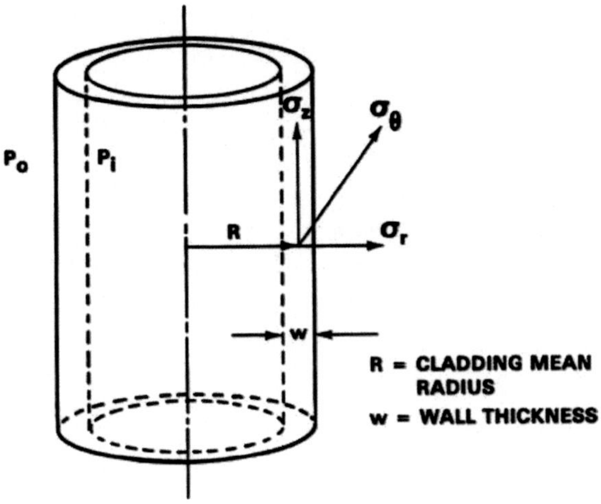

Fig. 8.11 Geometry for cladding stress analysis

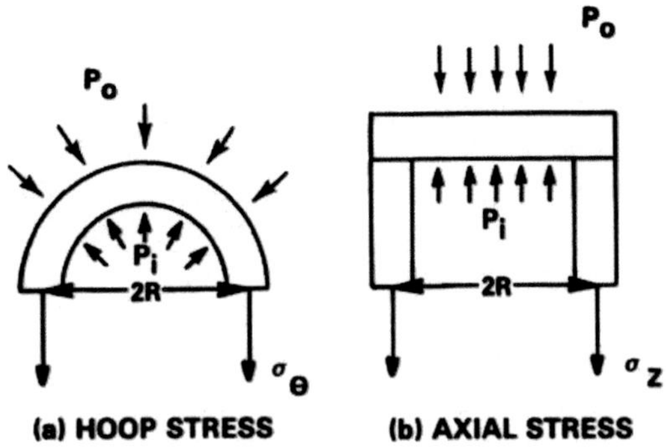

Fig. 8.12 Geometry for hoop stress and axial stress

$$(P_i - P_o) \cdot \pi R^2 = \sigma_z \cdot 2\pi R w,$$

or

$$\sigma_z = (P_i - P_o)\,\frac{R}{2w}. \tag{8.30}$$

This expression for σ_z is equal to $\sigma_\theta/2$ if P_i and P_o each are the same in Eqs. (8.29) and (8.30).

The radial stress, σ_r, is compressive (hence, negative). The maximum value of σ_r equals P_i at the inside of the cladding or P_o at the outside of the cladding. Some analysis codes assume that $\sigma_r = -P_i$; others ignore σ_r altogether, whereas others use the approximation

$$\sigma_r = -\frac{1}{2}\,(P_i - P_o). \tag{8.31}$$

The radial stress is always used with the hoop or axial stress and is much smaller than either; hence, the various approximations for σ_r have little effect on the final results.

For the calculation of deformation with multiaxial stresses, the various stresses combine to give a single *effective stress*, $\bar{\sigma}$, equal to

$$\bar{\sigma} = \frac{1}{\sqrt{2}}\left[(\sigma_\theta - \sigma_r)^2 + (\sigma_r - \sigma_z)^2 + (\sigma_z - \sigma_\theta)^2\right]^{1/2}. \tag{8.32}$$

However, for the determination of time-to-failure in Eqs. (8.20) and (8.23), the maximum principal stress is used (i.e., hoop stress). The effective stress from Eq. (8.32) will be used later in the Prandtl-Reuss flow rule Eq. (8.66).

For plenum gas loading of the cladding,

$$P_i = P_p, \text{ and } P_o = P_b, \tag{8.33}$$

where

P_p = fission gas plenum pressure
P_b = bulk coolant pressure at the axial position of interest.

8.3.4.3 Fuel-Cladding Mechanical Interaction Loading

When the gap is closed and the fuel-cladding mechanical interaction pressure is greater than the fission gas pressure, the interface pressure P_{fc} is the pressure that loads the cladding.

Neglecting the axial stress due to fission gas and fuel growth, the hoop and radial stresses in the cladding are again given by Eqs. (8.29) and (8.31), with

$$P_i = P_{fc}, \text{ and } P_o = P_b. \tag{8.34}$$

Given the assumption of zero axial stress, i.e.

$$\sigma_z = 0, \tag{8.35}$$

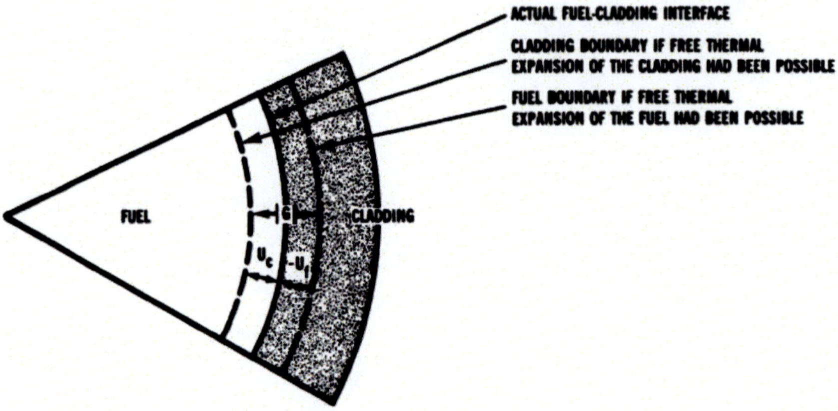

Fig. 8.13 Schematic of elastic fuel and cladding deformations that occur when gap closure leads to interface pressure loading

Equation (8.32) becomes

$$\bar{\sigma} = \frac{1}{\sqrt{2}}\left[(\sigma_\theta - \sigma_r)^2 + \sigma_r^2 + \sigma_\theta^2\right]^{1/2}. \tag{8.36}$$

When the gap computed from Eq. (8.25) is a negative quantity, the fuel applies a pressure $P_i = P_{fc}$ to the cladding. Defining u_c and u_f as the elastic deformations of the cladding and the fuel, as shown in Fig. 8.13, the magnitude $|G|$ is given as

$$|G| = u_c - u_f. \tag{8.37}$$

The fuel displacement u_f is negative (i.e., in the opposite direction from u_c) so that the magnitudes of the two deformations add together. In order to determine P_{fc}, we must express u_c and u_f in terms of P_{fc} and substitute the resulting expressions into Eq. (8.37).

Cladding Deformation (Thin-Walled Cylinder)

In order to obtain u_c in terms of P_{fc}, we write the hoop strain, ε_θ in terms of the hoop stress and radial stress as

$$\varepsilon_\theta = \frac{1}{E_c}\left(\sigma_\theta - \nu_c \sigma_r\right), \tag{8.38}$$

where

E_c = Young's modulus for the cladding
ν_c = Poisson's ratio for the cladding.

The hoop strain is defined as:

$$\varepsilon_\theta = \frac{\delta\,(\text{circumference})}{\text{circumference}} = \frac{\delta R}{R}, \tag{8.39}$$

where R is the radius at the fuel-cladding interface (essentially equal to R_{ci}). Combining Eqs. (8.38) and (8.39), and replacing δR by the radial cladding displacement u_c gives

$$u_c = \varepsilon_\theta R = \frac{R}{E_c}\left(\sigma_\theta - \nu_c \sigma_r\right). \tag{8.40}$$

Substituting Eqs. (8.29) and (8.31) for σ_θ and σ_r, together with Eq. (8.34), yields

$$u_c = \frac{R}{E_c}\left[\left(P_{fc} - P_b\right)\frac{R}{w} + \frac{\nu_c}{2}\left(P_{fc} + P_b\right)\right]. \tag{8.41}$$

Fuel Deformation (Thick-Walled Cylinder)

The next step is to express the fuel displacement, u_f, in terms of P_{fc}. The relation between u_f and P_{fc} involves the use of thick-walled vessel theory because the fuel must be considered as a thick cylinder. Since thick-walled theory may not be familiar to the reader, we shall divert our attention briefly from the flow chart of Fig. 8.10 to a derivation of stresses in a thick-walled cylinder needed to relate u_f to P_{fc}. The derivation appears in various textbooks, e.g., Ref. [25], which we follow closely here.

The geometry is shown in Fig. 8.14. We assume that $\sigma_z = 0$ and the axial strain in the fuel is constant. Using the nomenclature in Fig. 8.14, we can write the following force balance in the tangential direction on a ring of thickness dr (recalling that σ_r is the negative of pressure):

$$2\sigma_\theta dr + 2\sigma_r r = 2\left(\sigma_r + d\sigma_r\right)\left(r + dr\right). \tag{8.42}$$

Neglecting higher order terms gives the equilibrium equation for a thick walled cylinder:

$$r\frac{d\sigma_r}{dr} + \sigma_r - \sigma_\theta = 0. \tag{8.43}$$

Our objective is to solve this equation for the two unknowns, σ_r and σ_θ, in terms of the boundary condition, $P_o = P_{fc}$, and to relate these stresses at the fuel cladding interface to the fuel displacement u_f.

A second equation is needed to evaluate the two unknowns. This equation is supplied by the axial strain, ε_z which is independent of radius and is related to σ_θ and σ_r through Poisson's ratio, ν, and Young's modulus, E, for the thick-walled material:

$$\varepsilon_z = -\frac{\nu}{E}\left(\sigma_\theta + \sigma_r\right). \tag{8.44}$$

The negative sign is needed because σ_θ is a tensile stress and, hence, will be positive and will tend to decrease the axis length, while σ_r is compressive (negative) and will tend to increase the length. The

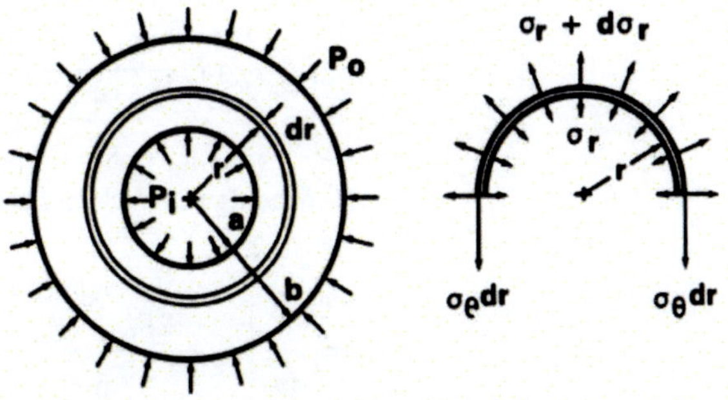

Fig. 8.14 Geometry for a thick-walled cylinder

axial strain, ε_z, is also unknown, but it will be evaluated in terms of the pressure boundary condition P_i at $r = a$ and P_o at $r = b$.

The hoop stress can be eliminated from Eq. (8.43) by means of Eq. (8.44) to give

$$r\frac{d\sigma_r}{dr} + 2\sigma_r = -\frac{E}{\nu_f}\varepsilon_z. \tag{8.45}$$

Multiplying this equation by r gives

$$\frac{d\left(r^2\sigma_r\right)}{dr} = -\frac{E}{\nu_f}\varepsilon_z r. \tag{8.46}$$

Integrating gives

$$\sigma_r = -\frac{E}{\nu_f}\frac{\varepsilon_z}{2} + \frac{C}{r^2}. \tag{8.47}$$

The first term on the right-hand side is a constant. This constant and C are evaluated simultaneously from the two boundary conditions:

$$\begin{aligned} \sigma_r &= -P_i \text{ at } r = a, \\ \sigma_r &= -P_o \text{ at } r = b. \end{aligned} \tag{8.48}$$

Hence,

$$C = \frac{a^2b^2\left(P_o - P_i\right)}{b^2 - a^2}, \tag{8.49}$$

$$-\frac{E}{\nu_f}\frac{\varepsilon_z}{2} = \frac{P_ia^2 - P_ob^2}{b^2 - a^2}. \tag{8.50}$$

From Eqs. (8.44) and (8.47),

$$\sigma_\theta = \frac{P_ia^2 - P_ob^2 - a^2b^2\left(P_o - P_i\right)/r^2}{b^2 - a^2}, \tag{8.51}$$

$$\sigma_r = \frac{P_ia^2 - P_ob^2 + a^2b^2\left(P_o - P_i\right)/r^2}{b^2 - a^2}. \tag{8.52}$$

The radial displacement of the outer surface of the thick-walled cylinder is

$$\begin{aligned} u\left(b\right) = \varepsilon_\theta b &= \frac{b}{E}\left[\sigma_\theta\left(b\right) - \nu\sigma_r\left(b\right)\right] \\ &= \frac{b}{E}\left[P_i\frac{2a^2}{b^2 - a^2} - P_o\left(\frac{b^2 + a^2}{b^2 - a^2} - \nu\right)\right]. \end{aligned} \tag{8.53}$$

Letting $P_i = P_p$, $P_o = P_{fc}$, $a = R_o$ (where R_o is the radius of the central void as shown in Figs. 8.5 and 9.4), and $b = R$ (R being satisfactorily approximated by R_{ci}), the fuel displacement can be obtained from Eq. (8.53) as

$$u_f = \frac{R}{E_f}\left[P_p\frac{2R_0^2}{R^2 - R_0^2} - P_{fc}\left(\frac{R^2 + R_0^2}{R^2 - R_0^2} - \nu_f\right)\right]. \tag{8.54}$$

Determination of P_{fc}

Substituting the expressions for u_f and u_c into Eq. (8.37) gives

$$\begin{aligned}|G| = &\frac{R}{E_c}\left[(P_{fc} - P_b)\frac{R}{w} + \frac{\nu_c}{2}(P_{fc} + P_b)\right] \\ &- \frac{R}{E_f}\left[P_p\frac{2R_0^2}{R^2 - R_0^2} - P_{fc}\left(\frac{R^2 + R_0^2}{R^2 - R_0^2} - \nu_f\right)\right].\end{aligned} \tag{8.55}$$

Solving for P_{fc} gives the desired result

$$P_{fc} = \frac{\dfrac{|G|}{R} + \dfrac{P_p}{E_f}\dfrac{2R_0^2}{R^2 - R_0^2} + \dfrac{P_b}{E_c}\left(\dfrac{R}{w} + \dfrac{\nu_c}{2}\right)}{\dfrac{1}{E_f}\left(\dfrac{R^2 + R_0^2}{R^2 - R_0^2} - \nu_f\right) + \dfrac{1}{E_c}\left(\dfrac{R}{w} + \dfrac{\nu_c}{2}\right)}, \tag{8.56}$$

where $|G|$ is the absolute value of the negative gap calculated by Eq. (8.25). We now have a value for the interface pressure that can be used to calculate cladding stresses when the gap is closed and $P_{fc} > P_p$.

8.3.4.4 Radial Deformation

Cladding

The *radial cladding deformation,* δR_c, is needed in order to determine if the cladding has expanded beyond the geometrical limits required for adequate cooling. [This deformation is different from the radial displacement, u_c, of Eq. (8.41).]

The radial cladding deformation is related to the creep strain, ε_θ^c, and the swelling strain, ε_θ^s, by

$$\delta R_c = R_c\left(\varepsilon_\theta^c + \varepsilon_\theta^s\right), \tag{8.57}$$

where R_c is the mid-wall cladding radius. The creep and swelling strains are calculated from a combination of the Prandtl-Reuss flow rule and correlations for swelling and thermal and irradiation creep. (In some structural codes, e.g., PECS, a radial velocity, or *deformation rate,* $\dot{R}_c$, is calculated from *strain rates,* and deformation is obtained by integration of $\dot{R}_c$.)

As will be discussed in Chapter 11, the swelling of 20% cold worked 316 stainless steel is a function of the fluence and irradiation temperature. Using the relationship for volumetric swelling $\Delta V/V$, given in Chapter 11, the swelling strain is

$$\varepsilon_\theta^s = \frac{1}{3}\frac{\Delta V}{V}. \tag{8.58}$$

The Prandtl-Reuss flow rule evaluates creep strain by means of *equivalent* stresses *and strains* (as described, for example, in Ref. [26]). For the case of interest here, the flow rule relates creep strain in the cladding to calculated stresses [as in Eqs. (8.29), (8.30) and (8.31)], an effective stress [Eq. (8.32) or (8.36)], and an empirical effective inelastic strain, $\bar{\varepsilon}$, as follows[9]:

[9]By comparison with the more complete Prandtl-Reuss flow rules of Eqs. (8.66), (8.67) and (8.68), it is observed that Eqs. (8.59) and (8.60) represent a special case for which only inelastic hoop strains are non-zero.

$$\varepsilon_\theta^c = \frac{\bar{\varepsilon}}{\bar{\sigma}}\left(\sigma_\theta - \frac{\sigma_r + \sigma_z}{2}\right), \tag{8.59}$$

where

$$\bar{\varepsilon} = \frac{2}{3}\left(\varepsilon_{\text{th creep}} + \varepsilon_{\text{ir creep}}\right), \tag{8.60}$$

and $\varepsilon_{th\ creep}$ and $\varepsilon_{ir\ creep}$ are thermal and irradiation creep strains.

Correlations for thermal and irradiation creep strain for 20% cold worked 316 stainless steel cladding are given below. These correlations are different for other material types and tend to change as more data become available. See Section 11.3 for a detailed discussion.

Thermal Creep [27]

$$\varepsilon_{\text{th creep}} = C_1\sigma_\theta \cosh^{-1}(1 + C_2 t) + C_3\sigma_\theta^n t^m + C_4\sigma_\theta^n t^{2.5}, \tag{8.61}$$

where

σ_θ = hoop stress (MPa)
t = time (h)

and the remaining constants (C_1, through C_4) and coefficients (n and m) are supplied by Ref. [27] for the particular temperature ranges of interest.

Irradiation Creep [28]

$$\varepsilon_{\text{ir creep}} = 10^{-6}\sigma_\theta\left[0.67F + 5.8 \times 10^4 e^{-8000/T}(F - 8.5\tanh F/8.5)\right], \tag{8.62}$$

where

σ_θ = hoop stress (MPa)
F = displacements per atom (dpa)
T = temperature (K).

Displacements per atom can be related to neutron fluence above 0.1 MeV for a particular reactor design. In EBR-II, where testing was done to obtain this equation, one dpa corresponds to $\sim 2 \times 10^{21} n/\text{cm}^2$ (above 0.1 MeV).

Fuel

Displacement of the fuel surface is also generally calculated in pin-design codes. One use of this calculation is to determine the amount of porosity required during fuel fabrication. A phenomenological treatment of fuel swelling and fuel thermal and irradiation creep (as, for example, in the LIFE code) is complex and will not be introduced in the present simplified model. It should be noted, however, that fuel deformation is just as important as cladding deformation as a cladding stress relief mechanism. Discussions of fuel swelling and fuel creep are given in Chapter 11.

8.3.4.5 Cladding Wastage

Corrosion of the cladding by sodium and chemical reactions of fuel and fission products with the cladding slowly decrease the thickness of the cladding. This phenomenon is called *wastage*. Wastage is accounted for in the stress-strain analysis through the calculation of the hoop and axial stresses. The reduction in cladding thickness due to wastage is calculated at each time step and the cladding thickness w in the equations for the stresses σ_θ and σ_z is modified accordingly.

8.3.5 *Life Code Approach*

8.3.5.1 Field Equations

The field equations for the stress analysis in the LIFE code [1] are listed below. They apply to both fuel and cladding.

(a) Equilibrium[10]

$$r\frac{d\sigma_r}{dr} + \sigma_r - \sigma_\theta = 0. \tag{8.63}$$

(b) Kinematics

$$\varepsilon_r^T = \frac{\partial u}{\partial r}, \varepsilon_\theta^T = u/r, \varepsilon_z^T = \text{constant}, \tag{8.64}$$

where

ε^T = total strain
u = radial displacement (m).

(c) Constitutive relations

$$\begin{aligned}
\varepsilon_r^T &= \frac{1}{E}\left[\sigma_r - \nu\left(\sigma_\theta + \sigma_z\right)\right] + \alpha T + \varepsilon_r^c + \varepsilon_r^s, \\
\varepsilon_\theta^T &= \frac{1}{E}\left[\sigma_\theta - \nu\left(\sigma_r + \sigma_z\right)\right] + \alpha T + \varepsilon_\theta^c + \varepsilon_\theta^s, \\
\varepsilon_z^T &= \frac{1}{E}\left[\sigma_z - \nu\left(\sigma_r + \sigma_\theta\right)\right] + \alpha T + \varepsilon_z^c + \varepsilon_z^s,
\end{aligned} \tag{8.65}$$

where

ε^c = creep strain
ε^s = swelling strain
$\alpha T = \int \alpha \, dT$, linear thermal expansion.

(d) Prandtl-Reuss flow rule

$$\begin{aligned}
\Delta\varepsilon_r^c &= \frac{\Delta\varepsilon^c}{\bar{\sigma}}\left(\sigma_r - \frac{\sigma_\theta + \sigma_z}{2}\right), \\
\Delta\varepsilon_\theta^c &= \frac{\Delta\varepsilon^c}{\bar{\sigma}}\left(\sigma_\theta - \frac{\sigma_z + \sigma_r}{2}\right), \\
\Delta\varepsilon_z^c &= \frac{\Delta\varepsilon^c}{\bar{\sigma}}\left(\sigma_z - \frac{\sigma_r + \sigma_\theta}{2}\right),
\end{aligned} \tag{8.66}$$

where $\Delta\varepsilon^c$ = change in ε^c in a time step, and $\bar{\sigma}$ is the same as Eq. (8.32)

$$\bar{\sigma} = \frac{1}{\sqrt{2}}\left[(\sigma_r - \sigma_\theta)^2 + (\sigma_r - \sigma_z)^2 + (\sigma_\theta - \sigma_z)^2\right]^{1/2}, \tag{8.67}$$

$$\Delta\varepsilon^c = \frac{\sqrt{2}}{3}\left[\left(\Delta\varepsilon_r^c - \Delta\varepsilon_\theta^c\right)^2 + \left(\Delta\varepsilon_r^c - \Delta\varepsilon_z^c\right)^2 + \left(\Delta\varepsilon_\theta^c - \Delta\varepsilon_z^c\right)^2\right]^{1/2}. \tag{8.68}$$

---

[10] Same as Eq. (8.43).

These equations are solved for several radial nodes in both the fuel and the cladding at various axial nodes. Cracking of fuel is accounted for in the fuel stress analysis. Attaining a solution for the pressure loading when the fuel-cladding gap is closed is considerably more complex than for the simplified model of Section 8.3.4.

8.3.5.2 Example of Calculated Pressure Loadings and Cladding Strains

A sample calculation of long-term fuel-pin deformation, as performed by the LIFE code, is shown in Fig. 8.15. Shown on the graph are the cladding swelling and the inelastic strain (creep) at the core midplane for a peak power SFR fuel pin irradiated at full power to design burnup, together with the cladding loading pressures.

Although the fuel and cladding are not initially in contact, the gap closes soon after operation begins and the fuel cladding mechanical interaction pressure is calculated by the LIFE code to load the cladding to 6 MPa. After 5,000 h, the gap is predicted to open again if the fuel maintained its structural integrity, i.e., did not crack. Beyond about 7,000 h, the plenum gas pressure provides the primary cladding loading. Also beyond this time, the two inelastic strain contributions control the radial deformation of the cladding.

In an actual fuel irradiation, the gap does not reopen as predicted in this calculation. Instead, fuel cracking due to repeated thermal stresses from shutdown and startup, together with fuel swelling, keeps the fuel pressed against the cladding. The fuel has little strength at this stage, however. Hence, the fuel-cladding interface pressure is essentially equal to the fission gas plenum pressure.

The complexity of fuel pin design, first illustrated in Fig. 8.1, is now summarized further in the diagram of Fig. 8.16. By now we have referred to most of the phenomena on the chart, although many will be discussed further in Chapters 9, 10 and 11. The relevant physical processes are shown in the ovals and the observable consequences on the fuel pin are shown in rectangles.

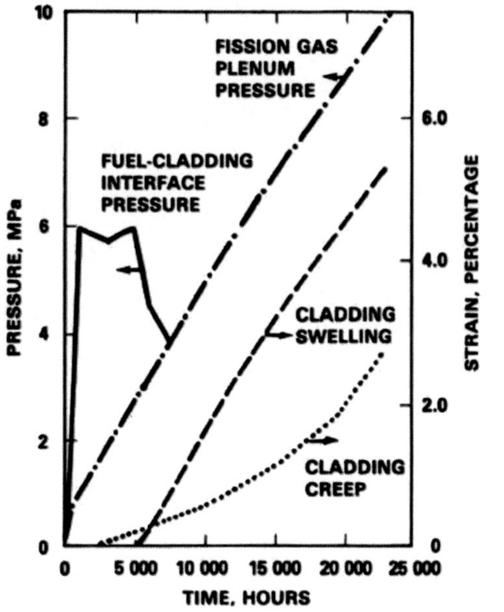

Fig. 8.15 LIFE analysis of a high power SFR fuel pin to design burnup. Compliments of D. S. Dutt, Hanford Engineering Development Laboratory, Richland, WA, 1978

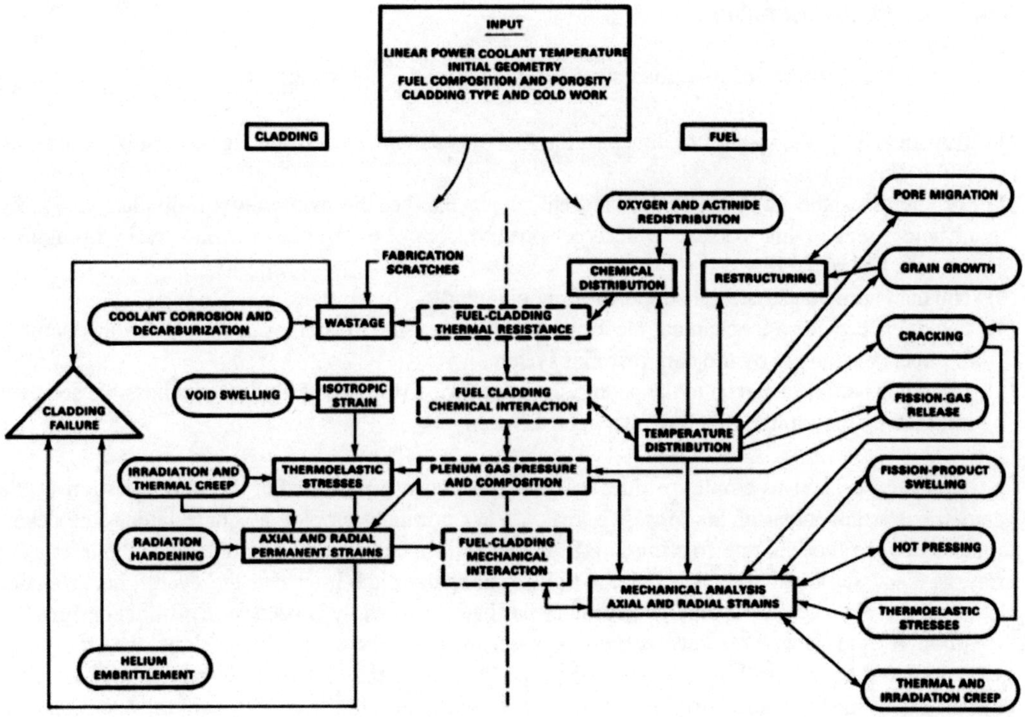

Fig. 8.16 Interrelation of mechanical, metallurgical, and chemical processes in fuel-element irradiation behavior [7]

8.4 Fuel Assembly Design

The method universally adopted[11] for assembling SFR fuel pins into manageable clusters is to collect them into a hexagonal duct, as depicted in Fig. 2.3. These ducts, together with the fuel pins and the associated end hardware, are referred to in the literature as both assemblies and subassemblies. They are assembled into an overall core according to the possible arrangements denoted by Fig. 2.2. Control assemblies are normally dispersed throughout the fueled core region whereas blanket assemblies surround the core.[12] Shielding assemblies are normally included outside the radial blanket. The purpose of the present section is to provide some background on the geometric and material considerations involved in designing such assemblies.

The bulk of the reactor power is generated in the fuel assemblies. A typical homogeneous SFR generates from 85 to 95% of the power in the active core, 3–8% in the radial blanket, and about 3–6% in the axial blanket. Hence, fuel assemblies constitute an area of very high power density, and the geometric arrangement of the pins within an assembly is of considerable importance.

[11] The only exception is Fermi-1, where the pins were enclosed in a square duct.

[12] Blanket assemblies are also included within the core for the heterogeneous design

8.4.1 Duct Considerations

The ducts serve a number of functions in a SFR, including the following:

(1) The ducts force the sodium to flow past the fuel pins and not bypass the high-flow-resistance path within the pin bundle.
(2) By enclosing the flow through an individual pin bundle, the ducts allow individual assembly orificing, thus giving the reactor designer positive control of the power-to-flow ratio throughout the core and radial blanket.
(3) The ducts provide structural support for the pin bundle.
(4) The ducts provide a mechanical means to load the pins into the core as a unit, with the assembly of ducts constrained by the core restraint system.
(5) The ducts provide a barrier to the potential propagation to the rest of the core of a possible accident initiated by the rupture of a few pins in an assembly.

Given the decision to employ a duct in which to house a pin bundle, the next question is the geometrical arrangement of the pins. Whereas LWR's normally employ a square lattice, the desire to maximize the fuel volume fraction in a SFR results in a considerable advantage to the triangular array, as discussed in Chapter 4. Also, the tighter triangular pitch improves the overall heat transfer characteristics. Hence, the hexagonal geometry has been universally chosen for SFR duct designs.

Although rigid ducts have become generally accepted for the early SFR designs, the amount of space occupied by the steel duct is well recognized to reduce the fuel volume fraction and, as such, directly reduce the breeding ratio. Three approaches to minimize the steel content in the duct walls are (1) to use a non-uniform duct wall thickness, (2) to slot the ducts to reduce the pressure differential across the duct wall, and (3) to use thinner duct walls. The first approach employs a mechanical deformation analysis to assess where duct thinning can be accomplished without an undue loss of structural strength. The second idea is to cut slots, as shown in Fig. 8.17, to reduce the internal duct pressure (and, hence, the stress). An axial vent location above the active core region results in a substantial drop for the in-core pressure without allowing coolant crossflow in the high heat generation regions. The crossflow that would occur could be minimized by a duct arrangement shown in the inset

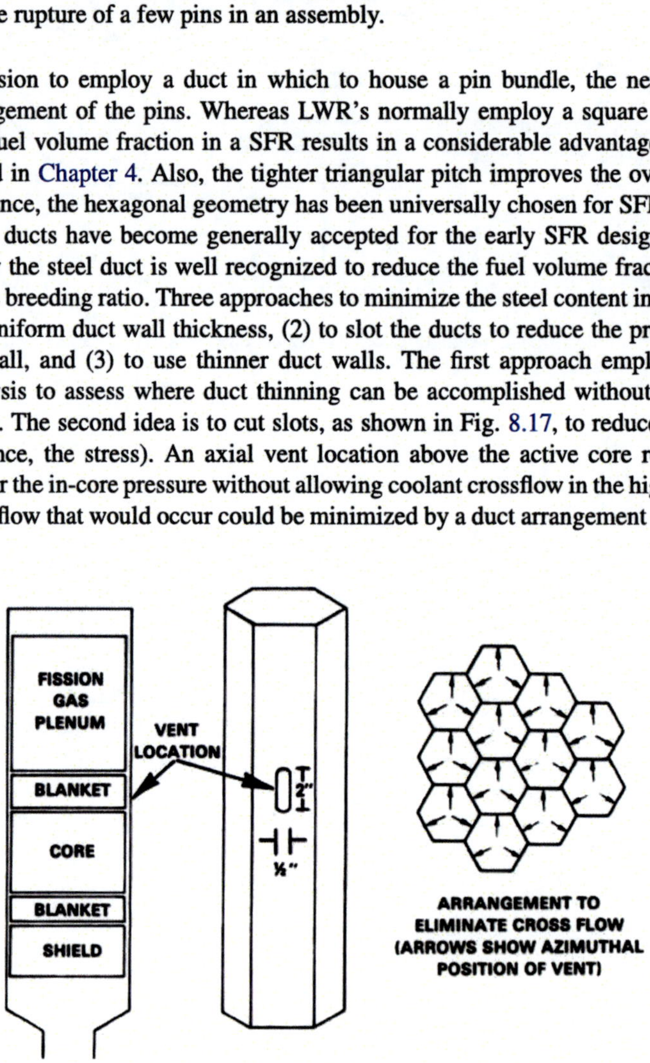

Fig. 8.17 Slotted duct design

to Fig. 8.17. The third approach, to use thinner duct walls, would allow the ducts to touch in the core region during operation. This would necessitate refueling at as low a temperature as practicable; in order to allow thermal contraction to reduce the loads required to withdraw the fuel assemblies.

Efforts to minimize the swelling characteristics of duct material are equally important to that of reducing wall thickness. In fact, it is quite possible that duct swelling may be more of a limitation to core lifetime than material degradation or stress effects in individual pins. The ducts cannot be allowed to swell to the point where they could stick to each other during times of refueling.

8.4.2 Fuel Assembly Size

An important design decision is to determine the number of pins per assembly. Included in the factors for consideration are the following [7]:

(1) Reactivity Worth per Assembly—There is a desire to keep this to a minimum for the peak worth assemblies to minimize the reactivity swing during refueling; an argument exists to keep the maximum worth less than one dollar to assure a mild transient response even in the extremely unlikely event of dropping a fuel assembly into a just critical core.
(2) Decay Heat Removal—Used fuel discharged from the reactor must be cooled, and the total heat removal capability required in the refueling equipment is directly proportional to the number of pins involved for a single movement.
(3) Assembly Weight—The heavier the assembly, the more difficult the handling procedures during fabrication and refueling.
(4) Mechanical Performance—Bowing and dilation problems become magnified with total size.
(5) Cost—Larger assemblies generally result in lower overall core cost.
(6) Refueling Time—Larger assemblies require fewer steps, hence less time, to refuel.
(7) Criticality During Shipment—Safety precautions normally require that the possibility of flooding with water be considered during ex-vessel shipments. Reactivity clearly increases with the size of the assembly.

Current practice is to cluster about 9 or 10 rows of fuel pins, which results in a total assembly size of 217 and 271 pins, respectively. The total number of pins as a function of pin rows is shown in tabular form in Fig. 2.3.

8.4.3 Fuel Pin Spacer Design

The two basic design choices commonly considered for radially separating the pins within an assembly are the use of wire wraps or grid spacers. The prototype and demonstration plants in the United Kingdom and Germany and the US Fermi reactor used grid spacers; other plants use wire wrap. In the wire wrap spacer design, a round steel wire is welded at one axial pin extremity and then spirally wound around the pin at a specified pitch or lead (the axial distance separating the repeat pattern) up to a weld at the other pin extremity. Grid spacers, on the other hand, consist of a cross-lacing of steel webbing, which is normally anchored to the assembly duct wall[13] at specified axial levels. The two principal variations of the grid concept are the staggered grid, wherein the grid mesh is offset at

[13] Some grid designs employ either three or six rods in duct corners to support and space the grids.

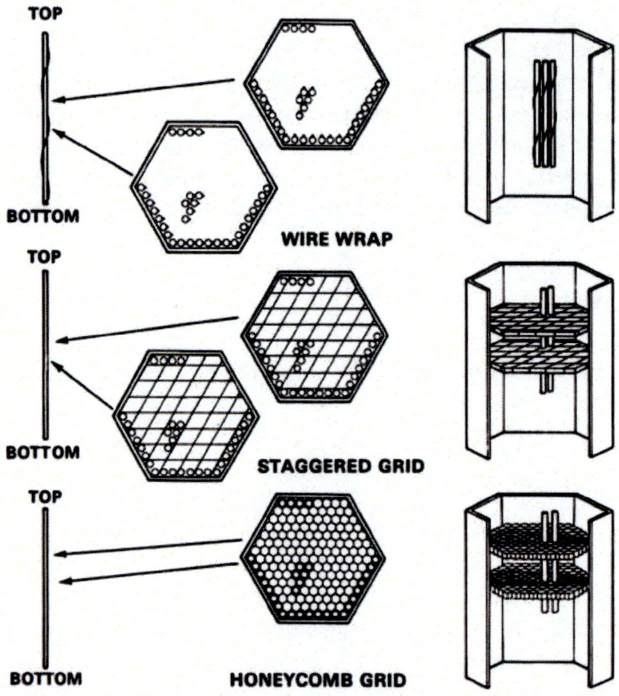

Fig. 8.18 Pictorial intercomparison of three primary pin separation concepts

alternate axial levels, and the honeycomb concept. Figure 8.18 contains a sketch of the wire wrap vs. the two grid pin spacer concepts.

Wire wraps have enjoyed the most popularity, mainly because fabrication is relatively easy and inexpensive. Mechanical vibration problems are minimized because even for a pitch of one foot, wire wrap contact with cladding occurs every two inches (because of the six nearest neighbors-each with their own wire wrap). It is possible, however, for the wire wrap to move slightly during handling. Thus, the precise location of support points for the pins is not known. This adds some uncertainty to the thermal hydraulic and fuel pin bundle structural calculations.

Perhaps the greatest stimulus for considering grid spacers is that such devices can be installed with appreciably less steel volume allocated to the pin spacer component. This potentially allows the fuel volume fraction of the core to be increased, which directly improves the breeding ratio and lowers the overall doubling time. The overall pressure drop tends to be lower[14] since the coolant has a straight-through pass (as opposed to swirling) and the required pump size is correspondingly lower. An additional advantage of the grid spacers is that the hot channel factor (the statistical factor which must be added to nominal temperature calculations to be assured that design limits are not exceeded, see Section 10.4) may be reduced because of reduced thermal hydraulic uncertainties. For a wire wrap design, the wire necessarily contacts the hottest cladding in the core (which occurs just below the top of the core); and local flow reduction, possibly induced by eddy currents behind the point of contact, could aggravate the local hot spot. When employing grid spacers, some design flexibility is available to locate the grids axially away from the position of the peak cladding temperature.

[14] Actual pressure drop is a function of (1) the number of grids required to maintain acceptable bundle deformation grid and (2) the grid design itself.

The principal concern with both grid concepts is the fabrication cost, since carefully aligned physical indentures must be made in the assembly duct wall to axially affix the grids. Perhaps the biggest uncertainty is simply how well such designs will perform for high burnup conditions. Very little long-term experience has been gained to fully assess overall design implications. Also, the safety considerations have yet to be fully assessed. One such concern is that associated with a low probability unprotected transient overpower accident. For a wire wrap design, molten fuel entering the coolant channel could be hydraulically removed from the active core region without mechanical obstruction (other than any blockage that might be formed by molten fuel or cladding refreezing), whereas the cross-lacing of a grid design would form a possible barrier to the flow of particulate material.

8.4.4 Fuel Assembly Length

The overall length of the fuel assembly is determined by the active fuel length, the upper and lower blankets, the fission gas plenum, the orifice-shielding block, the orifice assembly, and the handling socket. The considerations involved in specifying the fission gas plenum length were discussed earlier, and the blanket lengths are fixed from neutronics considerations.

There is not a unique criterion for establishing the length of the active fuel column. Several factors, including pressure drop, coolant temperature rise, and sodium void worth, are considered for determining an approximate height. Fine tuning of the axial core length is then generally accomplished by molding the total core power requirements with average linear heat ratings to arrive at a symmetric core layout.

Both the pressure drop and the rise in coolant temperature increase with fuel bundle length. Hence, there is a general incentive to keep the core height low, resulting in height-to-diameter (H/D) ratios below unity. As shown in Chapter 6, any design parameters that can be adjusted to increase the neutron leakage will reduce the magnitude of the positive reactivity effect resulting from sodium voiding. Hence, there have been attempts to flatten the core (greatly reduced H/D) or modularize it to enhance leakage. A heterogeneous core layout, as discussed briefly in Chapter 2, also increases neutron leakage and reduces the sodium void effect (cf. Chapter 6). Designs tend to have an active core height of about one meter.

Figure 8.19 shows a cutaway of a typical fuel pin assembly for a SFR. Coolant enters the assembly from the coolant inlet plenum via slotted ports that completely surround the assembly nozzle. Such a design is intended to prohibit accidental assembly blockage by allowing for a multitude of multidirectional coolant entry ports. From there, the coolant winds through an orifice-shielding block, sized to effect the degree of pressure drop (hence, flow reduction) desired. It then enters the pin bundle region.

As shown by the enlargement on the right side of the figure, the fuel pins are typically affixed to a lower rail by locking pins. Since the fuel pins are floating free at the top end, it is primarily this locking device that keeps the fuel bundle in position during its useful life.

The coolant then flows up past the lower fission gas plenum (if present below the core), through the lower blanket region, through the active core, past the upper blanket region and on past the upper fission gas plenum (if present above the core). Coolant then collects above the fuel bundle and is ejected into the upper coolant plenum via slots just below the assembly cap, which constitutes a mechanical grapple device for fuel handling purposes. The assembly shown in Fig. 8.19 is, of course, only typical of a SFR assembly. Several variations exist for particular designs, such as for the FFTF in which the axial blankets are replaced by reflectors and the coolant leaves the assembly out the top where it directly traverses an instrumentation tree prior to ultimate ejection into the upper coolant plenum region.

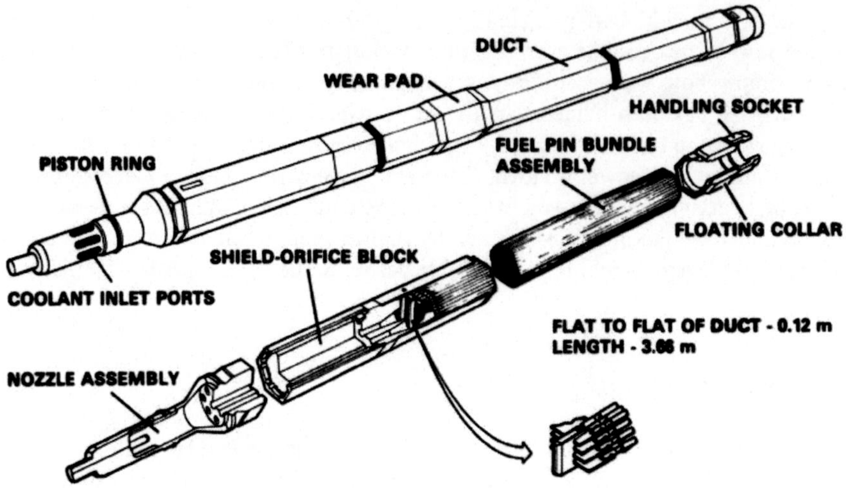

Fig. 8.19 Typical full length SFR assembly

8.5 Other Assemblies

The other principal components that make up a fast reactor core are the blanket, control, and shielding assemblies. The principal features of these key SFR components will be outlined below; a more complete discussion can be found in Ref. [29].

8.5.1 Blanket Assemblies

The primary function of the blanket assemblies is to allow the efficient conversion of fertile fuel (e.g., depleted UO_2) to fissile fuel via capture of stray neutrons. Blanket assemblies located at the radial periphery of the core also provide some shielding for structures beyond the blanket.

From their outward appearance, blanket assemblies look very similar to fuel assemblies. Both assemblies are hexagonally shaped, are the same approximate length, and are normally made of the same material. Since some fissioning occurs within the blanket pins, a fission gas plenum is normally present.

However, there are appreciable differences—in addition to the obvious change in fuel (a pure fertile loading in the blanket rather than a mixture of fertile and fissile fuel). Due both to low fissile fraction and low neutron flux,[15] the volumetric heat generation rate in the blanket fuel is much lower than in the core fuel. Hence, blanket pins can be larger in diameter than core pins while maintaining linear powers, χ, no greater than linear powers in the core. For this reason, and since it is more economical to incorporate the fertile fuel into large fuel pins, the blanket pins are substantially larger than core pins. Blanket pins are typically twice the diameter of fuel pins. As a result of the increased pin size, there are far fewer pins per assembly, and the coolant volume fraction is smaller. Blanket pin diameters are compared with core pin diameters in Table 8.1.

[15]The neutron flux can be quite high for in-core blankets in heterogeneous designs, but the volumetric heat generation rate is still low in comparison to fuel assemblies.

Because of the large pin diameter for blanket assemblies, blanket pins are considerably stiffer than core fuel pins. For blanket assemblies located in the radial periphery of the core, or for in-core positions (heterogeneous design), a pronounced power gradient within an assembly can occur. This combination of large temperature gradients and stiffer structure can lead to the development of higher stresses between pins—resulting in a potentially greater pin/duct interaction and stress problem than found in fuel assemblies.

Blanket assemblies located within the core proper, i.e., characteristic of the heterogeneous core design (Fig. 2.2), do not have the extreme radial power and temperature gradients typical of the radial blanket assemblies, but they tend to suffer more rapid power and temperature changes during normal operations. As a consequence, the lifetime of in-core blanket pins may be more limited by transient conditions than by cladding damage resulting from long-term steady-state irradiation.

8.5.2 Control Assemblies

Control assemblies are required to perform three primary functions: (1) reactivity compensation during the fuel burn cycle, (2) neutronic startup and shutdown capability for normal operations, and (3) rapid shutdown during off-normal conditions.

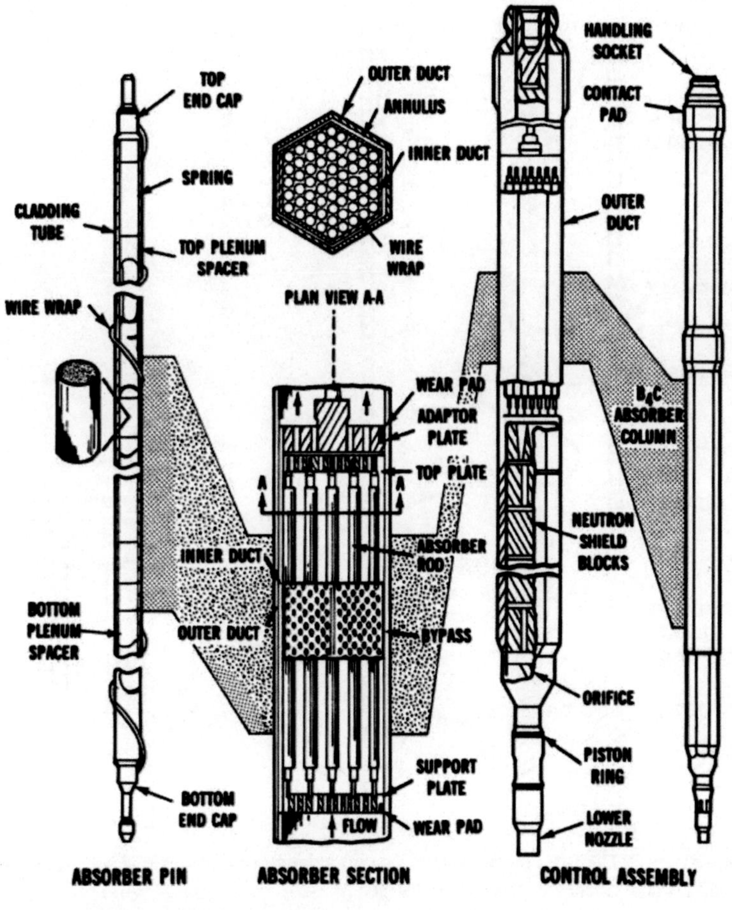

Fig. 8.20 Typical SFR control assembly composition

Whereas several control materials can be employed in a fast reactor (discussed in Chapter 11), boron carbide, B_4C, is the most commonly used substance. During reactor operation, the neutrons absorbed in ^{10}B results in an (n, α) reaction. This process both liberates heat and produces helium gas. Consequently, cooling must be provided for control rods, and a gas plenum is normally incorporated into the design to accommodate the pressure buildup inside the control pins.

The control material is usually fabricated in pellet form and assembled in steel tubing, as illustrated in Fig. 8.20. Pin diameters are typically about twice that of the fuel (i.e., comparable to the blanket pin diameter). The pins are clustered into the usual hexagonal array within a double-walled hexagonal duct assembly. When control rods are inserted or withdrawn, the inner ducted structure moves within the stationary outer duct. In some designs (e.g., Fig. 8.20), the inner hexagonal duct is perforated by thousands of holes, both to allow coolant crossflow within the control pins and to relieve excessive pressure buildup during a scram.

8.5.3 Shielding Assemblies

Radial shielding assemblies surround the core and blanket assemblies for all SFR designs. The principal function of these assemblies is to provide neutron and gamma shielding for the reactor vessel and

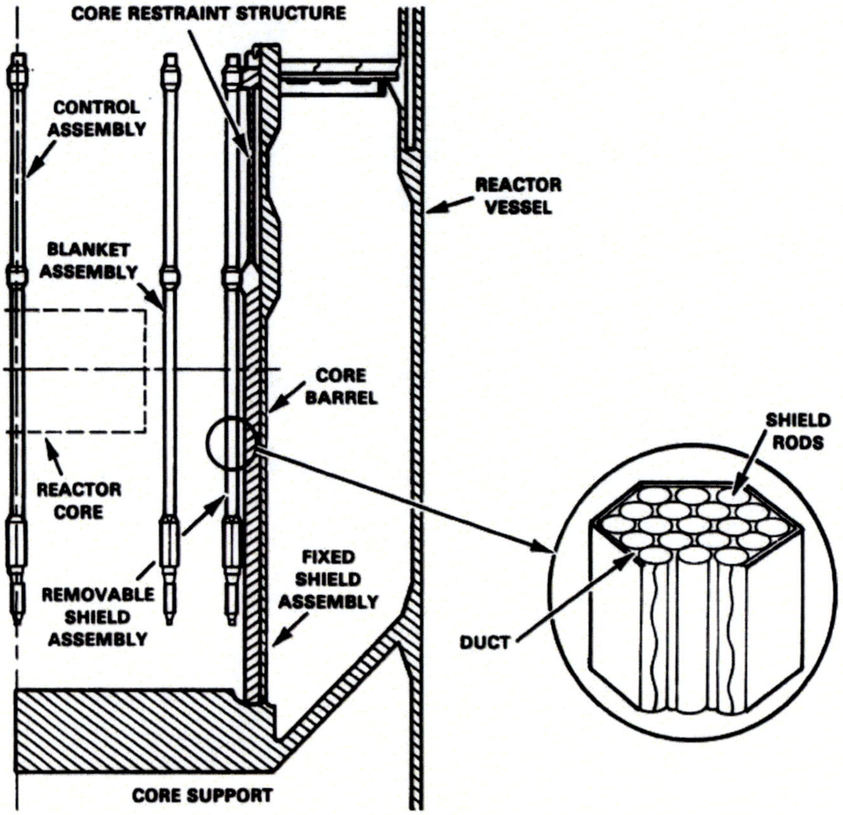

Fig. 8.21 Elevation view of CRBRP reactor system showing relation of removable shielding assemblies to the fixed shield assembly and the core barrel and core restraint structure

the major components within the vessel. An elevation view of the shielding assemblies for the CRBRP design is shown in Fig. 8.21. The assemblies fit into the regular hexagonal lattice with the core and blanket assemblies. They also serve to transmit the load between the core and the core restraint system. Similar removable radial shielding assemblies are used for pool designs, but there are several important differences. The height of the radial shielding is greater for the pool design in order to prevent activation of the secondary sodium in the IHX's, which are located in the reactor tank (e.g., see Fig. 12.7).

Shielding assemblies are normally designed in two types: removable and fixed. The removable assemblies, as the word implies, can be withdrawn and replaced during the core lifetime, whereas the fixed assemblies cannot. The removable positions are situated closest to the core, where the fluence levels are highest. Further discussion of both removable and fixed shielding appears in Section 12.4.

8.6 Clustered Assembly Behavior

Earlier sections of this chapter have been addressed to the principal design features and behavior associated with individual pins and pin assemblies. In this concluding section we turn our attention to duct assembly interaction effects that came into play as fuel, blanket, control and shielding assemblies are brought together to form a reactor. In this context we will discuss duct swelling, duct bowing, and core restraint systems. Duct swelling could have been treated in Section 8.4, in conjunction with fuel assembly design considerations, but it has been placed in the present context because (1) duct swelling is common to all types of ducts in the reactor, and (2) accommodation of duct swelling is an important function of the core restraint system.

8.6.1 Duct Swelling

The same mechanism that causes void swelling in the fuel pin cladding leads to appreciable swelling in the ducts. It is quite probable that assembly duct swelling may limit core lifetime rather than the onset of rupture in individual fuel pins.

Substantial swelling in both the axial and radial directions can take place, as illustrated in Figs. 8.22 and 8.23. Both fluence and temperature are responsible for the spatial problems depicted on the two illustrations. For instance, the relatively smooth drop in the axial elongation pattern noted from Fig. 8.22 clearly follows the fluence level; the peaks in the radial blanket and reflector result simply from the larger exposure time (hence, larger fluence even with lower flux). The magnitudes of the radial dilations can be better appreciated when one recognizes that the 1.3 mm separating the vertical lines in Fig. 8.23 would be quite adequate to satisfy mechanical requirements if there were no swelling. Because of swelling, initial gaps up to 7 mm may be required in this case. If these gaps were not supplied, the core would dilate radially during irradiation and would soon cause unacceptable misalignment of control rods and assembly handling heads. Refueling, which must be done in opaque sodium for a SFR, would then become very difficult.

Such gaps designed into the core to accommodate swelling result in an increase in fissile inventory and an appreciable reduction in the attainable breeding ratio. Hence, there is considerable incentive to reduce the swelling problem. Table 8.2 contains a qualitative assessment of potential design solutions to the duct swelling problem. The development of improved alloys to mitigate void swelling (see Section 11.3.2) would appear the most desirable.

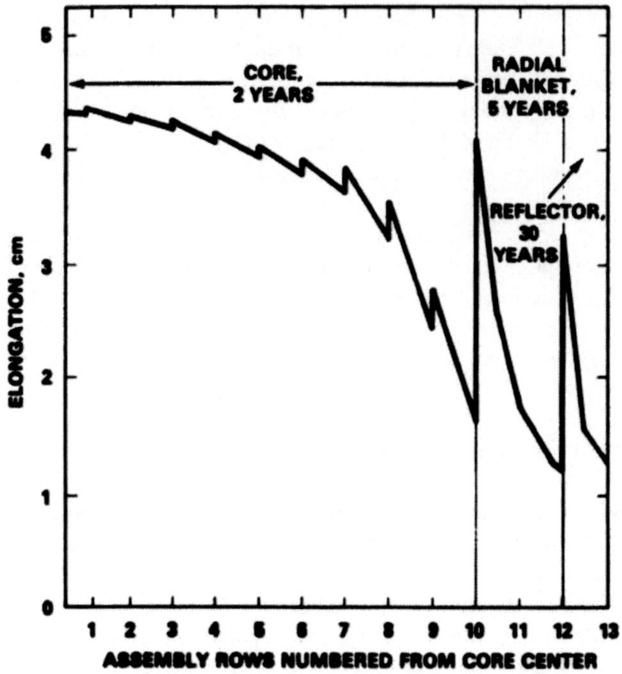

Fig. 8.22 Axial elongation of assembly duct walls due to void swelling at the nominal lifetime of each component [30]

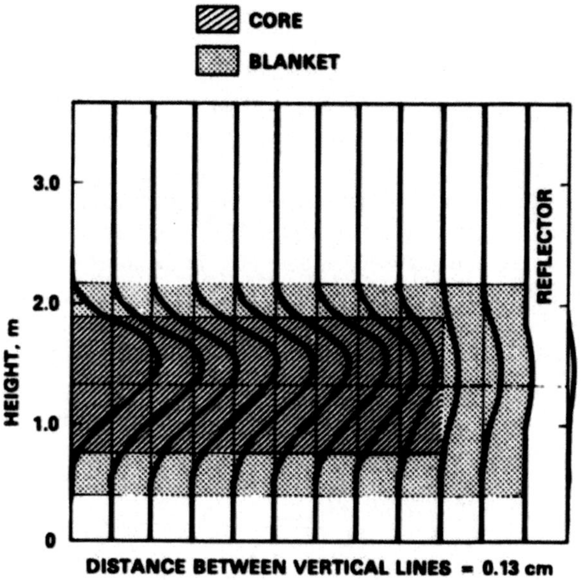

Fig. 8.23 Radial dilation of assembly duct due to void swelling at end-of-life [31]

Table 8.2 Possible design changes to accommodate assembly duct swelling

Design change	Implications
Mechanically open up grid	Lower breeding ratio; increased fissile inventory
Lower burnup	Increased fuel cycle cost
Reduce outlet temperature	Thermal efficiency reduced
Periodically rotate ducts by 180°	More down time
Remove ducts for high temperature annealing	Need more on-site fuel assemblies
Permit significant duct-to-duct contact	Requires refueling at lowest practical temperature
Develop alloys with improved swelling resistance	R&D effort

8.6.2 Duct Bowing

Whereas pure void swelling in ducts leads to a uniform dilation pattern of the type illustrated in Figs. 8.22 and 8.23, there are additional forces that lead to deformation in the duct walls. In particular, thermal expansion, irradiation creep, and mechanical interaction with neighboring assemblies occur simultaneously. The interaction of the above forces in combination with the mechanical tie-down points imposed by the core restraint system (to be discussed below) leads to a phenomenon called bowing.

Because of the complicated interties that can accrue due to the forces outlined, bowing is difficult to predict in a precise fashion. Each core configuration has to be individually assessed. Complex code systems such as CRASIB [32], AXICRP [33], and BOW-V [34] have been developed to analyze such systems.

Figure 8.24 contains a calculational sample for CRASIB, which qualitatively illustrates the bowing pattern emerging from various clamped and unclamped configurations (where clamping occurs at three axial levels: core support structure, top of the core, and top of duct). Figure 8.24 (a) depicts the effect of thermal bowing alone, and (b) represents pure swelling effects. Note that swelling results in a permanent deformation, even for the unclamped, power off condition. Case (c) illustrates the combined effects of thermal and swelling bowing. For the clamped core at power condition, this case results in a sharp duct curvature, which produces large bending stresses in the duct wall. However, these high stress points are effectively relaxed by irradiation creep, as illustrated in case (d). The importance of irradiation creep on eventual bowing patterns is dramatically illustrated by observing cases (c) and (d) for the unclamped, power off condition. Creep is seen to actually change the direction, as well as the magnitude, of duct bowing for unrestrained conditions. This underscores both the importance and complexity of determining duct bowing patterns for a particular reactor of interest.

In addition to the long-term structural problems arising from the bowing phenomena, it is also possible that bowing could cause a net positive power coefficient[16] over certain ranges of reactor operation. Since this generally represents an undesirable performance characteristic, considerable design analysis is normally given to assessing the bowing potential for reactor systems of interest.

[16]The power coefficient is defined as the change in reactivity resulting from an increase in core power, while holding all other variables fixed.

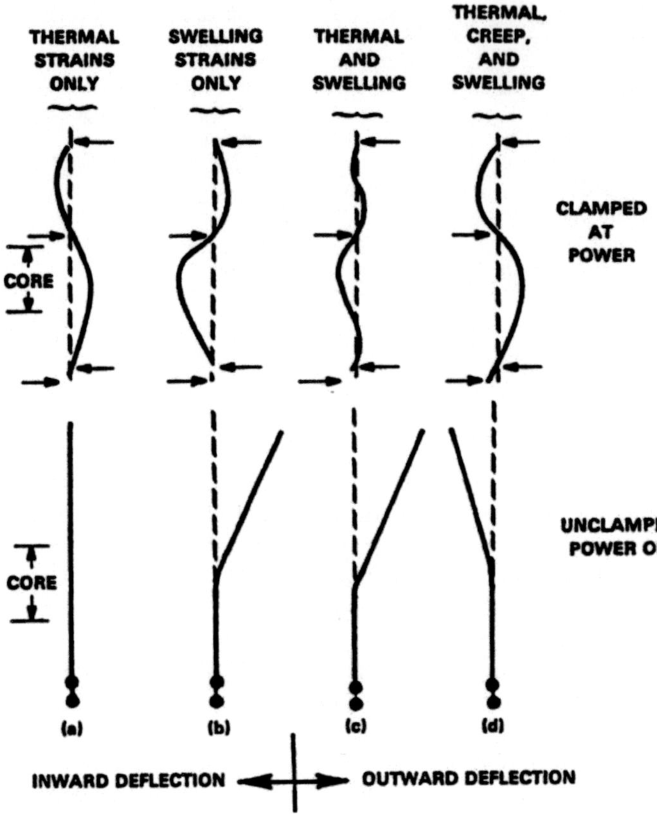

Fig. 8.24 Bowing of fuel assemblies with core clamps fixing the wrappers at the spacer-pad elevation [30]

8.6.3 Core Restraint

Given the requirements (a) to allow clearance between assembly ducts to accommodate swelling, and (b) to constrain the core to resist bowing (due to swelling and thermal gradients), it is necessary to provide a core restraint system. Such a system should fulfill three functions:

(1) provide a calculable and reproducible structural response of the core within the limits imposed by reactivity insertion considerations (during both long-term irradiation or transient conditions),
(2) maintain the tops of the reactor assemblies in a position such that handling heads can be remotely located and grappled by the refueling machine, and
(3) provide clearance for duct insertion and removal, with minimal vertical friction, during shutdown refueling conditions.

Uncertainties in areas such as (a) fast-flux irradiation effects on core structural materials, and (b) fluence and temperature conditions that affect duct bowing have led to several design concepts for core restraint devices. Early prototype systems employed an *active* device, wherein radial yokes could be repositioned to provide varying degrees of lateral force as conditions warranted. For commercial use, however, designers prefer to use *passive* devices, to minimize any maintenance problems that may be associated with hydraulic (or other) systems within the core barrel.

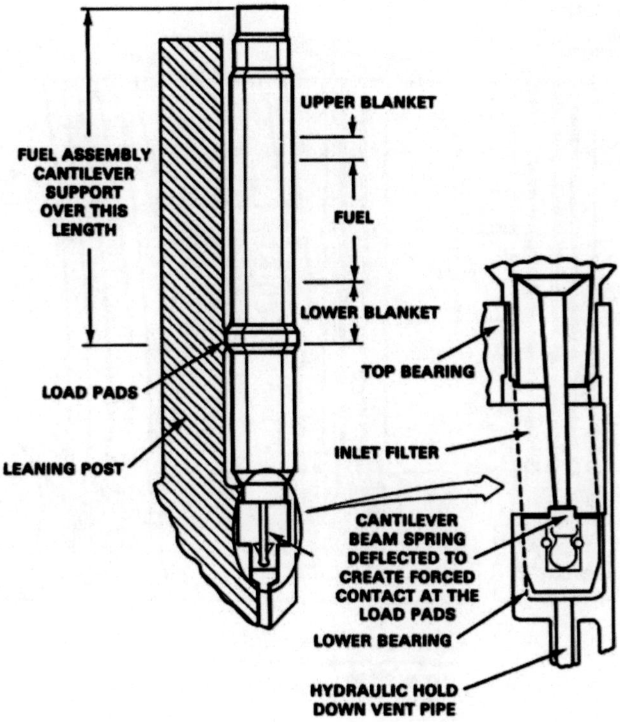

Fig. 8.25 The leaning post core restraint concept [29]

The three passive systems being used or under development are briefly described below [29].

Leaning Post. Figure 8.25 illustrates the leaning post concept, which was used in PFR. Transverse support pads are located in the lower axial blanket region of the fuel assembly. These support pads are forced into contact with a leaning post via a cantilever beam spring in the fuel assembly nozzle, which is forced into a deflected configuration during fuel assembly insertion. The fuel assemblies are supported in groups of six, each centered on a leaning post used as a control assembly. Thus, the large uncertainties associated with core restraint system design are more localized in a cluster of six assemblies, rather than being additive throughout an entire core. Principal limitations of the leaning post concept appear to be in the areas of seismic load capability and impact on core breeding capability.

Free Standing. Figure 8.26 illustrates the free standing core restraint concept, which allows the cantilever-supported fuel and blanket assemblies to bow freely outward until contact is made with shield assemblies at the core periphery. The large number of peripheral shield assemblies act as a soft spring, restraining further outward motion of the core assemblies. Pressed dimple-type spacer pads are located in or just above the active core region. Rapsodie, Phénix and EBR-II used this type of core restraint system, which features simplicity, but there are uncertainties associated with its extrapolation potential to larger cores and higher fluence levels. The free outward movement of assemblies in this restraint concept is considered as a potential contributing cause of decreasing reactivity anomalies [35] observed on four occasions at Phénix in 1989 and 1990.

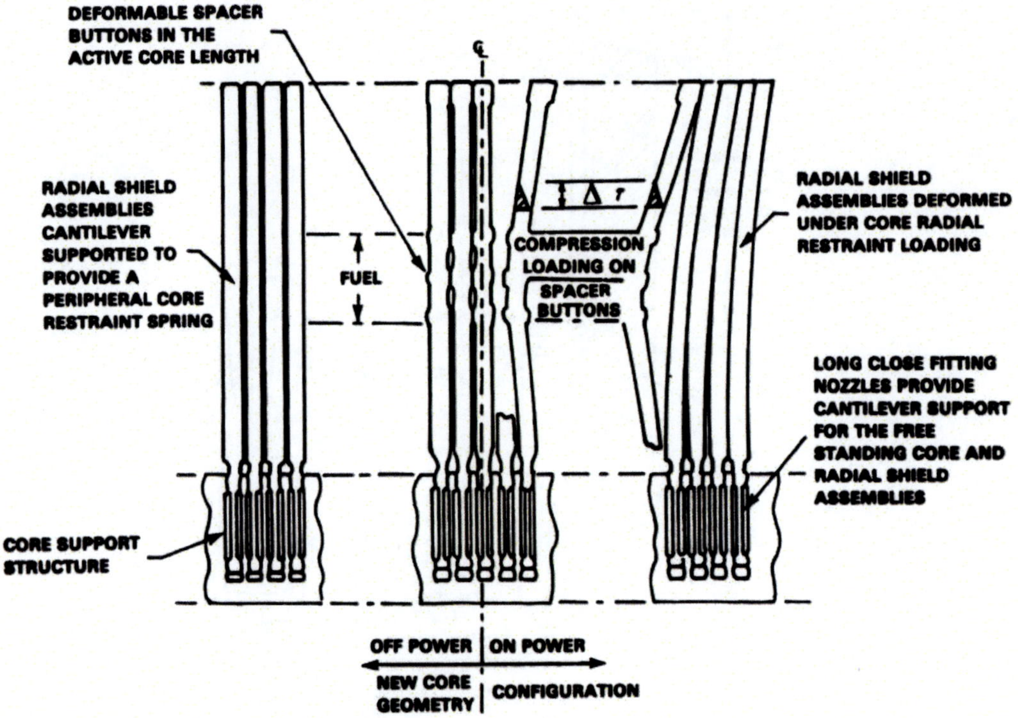

Fig. 8.26 Free-standing core restraint concept [29]

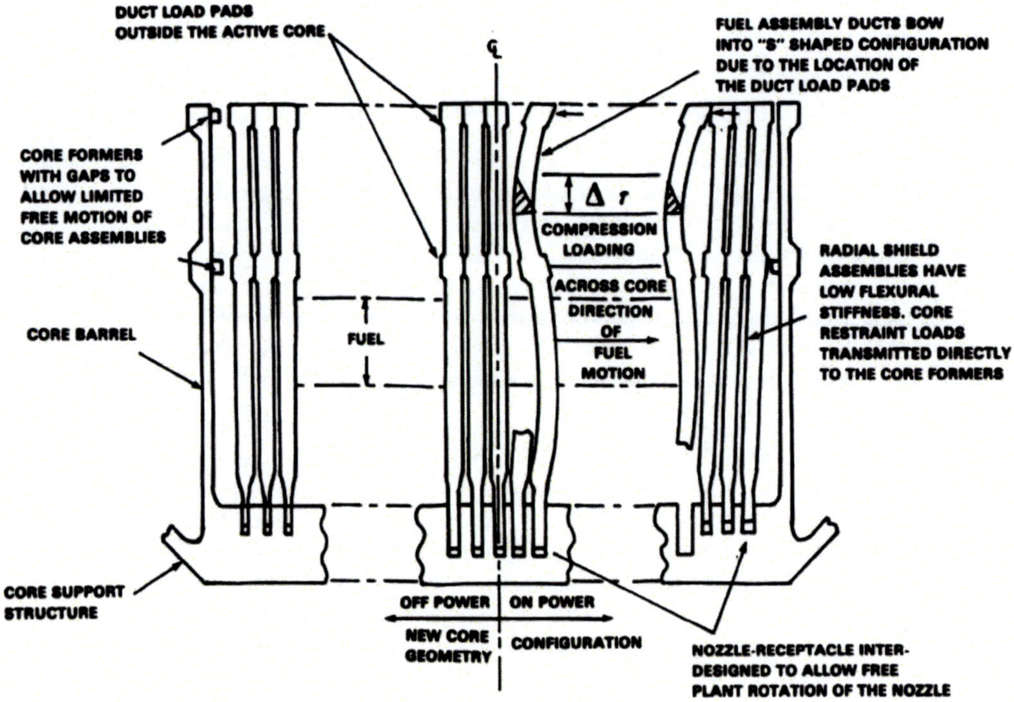

Fig. 8.27 Limited free bow core-restraint concept [29]

Limited Free Bow. Figure 8.27 illustrates the limited free bow core restraint system concept, wherein lateral restraint of the fuel assembly ducts is provided at three locations: one at either end of the assembly and a third in the low swelling region just above the active core. This corresponds to the clamping system depicted in Fig. 8.24. Such a support configuration results in the active core region of the fuel assemblies bowing radially outward from the core centerline in response to the temperature gradients generated in the ducts at power. Whereas this concept appears to have high burnup potential, it has only limited testing for reactor operating conditions. It was designed to accommodate the anticipated effects of irradiation swelling and creep as defined in Chapter 11. The limited free bow core restraint concept was adopted in the FFTF, CRBRP, and S-PRISM design.

References

1. V. Z. Jankus and R. W. Weeks, "LIFE-II – A Computer Analysis of Fast Reactor Fuel-Element Behavior as a Function of Reactor Operating History," *Nucl. Eng. Des.*, *18* (1972) 83–96.
2. *Clinch River Breeder Reactor Plant, Preliminary Safety Analysis Report*, Amendment 51, September 1979, pp. 4.2–4.78.
3. P. Delpeyroux, "Super-Phénix 1 Fuel Element Optimization-First Check of Overall Behavior," *Proc. from Conference on Optimisation of Sodium-Cooled Fast Reactors*, London, November 28–December 1, 1978.
4. G. Karsten, *The Fuel Element of the Sodium Breeder*, EURFNR-1418, Karlsruhe, Germany, October 1976.
5. L. H. Van Vlack, *Elements of Materials Science*, Addison-Wesley Publishing Co., Reading, MA, 1959, p. 99.
6. D. R. O'Boyle, F. L. Brown, and J. E. Sanecki, "Solid Fission Product Behavior in Uranium-Plutonium Oxide Fuel Irradiated in a Fast Neutron Flux," *J. Nucl. Mater.*, *29*, No. 1 (1969) 27.
7. D. R. Olander, *Fundamental Aspects of Nuclear Reactor Fuel Elements*, TID-26711-P1, U.S. ERDA, Springfield, VA, 1976.
8. D. S. Dutt and R. B. Baker, *SIEX-A Correlated Code for the Prediction of Liquid Metal Fast Breeder Reactor (LMFBR) Fuel Thermal Performance*, HEDL-TME 74-75, Hanford Engineering Development Laboratory, Richland, WA, 1975.
9. E. E. Gruber and L.W. Deitrich, "Dispersive Potential of Irradiated Breeder Reactor Fuel During a Thermal Transient," *Trans. ANS*, *27* (1977) 577.
10. J. R. Hofmann and C. C. Meek, "Internal Pressurization in Solid Mixed Oxide Fuel Due to Transient Fission Gas Release," *Nucl. Sci. Eng.*, *64* (1977) 713.
11. D. S. Dutt, D. C. Bullington, R. B. Baker, and L. A. Pember, "A Correlated Fission Gas Release Model for Fast Reactor Fuels," *Trans. ANS*, *15* (1972) 198–199.
12. G. D. Johnson, J. L. Straalsund, and G. L. Wire, "A New Approach to Stress-Rupture Data Correlation," *Mater. Sci. Eng.*, *28* (1977) 69–75.
13. J. Lovell, B. Y. Christensen, and B. A. Chin, "Observations of In-reactor Endurance and Rupture Life for Fueled and Unfueled FTR Cladding," *Trans. ANS*, *32* (1979) 217–218.
14. N. S. Cannon, F. H. Huang, and M. L. Hamilton, "Simulated Transient Behavior of HT9 Cladding," *Proc. from the 14th American Society for Testing and Materials International Conference*, Andover, MA, June 27–29, 1988.
15. N. S. Cannon, C. W. Hunter, K. L. Kear, and M. H. Wood, "PFR Fuel Cladding Transient Test Results and Analysis," *J. Nucl. Mater.*, *139* (1986) 60–69.
16. W. Hunter and G. D. Johnson, "Fuel Adjacency Effects on Fast Reactor Cladding Mechanical Properties," *Int. Conf. Fast Breeder Reactor Fuel Performance,* ANS/AIME, Monterey, CA, March 1979, pp. 478–488.
17. W. Hunter and G. D. Johnson, "Mechanical Properties of Fast Reactor Fuel Cladding for Transient Analysis," *ASTM Symposium on the Effects of Radiation on Structural Materials,* ASTM-STP-611, 1976.
18. G. D. Johnson and C. W. Hunter, "Mechanical Properties of Transient-Tested Irradiated Fast-Reactor Cladding," *Trans. ANS*, *30* (November 1978) 195–196.
19. F. E. Bard, G. L. Fox, D. F. Washburn, and J. E. Hanson, "Analytical Models for Fuel Pin Transient Performance," *Proceedings of International Meeting on Fast Reactor Safety and Related Physics,* CONF-761001, *III,* Chicago, IL, October 1976, US ERDA, 1977, p. 1007.
20. M. C. Bilione, "LIFE-METAL Analysis of U-Pu-Zr Fuel Performance," *89th Annual Meeting Abstracts*, American Ceramics Society, April 26–30, 1989, 15-N-87, p. 282.
21. M. C. Bilione, et al., "Status of Fuel Element Modeling Codes for Metallic Fuels," *Proceedings of the American Nuclear Society, International Conference on Reliable Fuels for Liquid Metal Reactors*, Tucson, AZ, September 7–11, 1986, pp. 5-77–5-92.

22. W. S. Lovejoy, M. R. Patel, D. G. Hoover, and F. J. Krommenhock, *PECS III: Probabilistic Evaluation of Cladding Lifetime in LMFBR Fuel Pins,* GEFR-00256, General Electric Co., Sunnyvale, CA, 1977.
23. D. S. Dutt, R. B. Baker, and S. A. Chastain, "Modeling of the Fuel Cladding Postirradiation Gap in Mixed-Oxide Fuel Pins," *Trans. ANS, 17* (1973) 175.
24. L. Leibowitz, E. C. Chang, M. G. Chasanov, R. L. Gibby, C. Kim, A. C. Millunzi, and D. Stahl, *Properties for LMFBR Safety Analysis*, ANL-CEN-RS-76-1, Argonne National Laboratory, Argonne, IL, March 1976.
25. J. E. Shigley, *Mechanical Engineering Design*, 2nd ed., McGraw-Hill Co., New York, NY, 1972, pp. 73–76.
26. A. Mendelson, *Plasticity: Theory and Applications*, MacMillan Co., New York, NY, 1968.
27. E. R. Gilbert and L. D. Blackburn, "Creep Deformation of 20% Cold Worked Type 316 Stainless Steel," *J. Eng. Mater. Technol., 99* (1977) 168–180.
28. E. R. Gilbert and J. F. Bates, "Dependence of Irradiation Creep on Temperature and Atom Displacements in 20% Cold Worked Type 316 Stainless Steel," *J. Nucl. Mater., 65* (1977) 204–209.
29. Y. S. Tang, R. D. Coffield, Jr., and R. A. Markley, *Thermal Analysis of Liquid Metal Fast Breeder Reactors,* American Nuclear Society, La Grange Park, IL, 1978, pp. 19–21, 47.
30. P. R. Huebotter, "Effects of Metal Swelling and Creep on Fast Reactor Design and Performance," *Reactor Technology, 15* (1972) 164.
31. P. R. Huebotter and T. R. Bump, "Implications of Metal Swelling in Fast Reactor Design." *Radiation-Induced Voids in Metals*, CONF-7 10601, June 9, 1971, AEC Symposium Series, No. 26, J. W. Corbett and L. C. Ianniello, eds., New York, NY, pp. 84–124.
32. W. H. Sutherland and V. B. Watwood, Jr., *Creep Analysis of Statistically Indeterminate Beams.* BNWL-1362, Battelle Northwest Laboratory, Richland, WA, 1970.
33. W. H. Sutherland, "AXICRP-Finite Element Computer Code for Creep Analysis of Plane Stress, Plane Strain and Axisymmetric Bodies," *Nucl. Eng. Des., 11* (1970) 269.
34. D. A. Kucera and D. Mohr, *BOW-V: A CDC-3600 Program to Calculate the Equilibrium Configurations of a Thermally-Bowed Reactor Core*, ANL/EBR-014, Argonne National Laboratory, Argonne, IL, 1970.
35. J. F. Sauvage, *Phènix-30 Years of History: The Heart of a Reactor*, Section 4.5, International Atomic Energy Agency, 2004. Available online at http://www.iaea.org/inisnkm/nkm/aws/fnss/phenix/book/index.html.

Chapter 9
Fuel Pin Thermal Performance

Alan Waltar and Donald Todd

9.1 Introduction

Fast reactor design requires the simultaneous application of mechanical and thermal-hydraulics analysis methods. We reviewed some aspects of mechanical analysis in Chapter 8; we will discuss mechanical design further in Chapter 12. Chapters 9 and 10 will deal with thermal-hydraulics analysis. In the present chapter we will investigate methods of determining temperature distributions within fuel pins. We will then extend these methods to assembly and core-wide temperature distributions in Chapter 10.

While the actual analysis of fuel pin thermal performance depends strongly on the material, the mathematics and concepts associated with determining temperature distributions within fuel pins is essentially independent of the material in a fast reactor. Throughout this chapter, the discussions of correlations and closure relationships necessary to complete the mathematical analyses, and to introduce important concepts, are generally based on oxide fuel with notes where other fuel types show different behavior. The behavior of carbide, nitride, and metallic fuel types are discussed from a materials perspective in detail in Section 11.2 in comparison to the oxide fuel.

There are several ways to approach thermal-hydraulics analysis. We could begin with coolant inlet flow and temperature conditions, work up through the core axially, and then go back to determine the pin temperatures for each axial level, working radially inward. This is the sequence followed by most thermal-hydraulic computational schemes and is the one we will follow in discussing hot channel factor applications in Chapter 10. For purposes of instruction, however, we have chosen to begin with the temperature distribution in the fuel itself, and then work radially out to the coolant. This allows us to begin with the heat source and physically follow the flow of heat in a decreasing temperature gradient through the system's several heat transfer barriers outward to the heat sink.

This chapter deals with steady-state fuel pin thermal performance, but a reactor designer is also interested in transient fuel pin behavior—both for anticipated transients (e.g., normal adjustments in power/flow operations and scrams) and off-normal or accident conditions. The same basic techniques apply to the analysis of transient conditions as to steady state, but the need to understand the interaction of a high-temperature environment with materials properties becomes more acute. These concerns are discussed in the chapters on core materials (Chapter 11) and on safety (Chapters 13, 14, 15 and 16).

Only liquid metal coolants are considered in the discussion of hydraulics and coolant heat transfer in Chapters 9 and 10. Discussion of the thermal hydraulics of gas-cooled fast reactors is briefly discussed in Chapter 17. A more detailed presentation on LMR thermal hydraulics, including the

---

A. Waltar (✉)
Pacific Northwest National Laboratory (PNNL), Retired, Richland, WA, USA
e-mail: alan.waltar@wildblue.net

A.E. Waltar et al. (eds.), *Fast Spectrum Reactors*, DOI 10.1007/978-1-4419-9572-8_9,

methods developed during the design of the FFTF and CRBRP, appears in the book by Tang, Coffield, and Markley, *Thermal Analysis of Liquid-Metal Fast Breeder Reactors* [1].

Information presented in this chapter relies heavily on the experience drawn from FFTF and CRBRP. This experience provides a wealth of information to introduce the key ideas and concepts that a student must consider in fuel pin thermal performance. However, the state of knowledge is continually expanding. Where appropriate, improvements in knowledge beyond the FFTF and CRBRP experience are summarized.

9.2 Fuel and Cladding Thermal Analysis

9.2.1 Radial Fuel Temperatures

The steady-state temperature distribution in cylindrical rods with an internal heat source is obtained from the heat conduction equation, which can be derived with the aid of Fig. 9.1. We assume that the heat source in the fuel is uniform, i.e., there is no neutronic spatial self-shielding within the pin.[1]

Neglecting heat transfer in the axial direction,[2] a one-dimensional heat balance on the cylindrical element dr gives, per unit length of the cylinder,

$$-2\pi rk\frac{dT}{dr} + Q2\pi rdr = -2\pi\left[rk\frac{dT}{dr} + \frac{d}{dr}\left(rk\frac{dT}{dr}\right)dr\right],$$

where

T = temperature (°C)
k = thermal conductivity (W/m·°C)
Q = uniform volumetric heat source (W/m^3)

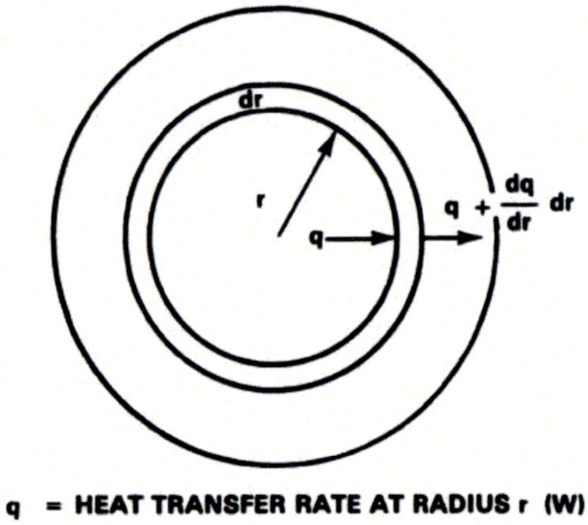

Fig. 9.1 Geometry for the cylindrical heat conduction equation

[1]Neglect of neutronics spatial self-shielding is an excellent assumption for fast reactors. The spatial variation of the heat source is complicated by restructuring of the fuel, however, as in Eq. (9.9).

[2]This is an acceptable approximation in most cases, since the radial temperature gradient is much greater than the gradient in the axial direction.

Simplifying gives the steady-state heat conduction equation:

$$\frac{1}{r}\frac{d}{dr}\left(rk\frac{dT}{dr}\right)+Q=0. \tag{9.1}$$

Note that the thermal conductivity remains in the argument for the derivative because thermal conductivity is a strong function of temperature and, hence, of radius.

The two boundary conditions needed to solve Eq. (9.1) are

(1) $\frac{dT}{dr}=0$ at $r=0$,

(2) $T=T_s$ at $r=R_f$,

where

R_f = radius at the outer surface of the fuel
T_s = temperature at the outer surface of the fuel.

Integrating Eq. (9.1) yields

$$rk\frac{dT}{dr}+Q\frac{r^2}{2}=C_1.$$

From boundary condition 1, $C_1=0$. We incorporate the second boundary condition directly into the second integration to give

$$\int_{T(r)}^{T_s} kdT+\frac{Q}{2}\int_r^{R_f} r\,dr=0.$$

Reversing the limits on the first integral, the result is

$$\int_{T_s}^{T(r)} kdT=\frac{Q}{4}\left(R_f^2-r^2\right). \tag{9.2}$$

Given the surface temperature T_s and thermal conductivity as a function of temperature, the fuel temperature at any radius r can be determined from the above expression.[3]

If we are interested only in the fuel centerline temperature T_{CL}, then $r=0$ and we have

$$\int_{T_s}^{T_{CL}} k\,dT=\frac{QR_f^2}{4}. \tag{9.3}$$

We can now evaluate the linear power, or linear heat rate, χ, originally introduced in Chapter 2, Eq. (2.2). Linear power and volumetric heat source are related by

$$\chi=Q\pi R_f^2. \tag{9.4}$$

[3] Alternatively, Eq. (9.2) can be obtained by defining; $\theta=\int_{T_s}^{T(r)} k\,dT$ so that $d\theta/dr=kdT/dr$. Equation (9.1) can then be written as $d^2\theta/dr^2+(1/r)(d\theta/dr)+Q=0$. The solution of this equation is $\theta(r)=-Qr^2/4+C\ln r$.

From Eqs. (9.3) and (9.4), one obtains

$$\chi = 4\pi \int_{T_s}^{T_{CL}} k\, dT. \tag{9.5}$$

We shall use the most logical SI unit for χ, kW/m. Units also used frequently for χ are W/cm and, in the U.S. literature, the hybrid unit kW/ft. Values for linear power in fast spectrum reactors were given in Table 8.1.

One method to evaluate $\int k\, dT$ is to use an analytical expression for k. A plot of k vs. T is given in Fig. 9.2 for mixed oxide fuel (80% U, 20% Pu) at 95% theoretical density and for an O/M ratio of 2.00, using the expression [2]

$$k = \left(0.042 + 2.71 \times 10^{-4} T\right)^{-1} + 6.9 \times 10^{-11} T^3 \quad (95\%\ TD), \tag{9.6}$$

where k is in W/m·K and T is in Kelvin.

The 95% *TD* case was chosen since that is the porosity of mixed oxide fuel typically employed in typical oxide fueled SFR reactor fuel systems. Porosity is the fraction of fuel pellet volume that is void. For fuel with porosity different from 5%, the thermal conductivity can be obtained by modifying the 95% *TD* value according to

$$k_p = k \frac{1 - 2.5p}{0.875} \text{ if } p \leq 0.1, \tag{9.7a}$$

$$k_p = k \frac{1 - p}{0.875\,(1 + 2p)} \text{ if } p > 0.1, \tag{9.7b}$$

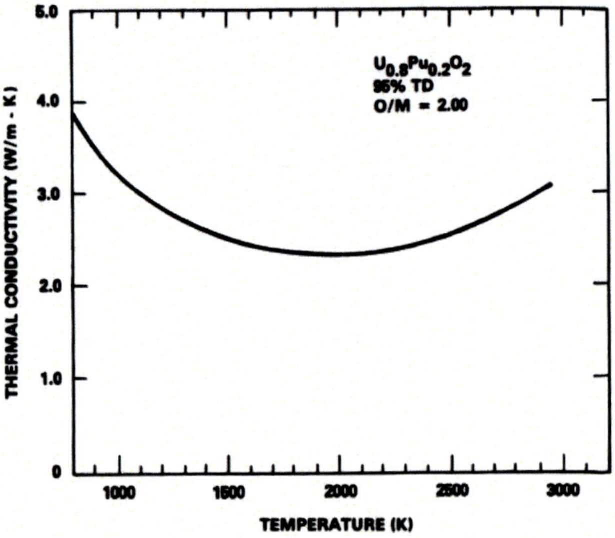

Fig. 9.2 Thermal conductivity for 80% U, 20% Pu-mixed oxide fuel [2]

where k_p = thermal conductivity of fuel with porosity p and k is determined from Eq. (9.6) for 95% TD (or 5% porosity).[4]

Another way to evaluate center temperature, given linear power and surface temperature, is to plot $\int_{T_{ref}}^{T} k_{ref} dT$ against fuel temperature T, where k_{ref} is the conductivity for a specified (or reference) fuel density or porosity, and T_{ref} is an arbitrary reference temperature. Such a plot is shown in Fig. 9.3 for mixed oxide fuel at 95% theoretical density. The first step is to find $\int_{T_S}^{T_{CL}} kdT$ from the linear power [by means of Eq. (9.5)]. If the actual density of the fuel is the same as that for which $\int k_{ref} dT$ has been provided, we start with the pin surface temperature and obtain the center temperature by using the calculated $\int_{T_S}^{T_{CL}} kdT$ as the increment in $\int kdT$ between T_s, and T_{CL} on the graph, i.e.,

$$\int_{T_S}^{T_{CL}} k_{ref}\, dT = \int_{T_{ref}}^{T_{CL}} k_{ref}\, dT - \int_{T_{ref}}^{T_s} k_{ref}\, dT. \tag{9.8}$$

(Conversely, if we know the surface temperature and the fuel melting temperature, we can use the graph to determine the linear power that would lead to centerline melting.)

If the density of the fuel is different from the value for which $\int k_{ref} dT$ has been provided, it is necessary to modify the increment in $\int k\, dT$ from the graph. Let k_p represent the fuel conductivity

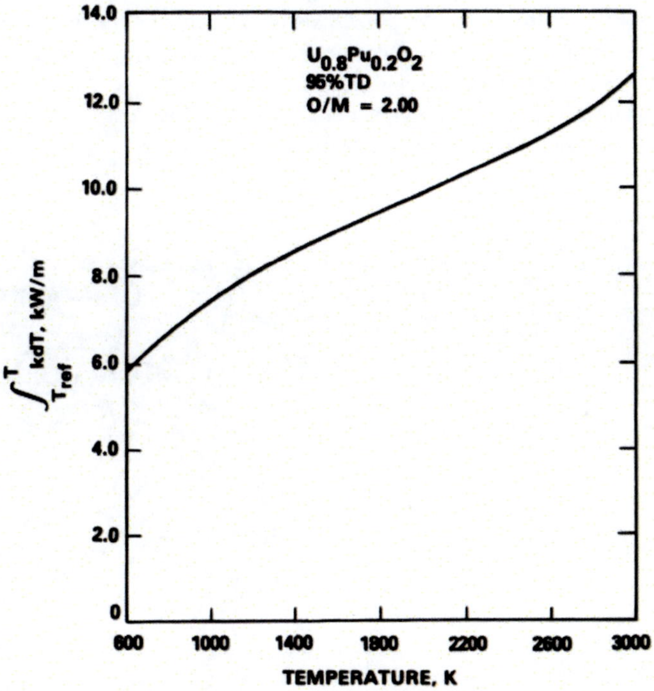

Fig. 9.3 Integral thermal conductivity data for (80% U, 20% Pu) mixed oxide fuel [constructed to be consistent with Eq. (9.6) and Fig. 9.2]

[4]Laboratory measurements indicate that the thermal conductivity of mixed oxide fuel varies with the oxygen-to-metal (O/M) ratio. Equation (9.6) and Fig. 9.2 correspond to a stoichiometric O/M ratio of 2.00. The precise effect of O/M ratio on fuel conductivity and temperature during actual reactor operation, however, is still in question.

at a given porosity, so that the integral for the fuel is $\int k_p dT$. One then multiplies $\int k_p dT$ by the ratio k_{ref}/k_p obtained from Eq. (9.7) and uses this result as the increment on the graph to find T_{CL}.

The problem for oxide fuel is actually more complicated because (as described in Section 8.2.3), the oxide fuel restructures, forming a void in the center of the pin.

Conductivity data for (80% U, 20% Pu) mixed oxide fuel is presented in this section to provide an illustration of the fuel behavior that must be accounted for in the analysis fuel pin thermal performance. Thermal conductivities along with other fuel properties of several uranium and plutonium fuels are summarized in Table 11.3 of Chapter 11. In addition, more refined data for material properties of various fuels can be found in Refs. [13, 30–37] of Chapter 11.

9.2.2 *Effects of Restructuring*

Metal, carbide, and nitride fuels do not restructure, so no voids form at their centers.[5] Hence, radial temperature distributions for these fuels can be determined by using the simple expressions derived above. Mixed oxide fuels, however, *do* restructure, as described in Section 8.2.3. With regard to fuel temperature distribution, the most important results of restructuring are the development of a central void and changes in thermal conductivity, density, and volumetric heat generation rates in the columnar and equiaxed regions.

Figure 9.4 is a sketch of the salient features of Fig. 9.5 and identifies the region and radius numbers used in the model for the restructured fuel pin. Given the density and the thermal conductivity of the three microstructure regions of Fig. 9.4, we can determine the temperature profile.

Due to porosity loss in the equiaxed and columnar grains, the local fuel density will increase. Hence, given the linear power χ (which does not change with restructuring), the actual volumetric heat generation rates in each of the three zones will be

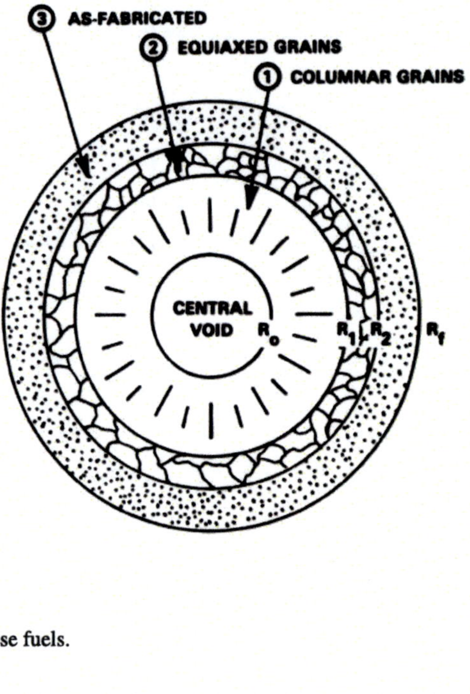

Fig. 9.4 Geometry for restructured fuel pin analysis

[5]See Chapter 11 for a discussion of the thermal properties of these fuels.

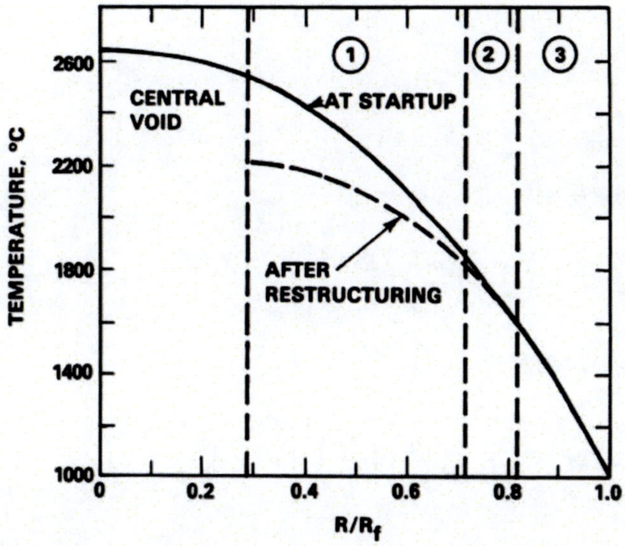

Fig. 9.5 Temperature distribution in a mixed-oxide fuel pin before and after restructuring. $\chi = 50$ kW/m; $T_s = 1{,}000$°C; Initial Density = 85% Theoretical Density; $T_1 = 1{,}800$°C, $\rho_1 = 98\%$ TD; $T_2 = 1{,}600$°C, $\rho_2 = 95\%$ *TD* (Ref. [3])

$$Q_3 = \frac{\chi}{\pi R_f^2}, \tag{9.9a}$$

$$Q_2 = \frac{\chi}{\pi R_f^2}\frac{\rho_2}{\rho_3}, \tag{9.9b}$$

$$Q_1 = \frac{\chi}{\pi R_f^2}\frac{\rho_1}{\rho_3}, \tag{9.9c}$$

where the densities, ρ, are straightforward reflections of the porosity loss.

With these new values of Q, the heat conduction equation is solved in each region, with the limits of $\int k\, dT$ being the surface temperatures bounding the different fuel structures. Between regions, e.g., between regions 1 and 2 with a boundary at R_1 continuity of heat flux is a boundary condition in addition to continuity of temperature, i.e.,

$$k_1\left(\frac{dT}{dr}\right)_1 = k_2\left(\frac{dT}{dr}\right)_2, \text{ and } T_1 = T_2, \text{ at } r = R_1. \tag{9.10}$$

Furthermore, the radii and densities are interrelated. For example, regions 1 and 2 contain material that was at density ρ_3 before restructuring. A mass balance inside radius R_2, gives

$$\rho_1\left(R_1^2 - R_0^2\right) + \rho_2\left(R_2^2 - R_1^2\right) = \rho_3 R_2^2. \tag{9.11}$$

Applying the boundary conditions [Eq. (9.10)] and the mass continuity relation, Eq. (9.11), a solution to the heat conduction equation Eq. (9.1) yields the following results for the three regions, where T_0, T_1, T_2, and T_s, refer to temperatures at R_0, R_1, R_2, and R_f[6]:

[6] Additional details in the solution are given in Ref. [3].

Region 3 (unrestructured fuel):

$$\int_{T_s}^{T_2} k_3 \, dT = \frac{\chi}{4\pi}\left[1 - \left(\frac{R_2}{R_f}\right)^2\right]. \tag{9.12}$$

Region 2 (equiaxed fuel):

$$\int_{T_2}^{T_1} k_2 \, dT = \left(\frac{\chi}{4\pi}\right)\left(\frac{\rho_2}{\rho_3}\right)\left(\frac{R_2}{R_f}\right)^2\left[1 - \left(\frac{R_1}{R_2}\right)^2 - 2\left(1 - \frac{\rho_3}{\rho_2}\right)\ln\frac{R_2}{R_1}\right]. \tag{9.13}$$

Region 1 (columnar fuel):

$$\int_{T_1}^{T_0} k_1 \, dT = \left(\frac{\chi}{4\pi}\right)\left(\frac{\rho_1}{\rho_3}\right)\left(\frac{R_1}{R_f}\right)^2\left[1 - \left(\frac{R_0}{R_1}\right)^2 - 2\left(\frac{R_0}{R_1}\right)^2\ln\frac{R_1}{R_0}\right]. \tag{9.14}$$

The center temperature, T_0, can be obtained from a knowledge of χ and using Fig. 9.3 and the above equations as follows. Start with Eq. (9.12) and Fig. 9.3 to find the temperature T_2. If the density of the unrestructured fuel is different from the density for which $\int k_{ref} dT$ is provided, the $\int k \, dT$ increment must be modified in the manner described in Section 9.2.1. Then, calculate T_1 using Eq. (9.13) and Fig. 9.3; and, finally, calculate T_0 using Eq. (9.14) and Fig. 9.3.

In Fig. 9.5, the radial fuel temperature distribution is plotted for a typical mixed-oxide fuel pin before and after restructuring. Note that the development of the central void and the enhanced thermal conductivity in the restructured regions decreases the hottest temperature by about 400°C, even though the linear power remains the same. Hence, restructuring in mixed oxide fuels partially offsets the disadvantages of low inherent thermal conductivity. A further reduction in center temperature occurs after the initial fuel-cladding gap closes, as explained below.

9.2.3 Fuel-Cladding Gap

Fuel temperatures determined in Sections 9.2.1 and 9.2.2 were based on a knowledge of the fuel surface temperature, T_s. Additional heat transfer barriers that must be traversed before reaching the bulk coolant include the fuel-cladding gap, the cladding, and the coolant film. Of these three barriers, the fuel-cladding gap provides the greatest resistance to heat flow. Initially, when the gap is open, the temperature drop across the gap is an order of magnitude greater than the drop across either the cladding or the coolant film. Even after the hot gap closes, the temperature drop across the contacting surfaces remains larger than the other two.

Most experimental evidence with irradiated oxide fuel pins indicates that the hot gap closes after a short time due to fuel swelling and thermal cracking in addition to the differential thermal expansion. At this time the gap resistance decreases significantly, which leads to a large reduction in fuel center temperature. Methods to calculate the thickness of the hot gap, G, as a function of irradiation were presented in Section 8.3.4.

In order to calculate the temperature drop across the gap, $(T_s - T_{ci})$, where T_{ci} is the inside cladding temperature, a gap conductance, h_g is defined such that

$$\chi = h_g 2\pi R_f (T_s - T_{ci}). \tag{9.15}$$

Methods are available to estimate gap conductances (equivalent to a heat transfer coefficient) for open and closed gaps [4]. The method generally used for fast reactor analyses follows the method developed for thermal reactors by Ross and Stoute [5]. Uncertainties in gap conductance are large because experimental measurements of this parameter must be unraveled from integral fuel pin experiments that involve linear power and $\int k\, dT$ values for the fuel in addition to gap conductance.

For an open gap, heat transfer can occur across the fuel-cladding gap by conduction, thermal radiation, and convection. However, surface temperatures are seldom high enough to allow the radiation component to be very appreciable and convection currents in the gap result in negligible heat transfer. Hence, the heat flow is mainly governed by conduction through the gas that fills the gap. The gap conductance for an open gap is given approximately by

$$h_g\,(\text{open}) \sim \frac{k_m}{G}, \tag{9.16}$$

where

k_m = thermal conductivity of the gas mixture in the gap (W/m · K)
G = gap (m).[7]

More accurate expressions for the open gap conductance take into account the roughness of the fuel and cladding surfaces and a quantity called the jump distance. Roughness is illustrated for an open gap in Fig. 9.6a. Reported values of roughness vary over a wide range of the order of 10^{-4} to 10^{-2} mm for both fuel and cladding [4]. These values are to be compared with initial hot gaps of the order of 0.1 mm.

The jump distance is an extrapolated distance used to account for incomplete energy exchange by gas molecules at each wall. The concept is illustrated schematically in Fig. 9.7.

By adding the jump distance to the gap, an effective gap thickness is obtained that provides the correct temperature drop across the gap when combined with the gas conductivity, k_m. Calculated values for jump distance are of the order of 10^{-4} mm or less, although higher values have been reported [5]; hence, they have little influence on the open gap.

Equation (9.16) can now be modified to include roughness and jump factors, as follows:

$$h_g\,(\text{open}) = \frac{k_m}{G + (\delta_f + \delta_c) + (g_f + g_c)}. \tag{9.17}$$

The gas thermal conductivity diminishes rapidly with burnup because the thermal conductivities of the fission product gases Xe and Kr are considerably less than that of the fill gas (normally helium).[8] On the other hand, the increasing gas pressure with burnup partially offsets this effect by increasing the molecular density. Moreover, gap closure after extended operation further increases the conductance.

For a closed gap, the gap coefficient is not reduced to zero because surface roughness remains, as illustrated in Fig. 9.6b. The gap conductance now becomes the sum of two parts, one due to the material contact surfaces and the other due to remaining gas caused by the roughness. The conductance due to the contact surfaces is proportional to the contact pressure, P_{fc} (defined earlier in Section 8.3.4) and

[7]Throughout the literature the word "gap" is used for the width or thickness of the gap in addition to referring to the region between the fuel and the cladding.

[8]The deleterious effect of Xe can be observed by comparing Figs. 9.10 and 9.11, which are included in the last part of this section in conjunction with linear-power-to-melting discussions.

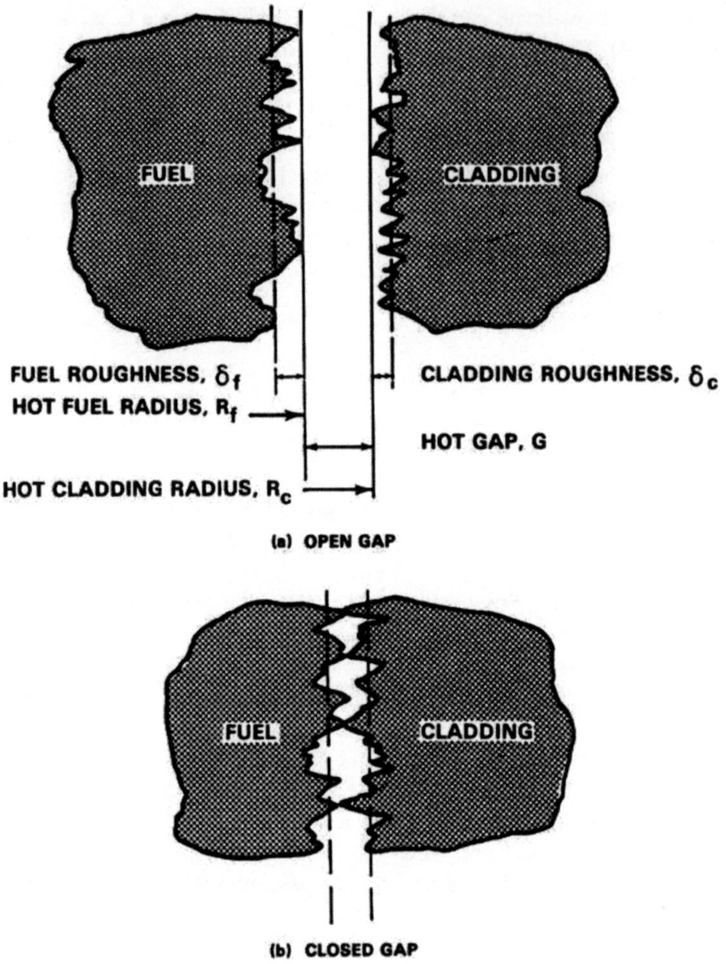

Fig. 9.6 Geometry for gap conductance for open and closed gaps. (**a**) open gap, (**b**) closed gap

an effective thermal conductivity of the fuel cladding surface materials. It is inversely proportional to the Meyer hardness of the softer material and the square root of an effective roughness. This effective roughness is defined as

$$\delta_{eff} = \sqrt{\left(\delta_f^2 + \delta_c^2\right)/2}.$$

The conductance due to gas conduction remains the same as for the open gap with $G = 0$ and with roughnesses assumed to be unchanged. Hence, the closed-gap conductance is

$$h_g\,(\text{closed}) = \frac{Ck_sP_{fc}}{H\sqrt{\delta_{eff}}} + \frac{k_m}{\left(\delta_f + \delta_c\right) + \left(g_f + g_c\right)}, \tag{9.18}$$

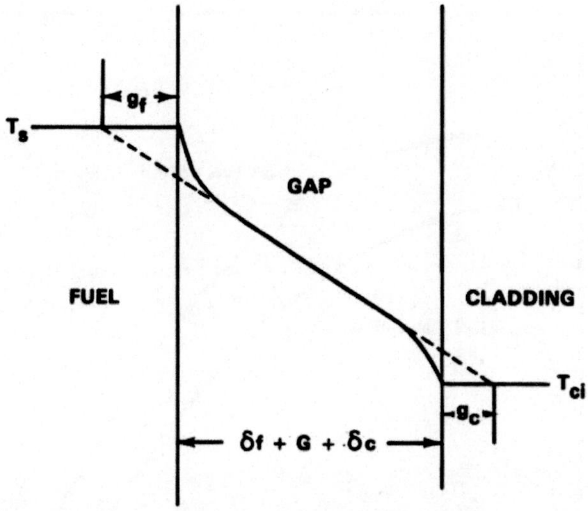

Fig. 9.7 Schematic temperature profile across an open gap to illustrate jump distances, g_f and g_c

where

$C =$ empirical constant ($m^{-1/2}$)

$k_s =$ effective conductivity of surface materials $\left(= \dfrac{2k_f k_c}{k_f + k_c}\right)$

$H =$ Meyer hardness of the softer material (Pa).

The net results of Eqs. (9.17) and (9.18) are qualitatively illustrated in Fig. 9.8, which shows the gradual increase in the effective heat transfer coefficient as the gap is reduced, and then the sharp increase upon physical contact. Further increases occur as the contact pressure becomes more intense up to the yield stress of one of the materials.

The influence of the fuel-cladding gap on the radial temperature distribution and the effect of uncertainties in the hot gap are illustrated in Fig. 9.9. Both curves represent peak temperatures (at

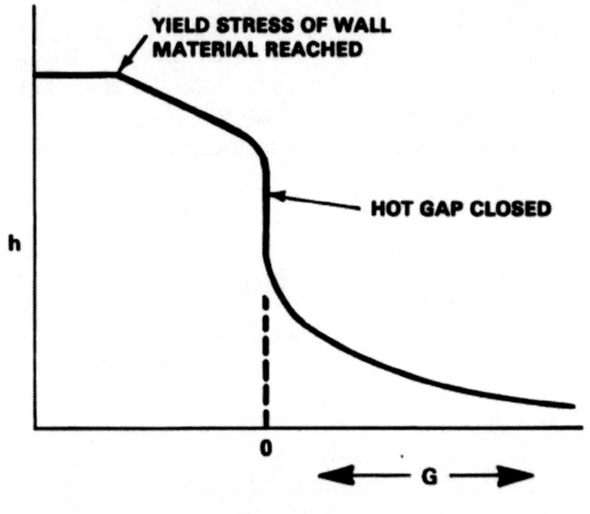

Fig. 9.8 Qualitative representation of the fuel-cladding coefficient

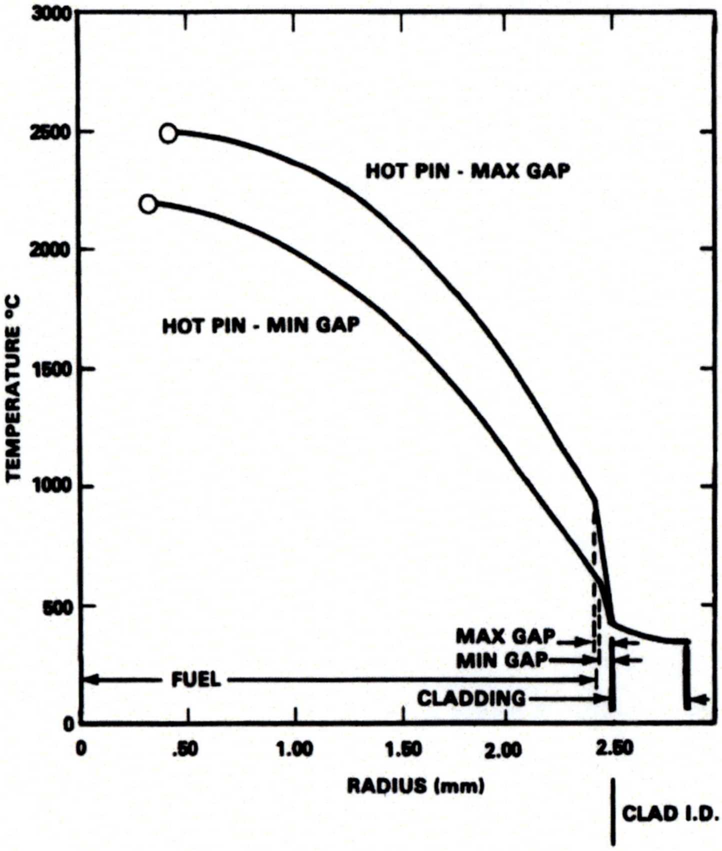

Fig. 9.9 Influence of a beginning-of-life gap on the hot pin temperature distribution. *Curves* are for the FFTF hot pin at the axial mid-plane at 115% overpower with hot channel factors included, one with the maximum gap and one with the minimum [6].

115% overpower and after the effect of hot channel factors have been included[9]) at beginning-of-life conditions at the axial midplane of the hottest FFTF fuel pin. The upper curve shows the effect of the maximum hot gap, G (max); the lower curve is for the minimum hot gap, G (min), but still at beginning-of-life. The 280°C difference in temperature drop across the gap results in virtually the same difference in center fuel temperature.

Further illustrations of the influence of gap on fuel temperature appear in Section 9.2.5 of this section and in Section 9.4 (Figs. 9.20 and 9.21).

9.2.4 Temperature Drop Across the Cladding

The temperature drop across the cladding is relatively small. The temperature difference between the inside and outside cladding surfaces, $(T_{ci} - T_{co})$, is obtained from Fourier's law for the heat flux, q (W/m^2):

[9]See Section 10.4 for a discussion of overpower and hot channel factors.

$$q = -k\frac{dT}{dr}.$$

In terms of the linear power, χ, this becomes

$$\chi = -k_c 2\pi r\frac{dT}{dr}.$$

Since the thermal conductivity of the cladding can normally be assumed constant, integrating across the cladding gives

$$T_{ci} - T_{co} = \frac{\chi}{2\pi k_c}\ln\frac{R_{co}}{R_{ci}}. \tag{9.19}$$

A note of caution should be added at this time. Throughout the thermal analysis so far we have tacitly assumed that all heat generation contributing to the linear power was in the fuel. This is not exactly correct, and detailed design codes must account for the deviation from this assumption. Gamma radiation, accounting for about 14% of the heat generated, is absorbed throughout the core—particularly in the more massive materials such as fuel, cladding, and structure. Also, a small amount of energy is transferred directly to the steel and the sodium through elastic scattering of neutrons and beta deposition. In all, about 2 or 3% of the fission energy is absorbed in the cladding.

9.2.5 Integral Measurements of Linear-Power-to-Melting

A major concern of the thermal analyst is to demonstrate that fuel does not melt. This is discussed at greater length in Section 10.4 on hot channel factors. At this point, however, it is useful to introduce the types of experimental results that are important for this demonstration.

From the preceding discussion of fuel thermal performance, it is evident that an elaborate calculation is needed to obtain the center temperature of the fuel as a function of linear power. This calculation has numerous potential sources of uncertainty. It would be desirable to take a fuel pin for a particular fast reactor and actually irradiate it in a fast test reactor to determine experimentally what linear power is required to cause centerline melting. This value is called *linear-power-to-melting* and is denoted by χ_m. The measurement of χ_m is called an integral measurement because it incorporates a series of phenomena into a single overall result. Such an experiment would provide a direct and independent demonstration of whether center melting would occur for the design peak linear power without having to depend on numerous intermediate calculations.

This type of experiment was performed for the FFTF and CRBRP design fuel pins by the Hanford Engineering Development Laboratory [7, 8] and in another experiment by General Electric [9]. In the HEDL-P-19 experiment, nineteen fresh mixed oxide (75% UO_2 -25% PuO_2) fuel pins with cold gap sizes varying from 0.09 to 0.25 mm (3.4–10 mils) and pin diameters of 5.84 mm (0.23 in.), corresponding to the FFTF driver fuel and the CRBRP fuel pin designs, and 6.35 mm (0.25 in.) diameter pins were irradiated in EBR-II. Peak linear powers varied from 55.4 to 57.7 kW/m (16.9–17.6 kW/ft) for the 5.84-mm pins and 65.6–70.2 kW/m (20.0–21.4 kW/ft) for the 6.35-mm pins. These linear powers were sufficient to produce centerline fuel melting in some, but not all, of the pins. Helium fill gas was used in all of the P-19 pins.

Later, a second similar experiment, called HEDL-P-20, was performed with preirradiated fuel pins. Preirradiation varied from 3.7 to 10.9 MWd/kg. In addition a few fresh pins were run in the P-20 experiment. Some of these pins were filled with 18% xenon and 82% helium to assess the effect of xenon tag gas.

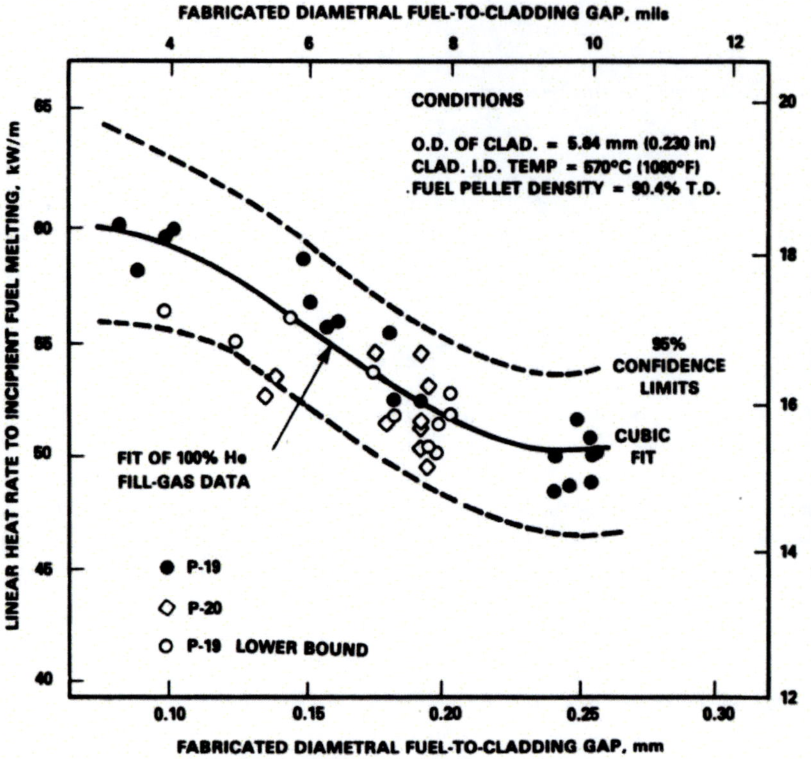

Fig. 9.10 Cubic fit of χ_m data from fresh pins normalized to FFTF conditions with helium fill-gas [8]. Lower bound data refer to pins in which no fuel melting occurred

Figure 9.10 shows the results of the P-19 experiment, together with results from the P-20 fresh fuel pins. Lower-bound points (Figs. 9.10 and 9.12) were obtained from values of pins that did not reach incipient melting (i.e., χ_m, lies above these values). Upper-bound points were obtained from pins that experienced some melting throughout the axial length of the pins; two such points are plotted on Fig. 9.11. For pins that experienced melting over part of the axial length, the χ for the axial position of incipient melting could be accurately determined.

The following correlation was found for linear-power-to-melting, χ_m, for fresh FFTF and CRBRP pins with 100% He fill gas [8]:

$$\chi_m = 53.51 + 199.3G_0 - 1.795 \times 10^3 G_0^2 + 3.764 \times 10^3 G_0^3, \quad (0.076 \leq G_0 \leq 0.254)\ (100\%\ \text{He fill gas}) \tag{9.20}$$

where G_0 is the cold gap in millimeters and χ_m is in kW/m. This fit and the corresponding 3σ level of confidence (discussed in Section 10.4) are plotted in Fig. 9.10.

A correlation was also developed for 10% Xe and 90% He fill gas [8]. The correlation is plotted on Fig. 9.11 together with the results of the two P-20 tests with 18% Xe—82% He fill gas in which melting occurred throughout the axial length of the pins. The correlation is

$$\chi_m = 56.7 + 100.7G_0 - 1.449 \times 10^3 G_0^2 + 3.424 \times 10^3 G_0^3 \quad (0.076 \leq G_0 \leq 0.254)\ (10\%\text{Xe},\ 90\%\ \text{He fill gas}). \tag{9.21}$$

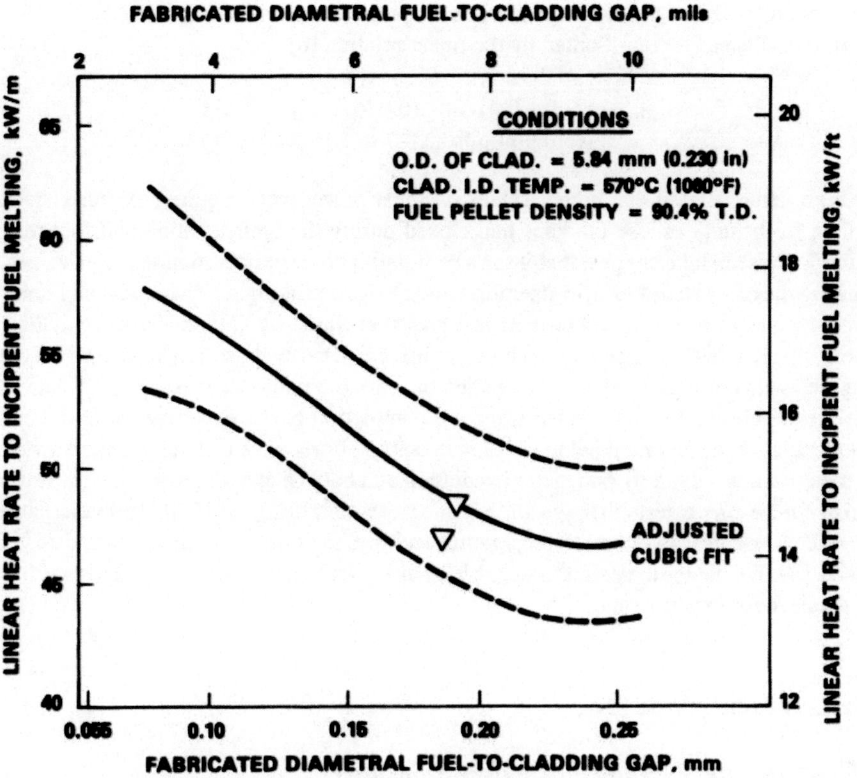

Fig. 9.11 Cubic fit of χ_m data from fresh pins adjusted to FFTF conditions with 10% Xe tag gas [8]. The two data points refer to χ_m for pins which contained 18% Xe and 82% He fill gas. Values of χ_m for 10% Xe, as in the FFTF pins, should be higher than for 18% Xe

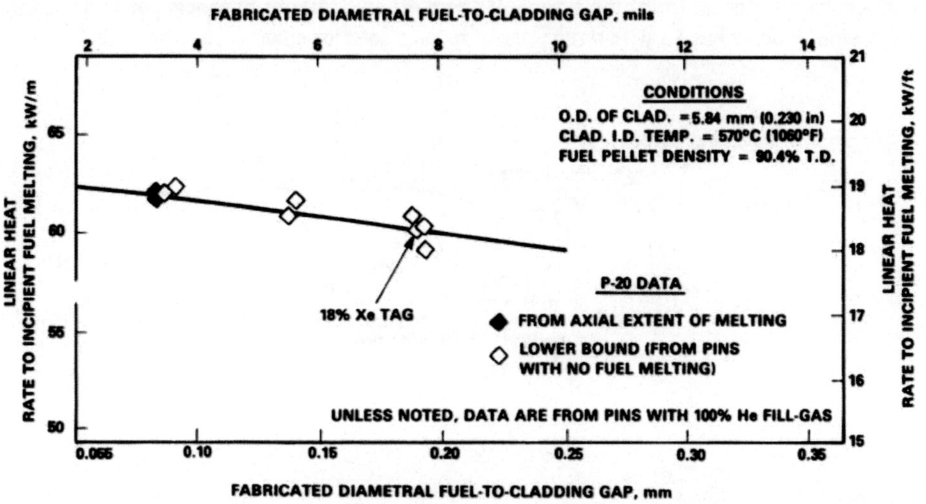

Fig. 9.12 Fit of lower bound χ_m data for fuel with burnups to 10.9 MWd/kg (1.12 at.%) [8]

Lower-bound results of the P-20 experiment for preirradiated pins are plotted in Fig. 9.12. These lower-bound results can be represented by the linear relation [8],

$$\chi_m = 66.3 - 0.41G_0 \quad (0.076 \leq G_0 \leq 0.254) \\ \text{(for burnup from 3.7 to 10.9 MWd/kg)}. \tag{9.22}$$

The P-20 test demonstrated that higher values of linear power were required to cause irradiated fuel to melt than fresh fuel because the gaps had closed during the preirradiation and fuel restructuring had matured. These results suggest that it may be possible to limit operation to lower values of linear power during the early stages of SFR operation and to operate at higher linear powers later.

A similar concept linear-power-to-melt test was performed by General Electric, called the F20 experiment. It included a range of variables complementary to those run in the HEDL tests. The results agreed with the HEDL tests and extended the variable range for these types of data.

Experimental efforts have continued since the completion of the experiments described above to refine the prediction of fuel melt behavior. However, the efforts have generally moved away from the integral measurements that rely on reactor irradiation and hot cell examinations, and towards carefully designed *separate effects* tests that specifically measure melt behavior [10]. Using the data from the separate effects tests to validate fuel temperature and fuel melt predictions in pin analysis codes like those discussed in Chapter 8, designers are able to analytically evaluate χ_m for different fuel designs without the need for experiments.

9.3 Coolant Heat Transfer

9.3.1 General Energy and Convection Relations

Energy is transferred from the fuel pins by convection to the coolant. The coolant, either liquid metal (usually sodium) or gas (usually He), flows between the pins in the assembly geometry (an intercomparison of several fast spectrum reactor coolants is provided in Table 11.11). Flow channels (sometimes called subchannels or cells) are designated as the spaces between pins and between pins and duct walls, as shown in Fig. 9.13. There are three types of channels—interior, edge, and corner. In the repeating interior lattice of the assembly, there are two interior channels per fuel rod. Designs with wire-wrap spacers have a wire wrap present in each interior channel half the time.[10] As noted in

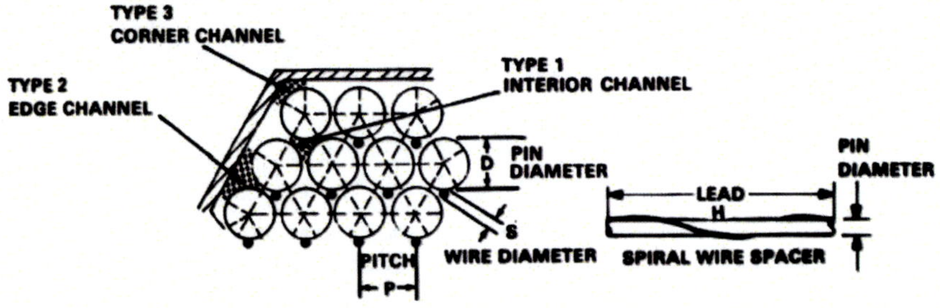

Fig. 9.13 Definition of flow channels and of wire spacer lead

[10]See Section 9.3.3 and Eqs. (9.31) and (9.32) for further details of geometry for each type of channel.

Fig. 9.13, the axial distance between each complete spiral of the wire is called the *lead* (sometimes referred to as the axial pitch).

The heat flux from the cladding to the coolant is given by

$$q = h\,(T_{co} - T_b) = \frac{\chi}{2\pi R_{co}}, \tag{9.23}$$

where

h = heat transfer coefficient (W/m^2·°C),
T_{co} = outer cladding temperature (°C),
T_b = coolant bulk temperature (°C), and
R_{co} = outer radius of cladding (m).

With the geometry defined by Fig. 9.14, this energy is convected away according to the following expression:

$$h\,(T_{co} - T_b)\,2\pi R_{co}dz = \chi dz = \dot{m}c_p dT_b, \tag{9.24}$$

where the coolant flow rate, $\dot{m}$, is the flow associated with one fuel pin. The model defined by Eq. (9.24) and Fig. 9.14 is highly simplified and will be used to describe the main features of the axial temperature distribution in Section 9.4.2. In Chapter 10, however, a more detailed analysis will be presented, which is required for actual SFR design. The mass flow rate in Eq. (9.24) is associated with one fuel pin, rather than an individual channel as in Fig. 9.13. The flow is shown upward for this calculation, as is the case for an SFR and modern gas-cooled fast reactor (GFR) designs. Though, as noted in Chapter 17, early GFR designs used downward flowing coolant, this practice is generally not applied today.

Before using Eqs. (9.23) and (9.24), we need to evaluate the heat transfer coefficient h. For this reason we shall digress from pin analysis in the remainder of this section in order to consider liquid metal thermal characteristics and heat transfer coefficient correlations. We will return to the use of Eqs. (9.23) and (9.24) in Section 9.4.

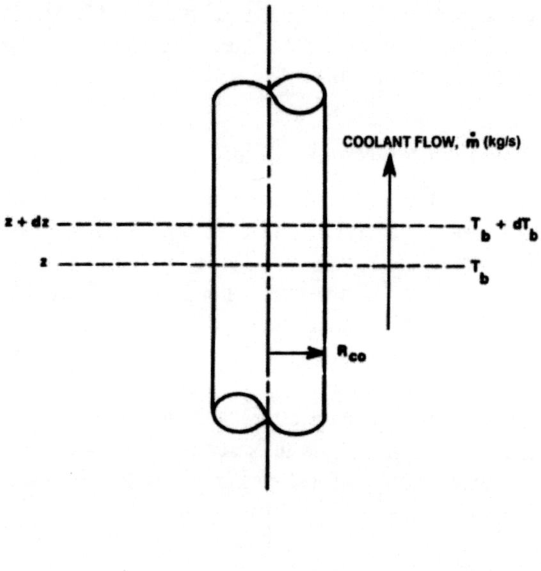

Fig. 9.14 Geometry for axial coolant flow (simplified)

9.3.2 Difference Between Liquid Metals and Other Fluids

As discussed in Section 8.2, a fast spectrum needs a high heat transfer coefficient in order to exploit the advantage of a small fuel-pin diameter. Heat transfer coefficients are higher for liquid metals than for other fluids at relatively low flow velocities and pressures; hence, liquid metals are prime candidates for use as fast spectrum reactor coolants. The high heat transfer coefficients for liquid metals result from their high thermal conductivities compared to other fluids.

Before presenting the correlations available for obtaining heat transfer coefficients for SFR core design, it is useful to consider some of the general differences between liquid metals and ordinary fluids that follow from the large difference in thermal conductivity.

The Prandtl number, Pr, is an important dimensionless grouping of properties that influence convection heat transfer:

$$\mathrm{Pr} = \frac{c_p\mu}{k} = \frac{\nu}{\alpha}, \tag{9.25}$$

where

ν = kinematic viscosity, μ/ρ
α = thermal diffusivity, $k/\rho c_p$.

Kinematic viscosity is related to the rate of momentum transfer in a fluid; thermal diffusivity is related to the rate of heat transfer by conduction. Hence, for a Prandtl number of unity, ν and α are equal, and the mechanism and rate of heat transfer are similar to those for momentum transfer. For many fluids, including water, Pr lies in the range from 1 to 10. For gases, Pr is generally about 0.7.

For liquid metals the Prandtl number is very small, generally in the range from 0.01 to 0.001. This means that the mechanisms of conductive heat transfer in liquid metals dominate over those of momentum transfer. The low Prandtl number results from the exceptionally high thermal conductivity of metals; the values for viscosity and specific heat for sodium are not appreciably different from those of water, whereas the thermal conductivity of sodium is about a factor of 100 greater than that of water. The high thermal conductivity for liquid metals results from the high mobility of free electrons in metal lattices. The Prandtl number for sodium at a typical mid-core temperature of 500°C is 0.0042.

Two classical calculations related to liquid metal heat transfer provide some insight into the physical consequences of a low Prandtl number. The first involves the difference in boundary layer behavior in laminar flow over a flat plate; the second is Martinelli's analysis of temperature distribution in turbulent flow in tubes.

9.3.2.1 Flow Over a Flat Plate

Laminar flow of liquid metals over a flat plate is analyzed in textbooks on heat transfer (e.g., Ref. [11]) and the analysis will not be reported here. Physically, though, the problem can be described as follows:

When a fluid flows over a flat plate and heat is simultaneously transferred between the fluid and the plate, two boundary layers develop along the plate surface: a hydrodynamic boundary layer and a thermal boundary layer. The velocity u and temperature T vary in the boundary layers and attain the free stream values u_∞ and T_∞ at the edges of the hydrodynamic and thermal boundary layers, respectively. The temperature at the wall is T_w. For fluids in which the Prandtl number does not differ greatly from unity, the thicknesses of the thermal and hydrodynamic boundary layers (δ_t and δ_h) are about equal, as illustrated in Fig. 9.15a. For liquid metals, however, the thickness of the thermal

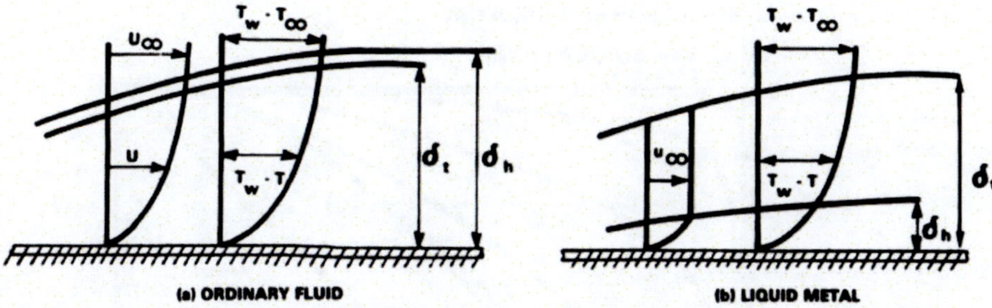

Fig. 9.15 Comparison of thermal (δ_t) and hydrodynamic (δ_h) boundary layers for ordinary fluids vs liquid metals. (**a**) ordinary fluid, (**b**) liquid metal

boundary layer is substantially larger than the thickness of the hydrodynamic boundary layer as a consequence of the large value of α relative to ν; the high thermal conductivity allows heat to be transported far out into the fluid with relatively little resistance, whereas the normal magnitude for viscosity allows the fluid to attain the free stream velocity relatively quickly. This situation is depicted in Fig. 9.15b.

The consequences of these differences is a difference in the heat transfer correlation for the Nusselt number (Nu $= hx/k$, where x is the distance down the flat plate). This difference also appears in correlations for turbulent flow in tubes. For liquid metals, the Reynolds (Re) and Prandtl (Pr) numbers in the correlations for both flat plates and tubes appear to the same power, so that the Nusselt number is correlated directly with their product, which is called the Peclet number, Pe. For a flat plate,

$$\mathrm{Pe} = \mathrm{Re}\,\mathrm{Pr} = \left(\frac{\rho V x}{\mu}\right)\left(\frac{\mu c_p}{k}\right) = \frac{\rho V x c_p}{k}. \tag{9.26}$$

Note that viscosity does not appear in the correlation at all. For ordinary fluids, Re and Pr appear to different powers, so that viscosity remains in the correlation and even becomes an important part of the correlation since viscosity is such a strong function of temperature.

9.3.2.2 Heated Turbulent Flow in Tubes

Turbulent flow is characterized by high Reynolds number (Re > 2,300 for tube flow), where

$$\mathrm{Re} = \frac{\rho V D}{\mu}. \tag{9.27}$$

In fully developed tube flow, the hydrodynamic boundary layer occupies the entire tube; one can imagine the boundary layers from the walls all meeting at the centerline of the tube. For turbulent flow, the hydrodynamic boundary layer consists of three sublayers: a laminar sublayer next to the wall, then a buffer layer, and finally a turbulent core that fills the center of the tube. Special equations for the velocity distributions in each sublayer are available for turbulent flow in tubes, known in fluid mechanics as the *law of the wall*.

Martinelli [12] analyzed turbulent flow through a tube with heat transfer by using velocity profiles through the three boundary layers but accounting for the differences in thermal conductivity and Prandtl number between liquid metals and ordinary fluids, and accounting for eddy diffusivity of heat and momentum. His results for a Reynolds number of 10,000 are reproduced in Fig. 9.16, in which the normalized temperature difference in the tube, $\left[T_w - T(y)\right] / [T_w - T_{CL}]$, is plotted against

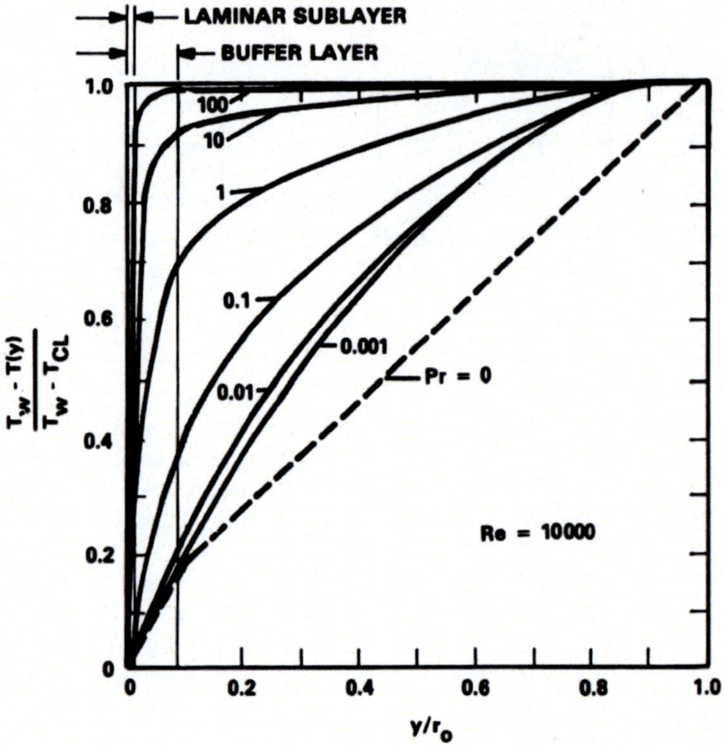

Fig. 9.16 Influence of Prandtl number on turbulent flow in tubes with heat transfer [12]

the normalized distance from the tube wall y/r_0, where r_0 is the tube radius, $y/r_0 = 1$ at the centerline, and T_w and T_{CL} are the wall and centerline temperatures. (Typical Reynolds numbers in liquid metal cooled cores are 50,000; Martinelli's results for Re = 100,000 are also similar to Fig. 9.16.) It is observed that for ordinary fluids (Pr~1) there is a high resistance to heat transfer in the laminar sublayer and the buffer layer, whereas for liquid metals (Pr < 0.01) the resistance to heat transfer is distributed throughout the tube, i.e., well into the turbulent core.

From this result one would expect that the temperature profile in a coolant channel in a liquid metal-cooled reactor would be similar to the curves in Fig. 9.16 for Pr = 0.01 and 0.001; hence, there would not be a thin film adjacent to the fuel pin across which most of the temperature drop would occur. This is indeed the case and represents one difference between liquid metal-cooled reactors and LWR's. In the water flow channels in an LWR, most of the temperature drop occurs across a thin film adjacent to the fuel rod. The liquid metal-cooled reactor literature continues to use the term "film heat transfer coefficient;" we, too, will use this term in Chapter 10, even referring to a region as a "film" region as if it physically existed between the coolant and the cladding. A hydrodynamic film of low-velocity flow adjacent to the tube wall does exist even in liquid metal tube flow, however.

Figure 9.16 also illustrates why viscosity plays a large role in correlations for heat transfer coefficients in turbulent flow in tubes for fluids with normal or high Prandtl numbers, whereas it does not enter into the liquid metal correlations. Although viscosity has a strong influence on velocity distributions in laminar and buffer layers in liquid metal tube flow, it does not affect temperature distributions; hence, heat transfer correlations are based on the viscosity-independent Peclet number.

9.3.3 Liquid-Metal Heat Transfer Correlations for Rod Bundles

Heat transfer coefficients for sodium flow through fuel-pin bundles with pitch-to-diameter ratios encountered in the SFR have been measured extensively, and numerous correlations have been proposed. Uncertainty limits on the heat transfer correlations are still rather large, but the importance of this uncertainty must be assessed with regard to the influence of the uncertainty on core and fuel pin design. This subject will be explored in Chapter 10 on hot channel factors. It will be observed that even a large uncertainty in heat transfer coefficients will have a relatively minor effect on maximum cladding or fuel temperatures. Hence, from the point of view of SFR design, knowledge of the sodium heat transfer coefficient is reasonably adequate.

Experimenters try to find a correlation that best fits existing data. The reactor designer must use a method of analysis that provides reasonable assurance that allowable cladding and fuel temperatures are not exceeded. One way to accomplish this is to use a correlation that gives heat transfer coefficients at some specified lower limit of uncertainty when compared to existing reliable experimental data. Another method is to use the best-fit correlation and provide for the uncertainties through appropriate hot-channel factors. Or, with the broad availability of high performance computers, methods based on statistical convolution concepts can be used to address uncertainties with sampling techniques that provide reasonable assurance the cladding and fuel temperatures are not exceeded at a specified statistical tolerance level.

For the FFTF and CRBRP analyses, a correlation was used that gives heat transfer coefficients below the best-fit values, and further allowance for uncertainties was added through hot channel factors. Both the FFTF/CRBRP correlations and several best-fit correlations are discussed in this section.

Liquid metal convection-heat-transfer correlations are generally presented in terms of Nusselt number versus Peclet number, where for tubes and rod bundles,

$$Nu = \frac{hD_e}{k}, \tag{9.28}$$

$$\mathrm{Pe} = \mathrm{Re}\,\mathrm{Pr} = \frac{\rho V D_e c_p}{k}, \tag{9.29}$$

and D_e is an effective hydraulic diameter [cf. Eq. (9.30)]. Once the Nusselt number is determined, the heat transfer coefficient, h, can be obtained and used in Eqs. (9.23) and (9.24).

When the experimental data for log(Nu) are plotted against log(Pe), the curve generally has a dog-leg[11] shape; the Nusselt number is constant up to some value of Peclet number called the critical Peclet number, $\mathrm{Pe_c}$ and gradually rises for values greater than $\mathrm{Pe_c}$. The value of $\mathrm{Pe_c}$ is about 200. Experimental data of this type are summarized in Fig. 9.17 for sodium flow through rod bundles of various pitch-to-diameter (*P*/*D*) ratios [13].

The difference in behavior above and below $\mathrm{Pe_c}$ results from eddy conduction. Below $\mathrm{Pe_c}$, heat transfer is by molecular conduction (which is proportional to the thermal conductivity k), while above $\mathrm{Pe_c}$ eddy conduction (which is proportional to effective eddy conductivity k_e), becomes important. Eddy conduction begins to make a contribution well above the laminar/turbulent boundary. For example, for sodium at 500°C (Pr = 0.0042) and Pe = 200, the Reynolds number is 50,000, which is well into the turbulent regime.

[11] A boomerang has a dog-leg shape. If the reader is still uncertain about this shape, he can ask anyone who plays golf, for there are always several dog-leg holes on a golf course.

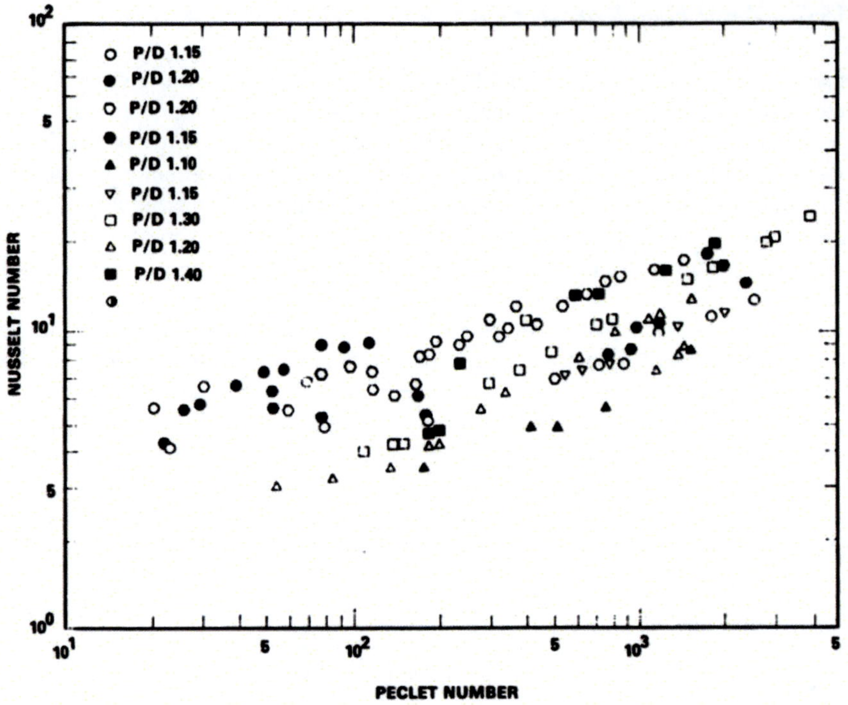

Fig. 9.17 Heat transfer coefficient correlation data for sodium in rod bundles [13]

For flow through rod bundles the diameter in the Re, Pe, and Nu numbers is an effective hydraulic diameter, D_e. For the channel types shown in Fig. 9.13, the value of D_e for channel i is

$$D_{ei} = \frac{4A_i}{P_{wi}}, \tag{9.30}$$

where

A_i = flow area of channel i
P_{wi} = the wetted perimeter.

For fuel assemblies with wire spacers, as shown in Fig. 9.13, the flow area in the basic channel ($i = 1$) is the area between the fuel pins less half the cross-sectional area of the wire wrap, because the wire wrap is in the channel for half the time on the average as the wires spiral around the three fuel pins adjacent to the channel.[12] In the edge channel ($i = 2$) the flow area is again the area between the pins and the wall less half the wire-wrap area. In the corner channel ($i = 3$), the flow area is the area between the pin and the corner less one sixth of the wire-wrap area. The wetted perimeter P_{wi} is the sum of the perimeters of the fuel-pin and wall surfaces in each channel plus one half the circumference of the wire wrap in channels 1 and 2 and one sixth of the wire-wrap circumference in channel 3. Therefore,

$$A_1 = \frac{\sqrt{3}}{4}P^2 - \frac{\pi D^2}{8} - \frac{\pi s^2}{8}, \tag{9.31a}$$

[12] This can be observed from Fig. 10.4 in Chapter 10.

$$A_2 = P\left(\frac{D}{2}+s\right) - \frac{\pi D^2}{8} - \frac{\pi s^2}{8}, \tag{9.31b}$$

$$A_3 = \frac{1}{\sqrt{3}}\left(\frac{D}{2}+s\right)^2 - \frac{\pi D^2}{24} - \frac{\pi s^2}{24}, \tag{9.31c}$$

$$P_{w1} = \frac{\pi D}{2} + \frac{\pi s}{2}, \tag{9.32a}$$

$$P_{w2} = P + \frac{\pi D}{2} + \frac{\pi s}{2}, \tag{9.32b}$$

$$P_{w3} = \frac{(D+2s)}{\sqrt{3}} + \frac{\pi D}{6} + \frac{\pi s}{6}, \tag{9.32c}$$

where D, P, and s are defined in Fig. 9.13.

The vertical flow velocity, V_i, in the Re and Pe numbers is governed by the average assembly velocity $\bar{V}$ and the channel flow area and hydraulic radius through the flow distribution factor, X_1, given in Section 10.2.1, Eq. (10.8).

An early review of liquid metal heat transfer correlations in rod bundles was given by Dwyer [14]. A useful later comparison of correlations and data was provided by Kazimi and Carelli [15]. This reference shows experimental data (Nu versus Pe) for pitch-to-diameter ratios from 1.04 to 1.30. These data were compared with four proposed correlations, which we will now discuss. The pitch-to-diameter ratios of principal interest for SFR liquid metal-cooled reactor designs are indicated by the valves listed in the table of Appendix A. These values are around 1.2 (varying from 1.15 to 1.32) for fuel, and lower (1.05–1.1) for control and radial blanket assemblies. Typical Peclet numbers for normal operation are from 150 to 300 in the fuel assemblies.

Borishanskii, Gotovskii, and Firsova proposed the following correlation [16]:

$$\text{Nu} = 24.15 \log_{10}\left[-8.12 + 12.76\,(P/D) - 3.65\,(P/D)^2\right] \tag{9.33a}$$

for $1.1 \le P/D \le 1.5$ and $\text{Pe} \le 200$, and

$$\begin{aligned}\text{Nu} = {} & 24.15 \log_{10}\left[-8.12 + 12.76\,(P/D) - 3.65\,(P/D)^2\right] \\ & +0.0174\left[1 - e^{-6\left(\frac{P}{D}-1\right)}\right](\text{Pe} - 200)^{0.9}\end{aligned} \tag{9.33b}$$

for $1.1 \le P/D \le 1.5$ and $200 \le \text{Pe} \le 2000$.

The critical Peclet number is 200. Reference [15] shows that this correlation agrees well with most of the experimental data over the ranges of *P/D* and Pe indicated.

Graber and Rieger proposed the following correlation [17]:

$$\text{Nu} = 0.25 + 6.2\,(P/D) + \left[0.32\,(P/D) - 0.007\right](\text{Pe})^{[0.8-0.024(P/D)]} \tag{9.34}$$

for $1.25 \le P/D \le 1.95$ and $150 \le \text{Pe} \le 3000$.

This correlation fits the data for Graber and Rieger's experiments ($P/D = 1.25$) and other data at $P/D = 1.2$, but overestimates values of Nu for *P/D* ratios below 1.2.

FFTF Analysis. The FFTF correlation is [6]

$$\text{Nu} = 4.0 + 0.16\,(P/D)^{5.0} + 0.33\,(P/D)^{3.8}\,(\text{Pe}/100)^{0.86} \tag{9.35}$$

for $20 \le \text{Pe} \le 1000$.

CRBRP Analysis. The correlation used for CRBRP [15] utilized the FFTF correlation in the *P/D* range of 1.2 and 1.3 and the Schad correlation as modified by Carelli [15] for the *P/D* range 1.05–1.15. The modified Schad correlation has two parts, one below Pe_c and one above Pe_c, where $Pe_c = 150$. The correlations used were

(a) $1.2 \leq P/D \leq 1.3$; the FFTF fuel correlation, Eq. (9.35).
(b) $1.05 \leq P/D \leq 1.15$; modified Shad:

$$\mathrm{Nu} = 4.496\left[-16.15 + 24.96\,(P/D) - 8.55\,(P/D)^2\right] \tag{9.36a}$$

for $\mathrm{Pe} \leq 150$, and

$$\mathrm{Nu} = 4.496\left[-16.15 + 24.96\,(P/D) - 8.55\,(P/D)^2\right]\left(\frac{\mathrm{Pe}}{150}\right)^{0.3} \tag{9.36b}$$

for $150 \leq \mathrm{Pe} \leq 1000$.

Values of Nu determined from experimental data are compared against values predicted by the various correlations in Figs. 9.18 and 9.19. Figure 9.19 also shows how the value of Nu can be affected by circumferential position within the core; data are from Dwyer and coworkers at Brookhaven National Laboratory [18]. The conservatism of the FFTF-CRBRP correlation can be readily discerned.

Development of new fast spectrum reactor designs has continued in the decades since the correlations and data described above were published. However, the experiments supporting the heat transfer correlations in triangular rod arrangements have generally been performed before 1975 and

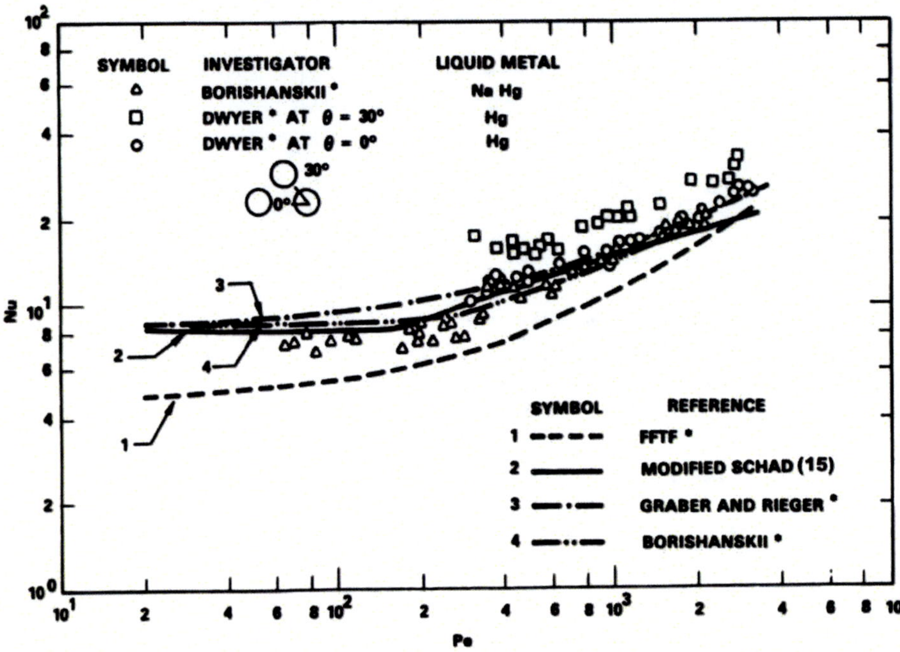

Fig. 9.18 Comparison of predicted and experimental results for *P/D* = 1.3 [15] (*References to investigators are found in Ref. [15].)

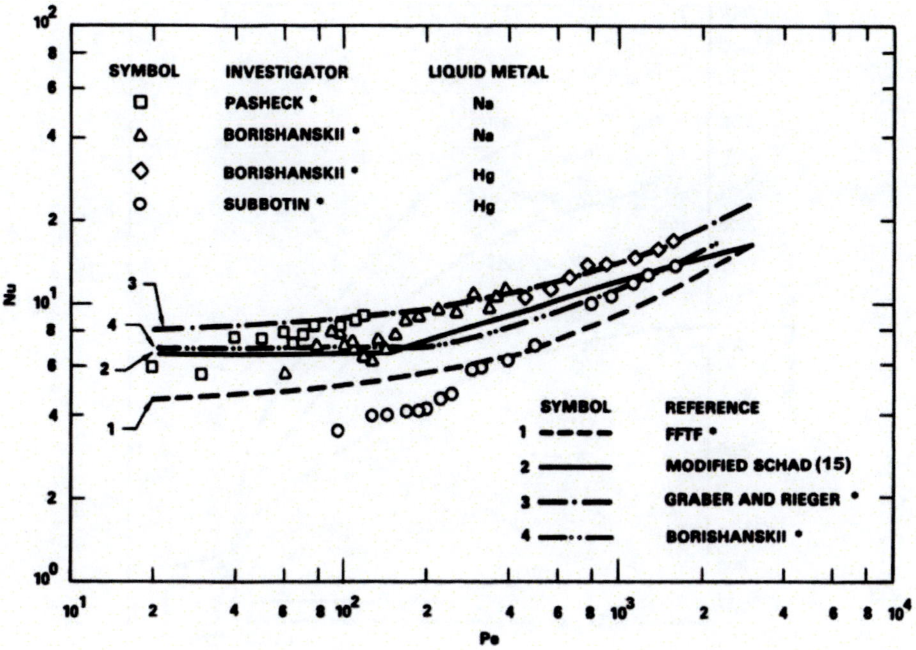

Fig. 9.19 Comparison of predicted and experimental results for *P/D*= 1.2 [15] (*References to investigators are found in Ref. [15].)

only limited effort to revise or recorrelate the data has taken place. A comparative review of several existing correlations is provided on Ref. [19] in the context of developing liquid metal cooled accelerator driven systems.

9.4 Fuel Pin Temperature Distribution

Having determined the means to compute the temperature distribution from the fuel center to the outer surface of cladding (Section 9.2) and to evaluate the coolant heat transfer coefficient (Section 9.3), we are now ready to complete the thermal analysis of the fuel pin.

9.4.1 Radial Temperature Distribution

Given the bulk coolant temperature at a specified core axial position, we can use Eq. (9.23) to determine the temperature rise to the outer surface of the cladding. This allows us to compute the absolute value of the radial temperature profile across the fuel pin.

Illustrative results of radial temperature calculations at the core midplane of the FFTF at beginning-of-life are shown in Fig. 9.20. The *peak* pin distribution is the nominal distribution for the peak power pin. The *hot* pin distribution is the distribution at 115% overpower and with the effects of all hot channel factors included. The *average* distribution is the nominal average-power fuel pin. Note that the center temperature for the average pin is below the value required for fuel restructuring so that no central void appears.

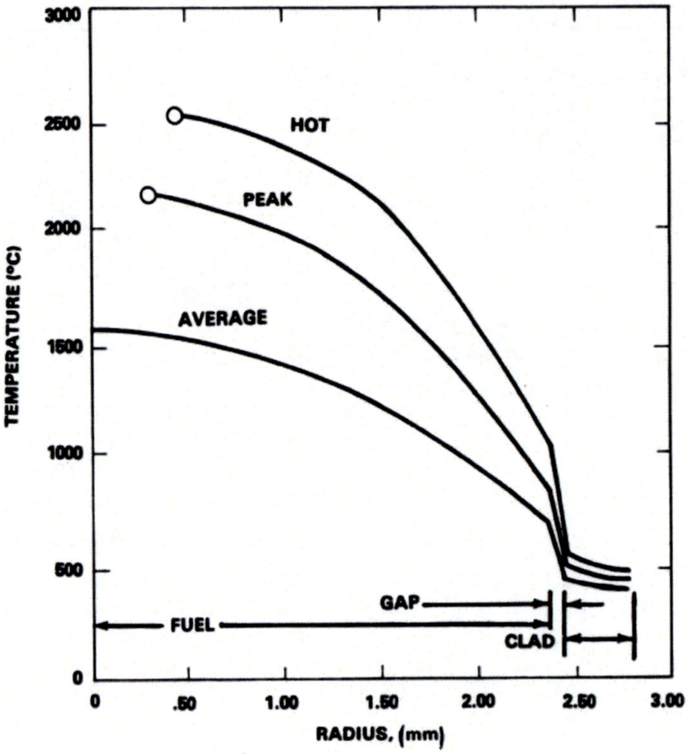

Fig. 9.20 Typical radial temperature distributions in SFR fuel pins [6]

9.4.2 Axial Temperature Distribution

In determining the axial temperature distribution in the coolant, it is tempting to simply integrate Eq. (9.24) across the assembly to obtain the coolant temperature rise in a single flow channel as the coolant flows through the assembly. Using average values for an assembly, one can obtain an average temperature rise, $\overline{\Delta T_b}$, for the assembly. As will be discussed in Chapter 10, however, there is actually a wide variation in ΔT_b between flow channels within an assembly. For the time being, defining the parameters as average values for an assembly, the average $\overline{\Delta T_b}$ for the assembly is

$$\overline{\Delta T_b} = \frac{1}{\dot{m}c_p} 2\pi R_{co} h \int \left[\bar{T}_{co}(z) - \bar{T}_b(z)\right] dz = \frac{1}{\dot{m}c_p} \int \chi(z)\, dz, \tag{9.37}$$

where the integration is over the assembly, including the axial blankets.

Note that the average coolant temperature rise across an assembly is directly proportional to linear power and inversely proportional to mass flow rate through the assembly. It is desirable to have a uniform $\overline{\Delta T_b}$ for all assemblies in the core (and to some extent in the radial blanket). Since χ varies radially in the core according to the radial power distribution, the achievement of a uniform $\overline{\Delta T_b}$ requires that $\dot{m}$ be varied between assemblies so that the ratio $\chi/\dot{m}$ is the same for each. This is done by orificing the flow rate through each assembly, meaning that an orifice is physically placed in each assembly near the coolant inlet. Exact $\overline{\Delta T_b}$ matching cannot be achieved since the radial power distribution varies with the fuel cycle, so the designer must optimize on a reasonable orificing scheme.

(Early GFR designs were even proposed with variable orificing during the fuel cycle, as discussed in Chapter 17.)

Axial temperature distributions for FFTF fuel, cladding, and coolant are plotted in Fig. 9.21 for fresh fuel and for fuel irradiated to 60 MWd/kg. Since χ (z) normally has a chopped cosine shape, the coolant and cladding temperatures are S-shaped curves and their shapes hardly change during the irradiation. The FFTF does not have axial blankets; otherwise there would be a coolant temperature rise in the blankets. Because of the large temperature drop across the fuel-cladding gap for gas-bonded pins and the closing of the gap with burnup, the axial temperature distribution in the fuel pin can change appreciably during irradiation. This effect is evident from Fig. 9.21. For the initial core conditions the fuel-cladding gap is open everywhere, and the smooth coolant temperature profile is reflected throughout the fuel pin. For later burnup conditions, however, fuel/cladding contact is made over the central two-thirds of the pin and the substantially enhanced gap conductance is responsible for the large dip observed in fuel surface temperature. The lower surface temperature leads to a drop in center fuel temperature as well.

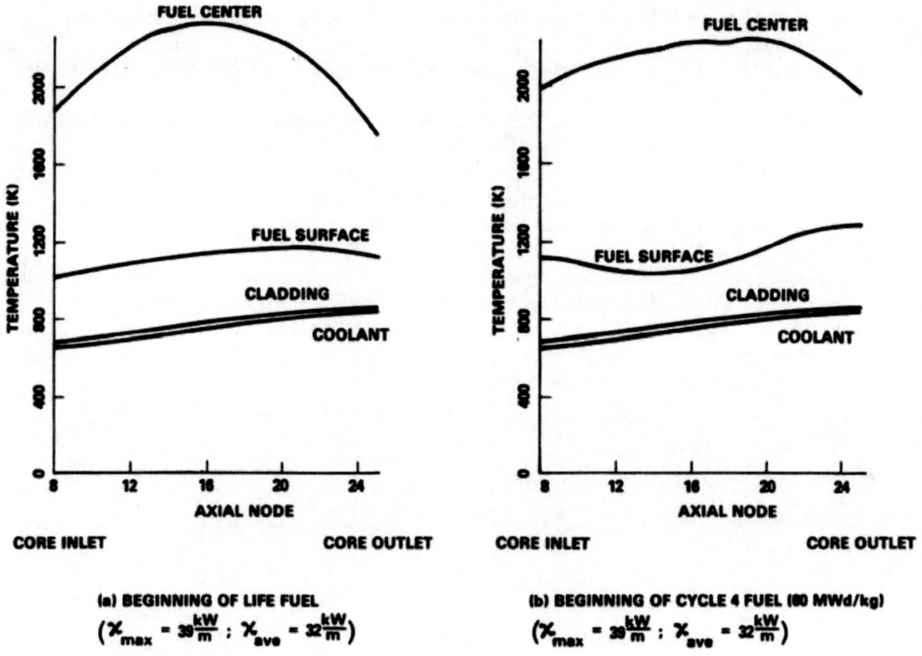

Fig. 9.21 Axial temperature distribution of FFTF fuel pins [20]

References

1. Y. S. Tang, R. D. Coffield, Jr., and R. A. Markley, *Thermal Analysis of Liquid Metal Fast Breeder Reactors*, The American Nuclear Society, La Grange Park, Illinois, 1978.
2. A. B. G. Washington, *Preferred Values for the Thermal Conductivity of Sintered Ceramic Fuel for Fast Reactor Use*, TRG-Report-2236, September 1973.
3. D. R. Olander, *Fundamental Aspects of Nuclear Reactor Fuel Elements*, Chapter 10, TID-26711-P1, Office of Public Affairs, U.S. ERDA, 1976.
4. R. B. Baker, *Calibration of a Fuel-to-Cladding Gap Conductance Model for Fast Reactor Fuel Pins*, HEDL-TME 77-86, Hanford Engineering Development Laboratory, Richland, WA, 1978.

5. A. M. Ross and R. L. Stoute, *Heat Transfer Coefficient between UO_2 and Zircalloy 2*, AECL-1552, Chalk River, Ontario, June 1962.
6. G. J. Calamai, R. D. Coffield, L. Jossens, J. L. Kerian, J. V. Miller, E. H. Novendstern, G. H. Ursim, H. West, and P. J. Wood, *Steady State Thermal and Hydraulic Characteristics of the FFTF Fuel Assemblies*, FRT-1582, June 1974.
7. R. D. Leggett, E. O. Ballard, R. B. Baker, G. R. Horn, and D. S. Dutt, "Linear Heat Rating for Incipient Fuel Melting in UO_2-PuO_2 Fuel," *Trans. ANS, 15*, 1972, 752.
8. R. B. Baker, *Integral Heat Rate-to-Incipient Melting in UO_2-PuO_2 Fast Reactor Fuel*, HEDL-TME 77-23. Hanford Engineering Development Laboratory, Richland, WA, 1978.
9. W. H. McCarthy, *Power to Melt Mixed-Oxide Fuel-A Progress Report on the GE F20 Experiment*, GEAP-14134, September 1976.
10. M. G. Adamson, E. A. Aitken, and R. W. Caputi, "Experimental and Thermodynamic Evaluation of the Melting Behavior of Irradiated Oxide Fuels," *J. Nucl. Mater., 130* (1985) 349–365.
11. J. P. Holman, *Heat Transfer*, McGraw Hill Co., New York, NY, 4th Edition, 1976.
12. R. C. Martinelli, "Heat Transfer to Molten Metals," *Trans. ASME, 69* (1947) 949–959.
13. J. Muraoka, R. E. Peterson, R. G. Brown, W. D. Yule, D. S. Dutt, and J. E. Hanson, *Assessment of FFTF Hot Channel Factors*, HEDL-TI-75226. Hanford Engineering Development Laboratory, Richland, WA, November 1976.
14. O. E. Dwyer, "Heat Transfer to Liquid Metals Flowing In-Line Through Unbaffled Rod Bundles: A Review," *Nucl. Eng. Des., 10* (1969) 3–20.
15. M. S. Kazimi and M. D. Carelli, *Heat Transfer Correlation for Analysis of CRBRP Assemblies*, CRBRP-ARD-0034, November 1976.
16. V. M. Borishanskii, M. A. Gotovskii and E. V. Firsova, "Heat Transfer to Liquid Metal Flowing Longitudinally in Wetted Bundles of Rods," *Sov. At. Energy 27* (1969) 1347–1350.
17. H. Graber and M. Reiger, "Experimental Study of Heat Transfer to Liquid Metals Flowing In-Line Through Tube Bundles," *Progress in Heat and Mass Transfer, 7*, Pergamon Press, New York, NY, 1973, 151–166.
18. O. E. Dwyer, H. Berry and P. Hlavac, "Heat Transfer to Liquid Metals Flowing Turbulently and Longitudinally Through Closely Spaced Rod Bundles," *Nucl. Eng. Des., 23* (1972) 295–308.
19. W. Pfrang and D. Struwe, *Assessment of Correlations for Heat Transfer to the Coolant for Heavy Liquid Metal Cooled Core Designs*, FZKA 7352.Forschungszentrum Karlsruhe GmbH, Karlsruhe, October 2007.
20. A. E. Waltar, N. P. Wilburn, D. C. Kolesar, L. D. O'Dell, A. Padilla, Jr., L. N. Stewart, and W. L. Partain, *An Analysis of the Unprotected Transient Overpower Accident In the FTR*, HEDL-TME 75-50., Hanford Engineering Development Laboratory, Richland, WA, June 1975.

Chapter 10
Core Thermal Hydraulics Design

Alan Waltar and Donald Todd

10.1 Introduction

In the previous chapter we explored the methodology for determining the temperature field for a single fuel pin. Since a typical fast reactor core comprises several thousand fuel pins clustered in groups of several hundred pins per assembly, a complete thermal-hydraulic analysis requires knowledge of coolant distributions and pressure losses throughout the core. This chapter will address these determinations.

We focus first on the coolant velocity and temperature distributions within an assembly. Then the discussion is broadened to include coolant velocity and pressure distributions throughout the reactor vessel. Having established these temperature and flow fields, the important design question of hot channel factor determination is then addressed.

10.2 Assembly Velocity and Temperature Distribution

In Chapter 9 we presented a simple relation for the average axial coolant temperature rise, based on an average bundle mass flow rate [Eq. (9.37)]. The problem is actually more complicated, for several reasons. First, the flow areas for the three types of channels in Fig. 9.13 are different, so the velocity and mass flow rates in each are different. Second, there is a radial power distribution, or *power skew*, across individual fuel assemblies. Third, there is *crossflow*, i.e., flow transverse to the axial direction, between assembly channels. In the first part of this section a simplified approximate velocity distribution will be derived that accounts for the variation in flow area between channels. In the second part the more complex set of equations accounting for crossflow and turbulent mixing between channels will be presented.

10.2.1 Approximate Velocity Distribution

An approximate velocity distribution, or flow split, between flow channels in Fig. 9.13 can be obtained from the physical fact that the axial pressure drop across each channel of a fuel assembly must be the same since the flow begins and ends in common inlet and outlet plenum regions. This technique was used by Novendstern [1] in his method for evaluating pressure drop in a fast reactor fuel assembly and

A. Waltar (✉)
Pacific Northwest National Laboratory (PNNL), Retired, Richland, WA, USA
e-mail: alan.waltar@wildblue.net

A.E. Waltar et al. (eds.), *Fast Spectrum Reactors*, DOI 10.1007/978-1-4419-9572-8_10,

is similar to the technique developed earlier by Sangster [2]. The approximate velocity distribution is useful for calculating assembly pressure drops and for use as a term in the Peclet number to calculate heat transfer coefficients. It is not sufficiently accurate for detailed design calculations of outlet temperature distribution or maximum cladding temperatures.

Novendstern's flow-split model was later improved by Chiu et al. [3]. The method of Novendstern is presented here, after which the final result for the Chiu-Rohsenow-Todreas model is given.

10.2.1.1 Novendstern Model

The pressure drop across channel i is

$$\Delta p_i = f_i \frac{L}{D_{ei}} \frac{\rho V_i^2}{2}, \tag{10.1}$$

where f_i is the effective friction factor in channel i (to be defined further in Section 10.3.3), V_i is the velocity, and D_{ei} is the effective hydraulic diameter [Eq. (9.30)]. Setting the three pressure drops equal gives

$$f_1 \frac{L}{D_{e1}} \frac{\rho V_1^2}{2} = f_2 \frac{L}{D_{e2}} \frac{\rho V_2^2}{2} = f_3 \frac{L}{D_{e3}} \frac{\rho V_3^2}{2}. \tag{10.2}$$

The velocities V_i must account for the way in which the mass flow is divided between the three types of channels. To relate V_i to the geometry, we first define an average axial velocity, $\overline{V}$, for the fuel assembly, which is related to the total mass flow rate $\dot{m}_T$ through the assembly by

$$\dot{m}_T = \rho A_T \overline{V}, \tag{10.3}$$

where A_T, is the sum of the flow areas of all the channels.

We let N represent the total number of channels in the assembly and N_i the number of channels of type i. The number of type-3 channels per assembly is always 6 (Fig. 9.13). For a 217-pin assembly there are 384 basic channels and 48 edge channels. For a 271-pin assembly there are 486 basic channels and 54 edge channels. Letting A_i represent the flow area in channel i averaged over one wire lead (where lead is defined in Fig. 9.13),

$$A_T = N_1 A_1 + N_2 A_2 + N_3 A_3. \tag{10.4}$$

The flow areas A_1, A_2 and A_3 are given by Eq. (9.31).

Assuming incompressible flow (i.e., constant coolant velocity), continuity gives

$$N_1 V_1 A_1 + N_2 V_2 A_2 + N_3 V_3 A_3 = \overline{V} A_T. \tag{10.5}$$

Flow distribution is governed by Eq. (10.2). The friction factor for each channel can be approximated by

$$f_i = \frac{C}{\mathrm{Re}_i^m} = \frac{C}{(\rho V_i D_{ei}/\mu)^m}, \tag{10.6}$$

where D_{ei} is given by Eq. (9.30), together with the associated Eqs. (9.31) and (9.32). For turbulent flow the recommended value for m is the exponent that appears in the Blasius relation for the friction factor [Eq. (10.36)], where $m = 0.25$.

Substituting Eq. (10.6) into Eq. (10.2) and assuming that C and m are the same for each channel gives

$$\frac{V_1^{(2-m)}}{D_{e1}^{(1+m)}} = \frac{V_2^{(2-m)}}{D_{e2}^{(1+m)}} = \frac{V_3^{(2-m)}}{D_{e3}^{(1+m)}}. \tag{10.7}$$

Using Eq. (10.7) to eliminate V_2, and V_3 from Eq. (10.5) gives

$$V_1\left[N_1A_1 + N_2A_2\left(\frac{D_{e2}}{D_{e1}}\right)^{\left(\frac{1+m}{2-m}\right)} + N_3A_3\left(\frac{D_{e3}}{D_{e1}}\right)^{\left(\frac{1+m}{2-m}\right)}\right] = \overline{V}A_T,$$

or

$$V_1 = \frac{\overline{V}A_T}{\sum\limits_{j=1}^{3} N_jA_j\left(\frac{D_{ej}}{D_{e1}}\right)^{\left(\frac{1+m}{2-m}\right)}} = X_1\overline{V}, \tag{10.8}$$

where X_1 is a *flow distribution factor*, or *flow-split parameter*, for channel 1.

For turbulent flow, for which $m = 0.25$,

$$V_1 = \frac{\overline{V}A_T}{N_1A_1 + N_2A_2\left(\frac{D_{e2}}{D_{e1}}\right)^{0.714} + N_3A_3\left(\frac{D_{e3}}{D_{e1}}\right)^{0.714}}. \tag{10.9}$$

In general, for the *i*th channel,

$$V_i = \frac{\overline{V}A_T}{\sum\limits_{j=1}^{3} N_jA_j\left(\frac{D_{ej}}{D_{e1}}\right)^{\left(\frac{1+m}{2-m}\right)}} = X_i\overline{V}. \tag{10.10}$$

The mass flow rates and flow velocities through the FFTF driver fuel assemblies are typical of SFR power reactors. The mass flow rate through the highest-power FFTF fuel assembly is 23.4 kg/s, and the average velocity, $\overline{V}$, is 6.4 m/s. The distribution according to Eq. (10.10) is plotted in Fig. 10.1 and compared to a more detailed computer prediction (which includes radial momentum exchange).

10.2.1.2 Chiu-Rohsenow-Todreas (CRT) Model

The CRT flow-split model represents an improvement over the simpler Novendstern model. The principal difference is that the CRT model divides the pressure drop across the channel into two

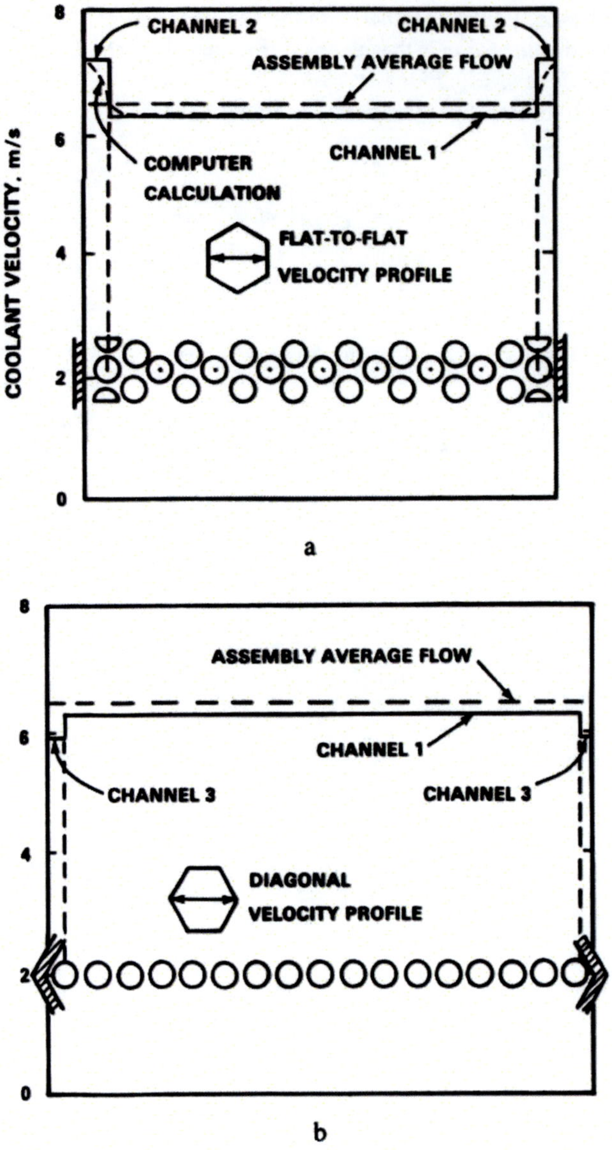

Fig. 10.1 Coolant velocity distribution in the maximum power FFTF assembly based on Eq. (10.10). **(a)** Assembly transverse (flat-to-flat), **(b)** Assembly transverse (diagonal)

components, one due to skin friction losses, and one due to form losses from flow perpendicular to the wire wrap. The bases for the CRT flow-split model, including the two pressure components, are given in Section 10.3.3. Flow split parameters are derived for channels 1 and 2 only; values for the corner channels (channel 3) are assumed to be the same as for the edge channels (channel 2) since the corner channels have so little influence on the flow. Results are given here for turbulent flow only. Results for laminar flow appear in Ref. [4]. The resulting velocities for channels 1 and 2 for turbulent flow are

$$V_1 = \frac{\overline{V}A_T}{N_1A_1 + (N_2A_2 + N_3A_3)\left(\frac{D_{e2}}{D_{e1}}\right)^{0.714}\left[\frac{C_1\frac{D_{e1}}{H}\frac{A_{r1}}{A_1'}\frac{P^2}{(\pi P)^2+H^2}+1}{C_3\left\{1+\left[C_2 n\left(\frac{V_T}{V_2}\right)_{gap}\right]^2\right\}^{1.375}}\right]^{0.571}}, \quad (10.11)$$

$$V_2 = \frac{\overline{V}A_T}{N_1A_1\left(\frac{D_{e1}}{D_{e2}}\right)^{0.714}\left[\frac{C_3\left\{1+\left[C_2 n\left(\frac{V_T}{V_2}\right)_{gap}\right]^2\right\}^{1.375}}{C_1\frac{D_{e1}}{H}\frac{A_{r1}}{A_1'}\frac{P^2}{(\pi P)^2+H^2}+1}\right]^{0.571} + N_2A_2 + N_3A_3}, \quad (10.12)$$

where C_1, C_2 and C_3 are dimensionless constants that must be determined experimentally; N_i, A_i, D_{ei}, A_T and $\overline{V}$ defined the same way as for the Novendstern method; V_T is a transverse velocity (defined in Fig. 10.14); and

P = pitch, as defined in Fig. 9.13
H = lead, as defined in Fig. 9.13
A_{r1} = projected area for one wire wrap in channel 1 over one lead,
$= \frac{1}{6}\left[\frac{\pi}{4}(D+2s)^2 - \frac{\pi D^2}{4}\right]$,

where D and s are defined in Fig. 9.13,

A'_1 = flow area in channel 1 without the wire wrap,
$= \frac{\sqrt{3}}{4}P^2 - \frac{\pi D^2}{8}$,

$$n = \frac{s}{\left[\left(\frac{D}{2}+s\right)\frac{P}{2} - \frac{\pi D^2}{16}\right]\Big/\frac{P}{2}},$$

$$\left(\frac{V_T}{V_2}\right)_{gap} = 10.5\left(\frac{s}{P}\right)^{0.35}\frac{P}{\sqrt{(\pi P)^2+H^2}}\left(\frac{A_{r2}}{A_2'}\right)^{0.5},$$

where A_{r2} = projected area for one wire wrap in channel 2 over one lead, i.e.,

$$A_{r2} = \frac{1}{4}\left[\frac{\pi}{4}(D+2s)^2 - \frac{\pi D^2}{4}\right],$$

$$A'_2 = P\left(\frac{D}{2}+s\right) - \frac{\pi D^2}{8}.$$

Based on experimental data from seven references, values for the constants C_1, C_2, and C_3, were determined to be [3]

$$C_1 = 2200$$
$$C_2 = 1.9$$
$$C_3 = 1.2.$$

Comparison of the CRT flow split with the Novendstern result, as given in Eq. (10.9), shows that the only significant difference is the large bracketed term in the denominators of the CRT values.

10.2.1.3 Circumferential Velocity and Temperature Distributions

Average flow velocities in the channels have been assumed throughout the above discussion. In reality, there is a circumferential variation of velocity around a fuel pin, with the axial component of velocity a minimum at the minimum gap between pins. This circumferential velocity variation leads to a circumferential cladding temperature variation, which in turn causes a hot spot in the cladding temperature at the minimum gap. For the CRBRP and FFTF designs, the ratio of the peak-to-average ΔT between the cladding temperature and the bulk coolant temperature due to the circumferential velocity distribution is ~1.5. This appears in the hot channel factors for the maximum cladding temperature calculation in the CRBRP design (discussed in Section 10.4.3). Obviously, circumferential velocity and temperature distributions for each SFR fuel assembly design must be carefully considered.

10.2.2 Effects of Crossflow on Velocity and Temperature Distributions

Although it is possible to obtain approximate values for the temperature and velocity distributions across a fuel assembly from the use of simple analytical models, the geometry of the reactor lattice is so complicated that it is virtually impossible to obtain exact solutions. This difficulty arises because the flow channels in a reactor lattice (defined in Fig. 9.13) are coupled to one another so tightly by lateral mass, momentum, and energy transport that it is not possible to analyze one channel independently of the others. The reactor designer, however, needs a better solution than is provided by the simple models because the fuel must be designed for a maximum cladding temperature, and this temperature is tied closely to the coolant temperature, which varies significantly between flow channels within a particular fuel assembly. A severe economic penalty would be imposed if the reactor designer were forced to set the maximum cladding temperature on the basis of the simple models. Accurate temperature distributions are also needed for the analysis of bowing behavior of assembly hexcans and fuel pins and, in safety-related transients, for the determination of the onset of coolant boiling.

For these reasons most SFR thermal hydraulics problems must be solved by complex numerical methods rather than either analytical means or simple models. Many computer programs have been developed to analyze the thermal hydraulic performance of SFR fuel assemblies. Some of these codes, such as COBRA, were originally developed for water reactor applications but have been modified for application to SFR's. SFR codes developed and applied in the U.S. include COBRA [5], THI3D [6], TRITON [7], and ORRIBLE [8]. A simple and faster code developed specifically for the SFR is ENERGY [9, 10] and later SUPERENERGY [11]. The SABRE code [12], developed in the U.K., is also widely used. The *control volume approach* described below is used in each of the codes listed.

Codes developed using the control volume approach are continually being developed for application to new transient conditions as well as for improvement in describing the steady state. However their use still relies heavily on experimentally determined coefficients, approximations, and validation against higher order mathematical models such as computational fluid dynamics (CFD).

10.2.2.1 The Control Volume Approach

The fuel lattice geometry imposes constraints on the flow field that cannot be handled conveniently with a conventional coordinate system. While reactor analysts continue to develop codes that provide a detailed resolution of the flow field within the boundaries of each individual flow channel by using the differential equations of fluid mechanics (e.g. CFD), the more useful approaches have historically involved solving systems of approximate equations that rely on simple physical arguments to describe the conservation of mass, energy, and momentum within each channel. These approximate equations are more readily solved for transient conditions and can more easily address scenarios in which phase-change occurs through correlations and approximations.

One such approach that has been particularly successful is the control volume approach developed by Meyer [13] and extended to three-dimensional reactor geometry by Rowe [5] (the author of the original COBRA series of codes) and later by others. In this approach the flow channels form control volumes that are allowed to communicate laterally with one another through the gaps between the fuel pins. A basic assumption of the approach is that any flow entering or leaving a control volume loses its sense of direction after crossing the control-volume boundary. Another important feature of the method is that the equations must ultimately be normalized to experiments or results from high order numerical methods for the particular lattice geometry and flow conditions, as noted earlier.

We present below the conservation equations used in the control volume approach. We shall limit this discussion to the steady-state case. Although transient solutions are needed for safety analysis, a satisfactory understanding of the physical problems can be obtained from the steady-state equations. We further limit the presentation to single-phase flow even though advanced codes like COBRA can, in some instances, be applied to two-phase flow to include boiling that may occur during transients. Finally, we will not discuss the numerical methods used to solve these equations since they are quite sophisticated and beyond the scope of this book. Moreover, new schemes are still being devised to reduce computer running time and simultaneously to provide more detailed information about coolant behavior, e.g., Ref. [14].

The objective of the calculation is to determine the velocities at both the inlet and outlet of each flow channel and the temperature (or enthalpy) and density at each channel outlet. The problem is usually defined in terms of four conservation equations, together with boundary conditions and an appropriate equation-of-state for the coolant. The four conservation equations generally used are conservation of mass, energy, axial momentum, and transverse momentum. It is usually assumed that the temperature (or enthalpy), density, and pressure at the inlet of the fuel assembly, and the pressure at the outlet of the assembly, are known and are the same for every flow channel in the assembly.

10.2.2.2 Conservation of Mass

Consider a control volume to be the space occupied by channel i in an axial section of length dz. Channel i is adjacent to J channels (with $J = 3$ for a repeating triangular lattice). The continuity equation is written for channel i adjacent to channel j, as defined by Fig. 10.2; a plan view of the geometry appears in Fig. 10.3.

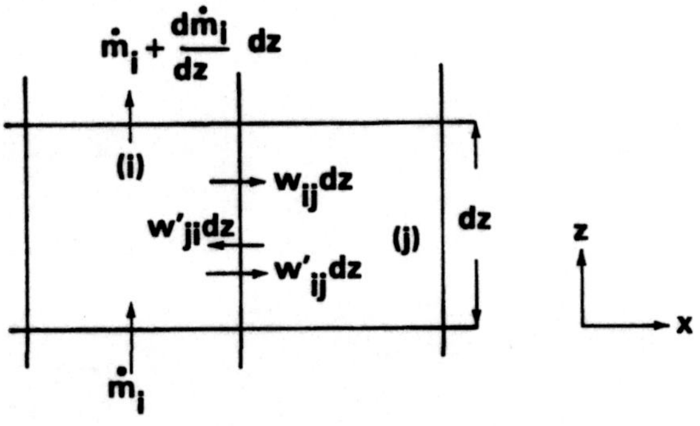

Fig. 10.2 Geometry for continuity equation

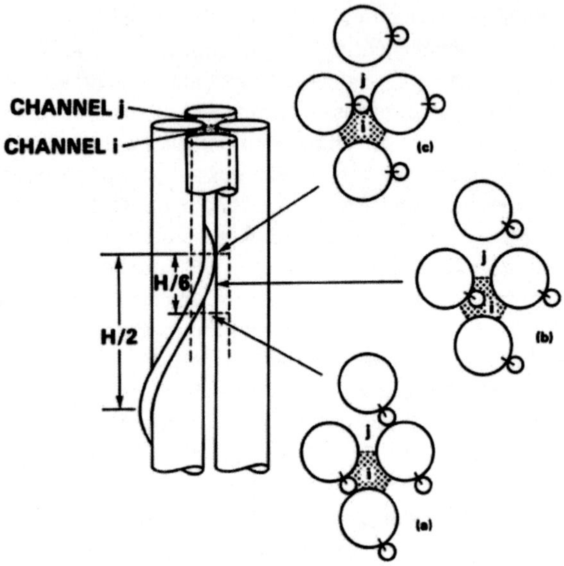

Fig. 10.3 Wire wrap orientation as it passes through channel *i* and into channel *j*

The upward flow rate $\dot{m}_i$ (kg/s) changes with axial length dz by $(d\dot{m}_i/dz)\,dz$.

The quantity $w_{ij}dz$ is the net *diversion crossflow*, i.e. the net transverse mass flow rate, from channel *i* to channel *j*. Consequently w_{ij} is the crossflow rate per unit length of channel (kg/m·s). Several sources of crossflow interact to produce the net diversion crossflow, including: (1) directional flow induced across the gap between fuel pins by the presence of the wire wrap, (2) non-directional flow induced by objects such as grid spacers, and (3) flow between channels caused by radial (transverse) pressure gradients.

The quantities w'_{ij} and w'_{ji} are *turbulent crossflow rates per unit length* (kg/m·s), and are associated with transport by eddy diffusion. These quantities are sometimes called the *turbulent mixing rates.* No net transport of mass occurs by this mechanism; hence, $w'_{ij} = w'_{ji}$. The phenomenon is important for energy and momentum transfer, however. For example, energy transport from turbulent mixing tends to reduce temperature gradients in the coolant.

Considering for the moment only one adjacent channel as shown in Fig. 10.2, a mass balance on the control volume gives, at steady state,

$$\dot{m}_i + w'_{ji}\,dz = \dot{m}_i + \frac{d\dot{m}_i}{dz}dz + w_{ij}\,dz + w'_{ij}\,dz. \tag{10.13}$$

Since $w'_{ij} = w'_{ji}$ this reduces to

$$\frac{d\dot{m}_i}{dz} = -w_{ij}. \tag{10.14}$$

Expanding the analysis to *J* channels around channel *i*,

$$\frac{d\dot{m}_i}{dz} = -\sum_{j=i}^{J} w_{ij}. \tag{10.14a}$$

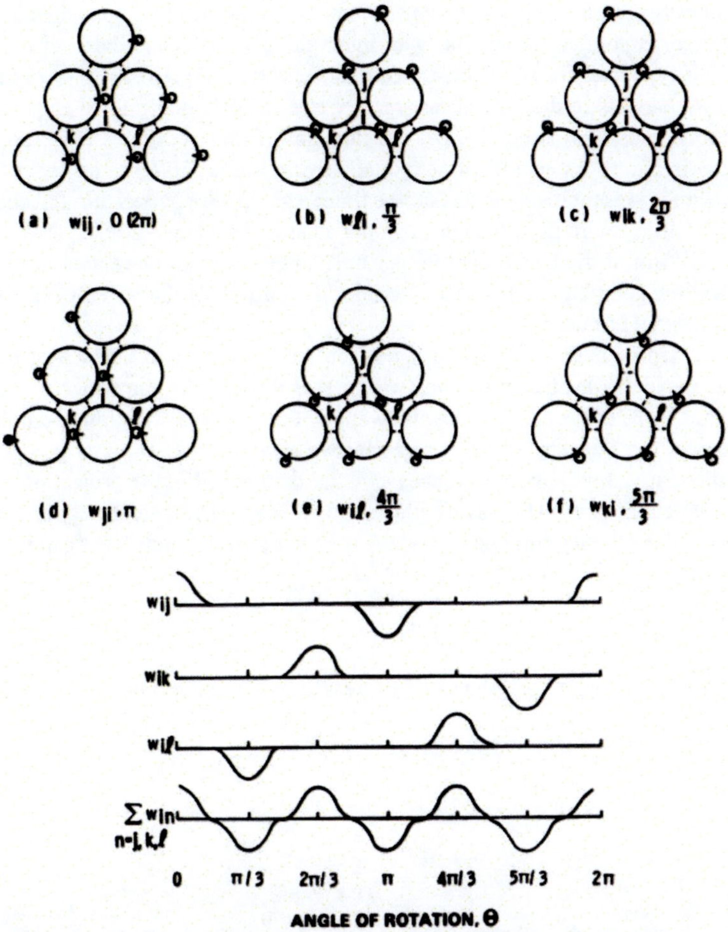

Fig. 10.4 Dependence of crossflow on orientation of the wire wrap. Listed below each configuration a through f is the positive crossflow which is maximum for the specific wire orientation followed by the angle of rotation of the wire, θ, relative to *a*. Plots of crossflows versus θ are qualitative; exact curves must be normalized to experiments. The *short lines* between the wires and the pins indicate the pin to which the wire is attached

Interchannel mixing has been described in several publications, e.g., Refs. [5, 15, 16]. Diversion crossflow induced by wire wrap can be described with the aid of Figs. 10.3 and 10.4. Flow channels i and j are illustrated in Fig. 10.3 and a wire is shown sweeping across channel i as it advances an axial distance $H/6$ and turns through an angle $\pi/3$ radians (60°). In the plan views the locations of the wire in channel i are illustrated at three axial positions. As the wire approaches channel j, then passes through the gap and on into channel j, it carries a fraction of the flow in channel i into channel j. The maximum crossflow occurs just below and above the wire as it passes through the gap (position c on Fig. 10.3). It is assumed that there is no crossflow between i and j when the wire enters channel i (position a), although crossflow enters channel i at that point from a channel other than j. At an axial height $H/2$ above and below position c, the wire from the pin on the right side of channel i will turn π radians around the pin and pass through the gap between channels i and j, and the crossflow between i and j will be a maximum in the reverse direction.

The crossflow between channels i and j consequently tends to pulse (in opposite directions) at axial intervals of $H/2$ (corresponding to angular wire intervals of π radians) when wire wrap spacers are used. This pulsing-type flow for w_{ij} is illustrated qualitatively in Fig. 10.4. Channel i is also adjacent to two other flow channels in the typical fast-reactor triangular lattice, shown as channels k and l in Fig. 10.4. The total crossflow from channel i is the sum of the crossflows to all adjacent channels and is oscillatory in nature, as also illustrated qualitatively in Fig. 10.4. It can be further noted from Fig. 10.4 that a wire is present in channel i half the time; e.g., no wire passes through channel i between positions a and b while a wire does pass through the channel between b and c.

The functional form of the crossflow rate per unit length, w_{ij} can be related to the channel mass flow rate, $\dot{m}_i$, and the geometry as follows. Consider the portion of the wire in Fig. 10.5 that carries crossflow from channel i to channel j.

Flow is carried from channel i to channel j as the wire advances a distance of roughly $H/6$, i.e. $H/12$ below and above the position where the wire passes through the gap which marks the boundary between the channels. Near the gap, the flow is swept by the wire in a direction approximately parallel to the wire. Hence, the flow through the gap has two components, an axial component u_i, and a transverse component v. As the wire advances an axial distance $H/6$, the projection of the crossflow follows an arc of length $(\pi/6)\,(D+s)$ as illustrated in Fig. 10.5. Therefore, for this approximate model, the ratio of the velocity components where the wire passes through the gap is

$$\frac{v}{u_i} = \frac{\pi\,(D+s)}{H}. \tag{10.15}$$

The diversion crossflow per unit length can be expressed by the following proportionality:

$$w_{ij} \propto \rho_i s v. \tag{10.16}$$

Replacing v by Eq. (10.15) and u_i by $\dot{m}_i/\rho_i A_i$, where A_i is the average channel flow area [defined by Eq. (9.31)], gives

$$w_{ij} = F\pi \frac{D+s}{H}\frac{s}{A_i}\dot{m}_i, \tag{10.17}$$

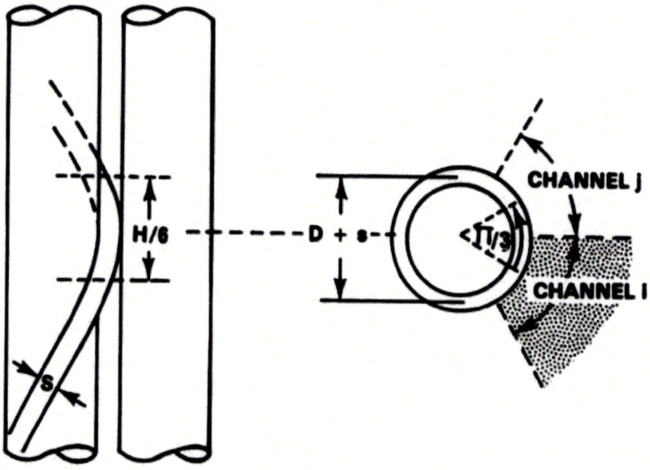

Fig. 10.5 Geometry for crossflow w_{ij}

where F is a proportionality constant that must be obtained from experiments and which may be a function of Re, *P/D*, and *H/D*.

10.2.2.3 Conservation of Energy

The energy flows are shown in Fig. 10.6. The quantity h^* refers to the enthalpy in the channel from which the diversion crossflow is coming. If the actual direction of the crossflow is from i to channel j (hence, w_{ij} positive), h^* is h_i; if the flow is from j to i, h^* is h_j. The χ_i dz term refers to energy from the fuel pins bounding the flow channel; χ_i, is the linear power per channel. The $-k_{ij}A_{ij}dT/dx$ term refers to heat conduction to adjacent channel j, where the x-direction is the direction perpendicular to the gap between flow channels i and j. This conduction term is more important in fast reactors than in water reactors because of the high conductivity of sodium and the tight spacing of the lattice.

The energy balance is

$$\begin{aligned} \dot{m}_i h_i + \chi_i dz + w'_{ji} h_j dz = \dot{m}_i h_i + \frac{d\left(\dot{m}_i h_i\right)}{dz} dz \\ -k_{ij} A_{ij} \frac{dT}{dx} + w_{ij} h^* dz + w'_{ij} h_i dz. \end{aligned} \tag{10.18}$$

We expand the derivative in the second term on the right as

$$\frac{d\left(\dot{m}_i\, h_i\right)}{dz} = \dot{m}_i \frac{dh_i}{dz} + h_i \frac{d\dot{m}_i}{dz} \tag{10.19}$$

and replace $d\dot{m}_i/dz$ by the continuity equation, so that

$$\frac{d\left(\dot{m}_i h_i\right)}{dz} = \dot{m}_i \frac{dh_i}{dz} - w_{ij} h_i. \tag{10.20}$$

The conduction term is written as

$$-k_{ij} A_{ij} \frac{dT}{dx} = -k_{ij} \left(\frac{s}{\Delta x}\right)_{ij} dz \left(T_j - T_i\right), \tag{10.21}$$

where s is the crossflow width ($A_{ij} = s_{ij} dz$) and Δx_{ij} is a representative distance between channels.

Recalling that $w'_{ji} = w'_{ij}$ substitution of the above relations gives for the energy equation

$$\frac{dh_i}{dz} = \frac{1}{\dot{m}_i} \left[\chi_i - k_{ij} \left(\frac{s}{\Delta x}\right)_{ij} \left(T_i - T_j\right) + w_{ij} \left(h_i - h^*\right) - w'_{ij} \left(h_i - h_j\right) \right]. \tag{10.22}$$

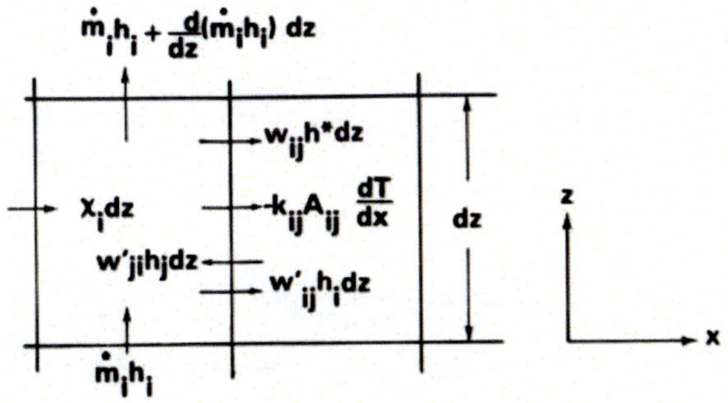

Fig. 10.6 Geometry for the energy equation

For J channels adjacent to channel i,

$$\frac{dh_i}{dz} = \frac{1}{\dot{m}_i}\left\{\chi_i + \sum_{j=1}^{J}\left[-k_{ij}\left(\frac{s}{\Delta x}\right)_{ij}(T_i - T_j) + w_{ij}\left(h_i - h^*\right) - w'_{ij}\left(h_i - h_j\right)\right]\right\}. \quad (10.22a)$$

10.2.2.4 Conservation of Axial Momentum

Axial forces acting on the control volume are shown in Fig. 10.7. The $F_i\ dz$ represents shear forces, $\rho_i g A_i\ dz$ is the gravitational force, $\dot{m}u_i$, is the axial momentum flux, $p_i A_i$ is the pressure force in the axial direction, and the $uw\ dz$ terms represent axial momentum exchanges between adjacent channels. The velocity u^* is the axial velocity in the channel from which the diversion crossflow is coming; thus, u^* is u_i for positive w_{ij} and u_j for negative w_{ij}. The flow area A_i is a function of z in the present context (varying with the changing position of the wire wrap) rather than being an average area as in Eqs. (9.31) and (10.17). Hence, A_i is included in the axial derivative of the pressure term.

The axial momentum equation[1] for one channel i adjacent to channel j is

$$\begin{aligned}\dot{m}_i u_i + p_i A_i + u_j w'_{ji} dz = \dot{m}_i u_i &+ \frac{d}{dz}(\dot{m}_i u_i)\,dz \\ &+ p_i A_i + \frac{d}{dz}(p_i A_i)\,dz + F_i dz + \rho_i g A_i\,dz \\ &+ u^* w_{ij} dz + u_i w'_{ij} dz.\end{aligned} \quad (10.23)$$

The shear term is divided into frictional and form loss terms as follows:

$$F_i dz = \left(f_i \frac{dz}{D_i}\frac{\rho_i u_i^2}{2} + K'_i\,dz \frac{\rho_i u_i^2}{2}\right) A_i, \quad (10.24)$$

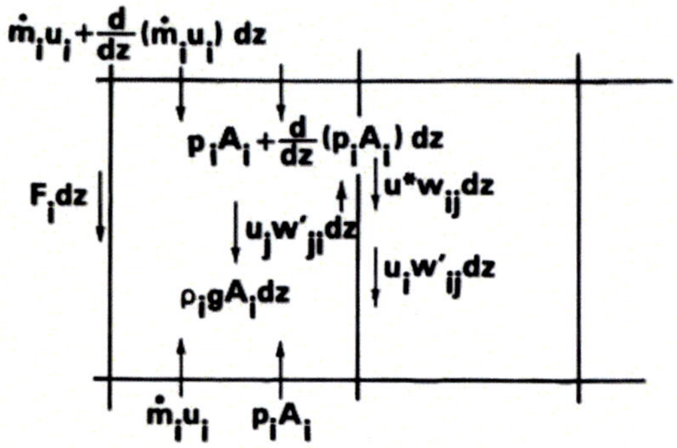

Fig. 10.7 Geometry for the axial momentum equation. The *arrows* indicate the direction of the force on the control volume

[1] A form of the momentum equation perhaps more familiar to some is $\vec{F}_S + \vec{F}_B = \int_{CS} \vec{V}\rho\vec{V}\cdot\vec{dA}$, where $\vec{F}_S$ and $\vec{F}_B$ are the surface and body forces acting on the control volume, and the momentum flux is integrated over the control surface (CS). The axial component equation [which is equivalent to Eq. (10.23)] is

$$F_{Sz} + F_{Bz} = -\dot{m}_i u_i + \dot{m}_i u_i + \frac{d}{dz}(\dot{m}_i u_i)\,dz + u^* w_{ij} dz + u_i w'_{ij} dz - u_j w'_{ji} dz.$$

where f_i is a friction factor and K_i' is a form loss coefficient per unit length. Since $\dot{m}_i = \rho_i u_i A_i$, this term can be written as

$$F_i\, dz = \left(\frac{f_i}{2D_i\rho_i} + \frac{K'_i}{2\rho_i}\right)\left(\frac{\dot{m}_i}{A_i}\right)^2 A_i\, dz. \tag{10.25}$$

The change in momentum flux can be expressed, using $\dot{m} = \rho u A$ and replacing $d\dot{m}/dz$ by the continuity equation, as

$$\begin{aligned}\frac{d}{dz}\left(\dot{m}_i u_i\right) dz &= \left(\dot{m}_i \frac{du_i}{dz} + u_i \frac{d\dot{m}_i}{dz}\right) dz \\ &= \left\{\dot{m}_i\left[\frac{1}{\rho_i A_i}\frac{d\dot{m}_i}{dz} - \frac{\dot{m}_i}{(\rho_i A_i)^2}\frac{d(\rho_i A_i)}{dz}\right] + u_i \frac{d\dot{m}_i}{dz}\right\} dz \\ &= \left[-2u_i w_{ij} - \left(\frac{\dot{m}_i}{\rho_i A_i}\right)^2 \frac{d(\rho_i A_i)}{dz}\right] dz.\end{aligned} \tag{10.26}$$

With these substitutions, the axial momentum equations can be written as

$$\left[-2u_i w_{ij} - \left(\frac{\dot{m}_i}{\rho_i A_i}\right)^2 \frac{d}{dz}(\rho_i A_i)\right] dz + \frac{d}{dz}(p_i A_i)\, dz$$

$$+ \left(\frac{f_i}{2D_i\rho_i} + \frac{K'_i}{2\rho_i}\right)\left(\frac{\dot{m}_i}{A_i}\right)^2 A_i\, dz + \rho_i g A_i\, dz + u^* w_{ij} dz + \left(u_i - u_j\right) w'_{ij} dz = 0$$

or

$$\begin{aligned}\frac{d}{dz}(p_i A_i) = &-\left(\frac{\dot{m}_i}{A_i}\right)^2\left[\frac{f_i A_i}{2D_i\rho_i} + \frac{K'_i A_i}{2\rho_i} - \frac{1}{\rho_i^2}\frac{d}{dz}(\rho_i A_i)\right] - \rho_i g A_i \\ &+ (2u_i - u^*)\, w_{ij} - \left(u_i - u_j\right) w'_{ij}.\end{aligned} \tag{10.27}$$

For J channels adjacent to channel i,

$$\begin{aligned}\frac{d}{dz}(p_i A_i) = &-\left(\frac{\dot{m}_i}{A_i}\right)^2\left[\frac{f_i A_i}{2D_i\rho_i} + \frac{K'_i A_i}{2\rho_i} - \frac{1}{\rho_i^2}\frac{d}{dz}(\rho_i A_i)\right] - \rho_i g A_i \\ &+ \sum_{j=1}^{J}\left[(2u_i - u^*)\, w_{ij} - \left(u_i - u_j\right) w'_{ij}\right].\end{aligned} \tag{10.27a}$$

10.2.2.5 Conservation of Transverse Momentum

The control volume for the equation to calculate momentum transfer between channels i and j is illustrated in Fig. 10.8. The control volume is the gap region between channel i and j in which s is the gap

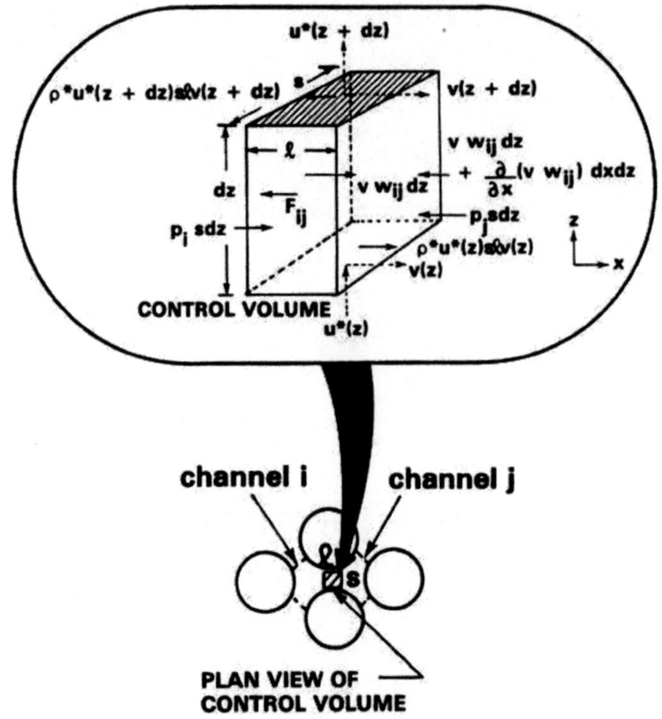

Fig. 10.8 Geometry for the transverse momentum equation. The *arrows* associated with the forces indicate the direction of the force on the control volume. The *dashed arrows* indicate the direction of the flow velocities, v and u^*

thickness (the minimum distance between fuel pins) and l is a characteristic distance between channels (a parameter which must be normalized to experiments or results from higher order numerical methods).

The transverse momentum entering the control volume with the flow upward through the bottom is $\rho^* u^*(z)\, slv(z)$, where ρ^* and u^* represent values in the channels from which the flow is coming. The momentum $\rho^* u^*(z+dz)\, slv(z+dz)$ is lost out the top. A second transverse momentum flux entering the control volume across the face of area $s\,dz$ is $vw_{ij}\,dz$, and the momentum flux leaving is $\left[vw_{ij} + \partial/\partial x\,(vw_{ij})\,dx\right] dz$. The force balance is

$$\rho^* u^*(z)\, slv(z) + vw_{ij}\,dz + p_i s\,dz = \rho^* u^*(z+dz)\, slv(z+dz) + \left[vw_{ij} + \frac{\partial}{\partial x}(vw_{ij})\,dx\right] dz + p_j s\,dz + K_{ij}\frac{\rho^* v^2}{2} s\,dz, \tag{10.28}$$

where K is a coefficient that accounts for both frictional and form losses for the crossflow. Since $\rho^* sv = w_{ij}$ in the gap, the equation can be written as

$$l\frac{\partial}{\partial z}(u^* w_{ij})\,dz + \frac{\partial}{\partial x}(vw_{ij})\,l\,dz = (p_i - p_j)\,s\,dz - K_{ij}\frac{w_{ij}^2}{2\rho s}\,dz,$$

or

$$\frac{\partial}{\partial z}\left(u^* w_{ij}\right) + \frac{\partial}{\partial x}\left(v w_{ij}\right) = \left(p_i - p_j\right)\frac{s}{l} - K_{ij}\frac{w_{ij}^2}{2\rho s l}. \tag{10.29}$$

The two momentum flux terms contribute little during normal steady-state flow. Both terms were omitted in early thermal-hydraulic codes, and the second one is sometimes still omitted; they are left in computer codes more for reasons of numerical stability for transient solutions than for the magnitude contributed to the momentum equation. Setting the left hand side of Eq. (10.29) equal to zero provides a simple relationship between $\left(p_i - p_j\right)$ and w_{ij} that is accurate for the steady-state case.

10.2.2.6 Calculated Results

Codes that solve these equations numerically have achieved considerable success in comparisons with experimental data. It is necessary, however, to obtain some of the coefficients in the equations from a combination of physical arguments and normalizations to experiments-quantities such as F, w'_{ij}, K_{ij}, l, $k\,(s/\Delta x)_{ij}$ and K'_i. The friction factor f_i in Eq. (10.24) can be determined by the methods described in Section 10.3.3.

Examples of detailed temperature distributions calculated with transverse mixing models are shown in Figs. 10.9 and 10.10. The curves in Fig. 10.9 were calculated for FFTF using the COBRA code. Outlet temperatures for a fuel assembly in Phénix are shown for calculations with and without accounting for transverse mixing. The asymmetry in Fig. 10.10 results from the radial gradient in the power distribution across the assembly.

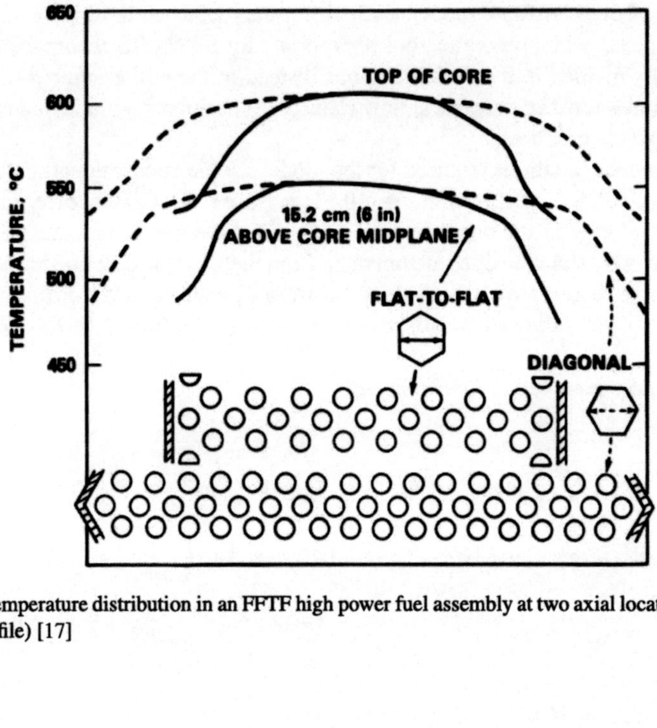

Fig. 10.9 Sodium temperature distribution in an FFTF high power fuel assembly at two axial locations (217-pin bundle; flat radial power profile) [17]

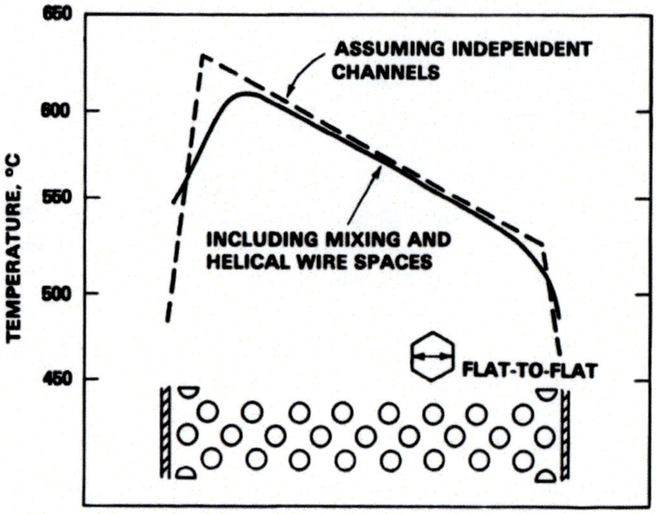

Fig. 10.10 Sodium temperature distribution in a Phénix fuel assembly at the top of the core (217-pin bundle; radial power skew) [18]

10.2.2.7 The ENERGY Model

The solution of the flow equations for every channel in an assembly is numerically intense and requires modern computers. A useful and simplified model called ENERGY, which calculates a reasonably good exit temperature distribution for forced convection problems in a short computer time, was developed specifically for the SFR when computer resources were more limited [9, 10]. Though no longer limited by computer resources (the need for simplified models is diminished), it is still a valuable exercise to review such simplified models because of the insight that they provide to identifying the most important terms for modeling the problem.

The simplicity of the ENERGY model results from the replacement of the exact momentum coupling between channels with approximations appropriate for SFR's. To accomplish this task, the fuel assembly is radially divided into only two regions instead of the entire number of subchannels. The central region includes most of the central flow channels. The outer region includes the flow channels near and adjacent to the hexcan.

The only differential equations required for the model are the energy equations in the two regions. Three velocities are defined, as shown in Fig. 10.11; these are the axial velocity in the central region $\left[(U_I)_z\right]$, the axial velocity in the outer region $\left[(U_{II})_z\right]$, and the circumferential velocity in the outer region $\left[(U)_s\right]$. The two axial velocities are obtained from the analysis described in Section 10.2.1. The effect of crossflow is represented by turbulent diffusion through an eddy diffusivity ε. The circumferential velocity $(U)_s$ and the eddy diffusivity ε are obtained by fitting the solution to experimental data.

The energy equations solved in ENERGY are

$$\rho c_p\,(U_I)_z\,\frac{\partial T}{\partial z} = \left(\rho c_p \varepsilon_I + \xi k\right)\left(\frac{\partial^2 T}{\partial x^2} + \frac{\partial^2 T}{\partial y^2}\right) + Q\,(x, y, z) \tag{10.30}$$

$$\begin{aligned}\rho c_p\,(U)_s\,\frac{\partial T}{\partial s} + \rho c_p\,(U_{II})_z\,\frac{\partial T}{\partial z} &= \left(\rho c_p \varepsilon_n + \xi k\right)\left(\frac{\partial^2 T}{\partial n^2}\right)\\ &\quad + \left(\rho c_p \varepsilon_s + \xi k\right)\left(\frac{\partial^2 T}{\partial s^2}\right) + Q\,(s, n, z)\,.\end{aligned} \tag{10.31}$$

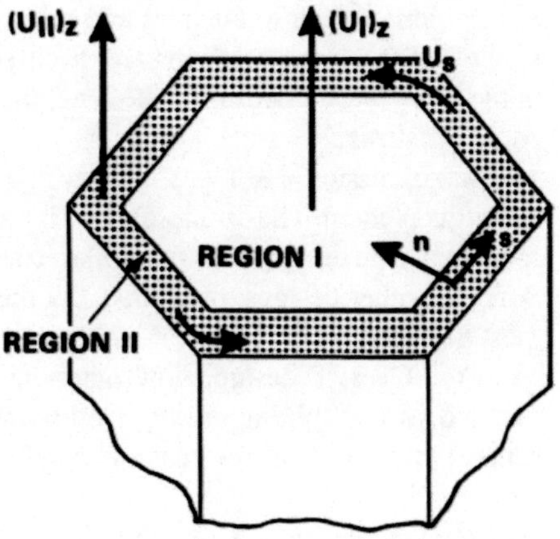

Fig. 10.11 Flow fields in the two regions of the ENERGY code (adapted from Ref. [9])

The terms on the left represent energy transport by convection. The first terms on the right side of the equations represent energy transport by eddy diffusion and conduction. The heat source from the fuel is given by Q. The factor ζ (< 1) that multiplies the conductivity k accounts for the winding (and hence lengthened) path between fuel pins followed by the sodium as energy is conducted in the direction transverse to the bulk flow; this factor is estimated from physical arguments. Good results are obtained by assuming that all eddy diffusivities are equal.

The boundary conditions for this system of equations include continuity of heat flux and temperature between zones, inlet temperature distribution, and a thermally insulated boundary at the hexcan wall. A finite difference method is used to solve the equations.

The model is normalized to experimental data through the two parameters ε and $(U)_s$. Nondimensional forms of these parameters are expressed as functions of Re, *P/D,* and *H/D* [19]. The authors of the ENERGY code report that temperature distributions obtained with that code for forced convection problems for rod bundles of normal geometry are as accurate as those obtained with larger codes, with a considerable savings in computation time. However, the model is not as effective as the more detailed codes for analyzing natural convection, transient effects, flow blockages, and other conditions out of the range of data to which the values of ε and $(U)_s$, have been normalized.

10.3 Coolant Flow Distributions and Pressure Losses

Having discussed coolant velocity and temperature fields within assembly bundles, we can now broaden our knowledge of core thermal-hydraulics by addressing the flow patterns and pressure drops within the reactor vessel as a whole. We will then conclude this section by focusing on the pressure drop across the fuel pins in a fuel bundle.

10.3.1 In-vessel Coolant Flow Distribution

Before discussing the flow patterns within the reactor vessel, it is important that we define the portion of the primary loop under discussion. We recall from Fig. 2.5 that for a loop-type system, the primary sodium is circulated through pumps and intermediate heat exchangers (IHX) outside the reactor vessel. For a pool-type system all the heavy primary components are contained within the main reactor vessel. As a consequence, the pump and IHX characteristics and the accompanying pressure distribution exterior to the inlet and outlet plenum are substantially different between these two basic systems. These considerations are discussed in Chapter 12. Between the inlet and outlet plenum, i.e., in the immediate environs of the core, the general flow characteristics are similar for both loop- and pool-type designs.

This discussion will focus on flow and pressure distributions as coolant is driven from the inlet to the outlet plenum. The example given is based on a loop-type design, but the methods would be similar for a pool-type design. This particular example is taken from the CRBRP design [20]. Numbers would differ for other designs, of course, but the flow division shown is instructive because it is typical of most SFR designs.

In the CRBRP design, flow enters an inlet plenum from three loops. The inlet flows must be directed so that mixing during a transient does not result in excessive thermal stresses in the core support structure and the reactor vessel when combined with other pressure and mechanical load stresses.

From the inlet plenum, the flow for the homogeneous CRBRP core was calculated to be distributed according to the following pattern:

Fuel assemblies	80%
Control assemblies	1.6%
Radial blanket assemblies	12%
Radial shielding (blanket modules)	0.3%
Periphery of core support structure	4.7%
Leakage through seals	1.4%

The flow through the fuel, control, and radial blanket assemblies passes from the inlet plenum through core support modules. Each module contains seven assemblies. The center inlet module has the highest flow rate; this flow dictates the core pressure drop. Flow distribution through the fuel and control assemblies is controlled by orificing in the individual assemblies rather than in the inlet module. For the radial blanket assemblies, the orificing is in the inlet module instead of the individual assemblies to allow radial blanket assemblies to be shuffled. There are some radial shielding assemblies adjacent to the radial blanket that form seven-assembly clusters with blanket assemblies, and hence are fed by inlet modules. These account for the 0.3% flow given in the list above.

From the inlet module, the sodium flows through the inlet nozzle at the bottom of each assembly. The sodium then flows through an orifice in the fuel and control assemblies. The orifice is designed to equalize maximum cladding midwall temperatures at the equilibrium cycle end-of-life conditions. In the fuel assembly, the sodium then flows past the neutron shielding below the core and upward into and through the rod bundle itself, i.e., the lower axial blanket, the core, the upper axial blanket, and the fission gas plenum. As described later in this section, a large part of the pressure drop occurs across the rod bundle.

The sodium exits from the assemblies into channels that guide the flow up to the vessel outlet plenum. From there it passes to the outlet pipes of the three primary system loops.

The maximum hydraulic load is on the core support plate because it is subjected to the core pressure drop. For the CRBRP the design pressure above the sodium pool, in the cover gas, is only 104 kPa absolute, or 2.5 kPa above atmospheric (0.36 psig). The maximum pressure in the inlet plenum below the core support plate is 950 kPa absolute (123 psig).

A hydraulic balance system is used to assure holddown of the fuel assemblies. A significant portion of the bottom of each assembly is exposed to the low pressure of the outlet plenum so that the net force downward on the fuel assembly is gravity and is equal to the buoyant weight of the assembly.

10.3.2 Pressure Losses

The pressure drop in the flow paths described above are considered to be either *form* or *friction* pressure losses. Form pressure losses are given by the equation

$$\Delta p = K\frac{\rho V^2}{2} \tag{10.32}$$

and friction pressure losses are given by

$$\Delta p = f\frac{L}{D}\frac{\rho V^2}{2}, \tag{10.33}$$

where the form factors K are either determined experimentally for the particular design or, like the friction factors, obtained from analysis.

Complete results are available for the CRBRP design [20]. It is instructive to see the detail with which the contributions to the pressure drop are calculated by studying the CRBRP results in Table 10.1 for the maximum-power fuel assembly. The *rod bundle* includes the active core region, the axial blankets, and the fission gas plenum.

All pressure drops except the rod bundle loss were reported to be known to ±20% uncertainty with 95% confidence. The rod-bundle pressure drop was known to ±14% uncertainty with 95% confidence. These uncertainties, when applied to the results in Table 10.1, lead to a maximum pressure drop of 850 kPa between the inlet and outlet plenums.

Table 10.1 Pressure losses in the CRBRP reactor vessel for maximum-power fuel assembly [20]

	Pressure drop		
Component sub-component	Component (kPa)	Sub-component (kPa)	Loss coefficients
Inlet Plenum	32		$K = 1.29$
Module	62		
Slots		39	$K = 2.01$
Strainer		3	$K = 0.363$
Expansion		8	$K = 0.41$
Manifold		6	$K + f\frac{L}{D} = 1.0$
Stalk		6	$K + f\frac{L}{D} = 1.0$
Inlet Nozzle	105		
Transition		19	$f\frac{L}{D} = 0.022$
Inlet loss (form)		59	$K = 0.098$
Inlet loss (frictional)		4	$f\frac{L}{D} = 0.0024$
Nozzle friction (1)		4	$f\frac{L}{D} = 0.000082$
Nozzle friction (2)		19	$f\frac{L}{D} = 0.00035$
Shield, Orifice	155		Not given
Rod Bundle Inlet	7		$K = 0.37$
Rod bundle	305		$f\frac{L}{D}$ From Novendstern correlation
Assembly exit	21		$K = 0.79$
Outlet Plenum	33		$K = 1.32$
Total	720		

10.3.3 Pressure Drop in the Fuel Assembly

Early United States, French, Russian, and Japanese prototype SFR designs utilized wire wrap to space the fuel pins in the hexagonal assembly channel, as discussed in Section 8.4.3. The pressure drop for this method of spacing is presented here. An alternate method that employs grid spacers at various heights along the fuel pins was used in early British and German designs and the U.S. Fermi Reactor, but the pressure drop caused by spacers is sensitive to the particular grid spacer design. Therefore, no general correlations for grid spacer pressure drop are given in this text.

Two pressure drop correlations for an SFR fuel assembly with wire wrap spacers are presented here, the first due to Novendstern [1] and the second due to Chiu et al. [3].

The earliest correlation used extensively in the U.S. for determining pressure losses in SFR pin bundles was developed by de Stordeur [21] for the Fermi Reactor in the early 1960s (a reactor which used grid spacers instead of wire wrap). In the late 1960s Sangster [2] presented a correlation for wire wrap spacers. Later Novendstern [1], adopting several of Sangster's principal features, developed a correlation that was used for the FFTF and CRBRP designs. Rehme [22] also contributed an improved method at about the same time as Sangster. Novendstern's analysis takes advantage of all the older methods, together with experimental data of Reihman [23], Rehme [22], and Baumann et al. [24], and his method agrees closely with pressure loss measurements for an FFTF fuel assembly [25]. Finally, Chiu, Rohsenow, and Todreas extended the analysis further to include drag across the wire wrap more explicitly [3].

Prior to the development of the Chiu-Rohsenow-Todreas model, the pressure drop was obtained from a modified form of the skin friction pressure drop equation, Eq. (10.33). Chiu et al., introduced separate expressions to account for the following two components of the pressure drop:

(1) form drag pressure drop induced by flow over the wires,
(2) skin friction pressure drop characterized by the resultant velocity of both the axial and transverse velocity components of the flow.

Hence, the Chiu-Rohsenow-Todreas model uses modified forms of both Eqs. (10.32) and (10.33).

10.3.3.1 Novendstern Model

In this model the influence of the wire wrap is accounted for by means of an effective friction factor in Eq. (10.33). A friction factor multiplier, *M*, is introduced such that the effective friction factor, f_1, for channel 1 (Fig. 9.13) is given by

$$f_1 = M f_{smooth}. \tag{10.34}$$

The multiplier *M* is obtained from a correlation based on the dimensionless parameters pitch-to-diameter (*P/D*), lead-to-diameter (*H/D*), and Reynolds number (Re); and f_{smooth} is a function of Re. Hence, *M* accounts for the increase in *f* above the smooth tube value caused by the wire spacer.

The correlation for *M* proposed by Novendstem is

$$M = \left[\frac{1.034}{(P/D)^{0.124}} + \frac{29.7\,(P/D)^{6.94}\,\mathrm{Re}_1^{0.086}}{(H/D)^{2.239}} \right]^{0.885}, \tag{10.35}$$

where $\mathrm{Re}_1 = \rho V_1 D_{e1}/\mu$.

In the expression for Re, and in the skin friction pressure drop equation, V_1 is given by Eq. (10.9) and D_{e1} by Eq. (9.30).

The Blasius relation for the friction factor, f_{smooth} is applicable for the flow conditions in SFRs; hence,

$$f_{smooth} = \frac{0.316}{\mathrm{Re}_1^{0.25}}. \tag{10.36}$$

Note that only the value of f_1 is given. It is sufficient that the final calculated pressure drop is based on channel type 1 because the definition of the flow distribution factor, X_i assures that Δp is the same for all channels (see Section 10.2.1).

The final value for Δp across the fuel pins comes from combining Eqs. (10.33) and (10.34) to obtain

$$\Delta p = M f_{smooth} \frac{L}{D_{e1}} \frac{\rho V_1^2}{2}. \tag{10.37}$$

Calculated results are compared [1] with experimental pressure drop data for a 217-pin FFTF fuel test assembly [25] in Fig. 10.12. These data were not used for the development of the correlation; hence, they provide an independent check of the correlation. The ±14% error curves (also from Ref. [1]) include most of the data points.

The friction factor multiplier, M, is plotted as a function of P/D and of H/D in Fig. 10.13 for a Reynolds number of 50,000, a fairly typical value for an SFR. For the FFTF the parameters are

$D = 5.84$ mm (0.230 in.)
$P = 7.31$ mm (0.288 in.) ($P/D = 1.25$)
$H = 305$ mm (12 in.) ($H/D = 52$)
$\mathrm{Re} = 58{,}000$

for which $M = 1.05$. Hence, the multiplier that accounts for the wire wrap deviates only slightly from unity for designs of most practical interest.

10.3.3.2 Chiu-Rohsenow-Todreas (CRT) Model

While the Novendstern model provides reasonably accurate results, as illustrated in Fig. 10.12, the CRT model treats the mechanisms responsible for the pressure drop in greater detail. Thus the CRT model has greater potential for application over a wider range of flow conditions.

In the CRT model the pressure drop is modeled as the sum of two terms:

$$\Delta p = \Delta p_s + \Delta p_r, \tag{10.38}$$

where

Δp_s = skin friction losses
Δp_r = form drag losses from the component of velocity in the direction perpendicular to the wire.

For the form drag term, the form coefficient K in Eq. (10.32) is expressed as $C_D A_r/A'$, where C_D, is a drag coefficient and A_r, and A' are defined in Eq. (10.11). The drag coefficient and the velocity

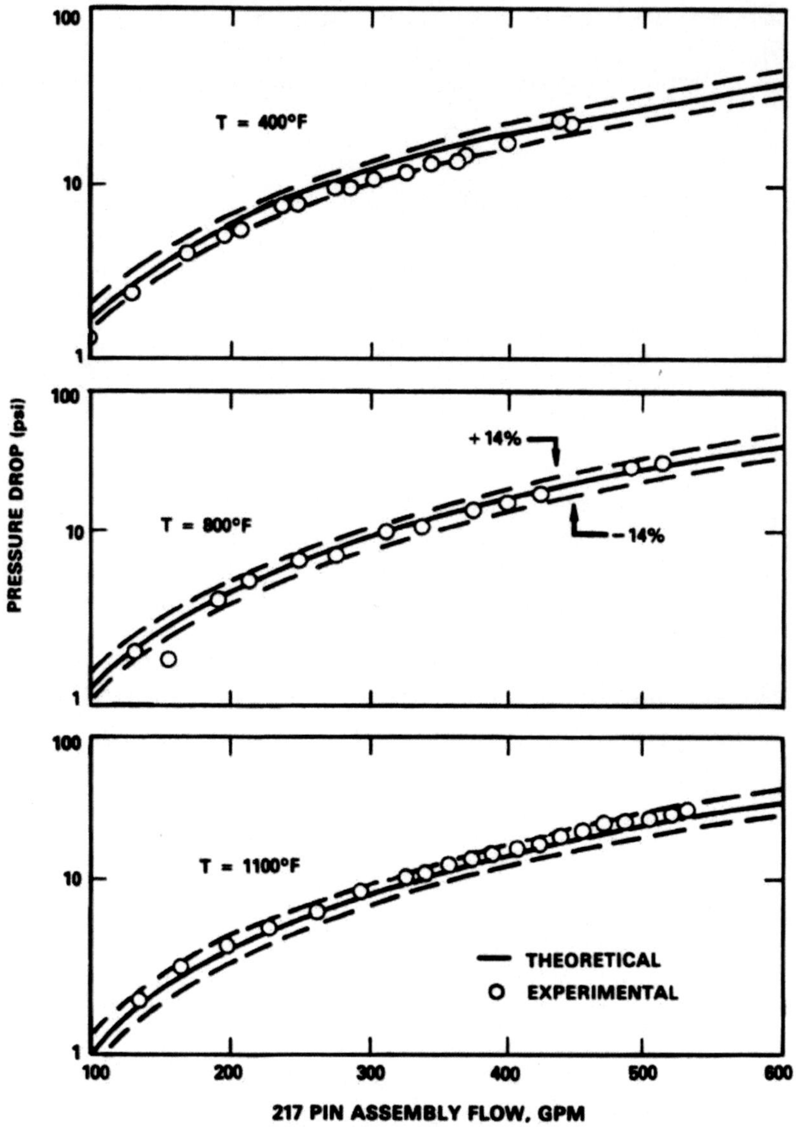

Fig. 10.12 Comparison of Novendstem model with experimental sodium pressure drop data [1]

component perpendicular to the wire, V_p, are related through a dimensionless constant, C'_1, and the skin friction factor based on the pressure drop in the flow channel without the wire, f_s, such that the drag pressure loss component over one lead of fuel pin is

$$\Delta p_r \text{ (one lead)} = C'_1 f_s \frac{A_r}{A'} \frac{\rho V_p^2}{2}. \tag{10.39}$$

For a fuel pin length L, there are L/H leads, so that Δp_r, for the entire pin is

$$\Delta p_r = C'_1 f_s \frac{A_r}{A'} \frac{L}{H} \frac{\rho V_p^2}{2}. \tag{10.40}$$

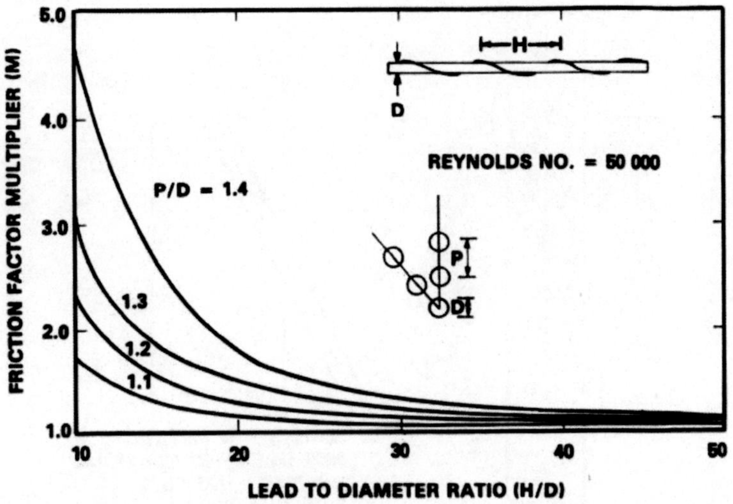

Fig. 10.13 Friction factor multiplier, M [1]

The velocity component V_p is expressed in terms of the more useful axial, transverse, and resultant velocities V_A, V_T, and V_R according to the geometrical considerations in Fig. 10.14. Thus,

$$V_p^2 = V_R^2 \sin^2\theta = V_A^2\left[1+\left(\frac{V_T}{V_A}\right)^2\right]\sin^2\theta. \tag{10.41}$$

The form drag pressure loss is, therefore,

$$\Delta p_r = C'_1 f_s \frac{A_r}{A'}\frac{L}{H}\frac{\rho V_A^2}{2}\left[1+\left(\frac{V_T}{V_A}\right)^2\right]\sin^2\theta. \tag{10.42}$$

The skin friction component of the pressure drop is

$$\Delta p_s = f_R \frac{L_R}{D_{eR}}\frac{\rho V_R^2}{2}, \tag{10.43}$$

where the subscript R denotes values related to the resultant velocity V_R. Geometrical arguments led Chiu et al., to replace the values related to V_R as follows:

$$\Delta p_s = f_s \frac{L}{D_e}\frac{\rho V_A^2}{2}\left[1+\left(C_2\frac{V_T}{V_A}\right)^2\right]^{1.375}, \tag{10.44}$$

where f_s, is the same as in Eq. (10.42) and C_2 is a constant to be determined empirically.

Chiu et al., next provide reasons why the skin friction component is controlling for the edge channel pressure drop to the extent that the Δp_r component can be dropped for channel 2. They argue further that $V_T/V_A << 1$ for the interior channels so that V_T/V_A can be made zero for channel 1. Moreover, when $V_T/V_A = 0$, θ and θ' on Fig. 10.14 become congruent so that $\sin\theta$ in Eq. (10.42) can be replaced by $\sin\theta'$ as given in Fig. 10.14. Finally the π^2 in the numerator of the expression for $\sin^2\theta'$ is absorbed into the constant C'_1 to give a new constant C_1 Thus, the form drag pressure drop in channel 1 is

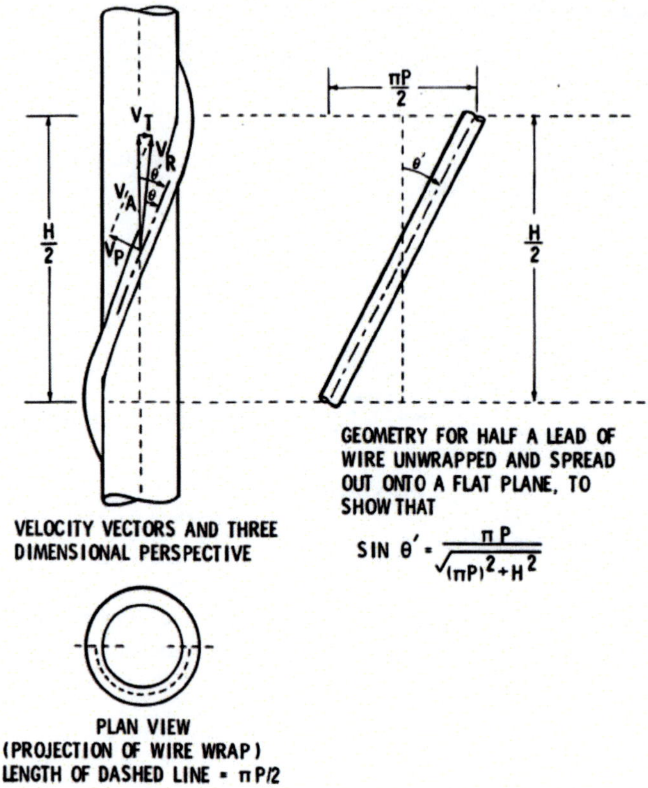

Fig. 10.14 Geometry for the Chiu-Rohsenow-Todreas model. The axial and transverse velocity components V_A, and V_T are equivalent to u and v in Section 10.2.2

$$\Delta p_{r1} = C_1 f_{s1} \frac{A_{r1}}{A'_1} \frac{L}{D_{e1}} \frac{D_{e1}}{H} \frac{P^2}{(\pi P)^2 + H^2} \frac{\rho V_1^2}{2}, \tag{10.45}$$

where V_A in channel 1 has been replaced by V_1.

The final expressions for the total pressure drop in channels 1 and 2 are

$$\Delta p_1 = f_{s1} \frac{L}{D_{e1}} \frac{\rho V_1^2}{2} \left[1 + C_1 \frac{A_{r1}}{A'_1} \frac{D_{e1}}{H} \frac{P^2}{(\pi P)^2 + H^2} \right], \tag{10.46}$$

$$\Delta p_2 = f_{s2} \frac{L}{D_{e2}} \frac{\rho V_2^2}{2} \left\{ 1 + \left[C_2 n \left(\frac{V_T}{V_2} \right)_{gap} \right]^2 \right\}^{1.375}, \tag{10.47}$$

where V_A in channel 2 has been replaced by V_2 and V_T/V_A has been replaced by $n\,(V_T/V_2)_{gap}$, where n and $(V_T/V_2)_{gap}$ are defined following Eq. (10.12).

The velocities V_1, and V_2 to be used in Eqs. (10.46) and (10.47) are those given by Eqs. (10.11) and (10.12). These velocities (and flow split parameters X_1 and X_2, equal to $V_1/\overline{V}$and $V_2/\overline{V}$) were obtained by forcing the above expressions for Δp_1 and Δp_2 to be equal, together with the continuity equation, Eq. (10.5).

10.3.3.3 Other Models

Numerous pressure drop correlations have been published in addition to the two described in this section. Generally, all such correlations are specific to a particular set of experimental data used by the authors to develop the correlation. Hence, the range of applicability of any such correlation is limited to the range of data used in developing it. Application of a correlation beyond the range of experimental conditions used in its development requires careful attention. This use should be justified by assessing the correlation performance using data that closely resembles the intended application in order to determine correlation biases and uncertainties in that range.

Several correlations, including Novendstern's [1], are described in Ref. [26]. These correlations are compared to a broad database of pressure drop data in order to evaluate which correlations have the best accuracy over a wide range of data and types of fluids. Such an evaluation helps to identify what correlations can be used in general design evaluations for fast reactor systems. However, for specific applications in which detailed reactor design and analysis is being performed, it is important to develop a new correlation or modify an existing one to predict design specific data with the best achievable accuracy.

10.4 Hot Channel Factors

The first three sections of this chapter have addressed the methods for calculating nominal temperatures, where "nominal" refers to values calculated for rated full-power operation without accounting for uncertainties. We noted in Section 10.2 that substantial variations in temperature can exist within an assembly at a given axial position. These are all predictable values, however, and these determinations constitute only the first part of the designer's job. The next part is the treatment of uncertainties in the analysis. The impacts of uncertainties in theoretical and experimental analyses, instrumentation accuracy, manufacturing tolerances, properties, and correlation uncertainties must be considered in order to assure safe and reliable reactor operation. Moreover, the influence of uncertainties on design will indicate the importance of increasingly sophisticated thermal-hydraulics experiments in the development of fast reactor technology. This section deals with the treatment of these uncertainties.

The thermal-hydraulics design of a fast reactor must conform to a set of *design bases*, or *design criteria*. Many of these relate to fuel, cladding, and coolant exit temperatures under various conditions. One fuel design criterion for FFTF and CRBRP, for example, was that no fuel melting should occur at some specified overpower; for both reactor designs this *overpower* was set at 115% of rated power. The maximum allowable cladding temperature must assure fuel pin integrity. The maximum allowable exit coolant temperature must assure structural integrity above the core. Other thermal-hydraulic design bases refer to other parameters such as allowable pressure drop and coolant flow velocities.

In order to satisfy the design bases, *design limits* must be specified. This is often done by establishing a statistical *level of confidence* with which selected parameters must be known. An uncertainty in a parameter, or a limiting value, is then associated with the specified level of confidence.

Uncertainties and limiting values are treated through the use of hot channel factors, or hot-spot factors. The hot channel factor, F, for a particular parameter is the ratio of the maximum value of that parameter to its nominal value. It is therefore a number greater than unity, and the decimal part (i.e., $F - 1$) represents the fractional uncertainty in the parameter. Hot channel factors must be based on a combination of experimental data and experimentally verified analytical methods; justification of these factors is one of the most important and challenging of the reactor designer's tasks.

The exact definitions and values of hot channel factors and the way they are applied to the analysis of uncertainties will change and evolve with each new fast reactor design and the complexities and speed of numerical analysis techniques. Methods used are strongly influenced by precedents used in the licensing process; indeed, LWR precedents have already influenced fast reactor methods. Since we have available the extensive uncertainty analyses for FFTF [27] and CRBRP [28–30], these are used here as guides to the methodology for calculating uncertainties. Analyses for later designs will continue to become more elaborate, particularly as more experimental data become available, more advanced numerical models are developed, and as incentives grow to decrease design margins. The important objective of this section, therefore, is to introduce the methodology involved in the use of hot channel factors, not to spell out the details of the particular analyses.

The parameter multiplied by a hot channel factor is generally a temperature difference. Consequently, the result is independent of the temperature units being used.

10.4.1 Statistical Methods

Two types of uncertainties might influence a particular parameter—*direct uncertainties*, or *biases*, and *random uncertainties*. A direct uncertainty in a parameter represents influence on the parameter by a variable that is not subject to random variation, but for which the exact value cannot be predicted in advance. A random uncertainty represents influence on the parameter by a variable that can be represented by a frequency distribution of occurrence. Several sources of each type of uncertainty might be present. Individual uncertainties propagate to yield an overall uncertainty for the parameter; the method of propagation differs for the two types of uncertainties, as will be described.

The general measure of uncertainty, or dispersion in data, is the *variance*, σ^2. If the statistical distribution of experimental data around a mean value (or of calculated values around the true value) is normal, the square root of the variance, σ, will represent a *standard deviation*. In this case, the probability is 67% that the measured or calculated value will lie within $\pm\sigma$ of the mean or true value, 95% that it will be within $\pm 2\sigma$, and 99.73% that it will lie within $\pm 3\sigma$. For the 3σ limit, this probability means that there is a 0.13% probability that the measured or calculated value will lie more than 3σ above the mean or true value and an equal probability that it will lie more than 3σ below the mean or true value. Hence, there is a 99.87% probability that the measured or calculated value will lie below the mean or true value plus 3σ.

Random uncertainties in thermal-hydraulics design are assumed to follow a normal distribution. If the design basis requires that the true value of a parameter for which the uncertainty is random be less than a specified design limit for that parameter with 99.9% confidence, then the nominal value of the parameter must be 3σ less than the design limits. This level of confidence is frequently required for SFR thermal-hydraulics designs; although for some parameters a 2σ level (97.5% confidence) is considered satisfactory.

If a 3σ level of confidence is required for a particular variable, the hot channel factor for that variable would be

$$F = 1 + 3\sigma. \tag{10.48}$$

It is not always possible to characterize an uncertainty by a standard deviation, especially in the case of direct uncertainties. In such a case it is necessary to find another justification to provide the appropriate level of confidence so that uncertainty factors will not allow design limits to be exceeded.

10.4.1.1 Combination of Hot Channel Factors

Numerous uncertainties influence each design parameter, and a hot channel factor is associated with each uncertainty. Individual hot channel factors combine to yield an overall hot channel factor for the design parameter.

Hot channel factors based on direct uncertainties multiply nominal values directly; hence, their combination is straightforward:

$$F_d = \prod_{k=1}^{D} F_{d,k}, \tag{10.49}$$

where

$F_{d,k}$ = value of a direct hot channel factor for hot channel effect k,
D = number of direct hot channel factors.

Hot channel factors based on random uncertainties are combined statistically before multiplying a nominal value. The propagated uncertainty is the square root of the sum of the squares of the individual random uncertainties. Each statistical hot channel factor, $F_{s,k}$, is related to the uncertainty, $n\sigma$, in a manner similar to Eq. (10.48), with the number of standard deviations n depending on the level of confidence required. Hence, the part of the statistical hot channel factor that is propagated as the sum of the squares $(F_{s,k} - 1)$. Therefore,

$$F_s = 1 + \left[\sum_{k=1}^{S} \left(F_{s,k} - 1\right)^2\right]^{1/2}, \tag{10.50}$$

where S = number of statistical hot channel factors.

The overall hot channel factor, F, is the product of the combined direct and statistical uncertainty factors:

$$F = F_d F_s. \tag{10.51}$$

10.4.2 CRBRP and FFTF Hot Channel Factors

The hot channel factors used for the early CRBRP homogeneous core design are listed in Table 10.2 [28]. Factors are listed for five regions (coolant, film, cladding, gap, and fuel) and for the heat flux. Hot channel factors are combined by columns in the table so that a total hot channel factor for each region and for the heat flux is obtained; the reason for this will become clear in Section 10.4.3 where the method for applying hot channel factors is illustrated.

The heat flux factor accounts for the uncertainties in power generation.[2] It is used in evaluating uncertainties in the film, cladding, gap, and fuel temperature differences. It does not apply to the coolant enthalpy rise (and, hence, to the temperature rise). Other factors, averaged along the flow channel, account for the coolant enthalpy uncertainties.

Hot channel factors presented in Table 10.2 are based on an early CRBRP homogeneous core design [28], which provides an adequate framework to present the basic concepts of the hot channel factors in the section. A more complete set of hot channel factors for the CRBRP heterogeneous core design is available [29], which takes into consideration the blanket assembly hot channel factors and other aspects not considered in the earlier work.

---

[2] In FFTF, the heat flux factor is included in the total hot channel factor for each region.

Table 10.2 Hot channel factors (early CRBRP homogeneous core design, 3σ level of confidence) [28]

	Coolant	Film	Cladding	Gap	Fuel	Heat Flux
A. Direct						
Power measurement and control system deadband	1.03					1.03
Inlet flow maldistribution	1.05	1.035				
Assembly flow maldistribution	1.08					
Cladding circumferential temperature variation		1.0 (1.7)[a]				
Direct combination (F_d)	**1.17**	**1.035 (1.76)[a]**				**1.03**
B. Statistical						
Inlet temperature variation	1.02					
Reactor ΔT variation	1.04					
Nuclear data (power distribution)	1.06					1.07
Fissile fuel maldistribution	1.01					1.04
Wire wrap orientation	1.01					
Coolant properties	1.01					
Subchannel flow area	1.03	1.0				
Film heat transfer coefficient		1.12				
Pellet-cladding eccentricity		1.15	1.15			
Cladding thickness and conductivity			1.12			
Gap conductance				1.48[b]		
Fuel thermal conductivity					1.10	
Statistical combination (F_s)	**1.08**	**1.19**	**1.19**	**1.48[b]**	**1.10**	**1.08**
Total (F)	1.26	1.23 (2.10)[a]	1.19	1.48	1.10	1.11

[a] This factor affects the maximum cladding temperature only; it does not affect the maximum fuel temperature.
[b] Used for fresh fuel only.

The FFTF hot channel factor analysis [27] was similar to the early CRBRP analysis but we focus here on the early CRBRP approach since it came later in design and applies to a power reactor design. Since both FFTF and CRBRP were based on oxide fuel, many of the values shown in Table 10.2 could change considerably for systems fueled with metal, carbide, or nitride fuel. One key change is that these higher thermal conductivity fuels would likely feature a fuel-cladding gap filled with sodium.

The individual direct and statistical factors for the CRBRP design are discussed briefly below.

10.4.2.1 Direct Factors

Power Level Measurement and Control System Dead Band

The calibration error in power measurement instrumentation is +2%, based mainly on water flow rate and feedwater temperature measurement uncertainties in the steam cycle.[3] An additional allowance of 1%, called a *dead band allowance,* is incorporated in the design of the control system to prevent excessive exercising of the control rod drives for small changes in reactor power. The combined hot channel factor for power measurement and dead band is 1.03; this was considered a direct uncertainty factor for the early CRBRP design.

Inlet Flow Maldistribution

Due to flow and pressure distributions within the inlet plenum, and due to the potential accumulation of fabricated dimensional tolerances in the orifice, fuel pin, wire wrap, and duct, the total flow to any one fuel assembly might be lower than nominal. Flow reduction affects the coolant enthalpy rise (and, hence, coolant temperature rise) directly. Allowance for a 5% flow reduction yields the hot channel factor of 1.05. The flow reduction influences the ΔT across the film, changing the Peclet number in the heat-transfer-coefficient correlation. The lower flow is also associated, however, with a lower hydraulic diameter for the flow, an effect that raises the heat transfer coefficient. The hot channel factor of 1.035 for inlet flow maldistribution for the film accounts for both reduced flow and reduced hydraulic diameter, as well as inlet assembly flow maldistributions.

Assembly Flow Maidistribution

Flow and coolant temperature distributions within an assembly due to differences between interior and edge flow channels were discussed in Section 10.2. The nominal value of the temperature rise in the hot channel of an assembly takes into account the ratio of the rise in the hot channel to the average value for the assembly as calculated by the methods of Section 10.2. The assembly flow maldistribution hot channel factor accounts for the uncertainty in this ratio.

Cladding Circumferential Temperature Variation

The coolant velocity and temperature distribution in the channel formed by three adjacent fuel pins is not uniform, but varies circumferentially around the rod, as discussed at the end of Section 10.2.1. The maximum cladding temperature occurs at the minimum gap between the rods. The presence of wire wrap in the gap further increases the cladding surface temperature. The calculated increase in ΔT across the film at the minimum gap was 1.5 times the average ΔT. Another 20% was added for calculation method uncertainties, to give a direct hot channel factor of 1.7. An alternate procedure would be to use the factor of 1.5 as a known design factor and include only the 20% uncertainty in the hot channel factor analysis.

This circumferential temperature factor is applied only to the cladding temperature limit. It does not affect the fuel-cladding gap or fuel temperature differences; hence, this hot channel factor is unity when calculating the film temperature rise for evaluation of maximum fuel temperature.

[3] One might suppose that this uncertainty is statistical and should be propagated as such rather than as a direct factor. This uncertainty is considered as a bias, however, in LWR licensing; hence, this is an example where LWR precedent has dictated SFR methodology

10.4.2.2 Statistical Factors

Inlet Temperature Variation

A statistical analysis of the integrated reactor and plant system, including primary and secondary sodium and steam systems, indicated an uncertainty in inlet coolant temperature of ±5°C. This uncertainty is included in the uncertainty in the coolant temperature rise across the core. For a nominal temperature rise of ~280°C, the ±5°C uncertainty corresponds to a ±2% variation in the temperature rise.

Reactor ΔT Variation

The same system analysis indicated a 3σ uncertainty in coolant flow of ±4%, which would lead to the same (±4%) uncertainty in ΔT across the core.

Nuclear Data (Power Distribution)

This factor represents uncertainties in the radial, axial, and local power peaking factors. It reflects uncertainties in nuclear data and nuclear design methods, after adjusting for comparisons with critical experiments. For the early CRBRP homogeneous core design, it was estimated that this uncertainty has a 6.5% influence on the heat flux factor and a 6% uncertainty on the coolant enthalpy rise.

Fissile Fuel Maldistribution

This factor is based on statistical measurements of fuel made for the FFTF. It is a 3.5% uncertainty in the fuel and cladding temperature limit analyses for an individual fuel pin, but only a 1% uncertainty for the coolant enthalpy rise since this rise is influenced by several pins.

Wire Wrap Orientation

This small contribution to the hot channel factor is associated with uncertainties due to swirl flow induced by the wire wrap in channels near the assembly ducts.

Coolant Properties

This reflects the effect of uncertainties in the specific heat and density of sodium.

Subchannel Flow Area

This factor represents the effect of propagated tolerances in fuel pin diameter, pitch-to-diameter ratio, and pin bowing. It does not affect the film hot channel factor because the decrease in hydraulic diameter associated with the 2.8% flow reduction overrides the influence of the Peclet number in calculating the film heat transfer coefficient.

Film Heat Transfer Coefficient

When calculating cladding and fuel temperature limits, one must be sure not to use too high a value for the heat transfer coefficient since the higher the coefficient, the lower the calculated cladding and fuel temperature. Comparison of the CRBRP (or FFTF) correlation with experimental data (e.g., Fig. 9.19) indicated that the minimum value of the actual heat transfer coefficient would not be more than 12% below the value calculated with the correlation.

Pellet-Cladding Eccentricity

An eccentric position of the fuel pellet within the cladding will result in an increased heat flux in the area of minimum fuel-cladding gap, with consequent increases in temperature rise through the coolant film and cladding. Calculations indicated that a hot channel factor of 1.15 is appropriate for the film and peak cladding temperatures. This factor is not applied to the peak fuel temperature because pellet eccentricity actually reduces this temperature.

Cladding Thickness and Conductivity

A 10% uncertainty in cladding conductivity (mostly to account for irradiation effects) was propagated statistically with a 6.7% uncertainty due to a 3σ cladding thickness tolerance of $\pm$1 mil (0.03 mm) to give a hot channel factor of 1.12.

Gap Conductance

Uncertainties in beginning-of-life (BOL) gap conductance are due to scatter in integral measurements (such as in the *P*-19 data, as illustrated earlier in Fig. 9.10), and to the difficulty of separating gap uncertainties from uncertainties in integral experiments resulting from fuel thermal conductivity and from tolerances on cladding inner diameter and fuel pellet outer diameter. For the fuel pin with the highest center fuel temperature, the gap hot channel factor was 1.48. (The cold gap nominal value for the early CRBRP design was 6.5 mils, with a tolerance of 2.5 mils on the pellet diameter and 0.5 mils on the cladding inner diameter.)

This hot channel factor is valid only for BOL fuel. After some period of irradiation and thermal cycling, fuel swelling and cracking cause the hot fuel-cladding gap to close; thus increasing the gap conductance, increasing the linear power-to-melting, and decreasing the peak fuel temperature. This was observed in the HEDL-P-20 experiments and can be seen by comparing Figs. 9.10 and 9.11 with Fig. 9.12.

Fuel Thermal Conductivity

Several sources of uncertainty in the fuel thermal conductivity propagate to an estimated uncertainty of $\pm$10% in the fuel temperature rise from the surface to the center of the fuel pellet, for any given value of linear power. The uncertainty in fuel thermal conductivity may actually be greater than this value, especially at high temperatures; but this uncertainty, when combined with the 1.48 factor for gap conductance, gives an appropriate overall uncertainty for the temperature change across gap plus fuel for BOL fuel.

Special attention must be given to applying the gap conductance and fuel thermal conductivity hot-spot factors, because the parameters to which they are applied are always interrelated in experimental measurements.

10.4.3 Application of Hot Channel Factors

The use of hot channel factors will be illustrated first by applying them to calculations of peak exit coolant temperature and maximum cladding temperature. Next, the method used to assure that fuel melting does not occur will be described.

The nomenclature used for the combined hot channel factors is listed below. Numerical values appearing in Table 10.2 for these factors are shown in parentheses.

Hot channel factor	Parameter affected
F_q	Heat flux (1.11)
F_b	Coolant enthalpy and temperature rise (1.26)
F_{film} (cladding)	Temperature rise across film at hot spot beneath the wire wrap, to be used in obtaining peak cladding temperatures (2.10)
F_{film} (fuel)	Temperature rise across film to be used in obtaining peak fuel temperature (1.23)
F_c	Temperature rise across cladding (1.19)
F_{gap}	Gap conductance (1.48)
F_{fuel}	Fuel conductivity (1.10)

Further nomenclature used includes[4]:

q_{OP} = Product of the nominal hot-channel heat flux at the outer surface of the pin cladding and the overpower factor (hence, heat flux at overpower conditions)
χ_{OP} = Product of nominal hot-channel linear power and overpower factor
h = Heat transfer coefficient
k = Thermal conductivity
D_i, D_o = Inside and outside diameter of the cladding
$T_{b,0}$ (inlet) = Nominal value of the inlet coolant temperature
$T_{ci,m}$ = Maximum value of the inner cladding temperature
$T_{co,m}$ = Maximum value of the outer cladding temperature
$\Delta T_{b,OP}$ = Product of nominal hot channel coolant temperature rise and overpower factor.

As the starting point for the temperature calculations we will use the nominal value of the inlet coolant temperature, $\Delta T_{b,0}$ (*inlet*), since the uncertainty in the inlet temperature is included in the temperature rise.

The maximum coolant temperature rise across a fuel assembly is

$$\Delta T_{b,m}\,(\text{exit}) = \Delta T_{b,OP}\,(\text{exit})\, F_b. \tag{10.52}$$

Hence, the maximum exit temperature is

$$T_{b,m}\,(\text{exit}) = T_{b,0}\,(\text{inlet}) + \Delta T_{b,m}\,(\text{exit})\,. \tag{10.53}$$

In order to calculate the peak temperature at the outer surface of the cladding, the first step is to locate the axial position of the peak nominal cladding temperature, $z_{\max}$. The peak outer surface temperature is the sum of the inlet coolant temperature, the maximum rise in coolant temperature to that axial position $\left[\Delta T_{b,m}\,(z_{\max})\right]$, and the maximum temperature rise across the film at that axial position, or

$$T_{co,m} = T_{b,0}\,(\text{inlet}) + \Delta T_{b,m}\,(z_{max}) + \Delta T_{film,m}\,(\text{cladding})\,, \tag{10.54}$$

where

$$\Delta T_{film,m}\,(\text{cladding}) = \frac{q_{OP} F_q}{h_{film}/F_{film}\,(\text{cladding})}, \tag{10.55}$$

[4]Subscript *m* refers to maximum values; subscript 0 refers to nominal values

and q_{OP}, and h_{film}, are calculated at the axial position of peak cladding temperature.

The maximum temperature rise across the cladding is

$$\Delta T_{c,m} = \frac{F_q \chi_{OP} \ln(D_o/D_i)}{2\pi k_c/F_c}. \tag{10.56}$$

Hence, the maximum inner surface cladding temperature would be the sum of Eqs. (10.54) and (10.56), with all terms evaluated at the axial level in question.

The third problem discussed here is to demonstrate that centerline fuel melting does not occur. Two approaches can be used to show this:

(1) the fuel centerline temperature can be calculated and compared with the fuel melting temperatures; or
(2) the operating linear power can be calculated and compared with experimentally observed values of linear-power-to-melting.

In the first approach, we find the axial location of nominal peak fuel temperature and then calculate the maximum coolant temperature at that position. Next, we calculate the maximum temperature rise across the film, but we use the film hot channel factor for the average cladding F_{film}(fuel), instead of the one that applies to the cladding hot spot beneath the wire wrap, or

$$\Delta T_{film,m}\,(\text{fuel}) = \frac{q_{OP} F_q}{h_{film}/F_{film}\,(\text{fuel})}. \tag{10.57}$$

From this point, the calculation of maximum fuel temperature becomes more complicated because of the experimental interrelation between gap conductance, fuel conductivity, and linear power. The procedure used for the early CRBRP homogeneous core is presented here.

The maximum temperature differences across the fuel and gap are calculated together. It is reasonable to propagate the uncertainties due to fuel conductivity and gap conductance randomly (i.e., as the sum of squares), since they cannot be isolated experimentally. This can be done as follows, recognizing that the uncertainty from the fuel conductivity is $\Delta T_{fuel,0}\left(F_{fuel} - 1\right)$ and the uncertainty due to gap conductance is $\Delta T_{gap,0}\left(F_{fuel} - 1\right)$.

The nominal temperature differences across the fuel and gap are both multiplied by the heat flux hot channel factor. To this is added the statistically propagated uncertainties due to gap conductance and fuel conductivity. Hence,

$$\Delta T_{(\text{fuel}+\text{gap}),m} = \left(\Delta T_{\text{fuel},0} + \Delta T_{\text{gap},0} +\right) F_q + \sqrt{\left[\Delta T_{\text{fuel},0}\left(F_{\text{fuel}} - 1\right)\right]^2 + \left[\Delta T_{\text{gap},0}\left(F_{\text{gap}} - 1\right)\right]^2}, \tag{10.58}$$

where $\Delta T_{(\text{fuel}+\text{gap}),m} = T_{\text{fuel},m} - T_{ci,m}$. Hence the maximum fuel temperature is

$$T_{\text{fuel},m}(z) = T_{\text{b},0}\,(\text{inlet}) + \Delta T_{\text{b},m}(z) + \Delta T_{\text{film},m}\,(\text{fuel}) + \Delta T_{c,m}(z) + \Delta T_{(\text{fuel}+\text{gap}),m}. \tag{10.59}$$

The second approach for showing that fuel melting will not occur is based on integral experiments of linear power required for incipient fuel melting, as described in Section 9.2.5. By comparing the operating linear power with the experimentally observed linear-power-to-melting, a *margin-to-melting*

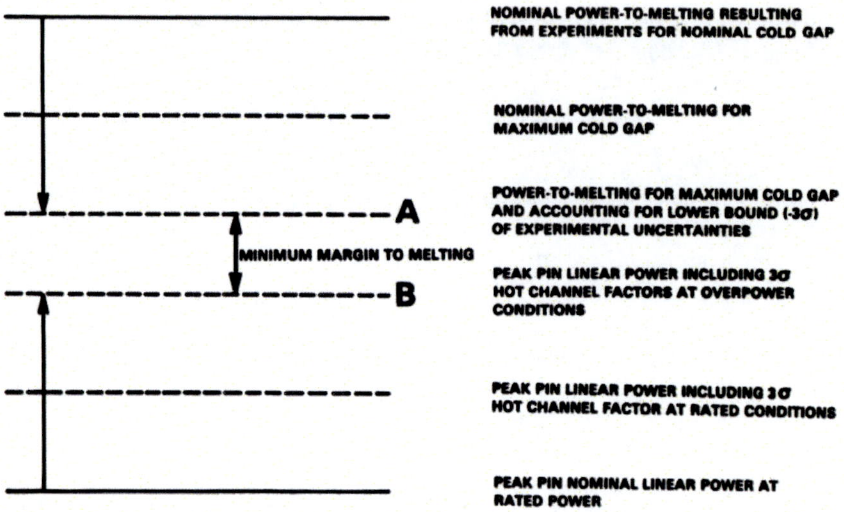

Fig. 10.15 Scheme of accounting for uncertainties in assessing margin-to-melting [30]

is obtained. The desired result is the margin after uncertainties have been taken into account. The procedure to do this is illustrated in Fig. 10.15 [30]. The *minimum linear-power-to-melting* is calculated by starting with the experimentally measured value for the nominal gap, decreasing this value to the result for the maximum (3σ level) cold gap, and decreasing it further by the experimental uncertainties at the -3σ confidence level. Such a procedure leads to line A on Fig. 10.15. This value must be greater than the *maximum operating linear power,* which is the nominal linear power, increased by the 3σ level of confidence (i.e., multiplied by F_q) and increased again by the overpower factor (i.e., 115% for FFTF and early CRBRP design). This result is line B on the figure. The amount by which line B is lower than line A is the design margin-to-melting.

10.4.4 Considerations for Other Fast Spectrum Reactor Designs

Development and application of hot channel factors is very specific to the design of the reactor, the fuel assembly/pin design and manufacture, the fuel material, and the accuracy of computer codes and correlations used to model the reactor. The examples provided earlier in this section were for SFRs using wire wrapped oxide fuel pins in a hexagonal assembly. Hot channel factors for other fast spectrum reactor designs must consider the specific features of the designs.

The single largest contributor to direct uncertainty listed in Table 10.2 is the cladding circumferential temperature variation, which is only applied to the cladding temperature limit. This direct uncertainty provides a clear example of how design affects hot channel factors. Assembly designs using a wider pin spacing (such as with lead coolant) or those with grid spacers instead of wire wraps can consider a smaller uncertainty in cladding circumferential temperature variation. The direct uncertainty is expected to decrease for lead coolant, which desires a very wide pin spacing to limit core pressure drop.

The pin design and the fuel material can have a large impact on the hot channel factors and the resulting margins—as they can affect items with the largest statistical uncertainties listed in Table 10.2. For example, sodium bonded pins (discussed in Chapter 11) can effectively eliminate the effect of pellet-cladding eccentricity by providing very high conductance across the gap [31] no matter the position of the fuel pellet within the cladding. Sodium bonding, bonding with another liquid metal, or very carefully controlled manufacturing processes can also improve the gap conductance uncertainty for fresh fuel for the same reason. In addition to affecting the pin design, the choice of fuel material also affects the fuel thermal conductivity uncertainty—potentially increasing the uncertainty for less well understood or less studied materials.

While generally not at the forefront of the reactor design process, consideration of design and material choices for their impact on hot channel factors can have an impact on the overall safety margins of a fast spectrum reactor design. Specifically, choosing a design or material that has outstanding nominal performance without consideration for the uncertainties, as they are addressed through hot channel factors, can degrade the overall benefit of the choice. However, this situation can be mitigated by recognizing early in the process what uncertainties are most important, and applying efforts to reduce these uncertainties through study or experimentation in order to reduce their impact on margins through the hot channel factors.

References

1. E. H. Novendstern, "Turbulent Flow Pressure Drop Model for Fuel Rod Assemblies Utilizing a Helical Wire-Wrap Spacer System," *Nucl. Eng. Des.*, *22* (1972) 19–27.
2. W. A. Sangster, "Calculation of Rod Bundle Pressure Loss," Paper 68-WA/HT-35, ASME, New York, NY, 1968.
3. C. Chiu, W. M. Rohsenow, and N. E. Todreas, *Flow Split Model for LMFBR Wire Wrapped Assemblies*, COO-2245-56TR, Massachusetts Institute of Technology, Cambridge, April 1978.
4. J. T. Hawley, C. Chiu, W. M. Rohsenow, and N. E. Todreas, "Parameters for Laminar, Transition, and Turbulent Longitudinal Flows in Wire Wrap Spaced Hexagonal Arrays," *Topical Meeting on Nuclear Reactor Thermal Hydraulics*, Saratoga, NY, 1980.
5. D. S. Rowe, *COBRA-IIIC: A Digital Computer Program for Steady Slate and Transient Thermal Hydraulic Analysis of Rod Bundle Nuclear Fuei Elements*, BNWL-1695, Battelle Pacific Northwest Laboratories, March 1973. See also T. L. George, K. L. Basehore, C. L. Wheeler, W. A. Prather, and R. E. Masterson, *COBRA-WC: A Version of COBRA for Single-Phase Multiassembly Thermal Hydraulic Transient Analysis*, Pacific Northwest Laboratory, PNL-3259, Richland, Washington, July 1980.
6. W. T. Sha, R. C. Schmitt, and P. R. Huebotter, "Boundary-Value Thermal Hydraulic Analysis of a Reactor Fuel Rod Bundle," *Nucl. Sci. Eng.*, *59* (1976) 140–160.
7. M. D. Carelli and C. W. Bach, "LMFBR Core Thermal Hydraulic Analysis Accounting for Interassembly Heat Transfer," *Trans. ANS*, 28 (June 1978) 560–562.
8. J. L. Wantland, "ORRIBLE—A Computer Program for Flow and Temperature Distribution in 19-Rod LMFBR Fuel Subassemblies," *Nucl. Tech.*, *24* (1974) 168–175.
9. E. U. Khan, W. M. Rohsenow, A. A. Sonein, and N. E. Todreas, "A Porous Body Model for Predicting Temperature Distribution in Wire-Wrapped Fuel Rod Assemblies," *Nucl. Eng. Des.*, *35* (1975) 1–12.
10. E. U. Khan, W. M. Rohsenow, A. A. Sonein, and N. E. Todreas, "A Porous Body Model for Predicting Temperature Distribution in Wire Wrapped Rod Assemblies in Combined Forced and Free Convection," *Nucl. Eng. Des.*, *35* (1975) 199–211.
11. B. Chen and N. E. Todreas, *Prediction of Coolant Temperature Field in a Breeder Reactor Including Interassembly Heat Transfer*, COO-2245-20TR, Massachusetts Institute of Technology, Cambridge, MA, 1975.
12. J. N. Lillington, *SABRE-3-A Computer Program for the Calculation of Steady State Boiling in Rod-Clusters*, AEEW-M-1647, United Kingdom Atomic Energy Authority, 1979.
13. J. E. Meyer, *Conservation Laws in One-Dimensional Hydrodynamics*, WAPD-BT-20, Westinghouse Electric Corp., Bettis Atomic Power Laboratory, Pittsburgh, PA, September 1960.
14. R. E. Masterson and L. Wolf, "An Efficient Multidimensional Numerical Method for the Thermal-Hydraulic Analysis of Nuclear Reactor Cores," *Nucl. Sci. Eng.*, *64* (1977) 222–236.
15. J. T. Rogers and N. E. Todreas, "Coolant Interchannel Mixing in Reactor Fuel Rod Bundles Single-Phase Coolants," *Symposium on Heat Transfer in Rod Bundles*, ASME, New York, NY, (1965), 1–56.
16. T. Ginsberg, "Forced-Flow Interchannel Miving Model for Fuel Rod Assemblies Utilizing a Helical Wire-Wrap Spacer System," *Nucl. Eng. Des.*, *22* (1972) 28–42.

17. M. W. Cappiello and T. F. Cillan, *Core Engineering Technical Program Progress Report, Jan-March 1977*, HEDL-TME 77-46 (July 1977), Hanford Engineering Development Laboratory, Richland, WA.
18. Chaumont, Clauzon, Delpeyroux, Estavoyer, Ginier, Marmonier, Mougniot, "Conception du Coeur et des Assemblages d'une Grande Centrale a Neutrons Rapides." *Conf. Nucleaire Europeene*, Paris, April 1975.
19. S. F. Wang and N. E. Todreas, *Input Parameters to Codes Which Analyze LMFBR Wire-Wrapped Bundles*, Rev. 1, COO-2245-17TR, Massachusetts Institute of Technology, Cambridge, MA, May 1979.
20. Preliminary Safety Analysis Report, Clinch River Breeder Reactor Plant, Project Management Corporation, 1974.
21. A. N. de Stordeur, "Drag Coefficients for Fuel-Element Spacers," *Nucleonics, 19* (1961) 74–79.
22. K. Rehme, The Measurement of Friction Factors for Axial Flow Through Rod Bundles with Different Spacers, Performed on the INR Test Rig, EURFNR-142P, November 1965.
23. T. C. Reihman, *An Experimental Study of Pressure Drop in Wire Wrapped FFTF Fuel Assemblies*, BNWL-1207, Richland, WA, September 1969.
24. W. Baumann, V. Casal, H. Hoffman, R. Moeller, and K. Rust, *Fuel Elements with Spiral Spacers for Fast Breeder Reactors*, EURFNR-571, April 1968.
25. R. A. Jaross and F. A. Smith, *Reactor Development Program Progress Report*, ANL-7742, Argonne National Laboratory, Argonne, IL, 1970, 30.
26. E. Bubelis and M. Schikorr, "Review and proposal for best fit of wire-wrapped fuel bundle friction factor and pressure drop predictions using various existing correlations," *Nucl. Eng. Des., 238* (2008) 3299–3320.
27. G. J. Calamai, R. D. Coffield, L. J. Ossens, J. L. Kerian, J. V. Miller, E. H. Novendstern, G. H. Ursin, H. West, and P. J. Wood, *Steady Stare Thermal and Hydraulic Characteristics of the FFTF Fuel Assemblies*, ARD-FRT-1582, Westinghouse Electric Corp., Sunnyvale, CA, June 1974.
28. M. D. Carelli and R. A. Markley, "Preliminary Thermal-Hydraulic Design and Predicted Performance of the Clinch River Breeder Reactor Core," *Nat. Heat Transfer Conf.*, ASME Paper 75-HT-71, ASME, New York (1975).
29. M. D. Carelli and A. J. Friedland, "Hot channel factors for rod temperature calculations in LMFBR assemblies," *Nucl. Eng. Des., 62* (1980) 155–180.
30. Y. S. Tang, R. K. Coffield, Jr., and R. A. Markley, *Thermal Analysis of Liquid Metal Fast Breeder Reactors*, The American Nuclear Society, La Grange Park, IL, 1978.
31. L. Walters, D. Wade, and G. Hofman, "An Innovative Particulate Metallic Fuel for Next Generation Nuclear Energy," Proceedings of ICAPP-10, San Diego, CA, June 13–17, 2010, Paper 10356.

Chapter 11
Core Materials

Baldev Raj

11.1 Introduction

The most hostile environment to be found in any nuclear reactor system is inside the core. Relative to a thermal reactor, the high flux, high burnup, and high temperature conditions encountered in a fast spectrum reactor place severe requirements on the materials selected for core design. Hence, considerable effort has been devoted to understanding and improving the performance of fuels and structural component candidates for fast spectrum reactor use.

This chapter is included to provide a more complete materials treatment of several of the general observations offered in Chapter 2 and of the design discussions included in Chapters 8, 9, and 10. However, because so much study has been given to the materials that comprise the primary building blocks of the fast spectrum reactor, it is not possible in an introductory text of this type to treat this subject with the degree of detail that a materials-oriented student would wish. Much of the information presented in this chapter, particularly that relating to stainless steel clad mixed oxide fuel, is taken from Olander [1] which represents an exceptionally clear and comprehensive treatment for the student interested in the materials phase of fast spectrum reactor technology.

This chapter is organized in four sections to deal with the four primary ingredients in the core; namely, the fuel, cladding and duct, coolant, and control material. Each section opens with general requirements and continues with general property considerations and the specific characteristics of the various candidate materials. A general intercomparison is then included as appropriate. Materials of direct interest to the sodium cooled fast spectrum reactors (SFR) are emphasized, in recognition of the central fast spectrum reactor role occupied by the SFR.

11.2 Fuel

11.2.1 Requirements

Perhaps the most demanding requirement placed on the fuel for a fast spectrum reactor is that of high burnup (≅150 MWd/kg or higher target exposure). Such an exposure level, about a factor of two to three times higher than for a modern light water reactor (LWR), leads directly to a high fission product generation. This requires particular design emphasis on accommodating larger fuel swelling, greater

B. Raj (✉)
Indira Gandhi Center for Atomic Research, Kalpakkam, Tamil Nadu, India
e-mail: dmg@igcar.gov.in

A.E. Waltar et al. (eds.), *Fast Spectrum Reactors*, DOI 10.1007/978-1-4419-9572-8_11,

fission gas release, and potentially higher consequences of fuel cladding chemical interaction relative to those associated with a LWR.

The fuel must also be capable of sustaining specific powers about a factor of four higher than for a LWR, as well as a higher temperature gradient (for oxide fuel), due to the smaller fuel pin diameter.

Finally, the inherent dynamic response of the core during accident conditions plays an important role in fuel selection. A desired feature of overall plant safety is that a temperature rise in the fuel should lead promptly and directly to a negative reactivity feedback effect, such that the accident condition causing the original temperature rise would be arrested rather than augmented. A negative Doppler coefficient and axial fuel expansion upon heating constitute two safety factors that could offer the desired response function.

11.2.2 Definitions

In searching for fuel materials capable of rendering high performance within the environment outlined above, the materials scientist is continually measuring and intercomparing certain characteristics. Before exploring the features of each of the major fuel candidates under study for fast spectrum reactor application, it is constructive to define some of the terminology commonly used to discuss nuclear fuel performance.

11.2.2.1 Burnup

As noted in Chapter 7, fuel *burnup* can be defined either in terms of energy yield, i.e., MWd/kg, or as the fraction of heavy atoms fissioned, i.e., atom percent (at.%) burned. The latter definition is usually the more meaningful to the fuel pin designer since it conveys most directly the potential damage done to the fuel matrix during irradiation. These two definitions can be quantitatively related to each other, as discussed in Section 7.3.

11.2.2.2 Dimensional Stability

Dimensional stability (or instability) is the term generally used to denote volumetric changes in the fuel matrix shape as a result of irradiation. The material distortion caused by radiation damage to fuel is extremely important because of the direct implications to pin and lattice design. There are two fundamental processes that contribute to dimensional instability. First, for every fission event that takes place, two fission fragments are formed. The resultant fission product compounds have a lower density than the parent matrix, which results in a fractional increase in matrix volume, about three times the burnup.

Second, since the atom undergoing fission no longer exists, a vacancy (the place originally occupied by that atom) is left in the fuel matrix. Additional vacancies are caused by fission fragments knocking other atoms away from their original sites. Both the fission products and the displaced atoms eventually come to rest at interstitial sites (i.e., at positions between those atomic sites that are arrayed in order).[1] The net effect of such vacancy/interstitial patterns on the dimensional stability of a particular fuel material depends heavily on the properties of the original, unirradiated lattice structure.

[1] It is possible, of course, that such slowing down atoms could relocate in a vacancy position. This is called vacancy annihilation.

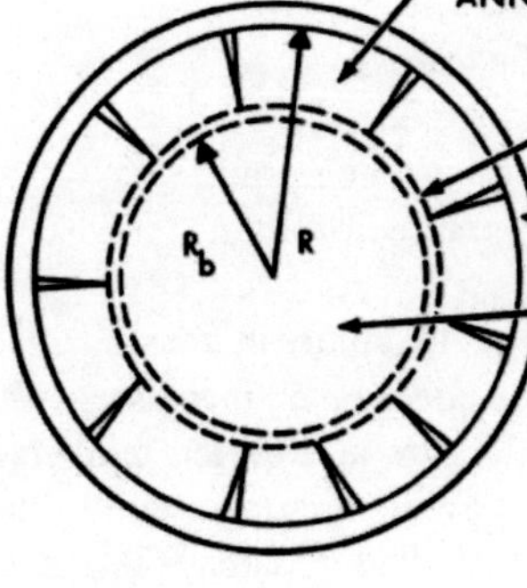

Fig. 11.1 Model of the mechanical state of a fuel pin cross section under irradiation

11.2.2.3 Microstructural Changes

Actual changes in grain size and orientation can be effected by even the slightest alloying of certain materials. Grain shapes and characteristics can also be grossly modified by the presence of strong temperature gradients, as noted in Section 8.2. Such structural changes can have a profound effect on subsequent material behavior. *Sintering* is a term used to describe the process of reduction of *porosity*[2] at high temperatures, where porosity is the fraction of void volume in the fuel. Thermal conductivity is an example of a property that could be altered significantly by changes in porosity.

11.2.2.4 Strength

Material mechanical properties such as hardness, yield and ultimate strength do not play nearly as significant a role in fuel selection as they do in the selection of cladding materials. This is because fuel breakup can be accommodated within the cladding jacket. An understanding of fuel cracking processes, however, is important in order to predict how fuel pins will perform under both steady-state and accident conditions. Figure 11.1 illustrates the possible mechanical state of a ceramic fuel material under early stages of irradiation. The hot interior exhibits plastic behavior whereas the cooler outer region, beyond radius R_b is relatively brittle and is characterized by radial cracks.

11.2.2.5 Creep

Creep is time-dependent strain occurring under constant stress over long periods of time. Whereas elastic or instantaneous plastic deformation can occur the moment that stress is applied, creep strain is a longer time-frame phenomenon. One source of constant stress that could cause creep in a fuel matrix is thermal stress resulting from a steep temperature gradient.

When a material is initially exposed to a stress field, the elastic and instantaneous plastic deformations that occur exhaust all the mechanisms readily available for allowing dislocation motion. Hence, any additional dislocations that allow creep to occur must result from thermal or irradiation activation. Since the probability per unit time of supplying the energy, E, needed to start a dislocation motion is proportional to the Boltzman factor, $\exp(-E/kT)$, the *thermal creep rate* $\dot{\varepsilon}$, can be generally expressed in the form:

$$\dot{\varepsilon} \propto \sigma^m \exp\left\{\frac{-E}{kT}\right\}. \qquad (11.1)$$

[2]This normally causes densification.

where

σ = applied stress,
m = stress exponent,
k = the Boltzman constant,
T = temperature in Kelvin.

Thus, the creep rate rapidly increases with stress level (since the exponent m is well in excess of unity) and, for a given stress level, increases with temperature.

Creep that occurs at elevated temperatures, but in the absence of an irradiation field, is called *thermal creep*. Creep occurring due to the presence of a radiation field where dislocations are the direct result of irradiation damage is called *irradiation creep*.

Figure 11.2 is included to illustrate the general deformation characteristics involved during a typical constant load creep test. In particular, the three general stages that occur prior to stress rupture are shown. Immediately after the occurrence of elastic and instantaneous plastic deformations, the material undergoes primary creep (assuming the temperature is at least one-third to one-half the melting temperature). This monotonic process becomes approximately constant (secondary creep) during the intermediate time frame and then accelerates (tertiary creep) just before failure.

11.2.2.6 Swelling

A target burnup of 15–17 at.% (~150 MWd/kg) in a fast reactor implies a total inventory of about 30% fission products somewhere in the fuel system. Most of these fission products get lodged within the fuel matrix as solid particles (as discussed in the above section on Dimensional Stability) and contribute to an overall volumetric increase, known as *fission product fuel swelling*. Swelling due to this effect can be defined as:

$$\left(\frac{\Delta V}{V}\right)_{\text{solid fp}} = \frac{V - V_0}{V_0}, \qquad (11.2)$$

where

V_0 = volume of a region of fresh fuel,
V = volume of the same region after irradiation.

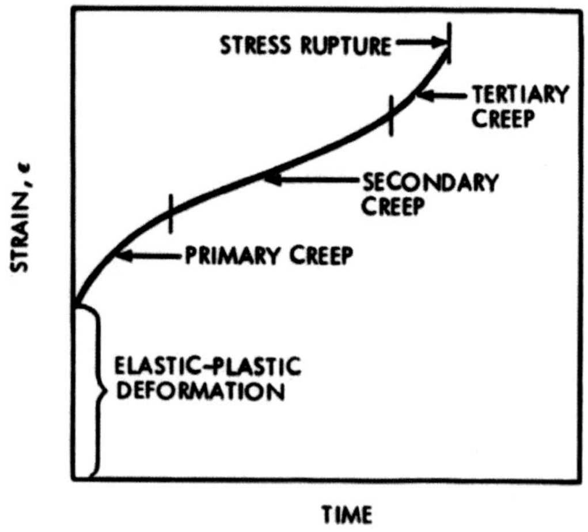

Fig. 11.2 Typical thermal creep curve

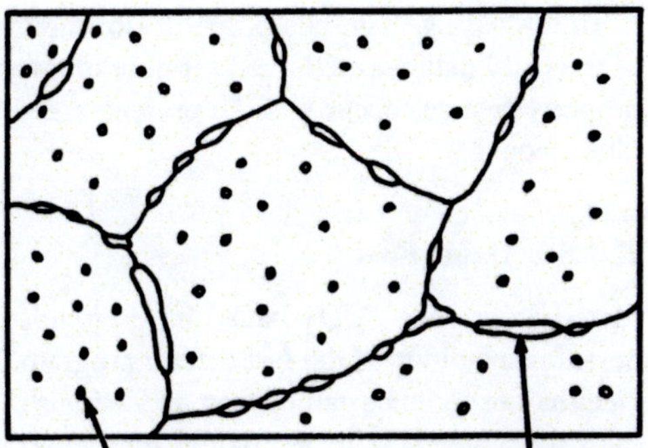

Fig. 11.3 A schematic distribution of fission gas bubbles in the fuel matrix

The unit cell volume being measured is assumed to envelop all solid fission products generated during the irradiation.

During the fission process, about 15% of the fission products occur as the rare gases xenon and krypton. Since these gases are insoluble in fuel, they tend to be rejected from the fuel matrix and contribute only indirectly to Eq. (11.2). These gases initially collect as small bubbles within the grains (intragranular) or as large bubbles along the grain boundaries (intergranular) as illustrated in Fig. 11.3. In either case, the net material density is appreciably less than if such space were filled by the original fuel. Hence, an additional volumetric increase, swelling due to fission gases, takes place.

11.2.2.7 Fission Gas Release

Not all of the fission product gases remain confined within the fuel. Depending largely on the nature of the fuel matrix structure and temperature, some of the gas can diffuse to the grain boundaries and from there much of it can flow via interconnected porosity to major cracks or open regions of the fuel. This released fission gas, unless intentionally bled off, then pressurizes the overall fuel system and applies stress to the cladding. A fission gas plenum is normally included in the fuel pin to keep the pressure buildup due to long-term irradiation within acceptable limits (discussed in Section 8.2). An alternate design option that has been considered is the use of vented pins, which releases filtered gas directly to the coolant system. The rate of fission gas release as a function of burnup for oxide fuel was shown in Fig. 8.8.

11.2.3 Uranium Based Fuels

The uranium-plutonium fuel system was shown in Chapter 1 to represent the best potential for high breeding performance in a fast spectrum reactor, compared to other combinations. Of the several U-Pu fuel matrix forms available, the most widely used and tested fuel to date is mixed oxide. Hence, this candidate will be given the most attention in the present section. Other fuels, however, have enjoyed certain success and have the potential to be employed in advanced fast reactor designs. These include uranium based metals, carbides and nitrides, and thorium based fuels. Cermets, a form in which fissile and fertile materials are dispersed in a matrix such as steel, will not be treated because the minimum amount of non-heavy metal required is normally high enough to render this fuel system unattractive for large-scale reactor application.

In discussing materials properties of the uranium based fuels, the reader is cautioned that much of the early data was obtained for pure uranium as opposed to the uranium-plutonium system. Some property changes occur with Pu present, and detailed fuel design procedures must take these changes into account.

11.2.3.1 Oxide Fuel

Mixed oxide fuels (UO_2-PuO_2) have provided the reference fuel material for most nations initially pursuing a buildup of the fast reactor program. Interest in this fuel system logically followed from the wealth of experience gained from oxide fuels in the water-cooled power reactors. Principal factors that motivated interest in mixed oxides for the fast spectrum reactor were the recognition of the high burnup potential, existence of an established industry for manufacturing oxide fuels, and rich operating experience.

For LWR applications, oxide fuels have demonstrated very satisfactory dimensional and radiation stability, as well as chemical compatibility with cladding and coolant materials. However, the environment in a fast spectrum reactor is considerably more hostile (higher temperatures and much larger burnup) and some alterations, beyond enrichment, must be made to the fuel to maintain satisfactory performance. For instance, whereas the oxygen-to-metal ratio (O/M) is maintained at almost precisely 2.00 for LWR oxide fuel, fast spectrum reactor fuel is purposely fabricated with a deficiency of oxygen (i.e., hypostoichiometric[3]). This is done primarily to reduce the propensity for irradiated fuel to oxidize the cladding (which leads to "thinning" or "wastage").

The principal disadvantages of mixed oxide fuel for fast spectrum reactor use are its low thermal conductivity and low fissile atom density. The former property leads to high temperature gradients in the fuel and low values for linear power, and the latter property is undesirable from a breeding-ratio point of view. The presence of oxygen atoms result in a softer neutron spectrum, which results in a lower breeding ratio. In addition, mixed oxide fuel has a lower fissile atom fraction, and a compatibility issue with liquid sodium (such that cladding rupture could lead to an enhanced risk of corrosion).

Fabrication. Because of the high melting temperature of mixed oxide fuel (~2,800°C for UO_2), fabrication is normally accomplished by powder metallurgy techniques. Uranium and plutonium oxides are mixed in the proportions required to achieve the desired fissile content and then the mixture is cold-compacted into a pellet. Sintering at temperatures around 1,600°C is then performed to achieve the desired degree of densification. Although a very high percentage of theoretical density (TD) can be attained by this method (and arc casting techniques can be employed to reach full TD), it is normally desired to fabricate pellets in the range of 85–95% TD. This allows sufficient internally distributed porosity to accommodate burnup and also provides for substantial melting accommodation under accident conditions.[4]

Microstructure. The principal alterations in the morphology of oxide fuel as it heats up are described in Section 8.2.3. Steep radial temperature profiles cause columnar and equiaxed grains to develop on the time scale of a few hours. Such restructuring patterns provide a strong interaction with fission gas retention patterns (discussed in Section 8.2.4) and overall thermal performance (discussed in Section 9.2.2).

Physical Properties. The thermal conductivity for UO_2 is presented in Section 9.2.1. Although the relatively low value for this physical property represents one of the principal disadvantages of UO_2 as a

[3]*A stoichiometric* material contains an atomic ratio exactly equal to its chemical formula. *Hypostoichiometric* refers to a deficiency in the non-heavy metal constituent, and *hyperstoichiometric* refers to an excess in the non-heavy metal constituent.

[4]Mixed oxide fuels undergo a volumetric increase of approximately 10% upon melting.

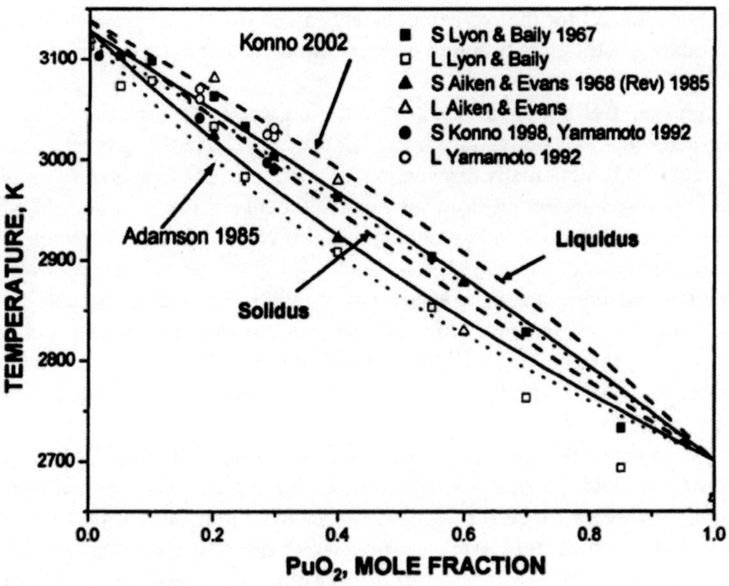

Fig. 11.4 Melting point of mixed uranium-plutonium oxide fuel [2]

fuel, the densification of the columnar grains up to nearly 100% TD improves the thermal conductivity in that region and allows a significant increase in linear power. Thermal conductivity is a function of temperature, density, O/M and Pu fraction in the fuel, which in turn depends on the fuel radius and the local morphology of the fuel matrix.

The mixed oxide fuel is a solid solution of UO_2 and PuO_2. For solid solutions the melting occurs over a temperature range. The lowest temperature at which melting starts is called *solidus* and the temperature at which the melting is complete is called *liquidus*. The solidus temperature of oxide fuel is high. Figure 11.4 contains the solidus and liquidus temperatures of mixed oxide fuel as a function of PuO_2 content. Note the substantial difference between the solidus and liquidus temperatures($\cong$50 K for 0.2 mole fraction PuO_2). This high melting point compensates somewhat for the low thermal conductivity in allowing adequate values of linear power [Eq. (2.2)].

Uranium dioxide has a fluorite crystalline structure, wherein the oxygen ions are arrayed in a simple cubic structure and the heavy metal ions form a face centered cubic sublattice. This arrangement results in an unoccupied interstitial position in the body center of the oxygen cubic structure, a situation that directly contributes to the relatively low density of mixed oxide fuel. Uranium carbide, on the other hand, exhibits an interlaced, face centered cubic structure, which leads to a higher concentration of heavy metal atoms and, thus, higher material density.

Mixed oxide, like most ceramics, is a relatively brittle material at temperatures less than half the melting point. Plastic deformation only occurs at the higher temperature levels. Consequently, the steep thermal gradients lead to rapid crack formation upon startup and cooldown. During steady-state power, cracks are believed to exist only near the radial periphery, as schematically illustrated in Fig. 11.1, the larger cracks having largely been annealed out. However, the thermal cycling that necessarily occurs in a power reactor leaves a cracked structure that is not well defined. As a result, it is difficult to know how much of a fuel continuum exists for thermal expansion during a thermal upset condition.

Swelling. As noted earlier, some porosity is intentionally incorporated into mixed oxide fuel to accommodate fuel swelling. Some of this swelling is due to solid fission products. Olander [1] estimates that swelling due to this source may range from 0.15 to 0.45% per at.% burnup. A number close to 0.2%

per at.% burnup is estimated for the oxygen deficient mixed oxide fuel generally used. This implies a solid fission product swelling contribution of approximately 2–3% for a typical target exposure of 100–150 MWd/kg.

Substantially greater swelling could result from fission gases. The situation is complicated, however, because the net swelling is derived from the balance between fission gas retention versus release that depend on grain structure, porosity distribution, temperature, and temperature gradient. For example, most of the fission gas generated from the relatively cool periphery of the pin is retained in the fuel mixture. Since the temperature is low, swelling in this region occurs in a manner similar to solid fission product swelling.

Further up the temperature gradient, in the equiaxed region,[5] even though a substantial fraction of the gas is still retained, the temperature is high enough to allow gas migration and collection along the grain boundaries; hence, substantial swelling results. In the still higher temperature columnar region, most of the fission gas is vented to the central void region. Consequently, very little swelling due to fission gas occurs in that locale.

In addition to built-in internal porosity, a physical gap between the fuel pellets and the cladding wall can be adjusted to yield the smear density desired for accommodating fuel swelling. Too large a gap is undesirable, however, since it greatly increases the temperature drop from the fuel surface to the cladding and also reduces the fission density (which directly lowers the breeding ratio). Pellet densities of approximately 90% TD and a smear density of 85% are sufficient to allow a typical mixed oxide fueled fast spectrum reactor to reach a burnup of 150 MWd/kg. As of now [3] over 300,000 mixed oxide fuel pins have been successfully irradiated (failure rates less than 1%) and, in some cases, burnup levels in excess of 150 MWd/kg have been achieved.

Fission Gas Release. Fission gas, once released from the fuel matrix and vented to open collecting zones (such as the central void or the fuel-cladding gap), does not re-enter the fuel matrix during continued irradiation. The normal procedure for collecting such gas is to provide a fission gas plenum immediately above or below the fuel column. It should be noted that since Xe and Kr are chemically insoluble in fuel, the buildup of pressure in the fission gas plenum (which is normally in approximate equilibrium with pressure through the fuel pin) does not affect the rate of gas released from the fuel matrix.

The processes by which fission gas bubbles nucleate, grow, diffuse, and eventually collect and condense at grain boundaries are extremely complex, though fuel temperature appears to provide a useful index for establishing the relative rates of fission gas mobility.[6] Below a temperature of about 1,300 K, fission gas mobility is very low and there is essentially no gas escape. Between 1,300 and 1,900 K, atomic motion allows some diffusion to take place such that an appreciable amount of gas can reach escape surfaces over a long period of time. Above 1,900 K, thermal gradients can drive gas bubbles to closed pores over distances comparable to grain sizes in days or months. Gas release occurs when the cavities reach a crack or other surface that communicates directly with free volume.

Approximate numerical values for fission gas release in the restructured and unrestructured regions as a function of burnup can be obtained using Eq. (8.6). Experience derived from Phénix operations in France [5] provided a somewhat different pattern, as illustrated in Fig. 11.5, although it is close to the higher linear heat rating illustrated in Fig. 8.8.

It should be noted that solid fission products tend to migrate toward the cooler region of the pin (radially toward the cladding and axially toward the ends of the pin). A particular concern is cesium

---

[5]See Section 8.2 for a discussion of equiaxed and columnar grain development.

[6]While most of the factors governing fission gas retention and redistribution are reasonably well understood, disagreement among ceramicists still remains over some of them [4]. Fuel temperature is only one of several governing factors.

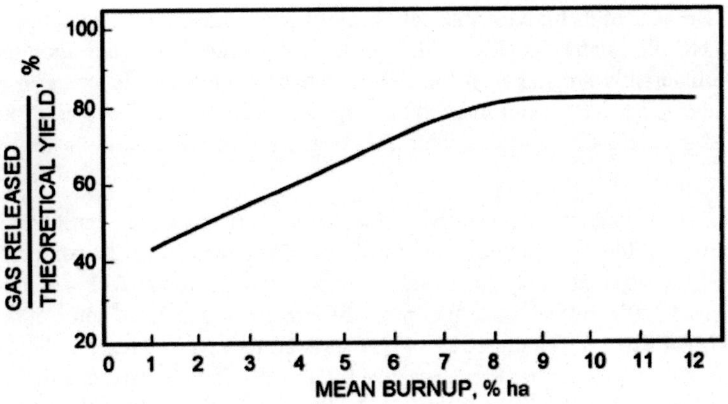

Fig. 11.5 Fission gas release in UO_2 vs. burnup [5]

which, if sufficiently concentrated near the pin periphery, can cause appreciable degradation of the cladding.

But there may be benefits as well. In particular, beyond about 7–8 at.% burnup, a certain oxide compound mixture has been identified [5] on the surface of the fuel and up to the cladding inner surface. This is described as *JOG "Joint Oxide Gain"* which is a solid fission product oxide layer and its major constituent has been identified essentially as Cs_2MoO_4. JOG has a higher thermal conductivity than the gas gap and this reduces the fuel surface temperature.

JOG is a soft material with a density of around 4.3 g/cm^3, formed in between fuel and cladding and plays a significant role in determining the integrity and lifetime of the fuel pin since it influences the heat transfer between fuel and cladding. Also, it reduces fuel-cladding mechanical interaction (FCMI). Since it induces uncertainties in power to melt determination at high burnup, a test was conducted in CABRI [6] to understand the behavior and develop parameters to be useful for predicting its behavior with burnup.

Fuel Pin Failure. Cladding failure is classified under four main categories: (i) manufacturing defects, (ii) mechanical interaction between pin, spacer and duct, (iii) FCMI and (iv) fuel-cladding chemical interaction (FCCI). Failure due to manufacturing defects might show up at the early stage of irradiation. The mechanical interaction is a complex phenomenon involving hot spot, etc. FCMI is generally overcome by suitable fuel pin design such as lower smear density, annular pellet, higher cladding thickness etc. However, this can be an issue at higher burnups (> 200 MWd/kg). Fuel cladding chemical interaction is caused by the migration of oxygen and accumulation of volatile fission products cesium and tellurium at the fuel surface. FCCI involves the oxidation of mainly chromium in the cladding in the presence of cesium, molybdenum, tellurium and iodine. In general the cladding wastage due to this at moderate burnups does not cause any concern. A majority of the cladding failures are generally traceable to fabrication defects. An analysis of the defect in advance is very important in enhancing quality assurance.

11.2.3.2 Carbide Fuel

Several compounds of uranium and carbon exist, but UC has been given the most attention because of its high uranium density. As indicated earlier, UC is a densely packed, face centered cubic structure and contains 4.8 wt.% carbon for a stoichiometric composition. Like UO_2, UC can be fabricated by compacting and sintering powders to achieve the degree of porosity desired. The primary factor of interest in UC as a fast reactor fuel, beyond its relatively high density, is its good thermal conductivity.

Microstructure. Because of its high thermal conductivity, peak temperature and temperature gradients are smaller in a UC pin relative to UO_2. Although some cracking does occur upon thermal cycling, such effects are inherently smaller than for UO_2. There is a tendency for pores to migrate up the temperature gradient, but the overall magnitude is far less than in UO_2 because of the much lower temperature gradients. Hence, no appreciable fuel restructuring occurs and no central void region develops.

Properties. Table 11.1 contains a qualitative comparison of mixed oxide, carbide, and nitride fuel materials properties. Table 11.2 indicates the principal differences in irradiation performances that follow from the properties of these three fuel systems. By comparison with the highly successful irradiation of over 300,000 mixed oxide fuel pins, the irradiation results of some 500 mixed carbide pins, up to 1977, indicated high failure rates [7]. However, recent experience indicates that a target exposure of 100 MWd/kg or higher can be realized with carbide fuel. Mixed carbide (U, Pu)C fuel with high Pu content used in the Indian Fast Breeder Test Reactor (FBTR) at Kalpakkam was found to be compatible with stainless steel clad materials [8]. Mark I fuel with a Pu/(U+Pu) fraction of 0.7 used in this reactor as driver fuel achieved a burnup of 155 MWd/kg without any pin failures [9].

Swelling. Control of stoichiometry during the fabrication of carbide fuel appears to be *very* important with regard to swelling characteristics. Hypostoichiometric mixtures, i.e., carbon deficient $UC_{(1-x)}$, leaves free uranium in the matrix. This material, during irradiation, tends to migrate and collect as

Table 11.1 Qualitative comparison of mixed oxide, carbide and nitride fuel materials properties (augmented from Refs. [7, 10])

Materials property	(U, Pu)O_2	(U, Pu)C	(U, Pu)N
Thermal conductivity	Low	Slightly higher than nitride	Much higher than oxide;
Thermal expansion	Higher	Lower than oxide	Lower than carbide
Thermal creep	Similar[a]	Similar[a]	Similar[a]
Irradiation creep	Higher	Lower than nitride	Lower than oxide
Swelling	Moderate	Higher than nitride	Higher than oxide
Powder	Not pyrophoric	Highly pyrophoric	Less pyrophoric
Handling	Inert atmosphere not needed	Ar, He, N_2 atmosphere (high purity)	Ar, He, N_2 atmosphere (commercial purity)
Dissolution in HNO_3 followed for reprocessing by PUREX process	Easy	Easy (may require destruction of organic complexes)	Easy; but risk of ^{14}C in reprocessing

[a] At equivalent temperatures, the thermal creep rates are similar although there are differences in the stress dependencies.

Table 11.2 Comparison of irradiation performance of mixed oxide, carbide and nitride fuels (adapted from Ref. [7])

Mixed oxide fuel	Mixed carbide and mixed nitride fuels
High centerline fuel temperature and low beginning-of-life (BOL) power capability (~50 kW/m)	Low centerline fuel temperature and high beginning-of-life (BOL) power capability (>100 kW/m)
Fuel restructuring	Little or no fuel restructuring
High fission gas release and significant fission gas pressure effects	Low fission gas release and significant fuel swelling
High plasticity fuel and low levels of fuel cladding mechanical interaction (FCMI)	Low plasticity fuel and high levels of fuel cladding mechanical interaction (FCMI)
Release of volatile fission products and fuel-cladding chemical interaction	Little or no release of volatile fission products and cladding carburization

a metallic phase at the grain boundaries. Hence, the swelling characteristics of metal fuel become manifest and lead to intolerable swelling rates [11]. In addition to this fabrication problem,[7] it is essential to ensure that no mechanism (such as fission product generation) could preferentially lead to a carbon deficiency. 10–20 wt.% sesquicarbide is typically added to take care of the carbon balance during irradiation. However, the carburization that is caused is only marginal. The excess carbon is present in the form of a two phase mixture of monocarbide and sequicarbide solid solutions in which the sesquicarbide is predominantly Pu_2C_3. The carbon deficiency in the hypostoichiometric carbide, though a single phase fuel at higher temperatures, is alleviated by using dispersed microdroplets of (U, Pu) alloy in the carbide matrix.

On the other hand, a carbon excess leads to precipitation of higher carbides, and these can be transferred to the cladding to cause cladding carburization. Cladding carburization is likely manageable for helium-bonded fuel, but this problem can be particularly acute with sodium bonding (sodium residing in the fuel-cladding gap). Sodium bonding is preferable to gas bonding from a thermal efficiency viewpoint, since the high temperature drop across a gas bond negates much of the advantage of the high-conductivity carbide fuel. One possible design solution to this problem is the inclusion of a sacrificial steel sheath, immersed within the sodium bond, to protect the cladding from chemical attack.

Fission Gas Release. For operating temperatures of interest, the fission gas released from stoichiometric UC is far less than that obtained from UO_2. Release rates in excess of 50% have been observed for specimens around 1,350°C, but such release rates are normally associated with hypostoichiometric mixtures, where the uranium phase has led to appreciable swelling. Figure 11.6 indicates that

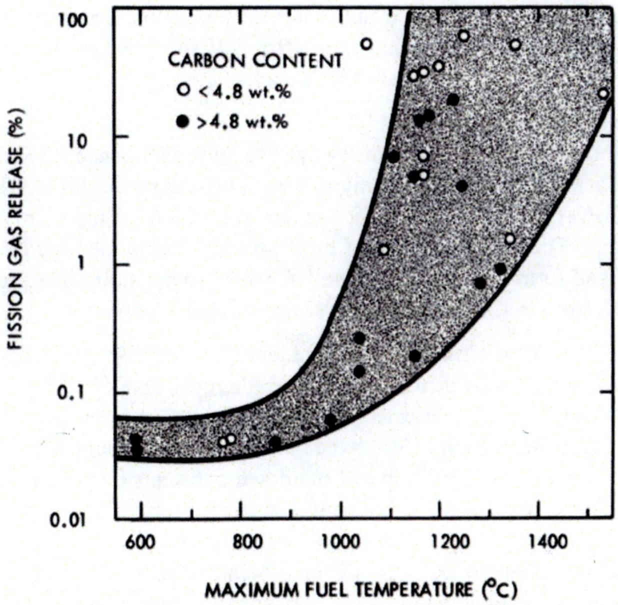

Fig. 11.6 The release of fission-product gases from UC as a function of the irradiation temperature [12]

[7] Another difficulty of the carbide fuel fabrication process is the necessity to keep oxygen concentrations at very low levels. Uranium will preferably oxidize, and the presence of oxygen could produce unwanted UO_2 as well as posing a fire hazard.

fission gas release, particularly for the carbon-deficient composition, can become significant above ~1,300°C.

Mixed Uranium-Plutonium Carbides. Whereas appreciably less in-pile experience has been accumulated for UC-PuC mixtures, data to date indicate that such mixtures have properties very similar to UC. Excess carbon tends to be precipitated as Pu_2C_3, and a carbon deficiency results in a U-Pu alloy deposition. The metallic alloy results in swelling characteristics very similar to hypostoichiometric UC. Therefore, mixed carbide fuels are chosen to be two phase mixtures containing monocarbide and a minor fraction of sesquicarbide to prevent the formation of the metal phase during irradiation. The Indian Fast Breeder Test Reactor (FBTR) employed mixed carbide fuel, which has undergone a burnup of 155 MWd/kg.

11.2.3.3 Nitride Fuel

Uranium plutonium mixed nitride has the same crystal structure as mixed monocarbide and has similar physical and chemical properties [10, 13, 14]. Along with mixed carbides, mixed nitrides are also considered advanced fast reactor fuels due to the higher fissile atom density, higher breeding ratio, higher thermal conductivity, and excellent compatibility with sodium coolant and cladding compared to the oxide fuel. Mixed nitrides also provide complete solid solubility. However, irradiation experience with nitride fuels is less than that with carbide fuels. Research and development programs on nitride fuels for fast spectrum reactor use were actively pursued in the USA, France, Germany, United Kingdom and the Russian Federation during 1960s and 1970s and a little later in Japan. But the investigations were restricted to UN, and (U, Pu)N fuels with a maximum plutonium content of 20%. Both helium-bonded and sodium-bonded nitride fuels have been developed and irradiated successfully to high burnup (up to 20 at.%). Nitride fuel has been accepted as the fuel of choice for the BREST-300 fast reactors in Russia (a reactor design that utilizes lead as the coolant).

Fabrication. The fabrication methods for nitride fuels are similar to that of mixed carbide fuels. Mixed nitride fuel is relatively easier to fabricate than mixed carbide fuel, since it is not as reactive and as pyrophoric as mixed carbides. Further, plutonium forms only the mononitride. The higher nitrides of uranium (UN_2 and U_2N_3) are unstable and decompose to form UN at high temperature ($\geq$1,673 K) under vacuum or argon. The major problem of mixed nitride fuel is the high parasitic absorption of fast neutrons by ^{14}N and formation of radioactive ^{14}C by (n, p) reactions. The problem of ^{14}C can be avoided by using ^{15}N. However, the process of ^{15}N enrichment is expensive.

Irradiation Performance. Mixed nitride fuels exhibit less fuel swelling, lower fission gas release, and are considerably easier to reprocess than the mixed carbide fuels. However, the production of biologically hazardous ^{14}C in nitride fuels fabricated using natural nitrogen poses a considerable concern for the reprocessing of mixed nitride fuels [15]. Mixed nitride fuels were found to exhibit extensive cracking and fragmentation during simple startup and shutdown transients if operated at high temperatures. This cracking phenomenon was considered to be the reason for the early fuel failures seen in US irradiation tests [16] and their peak temperature of operation was restricted to below 1,200°C to mitigate fragmentation. Since the mixed nitride fuels could dissociate at temperatures substantially lower than their congruent melting point, maintaining a nitrogen overpressure is necessary to prevent dissociation. Cladding carburization has not been studied to the same extent as the carbide fuels. However, carburization could be avoided by reducing the total amounts of oxygen and carbon impurities to less than 2,000 ppm.

11.2.3.4 Metal Fuel

All the early, low-power fast reactors employed metallic uranium fuel. It is relatively easy to fabricate, has excellent thermal conductivity (as shown in Fig. 11.13 in the section on general intercomparison), and has a high density (19.0 g/cm^3 at room temperature). The principal problem with metal fuel is its highly anisotropic growth patterns upon irradiation and the large associated dimensional change. If fully dense metallic fuel is used, along with a small initial fuel-cladding gap in order to achieve a high smear density, burnup is limited to the order of 10 MWd/kg, far too low to be economical. However, at least two techniques have been proposed to improve metal fuel burnup capability. One scheme [17] is to fabricate an axial hole in the fuel to accommodate fuel swelling internally. The second technique, developed in association with the EBR-I1 project [18], involves increasing the initial fuel-cladding gap to accommodate such swelling. Success in this approach with metal alloy fuels is described below in the write-up entitled EBR-II experience.

Metal fuels are known for their high breeding potential due to their high fissile metal atom density. A binary form of U-Pu alloy is the usual choice as fuel. These fuels have high thermal conductivity but have a low melting point. Also, they tend to form inter metallic compounds with the cladding, which have a lower melting point than the binary fuel. To overcome the deficiencies of a lower melting point and low melting eutectic formation, zirconium is added to the binary fuel matrix. The addition of Zr increases the melting point of the fuel, but reduces the thermal conductivity and also the breeding ratio. It also helps to stabilize the metallic phases of U and U–Pu, thereby limiting the swelling due to phase changes. As an alternate to meet the performance requirements, either the cladding inner tube is coated with some materials like V or W or a thin liner containing zirconium is kept between the fuel and cladding such that it acts as a barrier between them to prevent inter diffusion of metal constituents of fuel and cladding.

Smear Density. Since smear density of the fuel heavily influences the fission gas release and the swelling of the fuel, a judicious choice of smear density is required. A smear density of 75% is normally selected [19–21] for the fuel as it allows the pores in the fuel to get interconnected and facilitate the release of fission gases contained in the fuel (Fig. 11.7) such that fuel cladding mechanical interaction is less. A lower smear density also helps in accommodating swelling.

Crystalline Structure. Metallic uranium exhibits three crystalline structural phases (alpha, beta and gamma) over low, intermediate, and high temperature ranges. The alpha phase (orthorhombic

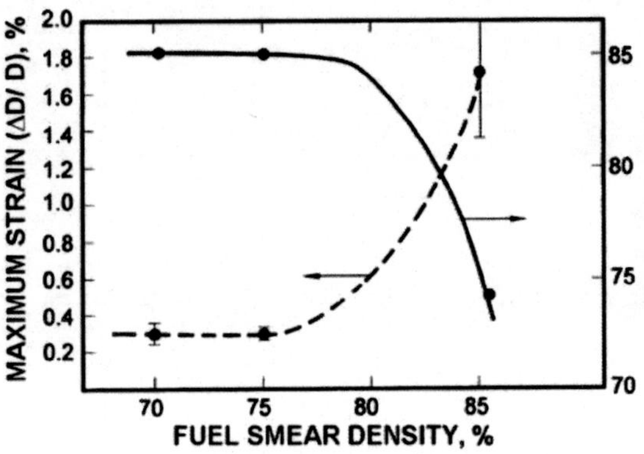

Fig. 11.7 Influence of smear density on gas release and diametral strain [19]

structure) is exceptionally anisotropic. Some of this anisotropy is suppressed in bulk material, wherein the grains are randomly oriented, but even the slightest amount of material cold working will preferentially orient the grains and allow highly nonuniform growth patterns. The gamma phase (body centered cubic) demonstrates the most isotropic behavior of the three phases.

Physical Properties. As in all fuel materials, thermal conductivity of metal fuel tends to decrease with burnup due to the material density loss. This is particularly aggravated when uranium swelling occurs. Both hardness and yield strength increase appreciably, even at low levels of irradiation, but there are conflicting data regarding the effects of radiation on ultimate tensile strength and ductility. The melting point of uranium metal is 1,135°C.

Swelling. Both uranium and uranium-plutonium metallic fuels have extremely low resistance to swelling[8] over the temperature ranges of interest, i.e., the 400–600°C range. Below 400°C, no appreciable swelling occurs. In the range from 400 to 500°C, tears occur along the grain boundaries and rapid swelling results. Deformation is believed to occur primarily due to grain-boundary sliding and cavitation, in preference to plastic flow within the grains. The fission gas diffuses to the grain boundaries and causes rapid grain boundary swelling, called "break-away" swelling. Beyond 500°C, a self-annealing process appears to take place that minimizes the effects of swelling.

Figure 11.8 shows radial averaged swelling of 85% smear density for U-19 wt.% Pu-10 wt.% Zr fuel plotted against burnup at an axial position 45% from the bottom of the core. These results are provided by the ALFUS code [22]—a code validated against experimental results. It is observed from Fig. 11.8a that for the case of 85% smear density, swelling due to the open pores is only around 5% at about 1 at.% burnup, when the initial slug-cladding gap is filled with a swollen slug. At a later stage

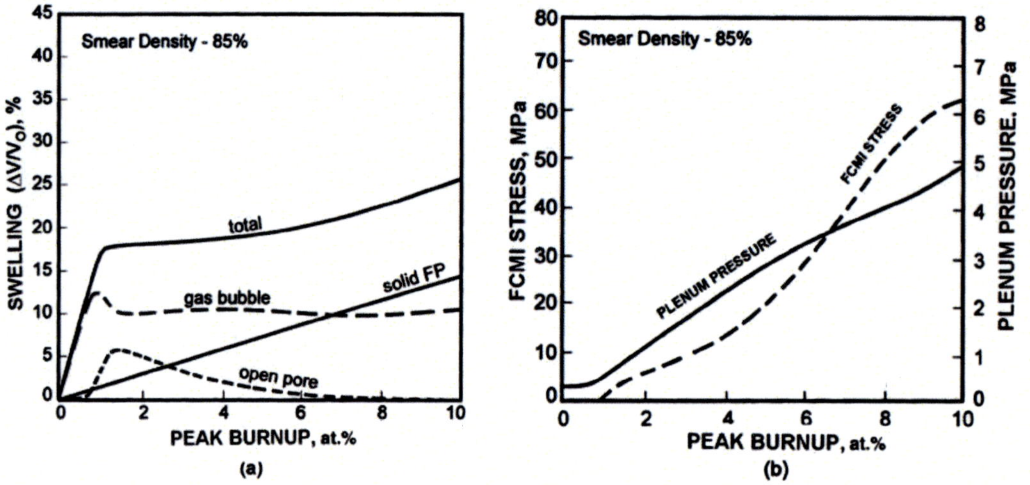

Fig. 11.8 Metal fuel swelling characteristics for 85% smear density (U-19% Pu-10% Zr)

[8]The term fuel "swelling" is somewhat ambiguous when applied to uranium metal because much of the observed growth over certain temperature ranges is due to the highly anisotropic behavior of uranium instead of fission product generation. However, the net effects of temperature and burnup are often referred to as swelling in the literature.

of irradiation, the open pore volume is replaced by the solid Fission Product (FP) swelling and this cannot be accommodated, and hence FCMI stress continues to grow as seen in Fig. 11.8b.

However, when the smear density is lowered to 75%, the open pore swelling amounts to about 20% and serves as a buffer to the solid FP swelling as seen from Fig. 11.9a. As a result, FCMI stress remains at a low level up to 10 at.% burnup as evident from Fig. 11.9b. It is observed that fuel makes contact with the cladding around 3 at.% burnup. Hence, 75% smear density fuel can accommodate about 35% swelling before fuel-cladding contact is made and this degree of swelling remains constant beyond 3 at.% burnup.

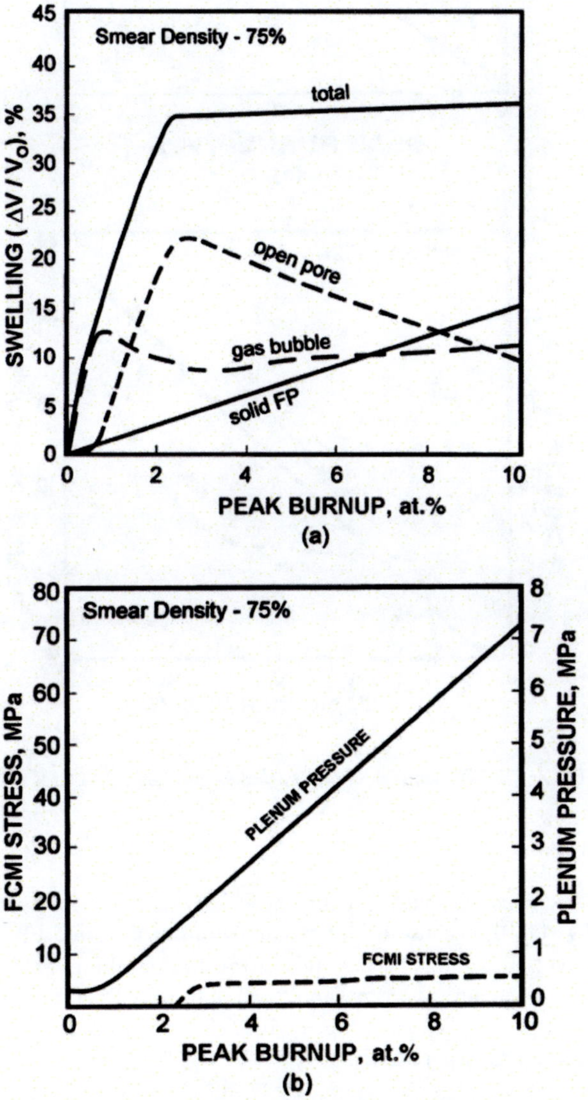

Fig. 11.9 U-19% Pu-10% Zr (75% TD) **(a)** Radial swelling; **(b)** FCMI and pressure build up in fuel pin [22]

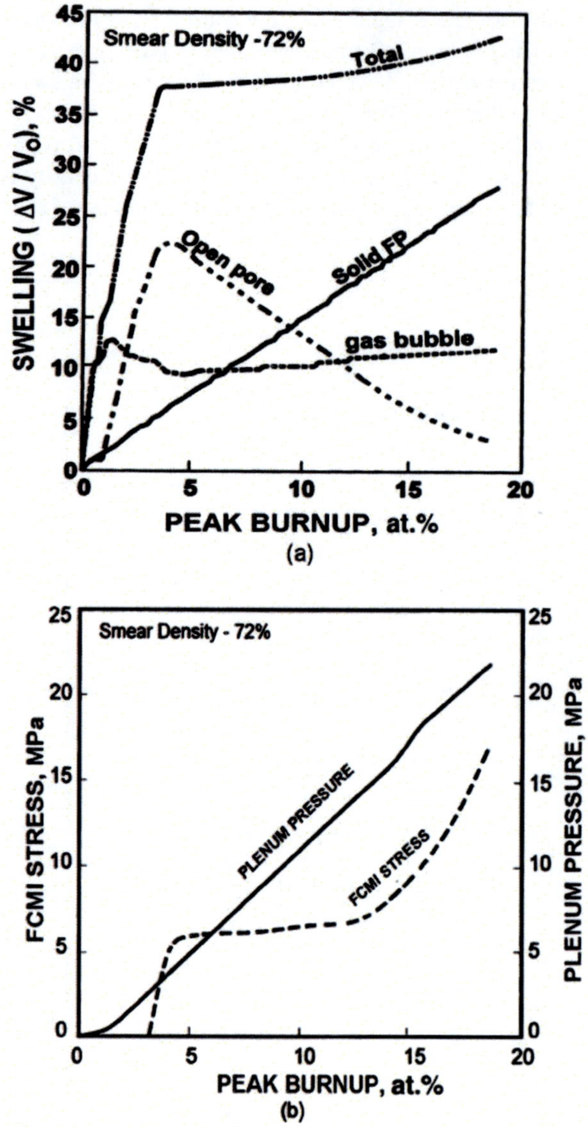

Fig. 11.10 U-19% Pu-10% Zr (72% TD) at high burnup (**a**) Radial swelling; (**b**) FCMI and pressure build up in fuel pin [22]

When the burnup becomes very high, significant FCMI can occur for even lower smear density fuel, as seen from Fig. 11.10, because of the accumulation of the solid FP. In this case, although the decrease in the open pore volume accommodates solid FP swelling up to ~13 at.% burnup, the remaining volume of the open pores becomes too small at the higher burnup. It leads to a significant increase in the FCMI stress as indicated in Fig. 11.10b. Hence, FCMI depends heavily on the smear density chosen for the fuel and target burnup.

Alloying. Because of the pronounced fuel swelling observed, much of the research effort on metallic fuels has focused on a search for alloying materials to mitigate the problem. In particular, efforts to stabilize the gamma phase, which has the most isotropic properties, have received primary attention.

Molybdenum has perhaps been the most successful material used in alloying uranium to retard the swelling problem. Both the Dounreay and Fermi reactors successfully employed a uranium molybdenum cermet fuel containing approximately 10 wt.% Mo. However, the burnup for these reactors was substantially less than desired for commercial plants. Achievement of high burnup with fissium and zirconium alloys is described in the section entitled EBR-II Experience.

Fission Gas Release. Figure 11.11 illustrates the fission gas released from metal fuel. These data indicate that fission gas release saturates at around 80%.

Axial Growth Rate. The axial expansion of the fuel is a major factor in the safety of metallic fuelled reactor (as discussed in the Chapters 13–16). The axial fuel growth due to fission gas retention is expected to be around 2–4% for the metal fuel composition most widely chosen. This is shown in Fig. 11.12, where this level is expected to be reached within 1–2 at.% burnup.

EBR-II Experience. The EBR-II reactor was fueled with a uranium-5 wt.% Fs[9] metal driver fuel, U-Fs alloy, which has operated with burnups as high as 10.5 at.%, for a full decade. The primary reason such high burnups have been achieved with metallic fuel is that substantial porosity has been designed into the fuel element. Since swelling cannot be suppressed, the fuel was fabricated to assure a fuel-cladding gap large enough to result in a cold smear density of approximately 75%. Initial swelling quickly expands the fuel to close the gap, but by the time contact is made with the cladding, sufficient swelling has occurred to open numerous paths for fission gas to escape to the fission gas plenum. Subsequent swelling is then restrained.

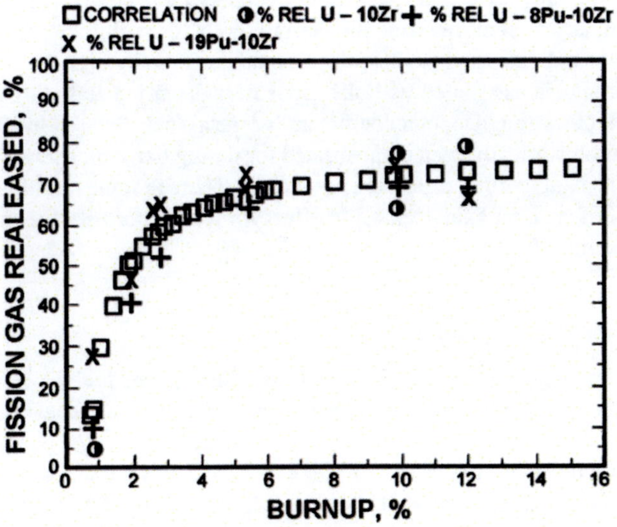

Fig. 11.11 Fission gas release for metal fuel [20, 23]

[9]Fissium (Fs) is an equilibrium concentration of fission product elements left by the pyrometallurgical reprocessing cycle designed specifically for EBR-II. Nominal 5 wt.% fissium consists of 2.4 wt.% Mo, 1.9 wt.% Ru, 0.3 wt.% Rh, 0.2 wt.% Pd, 0.1 wt.% Zr, and 0.01 wt.% Nb.

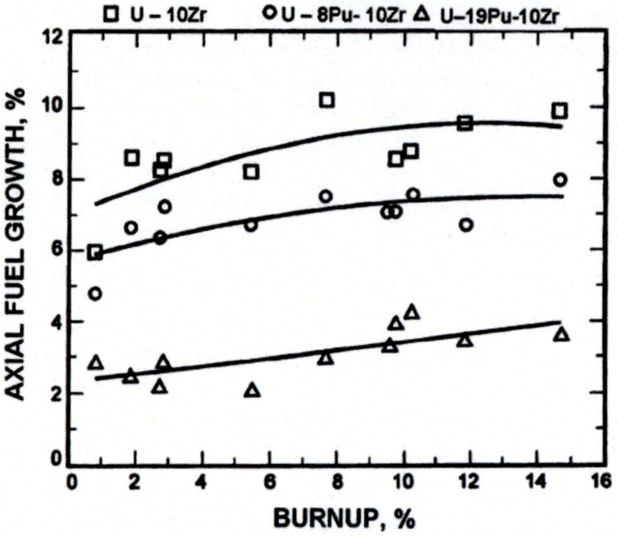

Fig. 11.12 Axial growth vs. burnup for various metal fuel compositions [23]

Research with a U-Pu-Zr alloy [18] suggests that performance even better than that achieved with the U-Fs alloy is possible. The principal improvement is the potential for higher temperature operation. A high eutectic temperature (~810°C) exists between this fuel and the stainless steel cladding, as well as a solidus temperature appreciably higher (~1,150°C) than previous metal alloys. (The comparable U-Fs alloy temperatures are 705°C and 1,000°C, respectively.) It is argued [18] that even with the low smear densities necessary to achieve economic burnup levels for such fuels, the fissile density is still better than other fuel candidates and the breeding potential is thereby high.

One of the uncertainties associated with the low smear density metal alloy fuels involves safety parameters. The lack of a strong Doppler coefficient for metal fuels (caused by the characteristic hard neutron spectrum) has historically been compensated for during transient overpower conditions by the presence of a reliable axial thermal expansion coefficient. There is some question, however, about the presence or efficacy of such an axial expansion effect for a high porosity fuel matrix. An update on this issue is provided in Ref. [24].

11.2.3.5 Other Uranium Compounds

In principle, many uranium compounds could serve as a fast reactor fuel. However, of the numerous other compounds, only uranium nitride (UN) has received appreciable attention, as noted above. It has physical properties [25] quite similar to UC. Uranium nitride is more compatible with cladding than UC (no carburization), but much less in-pile experience exists upon which to judge its overall merits. Fabrication problems are somewhat complicated, particularly if an arc casting process is used, because of the need for a nitrogen atmosphere to prevent nitrogen loss. Thermal decomposition of UN occurs above 2,000°C, which may raise certain safety questions but should not affect steady-state performance.

Uranium sulphide, US, likewise has many properties similar to UC, but its density is no better than that of UO_2. Hence, there is little incentive for active development. U_3Si has been studied as promising among the possible uranium-silicon systems because it yields a very high uranium density. Again, however, very little in-pile data are available upon which to judge its overall merits. A final uranium compound that has been studied to a limited extent as a potential fuel is the phosphide, UP.

11.2.4 Thorium Based Fuels

As noted from the discussion of Chapter 1, the ^{232}Th-^{233}U fuel system could be employed in a breeder reactor (thermal or fast), although with an inherently lower performance with regard to breeding ratio and doubling time. Because the potential to breed does exist for such a system, it is appropriate to provide some background [26] on the potential of thorium based fuel systems for fast reactor application.

Relative to uranium, thorium metal has both a higher thermal conductivity and a lower coefficient of thermal expansion. Both of these properties tend to reduce the thermal stress in the fuel elements, but the latter property would tend to reduce the negative reactivity feedback during a power excursion. Thorium has a considerably lower density than uranium.

Perhaps the most significant difference with regard to in-pile performance is that Th has an isotropic cubic crystalline structure (face centered cubic) and, therefore, undergoes appreciably less dimensional change upon thermal cycling and irradiation than does the anisotropic uranium. Thorium also has higher irradiation creep resistance, higher ductility, and a higher melting point (1,700°C). The latter property somewhat complicates the fabrication process since melting and casting are more difficult.

Cast thorium has a large grain size and easily cracks upon cold working. However, hot Th can actually be rolled to a thickness of 0.025 mm at room temperature without intermediate annealing. Overall, thorium metal is substantially easier to work with than uranium metal.

11.2.4.1 Th-U Oxides

For thorium to be used as a fuel, it must be mixed with uranium. In the case of a mixed thorium-uranium oxide system, it is possible to maintain a single phase structure over the entire mixture range of interest. The thorium-uranium oxide system can be fabricated by pressing and sintering in an air atmosphere, but sintering in a hydrogen atmosphere significantly reduces the oxygen excess.

Columnar grain growth, with associated lenticular pores, has been observed for thorium dioxide. The growth is very similar to that observed in UO_2. However, the vapor pressure of Th is less than for UO_2 and, as a result, such grain formation does not occur until higher temperatures (approximately 350°C higher) are reached. Equiaxed grain growth occurs in a manner very similar to UO_2. Thermal stress cracking patterns are also similar to UO_2. Pure ThO_2 has a thermal conductivity approximately 10% higher than pure UO_2. Irradiation experience that is available for ThO_2-UO_2 fuel yields little indication of fuel swelling. This fuel system should perform in a manner comparable to the UO_2-PuO_2 system. However, considerably more in-pile data needs to be collected for the ThO_2-UO_2 fuel system before a high degree of confidence can be attained regarding its long-term performance.

11.2.4.2 Th-U Carbides

ThC is a face centered cubic (FCC) structure with a melting point of 2,625°C and a density of 10.65 g/cm^3. One of the positive aspects of this carbide system is that the carbon content can vary from 3.8 to 4.9 wt.% without changing the crystalline structure (in sharp contrast to the UC system). Very little information has yet been gathered on the ThC-UC system.

11.2.4.3 Th-U Alloy

Metallic fuels of thorium and uranium can be readily obtained, but the uranium fraction must be kept below a certain level if reasonable dimensional stability is to be maintained. Only 1% of U can be dissolved into solid solution with thorium for temperatures up to 1,000°C. At higher content, U precipitates predominantly at the grain boundaries. For uranium additions up to 20 wt.%, the U can be regarded as a dispersion in the FCC thorium matrix; but beyond that point, the U may form a continuous metallic network.

This solubility characteristic is extremely important from the standpoint of dimensional stability. Good dimensional stability has been maintained in Th-U fuels for mixtures with low U content. Based on early in-pile work, it was reasoned [27] that the principal reasons for the encouraging results were (1) most fission fragments were deposited outside the U precipitates and directly in the Th metal, (2) the fine dispersion of U particles helped to anchor gas-filled pores, and (3) the higher Th melting point might have resulted in lower gas diffusion rates at a given temperature (hence, less propensity for fission gas induced swelling). Unless significant cracking occurs, fission gas retention is believed to be very high for Th-U fuel.

11.2.4.4 Th-Pu Alloys

Only limited data are available for Th-Pu and Th-U-Pu alloys [28]. Pu has been reported to have a very high solid solubility in thorium (48.5 at.%). However, even a small addition of uranium, even 2 at.%, was found to drastically reduce the solubility of plutonium in thorium. Also, the ternary Th-U-Pu alloys were found to form a liquid at low temperatures. Hence the potential power reactor Th-U-Pu ternary fuels could have a maximum 10 wt.% Pu and 25 wt.% U. Limited irradiation data are available for the ternary alloy fuels up to 5.6 at.% burnup. The fuel microstructure was not altered after irradiation, the swelling was low, and 70% of the fission gases were released to the plenum.

11.2.5 General Intercomparison

From the discussion above, it is apparent that several fuel systems exist that could be employed in fast reactors. Because of the hostile environment in which such fuels must perform, however, considerable high-burnup experience is critical to the confident prediction of long-term performance. Such data are generally quite expensive and time-consuming to accumulate; hence, the backlog of experience with mixed uranium-plutonium oxide fuels currently gives that fuel system a considerable advantage, although interest in metal fuel continues to develop.

In addition to in-pile performance, other considerations such as neutronics and safety are important in the fuel selection process. The reactor spectrum is one such factor since breeding ratio and transmutation effectiveness increase as the neutron spectrum is hardened. The high density metal reactor, having no moderator in the fuel matrix, yields the hardest neutron spectrum, but the flux in the principal Doppler resonance region (0.1–10 keV) is appreciably less than for the ceramic fueled reactor. Consequently, oxide, carbide, and nitride fueled fast reactors possess appreciably higher Doppler coefficients than metal fueled reactors.

Table 11.3 lists selected physical properties for uranium and plutonium fuels. Metals are clearly superior with regard to density but have substantially lower melting points than the ceramic fuels. Figure 11.13, which shows the thermal conductivities of stoichiometric fuels, indicates that full density metal fuel has substantially higher values than the carbide, nitride, or oxide fuel forms. On the other hand, as we saw from earlier discussions, fully dense metal U-Pu fuel cannot be used for even moderately high burnup applications in a fast spectrum reactor. A reduction in smear density

Table 11.3 Properties of uranium and plutonium fuels

	Melting point (K)	Heat of fusion (kJ/kg)	Density 100% TD (g/cm^3)	Thermal conductivity (W/mK) 500°C	Thermal conductivity (W/mK) 1,500°C	Specific heat (c_p) (kJ/kgK) 298 K	Specific heat at melting point Solid	Specific heat at melting point Liquid
Metal				30[a]				
U[b]	1,408	38.2	19.0			0.12	0.16	0.20
Pu	913	11.86	19.9			0.18[c]		
Oxide								
UO_2	3,138	277	10.97			0.24	0.77	0.50
$(U_{0.8}Pu_{0.2})O_2$[b]	3,023 (sol) 3,063 (liq)	277	11.08	4.0[d]	2.3[d]	0.26	0.64	0.50
PuO_2	2,670		11.46			0.35[e]		
Carbide				16[a]	17[a]			
UC	2,780	184	13.6			0.20	0.35	0.28
$(U_{0.8}Pu_{0.2})C$[b]	2,548 (sol) 2,780 (liq)	186	13.5			0.19	0.43	0.28
PuC[a]	1920		13.6			0.24[e]		
Nitride[a]				12	16			
UN	2,870		14.3			0.27[e]		
PuN	2,770		14.2			0.26[e]		
Phosphide[a]				17				
UP	2,880		10.2					
PuP	2,870		9.9					
Sulfide[a]				14				
US	2,750		10.9					
PuS	2,620		10.6					

[a]Adapted from page 169 of Ref. [29].
[b]Ref. [30].
[c]at 500°C.
[d]95% TD, O/M=2.00 (Cf. Fig. 9.2).
[e]at 1,500°C.

to approximately 75%, to allow reasonable burnup levels, directly reduces the effective thermal performance of the pin.

More refined data for both Table 11.3 and Fig. 11.13 may be found in Refs. [13, 30–37].

11.3 Cladding and Duct

The fuel cladding provides the basic structural integrity of the fuel element. It prevents fission gas from entering the primary coolant system[10] and provides physical separation of the fuel from the coolant. The duct is likewise a fundamental part of the fuel assembly design, as delineated in Section 8.4.1. Since both the cladding and duct provide structural support for the fuel and experience a similar temperature and radiation environment, the general materials background for both components will be provided in the same section (though specific references will generally be made to the cladding, since it tends to experience a more hostile environment).

[10]Controlled venting could be purposely designed into the system.

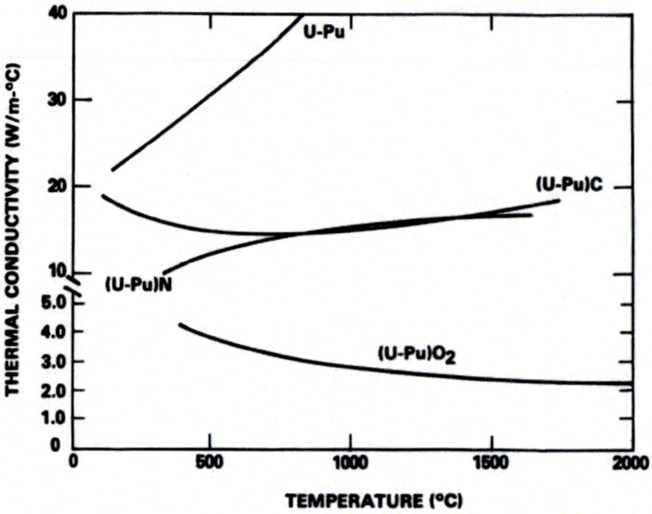

Fig. 11.13 Thermal conductivity of mixed oxide, nitride, carbide, and metal fuels (from page 33 of Ref. [29])

11.3.1 Requirements

High temperature strength and long exposure irradiation resistance provide the principal requirements for fast reactor cladding. For an economical fuel cycle, fast spectrum reactor cladding must be capable of residing in the core for up to three years or more at peak temperatures approaching 700°C. Whereas the peak exposure in a LWR may be around 10^{22} n/cm^2, the fluence requirements for fast spectrum reactor cladding are about 2×10^{23} n/cm^2 (a factor of 20 higher!). An important limitation to meeting the latter requirement is the phenomenon of void swelling, since accommodation of diametral cladding strains beyond about 3% begins to impose severe economic penalties. Also, cladding rupture or breach is an important life limiting phenomenon.

Since the cladding is a parasite from the standpoint of neutron economy, it is desirable to select a material with a low neutron capture cross section, $\sigma_{n,\gamma}$ (or σ_c). Table 11.4 contains values of $\sigma_{n,\gamma}$ for several metals in thermal and fast spectrum environments. When comparing these values, however, it should be kept in mind that neutron absorption in the cladding is considerably less important for a fast spectrum reactor than for a LWR. This situation arises because (1) the metal-to-fuel neutron cross section ratio is considerably lower in a fast spectrum, relative to a thermal spectrum, and (2) there is relatively less cladding material in a fast spectrum reactor. The latter fact is a consequence of a strong desire to maximize the fuel/cladding ratio to enhance the fast spectrum reactor internal conversion or breeding ratio.

It is noted from Table 11.4 that zirconium based alloys required in light water reactors for their low neutron capture cross section would not be attractive for fast spectrum reactor use due to their relatively high cross section at higher energies. Another reason for bypassing the standard LWR zircaloy cladding material in the fast spectrum reactor is that zircaloy does not possess sufficient strength at the high temperature operations required. Though only of secondary importance, neutronics considerations deter the use of high nickel bearing alloys in the fast spectrum reactor, even though some high nickel alloys may demonstrate generally superior irradiation properties.

Austenitic 316 stainless steel, normally cold worked to approximately 20%, was selected as the early reference cladding material for the SFR. Austenitic steel is a face centered cubic form of iron (as opposed to ferritic steel, which is body centered cubic). This austenitic form is the natural structure

Table 11.4 Absorption cross sections of the more important metals[a]

Group	Metal	$\sigma_{n,\gamma}$(100 keV) [millibarns]	$\sigma_{n,\gamma}$(thermal) [barns]
1	Al	4	0.230
	Ti	6	5.8
	Fe	6.1	2.53
	Cr	6.8	3.1
	V	9.5	15.1
2	Si	10.0	0.16
	Co	11.5	37.0
	Ni	12.6	4.8
	Zr	15.1	0.18
	Cu	24.9	3.77
	Mn	25.6	13.2
3	Mo	71.0	2.7
	Nb	100.0	1.15
	W	178.0	19.2
	Ta[b]	325.0	21.0

[a]From page 184 of Ref. [29].
[b]Ta is an absorber in control rods.

of iron between 910 and 1,400°C, and this structure is stabilized at room temperature by the addition of nickel. The austenitic steels have advantages over most ferritic steels in SFRs because of their better high-temperature creep strength and corrosion resistance. A few ferritic steels are being considered for cladding, however, and even more serious consideration is being given to ferritics for use in ducts.

11.3.2 Radiation Effects

On passage through a crystal lattice, neutrons interact with the lattice atoms and lose some of their energy to these atoms. These processes can be accompanied by damage to the crystal lattice in three general manners:

i. production of lattice atoms displaced from their regular lattice positions
ii. changes in the chemical composition by capture of neutrons followed by transmutation and
iii. excitation of electrons and ionization of atoms (although this process does not produce permanent damage in metals)

An energy transfer of only ~25 eV is required to displace an atom from its stable lattice position. A typical fast neutron-atom collision creates an energetic recoiling atom (the primary knock-on atom or PKA) which—if of sufficient energy (i.e. above ~1 keV and typically tens of keV)—then goes on to displace many neighboring atoms from their equilibrium lattice sites, generating vacancy and self-interstitial point defects in equal numbers. Many vacancy and interstitial defects are thus created from each single neutron collision event. The spatial distribution of each damage region is markedly heterogeneous and consists of (i) a dispersion of initially single point defects and (ii) a highly localized concentration of point defects known as a "collision cascade" at the end of the PKA track with dimensions of the order of 10 nm or less, and in the form of a central core of vacancies with the associated self interstitials ejected to the periphery. For such a defect configuration, there is a strong inherent tendency for the cascade core to collapse spontaneously, depending on the materials. This results in the creation of either a vacancy dislocation loop, or conceivably a microvoid or stacking fault tetrahedron.

The surrounding shell of interstitials also forms clusters or dislocation loops both ballistically and by subsequent rapid migration and coalescence. A large fraction of the point defect population is annihilated simply by mutual recombination, while a small proportion escape the cascade volume as freely migrating defects.

This classical picture of radiation damage has been supported in recent years by molecular dynamics based atomistic calculations that simulate cascade formation in a computer modeled crystal lattice. In metallurgically complex systems such as alloyed steels the ultimate defect configuration affecting materials behaviour under irradiation will depend on, in addition to neutron and PKA energies, parameters such as irradiation temperature, pre-existing point and line defect populations, and point defect-dislocation and point defect-solute interactions.

Since atomic displacements, as described above (rather than the neutron dose) are linked to the irradiation induced changes, the damage is most meaningfully characterized in terms of the average number of times that an individual atom is displaced from its lattice site. This leads to a damage/exposure unit called "*displacements per atom*" *(dpa),* which is generally accepted as a correlation parameter for mechanical property and microstructural changes in reactor materials irradiated in different neutron spectra. The dpa unit, calculated according to the widely accepted Norgett-Robinson-Torrens (NRT) model [38], includes both the neutron energy spectrum and the total number of neutrons per unit area passing through the material, and has been incorporated into an international standard. Early correlations of reactor data were based on the neutron dose unit "n/cm^2 (>1 MeV)" for thermal reactors or "n/cm^2 (>0.1 MeV)" for fast reactors, where a threshold energy of 1.0 or 0.1 MeV defines a "fast neutron" in the relevant neutron spectrum. Approximate conversion factors between the two units are: $1 \times 10^{22} n/\text{cm}^2$ (>0.1 MeV) $\cong$ 4.9 for fast reactor core components, and $1 \times 10^{19} n/\text{cm}^2$ (>MeV) $\cong$ 0.015 for thermal reactors. It should be noted that the dpa unit on the NRT model still has deficiencies insofar as it does not take into account damage annihilation during cascade formation; thus within a cascade, the fraction of surviving defects is strongly dependent on PKA energy and can be of the order of 20–30% of the NRT value. This fraction also depends on irradiation temperature.

Neutrons are also absorbed in nuclear reactions, resulting in changes in atomic nuclei and thereby creating new elements as impurity atoms. A large number of such transmutations is possible, involving both thermal and high energy neutrons, but the most important metallurgical implications are the (n, α) and (n, p) reactions, which generate helium and hydrogen gas, respectively. Helium is particularly important in irradiated steels and mainly arises from transmutation reactions involving nickel, boron, and iron as follows: (i) thermal neutrons and nickel are involved in the two step reaction: $^{58}\text{Ni}(n,\ \gamma)^{59}\,\text{Ni} \rightarrow\ ^{59}\text{Ni}(n,\ \alpha)^{56}\text{Fe}$; (ii) thermal neutrons and boron are involved in the reaction: $^{10}\text{B}(n,\ \alpha)^{7}\text{Li}$; and (iii) fast neutrons (>6 MeV) are involved in threshold type (n, α) reactions principally with nickel, but with significant contributions from other major alloying elements.

The core materials are, therefore, subjected to a demanding environment of high fast neutron flux coupled with high temperatures and high thermal gradients due to the high heat transfer property of sodium to extract heat efficiently (sodium is the preferred choice from efficient heat removal point of view, both under steady and transient conditions). A high flux of fast neutrons induces atomic displacements, as described above, in the core structural materials leading to phase instabilities, void swelling, irradiation creep and perceptible large changes in mechanical properties.

Neutron damage in reactor structural materials can be evaluated by testing or examination of samples machined (in radiation shielded facilities) from spent of failed components, or by using standard or miniature test pieces placed in rigs and irradiated in or close to the cores of material test reactors. The monitoring and control of experimental parameters like temperature, stress, flux, environment etc. under in-pile irradiations are difficult and complex. Furthermore, reactor irradiated specimens are highly radioactive and are handled in hot cells, which render many investigations rather difficult. Many of the critical damage effects are high dose effects (typically in the range of 100 dpa, corresponding to fast neutron fluence in excess of 10^{23} n/cm^2) and investigation of such effects requires very long irradiation times—up to a few years in material test reactors.

The number of such test reactors operating at high fast neutron fluxes is limited. These reasons have led to an alternative procedure, which is to irradiate samples with high energy ion beams from accelerators or cyclotrons, with the objective of simulating neutron radiation damage processes. An advantage with simulation irradiation damage at highly accelerated dose rates is that many years of reactor operation can, in principle, be simulated in several hours. With simultaneous multiple beam irradiation, it is possible to simulate synergistic effects of displacement and transmutation introduced by high-energy neutrons. Also, it is easy to set up in-situ experiments for investigating point defects dynamics and to explore long term kinetic pathways. Several laboratories, over the years, have developed simulation techniques using charge particle bombardment. This permits rapid screening of many candidate alloys and/or heat treatments as well as investigations of variables such as dose, dose rate, and irradiation temperature. However, since these variables interact in a complex manner in the evolution of the damage structure, ion beam simulation studies need to be evaluated with care to establish validity of the data; generally, benchmarking against reactor data.

The four major categories of mechanical behavior that must be investigated for cladding selection are summarized in this section. These include radiation hardening, embrittlement and fracture, void swelling, and irradiation creep.

11.3.2.1 Radiation Hardening

A material *hardens* when the matrix becomes entangled, i.e., when dislocations interact with each other or when interstitial atoms become lodged in glide planes to impede the motion of dislocations. Such a process, called *work hardening*, occurs during plastic deformation, as shown in Fig. 11.14.

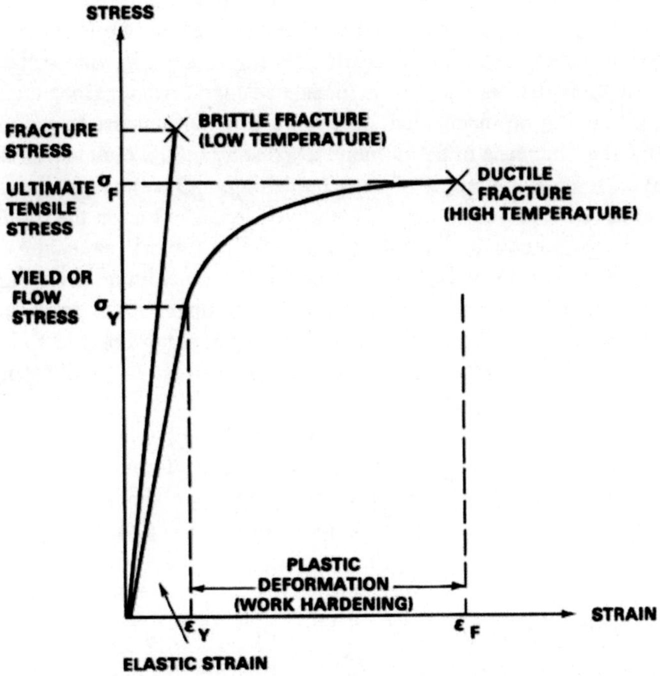

Fig. 11.14 Typical stress-strain curves at temperatures above and below the ductile-brittle transition temperature

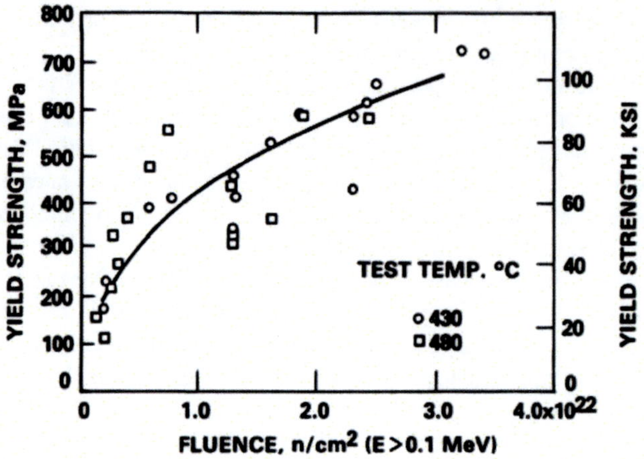

Fig. 11.15 Effect of fluence on the low temperature (430–480°C) yield strength (0.2% offset) of Type 316 stainless steel [39]

Once the elastic strain limit is passed, an increased stress causes permanent deformation. These increasing stress levels add to the deformation and entanglement process, leading to greater levels of hardening and material strength.

One of the standard techniques employed to modify the strength characteristics of a structural metal is to apply *cold working*. This is a process of reducing the cross-sectional area of the cladding by drawing the tubing under room temperature conditions. The 20% cold working used for SFR cladding and ducts results in a 20% reduction in area (which is nearly equivalent to a 20% reduction in cladding or duct thickness). A mechanical deformation process of this type introduces dislocation tangles into the matrix and leads to a greater yield strength and ultimate strength, while decreasing the ductility.

A similar hardening effect occurs in metals undergoing irradiation. Interstitial atoms, caused by neutron displacement damage from dislocation loops, considerably strengthen the metal. The general effects of neutron irradiation on mechanical behavior are (i) an increase in the yield strength (irradiation hardening); (ii) an increase in the ultimate tensile strength, which is less than the increase in yield strength; (iii) a decrease in the rate of work hardening; and (iv) a reduction in the uniform and total elongations. However, the overall effect is strongly dependent upon the irradiation temperature because if the lattice is hot enough, a substantial amount of thermal annealing can occur simultaneously to mitigate the hardening effect of neutron irradiation. Figures 11.15 and 11.16 compare typical yield strengths characteristics of irradiated and unirradiated metal at various temperatures.[11] At relatively low temperature ($T/T_m < 0.5$, where T_m is the melting point in Kelvin), the hardening effect is very pronounced, whereas thermal annealing removes most of the difference at temperatures more than $0.5T_m$.

Irradiation causes both source hardening and friction hardening to influence the motion of dislocations and hence the mechanical properties. *Source hardening* is a term used to describe the increased stress required to start a dislocation to move on its glide plane. *Friction hardening* measures the increased resistance to plastic deformation due to radiation-produced obstacles once movement along the slip plane occurs.

The relative dominance of the various mechanisms responsible for radiation hardening varies with the fluence level. For low fluence $\left(\phi t < 10^{21}\ n/\text{cm}^2\right)$ most of the resistance to plastic flow results

[11] Refer ahead to Fig. 11.19 for an explanation of the 0.2% offset concept.

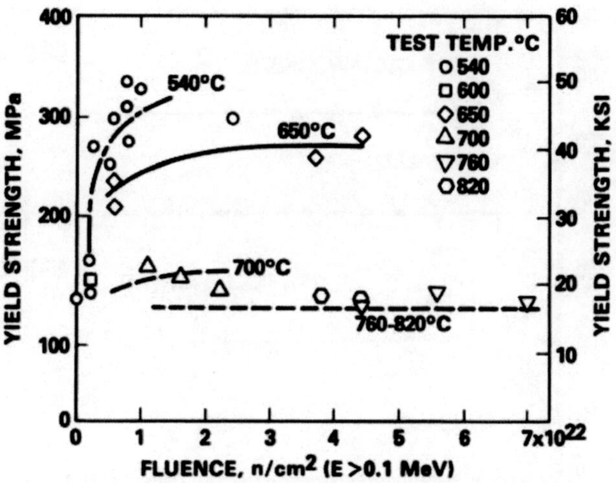

Fig. 11.16 Effect of fluence on the high temperature (>540°C) yield strength (0.2% offset) of Type 316 stainless steel [39]

from depleted gases. For the higher fluence levels of principal interest for fast spectrum reactors, dislocation loops constitute the dominant hardening mechanism.

It was mentioned earlier that radiation hardening is similar in net effect to cold working. In spite of this similarity, work hardening of irradiated materials is not nearly as effective as work hardening of unirradiated material. In irradiated material, there are already so many radiation-produced obstacles that any additional effect of work hardening causes only a small increase in frictional stress.

Cold working hardens the material and increases the yield strength of the material. When the cold worked material is irradiated in the reactor, irradiation softening occurs, unlike that occurring in the case of annealed material. This results in decrease in the yield strength of the material as shown in Fig. 11.17. The irradiation softening is a function of the degree of cold work and irradiation temperature. Figure 11.18 illustrates the lack of a self-annealing effect at lower irradiation temperatures.

11.3.2.2 Embrittlement and Fracture

When introducing the concept of embrittlement for cladding materials, it is useful to review the general stress-strain relationships involved. Figure 11.19 represents the relationship for ferritic steel. As stress is originally applied to a ferritic steel specimen, elastic strain occurs according to Hooke's law up to the proportional limit (PL), which is generally below the yield point. The yield point is attained at point U and the specimen will strain to point L, corresponding to the flow stress level. For such a material, an additional strain (called Lüders strain) will occur at this lower yield stress point. If the stress level is subsequently increased, additional strain will occur, and the work hardening that occurs in this area strengthens the material (by the mechanism described above). This process continues until the ultimate tensile strength (UTS) level is reached, beyond which necking predominates and fracture follows (unless the load is reduced). The dashed line in Fig. 11.19 represents the true stress-true strain curve, wherein actual area reductions and incremental strains in the specimen under test are properly taken into account.

Whereas Fig. 11.19 is representative of the behavior of ferritic steels, the austenitic steels behave somewhat differently, as depicted in Fig. 11.20. The principal difference in the stress-strain relationship is that there is no well defined yield point, i.e., there is no unique stress level at which plastic

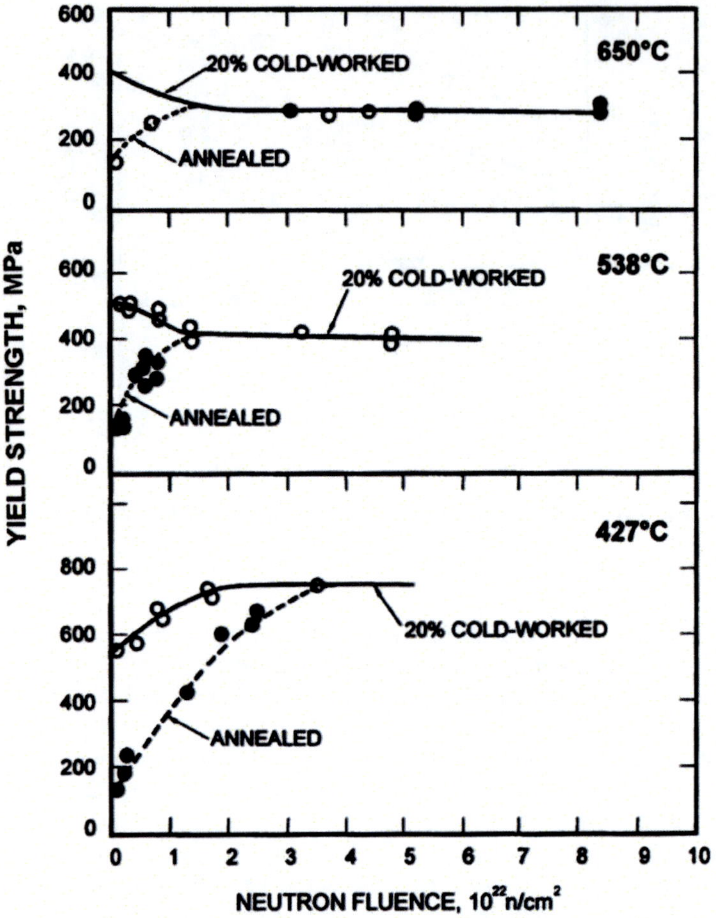

Fig. 11.17 Influence of temperature and neutron dose on evolution of yield strength in 20% CW AISI 316 irradiated in EBR-II [40]

flow begins. As a consequence, it is common practice in the case of such steels to arbitrarily label the yield stress level as that corresponding to a 0.2% permanent strain, i.e., where the 0.2% offset line on Fig. 11.20 intersects the stress-strain curve. This is called the 0.2% *offset yield strength* of the metal.

The term *total ductility* is defined as the strain of the specimen at fracture; whereas the strain between the yield strength and the ultimate tensile strength (UTS) ($\varepsilon_{UTS} - \varepsilon_Y$) is defined as the *uniform* strain. *Embrittlement* refers to the reduction in ductility. Fast neutron irradiation normally renders a metal less ductile than unirradiated metal. A brittle material fails prior to the yield point or before the 0.2% offset strain is reached.

The reduction in ductility caused by irradiation in normally ductile materials has been attributed to the phenomenon of dislocation channeling. The radiation-produced defect clusters are removed when dislocations are set in motion along the slip plane in the early stages of deformation. The slip process then continues by the motion of subsequent slip dislocations along the same channel in preference to a new slip band in an adjacent region of the crystal. As a result, deformation becomes considerably more localized than is the case for unirradiated metals. Owing to dislocation channeling, deformation takes place in the glide bands in a catastrophic manner and leads to plastic instability localized in the

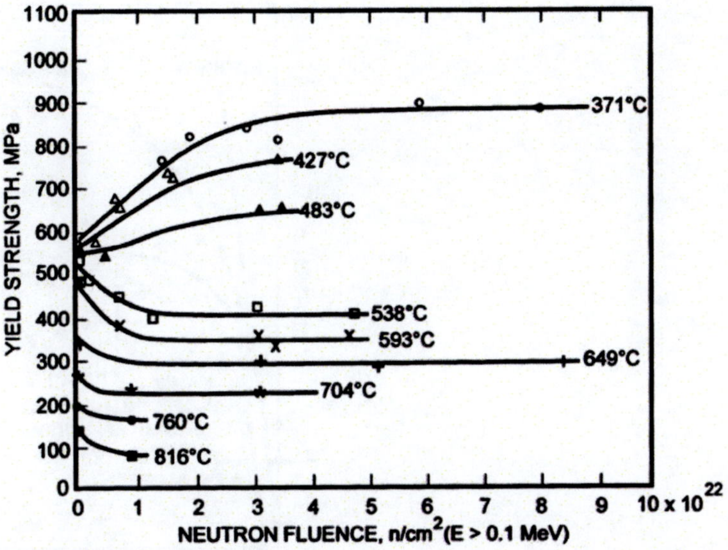

Fig. 11.18 Evolution of yield strength in AISI 316 SS irradiated in EBR-II at temperatures within the void swelling regime [40]

Fig. 11.19 Stress-strain curve for ferritic steel

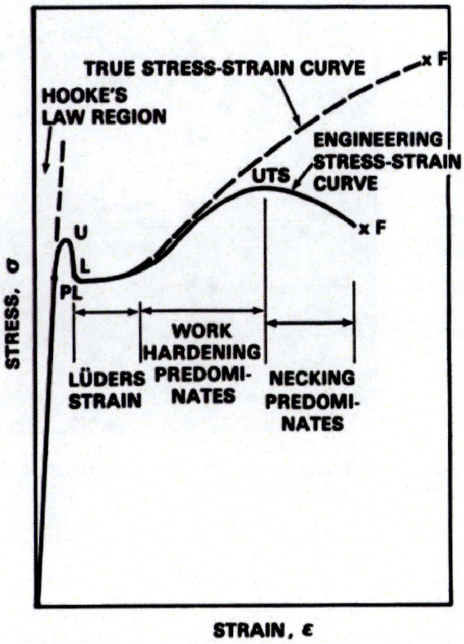

channels or slip bands. Figure 11.21 shows transmission electron microscope (TEM) images of dislocation channels in Fe18Cr-12Ni SS irradiated to 5.5 dpa at 360° C with 3 MeV protons and strained to 7% at 288°C. The phenomenon of dislocation channeling is also able to explain the reduction in work hardening rate owing to irradiation in ductile metals. The removal of irradiation-produced obstacles by the moving dislocations will result in a reduction in the enhancement of flow stress with strain

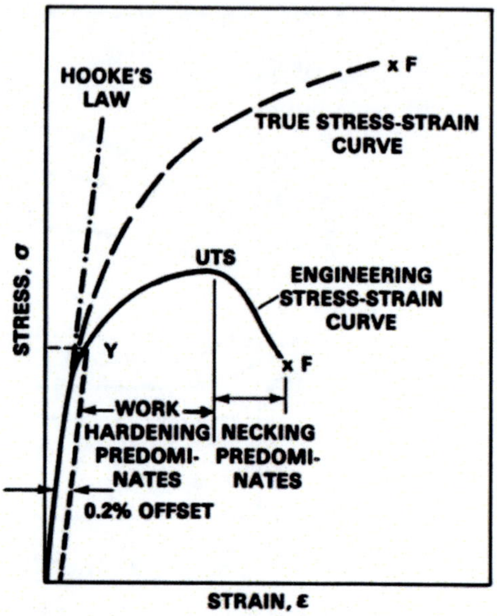

Fig. 11.20 Stress-strain curve for austenitic steel

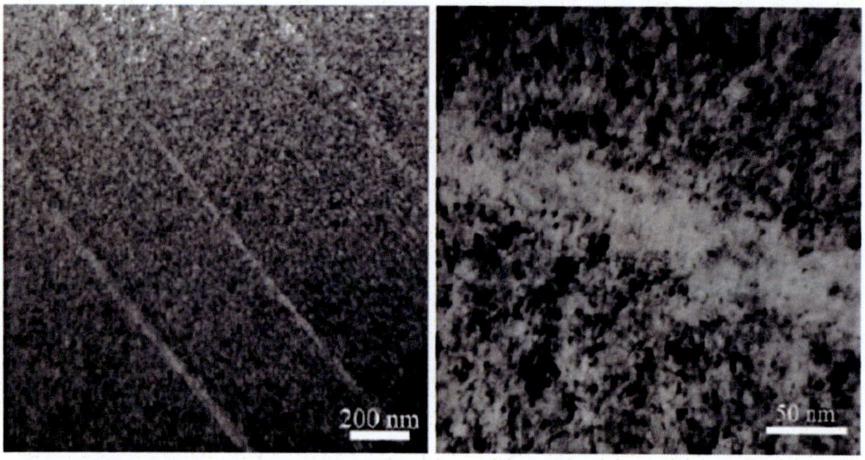

Fig. 11.21 Transmission electron micrographs of dislocation channels in Fe-18Cr-12Ni SS irradiated to 5.5 dpa with 3 MeV protons and tested at 288°C [38]

and consequently in a reduced work hardening rate. On a macroscopic scale, plastic instability and necking always set in when in a tensile test the increase in stress due to work hardening is too small to compensate for the reduction in cross-section so that the applied load remains constant or even decreases. Therefore, a general reduction in work hardening will always lead to premature necking with lower values of uniform and total elongations.

Whereas Figs. 11.19 and 11.20 represent typical stress-strain relationships for the common cladding material candidates, the actual magnitudes of the curves are strain rate dependent. The yield strength is strain rate dependent at higher temperatures, and at such elevated temperatures, time-dependent plastic deformation (creep) also occurs at low strain rates ($<10^{-4}$ /min). Such strain rates are characteristic of long term pressure on the cladding due to fuel swelling. When such tests are

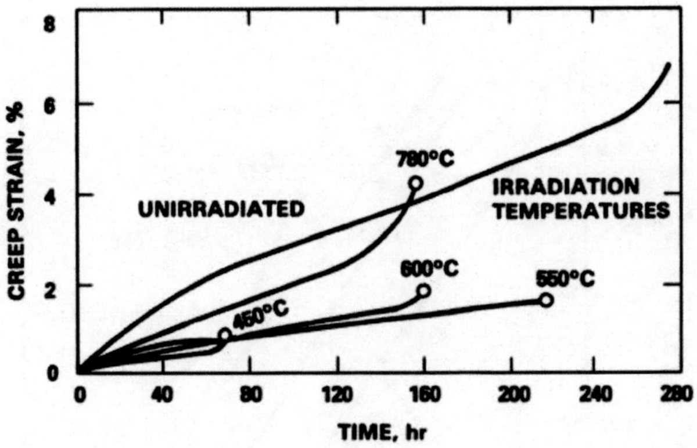

Fig. 11.22 Effect of irradiation temperature on the creep rupture of annealed type 304 stainless steel irradiated to $1.9 \times 10^{22}\ n/\text{cm}^2$ (> 0.1 MeV) and tested at 550°C under a stress of $3\times10^5\ \text{kN/m}^2$) [42]

conducted under high temperature conditions typical of an in-pile environment, they are called creep rupture tests. Data for higher strain rates ($\sim10^{-2}$ /min) are also collected to ascertain the cladding performance expected for typical startup, shutdown, and power cycling operations. Data are also collected under high strain rate and transient heating conditions for application to reactor transient conditions [41].

Due to the obvious necessity to ensure long-term structural integrity of cladding at high burnup and temperature levels, the effect of neutron irradiation on creep rupture strength is of considerable interest. Generally speaking, neutron irradiation leads to considerable matrix displacement damage and, at temperature conditions low enough to prevent thermal annealing, such damage impedes creep. Consequently, the strain rates at a given stress level are reduced for irradiated material as shown in Fig. 11.22. For the irradiated annealed type 304 stainless steel in the example, the irradiation temperature must be raised to nearly 800°C before the strain rate equals that of unirradiated material.

Another interesting factor revealed by Fig. 11.22 is that irradiation reduces the tensile elongation to fracture, i.e., irradiation significantly reduces the ductility of the metal. This embrittlement is graphically illustrated in Fig. 11.23 for the case of 304 stainless steel, where the elongation

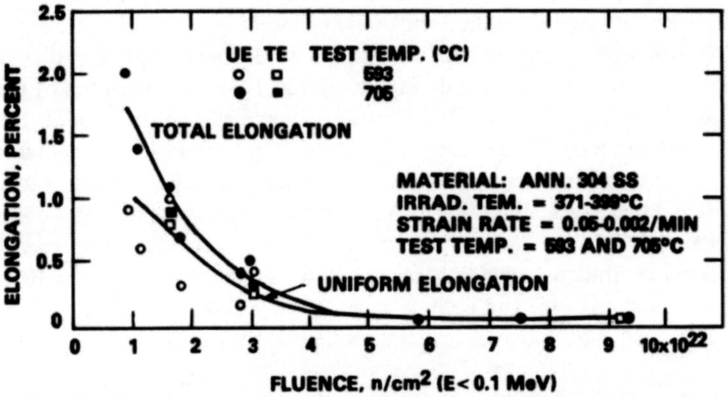

Fig. 11.23 Effect of fluence on high-temperature ductility of EBR-II type 304 stainless steel [43]

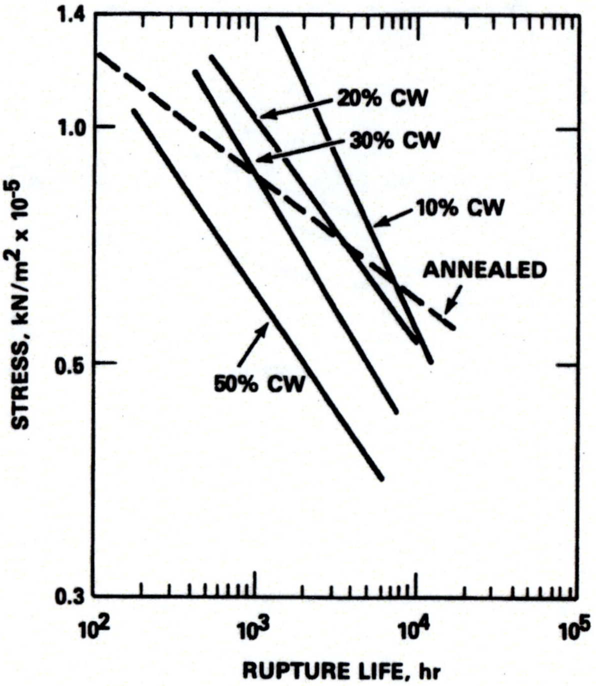

Fig. 11.24 Effect of cold working on the rupture life of type 304 stainless steel, tested at 700°C [44]

decreases exponentially to a saturation level at a fluence of $3–5 \times 10^{22}$ n/cm^2, and then does not further decrease.

It was mentioned earlier that cold working is often done to improve the strength characteristics of a metal. Figure 11.24 illustrates the effect of various degrees of cold working on the creep rupture strength for type 304 stainless steel. Cold working up to about 30% is seen to allow substantially higher stress levels for short-term rupture, but just the opposite is true for long-term strength. These data indicate that a fully annealed material would be preferred if the only criterion for cladding selection was long-term rupture life at low stress. However, as will be seen later, the principal reason for introducing cold work is to suppress void formation and swelling.

Other than cold working or irradiation induced matrix distortions, the primary phenomenon causing metal embrittlement is helium production during irradiation. Helium is produced by the (n, α) reaction occurring with the boron impurity that always exists to some amount in stainless steels. While this reaction tends to burn out most of the boron impurity relatively early in life, helium is continuously produced by (n, α) reactions with nickel. Boron is normally found interchanged with carbon at the grain boundaries, and it is precisely in this location that helium causes the most damage in terms of embrittlement potential.

Helium implantation into the cladding results from ternary fission in the fuel. As shown in Fig. 11.25, both calculations [46] and helium measurements show that helium from ternary fission can penetrate up to 0.13 mm into fuel cladding, which causes additional degradation of cladding adjacent to fuel.

Embrittlement due to helium is more damaging than that from cold work or displacement damage. With the latter, an increase in strength always occurs to compensate for the loss in ductility. Helium

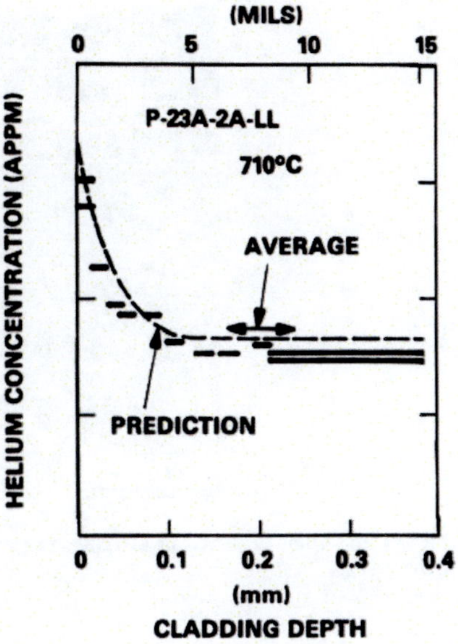

Fig. 11.25 Helium profile through the cladding wall in specimen from the fuel column region of fuel pin. The inner surface of the cladding is at wall position "zero" [45]

embrittlement, however, can decrease the strength and ductility simultaneously by producing premature fracture at the grain boundaries. Below 500°C, austenitic stainless steel cannot fracture at the grain boundaries and is therefore insensitive to helium embrittlement [44]. Above this temperature, however, the material can be severely embrittled, with strengths reduced to 50% of the unirradiated value and ductilities less than 0.1% [45]. A higher strain rate loading can prevent this grain boundary embrittlement until temperatures above 650°C are attained.

11.3.2.3 Void Swelling

During the early to mid-1960s, the key cladding problem for the SFR envisioned by designers was embrittlement. In 1967, however, designers were surprised by the discovery of another key problem. It was observed from irradiations at the Dounreay Fast Reactor that considerable *swelling* of the cladding took place under certain high fluence conditions [40]. The data indicated that such swelling only occurred for fast fluence levels above a threshold of about 10^{22} n/cm^2, but it increased rapidly beyond that point. Since it was known that fluence levels of about 2×10^{23} n/cm^2 would be required for successful cladding performance, considerable attention was drawn to these early results from Dounreay.

Close examination of the irradiated cladding specimens revealed that very small voids or cavities had been formed within the grains. They formed only within the temperature range of 350–700°C, but this is precisely the operating temperature domain for SFR application. Figure 11.26 shows voids observed in SS316 after irradiation in FBTR (India) for 40 dpa at 430°C [47].

The magnitude of volumetric increases observed for the high fluence levels was much higher than expected. Subsequent work demonstrated that most metals are susceptible to swelling by this mechanism for temperatures between 0.3 and 0.55 of their absolute melting temperature. As it turns out, stainless steel is one of the swelling resistant alloys. Nonetheless, the implications of having to provide for cladding growths of the magnitude implied by direct extrapolation of the existing data was sufficient to launch an extensive program towards understanding the basic phenomenon and searching for ways to mitigate the problem.

Fig. 11.26 Voids observed in 316 SS wrapper tube after irradiation in Fast Breeder Test Reactor (FBTR) for 40 dpa at 430°C [47]

Investigations revealed that four conditions must be satisfied for void swelling to occur [1]

1. Both interstitials and vacancies must be mobile in the solid. Whereas interstitials are always mobile in metals even at low temperatures, vacancies become mobile only at relatively high temperatures. Since immobile vacancies will be readily annihilated by a cloud of moving interstitials, void swelling cannot occur at low temperatures.
2. Point defects must be capable of being removed at sinks provided by structural defects in the solid, in addition to being destroyed by recombination. Furthermore, there must be a preference for interstitials at one sink in order for the vacancy population excess necessary for void formation to exist.
3. The supersaturation of vacancies must be large enough to permit voids and dislocation loops to nucleate and grow. At high enough temperatures, however, the thermal equilibrium concentration of vacancies at the void surfaces becomes comparable to that sustained within the matrix by irradiation. Hence, void nucleation and growth ceases at high temperatures.
4. Trace quantities of insoluble gas must be present to stabilize the embryo voids and prevent their collapse.[12] Helium provides this requirement, although O_2, N_2 or H_2 impurities can fill the same function. Since He is introduced into the cladding materials of interest only via neutron-induced transmutation, the reason for the fluence threshold for void swelling could be explained by the incubation period required to produce sufficient quantities of the He catalyst. Although such a He concentration could stabilize the growing voids, this concentration is far too low to classify these cavities as bubbles.

Theoretical treatments based on chemical reaction rate theory can explain the underlying physical processes and account quantitatively for the observed microstructures giving rise to swelling [38]. Vacancies and self-interstitials—created in equal numbers by irradiation—are lost either by mutual recombination or by absorption into sinks such as dislocations. Surviving self-interstitials aggregate rapidly into dislocations loops, which expand, coalesce and finally form a dislocation network. Surviving vacancies cluster in association with gas atoms (generally helium produced by transmutation, but also surface active gases already present) to form embryonic cavities. The dislocations

[12]This last requirement is a controversial point among materials specialists.

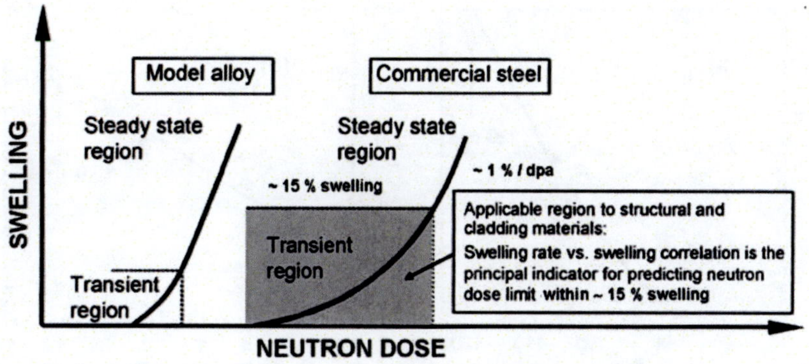

Fig. 11.27 Schematic representation of void swelling as a function of fluence

present in the system act as biased sinks for the preferential absorption of self-interstitials as a consequence of the differing strain fields associated with these point defects compared with vacancies. There is thus a net excess vacancy flux into neutral sinks such as the void embryos. When the latter contain a critical number of gas atoms (or, equivalently, reach a critical radius), biased driven void growth takes place, leading to breakaway or steady state swelling.

In most materials, void swelling is characterized by three regimes at a given dose rate and temperature: an incubation period (τ); a transient regime; and finally a steady state regime with monotonically increasing levels of swelling with increasing dose. The swelling behavior can be conveniently described as a low-swelling transient period followed by an acceleration to a region of nearly linear swelling of about 1%/dpa as shown in Fig. 11.27. An upper band of swelling acceptable for fuel cladding is generally less than 10–15%. Thus, designers require accurate swelling trends and values during the transient region. Such behavior is shown schematically in Fig. 11.27.

The commercial stainless steels exhibit a prolonged transient region before approaching a linear swelling rate. Thus, a curvature in the transient swelling region is significantly important from the design viewpoint. This information is, however, very limited. In the commercial stainless steels, nucleation and growth of voids concomitantly occur in the transient region, and significantly affect the swelling behavior.

Swelling is represented by means of power law swelling equation [48]

$$\frac{\Delta V}{V_0} = m_1 \left(\phi t - m_2\right)^n , \tag{11.3}$$

where $\Delta V \,/\, V_0$ is the volumetric swelling, ϕt is the neutron dose in dpa, m_1, m_2, and n are material coefficients. Taking the logarithm of both sides in Eq. (11.3), volumetric swelling versus neutron dose is expressed in a linear relationship, and the value of n can be derived from the slope of this relation. Figure 11.28 shows the log–log plot of the volumetric swelling versus neutron dose within the swelling transient region for 20% cold worked 316 stainless steel at 500°C. In this analysis, m_2 was selected as 83 dpa, which corresponds to the neutron dose just initiating void swelling, so as to minimize the standard deviation using a regression analysis of the least square method. The best-fit curve with the optimized values of n, m_1, m_2 is also denoted in Fig. 11.28. The two characteristic features can be derived from these analyses: the power law type swelling equation can be applied in a wide range of volumetric swelling from 0.58% (87 dpa) to 6.3% (107 dpa), and thus extrapolation

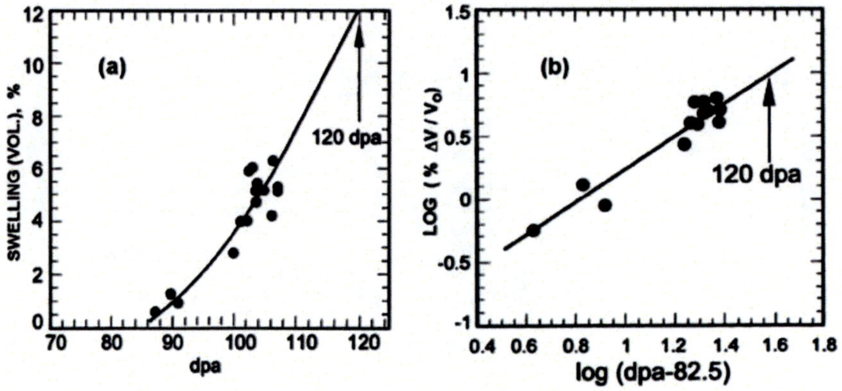

Fig. 11.28 Swelling vs. fluence for 20% cold worked 316 stainless steel at 500°C [48]

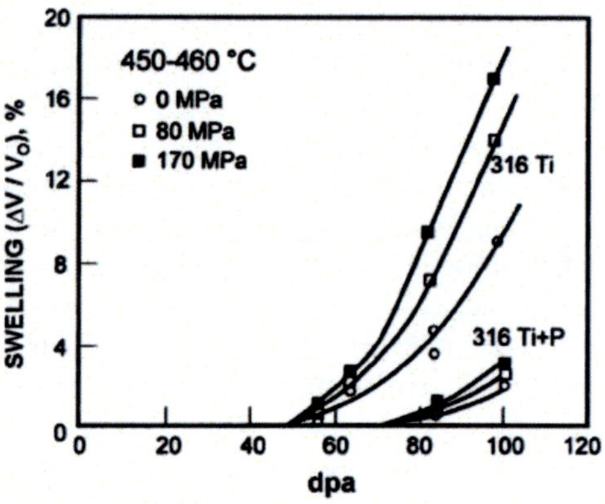

Fig. 11.29 Effect of stress on swelling of two modified 316 SS irradiation in the form of pressurized tubes in Phénix [49]

is made from 107 to 120 dpa using a linear log–log plot representation within the swelling transient region.

The results shown in Fig. 11.28 were obtained for a stress-free environment on the structural material. For most applications of cladding and duct material, however, a stress field will exist. Data for such stress fields indicate a reduction in the incubation period, τ, preceding the onset of steady-state swelling and higher volumetric change, as shown in Fig. 11.29 with the data derived from Phénix.

An extensive set of void swelling data vs. irradiation temperature is shown in Fig. 11.30. Reference [50] also contains a large data set for a variety of commercial alloys.

It is possible that saturation in the swelling effect may occur at high target exposures. Very little in-pile data are available to provide statistically meaningful results, but limited data from high energy ion bombardments [45] indicate that saturation may not occur until the fluence level is around 10^{24} n/cm^2. The present expectation is that commercial SFR cladding will swell from 0 to 15% at target fluence conditions.

It was mentioned earlier that swelling is quite sensitive to irradiation temperature. Figure 11.31 illustrates this behavior for type 316 stainless steel. For this case, maximum swelling occurs for a temperature of about 550°C.

The effect of cold working was mentioned previously in conjunction with improving rupture strength characteristics. Figure 11.32 indicates the improvement in swelling resistance offered by 20% cold-worked 316 stainless steel, as compared to the fully annealed material. Whereas some cold

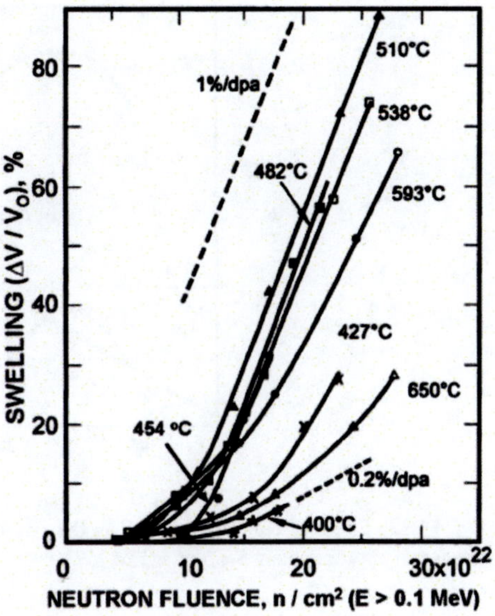

Fig. 11.30 Swelling vs. burnup for AISI 316 SS for a variety of irradiation temperatures [50]

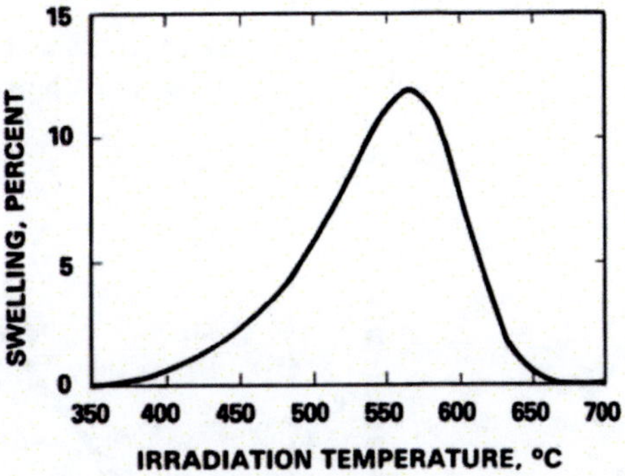

Fig. 11.31 Influence of temperature on void swelling for 20% cold worked 316 stainless steel irradiated to a fluence of 1×10^{23} n/cm^2 (EBR-II data, adapted from Ref. [51])

working is clearly advantageous with regard to inhibiting swelling, there is an apparent limit on the value of such a procedure. This is suggested by the 50% CW type 304 stainless steel swelling curves, also shown in Fig. 11.32. This material, differing from type 316 SS only by its reduced molybdenum content, exhibits two swelling humps. The second hump is believed to be the result of an instability of the dislocation network introduced by the large degree of cold work. At high temperatures, large segments of such a network become free of dislocations and provide a ready environment for the formation of voids and subsequent void growth.

The materials used for alloying steel can appreciably influence the swelling characteristics. Figure 11.33 illustrates the reduction in swelling affected by using a low purity nickel relative to high purity nickel. Apparently, the impurities present in low grade nickel precipitate and act as recombination sites for vacancies and interstitials, thus removing the swelling potential. As noted in Fig. 11.33, Inconel®[13] actually densifies slightly during irradiation. Unfortunately, it becomes severely embrittled because of its high nickel content.

[13] Inconel® is a registered trademark of International Nickel Company.

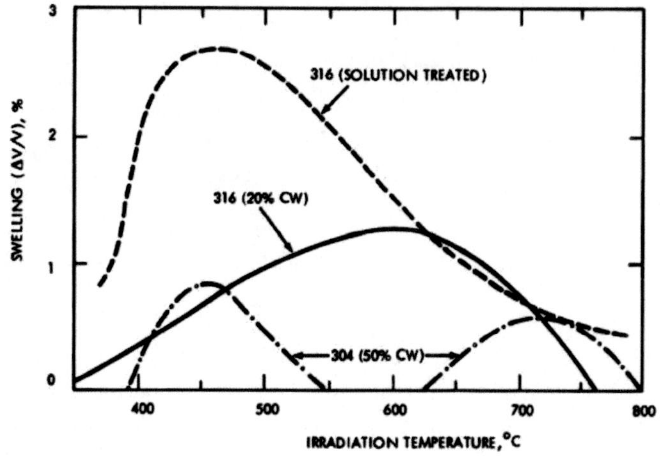

Fig. 11.32 Effect of cold work (CW) on the swelling behavior of austenitic stainless steels. The curves are all for a fluence of approximately 5×10^{22} n/cm^2 [52]

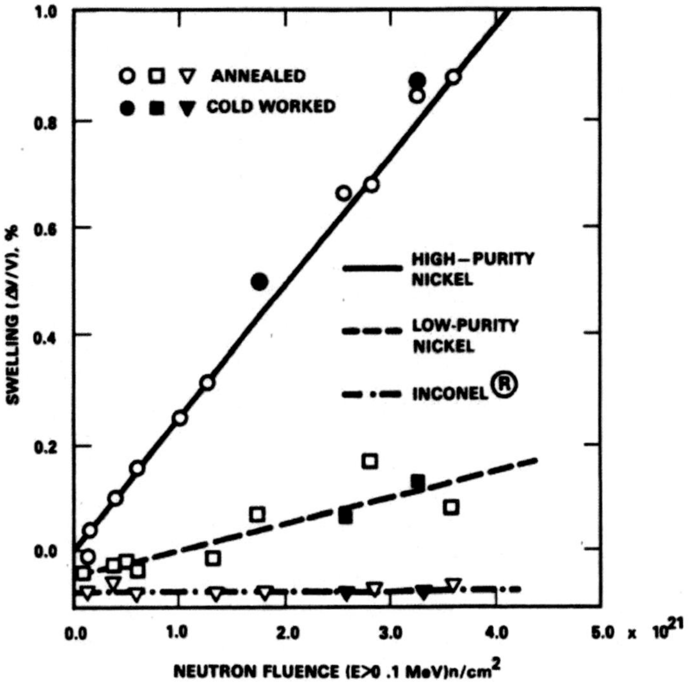

Fig. 11.33 Swelling of high-purity nickel, nickel of 99.6% purity, and Inconel® (73% Ni-17% Cr-8% Fe) at 425°C [53]. Values for swelling below zero indicate densification

In austenitic stainless steels, extensive studies have shown that void swelling varies systematically with the chromium and nickel contents. The swelling reduces with increasing nickel content and increases with increasing chromium. The resistance to void swelling can, therefore, be varied considerably by modifying alloy compositions with respect to major elements. Further, it is found that

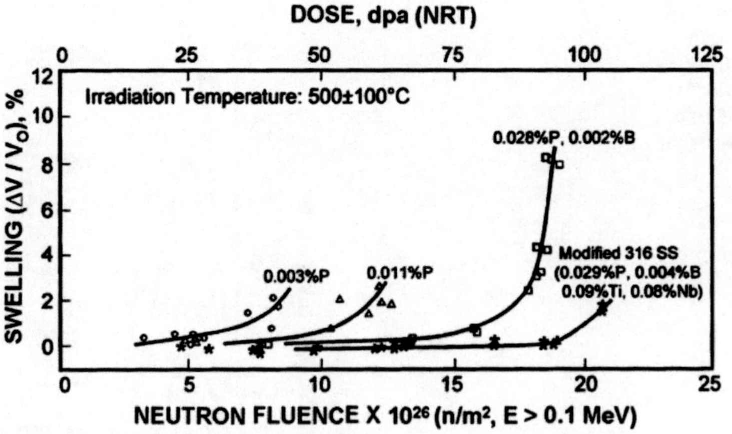

Fig. 11.34 Beneficial effect of phosphorus on swelling behaviour of 316 SS clad tube [54]

the commercial alloys swell less than the simple ternary alloys with the same chromium and nickel contents, suggesting that the minor element additions contribute to the swelling resistance. That minor elements exert considerable influence on swelling resistance has also been demonstrated by comparing the swelling behavior of pure metals with metals containing minor elements.

Minor elements that have been found particularly effective in increasing the resistance of stainless steels to swelling are titanium, silicon, and phosphorous. Among the minor elements, phosphorus is very effective in providing swelling resistance, particularly in combination with titanium as shown in Fig. 11.34. A drastic increase in the incubation period has been reported by adding 0.029% phosphorous and 0.09% Ti in 316 SS . Silicon also is known to reduce swelling drastically as shown in Fig. 11.35. The suppression of swelling by the addition of titanium is shown in Fig. 11.36. The beneficial effect of titanium tends to saturate above a Ti/C ratio of 2 for 20% cold work 316 SS. Based on these findings, a modified 316 SS was specified with typical composition of Fe-16Cr-14Ni-2.5Mo-0.06C- (0.7–0.8) Si-0.025P-0.004B-0.1Ti-0.1Nb. This material is designated as PNC316. Performance analyses have demonstrated satisfactory behavior of PNC316 SS fuel pin in the MONJU fast reactor core at a peak fuel burn-up of 131,000 MWd/t with fast neutron fluence of $2.3\times10^{23} n/\text{cm}^2$.

Recent irradiation experiments conducted on a variety of austenitic stainless steels have shown that void swelling appears to be increased when the dpa rate is decreased, primarily by a shortening of the transient regime of swelling as shown in Fig. 11.37.

Commercial ferritic-martensitic steels based on the 9–12%Cr composition exhibit the highest swelling resistance, and this low swelling response appears to be a generic property of ferritic alloys as a class. Typical of these is HT9 (12Cr-1Mo alloy). Such alloys therefore appeared ideal for fast breeder reactor applications. A variety of processes has been suggested to understand the generic high swelling resistance of ferritic steels. Interstitial solutes (C and N) with tetragonal strain fields in bcc iron interact strongly with both vacancy and self-interstitial point defects (trapping) and dislocations (Cottrell atmosphere formation) and furthermore can act in these roles conjointly with substitutional solutes in alloy steels. These processes enhance point defect recombination and reduce dislocation bias, respectively, to limit void nucleation and growth. In addition, the lath boundary structures in 10–12% Cr martensitic steels have been shown to act as significant point defect sinks with similar effects. Finally, the rather unique behavior of interstitial dislocation loops in bcc iron is postulated to suppress swelling, based on the observation of loops of predominantly a<100> after fast reactor irradiation.

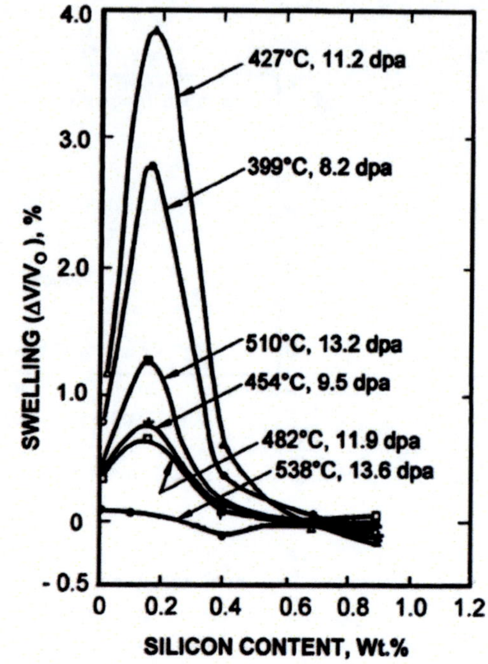

Fig. 11.35 Influence of silicon on swelling in annealed Fe-25Ni-15Cr at various combinations of temperature and fluence irradiated in EBR-II [38]

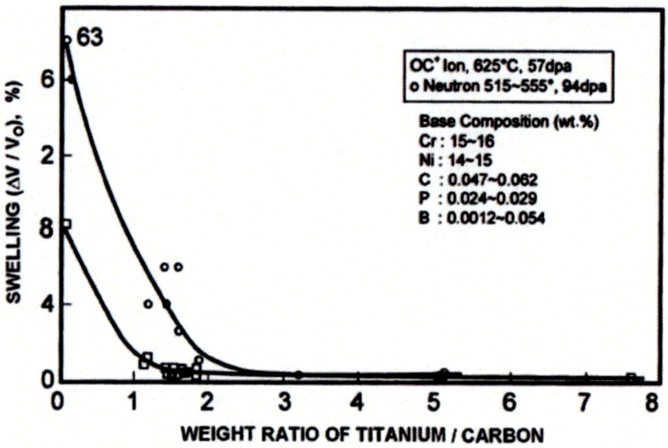

Fig. 11.36 Suppression of void swelling by Ti addition between 0.02 and 0.23 wt.% in 20% CW 316 SS [54]

11.3.2.4 Irradiation Creep

Creep deformation of core components under the influence of irradiation and stress, even at lower temperatures, has been one of the main concerns in SFRs [38, 40]. Several concerted studies have been oriented towards an understanding of the mechanisms of irradiation creep (IC) and theories developed invoke processes involving dislocation climb with point defect mobility as the rate-controlling process. IC normally involves interstitial related processes, as opposed to the vacancy related processes responsible for thermal creep. The mechanism of temperature independent stress induced preferential absorption (SIPA) of point defects at the network dislocations is found to be valid at

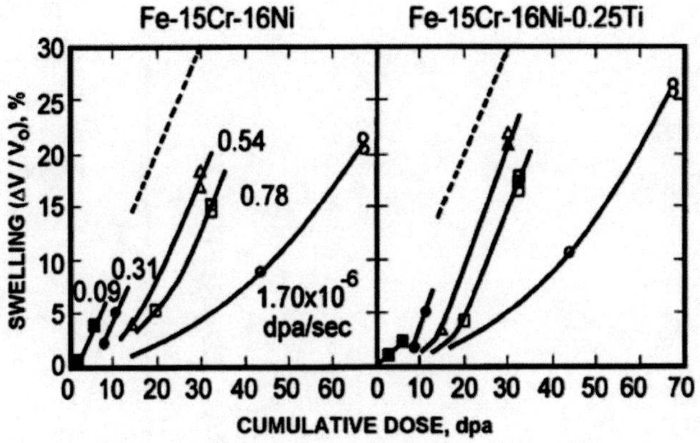

Fig. 11.37 Swelling of simple model fcc Fe–Cr–Ni alloys in FFTF–MOTA at 430°C, as observed by Okita and coworkers, showing the transient regime of swelling increases progressively as the dpa rate increases [55]

the doses, displacement rates, temperatures, and stress levels typical of SFR operating conditions. At higher stresses, a climb controlled glide (CCG) model with quadratic stress dependence works better. Complexity with commercial alloys arises out of the interplay of irradiation induced microchemical and microstructural evolution leading to combined effects of thermal and irradiation creep.

SFR design considerations usually avoid the stress-temperature regime of thermal creep and concentrate only on IC. Wassilew et al. [56] have developed a comprehensive model, the SIPA-AD model, to explain the experimental data on temperature, stress and dose rate dependences observed in austenitic steels including Ti-modified alloys. The SIPA-AD stress-induced preferential attraction due to anisotropic diffusion model is based on considering the point defect migration in an atomic lattice distorted by the dislocation and external stress fields; i.e. only the first-order size and shape interaction energy is considered, irrespective of the interaction with dislocations in different orientations, which is the basis of the normal SIPA model. It has been shown [57] earlier that this model yields a higher creep rate, compared to SIPA. In the Ti-modified alloys below 850 K, the migration of self-interstitials accounts for IC. Above this temperature, the migration of vacancies contributes substantially. Thus an effective diffusion coefficient is defined as [56]

$$\theta_c = a \exp\left(\frac{-\Delta H_v}{kT}\right) + b \exp\left(\frac{-\Delta H_i}{kT}\right), \qquad (11.4)$$

using the migration enthalpy changes ΔH_v and ΔH_i for vacancies and interstitials respectively and the corresponding frequency factors a and b. The temperature dependence of irradiation creep rate for austenitic and Ti-modified steels, as obtained from in-pile tests is shown in Fig. 11.38 and the role of relevant point defect processes in different temperature regimes is clear from the figure. The rate controlling defect varies with temperature, with activation energies for volume migration of 1.63 eV for vacancies and 0.09 eV for interstitials.

A dose rate $\dot{\kappa}$ dependence of creep rate as $\dot{\varepsilon} \sim \dot{\kappa}^{1/2}$ has been observed in austenitic stainless steels in the entire temperature range and this is shown in Fig. 11.39. The SIPA model predicts a linear behavior in the sink dominant regime. While recombination of vacancies and interstitials is typically a high temperature process, IC is of importance at lower temperatures as well. The observed departure from SIPA behavior seen in Fig. 11.39 is explained by the SIPA-AD model.

Inter-comparison and careful analysis of extensive experimental data on in-pile creep, out-of pile creep and pre-irradiated out-of-pile creep of austenitic steels and Ti-modified steels lead to the

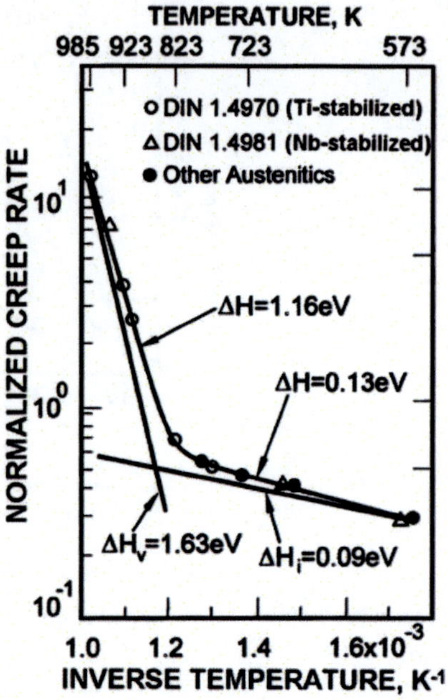

Fig. 11.38 Temperature dependence of irradiation creep rate in austenitic and Ti-modified steels. [The experimental data are from in-pile creep tests of pressurized tube specimens at EBR-II (up to 0.7×10^{22} n/cm^2) and FFTF (up to 1.01×10^{23} n/cm^2). The normalized creep rate is defined as $\theta_c = \varepsilon\, C_o/\acute{K}^{0.5}\sigma$, where C_o a constant, ε is creep strain, is irradiation dose rate and σ is stress.] [56]

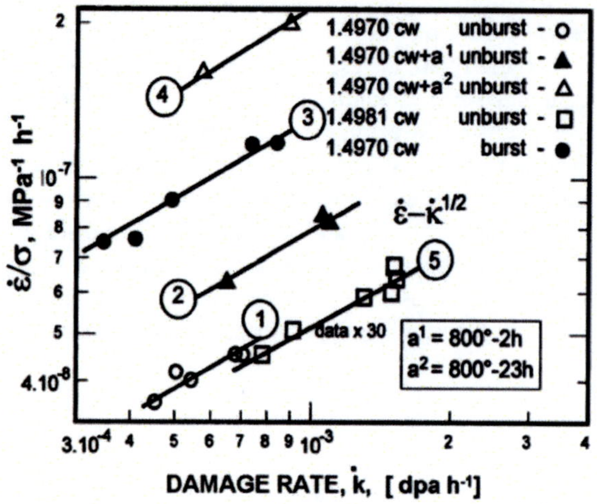

Fig. 11.39 Dose dependence of irradiation creep of Ti stabilized steel (1.4970) in cold worked (CW) and CW + aged condition. Data on Nb-stabilized steel (1.4981) is multiplied by 30 and plotted for comparison. The burst specimens have been tested till rupture (a^1 for ageing at 800°C for 2 h and a^2 for aging at 800°C for 23 h) [56]

following conclusions [56]: (1) While creep strains, in general, include contributions from different processes, the strain measured in un-irradiated and pre-irradiated and subsequently tested specimens possibly reflects only a single deformation process, giving rise to the steady state creep. (2) The in-pile creep data follow the elasto-diffusion SIPA creep mechanism (SIPA-AD). Even around $0.5T_m$, SIPA-AD operates independently and is not at all linked to thermal creep mechanisms like Nabarro-Herring creep or Coble creep. (3) There exists evidence to suggest a revision of the widely prevalent notion of treating IC as a temperature independent process. While IC at low temperatures is governed by migrating interstitials, there is contribution from migrating vacancies at higher temperatures. (4) IC deviates from linear stress dependence at stress levels above ∼180 MPa, showing a quadratic dependence as in climb controlled glide (CCG) creep [58].

11.3.2.5 Void Swelling and Irradiation Creep Interaction

The degree of enhancement of creep rate due to irradiation is greater at lower irradiation temperatures and decreases as the thermal creep regime is approached. This enhancement arises primarily due to the presence of the irradiation induced interstitials and vacancies in very large concentrations. The two main contributions to irradiation creep arise from a component independent of void swelling, and, secondly, a swelling driven creep component [40]. While swelling is quite sensitive to a variety of materials and irradiation variables, the instantaneous creep rate ($\dot{\varepsilon}$) is proportional to the applied stress (σ) and the instantaneous swelling rate ($\dot{S}$). Following the approach in Ref. [40], the instantaneous creep rate per unit stress and dose can be written as

$$B = \frac{\dot{\varepsilon}}{\sigma} = B_0 + D\dot{S},$$

where B_0 is the creep compliance independent of swelling, D is the creep—swelling coupling coefficient and $\dot{S}$ is the instantaneous volumetric swelling rate. B_0 is found to be largely independent of temperature over the temperature range of fast reactor interest (>400°C). Also, the value of D seems to be fairly independent of irradiation temperature and starting state for a variety of austenitic steels [40]. The above equation assumes that creep continues as long as swelling continues and that the coupling coefficient is invariant with respect to flux and stress. This assumption is not valid in the swelling dominated regime. As the swelling rate increases, the feedback effects of swelling and stress on irradiation creep become pronounced, and at some point, this feedback reverses sign resulting in the disappearance of creep. It has been shown [40, 59] that a diametral strain rate of 0.33%/dpa is the maximum attained value in 316 SS at all temperatures studied. It has also been demonstrated that unless creep and swelling predictions are generated from within the same experiment, the strong sensitivity of swelling to material and irradiation parameters would obscure the creep-swelling relationship and may produce incompatible results [40].

11.3.2.6 Engineering Consequences of Irradiation effects

Engineering consequences of irradiation-induced changes such as void swelling, irradiation creep and embrittlement are very important as these determine the residence time of core components and, hence, impact on the achievable burn up and consequent economic viability of fast spectrum reactors. Deformations of various components of the assemblies can occur due to void swelling, thermal creep and irradiation creep. Swelling and irradiation creep are very sensitive to environmental variables, temporal and spatial gradients that exist throughout the core. Limits have to be specified on deformations in view of the brittle nature of irradiated austenitic stainless steels as well as to avoid interactions between the core components. The latter aspect is very important to permit fuel loading and unloading with only moderate loads.

Differential swelling can occur because of gradients in flux and temperature at different locations in the core (Fig. 11.40). Differential deformations due to differential swelling should be limited between the pin and spacer wire. If the spacer wire deformation is higher than the pin deformation, the sodium coolant flow can be reduced and consequently the fuel pins can get overheated. On the other hand, when the pin deformation is higher than that of the spacer wire, there is a mechanical interaction and the fuel pin undergoes a helical deformation. If both fuel pin and spacer wire swell excessively, there would be interactions between the pins as well as between the fuel pins and the duct. Duct deformation should be limited; otherwise interactions between ducts would lead to excessive loads for fuel handling. Irradiation induced void swelling and creep can produce different types of duct deformation. At the centre of the core, assemblies are expected to remain straight with an axial elongation and an increase of distance across flats. But at the periphery, assemblies may bow outwards due to differential

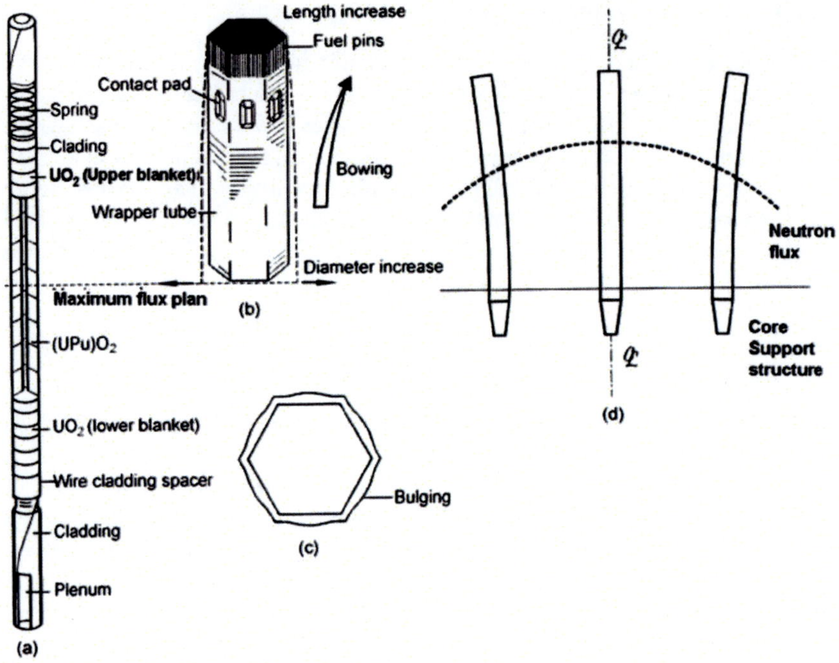

Fig. 11.40 Fuel pin assembly and in-pile deformations: (**a**) fuel pin, (**b**) deformation by swelling, (**c**) deformation by swelling and irradiation creep, and (**d**) bowing of assemblies due to differential swelling as a consequence of a flux gradient

void swelling on the opposite faces of the ducts as a consequence of the neutron flux gradient. The combination of void swelling and creep induced by internal (to hex can) sodium pressure produces dilation and rounding of duct faces.

The deformations due to swelling and irradiation creep are dependent on chemical composition and fabrication history—especially for austenitic stainless steels. Therefore it is important to fix precise specifications with narrow limits on chemical composition to avoid differences in deformations arising out of heat to heat variations for different core components. Gradients in swelling and creep can also affect the reactivity of the core, the movement of control rods and the flow of coolant. It has been noticed that irradiation creep and embrittlement are related to swelling. Thus, reduction of swelling would solve most of the engineering problems arising out of irradiation effects.

11.3.3 Cladding and Duct Materials

Given the intense high temperature and high fluence requirements imposed on fast reactor cladding and duct material candidates, much effort is being concentrated on the selection of special alloys that will provide the best overall balance of design [60]. As in any complex engineering application, optimal properties cannot be consistently obtained from any one material over the full range of parameters desired. Table 11.5 provides a composition listing for the stainless steels of most interest for fast reactor applications, as well as for the nickel base alloys.

Type 316, 20% cold worked stainless steel is the reference material that has been used for most SFR in-core cladding and structural applications. Table 11.6 lists the major thermophysical properties

Table 11.5 Nominal compositions of selected structural materials[a]

Type		C	Fe	Cr	Ni	Mn	Mo	Nb	Al	Ti	Si	W	P	S	Cu
Stainless steels[b]	304	0.08 max	base	18.0–20.0	8.0–12.0	2.0 max	–	–	–	–	1.0 max	–	0.045 max	0.030 max	–
	316	0.08 max	base	16.0–18.0	10.0–14.0	2.0 max	2.0–3.0	–	–	–	1.0 max	–	0.045 max	0.030 max	–
	321	0.08 max	base	17.0–19.0	9.0–12.0	2.0 max	–	–	–	5 X C (min)	1.0 max	–	0.045 max	0.030 max	–
	347	0.08 max	base	17.0–19.0	9.0–13.0	2.0 max	–	10 X C (min)	–	–	1.0 max	–	0.045 max	0.030 max	–
	Incoloy®[c] 800	0.04	46.0	20.5	32.0	0.75	–	–	–	–	0.35	–	–	–	0.30
Nickel base alloys	Inconel® 600	0.15	10.0	17.0	Base	1.0	–	–	–	–	0.5	–	–	–	–
	Inconel® X750	0.08	9.0	17.0	Base	1.0	–	1.2	1.0	2.75	0.5	–	–	–	–
	Hastel-loy X	0.1	18.5	22.0	Base	1.0	9.0	–	–	–	0.75	0.6	–	–	–
	Inconel® 718	0.1	base	21.0	55.0	0.35	3.3	5.5	0.8	1.15	0.35	–	–	–	–
	Inconel® 625	0.1	5.0	23.0	Base	0.5	10.0[a]	4.15	0.4	0.4	0.5	–	–	–	–

[a]Taken from pages 194 and 201 of Ref. [29].

[b]ASTM A271-64T specifies somewhat different compositions for heavy-walled stainless steel tubing and pipe. For example: 0.040% maximum phosphorous, 0.75% maximum silicon, and 8.0–11.0% and 9.0–13.0% for Type 304 and Types 321 and 347, respectively.

[c]Incoloy® is a registered trademark of the International Nickel Company.

Table 11.6 Thermophysical properties of 316 stainless [61] (all temperatures in Kelvin)

Melting point = 1,700 K		
Heat of fusion = 2.70×10^5 J/kg		
Boiling point = 3,090 K		
Heat of vaporization = 7.45×10^6 J/kg		
Specific heat:		
$c_p = 462 + 0.134\,T$	(J/kgK)	Solid region
$c_p = 775$	(J/kgK)	Liquid region
Thermal conductivity :		
$k = 9.248 + 0.01571\,T$	(W/mK)	Solid region
$k = 12.41 + 0.003279\,T$	(W/mK)	Solid region
Thermal expansion coefficient:		
$\alpha = 1.864 \times 10^{-5} + 3.917 \times 10^{-10} T + 2.833 \times 10^{-12} T^2$		Solid region
$\alpha = 1.864 \times 10^{-5} + 3.917 \times 10^{-10} T + 2.833 \times 10^{-12} T^2$		Liquid region
Viscosity (molten steel):		
$\log_{10}\mu = \frac{2385.2}{T} - 3.5958$	(μ in kg/ms)	
Vapor pressure (molten steel):		
$\log_{10}p = 11.1183 - \frac{18868}{T}$	(p in Pa)	
Density:		
$\rho = 8084 - 0.4209\,T - 3.894 \times 10^{-5} T^2$	(kg/m^3)	Solid region
$\rho = 7433 + 0.0393\,T - 1.801 \times 10^{-4} T^2$	(kg/m^3)	Liquid region

of type 316 stainless steel. The primary reasons for selecting this reference material are its excellent high temperature strength characteristics, its resistance to void swelling, its compatibility with mixed oxide fuels and sodium coolant, and it's relatively low cost.

Structural materials for fast reactor core components have evolved continuously in the quest to improve fuel element performance. Table 11.7 gives the different grades of materials used for cladding and duct tubes in various countries. Improvements in swelling resistance are by way of extended incubation and transient regimes. The first generation materials belonged to 304 SS and 316 SS grades. These steels quickly reached their limits because of unacceptable swelling at doses higher than 50 dpa. The trend in the development of radiation resistant 300 series austenitic stainless steels has been to increase nickel content and decrease chromium content in comparison to 304 SS and 316 SS. Solute elements like titanium, silicon, phosphorous, niobium, boron, and carbon play dominant roles in determining void swelling resistance. This observation has led to the development of advanced core structural materials such as 15Cr-15Ni-Ti SS (also called D9 SS). In D9, nickel is increased to 15% and chromium is lowered to 15% in comparison to that in 316 SS. Titanium is added in the range of 5–7 times of carbon. Silicon is limited to 0.75% maximum. Cold worked D9 SS variant has reached a record dose of 140 dpa without excessive deformation. In the light of recent international alloy development efforts, and in-reactor experience, a modified D9 SS designated as D9I SS is considered in the fast reactor community for still higher swelling resistance. The basic composition of D9I SS is similar to that of D9 SS, except that phosphorous is specified to be in the range of 0.025–0.04 wt.% and silicon in the range of 0.7–0.9 wt.%. Figure 11.41 shows the effect of phosphorous on decreasing peak swelling of D9I SS containing 0.75 wt.% silicon and 0.30 wt.% titanium [62]. The Ti/C ratio is known to play an important role in determining the creep strength of D9I SS (Fig. 11.42) [62].

The long-term solution for SFR core structural materials is considered to be the 9–12% Cr ferritic-martensitic (F-M) steels. These steels have excellent swelling resistance to doses even up to 200 dpa. However, the increase in ductile to brittle transition temperature (DBTT) due to irradiation is a cause

Table 11.7 Principal core structural materials used in fast reactors

Plant	Country	Peak target burnup (GWd/tHM)	Fuel cladding material	Duct material
Rapsodie	France	102	316 SS	316 SS
Phénix	France	150	Cr 17 Ni 13 Mo 2.5 Mn 1.5 Ti Si	316 SS
Super-Phénix	France	90	Cr 17 Ni 13 Mo 2.5 Mn 1.5 Ti Si	15-15Mo-Ti-Si
FBTR	India	50	316 M(CW)	316 L (CW)
PFBR	India	113	15Cr 15Ni Mo Ti (CW)	15Cr 15Ni Mo Ti
EBR-II	USA	80[a]	316 SS and HT9	HT9
FFTF	USA	155[b]	316 SS (20% CW)	316 SS & HT9
CRBRP	USA	97	316 SS (20% CW)	316 SS
KNK-II	Germany	172[a]	1.4970	SS
SNR-300	Germany	86	X10 Cr Ni Mo Ti B1515	X10CrNiMoTiB1515
SNR-2	Germany	150	1.4970	SS
JOYO	Japan	143	316 (20%CW)	316 (20%CW)
MONJU	Japan	94	Mod 316	Mod 316
DFBR	Japan	110	Advanced austenitic	PNC1520
BR-10	Russia	62[a]	Cr16 Ni15 Mo3 Nb	Cr16 Ni15 Mo3 Nb
BN-350	Kazakhstan	120	Cr16 Ni15 Mo2 MnTiSi (CW)	Cr 13 Mn Nb
BN-600	Russia	120	Cr16 Ni15 Mo2 MnTiSi (CW)	Cr 13 Mn Nb
BN-800	Russia	98	Cr16 Ni15 Mo2 MnTiSi (CW)	Cr 13 Mn Nb
BN-1600	Russia	170	Cr16 Ni15 Mo2 MnTiSi (CW)	Cr 13 Mn Nb
PFR	UK	250	316 SS, M 316 SS, PE 16	PE16/FV448

[a] Achieved peak burnup.
[b] 230 GWd/tHM was achieved with an advanced MOX annular fuel design, HT-9 clad, 169 pin/assembly

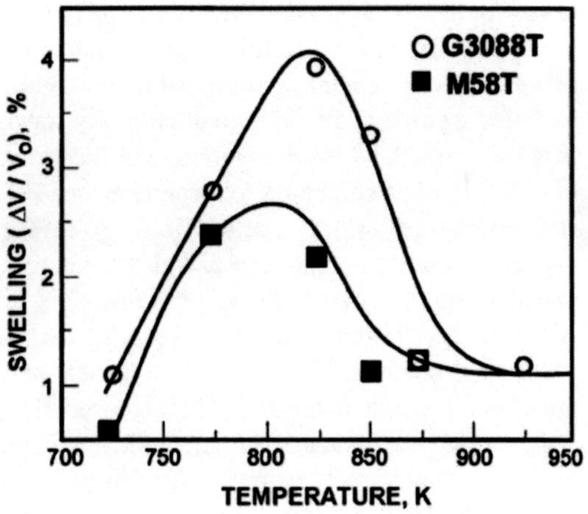

Fig. 11.41 Temperature dependence of void swelling measured by surface profilometry for the D9I SS alloys with two different phosphorous concentrations. The chemical composition of heats are G3088T: P = 0.026, Si = 0.74, Ti = 0.25; M58T: P = 0.047, Si = 0.77 and Ti = 0.31 [62]

of concern for F-M steels. Consequently, extensive studies involving modification of the composition and initial heat treatments have been carried out to improve the fracture toughness. The decrease in upper-shelf energy and shift in DBTT appear to saturate at about 10 dpa. A significant increase in toughness (i.e. low DBTT and high upper shelf energy) has been realised by (a) avoiding the formation of delta-ferrite and ensuring fully martensitic structure in 12% Cr steels by close control of nickel and chromium equivalent element concentration (delta-ferrite regions exhibit greater void formation and swelling than martensitic regions), (b) optimizing the austenitising temperature to refine the prior

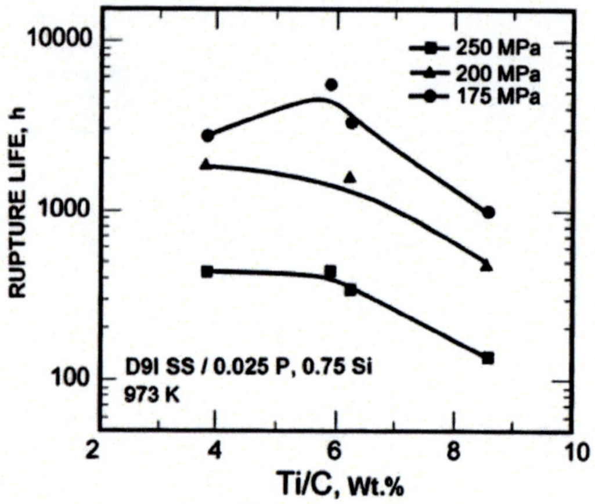

Fig. 11.42 Influence of titanium on creep rupture life of D9I SS (phosphorous = 0.025 wt.%, silicon = 0.75 wt.% and carbon = 0.04 wt.%) [62]

austenite grain size with strict control on embrittling tramp elements and (c) tempering treatments to reduce the strength of the martensite [63, 64]. 9Cr-1Mo grades of F-M steels have been reported to show the lowest increase in DBTT among the various grades of ferritic-martensitic steels [65]. These materials are very promising for duct applications.

The creep resistance of F-M steels decreases drastically above 550°C. Therefore, they are not generally suitable for cladding tube applications. Oxide dispersion strengthening is found to be a promising means of extending the creep resistance of F-M steels beyond 650°C without sacrificing the inherent advantages of high thermal conductivity and low swelling. This has led to the evolution of oxide dispersion strengthened F-M steels (ODS steels), with 0.3–0.4% yttria (Y_2O_3) particles as promising dispersiod. Chemical compositions of different ODS clad tubes are shown in Table 11.8 and the effect of chemical composition on creep rupture strength is shown in Fig. 11.43 [66]. Yttria particles are chemically inert and thermally stable up to around 1,250°C, and they provide high temperature strength by impeding mobile dislocations. Figure 11.44 shows improvement in creep strength of ODS steels as compared to dispersion free F-M steels [67]. The long term creep strength of ODS steels is also better than that of 316 SS [67]. The particle-matrix interface acts as trapping sites for irradiation induced point defects and thus retarding swelling. An addition of titanium has been found to refine the Y_2O_3 particles to ultra-fine (2–3 nm size) complex Y-Ti-O particles. Excess oxygen, which is the oxygen present in excess of that required for formation of Y_2O_3, has a very important role in controlling the formation of near-stiochiometric complex Y-Ti-O oxide particles and an associated increase in tensile and creep strength.

The precipitation hardened superalloys combine excellent swelling resistance with superior high temperature stress rupture properties. They are equally well suited to duct and cladding applications but their high temperature strength makes them an especially strong candidate to increase fuel pin life, compared to the other developmental alloys. But in nickel based superalloys, irradiation embrittlement is a major problem.

Unlike in the case of oxide fuels, for metallic fuels the operating temperature of cladding tubes is lower (< 625°C). The creep strength of modified 9Cr-1Mo steel is therefore adequate for it to be used as a cladding material.

To meet high burnup targets around 200 GWd/t, usage of advanced steels is inevitable. The creep rupture life of the ODS steels is superior to the conventional ferritic and austenitic steels. Hence, ODS steels development for advanced cladding continues in many countries as a priority area for materials development.

Table 11.8 Composition of ODS steels

	Chemical composition (wt.%)								
Cladding	C	Cr	W	Ni	Ti	Y_2O_3	Ex. O	N	Ar
M91	0.11	9.26	1.94	0.022	0.12	0.34	0.05	0.010	0.0033
M92	0.13	9.00	1.94	0.022	0.20	0.30	0.04	0.010	0.0034
M93	0.12	8.99	1.94	0.022	0.20	0.35	0.06	0.010	0.0033
M11	0.13	9.00	1.95	0.021	0.20	0.37	0.06	0.013	0.0025
MA957	0.02	14.0	–	0.13	1.0	0.27	–	–	–
F94	0.058	11.78	1.93	0.025	0.3	0.24	0.04	0.01	0.0003
F95	0.056	11.72	1.92	0.25	0.31	0.24	0.04	0.01	0.0038
1DS	0.09	10.98	2.67	0.15	0.4	0.4	0.119	0.014	–
1DK	0.045	12.87	2.81	0.16	0.52	0.34	–	0.0152	–
IGCAR	0.12	9.0	2.0	0.02 (max)	0.2	0.35	0.07	<0.01	0.002 (max)

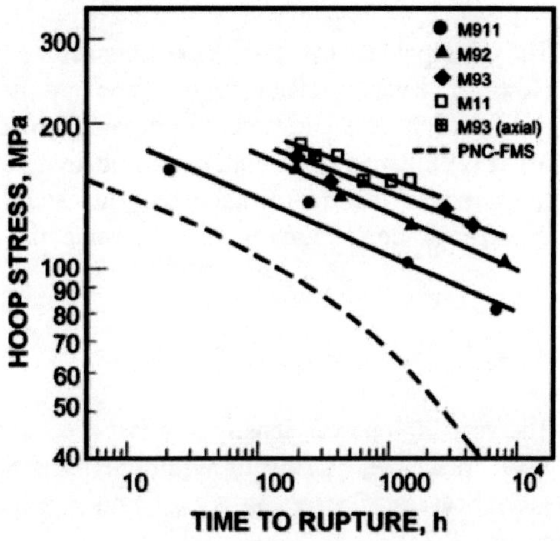

Fig. 11.43 Comparison of creep rupture strength of different ODS cladding steels in hoop direction at 973 K [66]

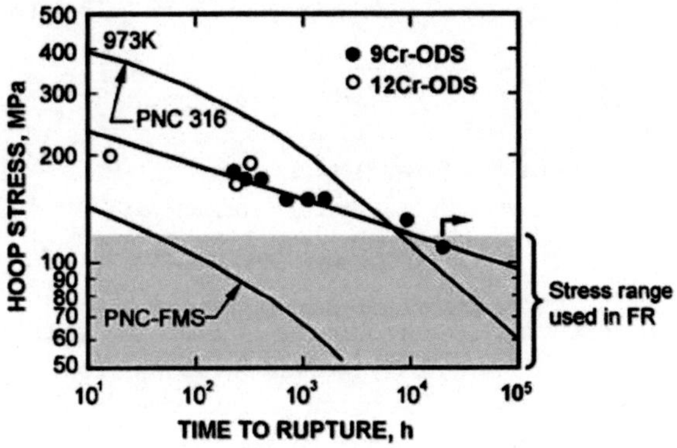

Fig. 11.44 Comparison of creep rupture strength of ODS steels with those of 316 SS and ferritic steel [67]

11.4 Coolant

The choice of coolant for a fast spectrum reactor probably has a greater effect on the overall physical plant layout than any other design selection. Whereas the coolant strongly influences the neutronic behavior of the core and provides a direct framework for the consideration of cladding materials, the most recognizable effect of the coolant selection is its effect on major components such as pumps and steam generators.

The liquid metal sodium has been selected as the coolant for all the major fast spectrum power reactor projects underway around the world. As a consequence, the properties of sodium and its implications in design will be given particular attention. The gas coolants helium and steam will be given limited attention, with brief reference to other metal coolants, to provide a framework for weighing the pros and cons of the options that have received the most attention to date. Chapters 17 and 18 respectively provide considerably more detail for gas and heavy metal coolants.

11.4.1 Requirements

The principal task of a fast spectrum reactor coolant is to remove heat from a high power density core. However, several considerations come into play in assessing the capability of the coolant to perform this function. It is convenient to organize the discussion of these facets along the lines suggested by Wirtz [29]; namely, thermal, neutronic, hydraulics, and compatibility considerations. Having covered the principal features of each coolant candidate, from the standpoint of those four categories, an intercomparison summary will be given at the end of this section.

11.4.2 Thermal Considerations

The very high power density in a fast spectrum reactor core (~400 W/l compared to ~100 W/l for a LWR) places very stringent requirements on the heat transfer properties of a coolant candidate. For a given heat transfer area, A (e.g., the outside cladding surface), the rate of heat removal, $\dot{Q}/A$, is given by Eq. (9.23):

$$q = \frac{\dot{Q}}{A} = h\,(T_{co} - T_b) \tag{11.5}$$

where

q = heal flux (W/m^2)
h = convection heat transfer coefficient (W/m$^2\cdot$°C)
$T_{co} - T_b$ = temperature drop from cladding surface to bulk coolant (°C).

Obviously, a large convection heat transfer coefficient is desirable for optimum heat removal.

Table 11.9 provides typical numbers for the convection heat transfer coefficient, as well as a typical coolant velocity and attendant pressure loss along the length of the core for the three cooling candidates initially considered of interest. Sodium is clearly superior from the standpoint of heat transfer characteristics, although He and steam can be pushed hard enough, given sufficient driving pressure, to accomplish the required task.

Table 11.9 Typical heat transfer data for coolant candidates[a]

Property	Coolant Na	He	Steam
h (W/m^2·°C)	85,000	2,300 (smooth surface) 10,000 (rough surface)	11,000
V (m/s)	6	115	25
p (MPa) (psi)	0.7 (100)	7.0 (1,000)	15 (2,200)
c_p (kJ/kg·°C)	1.3	5.2	2.6
ρ (kg/m^3)	815	4.2	41

[a]Agumented from page 28 of Ref. [29].

Table 11.10 is included to provide similar data for other liquid metals and, in particular, an inter-comparison with water. The numbers given are based upon a coolant flow of 3.3 m/s in a 25 mm diameter pipe. Water has the best heat capacity, but sodium is superior in all other categories. Water, of course, cannot be considered for fast spectrum reactor applications because of its major degradation of the fast neutron spectrum.

11.4.3 Neutronics Considerations

Whereas any coolant has some moderating effect on the neutron spectrum, the degree of moderation taking place is proportional to the atomic mass and the density. Sodium (mass 23) is clearly heavier than either helium or steam, and lead (mass 207) is substantially heavier than sodium. But when the density effect is taken into account, the helium-cooled reactor yields the hardest spectrum, steam yields the softest spectrum (because of the hydrogen content), and sodium provides an intermediate spectrum. Because of this spectrum difference, helium has a slight intrinsic edge regarding achievable breeding ratios.

Table 11.10 Comparison of heat transfer data for various fluids at high temperatures

Property	Coolant Na[a]	NaK[a]	Hg[a]	Pb[b]	Pb-Bi[b,c]	H_2O[a]
T_{melt} (°C)	98	18	−38	327.5	124.5	0
T_{boil} (°C)	880	826	357	1,743	1,670	100
c_p (kJ/kg·°C)	1.3	1.2	0.14	0.145	0.143	4.2
k (W/m·°C)	75	26	12	17.5	14.0	0.7
h (W/m^2·°C)[d]	36,000	20,000	32,000	24,900	22,500	17,000
Relative pumping power required (H_2O = 1)	0.93	0.93	13.1[e]	7.03[e]	6.29[e]	1.0

[a]Taken from page 29 Ref. [29].
[b]The Pb system data is taken from Ref. [68].
[c]Pb-Bi (45 at.% Pb—55 at.% Bi).
[d]For 3.3 m/s velocity in a 25 mm duct.
[e]The large pumping power for the heavy metal cooled cases can be reduced considerably by opening up the fuel pin lattice (cf. Chapter 18).

Another neutronics consideration is the effect of coolant activation. Helium, being inert, is unaffected by irradiation. Steam becomes somewhat radioactive (via ^{17}O), but the impact on design is minor. Sodium activation, however, attains a high degree of short half-life activation. The reaction that takes place is as follows:

$$^{23}Na + n \rightarrow ^{24}Na \xrightarrow{\beta(15\,hr)} ^{24}Mg$$

with an attendant release of 1.37 and 2.75 MeV gammas. This activation has led to the inclusion of an intermediate coolant loop for sodium cooled systems in order to ensure that all radiation is confined to the primary loop. Other reasons for including an intermediate loop in the SFR system are (1) to protect the core from possible pressure surges or positive reactivity effects due to hydrogen moderation should a steam generator leak occur and result in a sodium-water reaction, and (2) to protect the steam generator from radioactive corrosion products and fission products. Lead or lead-bismuth coolants are less reactive with neutrons, although the buildup of ^{210}Po (an alpha emitter with a half-life of 138 days) poses a challenge when Bi is included. Bismuth is added mainly to lower the melting temperature of the heavy eutectic metal. Both Pb and Pb-Bi coolants are inert to reactions with water or air.

A final neutronics consideration concerns the reactivity effect of coolant loss during an accident. As noted in Chapter 6, the large density change associated with sodium loss can cause an appreciable change in the neutron spectrum and an associated positive reactivity feedback. Neither of the gas cooling media exhibits such a characteristic, although a thorough safety analysis of such systems does cover the remote possibility of water entering the core as a result of a major break or depressurization. Under such conditions, similar variations in reactivity could occur. Both Pb and Pb-Bi systems have such high vaporization temperatures that coolant voiding is not considered a serious reality.

11.4.4 Hydraulics Considerations

In order for a reactor coolant to perform its heat removal function, a fraction of the total power output must be used to pump it through the core. Hence, it is of interest to compare the pumping requirements for the various coolant candidates.

For a given core power, $\dot{Q}$, the heat must be removed according to the relationship:

$$\dot{Q} = \dot{m}\, c_p\, \Delta T_{axial} \tag{11.6}$$

where

$\dot{Q}$ = core power (W)
c_p = specific heat (J/kg·s)
$\dot{m} = \rho VA$ = coolant mass flow rate (kg/s)
ρ = coolant .density (kg/m^3)
V = coolant velocity (m/s)
A = coolant flow area (m^2)
ΔT_{axial} = difference between outlet and inlet coolant temperature (°C).

Hence, for a fixed flow area and axial coolant temperature increase, the coolant velocity must be

$$V = \frac{\dot{Q}}{\rho c_p A \Delta T_{axial}}. \tag{11.7}$$

The pressure loss over the core can be written [cf. Eq. (10.33)] as

$$\Delta p = f \frac{L}{D_e} \frac{\rho V^2}{2} \tag{11.8}$$

where

Δp = pressure drop over core (N/m^2)
f = friction factor
L = length of core (m)
D_e = hydraulic diameter (m) = 4 A/P_w where P_w is the wetter perimeter.

Since the power required to move the coolant through the core is

$$\text{Pumping power} = \Delta p \cdot A \cdot V, \tag{11.9}$$

combining Eqs. (11.7) through (11.9) yields

$$\text{Pumping power} = \frac{LA}{2D_e}\left(f \rho V^3\right) = \left(\frac{L\dot{Q}}{2D_e\, \Delta T_{axial}}\right)\left(\frac{f V^2}{c_p}\right). \tag{11.10}$$

Using either expression, the strong dependence of pumping power on the coolant velocity is evident. While other parameters vary between sodium and gas (e.g., D_e, and c_p, are larger for helium designs than for sodium) the large velocity difference shown in Table 11.9 dominates. The numbers contained in the box below[14] indicate that, for typical sized plants, sodium coolant requires the least power and helium requires the most pumping power. Other liquid metals would appear to be ruled out, largely because of their much larger pumping requirements, as indicated in Table 11.10. However, as will be shown in Chapter 18, it is possible to open up the fuel pin lattice and allow lower flow velocities, thus removing much of this constraint.

	Sodium cooled	He-gas cooled	Steam cooled
Percent pumping power of electrical output	1.5	7	3

11.4.5 Compatibility Considerations

The coolant chosen for fast spectrum reactor applications must be compatible with the cladding and should also be at least reasonably compatible with the fuel so that an in-pile breach could be accommodated without adverse fuel–coolant interactions. Helium, being inert, is compatible with any metal or fuel.

[14] From page 28 of Ref. [29].

Steam, on the other hand, can be quite corrosive with the normal stainless steels. In fact, the dry-steam corrosion problem became sufficiently troublesome in the early test program [69] that a high nickel component steel, such as Incoloy® or Inconel® was required to provide acceptable compatibility conditions. As we noted from the discussions of Section 11.3.2, nickel has a deleterious neutronic effect, but even more importantly leads to considerable embrittlement from the (n, α) reaction. Steam has been dropped as an active fast spectrum reactor cooling candidate mainly because of the cladding corrosion problem.

Lead and lead-bismuth systems have generally been plagued with substantial oxidation corrosion. However, as covered in Chapter 18, there now appears to be ways to control the oxidation problem.

Sodium is quite compatible with the preferred cladding candidates discussed above. However, because sodium has been chosen as the base coolant, much work has been directed to every facet of its in-core environment in order to head off any long term surprises. Summarized below are the main categories [1] where efforts have been focused to isolate possible problem areas.

1. General Corrosion
   The principal metallic elements in the cladding (i.e., Fe, Cr and Ni) are very slowly dissolved in sodium from the hot core region and deposited in the cooler areas. This long-term uniform attack is called cladding *thinning* or *wastage.* Although this amounts to only tens of micrometers per year at 700°C, it must be considered in the overall load bearing capacity of the cladding, which is stressed by internal fission gas pressure. Oxygen concentrations must be kept low (<5 ppm) to minimize corrosion.
2. Selective Leaching
   Each of the principal metallic elements mentioned above has a different dissolution rate in sodium. Hence, selective leaching occurs and the composition of the steel at the outer surface is eventually altered. The effect, though small, is normally deleterious, since Cr and Ni are removed faster than Fe. Some of the more significant problems arising from the dissolution of minor constituents or activation productions from the steel in the core are discussed below.
3. Deposition
   The dissolved metal deposited in the colder region of the intermediate heat exchanger (IHX) can lead to higher heat transfer resistance (i.e., reduced heat exchanger efficiency). The deposits can become thick enough that the coolant driving pressure must be increased in order to maintain the desired flow. Some long-term Si deposits have even been observed. The silicon, originating as a minor constituent in the steel, forms a compound with sodium which, if precipitated in core assembly coolant channels, can influence pumping power or local flow conditions.
4. Transport of Radioactivity
   The major long-lived radioactive species produced in the cladding are ^{54}Mn, ^{58}Co and ^{60}Co. Though small in quantity, sodium transport and deposition of these radionuclides in the IHX and other primary system components can cause high enough local activity to affect routine maintenance.
5. Carbon Transport
   Carbon, included in the component steel alloys to improve strength characteristics, is quite mobile in solid cladding and can migrate fairly readily to the outer surface. Sodium can, therefore, transport carbon, as well as all the metallic elements, to colder surfaces. Both the hot steel ("decarburization") and the cold steel ("carburization") properties can be altered by these small changes in the carbon present. This is normally not a problem with cold worked 316 stainless steel, however, because rapid carbide precipitation in the steel lowers the carbon activity sufficiently to prevent decarburization. Oxygen and hydrogen concentration in sodium are controlled within the acceptable limits using cold traps. The cold trap is a crystallizer or precipitator chamber removing impurities that have a temperature dependent solubility in sodium.

6. Sodium Chemistry
   Because sodium is the transport medium for the processes described above, samples of sodium are usually extracted regularly and analyzed during normal operation. On-line instruments have been developed to monitor sodium purity for both the primary and secondary systems. In addition to determining oxygen and carbon content, it is often useful to monitor hydrogen levels, since this provides an indication of the tritium level in the cover gas (primary system) or steam generator leaks (secondary system).
7. Sodium–Fuel Interaction
   In the event of a small cladding breach for an oxide fuel system, a direct interaction of sodium and fuel could lead to the production of Na_3UO_4 or Na_3PuO_4. The concern is whether this substance could be swept into the coolant stream as a contaminant. There is also some concern about whether this material could lead to local fuel swelling aggravation and escalation of the cladding leak into a fuel rupture. If the fuel should be molten, the thermal aspect of the interaction is the principal concern. This problem is discussed in Chapter 16.

11.4.6 Sodium Properties

A consistent set of thermodynamic properties of sodium is reported by Fink and Leibowitz [70]. Thermodynamic properties listed in this section were obtained from that reference.

Vapor pressure of sodium as a function of temperature is plotted in Fig. 11.45. Note that the hottest sodium in a typical SFR system (800 K) is approximately 350°C below the boiling point at atmospheric pressure. An equation for the vapor pressure of sodium [70] developed by and consistent with the high temperature data of Bhise and Bonilla [71], and Das Gupta [72] is

$$P = \exp\left[18.832 - \frac{13113}{T} - 1.0948 \ln T + 1.9777 \times 10^{-4} T\right],$$

where P is in atmosphere and T is in Kelvin.

The critical properties, as determined by Bonilla et al., are

$P_c = 25.6$ MPa (253 atm)
$T_c = 2{,}509$ K
$\rho_c = 214$ kg/m^3.

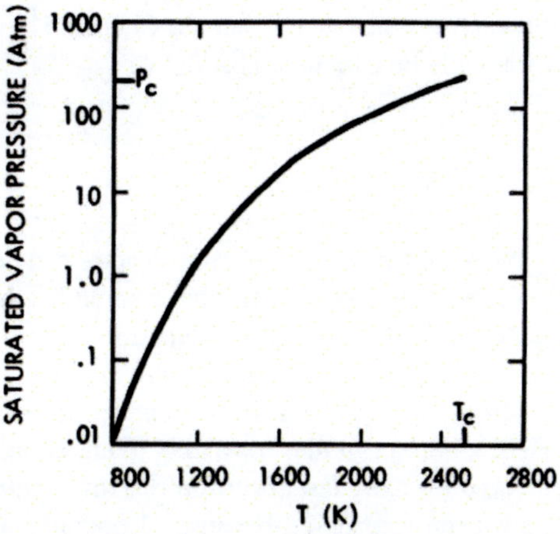

Fig. 11.45 Saturated vapor pressure of sodium

The specific heat, c_p for sodium at typical SFR temperatures (e.g., from 400°C to 550°C) is 1.3 kJ/kgK.

The density of liquid saturated sodium up to 1,644 K is

$$\rho_l\left(kg/m^3\right) = 1011.8 - 0.22054T - 1.9226 \times 10^{-5}T^2 + 5.6371 \times 10^{-9}T^3.$$

where T is in Kelvin.

Expressions for ρ_l, from 1,644 K to the critical point, together with h_l, h_g, ρ_g, and other saturation properties, and superheat and subcooled properties, are given in Ref. [70].

Expressions for the physical properties thermal conductivity, k [73], and viscosity, μ [74], as a function of temperature (Kelvin) are given below:

$$k(W/m \cdot K) = 93.0 - 0.0581\,(T - 273) + 1.173 \times 10^{-5}\,(T - 273)^2$$
$$\mu(kg/m \cdot s) = A\rho^{1/3}e^{B\rho/T}\left(\rho \text{ in } kg/m^3\right)$$

where $A = 1.235 \times 10^{-5}$ and $B = 0.697$ for $T < 773\ K$,

$A = 0.851 \times 10^{-5}$ and $B = 1.040$ for $T > 773\ K$.

11.4.7 General Intercomparison of Coolants

This section summarizes the pros and cons of each coolant discussed previously, and refers to other items that bear on coolant selection. Table 11.11 lists the advantages and disadvantages of each coolant.

A principal advantage of sodium, beyond its excellent heat transfer properties, is that it does not require pressurization to prevent coolant boiling. This feature contrasts sharply with the thick-walled high-pressure systems required for He- or steam-cooled systems.

Helium is still receiving attention in the gas-cooled fast reactor program and is discussed in Chapter 17. Steam has been essentially eliminated as a serious contender, mainly because of cladding corrosion problems.

Heavy metal coolants (Pb and Pb-Bi systems) avoid coolant voiding issues (due to the very high vaporization temperatures) and they do not react with air or water. However, corrosion issues remain and the technology is sufficiently new (at least for civilian reactor applications) that the data base for such systems is quite small. Chapter 18 provides a perspective on how such coolant systems may eventually be commercialized.

11.5 Control

Control of a large fast spectrum reactor can be provided by either removing fuel or inserting a neutron absorber. Although the former method has been used (e.g., in EBR-II), the latter is by far the most widely employed technique. Neutron cross sections of all absorber materials are much smaller in the high energy range than in the thermal neutron domain, but materials widely employed for thermal reactor control purposes also tend to be good absorbers in the fast spectrum reactor energy range. Hence, boron carbide, consisting of either natural boron or enriched with ^{10}B, is the absorber material favored by most fast spectrum reactor designers. Other candidates, including tantalum and europia, have been considered because of potential gas release and swelling problems with boron carbide.

Table 11.11 Intercomparison of fast spectrum reactor coolants[a]

Coolant	Advantages	Disadvantages
Sodium	– Excellent heat transport properties – Low pressure system – Low pumping power requirements – Lowest fuel cladding temperature – Potentially high breeding ratio – Inherent emergency cooling of fuel – Extensive sodium reactor experience – Potential for vented fuel	– Radioactivity (intermediate sodium loop employed) – Unfavorable coolant reactivity void coefficient – Chemical reactions with air and water – Nonvisible refueling procedure – Solid at room temperature – Maintenance on primary system impeded by radioactivity
Helium	– No intermediate loop – Coolant not activated – Potentially high breeding ratio – Visible refueling/maintenance – Minimal void coefficient – Most compatible with materials – Utilization of thermal GCR technology – Potential for vented fuel – Potential for direct cycle – Flooding with H_2O tolerable	– High pressure system – High pumping power requirements – Cladding roughening required – Emergency cooling provisions not established – Unproven high power density capability – Lack of fast spectrum reactor technology – Gas leakage difficult to control – High performance demands on pumps and valves
Steam	– Direct cycle – Visible refueling/maintenance – Industrial capability available for components – Minimum chemical reactions – Fluid at room temperature	– High pressure system – High pumping power – Cladding corrosion – Lack of fast spectrum reactor technology – Emergency cooling provision not established – Low breeding ratio – Fission product carryover to turbine – Unfavorable coolant reactivity coefficient
Lead	– Does not have the combustibility of sodium or other alkali metal coolants – Does not react vigorously with water, steam, air, or carbon dioxide – High boiling temperature – Excellent heat transport properties – Low pressure system – Slightly lower absorption of neutrons relative to sodium (permits derated core coolant volume to be increased, lowering pressure drop) – Acceptably low pumping power for derated core with opened lattice – High density provides sufficient net positive suction head to install primary coolant pumps in primary circuit hot leg in pool configuration – Significantly lesser amounts of ^{210}Po by two to four orders of magnitude compared with Pb-Bi – High density limits coolant void growth following steam generator or heat exchanger failure and blowdown of steam or carbon dioxide into Pb	– Higher melting temperature relative to sodium or Pb-Bi – Thicker structures are required to support coolant weight – Thicker structures are required to resist forces during seismic events – Size of a pool reactor may be limited by need to accommodate seismic events – Opaque coolant – Solid at room temperature – Radioactive coolant (Dominant radiation is gammas from ^{207m}Pb having a half-life of 0.806 s)

Table 11.11 (continued)

Coolant	Advantages	Disadvantages
Lead-bismuth eutectic	– Does not have the combustibility of sodium or other alkali metal coolants – Does not react vigorously with water, steam, or air – Reacts only weakly with carbon dioxide – Lower melting temperature compared with Pb – High boiling temperature – Excellent heat transport properties – Low pressure system – Slightly lower absorption of neutrons relative to sodium (permits derated core coolant volume to be increased, lowering pressure drop) – Acceptably low pumping power for derated core with opened lattice – High density provides sufficient net positive suction head to install primary coolant pumps in primary circuit hot leg in pool configuration – High density limits coolant void growth following steam generator or heat exchanger failure and blowdown of steam or carbon dioxide into Pb	– Neutron capture in ^{209}Bi results in ^{210}Po, which is an alpha emitter with a half-life of 138 days; exposure to moisture can result in volatile polonium hydride resulting in an airborn hazard – Thicker structures are required to support coolant weight – Thicker structures are required to resist forces during seismic events – Size of a pool reactor may be limited by need to accommodate seismic events – Opaque coolant – Solid at room temperature – Radioactive coolant (Dominant radiation is gamma from ^{207m}Pb with half-life of 0.806 s)

[a] Adapted from page 27 of Ref. [29].

Silver and silver alloys have enjoyed considerable success for pressurized water reactor applications but have not been seriously considered for fast spectrum reactor applications due to their relatively low reactivity worth and high cost.

11.5.1 Requirements

The basic requirements of the control rod systems are (1) to compensate for built-in burnup reactivity, and (2) to provide neutronic shutdown for routine operations and safety measures. As such, enough absorber material must be provided to meet the reactivity worth requirements outlined in Chapter 6. The control system installed must be capable of long life (e.g., 3 years or longer), which means the absorber material must be compatible with its cladding, and dimensional instability must be well characterized. Finally, overall plant economics favor absorber materials that are readily available in large quantities, i.e., an industrial product.

11.5.2 Boron

Boron, present in the form of boron carbide, B_4C, is the absorber material generally used in fast spectrum reactor systems. The principal advantages of boron carbide are (1) its relatively high neutron absorption cross section, (2) its availability and low cost, (3) its comparative ease of fabrication, and (4) its low radioactivity after irradiation. The ^{10}B neutron absorption mechanism is the (n, α) reaction, and the He atoms produced lead to both matrix swelling and gas release difficulties. This problem can be accommodated by operating at relatively high temperatures (cf. Fig. 11.49) and/or by venting the pins.[15]

11.5.2.1 Neutronics

Figure 11.46 shows the $^{10}B(n, \alpha)^{7}Li$ absorption cross section over the complete thermal to fast spectrum range, as well as the ^{239}Pu fission cross section from Fig. 2.7 (somewhat smoothed out) for purposes of comparison. As clearly seen from this figure, the boron-10 cross section is only of the order of one barn for the fast neutron spectrum, whereas it is of the order of 1,000 barns for a thermal reactor. Furthermore, the absorption/fission cross section ratio drops with a harder spectrum, leading to a requirement for more absorber material in a fast spectrum reactor than an LWR. However, there is a compensating effect in that much of the neutron absorption occurs near the absorber material surface in an LWR, due to self-shielding, whereas absorptions are more uniformly distributed in the absorber material for a fast spectrum reactor environment.

Also contained in Fig. 11.46 is the $^{10}B(n, t)\ 2\alpha$ cross section, which is appreciable only for the very high energy tail. The latter reaction, while favorable from the standpoint of control rod reactivity worth density, is basically undesirable because the tritium therein produced must be contained to prevent leakage from the reactor systems. This reaction is responsible for about one-half of the low-level tritium radiation production in the plant. Natural boron consists of 19.8% ^{10}B, with the remainder being ^{11}B (the latter possessing a very small neutron absorption cross section). Hence, the reactivity worth of a given dilute sample of B_4C can be increased by nearly a factor of five by enriching it with the lighter ^{10}B isotope. In a fast spectrum reactor absorber assembly, self-shielding limits this worth increase to about 60% of the change in enrichment.

11.5.2.2 Physical Properties

Boron carbide has a rhombohedral crystalline structure. Although hot pressed pellets of 92% theoretical density are brittle, the thermophysical properties [75] of this material are generally compatible with

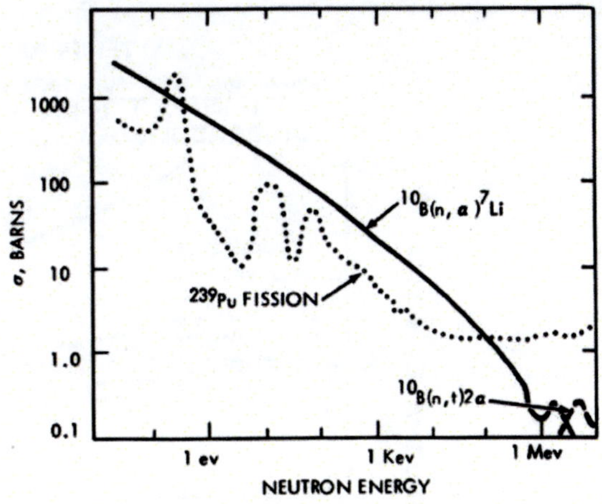

Fig. 11.46 Absorption cross section of ^{10}B compared with the fission cross section of ^{239}Pu

[15]Considerable interest exists in venting the absorber pins in order to avoid the necessity of a plenum to contain the He buildup.

control system requirements. Structural integrity of absorber pins is provided by cladding material, thereby allowing B_4C to be employed either in the form of powder or pellets. Below 800°C, boron carbide retains about 80% of the tritium produced. Unfortunately, cladding will not contain the tritium gas that escapes the boron carbide matrix.

Thermal conductivity is perhaps the most important thermophysical property of present interest to the boron carbide system. The (n, α) reaction liberates 2.78 MeV per event, and most of this energy is deposited directly within the B_4C matrix (~75 W/cm^3 for naturally occurring B_4C in a fast spectrum reactor). Hence, the thermal gradient in the absorber material is a direct function of thermal conductivity. Figure 11.47 illustrates the thermal conductivity of fully dense B_4C. It is of particular interest to note from Fig. 11.47 that the thermal conductivity drops significantly upon irradiation. The lower curves represent a large collection of irradiation data, including results for differing irradiation conditions.

11.5.2.3 Irradiation Behavior

The principal effects of irradiation on B_4C are helium generation and matrix swelling. For unenriched B_4C, a complete burnup of all the ^{10}B atoms (0.22×10^{23} captures/cm^3 of B_4C) would yield 814 cm^3 of He at standard pressure and temperature. The amount of this gas released during burnup is sensitive to both irradiation temperature and to exposure level. Figure 11.48 illustrates this effect with a three-dimensional plot. The percent gas release builds up with irradiation temperature, with a maximum occurring around 1,100 K. This gas released must either be vented to the coolant or sufficient plenum volume must be provided in order to prevent excessive loading on the control cladding material. In addition to temperature, the stoichiometric composition also affects the degree of gas release.

Both the He and Li atoms resulting from the neutron absorption process are larger than the original boron atoms. Hence, swelling occurs due to both the Li and the portion of He atoms remaining in the matrix. Figure 11.49 indicates that the resulting swelling is approximately linear with burnup and tends to drop as the irradiation temperature increases.

The useful life of absorber pins containing boron carbide could be enhanced by the ability of the cladding to accommodate the stresses imposed by B_4C swelling. Hence, there is considerable interest in providing cladding material for absorber pins that displays good ductility characteristics.

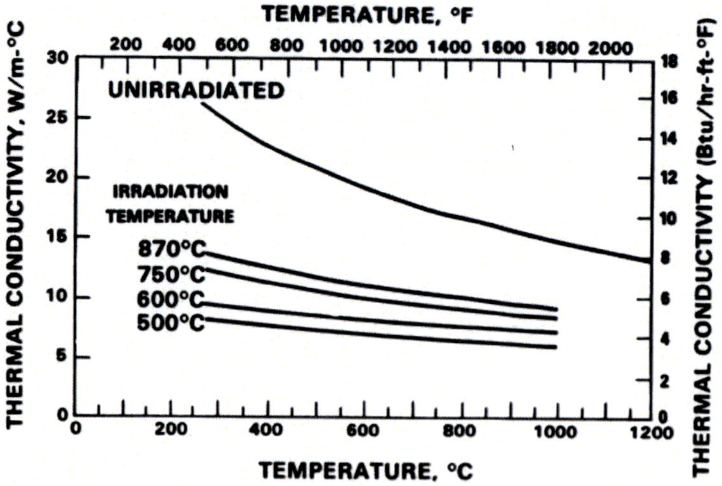

Fig. 11.47 Effect of irradiation on thermal conductivity of boron carbide [76]

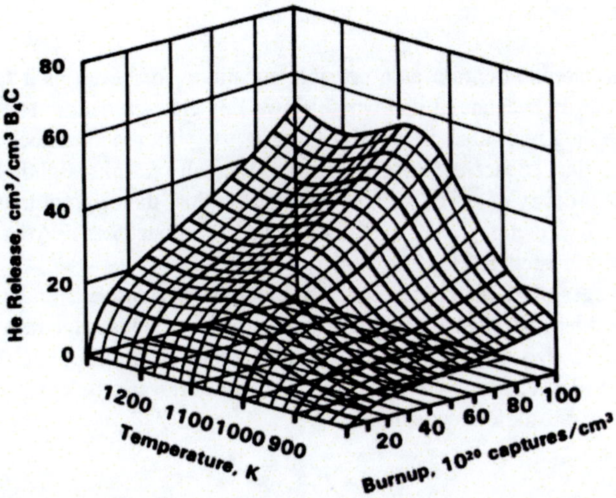

Fig. 11.48 Helium gas release as a function of temperature and burnup [77]

11.5.2.4 Compatibility

Boron carbide is quite compatible with most steel alloy cladding materials, with no appreciable metal attack below 700°C. Above that temperature, however, a layer of Fe_2B tends to build up, and the attack that occurs is about a factor of three greater in the presence of sodium. A eutectic is formed if the temperature should ever climb to 1,226°C. Excess boron in the B_4C will enhance the steel attack characteristics, and the presence of lithium from the (n, α) reaction has been shown to enhance the diffusion of boron and carbon into the steel.

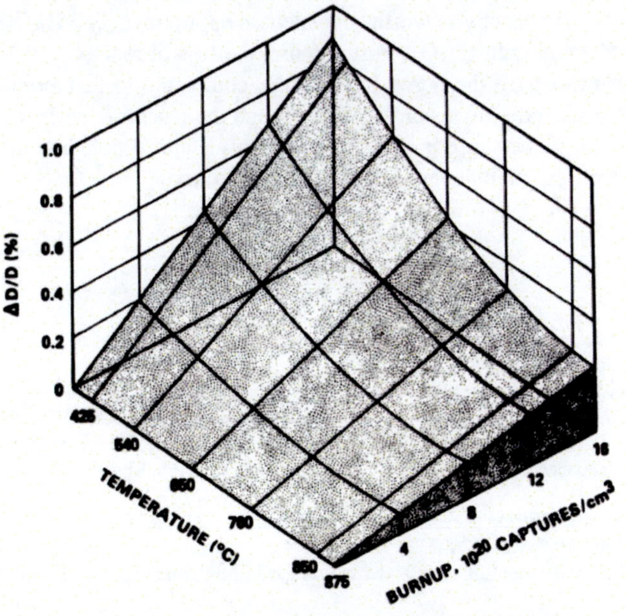

Fig. 11.49 Boron carbide linear swelling rate ($\Delta D/D$) during fast reactor exposure [78]

11.5.3 Tantalum

Tantalum is receiving some attention as a possible substitute for boron as a fast spectrum reactor control material, primarily because of its favorable swelling characteristics. The fundamental reason for the improved swelling resistance is that the $^{181}Ta(n, \gamma)$ ^{182}Ta reaction does not yield helium, as does the boron-10 reaction. In fact, at 530°C the material actually shrinks slightly because the tungsten atoms produced by a β-decay of ^{182}Ta to ^{182}W are smaller than the original tantalum atoms. On the other hand, since Ta is a metal, it will eventually swell to about 1%$\Delta V/V$ (by the void formation mechanisms discussed in Section 11.3.2).

Additional advantages of Ta as a control material include its relative abundance (fairly low cost), ease of fabrication, moderately high absorption cross section in the fast spectrum domain, and daughter products (which are also good absorbers). The principal disadvantage is the 115-day half-life γ-decay, from ^{182}Ta to ^{182}W, which causes long-term heat removal problems. Also, Ta is soluble in Na, making cladding an absolute requirement.

11.5.4 Europium

Europium, in the sesquioxide form Eu_2O_3, has received considerable attention in recent years because of its high fast spectrum absorption cross section. Both naturally occurring isotopes (47.8% ^{151}Eu and 52.2% ^{153}Eu) have dilute sample neutron absorption cross sections over twice that of ^{10}B averaged over the SFR spectrum. Like tantalum, the daughter products are also good absorbers and the (n, γ) reaction avoids the need for a gas plenum.

Unfortunately, self-shielding is such that a full assembly has a reactivity worth equivalent only to that for an unenriched B_4C assembly in a fast spectrum reactor. The other disadvantages of europia include (1) the high level of radioactivity induced into the europium decay chain, (2) the low supply (as a rare-earth), and (3) low thermal conductivity (which requires fairly small diameter pins).

An interesting attempt to combine the attributes of both Eu and B is the europium compound EuB_6. This compound has a reactivity worth equivalent to approximately 25% ^{10}B-enriched B_4C, and it is worth about 10% more than Eu_2O_3. Furthermore, the loss of reactivity is lower than in B_4C. It is, of course, possible to enrich the boron in the EuB_6 compound to further enhance its absorption capability, but the expense goes up accordingly. Although the reactor experience is quite-limited, the dimensional stability properties appear attractive. The only major problem is that the He gas release is much larger than for B_4C, so that venting is probably required.

References

1. D. R. Olander, "Fundamental Aspects of Nuclear Reactor Fuel Elements," TID-26711-P1, Office of Public Affairs, U. S. ERDA, Washington, DC, 1976.
2. S. Vana Varamban, "Estimation of solidus and liquidus temperature for UO_2 and PuO_2 psuedo binary system," *Proceedings of Sixteenth National Symposium on Thermal Analysis*, THERMANS-2008, February 4–6, 2008, Kalpakkam, India. (Eds. Salil Varma, K.V. Govindan Kutty, S.K, Mukerjee, T. Gnanasekaran, Mrs. S.R. Bharadwaj, and V. Vengopal). Published by Scientific Information Resource Division, Bhabaha Atomic Research Centre, India, 2008, pp. 434–436.
3. Nuclear Energy Agency, OECD, "Accelerator-driven systems (ADS) and fast reactors (FR) in advanced nuclear fuel cycles." NEA No. 3109, Paris, 2002.
4. E. H. Randklev, "Radial distribution of retained fission gas in irradiated mixed oxide fuel," *Trans. ANS*, *28* (June 1978) 234–236.

5. M. Tourasse, M. Boidron, and B. Pasquet, "Fission products behaviour in Phénix fuel pins at high burnup," *J. Nucl. Mater., 188* (1992) 49.
6. J. C. Melis, H. Plitz, and R. Thetford, "Highly irradiated fuel behaviour up to melting: JOG tests in CABRI," *J. Nucl. Mater., 204* (1993) 212.
7. U. P. Nayak, A. Boltax, R. J. Skalka, and A. Biancheria, "An analytical comparison of the irradiation behavior of fast reactor carbide and oxide fuel pins," *Proceedings of the Topical Meeting Advanced LMFBR Fuels,* Tucson, AZ, ERDA 4455, October 10–13, 1977, 537.
8. C. Ganguly and A. K. Sengupta, "Out-of-pile chemical compatibility of hyperstoichiometric $(Pu_{0.7}U_{0.3})C$ with stainless steel cladding and sodium coolant," *J. Nucl. Mater., 158* (1988) 159.
9. B. Raj, "Plutonium and the Indian atomic energy programme," *J. Nucl. Mater., 385* (2009) 142.
10. IAEA-TECDOC-1374, "Development status of metallic, dispersion and non-oxide advanced and alternative fuels for power and research reactors," September, 2003.
11. T. N. Washburn and J. L. Scott, "Performance capability of advanced fuels for fast breeder reactors," *Proceedings of the Conference on Fast Reactor Fuel Element Technology,* New Orleans, LA, April 13–15, 1971, 741–752.
12. J. A. L. Robertson, "Irradiation Effects in Nuclear Fuels," Gordon and Breach, New York, NY, 1969, 223. (Copyright held by American Nuclear Society, LaGrange Park, IL.)
13. H. Blank, "Nonoxide Ceramic Nuclear Fuels," from "Nuclear Materials," Part I Volume 10 A, edited by B. R. T. Frost, appearing in "Materials Science and Technology, A Comprehensive Treatment," edited by R. W. Cahn, P. Haasen, and E. J. Kramer, VCH Verlagsgesellschaft GmbH, Weinheim, 1994, 191.
14. H.J. Matzke, "Science of Advanced LMFBR Fuels," chapter 4, "Physical properties" North-Holland, Amsterdam, 1986, 176.
15. W. F. Lyon, R. B. Baker, and R. D. Leggett, "Performance analysis of a mixed nitride fuel system for an advanced liquid metal reactor," in *LMR: A Decade of LMR Progress and Promise*, Washington, DC, American Nuclear Society, La Grange Park, IL, November 11–15, 1990, 236–241.
16. A. A. Bauer, P. Cybulskis, and J. L. Green, "Mixed nitride fuel performance in EBR-II," *Proceedings of the Symposium on Advanced LMFBR Fuels*, Tucson, AZ, October 10–13, 1977, pp. 299–312.
17. R. D. Leggett, R. K. Marshall, C. R. Hanu, and C. H. McGilton, "Achieving high exposure in metallic uranium fuel elements" *Nucl. Appl. Tech., 9* (1970) 673.
18. C. W. Walter, G. H. Golden, and N. J. Olson, "U-Pu-Zr Metal Alloy: A Potential Fuel for LMFBR's," ANL 76-28, Argonne National Laboratory, Argonne, IL, November 1975.
19. G. L. Hofman, L. C. Walters, and T. H. Bauer, " Metallic fast reactor fuels," *Progress in Nuclear Energy, 31* (1997) 83.
20. R. G. Pahl, D. L. Porter, D. C. Crawford, and L. C. Walters, "Irradiation Behaviour of Metallic Fast Reactor Fuels," ANL/CP-73323, Argonne National Laboratory, Argonne, IL, 1991.
21. B. R. Seidel and L. C. Walters, "Performance of metallic fuel in liquid-metal fast reactors," ANS 1984, International meeting November 11–16, 1984, CONF-841105-2.
22. T. Ogata and T. Yokoo, "Development and validation of ALFUS: An irradiation behaviour analysis code for metallic fast reactor fuels," *Nucl. Tech., 128* (1999) 113.
23. R. G. Pahl, D. L. Porter, C. E. Lahm, and G. L. Hofman, "Experimental studies of U-Pu-Zr fast reactor fuel pins in EBR-II," in AIME Meeting, *Symposium on Irradiation Enhanced Materials Science*, Chicago, IL, Fall 1988, CONF 8809202-2.
24. D. C. Crawford, D. L. Porter, and S. L. Hayes, "Fuels for sodium-cooled fast reactors: US perspective," *J. Nucl. Mater., 371* (2007) 202.
25. A. Bauer, "Nitride fuels: properties and potential," *Reactor Tech., 15* (2) (1972).
26. V. S. Yemel'yanov and A. I. Yebstyukhin, "The Metallurgy of Nuclear Fuel-Properties and Principles of the Technology of Uranium, Thorium and Plutonium" (trans. Anne Foster), Pergamon Press, New York, NY, 1969.
27. J. H. Kittel, J. A. Horak, W. F. Murphy, and S. H. Paine, *Effects of Irradiation on Thorium and Thorium-Uranium Alloys*, ANL-5674, Argonne National Laboratory, Argonne, IL, 1963.
28. B. Blumenthal, J. E. Saneki, and D. E. Busch, "Thorium-Uranium-Plutonium Alloys as Potential Fast Power Reactor Fuels, Part II. Properties and Irradiation Behaviour of Thorium-Uranium-Plutonium Alloys," ANL-7259, Argonne National Laboratory, Argonne, IL, 1969.
29. K. Wirtz, "Lectures on Fast Reactors," Kernforschungszentrum, Karlsruhe, 1973 (Published by Gesellschaft fur Kernforschungszentrum).
30. J. K. Fink, M. G. Chasanov, and L. Leibowitz, "Thermophysical Properties of Thorium and Uranium System for Use in Reactor Safety Analysis," ANL-CEN-RSD-77-1, Argonne National Laboratory, Argonne, IL, June 1977.
31. J. K. Fink, "Thermophysical properties of uranium dioxide," *J. Nucl. Mater., 279* (2000) 1.
32. D. G. B. Martin, "The thermal expansion of solid UO_2 and (U, Pu) mixed oxides — a review and recommendations," *J. Nucl. Mater., 152* (1988) 94.

33. J. J. Carbajo, G. L. Yoder , S. G. Popov, and V. K. Ivanov, "A review of the thermophysical properties of MOX and UO_2 fuels," *J. Nucl. Mater., 299* (2001) 181.
34. Y. Philipponneau, "Thermal conductivity of (U, Pu)O_{2-x} mixed oxide fuel," *J. Nucl. Mater., 188* (1992) 194.
35. C. Duriez, J. P. Alessandri, T. Gervais, and Y. Philipponneau, "Thermal conductivity of hypostoichiometric low Pu content (U, Pu)O_{2-x} mixed oxide," *J. Nucl. Mater., 277* (2000) 143.
36. M. Inoue, "Thermal conductivity of uranium–plutonium oxide fuel for fast reactors," *J. Nucl. Mater., 282* (2000) 186.
37. S. Majumdar, A. K. Sengupta, and H. S. Kamath, "Fabrication, characterization and property evaluation of mixed carbide fuels for a test fast breeder reactor," *J. Nucl. Mater., 352* (2006) 165.
38. G. S. Was, "Fundamentals of Radiation Materials Science: Metals and Alloys," Springer, Berlin, Heidelberg, 2007.
39. R. L. Fish and J. J. Holmes, "Tensile properties of annealed type 316 stainless steel after EBR-II irradiation," *J. Nucl. Mater., 46* (1973) 113.
40. F. A. Garner, "Irradiation performance of cladding and structural steels in liquid metal reactors," *Material Science and Technology*, edited by R. W. Cahn et al. Vol. 10A, VCH, Weinheim, 1994, 419–543.
41. C. W. Hunter, R. L. Fish, and J. J. Holmes, "Mechanical properties of unirradiated fast reactor cladding during simulated overpower transients," *Nucl. Tech., 27* (1975) 367.
42. E. E. Bloom and J. R. Weir, "Effect of neutron irradiation on the ductility of austenitic stainless steel," *Nucl. Tech., 16* (1972) 45.
43. R. L. Fish and C. W. Hunter, "Tensile properties of fast reactor irradiated type 304 stainless steel," *ASTM Symposium on the Effects of Radiation on Structural Materials*, ASTM STP 611, 1976, 119.
44. T. Lauritzen, *Stress-Rupture Behavior of Austenitic Steel Tubing: Influence of Cold Work and Effect of Surface Defects*, USAEC Report GEAP-13897, 1972.
45. C. W. Hunter and G. D. Johnson, "Fuel adjacency effects on fast reactor cladding mechanical properties," *International Conference on Fast Breeder Reactor Fuel Performance*, ANS/AIME, Monterey, CA, March 1979.
46. H. Farrar, C. W. Hunter, G. D. Johnson, and E. P. Lippincott, "Helium profiles across fast reactor fuel pin cladding" *Trans. ANS, 23* (1976).
47. B. Raj, M. Vijayalakshmi, P. V. Sivaprasad, B. K. Panigrahi, and G. Amarendra, "Approaches to development of steel and manufacturing technology for fusion reactors," *Proceedings of the 29th Risø International Symposium on Energy Materials* – "Advances in Characterization, Modelling and Application," Editors Andersen, N. H.; Eldrup, M. Hansen, N. Juul Jensen, D. Nielsen, E. M. Nielsen, S. F. Sørensen, B. F. Pedersen, A. S. Vegge, T. West, S. S. Risø DTU, 2008.
48. S. Ukai, and T. Uwaba, "Swelling rate versus swelling correlation in 20% cold-worked 316 stainless steels," *J. Nucl. Mater., 317* (2003) 93.
49. P. Dubuisson, A. Maillard, C. Delalande, D. Gilbon, and J. L. Seran, "The effect of phosphorus on the radiation-induced microstructure of stabilized austenitic stainless steel," in *Effects of Radiation on Materials, 15th International Symposium*, STP 1125, ASTM, Philadelphia, PA, 1992, 995–1014.
50. F. A. Garner and D.S. Gelles, "Neutron-induced swelling of commercial alloys at very high exposures," *Effects of Radiation on Materials, 14th International Symposium* (Vol II), STP 1046, ASTM, Philadelphia, 1990, 673–683.
51. J. L. Bates and M. K. Korenko, *Updated Design Equations for Swelling of 20% CW AISI 316 Stainless Steel*, HEDL-TME 78-3, Hanford Engineering Development Laboratory, Richland, WA, January 1978.
52. J. L. Straalsund, H. R. Brager, and J. J. Holmes, "Radiation-Induced Voids in Metals," J. W. Corbett and L. C. Ianniello, eds. AEC Symposium Series No. 26, CONF-710601, 1972, 142.
53. J. J. Holmes, "Irradiation damage to cladding and structural materials – I," *Trans. ANS, 12* (1969) 117.
54. Y. Tateishi, "Development of long life FBR fuels with particular emphasis on cladding material improvement and fuel fabrication," *J. Nucl. Sci. Tech., 26* (1989) 132.
55. N. Sekimura, T. Okita, and F. A. Garner, "Influence of carbon addition on neutron-induced void swelling of Fe–15Cr–16Ni–0.25Ti model alloy," *J. Nucl. Mater., 367–370* (2007) 897.
56. C. Wassilew, K. Ehrlich, and J. J. Bergman, "Analysis of the in-reactor creep and rupture life behavior of stabilized austenitic stainless steels and nickel-based alloy hastellay-x," *Effects of Radiation on Materials*, ASTM STP 956, ASTM, Philadelphia, 1987, 30.
57. C. H. Woo, "Irradiation creep due to elastodiffusion," *J. Nucl. Mater., 120* (1984) 55.
58. K. Mansur, "Irradiation creep by climb-enabled glide of dislocations resulting from preferred absorption of point defects," *Phil. Mag. A, 39* (1979) 497.
59. M. B. Toloczko and F. A. Garner, "Relationship between swelling and irradiation creep in cold-worked PCA stainless steel irradiated to ~178 dpa at ~400°C," *J. Nucl. Mater., 212–215* (1994) 509.
60. B. Raj, "Materials science research for sodium cooled fast reactors," *Bull. Mater. Sci., 32* (2009) 271.
61. L. Leibowitz, E. C. Chang, M. G. Chasanov, R. L. Gibby, C. Kim, A. C. Millunzi, and D. Stahl, "Properties for LMFBR Safety Analysis," ANL-CEN-RS-76-1, Argonne National Laboratory, Argonne , IL, March 1976.

62. M. D. Mathew, "Towards developing an improved alloy D9 SS for clad and wrapper tubes of PFBR," *IGC News Letter*, *75* (2008) 4.
63. P. Dubuisson, D. Gilbon, and J. L. Seran, "Microstructural evolution of ferritic-martensitic steels irradiated in the fast breeder reactor Phénix," *J. Nucl. Mater.*, *205* (1993) 178.
64. R. L. Klueh and D. R. Harries, eds., "High Chromium Ferritic and Martensitic Steels for Nuclear Applications," ASTM, Philadelphia, PA, 2001, 90.
65. A. Kohyama, A. Hishinuma, D. S. Gelles, R. L. Klueh, W. Dietz, and K. Elrich, "Low-activation ferritic and martensitic steels for fusion application," *J. Nucl. Mater.*, *233–237* (1996) 138.
66. S. Ukai, S. Mizuta, M. Fujiwara, T. Okuda, and T. Kobayashi, "Development of 9Cr-ODS martensitic steel cladding for fuel pins by means of ferrite to austenite phase transformation," *J. Nucl. Sci. Tech.*, *39* (2002) 778.
67. S. Ohtsuka, S. Ukai, and M. Fujiwara, "Nano-mesoscopic structural control in 9CrODS ferritic/martensitic steels," *J. Nucl. Mater.*, *351* (2006) 241.
68. *Handbook on Lead-bismuth Eutectic Alloy and Lead Properties, Materials Compatibility, Thermal-hydraulics and Technologies*, 2007 Edition, OECD NEA No. 6195, ISBN 978-64-99002-9. Available on line at http://www.nea.fr/html/science/reports/2007/nea6195-handbook.htm.
69. W. Hăfele, D. Faude, E. A. Fischer, and H. J. Laue, "Fast breeder reactors," *Annual Review of Nuclear Science*, Annual Reviews, Inc., Palo Alto, CA, 1970.
70. J. K. Fink and L. Leibowitz, "Thermophysical Properties of Sodium," ANL-CEN-RSD-79-1, Argonne National Laboratory, Argonne, IL, May 1979.
71. V. S. Bhise and C. F. Bonilla, "The experimental pressure and critical point of sodium," *Proceedings of the International Conference on Liquid Metal Technology in Energy Production*, Seven Springs, PA, May 1977. (Also COO-3027-21, NTIS [1976].)
72. S. Das Gupta, "Experimental high temperature coefficients of compressibility and expansivity of liquid sodium and other related properties," D.E.S. Dissertation with C. F. Bonilla, Dept. of Chemical Engineering and Applied Chemistry, Columbia University, Xerox-University Microfilms (1977). Also COO-3027-27, NTIS (1977).
73. G. H. Golden and J. D. Tokar, "Thermophysical Properties of Sodium," ANL-7323, Argonne National Laboratory, Argonne, IL, 1967.
74. O. J. Foust, ed., "Sodium-NaK Engineering Handbook," Vol. 1, Gordon and Breach, New York, NY, 1972, 23.
75. W. K. Anderson and J. S. Theilacker, eds., *Neutron Absorber Materials for Reactor Control*, U.S. Government Printing Office, Washington, DC, 1962.
76. Neutron Absorber Technology Staff, "A Compilation of Boron Carbide Design Support Data for LMFBR Control Elements," HEDL-TME 75-19, Hanford Engineering Development Laboratory, Richland, WA, February 1975.
77. J. A. Basmajian and A. L. Pitner, "A correlation of boron carbide helium release in fast reactors," *Trans. ANS*, *26* (1977) 174.
78. D. E. Mahagin and R. E. Dahl, "Nuclear Applications of Boron and the Borides," HEDL-SA-713, Hanford Engineering Development Laboratory, Richland, WA, April 1974. (See also Ref. [70].)

Part IV
Safety

John Sackett

Safety analysis constitutes an integral part of a fast reactor system, just as it does for any nuclear reactor or other industrial system. The discussion is divided two parts, one in which the plant protection system (PPS) operates as designed (Protected Transients) and one in which it is postulated to fail (Unprotected Transients). The first part of the safety analysis presented here concerns basic safety design philosophy and approaches, addressed in Chapter 13. A more detailed discussion of Protected Accidents is addressed in Chapter 14. The discussion of Unprotected Transients begins in Chapter 15 where opportunities for incorporating self-protecting features are discussed. Hypothetical events leading to energetic core disruption are considered in Chapter 16 with a more review of the steps historically taken to understand the complex interactions involved in these hypothetical events contained in Appendix G. Containment design considerations are also discussed in Chapter 16.

These four chapters are focused primarily on sodium-cooled systems due to the nearly universal acceptance of sodium cooling for present fast spectrum reactors. Special safety considerations for gas-cooled and lead-cooled fast reactors are covered in Part V.

Chapter 12
Reactor Plant Systems

Pavel Tsvetkov, Alan Waltar, and Donald Todd

12.1 Introduction

The principal objective of the sodium-cooled fast reactor (SFR) power plant is to generate electricity.[1] This is accomplished by transferring energy from nuclear fission to a steam system to run a turbine-generator. In this chapter we describe the SFR systems outside the core that are needed to meet this objective. The main emphasis, discussed in Section 12.2, is on the heat transport system, focusing on the design problems unique to SFRs. First, the overall heat transport system is described, including the primary and secondary sodium systems and the various steam cycles in use and proposed. Discussions then follow in Section 12.3 for the main components in the sodium system—the reactor vessel and reactor tank, sodium pumps, intermediate heat exchangers, and steam generators.[2]

The four remaining sections address shielding problems peculiar to the SFR (Section 12.4), refueling considerations (Section 12.5), core and coolant system instrumentation (Section 12.6), and auxiliary systems necessary for sodium-cooled plants (Section 12.7). A few words are in order to place the topics of this chapter in perspective relative to the other subjects covered in this book. Our treatment of plant systems is very brief; we devote only one chapter to design areas that require an extraordinary share of the development effort in order to bring the SFR to commercial status. In the long run, the successful commercialization of the SFR will depend to a large degree on the successful development of components that operate safely and reliably and are built at a reasonable capital cost.

12.2 Heat Transport System

12.2.1 SFR System Layout

The SFR heat transport system consists of a primary sodium system, a secondary (or intermediate) sodium system, and a steam system. The secondary system is unique to sodium cooled reactors. It is employed to prevent the possibility of contact between radioactive sodium in the primary system

[1] As noted in many other parts of this book, another key objective of fast spectrum reactors is to transmute objectionable higher actinides into more benign isotopes to minimize nuclear waste issues. Accordingly, there may be cases where special SFRs would be built exclusively for that purpose, in which case it is possible that the heat generated in the fission process would be dumped without attempts to recover it for the production of electricity.

[2] A more complete discussion of SFR systems can be found in Agrawal and Khatib-Rahbar [1].

P. Tsvetkov (✉)
Zachry Engineering Center, Department of Nuclear Engineering, Texas A&M University, College Station, TX, USA
e-mail: tsvetkov@tamu.edu

A.E. Waltar et al. (eds.), *Fast Spectrum Reactors*, DOI 10.1007/978-1-4419-9572-8_12,

(mainly the 15 h half-life ^{24}Na) and water in the steam generator. The main components in the sodium systems are the reactor vessel or tank, primary pumps, intermediate heat exchangers (IHX), secondary pumps, and steam generators.

12.2.1.1 Pool vs. Loop Systems

Two types of primary systems are being used, as described in Section 2.3.5, the pool and the loop system. In the pool system the entire primary system (i.e., reactor, primary pumps, and IHX's) is located in a large sodium pool in the reactor tank. In the loop system the primary pumps and IHX's are located in cells outside the reactor vessel with interconnecting piping. The heat transport systems for both the loop and pool designs are shown schematically in Fig. 12.1. The system selected for each prototype or demonstration plant design is listed in Table 12.1, as well as the main design parameters for the heat

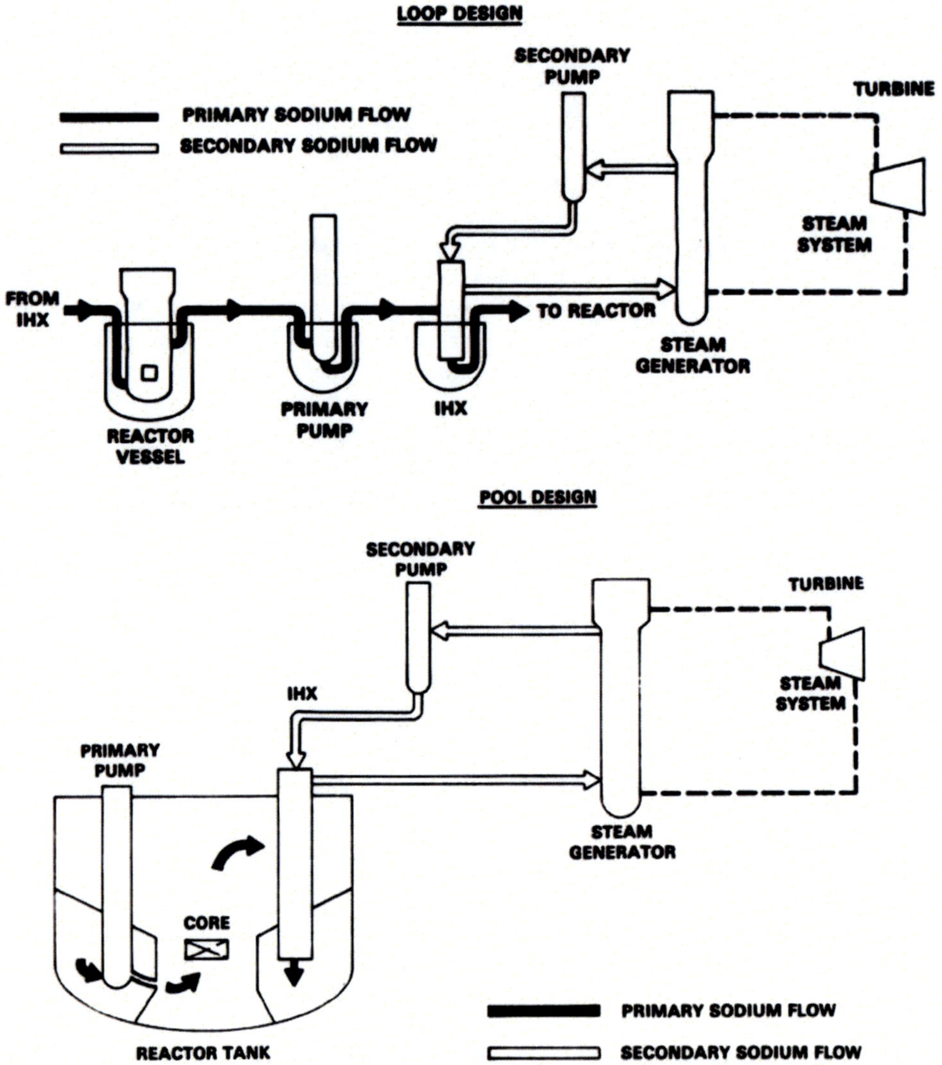

Fig. 12.1 Heat transport systems

Table 12.1 Main plant parameters for a range of fast spectrum reactors

	BN-350 (Kazakhstan)	Phénix (France)	PFR (U.K.)	SNR-300 (Germany)	MONJU (Japan)	CRBRP (U.S.)	BN-600 (Russia)	Super Phénix (France)	CDFR (UK.)	SNR-2 (Germany)	BN-1600 (Russia)	EFR	PFBR (India)	KALI-MER (Korea)	DFBR (Japan)	CEFR (China)
Electrical rating. MW_e	130	255	250	327	280	380	600	1,242	1,500	1,497	1,600	1,580	500	162.2	660	23.4
Thermal power MW_{th}	750	563	650	762	714	975	1,470	2,990	3,800	3,420	4,200	3,600	1,250	392.2	1,600	65
System	Loop	Pool	Pool	Loop	Loop	Loop	Pool	Pool	Pool	Loop	Pool	Pool	Pool	Pool	Loop	Pool
Number of loops	6	3	3	3	3	3	3	4	6	4	4	6	2	2	3	2
Primary pump location	Cold leg	Cold pool	Cold pool	Hot leg	Cold leg	Hot leg	Cold pool	Cold pool	Cold pool	Hot leg	Cold pool	Cold pool	Cold pool	Cold pool	Cold leg	
Number of IHX's per loop	2	2	2	3	1	1	2	2	2	2	1	1	2		1	
IHX temperatures :																
Reactor outlet, °C	430	560	550	546	529	535	550	542	540	540	550	545	544	530	550	516
Reactor inlet, °C	280	395	399	377	397	388	365	395	370	390	395	395	316	316	395	360
Secondary outlet. °C	415	550	540	520	505	502	510	525	510	510	515	525	525	511	520	495
Secondary inlet, °C	260	343	370	335	325	344	315	345	335	340	345	340	355	340	335	310
Steam generator	Integral; bayonet tube evaporator. U-tube superheater	Modular: "S"-shaped	Separate; U-tube	Separate: 2-straight I-helical	Separate; helical coil	Separate: hockey stick	Separate: straight tube	Integral: helical coil	Integral: helical coil	Integral: helical coil or straight tube		Inte-grated	Inte-grated		Inte-grated	
Steam cycle	Recirculating	Once through (Benson)	Recir-culating	Once through (Sulzer)	Once through (Benson)	Recircu-lating	Once through (Benson)	Once through (Benson)	Once through	Once through		Once through	Once through		Once through	
Number of units per loop:																
Integral steam generators	–	–	–	–	–	–	–	1	–	2		1	4		–	
Separate evaporators	2	12	1	3	1	2	1	–		–		–	–		–	

Table 12.1 (continued)

	BN-350 (Kazakhstan)	Phénix (France)	PFR (U.K.)	SNR-300 (Germany)	MONJU (Japan)	CRBRP (U.S.)	BN-600 (Russia)	Super Phénix (France)	CDFR (UK.)	SNR-2 (Germany)	BN-1600 (Russia)	EFR	PFBR (India)	KALI-MER (Korea)	DFBR (Japan)	CEFR (China)
Separate superheaters	1	12	1	3	1	1	1	–		–		–	–		–	
Steam drums	0	1	1	0	0	1	–	–		–		–	–		–	
Moisture separators	–	1	–	1	1	–	1	–	–	–		–	–		–	
Reheaters	0	12	1	0	0	0	1	0		–		–	–			
Turbine																
Inlet pressure, MPa	4.9	16.3	12.8	16.0	12.5	10.0	14.2	18.4	16.0	16.5	14.0	18.5 @SG	16.7	15.0		16.9
Inlet temperature. °C	435	510	513	495	483	482	505	490	490	495	500	490 @SG	490	480		495
Number/Rating, MWe	1	1	1/250	1	1	1/434		2/600	2/660	1/300	2/800		1/500			
Type	K-I00-45	Conden-sing	Tandem com-pound	Conden-sing	Tandem com-pound	Tandem com-pound	K-200-130	Conden-sing	Tandem com-pound	Single shaft			Tandem com-pound con-densing C			

transport systems.[3] The only test reactor to use the pool design is EBR-II; the rest employed the loop system, including FFTF, BOR-60, Rapsodie, JOYO, KNK-II, Fermi, Dounreay, FBTR, and PEC.

The term loop applies to a sequential series of components in the heat transport system between the reactor and the turbine, each operating independently of the other loops in the system. In a SFR loop system, each loop consists of a single primary and secondary pump and one or more IHX's and steam generators. For the pool system, the term loop refers to the IHX's and secondary sodium system; there is no requirement that the number of primary pumps and loops be the same, though they frequently are. Most commercial size demonstration plants have four loops; most of the smaller prototype plants have three loops. In Fig. 12.1, only one loop is illustrated. The four loops in the pool reactor Super Phénix (for which there are also four primary pumps) are shown in the plan view in Fig. 12.2. This figure shows two IHX's in the pool for each loop and each primary pump.

Both pool and loop designs have particular advantages, and there are advocates of both concepts. Several *advantages* of *the pool system* can be identified as follows: Leakage in the primary system components and piping does not result in leakage from the primary system and there may be a lower chance that a primary system pipe could be broken. The mass of sodium in the primary system is of

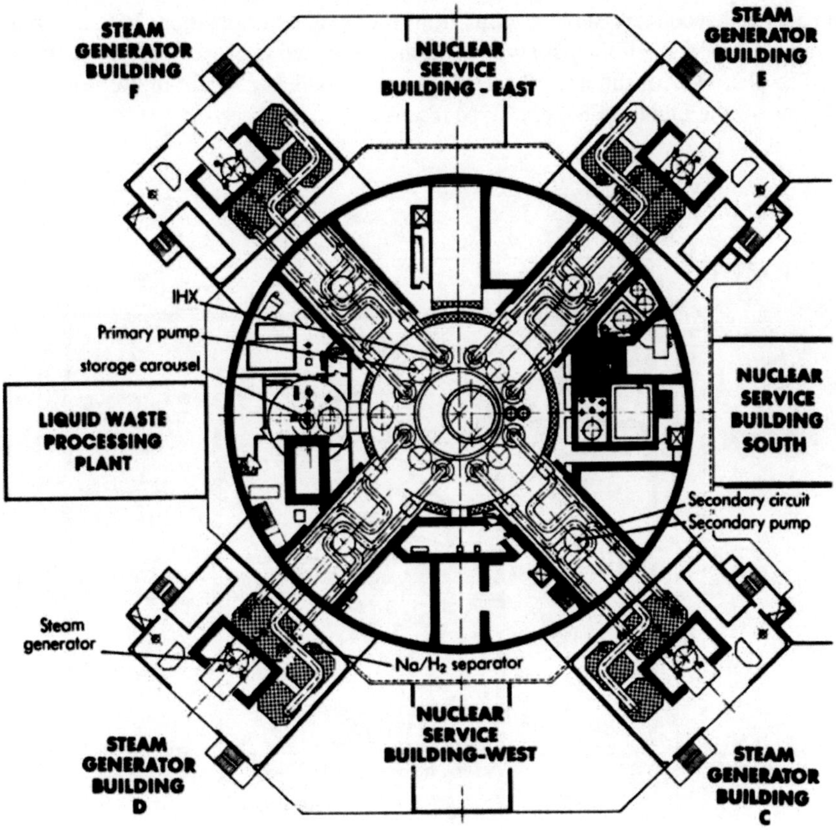

Fig. 12.2 Plan view of Super Phénix pool, showing the 4 loops, 4 primary pumps, and 8 IHX's

[3] Additional details for the plants, including references, are provided in Appendix A.

the order of three times that of a loop system, thus providing three times the heat capacity. This results in a lower temperature rise in off-normal transients or a longer time to reach boiling if heat sinks are isolated. The large thermal inertia of the pool tends to dampen transient thermal effects in other parts of the system. The cover gas system can be simpler since the only free surface needed is the free surface in the tank unless an actively controlled cover gas is used, as in Super Phénix, to reduce reactor vessel wall thermal stresses.

Advantages of the loop system include the following: Maintenance is simpler for the loop system since components can be isolated in cells. This feature also provides for greater flexibility in making system modifications and major maintenance during reactor operation. Less neutron shielding is required to prevent activation of the secondary sodium. The structural design of the vessel head is simpler than the large roof deck of a pool reactor. Greater difference in vertical elevation of the IHX relative to the core enhances natural circulation of a loop relative to that of a pool, and the well defined primary coolant flow path allows a more reliable prediction of natural circulation for a loop-type SFR. Tighter coupling (i.e., quicker response to changes) of the steam and secondary sodium systems to the primary sodium system and reactor in a loop design, due to the smaller mass of sodium in the primary system, influences the control and load following characteristics of the overall heat transport and steam system, but it is unclear whether this involves a net advantage relative to a pool design.

For smaller reactor capacities, there is distinctive advantage in optimizing the number and sizes of the components and secondary sodium loops from economic considerations without compromising on the overall plant safety. For the pool type SFR, the optimization process has led to the adoption of two loops. The plan view of the nuclear island connected building layout of the reactor is shown in Fig. 12.3 below for the Indian PFBR pool-type reactor.

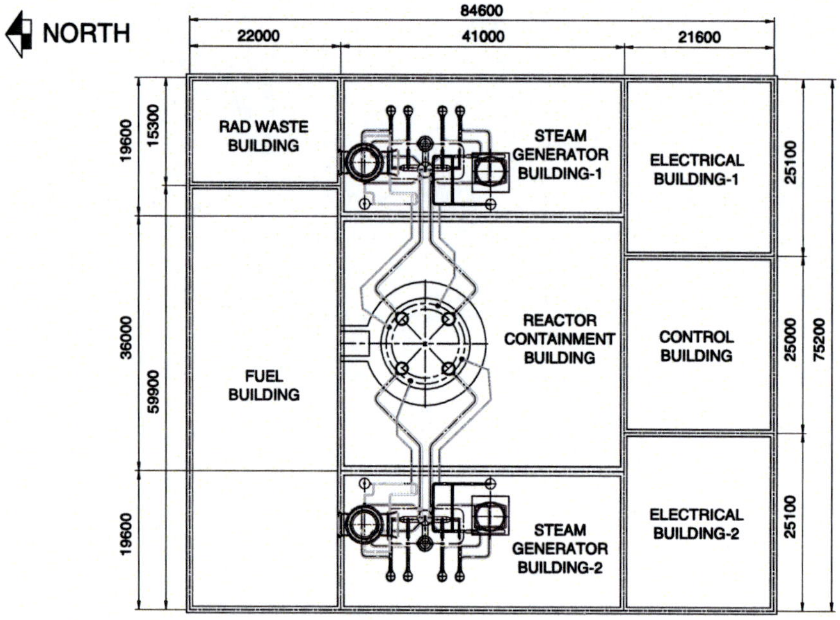

Fig. 12.3 The PFBR nuclear island as an example of a possible pool layout

12.2.1.2 Heat Exchanger Configurations

An important aspect of the heat transport system layout, which is illustrated in Fig. 12.1, is the relative axial elevation of all the components. The main components of both the primary and secondary systems are arranged so that the thermal center of each component is above the thermal center of the previous one in the flow cycle. This profile provides natural circulation of the sodium in order to transfer heat from the core to an ultimate heat sink in the event of a loss of pumping capability. Examples of this layout are shown for the S-PRISM pool-type reactor in Fig. B.2 of Appendix B and for the JSFR loop-type reactor in Fig. C.9 of Appendix C.

Reactor inlet and outlet temperatures of SFRs are generally of the order of 400 and 550°C, respectively, although these values will vary as designs become more or less conservative. These temperatures then correspond, except for small heat losses, to the temperatures across the primary side of the IHX.

Primary and secondary flows in the IHX are generally countercurrent, with log mean temperature differences between the primary and secondary sodium of the order of 30–40°C. With the single exception of PFR, secondary flow is on the tube side in order to facilitate cleanup of sodium water reaction products from the steam generator should a major leak occur there. The pressure on the secondary side is higher than on the primary side to avoid leakage of radioactive sodium to the secondary in case of tube leaks (which is another reason for putting the secondary on the tube side).

Steam generators can be categorized as *integral* or *separate.*[4] In an integral system, evaporation and superheating occur without separation of steam and water between the two processes. For most integral steam generators, both evaporation and superheating take place in an integral unit (i.e., within the same shell), as in Super Phénix. Other integral steam generator systems employ separate components, as in BN-350. In a separate steam generator, evaporation and superheating occur in different units, with steam separation between the two processes. Steam separation usually takes place in a steam drum (or steam separator) or moisture separator, which are separate components between the evaporator and super heater. Steam separation, however, can be incorporated as an integral part of the evaporator unit. In some designs more than one evaporator feeds into a single super heater. Steam-water is always on the tube side with sodium on the shell side. Since steam-water pressure is higher than the sodium pressure, in the event of tube leak, steam or water will flow into the sodium instead of the reverse in order to prevent the contamination of the turbine with sodium oxide.

12.2.2 Steam Cycle

The high temperature of the sodium coolant allows the SFR to take advantage of a superheated steam cycle. This allows thermal efficiencies close to 40%, near those of modern day fossil plants and considerably higher than efficiencies of the order of 32% for light water reactors (LWRs). The use of a saturated steam cycle (as used in LWR's) is also being considered for SFRs, perhaps in order to achieve higher reliability at the expense of thermal efficiency.

Four steam cycles being used and suggested for SFRs are illustrated in Fig. 12.4. All are, of course, modifications of the basic Rankine cycle.[5]

---

[4]The term "modular" has also been used for systems consisting of several separate modules that are interconnected to perform a single function, as in Phénix. The meaning of the terms integral and separate as applied to steam generator systems is not entirely consistent throughout the SFR industry. The definitions described here appear to prevail at the time of publication, but further redefinition may be expected as new variations in SFR plant designs evolve.

[5]As an alternate approach, some designers are proposing to use super-critical CO_2 in a Brayton Cycle to replace the intermediate loop and directly drive the turbines to produce electricity,[2] but this approach is yet to be fully developed.

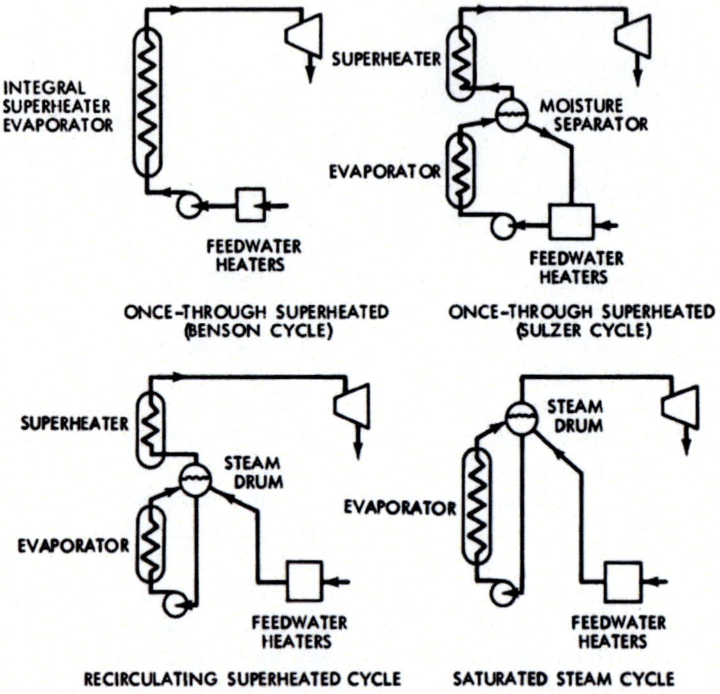

Fig. 12.4 SFR steam cycles

Two classes of superheat cycles are represented—the once-through cycle and the recirculating cycle. The once-through cycle is further subdivided into the Benson and Sulzer cycles. The Benson cycle employs an integral steam generator, whereas the Sulzer cycle uses separate evaporators and super heaters with a moisture separator in between. The exit quality from the evaporator in the Sulzer cycle is high, e.g., 95% in SNR-300 for which this cycle was designed. In contrast, the exit quality from the evaporators in a recirculation super heater cycle is considerably lower, e.g., 50% in the CRBRP design. The moisture separator and the steam drums are shown in Fig. 12.4 as separate units, but they can also be incorporated as an integral part of the evaporator.

12.2.3 Plant Control

Plant control is a complex process that varies with each plant and combination of systems selected. As a consequence, we will introduce only the basic concepts in this overview book. A sense of the factors involved in the control of a SFR heat transport system can be derived from outlining the plant control developed for the CRBRP design, as written by members of the CRBRP staff. This description, together with Fig. 12.5, follows.

Plant control is achieved by a two-level system. It provides automatic and manual control of the reactor, heat transfer systems, turbine and auxiliary systems for normal and off-normal operation. The first level of control is load demand established by either an automatic signal from the grid load dispatching system or a plant-operator-determined set point. The supervisory controller receives the load demand signal and also input of steam temperature, steam pressure and generator output. The controller compares the steam side conditions to the load demand and, in the case of any variation,

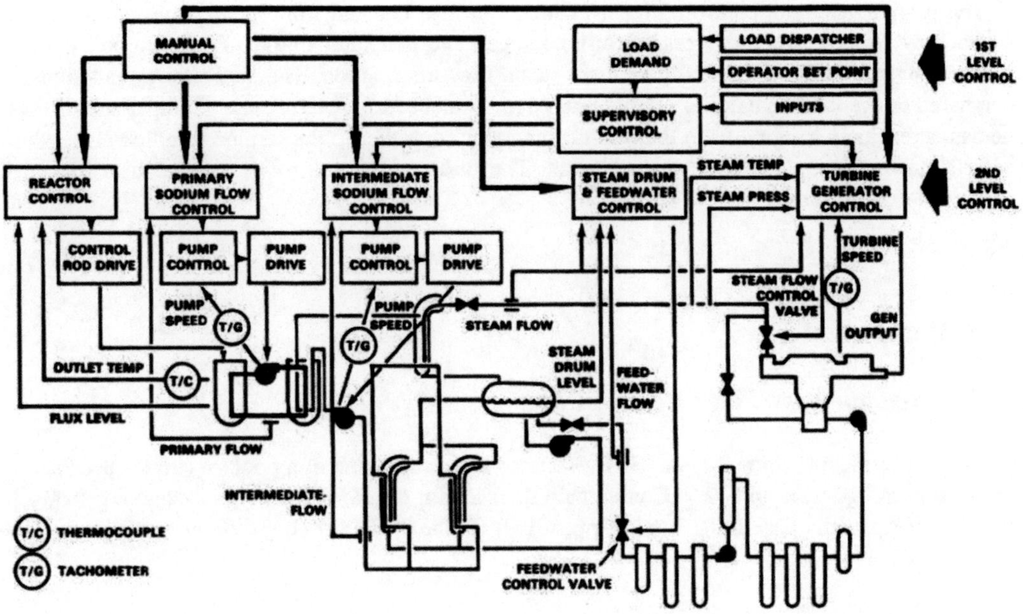

Fig. 12.5 Plant control (CRBRP design)

sends demands to each of the second-level controllers that adjust the reactor power level, system flows and steam supply to meet the new load demand.

Each of the second-level controllers can be operated manually. Manual control is used while the plant is shut down and during startup to bring the plant to 40% power level after which automatic control may be initiated.

The reactor controller receives a demand input of reactor power level from the supervisory controller and input of sodium temperature and flux level in the reactor. The controller compares the flux level with the power demand signal to determine the change required to meet the demand. A trim signal is provided from the steam temperature to the turbine throttle to maintain throttle conditions at a preset temperature value. At the same time, core outlet temperature information is used to modify the controller output so that core outlet temperature is maintained within predetermined limits. The resulting signal is a demand to the control rod drive mechanism controller to adjust rod position to a point where the difference between power output and power demand is zero.

The primary and intermediate sodium flow controllers receive inputs of flow demand from the supervisory controller and sodium flow in the cold leg of each loop. The controller compares the demand with the flow signal to determine any change required. A trim signal is provided to the intermediate pump flow controller from the turbine throttle to maintain throttle pressure at the preset value. The resulting signal is a demand to the pump controllers, which also receive input from tachometers on each of the pumps. Comparison of these inputs results in a demand signal from the pump controllers to the pump drive controls for any changes in pump speed.

The steam drum and feed water controller does not receive input from the supervisory controller because the steam drum level is kept constant for all power levels. The controller receives inputs of flow in the main steam line, steam drum level, and flow in the feed water line. Comparison of these inputs results in a demand signal to the feed water control valves, which adjusts flow to maintain steam drum level.

The turbine generator controller receives inputs of load demand from the supervisory controller, steam flow temperature and pressure, turbine speed and generator output. Flow, temperature and pressure signals are combined to give mass steam flow information. The generator output signal is compared to the load demand to establish any requirement for power change. Comparison of these demands results in a signal from the turbine generator controller to the turbine throttle valve, which adjusts steam flow to meet the load demand. The turbine speed signal is used to trim for small variations in turbine speed.

12.3 Components

12.3.1 Reactor Vessel and Reactor Tank

The reactor vessel of a loop system (SNR-300) and the reactor tank of a pool system (Super Phénix) are shown in Figs. 12.6 and 12.7. Comparable sketches for the JSFR loop-type system are shown in Fig. C.6 of Appendix C and Fig. B.6 in Appendix B for the S-PRISM pool-type system.

12.3.1.1 Reactor Vessel (Loop System)

The reactor vessel in a loop design is a vertical, cylindrical shell with a dome shaped bottom. The vessel is hung at the top from a support ring.

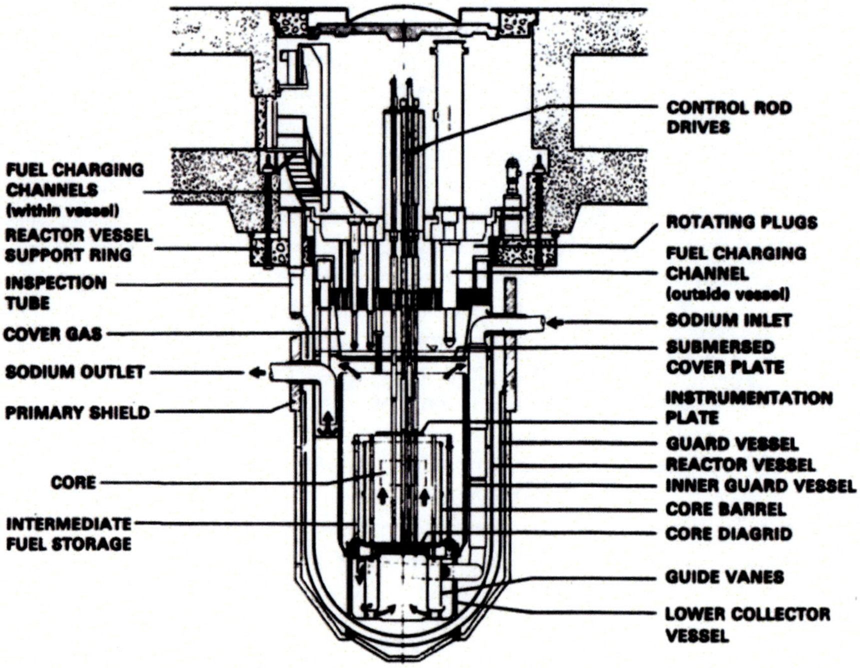

Fig. 12.6 Reactor vessel for the loop system, SNR-300

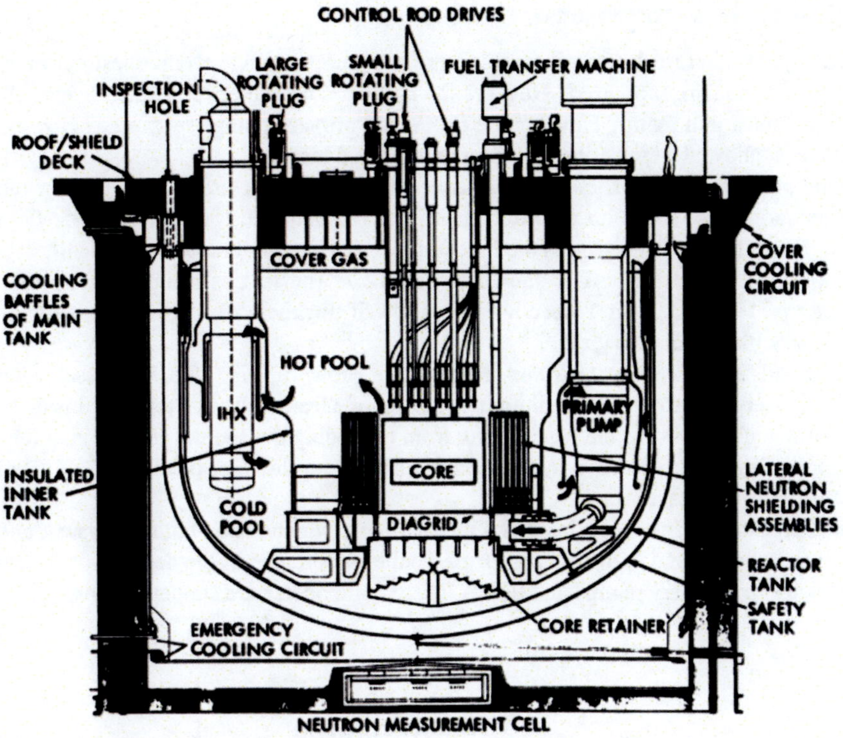

Fig. 12.7 Reactor tank for the pool system, Super Phénix

The fuel assemblies rest on a core support structure (also called a plate, grid, or diagrid). In SNR-300 the core support structure is attached to an inner guard vessel (not present in CRBRP), which is in turn supported at the bottom by the reactor vessel. In the CRBRP design the core support structure is attached to a core support ledge, which is attached to the side of the vessel. Joined to the core support structure (in both SNR-300 and CRBRP) is a core barrel or jacket that separates the sodium flowing through the core, blankets, and radial shielding from the surrounding sodium pool. An inlet flow structure guides the sodium from the inlet plenum to the fuel assemblies; an upper internals structure guides the flow from the assemblies into the outlet plenum. Containers for interim fuel storage during refueling are located outside the core barrel.

A guard vessel is located outside the reactor vessel to protect against any potential loss of sodium from the vessel. The reactor and guard vessel sit in a reactor cavity. Both inlet and outlet sodium pipes enter above the guard vessel so that any pipe rupture within the reactor cavity will not result in sodium loss to the core. In SNR-300 the inlet pipes penetrate the vessel above the outlet pipes, and the inlet pipes run downward between the inner guard vessel and the reactor vessel to the inlet plenum below the core. In the FFTF and the CRBRP design, the inlet pipes run downward between the reactor vessel and the guard vessel and enter the vessel near the bottom. In CRBRP, the inlet and outlet pipes are stainless steel with diameters of 0.6 and 0.9 m, respectively, and wall thicknesses of 13 mm.

Several centimeters of argon gas cover the sodium (hence, the common name, cover gas), separating the sodium pool from the reactor head (also called closure head, or cover). The head provides access for both control rods and refueling ports and is usually fabricated with several rotating plugs for refueling (discussed in Section 12.5).

12.3.1.2 Reactor Tank (Pool System)

In considering a structure as large and complex as a pool reactor tank, several questions quickly come to mind. How is the tank supported? How are the hot and cold sodium separated (especially around the IHX's that must join them)? How are the IHX's and primary pumps supported? How is the large roof/shield deck above the tank constructed, supported and insulated?

Since the answers vary to some degree between every plant, a useful way to gain insight into proposed solutions is to focus on a particular design. We shall illustrate the basic design solutions adopted for Super Phénix, since it was the first commercial-size pool reactor to be built.

The Super Phénix reactor tank is shown in Fig. 12.7. The tank is stainless steel with a height and diameter of 19.5 and 21 m, respectively, and a wall thickness of ~50 mm. It is hung from the roof/shield deck that covers the tank.

Further details of the reactor tank and internals are shown in Fig. 12.8. The reactor tank is surrounded by a safety tank to contain sodium in the event of a leak in the reactor tank. Inside the reactor tank is a baffle tank to prevent the reactor tank from exceeding the sodium reactor inlet temperature. This is accomplished by providing bypass sodium flow from the inlet plenum to the annular space between the baffle and reactor tanks.

Yet another internal tank is the insulated internal tank, which forms both a physical and thermal barrier between the hot and cold sodium. Of particular importance in the design of this tank are the penetrations for the primary pumps and IHX's. These components are supported from the roof/shield

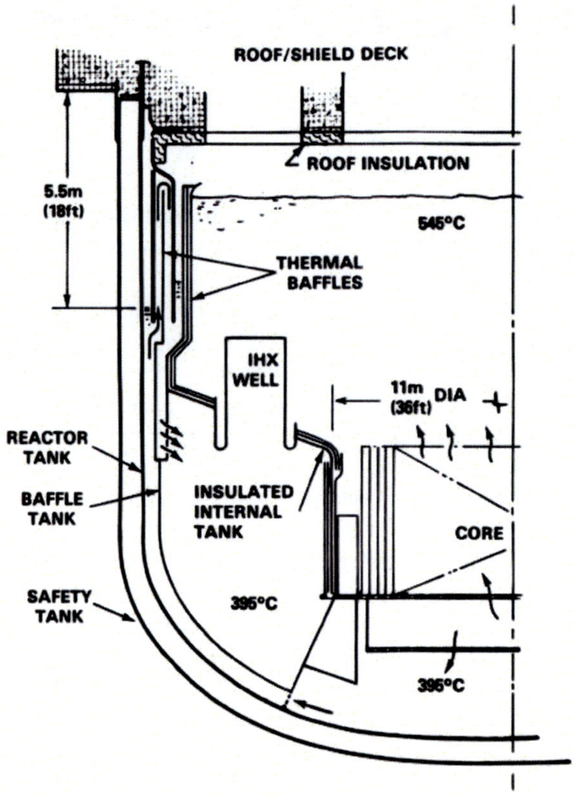

Fig. 12.8 Primary tank and internals (Super Phénix)

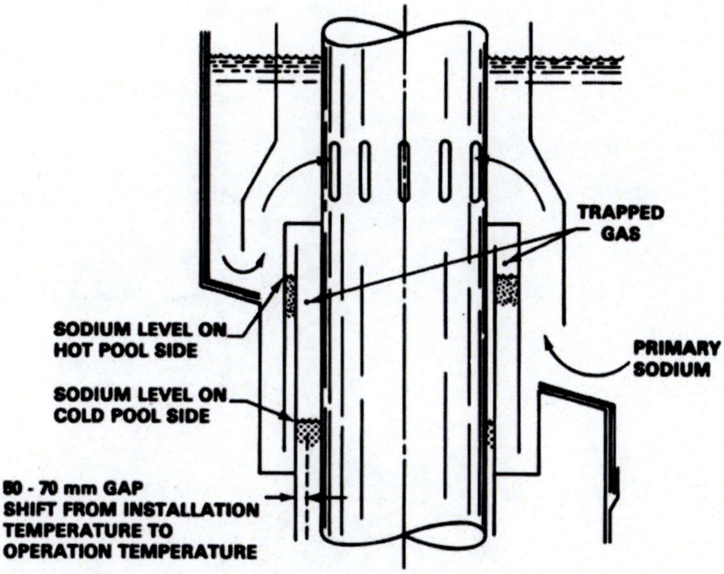

Fig. 12.9 IHX penetration in insulated internal tank (Super Phénix)

deck. Because of the large temperature variations, from room temperature above the deck to the sodium pool temperatures below, the IHX penetrations of the insulated internal tank must accommodate 50–70 mm of axial expansion. A bell-jar type seal is used to accomplish this, as shown in Fig. 12.9. The pump fits loosely into a well which extends from the cold sodium to the roof/shield deck so that no seal is required. An added problem for the pumps, however, involves the pipe connection to the inlet plenum; a flexible joint must be designed to accommodate the roughly 50 mm radial and 100 mm axial thermal expansions due to differences between the deck and pump operating temperatures.

The roof/shield deck represents one of the most difficult design problems for the pool concept. The construction of the roof/shield deck is illustrated schematically in Fig. 12.10; it is composed of a steel web structure filled with concrete. The deck is supported by the concrete vault that surrounds the reactor tank, and the deck in turn supports the tank. The deck must also support the primary pumps, IHX's, control rod drives, and fuel handling equipment, and must in addition be designed to remain leak tight following potential deformation from a hypothetical core disruptive accident. As in the case of a loop system, the deck is separated from the sodium pool by a cover gas. In addition, the deck must be insulated from the high temperature sodium, and even with insulation some cooling of the deck is required. The Super Phénix insulation is stainless steel gauze sandwiched between stainless steel foil (Similar insulation is also required to insulate the concrete walls surrounding the reactor tank.)

12.3.2 Sodium Pumps

Primary and secondary pumps in SFRs generally fall into the category of mechanical, vertical-shaft, single-stage, double suction impeller, free-surface centrifugal pumps. Figure 12.11 shows a diagram of the CRBRP primary pump to illustrate the nature of the pumps in general use. The sodium level is just below the thermal shield. Argon cover gas is over the sodium, and this gas is connected to the reactor vessel cover gas through a pressure equalization line.

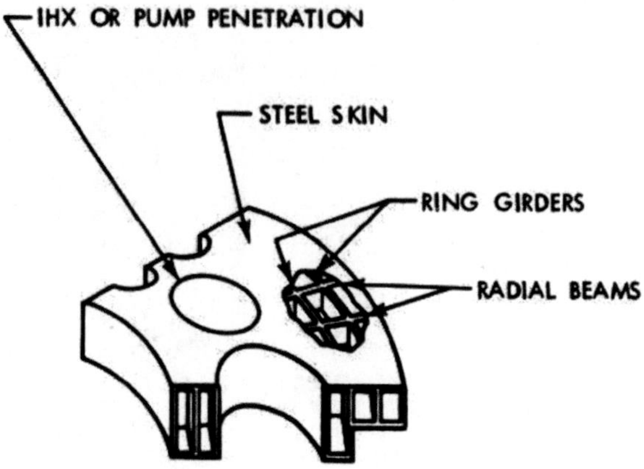

Fig. 12.10 Section showing construction of roof/shield deck (Super Phénix)

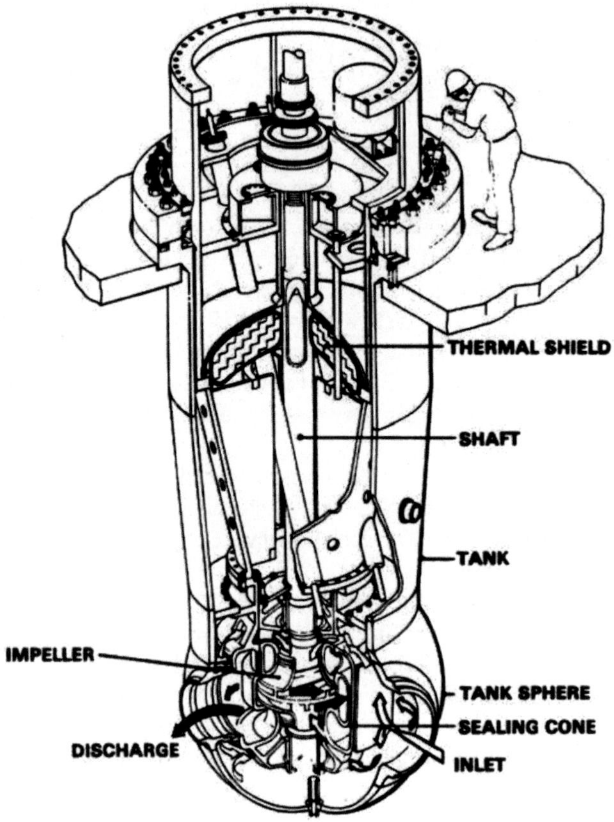

Fig. 12.11 Diagram of primary system sodium pump (CRBRP)

Choices affecting pump design involve differences between primary pumps for the loop versus pool system, differences between primary and secondary pumps (often minor), and pump location. Important design choices (not discussed here) include seals and bearings and impeller and bypass flow arrangements.

Electromagnetic pumps can also be used for SFRs since sodium is an excellent electrical conductor. An electromagnetic pump was used for the secondary sodium in EBR-II, and they were used in SEFOR and in the primary circuit of the Dounreay Fast Reactor. They are not used in the main loops in large SFR power plants, but they are being used in some backup decay heat removal systems, e.g., in SNR-300 and Super Phénix (see Section 14.3.3).

Only mechanical pumps are being used in large SFRs. Some testing was done in the US on the use of a relatively small inducer-type mechanical pumping element in series with a large centrifugal pump to assist in meeting the strict suction requirements of SFR centrifugal pumps (discussed below on net positive suction head).

12.3.2.1 Pump Location

A classic selection problem for the loop design SFR has been the location of the pumps. For the pool design, the primary pump is always located in the cold sodium. For the loop system, however, the location of the primary pump can be either on the hot leg or the cold leg. In all secondary systems, the pump is located in the cold leg both because the advantages of hot-leg location existing for the primary loop are not present in the secondary loop and also because it is important to pressurize the secondary sodium in the IHX in order to force flow from leaking tubes in the direction from the nonradioactive secondary to the radioactive primary.

In the loop design, despite the obvious advantage (e.g. to seals and bearings) of placing the primary pump in the colder sodium environment of the cold leg (as in MONJU, BN-350, and most of the early experimental loop reactors), the hot leg has been selected for several loop designs (SNR-300, SNR-2, FFTF, and CRBRP). The main reason for the hot-leg choice involves suction requirements for primary system pumps. These requirements are stringent in order to accommodate the full range of anticipated transients, a subject treated in Ref. [3]. The basic argument can be explained briefly by examining conditions during normal steady-state operation.

The *net positive suction head* (NPSH) of a pump is the difference between the absolute pressure at the pump suction and the vapor pressure of the fluid pumped. Any pump has, inherent in its design, a minimum required NPSH at any given flow rate to prevent cavitation. The available NPSH must, therefore, always be greater than this minimum value. The available NPSH can be obtained from the relation

$$\text{NPSH} = H_p + H_z + H_l + H_v,$$

where

H_p = pressure at the liquid surface of the source from which the pump takes suction (hence, the cover gas pressure for the present application),
H_z = hydrostatic head of the liquid source above the impeller,
H_l = pressure drop (losses) in the piping and equipment upstream of the pump,
H_v = vapor pressure of the fluid at the suction.

The pressure above the sodium surface in the SFR primary pumps is equalized with the reactor cover gas pressure, H_p, and need be only slightly above atmospheric in a SFR—just enough to prevent in-leakage of air or inert gas. The hydrostatic head, H_z controls the length of the pump shaft. The sodium vapor pressure, H_v is small, on the order of only 1 kPa for the hot leg sodium.

If the pump is on the hot leg, the pressure drop H_l includes only the losses through a short distance of piping plus vessel exit and pump entrance losses. With the pump on the cold leg, however, H_l, must include the pressure drop through the IHX plus additional piping. The IHX pressure drop is generally of the order of 50–100 kPa. The only way, therefore, to obtain a comparable NPSH in the cold leg to that in the hot leg is to increase the length of the pump shaft (hence H_z) or to pressurize the argon cover gas in the reactor (H_p),[6] or a combination. Another option is to develop a more advanced pump that can operate with a lower NPSH. The difficulty of achieving these options provides the incentive for the hot leg location. The problem is further aggravated as SFR plant size increases since the required NPSH for a particular pump increases with increasing rated capacity.

12.3.3 Intermediate Heat Exchanger

Most SFR intermediate heat exchangers are of similar design. Illustrations of an IHX for both a pool design (Super Phénix) and a loop design (CRBRP) are provided in Fig. 12.12. A novel approach in which the primary pump and IHX are combined into a single unit is illustrated for the JSFR in Fig. C.11 of Appendix C.

Except for BN-350,[7] the IHX's in all of the prototype and demonstration plants are vertical, counter flow, shell-and-tube heat exchangers with basically straight tubes. With the exception of PFR, the secondary sodium enters at the top and flows to the bottom through a central down comer; flow is then reversed and returns upward through the tubes. Primary sodium generally flows downward on the

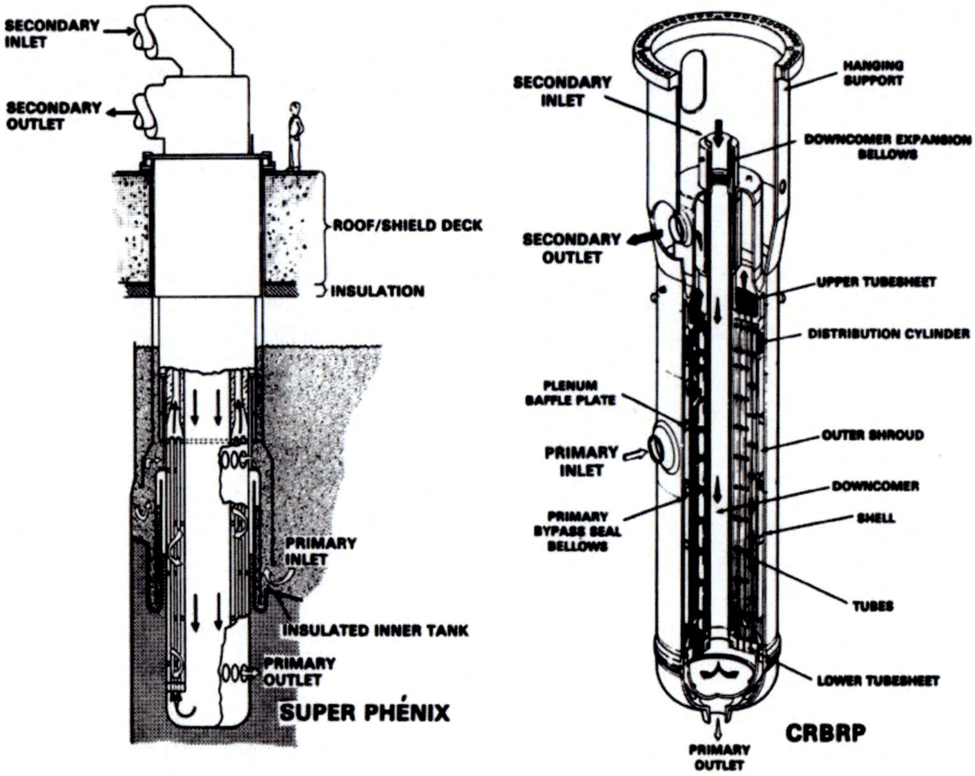

Fig. 12.12 Intermediate heat exchanger designs for a pool system (Super Phénix) and for a loop system (CRBRP)

[6]Pressurizing the cover gas increases the potential for leakage of radioactive gas through the cover seals.
[7]BN350 uses U-tubes.

shell side and exits at the bottom. Principal differences in IHX's between the loop and pool systems appear at the entrance and exit flow nozzles.

The tube bundle must be mounted to allow for differential thermal expansion between the tubes and the shell. To accomplish this, the lower tube sheet is allowed to float and, therefore, is supported by the tube bundle. The tubes are supported by the upper tube sheet. The CRBRP design shows a flexible bellows at the top of the down comer to allow differential expansion between the down comer and the tube bundle. An alternate method, used in FFTF, for accommodating this differential expansion is to design a bend in part of the tube bundle. A sine wave bend is used in PFR and CDFR.

Each of the eight IHX's in Super Phénix includes 5,380 tubes, 14 mm outside diameter, 12 mm inside diameter, and 6.5 m length. A remotely operated machine has been developed to seal off tubes that leak. Both 316 and 304 stainless steels are generally used in the IHX components.

Uniformity of temperature along the axial direction of the tubes is necessary to minimize the thermal stresses in both tube and tube sheet. For this purpose, incorporation of a flow distribution device below the bottom tube-sheet facilitating higher flow through outer row of tubes as compared to the inner rows can be considered. Additionally, a flow mixing device as shown in Fig. 12.13 at the secondary outlet over the top tube sheet will offer uniform mixed mean temperature, thereby overcoming the problems of differential thermal expansion between the hot outer shell and inner parts of IHX. Further, as an additional measure, bellows are also incorporated to accommodate the differential thermal expansion leading to reduced stress levels as shown in Fig. 12.13.

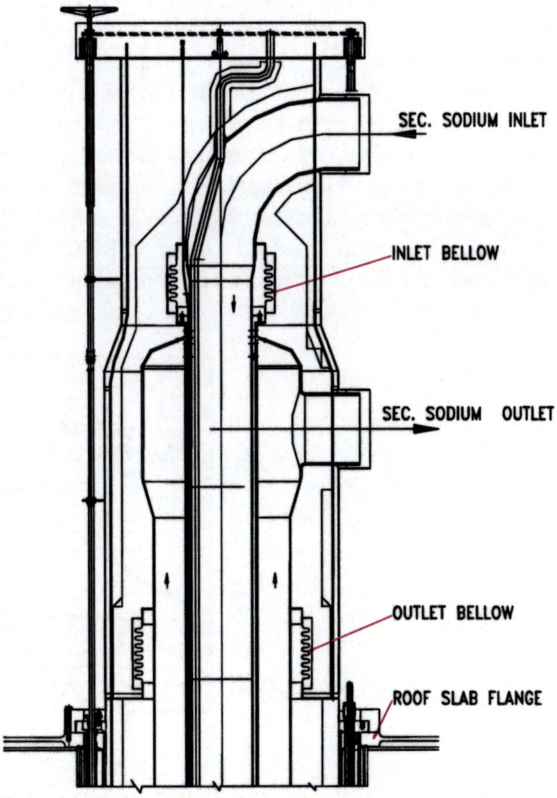

Fig. 12.13 Stress reduction approach

In the pool-system configurations, the interface between the internal tank (inner vessel) and IHX could be sealed through a mechanical arrangement. This would preclude the use of gas for sealing, the accidental leaking of which is considered as one of the sources of reactivity oscillations in the core.

12.3.4 Steam Generators

Steam generators can be integral or separate, as discussed in Section 12.2.1. They can use tubes that are straight, helical, U-tube, or hockey-stick shaped. Either single wall or double wall tubes can be used. Steam generator characteristics for the prototype and demonstration plants were given in Table 12.1.

The choice of integral versus separate involves the steam cycle selection, as discussed in Section 12.2.2. The general shapes of the temperature distributions in an integral unit are shown in Fig. 12.14. After boiling is complete, the steam can be superheated close to the inlet sodium temperature.

In separate steam generators the sub cooled heating and boiling take place in the evaporator, while the superheating occurs in the super heater. In the separate evaporator, complete evaporation does not occur. The CRBRP evaporators, as an example of a recirculating cycle, were designed for 50% exit quality. The exit quality for the SNR-300 evaporators, which operate on the Sulzer once-through cycle, is 95%.

In both cases the steam is separated from the liquid before entering the super heater. As can be observed from Table 12.1, both integral and separate systems have been widely used.

Several designs for steam generators have evolved—all having particular advantages and disadvantages. A fundamental consideration in each design is the method for accommodating thermal expansion. Figure 12.15 contains examples of the basic configurations capable of providing such accommodation: helical coil, U-tube, and hockey sticks. The straight tube configuration, also shown in Fig. 12.15, requires special provisions to accommodate thermal expansion-similar to those described for the IHX. Helical coils constituted the design for Super Phénix (France), CDFR (U.K.), SNR-2 (Germany), and MONJU (Japan). U-tubes were used in PFR (UK) and in the BN-350 (Kazakhstan) super heaters. Straight tubes are employed in BN-600 (Russia) as well as in two loops of the SNR-300 design (Germany). A third loop of SNR-300 design incorporated a helical coil steam generator. Either helical coils or straight tubes were planned for SNR-2. In the United States, the hockey stick

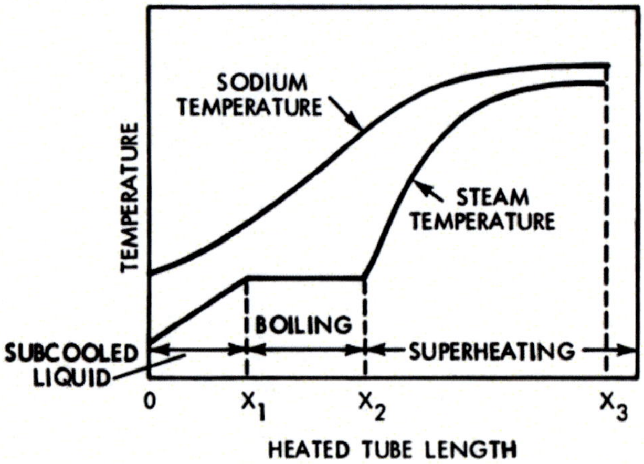

Fig. 12.14 Temperature distributions in an integral super heater (based on Figures 7 and 8, Ref. [4])

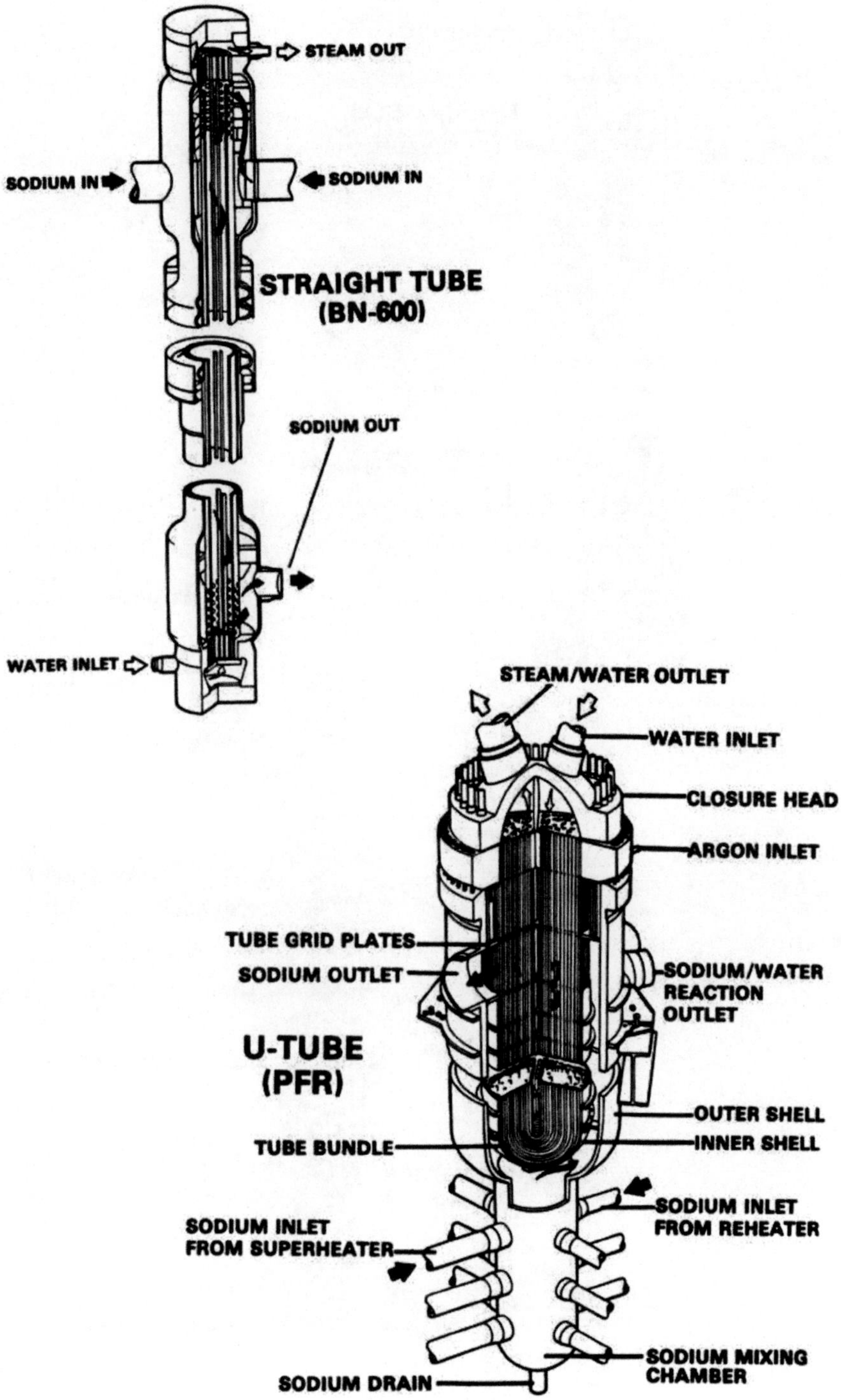

Fig. 12.15 Steam generator designs, showing different tube configurations

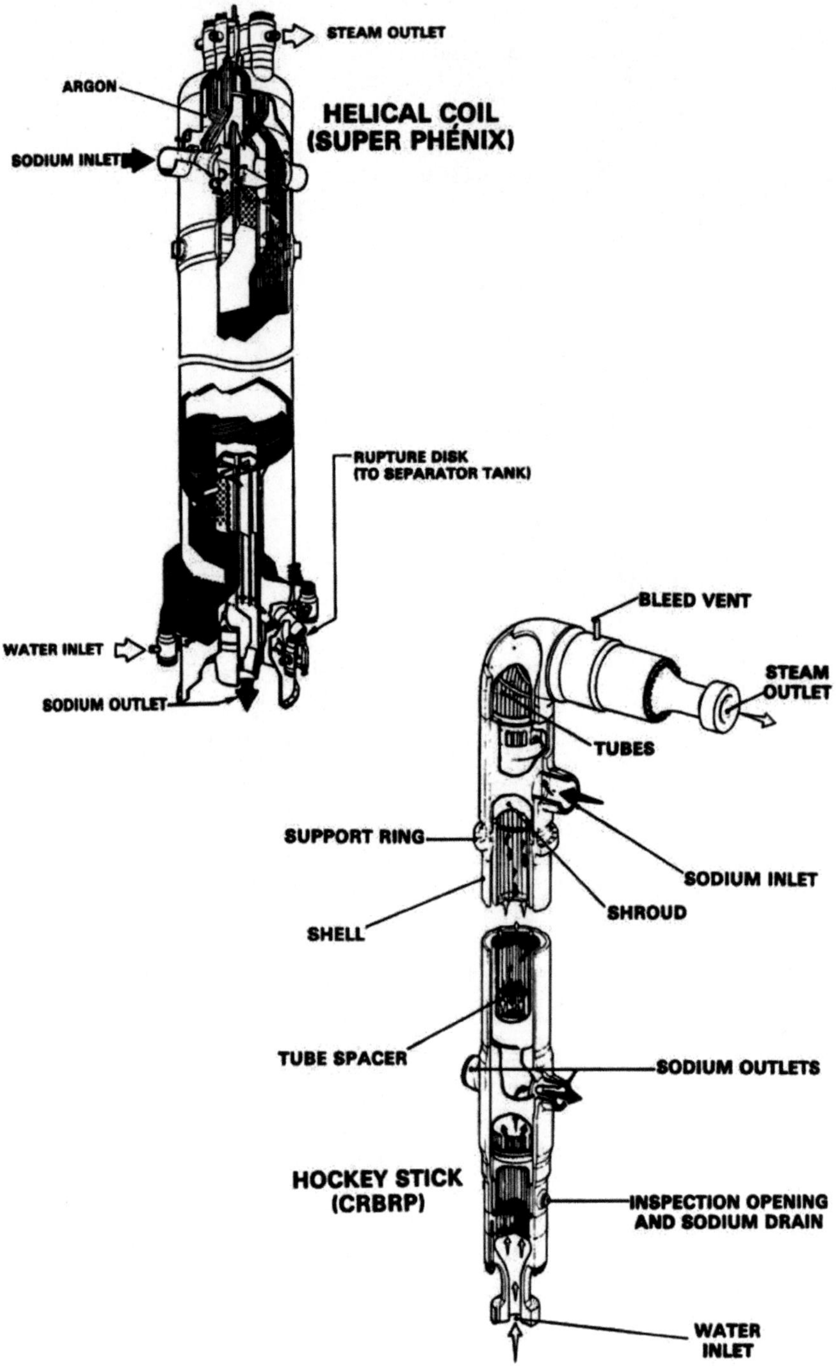

Fig. 12.15 (continued)

was selected and tested for the CRBRP design, and development of both helical coil and straight tube designs continue. Another type of tube was used in the BN-350 evaporators called bayonet tubes. These tubes have a central tube surrounded by an annulus. Water flows down the central tube and a water-steam mixture flows back up the annulus.

Tube integrity is far more important for the SFR steam generator than for the LWR due to the potential chemical reaction between sodium and water. Consideration has been given to the use of double wall tubes in which leaks could be detected from gas conditions between the tube walls.

Double wall tubes are used, for example, in EBR-II and Dounreay fast reactor steam generators. When experiments demonstrated that sodium-water reactions could be adequately contained, however, the emphasis shifted to the simpler single wall tube design, and all prototype and demonstration plants now use single wall tubes. Some interest in double wall tubes is being revived with the argument that they might lead to greater steam generator reliability, which might be especially important for the early generation plants. While experience with SFR steam generators has been reasonably encouraging, BN-350, PFR, and the Fermi Reactor all experienced difficulties with steam generator leaks.

Pressure relief systems sealed by rupture discs are incorporated in the steam generator design to accommodate pressures resulting from potential sodium-water reactions. A sodium-water reaction outlet to a separator tank is shown for the Super Phénix and PFR steam generators in Fig. 12.15; relief lines exist on the other designs but are not shown.

Most steam generators, both tubes and shell, are made of ferritic steel containing 2+% chromium and 1% molybdenum, a material selected to minimize chloride stress corrosion. In some cases this is stabilized with 1% niobium to reduce carbon loss to the sodium. Exceptions to this are the selection of Incoloy® 800 for Super Phénix and the use of austenitic stainless steel for the original PFR super heaters and reheaters.

An important condition in steam generator design is the transition between nucleate boiling and film boiling, or the departure from nucleate boiling (DNB). At this point the temperature of the tube wall rises sharply and an instability in tube wall temperature occurs in the transition zone. This temperature behavior is illustrated schematically in Fig. 12.16. The transition occurs between x_1 and x_2; in this region the tube wall is intermittently in contact with water or steam, and the wall temperature oscillates rapidly. Fluctuations of this kind, if too large a magnitude, could cause thermal fatigue of the tube or structural changes that might enhance waterside corrosion. These and other considerations regarding temperature distributions, dry out, and heat transfer correlations are discussed in Ref. [4].

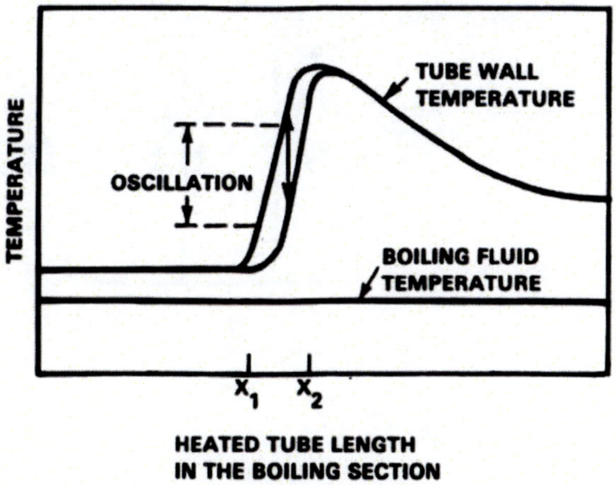

Fig. 12.16 Instability and rise in wall temperature at the transition from nucleate to film boiling (based on Figures 7–15, Ref. [4])

To absorb differential thermal expansion of tubes in designs with straight tubes, an expansion bend is provided as in PFBR. The expansion bends are located preferably in the single phase flow regime, at the bottom, in order to permit higher allowable stress and avoid creep damage.

Since integrity of steam generator tubes during life of the plant is very important, incorporation of in-service inspection provision for tubes ensures monitoring of the steam generator through assessing degradation in tubes. Further, the operating experience of worldwide SFR has highlighted the junction between tube and tube sheet as the most critical zone. A raised spigot butt state-of-art weld (Fig. 12.17) as in the PFBR steam generator offers a potential solution for this issue.

Modified 9Cr-1Mo (Gr.91) is the preferred material of construction for steam generators, mainly due to its good high temperature mechanical properties, resistance to decarburization and chloride stress corrosion cracking, and resistant to the environment resulting from the sodium water reaction.

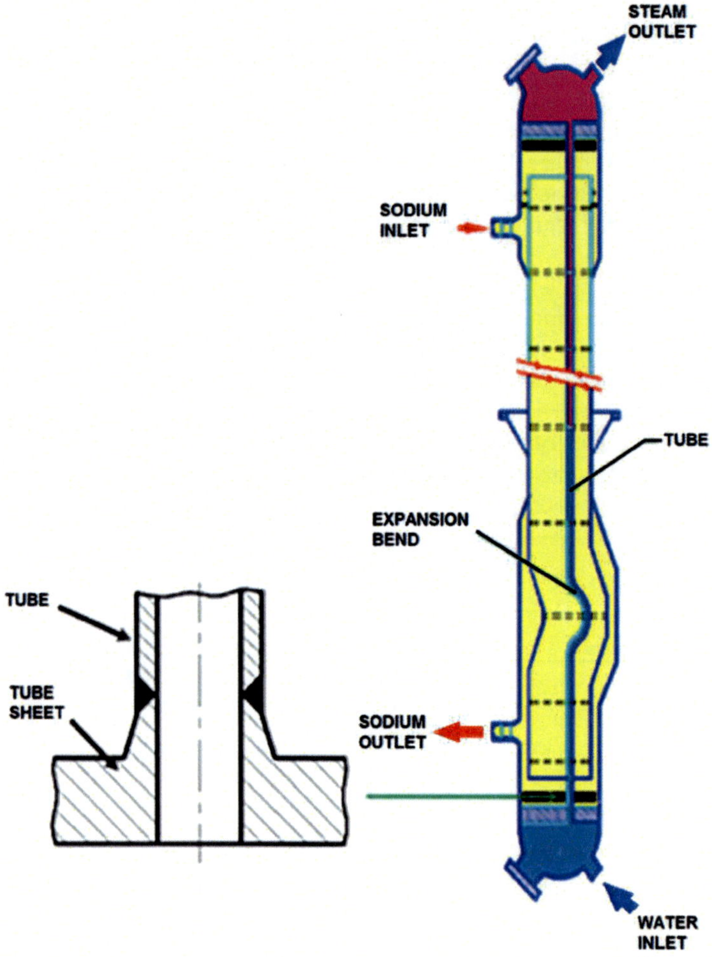

Fig. 12.17 Vertical steam generator

12.4 Shielding

Greater attention must be given to shielding design and analysis for fast reactors than for thermal reactors. The high energy neutron flux in a SFR is considerably higher than in an LWR, and high energy neutron leakage from the core is greater. Although the neutron production rates for a SFR and LWR of the same power level are comparable, the power density (kW/liter) is higher for the SFR. Also, the neutrons in an LWR are slowed down to thermal energies close to the fission source so that the high energy leakage source for irradiation of surrounding structures is relatively low for an LWR compared to that for an SFR.

SFR shielding can be discussed according to areas that require extensive shielding design and analysis. For both pool and loop designs, these areas include in-vessel radial shielding, the closure head assembly and its penetrations, and neutron flux monitors. For pool designs, special attention is required for the intermediate heat exchangers where secondary sodium can be activated by neutrons. For loop designs, other key areas include the reactor vessel support area and the primary heat transport system pipe ways.

Shielding in these areas will be discussed briefly here. Other areas that require shielding, but which will not be discussed, include auxiliary piping penetrations, heating and venting system penetrations, shielding for cover gas and coolant purification systems, fuel handling equipment shielding, and biological shielding for areas that require personnel access during normal, full power operation.

Information on FFTF and CRBRP shielding is reported in Refs. [5, 6].

12.4.1 Coolant and Cover Gas Radioactivity

Before describing the shielding systems in a SFR, radioactivity of the coolant and cover gas will be discussed. The use of sodium as the coolant in the SFR introduces shielding problems different from those in the LWR as a result of neutron activation of the sodium. Sodium in nature is composed entirely of sodium-23. The (n, γ) reaction in sodium produces radioactive ^{24}Na, which has a 15.0-h half life and emits both a 1.4 and 2.8-MeV gamma with decay. An $(n, 2n)$ threshold reaction also occurs, producing ^{22}Na, which has a 2.6-year half life and emits a 1.3 MeV gamma. During operation ^{24}Na is the dominant activation product, and shielding against gammas from this source in the primary sodium is one of the important shielding problems in SFR design. The ^{22}Na activity becomes the dominant activity in the sodium ~10 days after shutdown. For maintenance of primary pumps and IHX's, however, radioactivity from corrosion products (discussed in Section 11.4.5) becomes the most important radioactive source. The calculated ^{24}Na specific activity in the CRBRP primary sodium is 30 Ci/kg (based on a primary system sodium inventory of 6.4×10^5 kg). For FFTF, the calculated ^{24}Na activity was 11 Ci/kg. For an early (1968) General Electric pool design [7], the calculated ^{24}Na activity was 18 Ci/kg (based on a primary system sodium inventory of 1.3×10^6 kg). The calculated CRBRP ^{22}Na specific activity after 30 years of operation is 3.5 mCi/kg. The corresponding FFTF value for ^{22}Na is 1 mCi/kg.

Another radiation source in a SFR is the reactor cover gas. Activation of impurities in the sodium and direct activation of ^{40}A and ^{41}A contribute to activity in the cover gas.[8] Even ^{23}Ne, appears from an (n, p) reaction with ^{23}Na, but its half life is short (38 s).

The main design requirement, however, is to permit reactor operation with leakage occurring in a specified fraction of the fuel pins. For FFTF, this fraction of defective pins was set at 1%. Calculated activity due to fission gas in the reactor cover gas for these design-basis conditions is given in Table 12.2. Since failed fuel never approached 1%, actual activities were far below these levels.

---

[8] Activity in FFTF was in the range of 0.15–0.4 Ci/m^3; not a major problem.

Table 12.2 FFTF design basis cover gas activity

Isotope	Activity (Ci/m^3)
Xe-131m	5.43×10^{-1}
Xe-133m	1.47×10^{1}
Xe-133	2.67×10^{2}
Xe-135	1.26×10^{3}
Kr-83m	6.82×10^{1}
Kr-85m	1.34×10^{2}
Kr-85	9.30×10^{-9}
Kr-87	1.80×10^{2}
Kr-88	2.64×10^{2}

Note: The "m" designates the metastable state

12.4.2 In-vessel Shielding

In-vessel radial shielding is required (1) to prevent excessive radiation damage to structural materials that must remain in the reactor for the lifetime of the plant, and (2) to protect the vessel itself. Examples of in-vessel structures are the core barrel and the core restraint systems. Axial shielding below the core is needed to protect the core support structure. The shielding design must assure that permanent components have an end-of-life ductility consistent with a threshold criterion for brittle fracture. In both the FFTF and the CRBRP design, the threshold of ductility was chosen to be 10% total elongation, a level that assures a ductile mode of deformation up to failure and permits conventional structural analysis methods and criteria to be used in design.

The radial blanket serves as the first shield between the core and the radial structures.[9] Beyond the blanket are located removable radial shielding (RRS) assemblies. The arrangement of these assemblies for an early CRBRP homogeneous core design is shown in Fig. 12.18.[10] A corresponding elevation view was shown in Fig. 8.21. The RRS assemblies contain rods of stainless steel or of nickel based alloys compatible with sodium, such as Inconel®, as illustrated in Fig. 8.21. Nickel has a high inelastic scattering cross section and is particularly effective in degrading the energy of fast neutrons. Iron is also effective, though less effective than nickel. However, stainless steel is less expensive than Inconel®. The material choice for CRBRP was narrowed to 316 stainless steel and Inconel 600®. The shielding rods in the CRBRP design extend from the bottom of the lower axial blanket to the approximate top of the upper axial blanket. The shielding is in the form of rod bundles in order to allow cooling of the shield by sodium. The main heat sources in the shield are gammas from the core and blanket and gammas generated in the shield itself from both inelastic scattering and neutron capture reactions. In FFTF, the radial reflectors are made up of hexagonal blocks of Inconel® that are bolted together. Holes through the blocks provide cooling passages.

In the CRBRP design, the removable radial shield assemblies are surrounded by a fixed radial shield. The fixed radial shield is an annulus of 316 stainless steel 0.146 m thick, which can experience relatively high fluence since it is not a load bearing component. The FFTF fixed radial shield is formed from flat plates to create a 12-sided shield between the core and the core barrel.

[9] FFTF has no blanket but uses removable radial reflectors to protect the fixed radial shield and the radial support structure and core barrel.

[10] Core layouts for the S-PRISM pool-type design are illustrated in Fig. B.7 of Appendix B and for the JSFR loop-type design in Fig. C.4 of Appendix C.

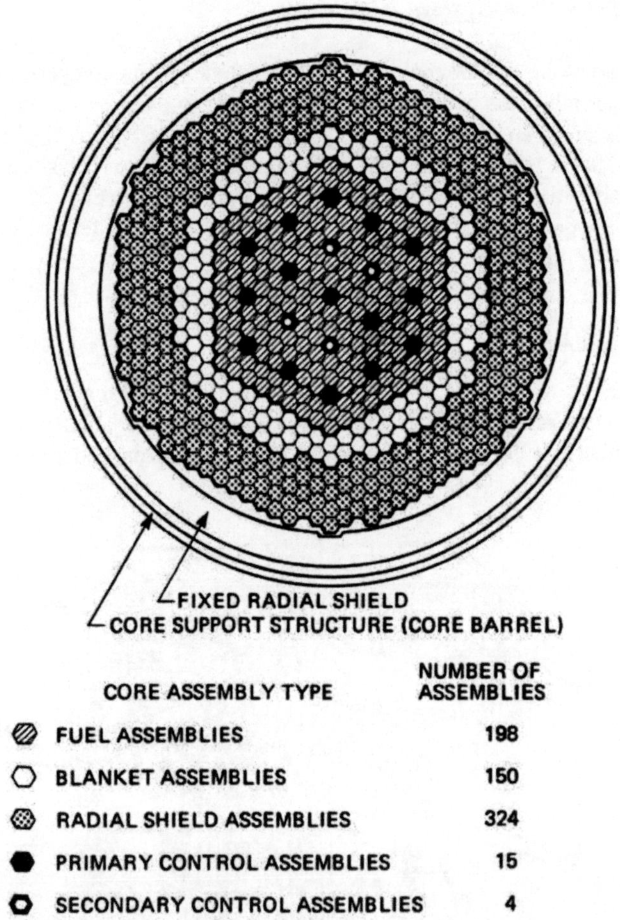

Fig. 12.18 Plan view of CRBRP reactor core layout (early homogeneous core design)

In the pool design, graphite is usually incorporated into the radial shielding to moderate neutrons and allow them to be absorbed before reaching the IHX. Local B_4C shielding is used near each IHX and primary pump in order to reduce activation of the structural materials, thereby allowing maintenance as well as reducing secondary sodium activation by thermal neutrons in the IHX's.

An axial shield is located below the lower axial blanket to protect the core support structure and lower inlet modules. In CRBRP, this shielding consists of a 0.51 m long 316 stainless steel shield block in each fuel, blanket, and control assembly. In FFTF, the lower shield and inlet orifice assembly are 0.54 m long. No special upper axial shielding is required for either FFTF or CRBRP because of the shielding provided by the upper sodium pool.

The most challenging problem to be resolved in the CRBRP in-vessel shielding analysis was the prediction of neutron streaming. Pathways for streaming, which influence the analysis, include the clearance gaps required in the design of the fixed radial shield, the cooling channels of the axial shielding provided in each core assembly, the fission gas plenum of each core assembly, and interfaces between in-vessel components.

12.4.3 Reactor Enclosure System Shielding

Key areas for shielding of the reactor enclosure system include component penetrations and interfaces in the closure head assembly, the reactor vessel support area, and the ex-vessel flux monitors (in the reactor cavity). These areas are illustrated in Fig. 12.19 for CRBRP, typical of a loop design. For a pool design, neutron flux levels at the tank support area and reactor cavity walls are lower than for the loop design. Hence, the shielding problems in these areas are less severe.

We describe here the design solutions in these areas for the CRBRP design as being representative of the types of shielding problems encountered.

12.4.3.1 Closure Head Assembly Shielding

The CRBRP closure head assembly (CHA) is shown in Fig. 12.20. The main penetrations involve refueling components, control rod drive mechanisms, and upper internals jacking mechanisms. Radiation source terms that affect the CHA shielding design include neutron and gamma streaming up the

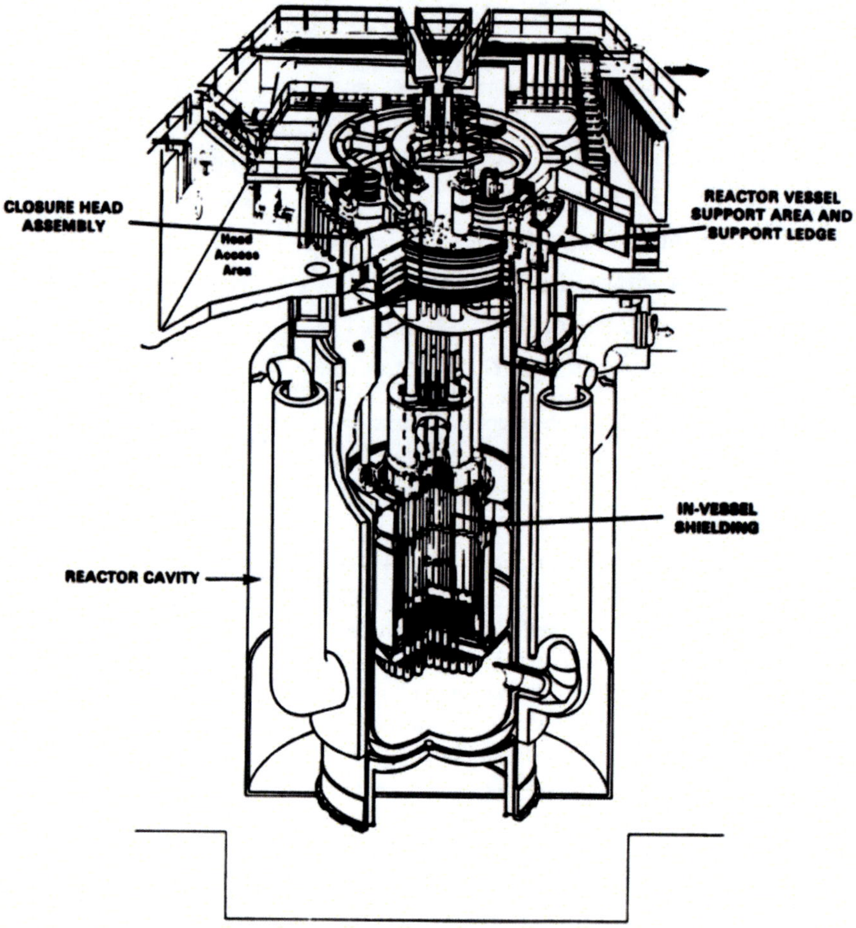

Fig. 12.19 Key shielding areas in CRBRP reactor and reactor enclosure system

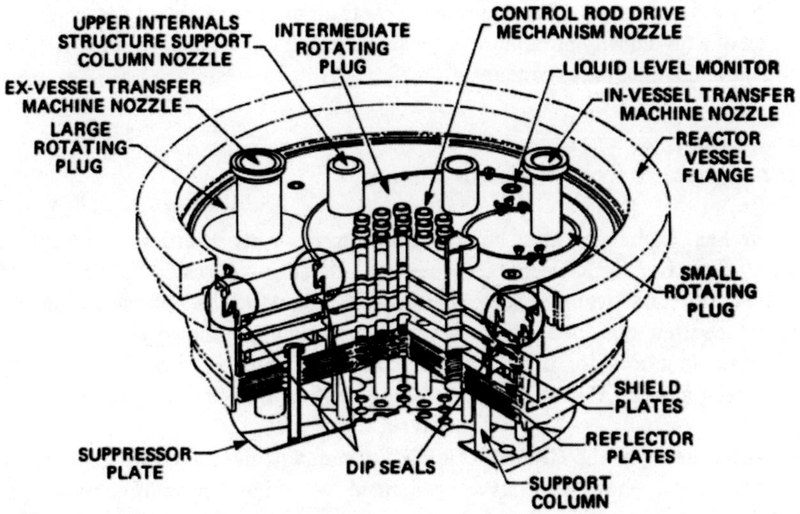

Fig. 12.20 Closure head assembly configuration (CRBRP)

stepped annuli of the CHA penetrations or component interfaces, the radioactive cover gas below and in the CHA penetrations, and neutron and gamma penetration through the CHA bulk shielding.

The sodium filled dip seals around the refueling plugs shown in Fig. 12.20 present the major shield design problem for the closure head assembly. These seals form the barrier for the cover gas in the CHA rotating plug annuli. CRBRP was designed to operate with 1% failed fuel. Radioactive fission product gases in the cover gas from this failed fuel require ~0.3 m of steel shielding for personnel access to the head access area. Dip-seal tradeoff studies resulted in the location of the seals in the closure head as shown in Fig. 12.20. Radiation through these dip seals is the largest contributor to the dose rate in the head access area.

12.4.3.2 Reactor Vessel Support Area Shielding

Shielding the reactor vessel support area created a significant design problem for both the FFTF and CRBRP. The radiation source is predominantly due to streaming in the reactor cavity. In both the FFTF and the CRBRP design, a canned B_4C radiological shield is placed at the lower elevation of the support ledge to stop thermal neutrons, and a carbon steel collar reduces the streaming gap at the vessel flange elevation. A concrete shield ring above the support ledge reduces radiation streaming into the head access area.

12.4.3.3 Ex-Vessel Source Range Flux Monitor

The Source Range Flux Monitor (SRFM), which monitors the core during shutdown and refueling, is located in the reactor cavity for CRBRP. The neutron flux from the core that reaches the flux monitors must be great enough to allow monitoring of changes in core sub criticality, while neutron fluxes from extraneous sources and gamma dose rates must be sufficiently low. These objectives are accomplished by shielding at the SRFM, which consists of a graphite moderator block of 0.51 m by 0.63 m surrounded by lead and B_4C background shields. Neutron background, due to fuel-in-transfer or fuel storage in the fuel transfer and storage assembly, is reduced by means of B_4C shields in the reactor cavity. The gamma background at the SRFM is reduced to acceptable levels by surrounding the moderator block with lead to reduce gamma levels from the vessel, guard vessel, and sodium, and by the utilization of a high purity aluminum alloy as the structural material for the SRFM to minimize its neutron activation gamma background.

12.4.4 Heat Transport System (Loop Design)

The intermediate heat exchanger must be shielded from neutrons in order to prevent activation of the secondary sodium. In CRBRP, secondary sodium activation was to be held below 0.07 μCi/kg. Ordinary concrete structural walls of the equipment cells provide the bulk shielding. Considerable design effort was required to reduce neutron streaming through the pipe ways into the cells.

Delayed neutron monitors for detecting fuel cladding failure are also located in the CRBRP heat transport piping system. Hence, the background neutron flux levels at these monitors must be minimized.

A problem encountered in the CRBRP shielding design was the photo neutron production in the concrete cell walls. Those photo neutrons are generated by the interaction of gamma rays from ^{24}Na in the primary coolant pipes with deuterium in the concrete. Since more than 80% of the neutron background at the delayed neutron monitors was attributed to photo neutrons from the concrete, non-hydrogenous materials were specified for neutron background shielding around the monitors.

12.4.5 Shielding Methods

Neutron flux distributions and radiation sources near the reactor core can be obtained from one, two, and three-dimensional diffusion theory calculations. For neutron and gamma dose rates at large distances from the core and external to the vessel, however, transport theory solutions are required. Special complex techniques, combined with transport theory, are required for streaming calculations; and it is often required that streaming calculations be verified by experiment when design margins are low.

For the CRBRP and FFTF shielding analysis, 40–60 neutron groups were used in discrete-ordinate transport theory calculations. Forward biased quadrature sets containing 100–166 angles were used to solve streaming problems, and Monte Carlo transport methods using albedo scatter data, together with experiments, were developed to verify the design calculations.

Fast reactor shield design involves computation of neutron transport through shield materials with attenuations of the order 10^8–10^{10}. Shielding designs for these deep penetration problems is difficult and is associated with large uncertainties. In line with the international practice to obtain the bias factors for shield designs, mockup experiments have been carried out in the shielding corner cavity of a Indian Experimental water reactor called APSARA. The neutron flux was enhanced to 1.03×10^{10} n/cm^2·s by displacing most of the water between the core edge and stainless steel liner of APSARA pool on the shielding corner-side with an air-filled aluminum box. Converter assemblies made of depleted uranium were used. The following neutron experiments were conducted in phases with various shielding models simulating radial and axial shields[8]:

1. Steel—sodium models, steel-borated graphite-sodium models and steel-boron carbide-sodium models for simulating similar shielding in the radial direction;
2. Steel-graphite-boron carbide-sodium models to study the effect of replacing boron carbide by graphite in radial shields;

3. Steel-sodium models simulating upper axial shields to obtain bias factors for flux at detector locations;
4. Radiation streaming through the gas plenum model simulating lower axial gas plenum to obtain bias factors for the estimation of dpa at the grid plate; and
5. Scaled down models of top shield and transfer arm to validate the radiation streaming calculations.

In the case of transport through steel-sodium shields, calculations predict fluxes within a factor of two. For shield models with borated graphite and boron carbide, the calculations generally under predict the neutron fluxes by factors in the range 3–5. The measurements for the plenum model show that the fast neutron fluence (above 0.1 MeV) and, hence, dpa are under predicted by a factor of 1.5–3.5. Radiation streaming through top shield models is predicted within a factor of 2 and, hence, a bias factor of two in gamma shielding calculations is used. However, the fast neutron flux streaming is underestimated by a factor 2. Similarly, in all the configurations of transfer arm model, fast neutron flux is under predicted by a factor 2.

For the shielding applications, the 121 (100 neutron, 21 gamma) group coupled cross-section set based on ENDF-IV, called DLC-37 of US origin, had been in use. In view of non availability of cross sections of many important nuclides in DLC-37, a new 121 group set, viz. IGC-S2, in the same format as the DLC-37, was created[9] from ENDF/B-VI with anisotropy considered up to P5. In the long run, in order to reduce the uncertainties to less than a factor of two, a new 175 neutron +42 gamma group cross-section library has been derived from ENDF/B-VI. Use of this library with self shielded cross-sections and larger number of thermal groups led to better predictions[10]. This 175 group library is used for in-vessel shield design and local shielding of the intermediate heat exchanger.

The top shield/roof slab of a SFR has many openings that are required for various components penetrating through it. In order to avoid radiation streaming through the annular gaps between the openings and the components passing through them, complementary shielding must be provided above the top shield/roof slab. This shielding is required to provide shielding for gamma radiation alone from the primary sodium, since core neutrons are shielded by the bulk primary sodium itself available above the core assemblies.

Both steel as well as lead are the primary candidate materials for complementary shielding and the selection is based on the operating temperature, availability of space and ease of fabrication. The importance of complementary shielding design stems from the fact that the weight of complementary shielding constitutes a significant portion of the weight of Reactor Assembly and it adds loading on the top shield/roof slab.

12.5 Refueling

Fuel handling involves two phases: (1) exchanging used fuel assemblies with fresh assemblies in the reactor, and (2) receiving new fuel at the power plant and shipping used fuel from the plant. Both phases will be discussed in this section.

Before discussing the mechanical processes associated with fuel loading and unloading, it is instructive to briefly review the major steps in the overall flow, as noted in Fig. 12.21. Such steps must be taken irrespective of the particular mechanics employed to move the fuel assemblies.

The principal components and building arrangement for both pool and loop SFR designs are shown in Figs. 12.22 and 12.23. The pool design is for Super Phénix. The loop design is one developed by Atomics International (AI) for a demonstration size plant[11]. Refueling designs for liquid metal

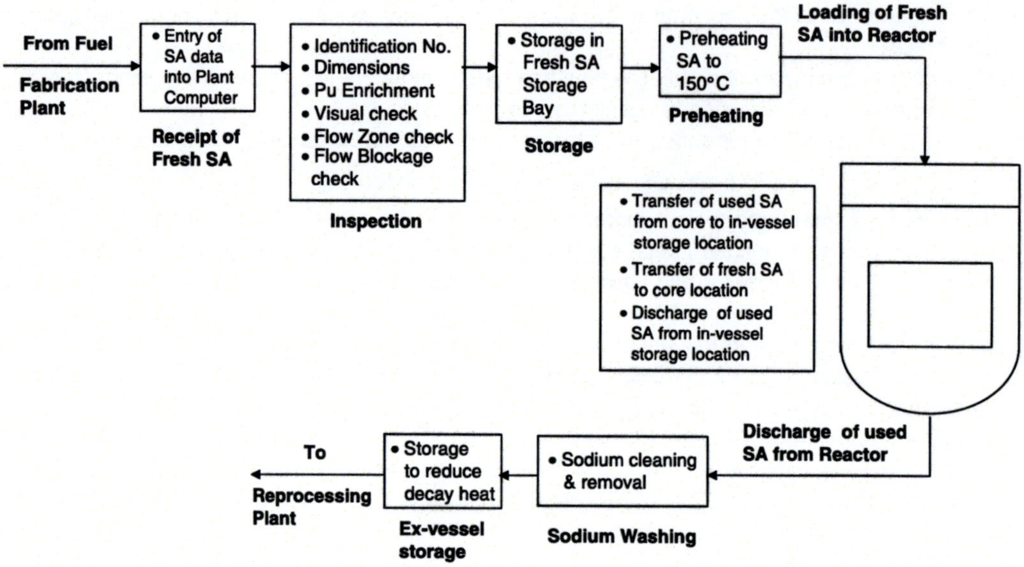

Fig. 12.21 Sequence of core SA handling

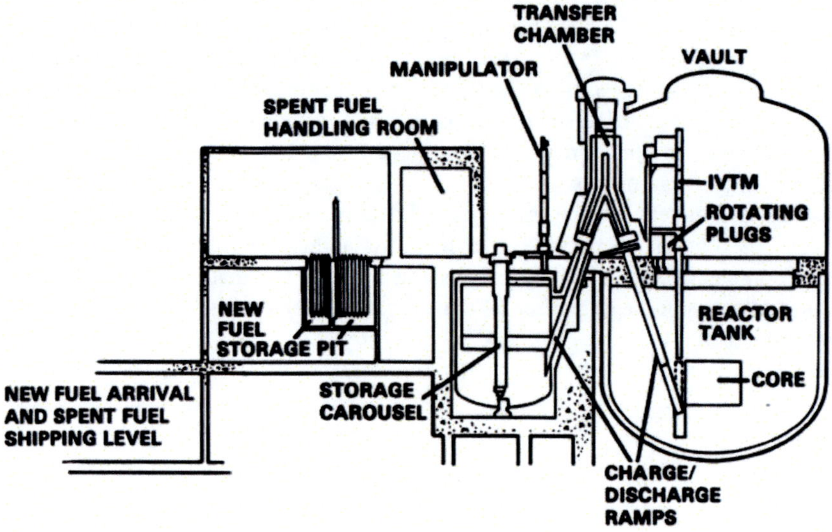

Fig. 12.22 Refueling system for Super Phénix (pool design)

reactors are reviewed in Ref. [12]. The Phénix refueling system is described in Ref. [13]. The refueling steps for the JSFR loop-type reactor are described in Appendix C and illustrated in Fig. C.13.

Reactor refueling (phase 1) is done with the reactor shut down. The components and cells are arranged with ex-vessel storage outside of the Reactor Containment Building so that receiving and shipping (phase 2) can be done while the reactor is operating.

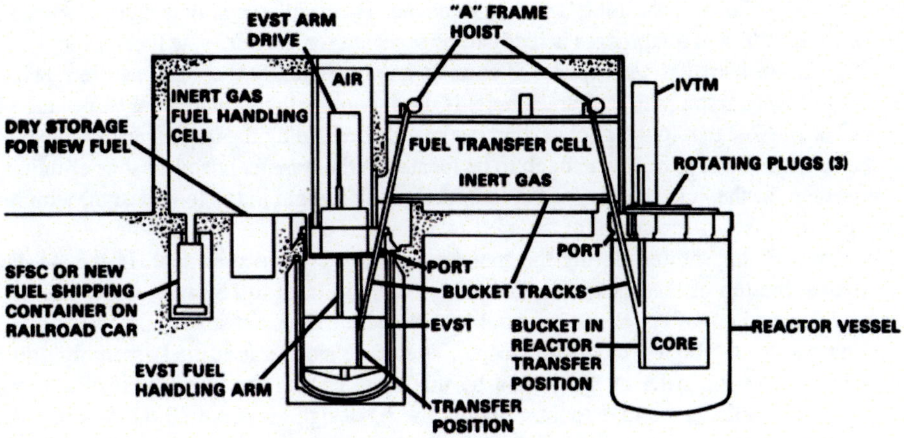

Fig. 12.23 Refueling system for a loop design (Atomics International Design, Ref. [11])

12.5.1 Reactor Refueling

Refueling normally occurs once a year (although it can be a much longer interval) and usually consists of replacing one third of the core fuel at each refueling shutdown. Refueling systems are designed to complete normal refueling in less than 2 weeks.

Assemblies that must be replaced periodically include core fuel, radial blanket, control, and shielding assemblies, In contrast to LWR's, refueling of SFRs is done without removing the head of the reactor vessel. This technique is called *under-the-plug-refueling*.

As noted in Figs. 12.22 and 12.23, there are three areas involved in the refueling process: the reactor vessel or tank, the fuel transfer cell (FTC) or transfer chamber, and the ex-vessel storage tank (EVST) or storage carousel. An In-Vessel Transfer Machine (IVTM) transfers fuel inside the reactor vessel. For these two designs, fuel is transferred between the vessel and the EVST in a transfer bucket by means of an "A-frame" hoist. Transfer ports are located between the reactor vessel and the FTC and between the EVST and the FTC. A fuel handling arm, or manipulator, transfers fuel in the EVST. In the AI design, the EVST can store up to one refueling batch of new fuel plus an entire core load of used fuel.

Fuel remains under sodium throughout the fuel transfer process. Consider the sequence of replacing a used fuel assembly with a new assembly, starting with the new assembly in the EVST. We will illustrate this sequence with details from the AI design; different designs will vary in detail but will be similar in general.

The new assembly is lifted out of its storage position in the EVST by the EVST fuel handling arm. The assembly (under sodium) is transferred to and placed in the transfer bucket. The bucket has space for two assemblies, one new and one used. At this point the bucket contains only the one new fuel assembly, in a vertical position. The bucket is then hoisted at an angle through a fuel transfer port and guided by tracks up into the FTC. It is then guided to the reactor vessel and down through a second fuel transfer port into the reactor vessel and placed in a vertical position in the shielding region outside the reactor core.

The IVTM is then moved to a position directly above the used fuel assembly to be replaced and the assembly is grappled. The used assembly is raised above the remaining assemblies and transferred through the sodium pool to the open space in the transfer bucket, into which it is then lowered. The fresh assembly is next withdrawn from the bucket by the IVTM and transferred to the position in the reactor from which the used assembly was just removed. The used assembly is then returned through the FTC to the EVST, and the process is ready to be repeated for the next new fuel assembly.

The FTC has an inert gas atmosphere. The inert gas is generally the same as the reactor cover gas; hence, generally argon. An "A-frame" hoist is used to guide the transfer bucket into the FTC, or transfer chamber. Note that in the pool design, the space provided by the large reactor tank allows the apex of the A-frame hoist to be low enough to be located in the transfer chamber, thus eliminating the lateral movement of the hoist shown for the loop design. The angle in the loop design, with its small diameter reactor vessel, is too steep for the apex to be in the FTC.

The bucket contains sodium during the transfer of both new and used fuel. In the AI design, a siphon exists at the top of the bucket that lowers the sodium level to 150 mm below the top of the bucket as it is hoisted out of either sodium pool so that sodium does not spill in the FTC.

Refueling can begin about 2 days after reactor shutdown since a used fuel assembly will have a decay power level of the order of 30–40 kW by this time. During the transfer process through the FTC, the transfer bucket is out of a sodium pool in the AI design for about 10 min. During this time the sodium temperature in the bucket will be increased by heat from a used fuel assembly by about 20°C. If the bucket becomes stuck in the FTC, the decay heat will be transferred from the bucket surface to the cell walls by radiation. The bucket temperature will rise to a maximum of ~500°C in 3 h or more. The heat is then transferred through the cell walls to the air in the building. If the bucket gets stuck in a transfer port, forced convection is required to maintain the bucket temperature at 500°C.

The large SFR designs use a rotating plug concept for refueling, in which several rotating plugs are located in the reactor head, and the IVTM is mounted on the smallest plug. Usually three rotating plugs are used. These plugs are illustrated in Fig. 12.20 for the CRBRP design. Note that the largest plug is concentric with the reactor vessel flange but that the smaller two plugs are eccentric. Each plug can rotate independently so that the IVTM can move to a position directly above any assembly in the reactor.

In the EVST, only one rotating plug is used in the AI design, and an arm under the plug is used to transfer assemblies. In Super Phénix the assemblies are rotated on a carousel rather than having a rotating plug in the cover.

The lucid logic of the refueling designs described should not give the reader the impression that these designs were easy to develop and are the obvious or only natural way to accomplish refueling in a SFR. These designs are the result of long-term and still on-going evolution in design, with frequent ingenious mechanical innovations. Other refueling schemes are possible for the SFR and, in fact, have been used. For example, FFTF does not use three rotating plugs of varying sizes, but instead uses three small independent rotating plugs with under-the-head arms to transfer fuel, each of which can cover one third of the assemblies. In SEFOR an entirely different, and simpler, system was used. SEFOR had an inerted (argon) refueling cell over the reactor vessel. During refueling the vessel cover was removed (as in LWR refueling) and fuel was withdrawn up into the refueling cell and transferred to a storage tank. Radiation levels were low enough in SEFOR that natural convection cooling by argon was sufficient, which made this simple refueling scheme particularly attractive for this case.

Subsequent to the problems experienced with the ex-vessel storage carousel in Super Phénix, some designers are moving toward in-vessel storage of fresh and used fuel. Whereas this requires a somewhat larger primary vessel, it avoids the extra heating and cooling systems required for the ex-vessel storage system, plus other items such as another cover gas system, leak detection systems, emergency cooling systems, etc.—all of which add to the cost of the plant. The Indian PFBR represents an example of the new designs that eliminate the ex-vessel storage system.

12.5.2 Receiving and Shipping

Receiving and shipping is accomplished in a Fuel Handling Building that houses a new fuel arrival-used fuel shipping area, a new fuel handling area, and a used fuel handling area. All of these areas cannot be illustrated adequately in Figs. 12.22 and 12.23 because they are located at different azimuthal positions around the reactor.

New fuel can be received and safely stored in a dry storage tank while the reactor is operating. This fuel can then be moved to the EVST before the reactor is shut down for refueling.

Used fuel is moved from the EVST during reactor operation into the used fuel handling area and later placed in a shielded used fuel shipping cask (SFSC). Low power fuel might be placed in a gas atmosphere in the SFSC for shipping. Several options are available for the higher power fuel assemblies. They could be shipped in sodium if reprocessing is to occur on site. For reprocessing away from the site, they could be stored long enough to allow shipment in a gas atmosphere SFSC, or they could be shipped in a forced convection gas-cooled SFSC.

12.6 Instrumentation

All nuclear reactor systems incorporate a high degree of instrumentation in order to provide the continued monitoring features necessary for both plant control and scientific data collection. Much of this instrumentation, such as radiation monitoring, is common to all reactor types and need not be considered in this text. However, the presence of a liquid metal poses a few instrumentation challenges that are unique to the SFR system. These challenges constitute the basis for the present section.

12.6.1 Core Parameter Monitoring

Variables such as flux, temperature, flow, and pressure must be determined in any nuclear reactor, but the sodium environment requires measurement techniques somewhat different from those of light water systems.

12.6.1.1 Flux

It was noted in Section 12.4.3 that the flux monitoring for a typical SFR system consists of a group of neutron detectors located in the reactor cavity external to the reactor vessel. In-core or in-vessel detectors may be used for initial startup operation, but the neutron source from spontaneous fission of ^{240}Pu is normally present, which is strong enough to activate such remotely located detectors. An appreciable neutron source is also possible from ^{242}Am, especially if recycle fuel is used.

Figure 12.24 shows a flux monitoring set for the CRBRP design. For this design, the highly sensitive BF_3 detectors are used for low power operation (startup, refueling), U-235 fission chambers are employed for mid-range operations, and compensated ion chambers are used for power range measurements. Three sets, identical to that shown in Fig. 12.24, provide an overlapping range of flux monitoring to continuously record neutron flux from shutdown to more than full power. Electrical signals from these detectors, which are proportional to reactor power, are used for both reactor control and the plant protection system (PPS). Such signals also feed the data logging system and provide annunciator trips in the control room for out-of-limit conditions.

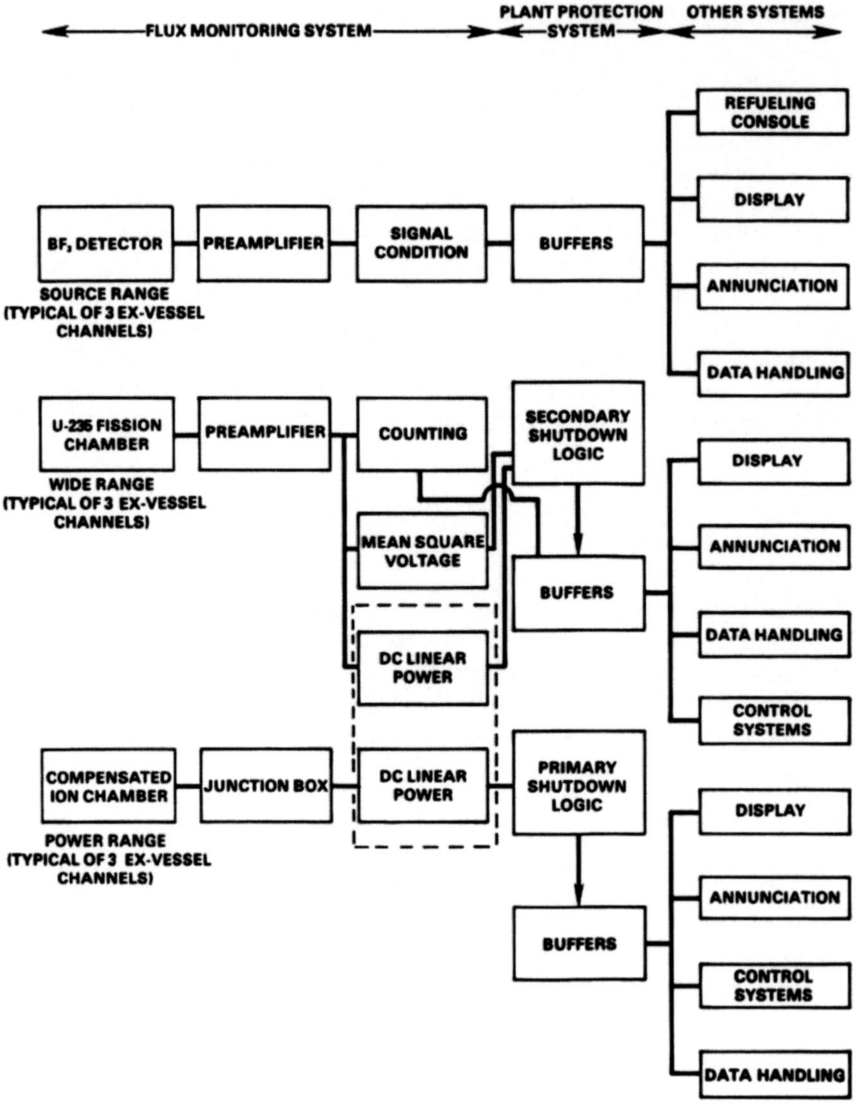

Fig. 12.24 Flux monitoring set for CRBRP design

12.6.1.2 Temperature

Sodium temperatures must be measured routinely throughout the primary and secondary circuits to calculate thermal power and determine loop operating conditions. Two types of detectors are in common use: Resistance Temperature Detectors (RTD's) and thermocouples. The RTD's provide a highly accurate and reliable measurement to ensure that the plant is operating within design limits. The sensor for such a device typically consists of a double element of platinum contained within a sheath that is spring loaded against the bottom of a thermowell, as illustrated in Fig. 12.25. Insertion of this sensor raises the possibility of sodium leakage from the penetration in the event of thermowell failure.

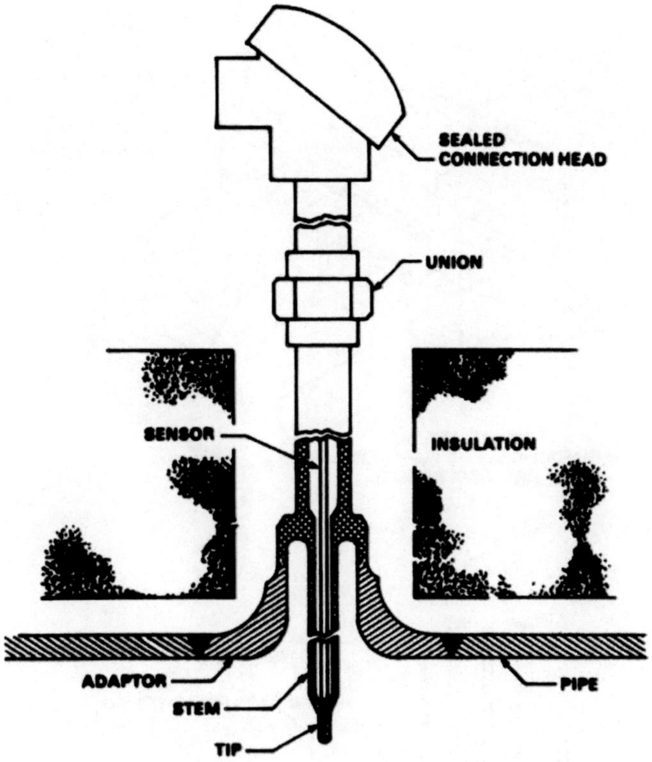

Fig. 12.25 A typical thermowell assembly incorporating a Resistance Temperature Detector (RTD) for measuring sodium temperatures

Whereas well failures in such devices appear to be rare, a backup cable penetration seal (as shown by the sealed connection head in Fig. 12.25) can be provided to prevent such leakage.

12.6.1.3 Flow

Coolant flow measurements must be made to complete the thermal power calculations and loop operating characteristics. Both the standard Venturi flow meter and a magnetic flow meter are often used in liquid metal systems. The Venturi meter is highly accurate, but suffers from a response time that is often too slow for control system and PPS use. The magnetic flow meter, on the other hand, tends to be less accurate but exhibits a rapid response time. When the two are used in series, the Venturi meter can be used to provide in-place calibration[11] of the rapid-response magnetic flow meter.

The magnetic flow meter is unique to liquid metal systems. It is a direct result of the electric properties of the liquid metal coolant. Figure 12.26 contains a schematic of a permanent magnet flow meter.

Another type of flow meter made possible by the unique properties of the liquid metal is an eddy current flow meter. Figure 12.27 contains a simplified diagram of such a device used in measuring the rate of coolant discharge from an assembly into the upper sodium pool.

[11] It is possible to calibrate both types of meters by activating the sodium with a pulsed neutron device and using time-of-flight recording techniques. This procedure was successfully employed in FFTF.

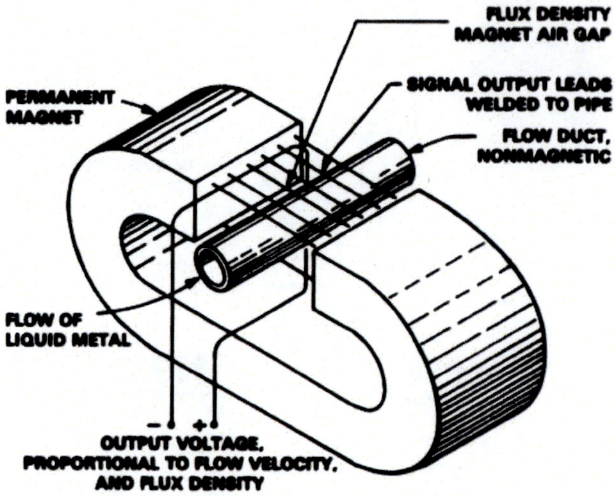

Fig. 12.26 Schematic of permanent magnet flow meter

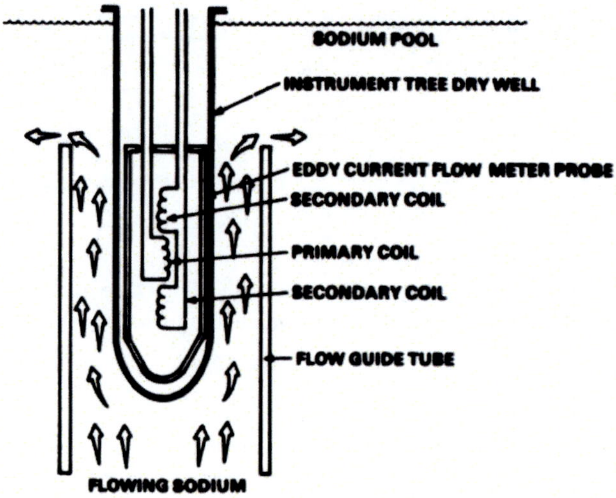

Fig. 12.27 Simplified diagram of eddy current flow meter

12.6.1.4 Pressure

Liquid pressure measurements are normally made by routing a small column of the high pressure liquid onto one side of a sensing diaphragm. This causes a complication when measuring sodium pressure because sodium solidifies well above room temperature. Trace heating could be provided to assure liquid sodium conditions, but this becomes unreliable for many applications. An alternate method often employed is to interface the sodium with NaK via a bellows system, as illustrated in Fig. 12.28, taking advantage of the fact that NaK is a liquid at room temperature in the close-tolerance pressure transducer assembly.

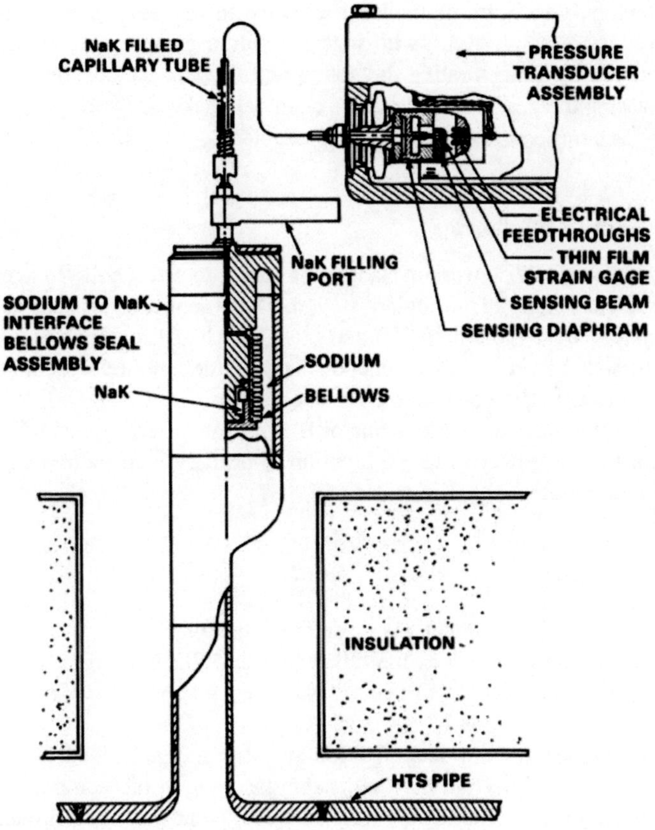

Fig. 12.28 Typical pressure sensor installation for an SFR system

12.6.2 Fuel Failure Detection

The detection of a cladding breach in a fuel element can be normally accomplished by monitoring increased cover gas activity or by detecting the presence of delayed neutrons in the sodium leaving the reactor. Locating the fuel assembly containing the leaking fuel element is more difficult; the use of gas tagging provides one means to provide such identification.

12.6.2.1 Cover Gas Monitoring

A cover gas monitoring system normally exists to allow the presence of fission products that escape the fuel pin to be discovered. The most abundant fission products that escape are isotopes of the noble gases xenon and krypton. Many of the fission products emit gamma rays with relatively low energy (~100 keV) and detection of such activity in a prevailing background of high energy ^{23}Ne (440 keV) and ^{41}A (1,300 keV) gammas could present a difficulty.[12] However, numerous Xe isotopes emit relatively high energy gammas and, as such, experience indicates that detection of cladding failure for a SFR system is readily attainable.

[12]The ^{23}Ne activity is considerably less in a pool type reactor, relative to a loop type system, due to the larger holdup time (which allows the 38 s half-life ^{23}Ne to decay) and less turbulence in the pool level.

Germanium detectors, which are particularly sensitive to low level gamma detection, are often utilized in the cover gas system, along with high resolution gamma spectrometers. Detector efficiency can be improved by concentrating the xenon and krypton isotopes in the cover gas; this is accomplished by passing the gas stream through a charcoal-packed column. The time for fuel failure detection by cover gas monitoring is of the order of minutes.

12.6.2.2 Delayed Neutron Monitoring

A second system often used to discover pin cladding failure is to detect delayed neutrons that emanate from fission products circulating in the coolant stream. These neutron emissions are produced chiefly by two fission products, ^{87}Br (56 s half life) and ^{137}I (25 s half life). Both isotopes are soluble in sodium and enter the sodium via sodium contact on exposed fuel, by fuel washout through a cladding breach, or by fission gas expulsion of fuel into the coolant.

Delayed neutron detectors, usually consisting of BF_3 chambers, are typically located near primary pumps. Overall detection time is obviously a function of primary loop sodium transport time; this is typically the order of a minute.

12.6.2.3 Location Indicators

Detection of a fuel failure via the cover gas or delayed neutron detector system provides no information on which fuel assembly contains the failed fuel pin. Since a large SFR may be comprised of the order of 300 fuel assemblies, it is important to have some techniques available to identify the offending assembly.

One technique developed for this identification is gas tagging. Unique blends of stable xenon and krypton gas isotopes are injected into the fission gas plenum of each pin during final fabrication; all pins within the same fuel assembly have identical blends. A three-dimensional network of $^{126}Xe/^{129}Xe$, $^{78}Kr/^{80}Kr$, and $^{82}Kr/^{80}Kr$ was used to provide over 100 unique gas tags for the fuel and absorber assemblies in the FFTF[14]. As illustrated in Fig. 12.29, the failed assembly is identified by

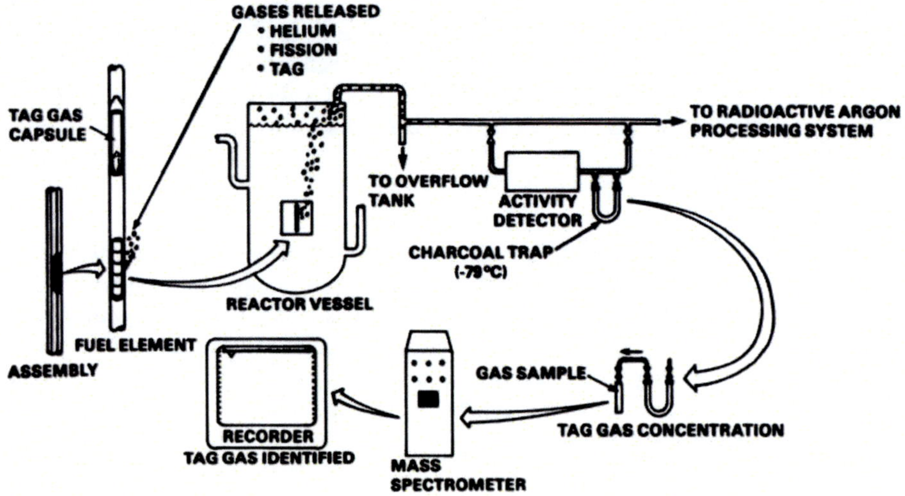

Fig. 12.29 Fuel failure location determination by gas tagging

matching the results of the cover gas mass spectrometer analysis with previously determined analysis of all gas tags in the reactor, suitably corrected for burn up and background.

12.6.3 Sodium Leaks and Level Measurement

The detection of sodium leaks is important for at least three reasons: (1) sodium will burn in an air atmosphere (2) the primary loop sodium is radioactive, and (3) loss of substantial amount of sodium could impair cooling capacity. Sodium levels must be measured in all vessels containing sodium.

12.6.3.1 Leak Detection

One method used to detect the presence of sodium exploits the electrical conductivity properties of the liquid metal. Contact type sensors consist of two electrodes extended to a gap at a location where leaking sodium may be expected to collect. The presence of sodium shorts out the electrode gap and allows a signal to be delivered to the control room. Such detectors are often placed under the reactor vessel and at any low point surrounding piping systems that contain flowing sodium. The principal difficulties of such detectors are (1) oxidation of the electrodes, and (2) providing assurance that leaking sodium will actually reach the detector (a major concern in the case of small leaks).

A second type of sodium leak detector checks for the presence of sodium aerosol. Atmospheric gas samples from the area in question can be analyzed by either ionization detector or filter examination. For the ionization technique the gas stream is passed over a heated filament that produces sodium ions. Collector electrodes will then induce an ion current if sodium is present. In the case of the filter technique, a replaceable submicron filter is placed in the gas stream and the filter is then periodically removed and analyzed chemically for sodium deposits.

12.6.3.2 Sodium Level

Induction level probes can be used to measure the liquid level of sodium, as illustrated schematically in Fig. 12.30. The induction field established by the primary coil is modified by the electrical properties of the liquid coolant, thereby leading to a secondary signal that is directly proportional to the sodium level. Though useful for maintaining inventory records in all sodium repositories, this is especially important within the reactor vessel where a fundamental safety concern is to guarantee a coolant level well above the top of the core at all times.

12.7 Auxiliary Systems

Beyond the major systems mentioned earlier in this chapter, numerous auxiliary systems exist which are necessary to support overall plant operation. Many of these are large systems, such as heating and ventilation but they are not unique to the SFR (except for the general desire to minimize the presence of water in the immediate vicinity of sodium systems). Our present purpose is to describe briefly auxiliary systems that are unique to a sodium cooled plant.

There are numerous ways to categorize such systems. We have chosen only three: (1) inert gas (which recognizes the need for an inert atmosphere surrounding a combustible coolant) (2) trace heating (necessary to keep sodium in the liquid phase for low core power levels), and (3) sodium purification (clearly unique to sodium coolant).

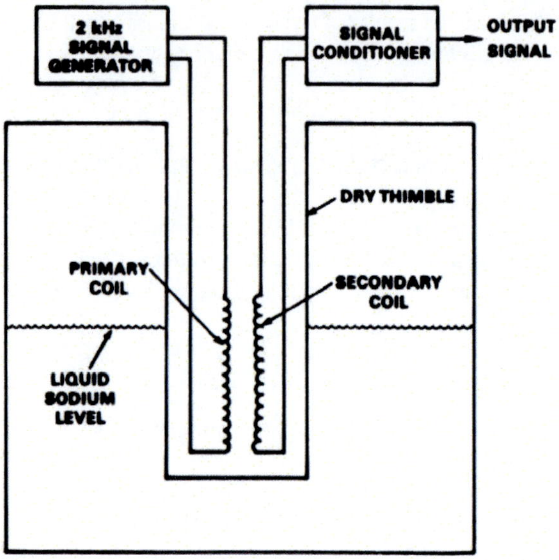

Fig. 12.30 Induction sodium level probe schematic

12.7.1 Inert Gas

An inert cover gas is a requirement for any part of a SFR system where a free liquid sodium surface can exist (i.e., vessel, pumps, IHX). Prudent design also normally requires an inert atmosphere in any of the cells that house sodium piping systems. Whereas the word "inert" normally implies a noble gas, the essential characteristic of the atmosphere desired is that it be chemically inert to sodium. Nitrogen satisfies this requirement and is both abundant and relatively inexpensive. Hence, it has been almost universally employed as the inert atmosphere for equipment cells. Unfortunately, it cannot be used for high temperature application (≥400°C) because of nitriding problems in the steel enclosures. Consequently, argon has been selected as the cover gas *within* the vessel, the piping systems, and the refueling transfer chamber for all major SFR projects to date. Helium is a potential alternative.

Because of this normal split of inert gases employed, and the natural associated split in physical layout of the two subsystems, it is appropriate to discuss the key features of each separately.

12.7.1.1 Argon Cover Gas Subsystem

The argon cover gas subsystem provides an inert atmosphere and pressure control for all liquid metal-gas interfaces. Chemical purification features include both sodium vapor and oil vapor traps. Compressors and storage facilities are necessary ingredients of this subsystem as well as pressure equalization lines to keep all cover gases at the same pressure. The purging system needs to have a high enough capacity to allow complete changes of atmosphere to accommodate maintenance operations.

Because of the possibility of radioactive contamination (discussed in Section 12.4.1) a key feature of this subsystem is the radioactive argon processing subsystem (RAPS), which removes krypton and xenon radioisotopes. A charcoal bed, which utilizes a cryogenic still, provides an effective way to remove the krypton and xenon isotopes from the argon stream. Surge tanks are useful to allow short-lived isotopes in the contaminated argon to decay.

12.7.1.2 Nitrogen Subsystem

A nitrogen subsystem is normally incorporated to supply and control atmospheric pressure and purity in the inerted equipment cells. This is done by means of a feed-and-bleed system to regulate pressure and fresh nitrogen purging to minimize contamination, using measurements of the oxygen or water-vapor in the cells as a control signal. Nitrogen is also normally supplied for the sodium/water reaction pressure relief system in the steam generator, for cleaning operations, and for valve actuation in inerted cells.

A key feature employed to remove radioactive contaminants is the cell atmosphere processing subsystem (CAPS). It works on the same principle as the RAPS System, complete with cryogenic features, but usually has substantially larger capacity. One reason for the large capacity requirement is to allow pressurization of equipment cells during pressure testing of plant containment.

12.7.2 Trace Heating

Sodium melts at 98°C; hence, it must be heated at low reactor power levels to remain in the liquid form. The method normally employed to provide such heating is electrical trace heaters. A typical trace heating assembly, as illustrated in Fig. 12.31, consists of a nickel-chromium resistance element insulated with magnesia, covered with a nickel-iron-chromium alloy heater sheath, and surrounded by a large thickness of thermal insulation.

Such heaters provide a heat flux around 10–20 kW/m^2. For a large plant, the trace heating system may consume on the order of 10 MW during initial startup (cold core) conditions. The requirements for trace heating drop appreciably when primary and secondary pumps are activated—due to frictional heating resulting from pumping action.

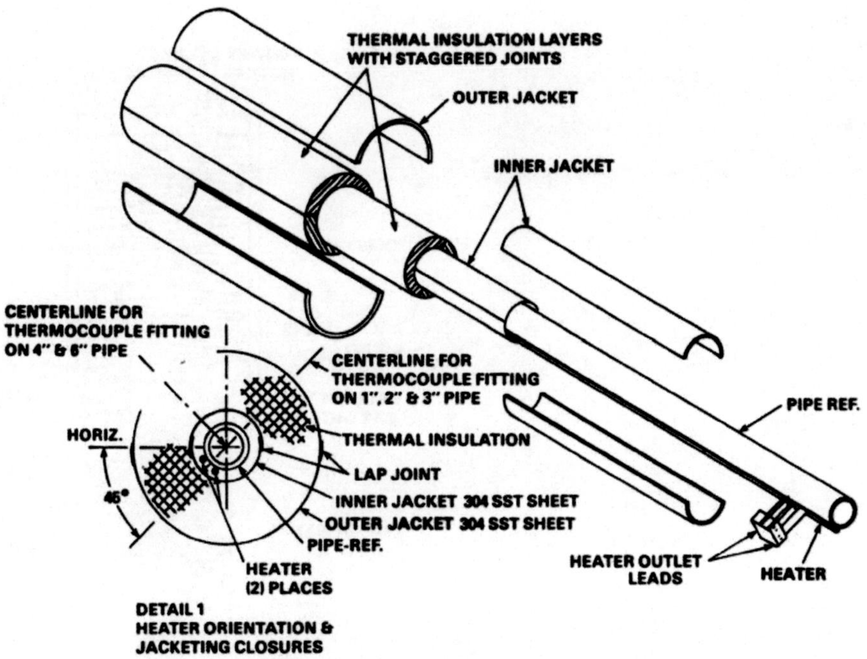

Fig. 12.31 Trace heating and insulation for one-to-six inch pipe

12.7.3 Sodium Purification

The principal objective of the sodium purification system is to maintain the sodium clean from chemical or radioactive particulate contaminants. As noted in Chapter 11, several trace elements from in-core structural materials dissolve into the flowing sodium coolant during normal operation. Table 12.3 provides a listing of elements that a typical sodium purification system may be designed to monitor.

The main component incorporated in such a system to remove such impurities is the cold trap. This device, which is connected to a bypass line from the main sodium loop, removes impurities by crystallization at a temperature (~150°C) significantly below the main-stream sodium temperature. Figure 12.32 contains an illustration of a typical cold trap. Sodium oxide crystallizes on the packing. This packing is replaced when it begins to plug.

An interesting feature of the cold trap shown is the economizer. Inlet sodium must be cooled to low temperatures prior to entering the crystallizer, but returning (purified) sodium must be reheated

Table 12.3 Contaminants monitored by a typical sodium purification system

Boron	Iron
Carbon activity	Lithium
Cesium-137	Manganese
Chlorine	Molybdenum
Chromium	Nickel
Hydrogen	Nitrogen
Iodine-131	Oxygen
Iodine-132	Plutonium
Iodine-133	Tritium
Iodine-135	Uranium

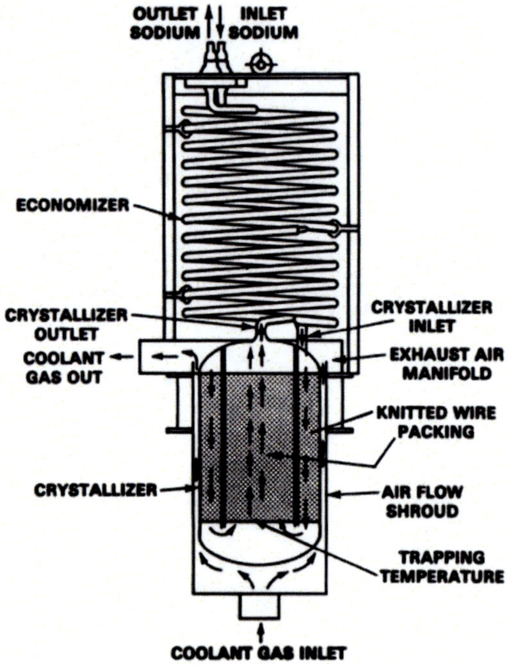

Fig. 12.32 A typical sodium cold trap

to nearly the bulk coolant temperature prior to returning to the main coolant system. Both of these functions can be performed by bringing the inlet sodium through a tube that is concentrically enclosed by an outer tube containing the counter-flowing purified sodium. A much smaller auxiliary cooling and heating system is then required to allow satisfactory cold trap performance than would be the case without the economizer feature.

References

1. A. K. Agrawal and M. Khatib-Rahbar, "Dynamic Simulation of LMFBR Systems," *Atomic Energy Review, 18*, 2 (1980), IAEA, Vienna.
2. V. Dostal, "A Supercritical Carbon Dioxide Cycle for Next Generation Nuclear Reactors," Ph.D. Dissertation, MIT (2004).
3. J. Graham, *Fast Reactor Safely,* Academic, New York, NY (1971).
4. Y. S. Tang, R. D. Coffield, Jr., and R. A. Merkley, *Thermal Analysis of Liquid Metal Fast Breeder Reactors,* American Nuclear Society, La Grange Park, IL (1978).
5. W. L. Bunch, J. L. Rathbun, and L. D. Swenson, *Design Experience-FFTF Shielding,* S/A-1634, Hanford Engineering Development Laboratory (presented at the US/USSR Joint Fast Breeder Reactor Shielding Seminar, Obninsk, USSR, November 1978).
6. R. K. Disney, T. C. Chen, F. G. Galle, L. R. Hedgecock, C. A. McGinnis, and G. N. Wright, *Design Experience-CRBRP Radiation Shielding,* CRBRP-PMC 79-02, CRBRP Technical Review (April 1979) 7–28.
7. A. S. Gibson, P. M. Murphy, and W. R. Gee, Jr., *Conceptual Plan! Design,* System Descriptions, and Costs for a 1000 MWe Sodium Cooled Fast Reactor, Task I1 Report, AEC Follow-On Study, GEAP 5678, General Electric Company (December 1968).
8. R. Indira et al., "Fast Reactor Bulk Shielding Experiments for Validation of Shielding Computational Techniques," *Proceedings of Conference on Nuclear Mathematical and Computational Sciences: A Century in Review, A Century Anew*, ANS, Gatlinburg, TN (2003).
9. K. Devan et al., "Generation and Validation of a New 121 Group Coupled (n, γ) Cross-section Set for Fast Reactor Applications," *Annals of Nuclear Energy, 23* (1996) 791.
10. K. Devan et al., "Effects of Cross-Section Sets and Quadrature Orders on Neutron Fluxes and on Secondary ^{24}Na Activation Rate of a Pool Type 500 MWe FBR," *Annals of Nuclear Energy, 30* (2003) 1181.
11. J. S. McDonald (AX), C. L. Storrs (CE), R. A. Johnson (AI), and W. P. Stoker (CE), "LMFBR Development Plant Reactor Assembly and Refueling Systems," *Presented at ASME Meeting*, San Francisco, CA (August 18–21, 1980).
12. K. W. Foster, "Fuel Handling Experience with Liquid Metal Reactors," *Proceedings of the International Symposium on Design, Construction and Operating Experience of Demonstration Liquid Metal Fast Breeder Reactors*, Bologna, Italy (April 1978).
13. E. Benoist and C. Bouliner, "Fuel and Special Handling Facilities for Phénix," *Nuclear Engineering International, 7* (1971), 571–576.
14. N. J. McCormick and R. E. Schenter, "Gas Tag Identification of Failed Fuel I: Synergistic Use of Inert Gases," *Nuclear Technology, 24* (1974), 149–155. See also Part II, "Gas Tag Identification of Failed Fuel II: Resolution Between Single and Multiple Failures," *Nuclear Technology, 24* (1974), 156–167.

Chapter 13
General Safety Considerations

John Sackett

13.1 Introduction

Fast reactors exhibit some unique characteristics related to safety in comparison to thermal reactors. At first glance, it might appear that achieving exceptional safety in a fast reactor might be more challenging than in a thermal reactor, considering that sodium-cooled fast reactors (SFR) have a higher core power density, the neutron lifetime is shorter, the effective delayed neutron fraction is less, the core is not arranged in its most reactive configuration, the sodium void effect is usually positive, and sodium interacts rather violently with air or water. On the other hand, the boiling point of sodium is sufficiently high that the reactor can be operated near atmospheric conditions (eliminating the massive pressures required for water-cooled systems), sodium has a very high heat capacity and thermal conductivity, the neuron mean free path is sufficiently long that spatial power shifts are negligible, and xenon poisoning is a non-issue. Furthermore, it has been demonstrated that passive safety features can be more easily incorporated than in the thermal reactor systems.

Whereas oxide fuel is almost universally employed in thermal reactors, there are several viable choices of fuel for fast reactors. Oxide, or uranium-plutonium mixed oxide (MOX), fuels possess a strong negative fuel Doppler feedback effect, which is a positive factor to arrest unprotected overpower transients, but it is a liability during unprotected loss of flow conditions—where the goal is to avoid sodium boiling. Metal-fueled fast reactors exhibit a much smaller Doppler effect (due to the harder neutron spectrum) and, therefore, perform considerably better during an unprotected loss of flow event. Favorable response to unprotected overpower events relies more on axial expansion of the core than on Doppler feedback. Carbide and nitride-fueled fast reactors respond more like metal-fueled reactors, but with an intermediate Doppler effect.

As long as sodium voiding can be prevented, the transient response of a SFR is very manageable, even for unprotected conditions (i.e. very low probability events where the entire plant protection system fails). Should sodium voiding occur, however, substantial core damage could result. Dealing with such a situation greatly complicates the analyses and has provided the stimulation for a great deal of R&D (covered in Chapters 15 and 16 and Appendix G).

In most respects, the approach to fast reactor safety analysis is the same as for thermal reactor systems, with emphasis on multiple barriers to accident progression, risk analysis and ultimate licensing. An interesting result of this work has been the discovery that SFRs can be made self-protecting to a large degree, an aspect of design confirmed by formal risk assessments conducted for fast reactors.

Another aspect of fast reactor core design important to safety, given the sensitivity of reactivity changes to changes in core geometry, is bowing of fuel assemblies due to radial temperature gradients

J. Sackett (✉)
Argonne National Laboratory (Retired), Argonne, IL, USA
e-mail: johnflyer@3riversdbs.net

A.E. Waltar et al. (eds.), *Fast Spectrum Reactors*, DOI 10.1007/978-1-4419-9572-8_13,

across the core. Attention must be paid to core restraint system design (discussed in Section 8.6.3), given the variations in power and temperature gradients across the core during all operating conditions.

13.2 Multiple Barrier Approach to Safety

Since sodium will interact chemically with air or water, the first barrier important to SFR safety is maintaining the integrity of the primary coolant system. This is accomplished in several ways and there are two different philosophies for achieving it. The first is to locate all components of the primary coolant system (pumps, heat exchangers, piping) in a double-walled tank. The second is to design a more traditional piping system that does not require a tank, making components more accessible for maintenance. The piping is doubly contained, which provides an additional barrier to leakage. In both systems, leak detection is provided between the primary barrier and its guard vessel. These philosophies are the pool- and loop-type SFR designs illustrated in Fig. 2.5 in Chapter 2 and described in detail in Chapter 12.

Another important boundary associated with the primary coolant system is the cover-gas, which isolates the sodium from the atmosphere. Generally this is the inert gas Argon and much attention needs to be given to ensure its purity. Ingress of air (specifically oxygen) into this cover gas will result in the formation of sodium oxide, which will precipitate as a solid below certain temperatures. (This behavior facilitates cleanup of sodium using cold traps, which is also important to ensuring integrity of the coolant flow paths).

The third barrier related to the sodium coolant is the sodium–water interface in the steam generating system. To ensure protection of the primary coolant systems, an intermediate loop of sodium is normally installed that takes heat from the primary sodium and transfers it to the steam generating system. Design of the steam generators is also an important consideration, but in this arrangement, failure of the sodium–water interface in the intermediate loop will not endanger the reactor. As an alternate approach, some designers are proposing to use super-critical CO_2 in a Brayton Cycle to replace the intermediate loop and directly drive the turbines to produce electricity [1].

Safety awareness, which consciously permeates the approach from conceptual design through the licensing process and long-term operation of nuclear power plants, has been implemented to a far greater extent in the nuclear industry than normally present in other areas of human endeavor. The result has been an exceptionally good safety record for modern nuclear power plants. The accident at the Three Mile Reactor Unit 2 in March 1979, which involved loss of cooling and melting of the core, resulted in minimal release of radioactivity and no injury to the public. The accident at Chernobyl, which involved a more violent disruption of the core, resulted in significant release of radioactivity and fatalities. However, this was a reactor that did not incorporate a containment building or modern approaches to protection of the reactor core. The massive earthquake and highly destructive tsunami at the Fukashima site in Japan caused extensive damage to several nuclear plants, although the radiation releases resulted in no fatalities. This event emphasizes the need to consider combinations of events that are normally unexpected and the value of passive safety features for modern plant designs, namely safety response that does not require electric power or operator intervention.

13.3 International Experience Related to Safety

Fast reactor experience is extensive, and is summarized in Table 13.1 (See the complete listing, including references in Appendix A). Fast reactors have been operated in the US, France, UK, Germany, Japan, India, China, Russia and Kazakhstan. In each of these countries, small reactors of similar

Table 13.1 Summary of fast reactors world-wide

Reactor	Country	Dates of operation	Power (MW_{th})
EBR-I	US	1951–1963	1.2
EBR-II	US	1961–1994	62.5
Fermi-1	US	1963–1972	200
FFTF	US	1980–1992	400
CRBRP	US	Cancelled (1983)	975
Rapsodie	France	1967–1983	40
Phénix	France	1973–2009	563
Super-Phénix	France	1985–1997	2,990
BR-5/BR-10	Russia	1958–2002	8
BOR-60	Russia	1969–	55
BN-350	Kazakhstan	1972–1999	750
BN-600	Russia	1980–	1,470
BN-800	Russia	Construction phase	2,100
JOYO	Japan	1977–	140[a]
MONJU	Japan	1994–1995 and 2010–	714
DFR	UK	1959–1977	60
PFR	UK	1974–1994	650
KNK-II	Germany	1972–1991	58
SNR-300	Germany	Cancelled (1990)	762
FBTR	India	1985–	40
PFBR	India	Construction phase	1,250
CEFR	China	2010–	65

[a]Mark-III core.

size to Experimental Breeder Reactor-II (EBR-II) were operated to develop and test the technology. The early test reactors were, EBR-II (US), Rapsodie (France), DFR (UK), KNK (Germany), JOYO (Japan), and BOR-60 (Russia). Of these, EBR-II, KNK, DFR were complete power plants. In the US and Russia, small, specialized fast-spectrum test reactors were operated to address questions of physics; namely, SEFOR and EBR-I (US), and BR-2, BR-5/BR-10 (Russia).

The next generation of fast reactors was made up primarily of complete power plants that incrementally increased power levels over the test reactors that preceded them. These reactors were Fermi-1 (US), Phénix (France), PFR (UK), SNR-300 (Germany), MONJU (Japan) and BN-350 (Kazakhstan).

France and Russia operated larger commercial plants, Super-Phénix (France) and BN-600 (Russia). In addition the US constructed and successfully operated a second research reactor, FFTF (but without an electricity generating system). The US also pursed design of the CRBRP, which was cancelled before construction was completed. A similar fate befell the German fast reactor, SNR-300.

All of the reactors listed in Table 13.1 were/are cooled with sodium. As described in Chapter 18, lead-bismuth eutectic-cooled reactors were developed and built in the former Soviet Union as naval submarine propulsion reactors and supporting land prototypes. However, those reactors were strictly for military purposes and are not addressed here. Sodium supports a fast-neutron spectrum because of low neutron moderation and absorption. It has excellent thermal conductivity and high heat capacity, which allows for high power density in the core. Its relatively low density reduces pumping power requirements and its large margin to boiling allows for operation at atmospheric pressure. The coolant is also chemically compatible with structural materials, which minimizes corrosion in plant cooling systems. However, an inert atmosphere covering the sodium is needed because it is reactive with air. Sodium will react if exposed to air and special fire-suppression systems are an important part of design.

This extensive experience with sodium-cooled reactor operation and testing has resulted in the following major conclusions:

Positives

- Fast reactor fuel is reliable and safe, with substantial success with oxide, metal, and carbide fuels. Cladding failure does not lead to progressive fuel failure during normal or off-normal reactor operation [2–4].
- High burn up of fast reactor fuel is achievable, whether metal or oxide. Acceptable conversion ratios (either as breeders or burners) are also achievable with either fuel type [4, 5].
- Sodium is not corrosive to stainless steel or components immersed within it [6].
- Leakage in steam generating systems with resultant sodium-water reaction does not lead to serious safety problems. Such reactions are not catastrophic, as initially believed, and can be detected, contained and isolated [6].
- Leakage of high-temperature sodium coolant, leading to a sodium reaction, is not catastrophic and can be contained, suppressed and extinguished. There have been no injuries from sodium leakage and fire (operation at near atmospheric pressure is an advantage to safety) [6].
- Fast reactors can be self-protecting against Anticipated Transients without Scram when fueled with metal fuel. Load-following is also straightforward [7–9].
- Passive transition to natural convective core-cooling and passive rejection of decay heat has been demonstrated [8–10].
- Reliable control and safety-system response has been demonstrated [6, 11].
- Effective systems for purity control of sodium and cleanup have been demonstrated [6].
- Efficient reprocessing of metal fuel, including remote fabrication, has been demonstrated [6].
- Low radiation exposures are the norm for operating and plant maintenance personnel, less than 10% of that typical for light water reactors (LWRs) [6].
- Emissions are quite low, in part because sodium reacts chemically with many fission products if fuel cladding is breached.
- Maintenance and repair techniques are well developed and straightforward [6].
- Electromagnetic pumps for moving sodium operate reliably.

Negatives

- Steam generators have not been reliable and are expensive to design and fabricate.
- Sodium heat-transport systems have experienced a significant number of leaks because of poor manufacturing quality control and difficulty with welds. Also, because of sodium's high thermal conductivity, many designs did not adequately anticipate the potential for high thermal stress in transients.
- Many problems with handling fuel in sodium systems have occurred, primarily because of the inability to visually monitor operations.
- Failure of in-sodium components without adequate means for removal and repair has resulted in costly and time-consuming recovery.
- Sodium-cooled fast reactors have been more expensive to build than water-cooled-reactor systems.
- Reactivity anomalies have occurred in a number of fast reactors, requiring careful attention to core restraint systems and potential for gas entrainment in sodium flowing through the core.
- Operational problems have been encountered at the sodium/cover–gas interface, resulting from formation of sodium-oxide that can lead to binding of rotating machinery, control-rod drives and contamination of the sodium coolant.

13.4 Approaches to Safety

13.4.1 Basic Safety Objectives

The first objective of nuclear reactor safety is to *protect public health* by reducing the risk from releases of radioactivity to acceptable levels. This is accomplished by providing a design that complies with established regulatory requirements, and also provides additional margins beyond the requirements.

The second safety objective is to *protect plant worker health and safety* by providing a safe working environment that reduces the risk of injury and complies with the regulatory requirements for worker protection.

The next safety objective is to build, operate, and decommission the plant in a manner that *protects environmental quality*.

Finally, the plant must be designed, equipped, operated, and maintained to *preserve investor equity* and produce electricity reliably.

The design basis for radiation protection consists of (a) barriers to contain radioactive materials and prevent human exposure or release to the environment, (b) distance to provide spatial margins to reduce the intensity of radiation exposure, and (c) time to isolate radioactive material until it has decayed to a stable or less harmful state.

13.4.2 Safety Design and Defense-in-Depth

By design, the plant and all its systems are configured and constructed in a manner that assures safe, stable, and reliable operation while preserving protection mechanisms. Engineered safety systems are provided to prevent conditions that can defeat the basic radiation protection mechanisms, and to mitigate the consequences of equipment failure or inappropriate operator actions. Inherent protection margins can be provided by selection of materials and arrangement of components in a manner that mitigates the consequences of an event without active systems or operator actions. The design principle of defense-in-depth is applied for important safety-related functions: containment, reactor shutdown, and residual heat removal. In addition, safety margins beyond the design basis are provided to minimize public risk.

As described by the former Chairman of the United States Nuclear Regulatory Commission (NRC), Nil Diaz, in an address in 2004, "the concept of defense-in-depth is a centerpiece of our approach to ensuring public health and safety, and it goes beyond pieces of equipment. It calls for, among other things, high quality design, fabrication, construction, inspection, and testing; plus multiple barriers to fission product release; plus redundancy and diversity in safety equipment; plus procedures and strategies; and lastly, emergency preparedness, which includes coordination with local authorities, sheltering, evacuation, and/or administration of prophylactics (for example, potassium iodide tablets). This approach addresses the expected as well as the unexpected." [12].

Plant functions that are important to the preservation of safety protection mechanisms are designed according to the defense-in-depth principle, which provides multiple layers of safety assurance. At the first level, the aim is to provide a conservative design with large safety margins that can be constructed and operated *normally* without challenges to safety limits. The next level provides additional design features that protect against a single, *unlikely* fault (~once in the plant lifetime) resulting in limited damage that requires minor repair. The third level provides additional design features that protect against a single, *extremely unlikely* fault (not expected in the plant lifetime) that may cause extensive damage requiring major repairs. These foreseen events are within the plant safety design basis, and the most demanding events are usually identified as *Design Basis Accidents* (DBAs). DBAs are analyzed to demonstrate compliance with regulatory requirements, and the analysis results are documented in safety reports to verify safety margins for licensing.

Key safety functions are met by employing multiple, diverse, and independent structures, systems, or components—each of which is capable of achieving the defined safety function. Redundancy, diversity, and independence assure that all safety functions cannot be lost due to a single failure, either internal (equipment failure, operator action) or external (earthquake, fire, flood). *Safety grade systems*, components, and structures are designed and maintained to criteria that assure their reliable operation, with special attention to quality assurance and provisions for inspection, testing, and repair. The key safety functions are as follows:

Containment: Containment of radioactive material is assured by multiple physical barriers: the fuel cladding, the primary coolant system boundary, and the containment structure.

Reactor shutdown: Reactor shutdown is assured by multiple reactor control and protection systems: a primary system for startup, shutdown, reactor power changes, and power distribution management, and a secondary shutdown system, always available for activation.

Residual heat removal: Residual heat removal is assured by multiple heat transport paths and systems: the normal heat removal system (steam generator, condenser), and dedicated emergency shutdown heat removal systems.

In addition to safety margins provided by design features incorporated into the design basis, it is customary to analyze the performance of the design in events that exceed the normal safety design envelope. These events are assumed to involve an accident initiator *and* failure of one or more safety-grade systems. As double- or multiple-fault events, these accidents are beyond the single-fault criterion for design-basis accidents, and are classified as *Beyond Design Basis Accidents* (BDBA). BDBA analyses provide identification of mechanisms, features, and phenomena that are relevant to the overall safety capability of the nuclear reactor plant design. In addition, BDBA analyses provide quantification of margins beyond the safety basis, and an indication of design areas where safety margins could be enhanced to balance the protection spectrum.

One type of BDBA that has received special attention is the Anticipated Transient without Scram (ATWS), in which it is assumed that all of the safety-grade reactor shutdown systems fail when called upon to act in response to an operational or DBA initiator. For a SFR in a normal operating condition, there exist only three events that can perturb the reactor state: (1) a change in the coolant flow rate, (2) a change in the coolant inlet temperature, or (3) a change in the reactivity. Consequently, the three limiting ATWS events are:

(1) an unprotected loss of coolant flow (ULOF) accident in which all reactor coolant pumps cease operation *and* the primary and backup reactor shutdown systems fail,
(2) an unprotected loss of heat sink (ULOHS) accident in which normal heat rejection capability is lost *and* the primary and backup reactor shutdown systems fail, and
(3) an unprotected transient overpower (UTOP) in which an unexpected positive reactivity is inserted to the core (such as a rapid removal of one or more control rods) *and* the primary and backup reactor shutdown systems fail.

In past US regulatory reviews of the Fast Flux Test Facility (FFTF) and the Clinch River Breeder Reactor Plant (CRBRP), such limiting ATWS events have been identified as precursors for severe accident progression (core melting) [13–16]. In particular, the ULOF BDBA sequence was identified as an enveloping event for containment margin assessment.

Following the FFTF/CRBRP era, further research and development in the US led to design safety enhancements [17–20]. The approach was developed from findings of the Advanced Liquid Metal Reactor (ALMR) program in the 1980s, and in particular the Integral Fast Reactor (IFR) program at Argonne National Laboratory (ANL). Testing performed at EBR-II in 1986 demonstrated benign reactor behavior (no cladding failures, no coolant boiling, and no fuel melting) in full-scale and full-scope ULOF and ULOHS BDBA simulations [9]. These tests showed that the key design features for *prevention* of severe accident progression are (1) inherent reactivity feedbacks to shut down fission power in unprotected accidents, and (2) natural circulation shutdown heat removal. The EBR-II tests demonstrated that, with appropriate design selections, the consequences of accident initiators that led to calculated core melting in FFTF and CRBRP could be limited to an elevated coolant temperature only slightly above the normal operating temperature. Subsequent natural convection tests at FFTF demonstrated adequacy of natural convection cooling and the ability to accommodate the ULOF and ULOHS BDBA events from partial power [21, 22]. Inherent negative reactivity feedback for fission power shutdown can be most readily assured with metallic fuel, with its high thermal conductivity and low operating temperature. These design features were employed in the Rockwell International SAFR and General Electric PRISM (described in Appendix B) concepts developed in the ALMR program.

13.4.3 Classification of Accidents

Whereas the defense-in-depth approach provides the basic philosophy in all fast reactor safety studies, reactor designers and safety analysts need more specific guidance regarding (1) the treatment of specific accidents, and (2) the specification of equipment requirements in order to achieve the safety margins desired in the overall plant.

The NRC has provided a categorization of accident types as shown in Table 13.2. The examples shown for LWR systems have undergone considerable cross examination and have certain regulatory status. The fast reactor examples, however, are offered simply as possible candidates to provide a framework for many of the accidents discussed in this text; they are without any regulatory status.

Another accident categorization, which focuses on the classification of safety equipment, is contained in nuclear standards developed under the sponsorship of the American Nuclear Society (ANS). These standards were developed in accordance with procedures provided by the American National Standards Institute (ANSI), which ensure a consensus of affected interests to permit their approval as American National Standards. Because of the nature of the ANSI procedures, which require industry, government and public acceptance before any proposed guidelines can attain the status of approved American National Standards, many of the fast reactor standards are yet to be finalized. Several of the proposed standards were published as draft standards by the ANS for trial use and comment. However, much of this work was discontinued in the 1980s following the cancellation of the CRBRP and activity has only recently been restarted due to renewed interest in fast reactors, especially small modular systems. It is instructive to sketch out the classifications and indicate the general guidance available to reactor designers. An example of classifications found in those standards is shown in Table 13.3.

13.4.4 Use of Probabilistic Risk Assessment in Fast Spectrum Reactor Design for Safety

As previously noted, the evolution of the design of a fast reactor must incorporate the sound and well-practiced safety concepts of defense in depth and adequate safety margins. These goals can be achieved through the use of a probabilistic risk assessment (PRA) [22–24].

Table 13.2 Classification of postulated accidents

Description	LWR examples	Fast reactor examples
1 Trivial incidents	Small spills: Small leaks outside containment	Single seal failures: minor sodium leaks
2 Small releases	Spills, leaks and pipe breaks	IHTS valve, seal leaks; condensate storage tank valve leak; turbine trip/steam venting
3 Radwaste system failures	Equipment failure: release of waste storage tank contents	RAPS/CAPS valve leaks; RAPS surge tank failure: cover gas diversion to CAPS: liquid tank leaks
4 Events that release radioactivity into the primary system	Fuel failures during normal operation; transients outside expected range of variables	Fuel failures during normal operations: transients outside expected range of variables
5 Events that release radioactivity into the secondary system	Class 4 and heat exchanger and SG leak	Class 4 and heat exchanger leak
6 Refueling accidents inside containment	Drop fuel element: drop heavy object onto fuel: mechanical malfunction or loss of cooling in transfer tube	Drop of fuel element: crane impact on head; inadvertent floor valve opening: leak in fuel transfer cell/chamber
7 Accidents to spent fuel outside containment	Drop fuel element. Drop heavy object onto fuel: drop shielding cask. Loss of cooling to cask: transportation incident *on site*	Shipping cask drop; EVST/FHC system leaks; loss of forced cooling to EVST
8 Accident initiation events considered in design-basis evaluation in the safety analysis report	Reactivity transient; rupture of primary piping: flow decrease. Steam-line break	SG leaks: Na/Water reaction; fuel failure propagation; rupture of primary piping; pump failure or reactivity transient (with PPS operating)[a]
9 Hypothetical sequences of failures more severe than class 8	Successive failures of multiple barters normally provided and maintained	Successive failures of multiple barriers normally provided and maintained

CAPS: Cell atmosphere processing system, IHTS: Intermediate heat transport system, EVST: Ex-vessel storage tank (in spent fuel), RAPS: Radioactive argon processing system (purifies contaminated core gas), FHC: Fuel handling cell, SG: Steam generator

[a]Such events with PPS failure have often been assessed to establish the safety margin in the containment design and the long term decay heat removal capability

The safety of reactors is traditionally based on deterministic approaches, which consider a set of challenges to safety and determine how they can be mitigated. A probabilistic approach to safety enhances and extends the deterministic approach by providing a logical means for prioritizing these challenges based on risk significance, and by allowing consideration of a broader set of resources to defend against them. A PRA, because it considers both likelihoods and consequences of events, can very readily be used to identify and study those events that would fall within a licensing basis.

Events that have large consequences, beyond the capability of the reactor plant to safely mitigate, are reduced in risk significance by requiring that their likelihoods are acceptably small such that they do not introduce risk beyond an accepted norm. A risk profile is then developed for the design of the plant that preferentially limits the more likely events to be of less consequence such that the integrated risk is acceptably low. Significance is determined by the policy of the regulator. For example, the NRC policy statement on safety goals of 1986 has been used in conjunction with risk-informing many safety issues for LWRs. For advanced plants, such as the SFRs, even higher aspirations for safety may be

Table 13.3 Conditions categorized by ANSI standards

PWR and BWR		Fast reactor	
Category	Examples	Category	Examples
I. Normal operation (expected frequently during normal course of operation and maintenance)	Startup, shutdown, standby, power ascension from partial load to full power cladding defects within Tech Specs; refueling	I. Normal operation (expected frequently during normal course of operation and maintenance)	Startup, normal shutdown, standby, load following; cladding defect within Tech Specs; refueling
II. Incidents of moderate frequency (any one of which may occur during a *calendar year* for a particular plant)	Inadvertent control rod group withdrawal; partial loss of core cooling; moderate cool-down; loss of off-site power; single error of operator	II. Anticipated operational occurrences (may individually occur one or more times during the *lifetime* of the plant)	Tripping of Na pumps; failure of all off-site power; tripping of Turbine-Generator set; inadvertent control rod withdrawal
III. Infrequent events (may occur during the lifetime of a particular plant)	Loss of reactor coolant (with normal coolant makeup system only); secondary pipe break; fuel assembly in violation of Tech Specs; control rod withdrawal in violation of Tech Specs; unexplained reactivity insertion; complete loss of core flow (excluding pump locked rotor)	III. Postulated accidents (not expected to occur-but included in the design basis to provide additional margins for assuring no undue risk to the health and safety of the public)	Spectrum of events appropriate to a specific design considering both the probability and consequences of the events (e.g. pipe rupture, large Na fire, large Na-water reaction, rupture of radwaste system tank)
IV. Limiting faults (not expected to occur-but postulated because of potential for significant radioactive release; most drastic faults that must be designed against)	Major pipe rupture (up to and including double-ended rupture of largest pipe); fuel or structure movement due to core damage; ejection of single control rod; major secondary system pipe rupture (double-ended): coolant pump locked rotor		

expected because the advanced plants incorporate improved design features and may benefit from operational experience with the current fleet of LWRs. It is for this reason that their ability to provide inherent protection against ATWS events is especially relevant and the use of PRA can help quantify reduction of risk.

The design evolution of SFRs can be an iterative process in which functional design characteristics are put forth and then tested against a PRA through a quantitative evaluation. The design is then further refined and developed as needed to produce an acceptable risk profile for the plant. As more specific features are incorporated in the design, the risk models given in the PRA are also commensurably refined. This process of progressive refinement continues until the plant is mature enough to receive a license to operate. Beyond this point in time, the PRA is used as an operational tool to aid in the decision making of the operator/owner as power is produced by the plant in a safe, economic, and reliable way. It is important to maintain a living PRA model of the plant throughout its life history, especially in anticipation of new safety issues that might arise. PRAs have also been shown to be

Table 13.4 Classification of events and consequences for reactor licensing

Events	Frequency	Current NRC allowable consequences
Anticipated operational occurrences (AOOs)		
Operational events	Expected during the lifetime of the plant (>10^{-2} per reactor year)	None; maintain margin to fuel damage
Postulated accidents		
Design Basis Accidents (DBAs), typically failure of one safety grade system	Not expected to occur during the lifetime of the plant, but anticipated in the design (>10^{-5} per reactor year)	Minor fuel damage permissible at lower probability (<10^{-4} per reactor year); allowable individual exposure <25Rem (2.5mSv)
Beyond Design Basis Accidents (BDBAs), e.g., multiple failures of safety grade systems, including ATWS events	Accidents of very low probability not considered as part of the design basis for the plant (<10^{-5} per reactor year)	Substantial fuel damage permissible; allowable exposure >25Rem (2.5mSv) to public at lower probability (<10^{-6} per reactor year)

useful for training programs associated with safe operation and in communicating a perspective on plant safety to various stakeholders.

In summary, a PRA can be a beneficial tool to overall plant design and operation and, in particular, to the early safety-related design considerations that would be part of a gap analysis on accident phenomenology [25–27]. Shown in Table 13.4 is one approach to accident classification that takes advantage of a PRA formulation.

13.4.5 NRC Procedural Requirements

The NRC has established a process for achieving design certification and/or licensing [28, 29]. Design certification involves extensive review and once approved, obtaining a combined license (COL) for construction and operating of the reactor design at a specific site can be greatly expedited. For the NRC to grant a COL, assurance must be given that the safety characteristics of the certified design are compatible with characteristics of the site.

Obtaining a design certification involves a comprehensive review along the following schedule:

- 1–3 years: Pre-application review to address major technical/safety issues important in the license application.
- 1–2 years: Preparation of the license application.
- 1–4 years: NRC staff review of the license application and completion of a Safety Evaluation Report; if all questions have been successfully resolved, a recommendation to the commission for design certification.
- 1–2 years: Review by the commission for design certification, aided by independent review and if questions are satisfactorily resolved, issuance of a design certification.

The legal framework for these activities and definitions of what must be addressed by the applicant are defined in Title 10 of the Code of Federal Regulations under Part 50 and Part 52 (10 CFR Part 50) and (10 CFR Part 52). Under 10 CFR Part 50, the applicant must provide (1) a preliminary safety analysis report, (2) an environmental impact report, and (3) documentation of financial competence and compliance with antitrust regulations. The NRC review includes not only a safety evaluation but also compliance assessment with applicable regulations for site characteristics, plant design, plant response to hypothetical accidents, plant operations, environmental impacts and emergency response. The NRC review is documented in a Safety Evaluation Report and a Draft Environmental Impact Statement. The Advisory Committee on Reactor Safeguards (ACRS) also reviews the application and reports its findings to the chairman of the NRC.

When a construction permit is issued, initial construction can begin and the detailed design and operating procedures completed. An operating license application is submitted, which includes a Final Safety Analysis Report (FSAR). Following review, the NRC documents its review in a Final Safety Evaluation Report and the ACRS performs an independent evaluation of the FSAR and the NRC staff review.

The process described above as defined in 10 CFR Part 50 is tailored for current LWR designs. Non LWR applications of 10 CFR Part 50 are accomplished in a case-by-case review pursuant to agreements reached between the applicant and the NRC.

In 1989, the NRC established an alternative licensing process under 10 CFR Part 52, which defines both design certification and granting of a combined construction and operating licensing process. The intent is to facilitate licensing by defining a one-step licensing process and emphasizing up-front review and approval of both the site and the reactor design. The early site permit is intended to resolve site safety, environmental protection, and emergency preparedness issues independent of a specific nuclear plant design. The application must provide information on site boundaries, site characteristics, proximate facilities, proximate populations, site plant locations, maximum thermal and radiological discharges, hypothetical accident considerations, and emergency plans.

13.5 Accident Analysis Perspective

The approach to fast reactor safety considerations has benefited from the large reservoir of experience developed from LWR systems even though the precise accident paths differ. A discussion of important differences is provided below.

SFRs enjoy a safety advantage since the sodium cooled reactor is a low pressure system in which the single phase primary loop flow would continue operating independent of the pressurization requirements evident in a LWR.

Perhaps the most significant difference between fast reactors and LWRs is the core configuration. In a LWR, the core components (fuel, cladding, coolant) are arranged to maximize k_{eff}. Any rearrangement, due to fuel melting or coolant loss, tends to result in significant negative reactivity feedback. A fast reactor, on the other hand, needs no moderator and is not in its most reactive configuration. Consequently, if fuel melting should occur, a core compaction process could lead to a reactivity gain. Also, complete voiding of sodium in a large core could result in a reactivity gain [30]. These possibilities provided the underlying concern for the majority of the large-accident safety studies done for the early fast reactor systems.

As will be noted in Chapter 15, there is now considerable evidence that the inherent forces within a SFR core in the event of a postulated fuel meltdown would more likely disperse the fuel early in the transient, thus leading to a reactivity loss rather than a reactivity gain [31, 32]. Nevertheless, the potential for sodium voiding and a major increase in reactor power followed by fuel melting provides a fundamental departure in behavior relative to the LWR system.

The LWR water coolant needs to be pressurized [~7 MPa (1,000 psi) for a BWR and 15 MPa (2,200 psi) for a PWR] in order to obtain reasonable coolant temperatures. Hence, any breach in the primary system directly leads to coolant "blowdown." This concern is absent in fast reactor systems, since the sodium is highly sub-cooled (by about 350°C) at normal operating conditions. The maximum system pressure [~1 MPa (150 psi)] is just high enough to insure proper coolant flow through the core. A breach in the primary system of a fast reactor would not result in coolant boiling. The primary concern would be eventual loss of coolant and any chemical reactions that might occur. The chemical reactions of concern are primarily sodium-air and sodium-water reactions. A barrier to such reactions for normal operation is provided by covering all sodium systems with an inert cover gas.

The nuclear properties of a LWR would seem to imply easier dynamic control, relative to a Pu-fueled fast reactor, because the delayed neutron fraction is about twice the SFR value (due to the presence of ^{235}U rather than ^{239}Pu), and the prompt neutron generation time is about two to three orders of magnitude larger. However, these differences come into play only under major accident conditions, and concerns about the short neutron lifetime of the fast reactor, which were expressed in the early days of safety studies, have been largely allayed due to inherent feedback characteristics that limit the rise of power upon attaining prompt criticality (discussed in Chapter 15).

The thermal time constants of a SFR tend to be shorter than those of an LWR, due mainly to the smaller fuel pin size and smaller coolant flow area per pin. Such factors are routinely taken into account in performing transient studies. This is an important consideration because of the potential for thermal shock due to the relatively high heat capacity and thermal conductivity of sodium.

13.5.1 Accident Initiators

Although there are substantial differences in the systems associated with LWR's and fast reactors, which necessarily lead safety analysts to consider several physically different mechanisms as potential accident initiators, the neutronic response of each reactor type is very similar for the more probable accidents. The major low probability-large consequence concern in the LWR is a massive rupture in the primary piping system, which would allow coolant flashing and subsequent blowdown. Should such an event occur, operation of the emergency core cooling system (ECCS) is required to prevent fuel melting. Hence, considerable efforts have been and continue to be expended to assure the function-ability and reliability of the ECCS to provide means of cooling the core over a long enough time frame to remove the decay heat. As an upper limit to the public consequences possible from an LWR accident, a double-ended pipe break has historically served as the initiating event for LWR accident consequences evaluation.

A major pipe break in a fast reactor does not lead to the same consequences as in an LWR because rapid system depressurization would not result in coolant flashing. Hence, the accidents analyzed for public consequence considerations have been traditionally associated with core rearrangement. In order for such a rearrangement to occur, fuel melting is necessary and, consequently, bounding accidents for fast reactors have been historically associated with a postulated power/flow imbalance-coupled with failure of the PPS to terminate the excursion. The heat generation/heat removal imbalance can come about due to either a *transient overpower* (TOP) condition, in which a reactivity insertion causes the power to rise, or a *transient undercooling* (TUC) condition, in which the power remains essentially constant, but primary coolant flow is lost.

The major advance in fast reactor safety has been developing designs to provide inherently safe response to these events. Aggressive operational-safety tests conducted in EBR-II in the 1980s to address these issues are reported in Chapter 15.

13.5.2 Historical Perspective of Fast Reactor Accidents

Prior to launching into the various approaches under current use in analyzing fast reactor accidents, it is appropriate to review briefly the early approaches to fast reactor safety. The reason for this backward glance is that much of the present work—especially the efforts focused on major accident considerations—is strongly influenced by safety studies conducted some two to three decades ago.

13.5.2.1 Early Concerns

Certainly the strongest factor that influenced the early safety analysts was the question of possible core compaction. The fact that fuel densification would increase the system reactivity, in contrast to thermal reactor systems, dominated the types of accidents postulated for analysis. In addition to the core compaction question, the small *effective beta*, $\bar{\beta}$, described in Section 6.4 (hence, small value for the dollar unit of reactivity) and short prompt neutron lifetime provided concern that a core meltdown may rapidly lead to a prompt critical condition and that the power rise beyond that point could be extremely rapid. Although fuel vaporization, an inherent shutdown mechanism, would ultimately limit the energy generated by physically moving the core apart, the concern was that the potential work energy released might be larger than practically containable.

Hypothetical Core Disruptive Accident (HCDA) studies have been conducted for nearly all of the fast reactors in the United States and Western Europe [33–35]. A major result of each study has been the calculation of a maximum HCDA work energy, defined as an upper limit potential for work done on the reactor environs by expanding core debris.

The original Bethe-Tait analysis (briefly described in Chapter 16) would predict increasing values for maximum HCDA work energy for the larger core sizes. However, as discussed earlier, major improvements have been made to analytical techniques, most of which demonstrate reduced consequences. It is clear that the efforts expended in better understanding the sequence and ultimate consequence of even an unprotected hypothetical core disruptive accident have made a favorable impact on containment demands.

13.6 Risk and Accident Analysis Approaches

As to be expected, the relative emphasis given to the concept of risk, as well as to the degree to which the basic elements of risk are blended into an overall safety analysis effort, varies somewhat within the international fast reactor community. Despite the particular approaches used, however, it should be clear from the preceding section that substantial progress has been made over the past decades in reducing the requirements for fast reactor accident containment. Much of this progress has resulted from an increased attempt to analyze postulated accident sequences mechanistically, i.e., computationally tracking through cause and effect relationships—rather than relying on upper bound calculations that contain an undetermined number of safety margin layers.

The discussion below summarizes the basic ingredients required to determine overall risk; namely, the mechanistic and probabilistic approaches. A principal task is to blend these approaches in an optimal manner to allow an overall system risk assessment to be confidently attained.

13.6.1 Mechanistic

The mechanistic (or deterministic) approach is conceptually straightforward. It represents an attempt, by the analyst, to track a scenario from the postulated inception of the accident all the way through the various material motions and component failure sequences until the system has reached a long term steady-state condition. All of the governing inherent system response features are taken into account in performing this tracking procedure.

This approach is popular because it forces both the analyst and experimenter to determine the accident sequence resulting from direct cause and effect relationships. It is logically appealing because it allows one to visualize the factors of governing importance and also allows sensitivity analyses to be rationally conducted to allow for areas of greatest uncertainty. Numerous computer code systems have evolved and such computational capability has been highly successful in both reducing containment requirements for major-accident calculations and providing a focus on those areas of uncertainty where both analytical and experimental effects are most urgently needed. Examples of employing the mechanistic approach are given in Chapters 15 and 16.

The principal disadvantage of this approach is that large and complex computational systems must be developed and continually improved to provide a systematic tracking of the numerous combinations of interactions that could occur—especially for analyzing whole-core transients. An additional problem is that of developing computational capability for adequate simulation beyond the point where gross core geometric changes occur. However, significant advances in computation capability for modeling and simulation have enhanced the value of this approach. Their usefulness is dependent upon recovering important data from previous tests and experiments, largely conducted in the decades of the 1980s and 1990s.

13.6.2 Probabilistic

The probabilistic approach takes into account a systematic evaluation of event trees, fault trees, and the appropriate mathematical techniques to evaluate overall probability for system failures. With regard to the latter area, the approach quantitatively takes into account variations in physical parameters in the reactor system and variations in the models used to describe certain complex physical phenomena (e.g., fuel pin failure models).

As inferred from this discussion, it is clear that the probabilistic approach requires some consequence evaluations to be made in order to assess the implications of probabilistic variations in various parameters in the overall analysis. By including distribution functions for reactor parameters (e.g., fuel Doppler coefficients) into a fast running code, accident spectra useful for probabilistic predictions can be obtained. A major difficulty in employing the probabilistic approach is that the distribution functions for the numerous parameters of uncertainty are not well known. Little failure data are available for systems that have low failure rates. Also, there are many facets of fast reactor accident analysis—particularly for unprotected accidents—where simplified consequence models are not available for use in probabilistic assessments.

References

1. V. Dostal, "A Supercritical Carbon Dioxide Cycle for Next Generation Nuclear Reactors," Ph.D. Dissertation, MIT (2004).
2. J. D. B. Lambert, et al., "Performance of Breached LMFBR Fuel Pins During Continued Operation," BNES Conference on Nuclear Fuel Performance (March 25–29, 1985).
3. L. C. Walters, B. R. Seidel, and J. S. Kittel, "Performance of Metallic Fuels and Blankets in Liquid-Metal Fast Breeder Reactors," *Nuclear Technology*, *65* (1984).
4. L. C. Walters and G. L. Hoffman, "Metallic Fast Reactor Fuels: A Comprehensive Treatment," in *Material Science and Technology*, Vol. 10A, VCH, eds. R. W. Cahn, et al. (1994).
5. C. M. Cox and R. F. Hilbert, "U.S. Experience in Irradiation Testing of Advanced Oxide Fuels," *Proceedings of the ANS Topical Meeting on Advanced LMFBR Fuels* (1977).

6. J. I. Sackett and H. W. Buschman, "Safety Implications from 20 Years of Operation at EBR II," *International Topical Meeting on Fast Reactor Safety*, Knoxville, TN (April 21–25, 1985).
7. J. I. Sackett, R. M. Singer, and A. Amorosi, "Design Features to Maximize Simplicity, Operability and Inherent Safety of LMFBRs," *Transactions of the American Nuclear Society, 38* (1982).
8. R. M. Singer, et al., "Decay Heat Removal and Dynamic Plant Testing at EBR II," Second Specialists' *Meeting on Decay Heat Removal and Natural Convection in LMFBRs*, BNL, Upton, New York, NY (April 17–19, 1985).
9. H. P. Planchon, et al., "The Experimental Breeder Reactor II Inherent Shutdown and Heat Removal Tests – Results and Analyses," *Nuclear Engineering and Design, 91* (1986), 287–296.
10. T. R. Beaver, et al., "Transient Testing of the FFTF for Decay Heat Removal by Natural Convection," *Proceedings of the LMFBR Safety Topical Meeting*, Vol. II, pp. 525–534, European Nuclear Society, Lyon, France (July 19–23, 1982).
11. J. I. Sackett, "Safety Philosophy in Upgrading the EBR II Protection System," *Proceedings of the International Meeting on Fast Reactor Safety and Related Physics*, American Nuclear society, Chicago, IL (November, 1976).
12. N. Diaz "The 3rd Annual Homeland Security Summit, The NRC's Defense-in Depth Philosophy," (June 3, 2004).
13. U.S. Nuclear Regulatory Commission, *Safety Evaluation Report Related to the Construction of the Clinch River Breeder Reactor Plant*, NUREG-0968, Vols. 1 and 2 (March, 1983).
14. D. E. Simpson, "Resolution of Key Safety-Related Issues in FFTF Regulatory Review," *Proceedings of the International Meeting on Fast Reactor Safety and Related Physics*, CONF-761011, Vol. II, pp. 400–441, American Nuclear Society, Chicago IL (October 5–8, 1976).
15. R. J. Slember, "Safety-Related Design Considerations for the Clinch River Breeder Reactor Plant," *Proceedings of the International Meeting on Fast Reactor Safety and Related Physics*, CONF-761001, Vol. I, pp. 112–125, American Nuclear Society, Chicago, IL (October 5–8, 1976).
16. T. G. Theofanous and C. R. Bell, "An Assessment of CRBR Core Disruptive Accident Energetics," *Proceedings of the International Meeting on Fast Reactor Safety*, Vol. 1, pp. 471–480, American Nuclear Society, Knoxville, TN (April 21–25, 1985).
17. Y. I. Chang, "Advanced Nuclear Energy System for the Twenty-First Century," *PHYSOR 2002*, Seoul, Korea (October 7–10, 2002).
18. G. J. Van Tuyle, et al., *Summary of Advanced LMR Evaluations – PRISM and SAFR*, NUREG/CR-5364, BNL-NUREG-52197 (1989).
19. U.S. Nuclear Regulatory Commission, *Preapplication Safety Evaluation Report for the Sodium Advanced Fast Reactor(SAFR) Liquid Metal Reactor*, NUREG-1369 (December 1991).
20. U.S. Nuclear Regulatory Commission, *Preapplication Safety Evaluation Report for the Power Reactor Innovative Small Module (PRISM) Liquid Metal Reactor*, NUREG-1368 (February 1994).
21. T. R. Beaver, et al., "Transient Testing of the FFTF for Decay Heat Removal by Natural Convection," *Proceedingsof the LMFBR Safety Topical Meeting*, Vol. II, pp. 525–534, European Nuclear Society, Lyon, France (July 19–23, 1982).
22. A. Padilla and D. J. Hill, "Comparison of Reactivity Feedback Models for the FFTF Passive Safety Tests," *Proceedings of the Safety of Next Generation Power Reactors Conference*, Seattle, WA (May 1988).
23. U.S. Nuclear Regulatory Commission, *Staff Plan to Make a Risk-Informed and Performance-Based Revision to 10 CFR Part 50*, SECY-06-0007 (January 9, 2006).
24. U.S. Nuclear Regulatory Commission, *Approaches to Risk-Informed and Performance-Based Requirements for Nuclear Power Reactors,10 CFR Parts 50 and 53*, RIN 3150-AH-81 (May 4, 2006).
25. D. J. Hill, "An Overview of the EBR II PRA," *Proceedingsof the 1990 International Fast Reactor Safety Meeting*, Snowbird, UT, Vol. IV, p. 33, American Nuclear Society (August 12–16, 1990).
26. R. S. May, et al., "Probabilistic Methods for Evaluating Operational Transient Margins and Uncertainties," *Nuclear Science and Engineering, 103* (1989) 81.
27. R. W. Schaefer, "A Probabilistic Method for Evaluating Reactivity Feedbacks and Its Application to EBR II," *Advances in Mathematics, Computations, and Reactor Physics*, Pittsburgh, PA (April 28–May 1, 1991).
28. U.S. Nuclear Regulatory Commission, *Nuclear Power Plant Licensing Process*, NUREG/BR-0298, Rev. 2 (July 2004).
29. U.S. Nuclear Regulatory Commission, World Wide Web site: www.nrc.gov.
30. R. A. Wigeland, R. B. Turski, and P. A. Pizzica, "Impact of Reducing Sodium Void Worth on the Severe Accident Response of Metallic-Fueled Sodium-Cooled Reactors," *Proceedings of the International Topical Meeting on Advanced Reactors Safety*, Pittsburgh, PA, Vol. II, p. 1062 (April 1994).
31. J. E. Cahalan, R. A. Wigeland, G. Friedel, G. Kussmaul, J. Moreau, M. Perks, and P. Royl, "Performance of Metal and Oxide Fuels during Accidents in a Large Liquid Metal Cooled Reactors," *Proceedings of the International Fast Reactor Safety Meeting*, Snowbird, UT, Vol. IV, p. 73 (August 1990).

32. L. R. Campbell, et al., *FFTF Loss of Flow Without Scram Experiments with GEMS*, HEDL-TC-2947, Hanford Engineering Development Laboratory, Richland, WA (June 1987).
33. L. E. Strawbridge and G. H. Clare, "Exclusion of Core Disruptive Accidents from the Design Basis Accident Envelope in CRBRP," *Proceedings of the International Meeting on Fast Reactor Safety*, Vol. 1, pp. 317–327, American Nuclear Society, Knoxville, TN (April 21–25, 1985).
34. L. W. Deitrich, et al., "A Review of Experiments and Results from the Transient Reactor Test (TREAT) Facility," ANS Winter Meeting, Washington, D.C. (November 15–19, 1998).
35. C. L. Beck, et al., "Sodium Void Reactivity in LMFBRS: A physics Assessment," *Proceedings of the Topical Meeting on Advances in Reactor Physics and Core Thermal Hydraulics*, Kiamesha Lake, NY, NUREG/CP-0034, Vol. 1, p. 78 (September 22–24, 1982).

Chapter 14
Protected Transients

John Sackett

14.1 Introduction

As one of the fundamental principles of defense-in-depth, there are multiple physical barriers to prevent radiation release and redundant defenses to protect them. In this way, the failure of any single physical barrier does not result in a risk to public health and safety. A typical example of the multiple barrier approach for previous sodium-cooled fast reactor designs such as EBR-II, FFTF, and CRBRP has been the use of cladding on the fuel as the first barrier, the closed primary coolant loop as the second barrier, and the containment building as the third barrier. There are other examples of fast reactors with both fewer and greater numbers of such barriers, and the adequacy of any specific approach is one of the evaluations performed by the appropriate national nuclear regulating body as part of reviewing a license application, such as the United States Nuclear Regulatory Commission (NRC).

A Plant Protection System (PPS) is routinely provided in a Sodium Fast Reactor (SFR), as in any nuclear reactor system, to provide a rapid and safe shutdown for off-normal situations and to protect all barriers to radiation release. The prime objective of such a PPS is to assure protection of the general public, the workers, and the plant investment. In this sense, *protected transients* are those for which an engineered, active plant protection system operates as designed. Events for which the PPS fails to function are considered in the next chapter.

An effective PPS must be able to arrest any undesirable increase in heat generation or loss of heat removal capability. This is reflected in the two major parts of the PPS: (1) a reactivity shutdown system, and (2) a decay, or residual, heat removal system. Other engineered safeguards features may be included in the overall plant system, such as isolating the containment building, fission product removal from containment, or post-accident heat removal devices, but such systems will not be discussed in this chapter since they are not needed or activated if the PPS successfully performs its function.

It is important to recognize that successful action by the PPS to off-normal conditions renders the SFR an exceptionally safe system. The low pressure operation of the system (at temperatures far below boiling), excellent heat transfer characteristics, the large heat capacity of the coolant, and strong natural convective cooling ensure this result.

The purpose of the present chapter is to focus attention on protected transients and to address the ability of the PPS to detect an off-normal condition, to respond appropriately and in a timely manner, and to do so with very high reliability. The entire range of possible transient events is considered, from local fault propagation to events that affect the entire reactor plant.

J. Sackett (✉)
Argonne National Laboratory (Retired), Argonne, IL, USA
e-mail: johnflyer@3riversdbs.net

A.E. Waltar et al. (eds.), *Fast Spectrum Reactors*, DOI 10.1007/978-1-4419-9572-8_14,

14.2 Plant Protection System

14.2.1 Transient Condition Envelope

An imbalance in the heat generation-to-heat removal ratio constitutes the primary concern for in-core transient analyses. As long as the nuclear heat generated is adequately removed, intrinsic barriers to radiation release remain intact. It is only when an over-heated condition exists that a potentially serious safety question arises. Consequently, the fundamental questions that must be posed in designing an effective PPS are focused on malfunctions of controls that could lead to overheating due to higher than normal power generation (i.e., reactivity insertion mechanisms) or under cooling due to reduced reactor cooling (i.e., impairment of heat removal).

14.2.1.1 Reactivity Addition Mechanisms

The most obvious area for scrutinizing possible reactivity additions is the potential relative motion or displacement of core materials that involve significant reactivity worth, such as control rods, fuel, and sodium. Control rod movement constitutes the dominant operational mode for purposely increasing or decreasing reactivity into the reactor. Hence, it is natural to ask what the core response would be if it were possible to inadvertently withdraw control rods too fast or too far. The design of the control-rod system contains interlocks that normally prevent more than one rod from being withdrawn at one time, and a maximum speed of withdrawal (runaway speed) is purposely designed into the system to prevent very rapid reactivity additions. Even given the worst combinations of failures leading to the rapid withdrawal of one or more control rods, with proper design the maximum reactivity insertion rates are typically on the order of only a few cents per second, and the amount of reactivity added to the core can be correspondingly limited to prevent damage.

Reactor designs are developed to eliminate other possible mechanisms that can add reactivity to a reactor core, including possible control assembly or fuel assembly meltdown, loss of hydraulic balance to the fuel assemblies, inward radial movement of the core assemblies (core compaction), boiling of the sodium coolant in the central regions of the reactor core, and the introduction of colder sodium into the reactor core. Each of these mechanisms has been the subject of extensive analysis and testing over several decades of fast reactor development, backed up by considerable experience in SFR operation. Modern SFR designs render each of them either physically not credible, or to extremely low levels of probability [1–5]. In any event, if they were to occur, protection would be provided by the PPS.

14.2.1.2 Impairment of Heat Removal

The inadvertent reduction or loss of primary-loop coolant flow constitutes the most serious concern from the standpoint of the heat transport system, although any event that could allow the inlet coolant temperature to rise is also to be avoided since this erodes the design margins for core temperatures that have been established for safe operation.

Primary and intermediate coolant pumps, steam generator feed water pumps, and condenser cooling water pumps are typically operated from an external or off-site power supply, e.g., the power grid. The loss of off-site electrical power leading to loss of power to coolant pumps is normally included as an anticipated event (expected to occur once or more during the plant lifetime) and, consequently, it is particularly important to demonstrate that the PPS system will function properly in this event. A safe and non-damaging shutdown must also be assured even if the backup electrical supply (supplied for all nuclear plants as part of the defense in depth philosophy) fails to come on line. Failures in the secondary sodium or steam loop systems are more easily protected since they tend to provide less severe consequences to the core and some thermal time delay is inherently present. However, the consequence of all credible events must be assessed to assure that the propagation of the initiating event could not lead to unacceptable shutdown conditions and that the plant protection systems respond appropriately.

14.2.2 PPS Effectiveness Considerations

A properly functioning PPS should arrest the types of initiating events discussed above and affect a safe, cold shutdown with no core damage. Typical transient initiators considered for PPS design are shown in Table 14.1.

14.2.3 Reactor System Integrity Limits

Given the spectrum of event initiators listed in Table 14.1 it is essential to ensure that a proper response of the PPS will provide adequate protection to the reactor system and barriers to radiation release. To be assured of such protection, it is necessary to determine the fuel pin and heat-transport system limits that must not be exceeded during such conditions.

14.2.3.1 Fuel Damage Limits

In establishing transient design limits for reactor core structural components, the fuel pin and control rod absorber pin cladding receive particular scrutiny due to their lead role as the first barrier to release of radioactive materials. The objective of such analysis is to define realistic limits to establish the capability of the cladding under steady-state and transient conditions for the fuel system employed. Approaches to perform the analysis are described in Section 8.3.

Table 14.1 Typical initiators considered for PPS design

Event initiator	Equipment failure	Operator error	External events
Reactivity insertion	Control rod uncommanded motion; Overcooling from pump run-up; BOP system blow down; Gas bubble entrainment	Control-rod controller error; Coolant-pump control error; Actuation of pressure relief valve	Seismic
Loss of flow	Electrical faults; Loss of off-site power; Controller failures; Mechanical faults; Pump mechanical failure; Loss of piping integrity	Turning off power to pumps; Opening breakers to power supplies	Seismic
Loss of heat sink	Steam generator failure; Intermediate heat transport failure; Supercritical CO_2 system failure; Loss of electric grid load; Flow blockage in heat transfer loop	Stopping flow in intermediate loop; Blow-down of stream generators; Isolating plant from the grid	Seismic
Primary sodium leak	Primary system overpressure; Contamination or loss of inert cover gas; Leak at penetrations of piping/vessel; Leak in auxiliary systems	Overpressure of cover gas; Open valve in auxiliary systems	Seismic

Cladding damage can occur by metallurgical changes or by mechanical loading. Both mechanisms tend to be very strong functions of temperature and fuel composition. Metallurgical damage includes micro-structural changes, alloy depletion, fission product attack and fuel-cladding metallurgical interaction. Mechanical damage occurs primarily by creep deformation resulting in grain boundary cavitation and local instabilities in the plastic deformation.

Metallic fuels can also interact metallurgically with the cladding. A diffusion zone between the fuel and the cladding may form during steady-state operation involving formation of a eutectic mixture of iron and uranium. Additional metallurgical attack during transients may occur depending on the temperature at the fuel/cladding interface during the transient. If this temperature is high enough, liquids can form at the interface of fuel and cladding, and the eutectic penetration rates that attack the cladding can be quite rapid. Significant transient thinning of the cladding wall may occur.

Fuel melting criteria are very important for judging the response to reactor transients. Since metal fuel (with high thermal conductivity and low fuel centerline temperature but low melting temperature) and oxide fuel (with low thermal conductivity and high fuel centerline temperature but high melting temperature) behave in very different ways to overpower or under cooling transients, this becomes an important distinction in approaches to PPS design and safety analysis.

One example for establishing fuel damage limits is the cumulative damage function (CDF) approach (see Section 8.3.2). This approach was developed in support of both FFTF and CRBR safety analyses to describe behavior of the cladding during steady-state and transient operation—in this case 20% cold-worked, 316 stainless steel and oxide fuel. (Such mechanical stress does not occur in metal fuel if sufficient porosity is designed into the fuel).

The conservatism of this approach has been demonstrated by comparison of predicted steady-state design lifetime with test data from approximately 600 FFTF prototypic fuel pins (including several transient tests in capsules) with 20% cold worked 316 stainless steel cladding irradiated in EBR-II [6, 7] and about 50 pins transient-tested in the Transient Reactor Test Facility (TREAT) [8].

A similar result has been found with metal fuel pins for both steady-state and transient conditions. Transient tests in TREAT have demonstrated that both oxide and metal fuel experience cladding failure at about 4 times overpower [9]. Mild transients also demonstrate considerable margin to cladding failure as was demonstrated in support of the Anticipated Transients without Scram (ATWS) tests in EBR-II [10].

14.2.3.2 Reactor Heat Transport System Damage Limits

The Reactor Heat Transport System (HTS) and connected systems provide the necessary functions to remove reactor heat safely under all plant operating conditions. Boundary materials and service conditions of heat transport systems should be selected and maintained to provide materials properties that will assure non-brittle behavior and avoid conditions that could lead to rapidly propagating failures. The reactor coolant pressure boundary components are generally designed and analyzed in accordance with the ASME Boiler and Pressure Vessel Code, Section III, Class 1 requirements (Section III, Nuclear Power Plant Components) [11] or equivalent code in other regions. Conformance with these requirements will assure HTS structural adequacy at operating temperatures under specified loading conditions, both normal and off-normal, including internal pressure, dead weight, seismic events, design thermal events, and design basis accidents. The PPS contributes to assuring conformance with the requirements by monitoring the HTS pressure and sodium temperatures to assure they satisfy defined limits.

An important consideration for sodium heat-transport systems is the potential for significant thermal shock during transients [12]. The high thermal conductivity of sodium and large heat capacity underlay this potential. Particular attention must be paid to welds and materials used in the heat transport system. Of special interest has been the degradation of mechanical structures from thermal striping, namely very rapid and repeated changes in sodium temperature due to the unsteady motion of the colder and hotter sodium streams exiting the assemblies impinging on reactor structures [13].

14.3 Reliability Considerations

Probabilistic risk assessment (PRA) with modern PPS designs suggest that a very high level of reliability can be achieved, perhaps as low as 10^{-5} probability of PPS failure per reactor year of operation. However, to achieve this level of reliability requires attention to several fundamental principles of design.

14.3.1 Basic Elements of Reliability

The five basic elements that contribute to overall system reliability include (1) functionality, (2) redundancy, (3) diversity, (4) independence, and (5) operational reliability.

Functionality. By functionality, we mean the design capability of the PPS to perform its assigned task if called upon to function. It must, for example, contain sufficient negative reactivity worth to overcome any reactivity that might be inserted by the initiating mechanism. Furthermore, the system must be capable of inserting sufficient negative reactivity into the reactor fast enough to overcome the transient reactivity effects that are an outcome of the initiating mechanism. Fortunately, for any of the credible initiators that have been identified, the functionality requirements of the PPS are relatively easy to meet. For a control assembly in a commercial size plant, a reactivity worth of a few dollars and an insertion time of the order of a second are quite sufficient. Spring assisted rod insertion drives are often used for certain SFR shutdown rods, but this is more a matter of system diversity than of need for rapid response. Hence, systems similar to those routinely employed in the thermal reactors are generally adequate for fast reactor application.

Redundancy. Redundancy implies that multiple systems exist, each fully capable of providing the required safety functions. This feature extends to the instrumentation needed to detect an off-normal situation and the physical systems (control rods, pumps, etc) that protect barriers to release of radioactive material.

Diversity. Diversity means that different principles are employed in the multiple safety systems utilized. Diversity is important if the reactivity shutdown and heat removal systems are to be as free as possible from common cause failures. A considerable incentive exists with regard to enhancing overall PPS reliability to employ diverse systems all the way from the initiating event sensing devices to the activation of the power reducing or heat removal component itself. For example, diversity can be assured by deployment of active shutdown systems (responding to electronic signals) and self-actuated shutdown systems, any one of which provides protection.

Independence. Independence denotes complete separation of the multiple safety systems so that the successful operation of one system in no way is affected by the success or failure of another system. Elements of system independence have been shown in some of the previous examples, but there must be a conscious effort in the design and construction of an effective PPS to assure that the separate safety systems are indeed independent.

Operational Reliability. Operational reliability refers to the absence of spurious trips of the PPS that inappropriately impact reactor operation. It would be fairly easy to design a PPS with so much redundancy, diversity, and independence that the probability of failure when called upon to function would be exceedingly low. The challenge is to design such a system that retains this attribute without spuriously initiating scrams during periods of normal operation.

14.3.1.1 Examples of Reactivity Shutdown System Reliability

Modern examples of PPS design can be found in the PRISM [14] and SAFR [15] reactors, which were commercial designs developed as part of the Advanced Liquid Metal Reactor (ALMR) program conducted by the US. DOE during the 1980s and early 1990s. These designs employ both active and passive shutdown systems for achieving the desired level of protection. Several basic elements of reliability for active safety systems are common to all modern PPS designs. Independence and redundancy are achieved by ensuring separation between two (primary and secondary) shutdown systems, each capable of independently protecting the barriers to release of radioactive material.

With such a design, *both* the primary and the secondary system have to fail completely in order to render the reactivity shutdown system ineffective. Since the secondary system is typically a mirror image of the primary (but with different input signals[1]), the numerical probability of reactivity shutdown system failure is the product of the failure probability for these two independent and redundant systems. A prime consideration with regard to independence is that of providing physical barriers between the systems and/or maximizing the separation of system electrical cables.

Diversity also constitutes a very important part of the typical PPS system. The degree to which diverse driving signals and diverse mechanical hardware is present in such a design is discussed in the following section.

Modern PRAs establish a formal methodology for evaluating reliability of the PPS and risks associated with abnormal events that may occur in a fast reactor. An example of a level-1 PRA for a SFR was done for EBR-II, with the result that risk levels were ~10% of those for typical PWRs [16].

14.3.2 Illustrations of Diversity in the Reactivity Shutdown System

A few examples of system diversity in the primary and secondary reactivity shutdown systems are included in this section to illustrate methods that have been used to minimize the potential for common cause failure.

Table 14.2 lists the scram signals that actuated the primary and the secondary shutdown systems for FFTF. An immediate observation is that there are 11 or 12 signals to initiate either shutdown system. Of perhaps more importance, the input signals are different for each system. Several terms are important in making this distinction. Namely: *diversity*, which implies different mechanisms, design and/or approach to detection, *redundancy,* which implies multiple detectors, and *independence*, which implies that no common failure mode exists between them.

Table 14.3 shows the diversity incorporated into the early design of the reactivity shutdown system of the CRBRP. The conscious effort to include diversity in the design wherever possible is clearly shown [17], although achieving the goal of diversity in the design concept must always be considered with the desire for extremely high reliability in mind.

Although this example demonstrates how diversity can be included in the PPS design, the physical reactivity shutdown devices all used pin bundle configurations for the absorber elements. In the past, there have been concerns that an accident sufficiently severe to grossly distort the control rod or safety rod assembly guide tubes might disable both systems for driving absorber material into the core, representing an example where a single cause could undermine the intended redundancy. Hence,

---

[1] See Table 14.2.

Table 14.2 Scram signals for actuating reactivity shutdown systems in the FFTF

Primary shutdown system	Secondary shutdown system
1. Power range nuclear high flux	1. Flux/total flow
2. Power range nuclear low setting	2. Flux-increasing delayed flux
3. High startup flux	3. Flux-decreasing delayed flux
4. Flux-decreasing delayed flux	4. Low primary loop flow
5. Flux-increasing delayed flux	5. High primary loop flow
6. $Flux^2$/loop pressure	6. Low secondary flow
7. Intermediate heat exchanger (IHX) primary outlet temperature	7. Loss of off-site power
8. Reactor vessel coolant level	8. Reactor outlet plenum temperature
9. Flux/closed loop flow	9. Closed loop outlet temperature
10. Closed loop IHX primary outlet temperature	10. Experiment-associated trip functions
11. Experiment-associated trip functions	11. Flux/closed loop flow
	12. High closed loop flow

there have been efforts to devise an effective shutdown system that uses a principle for core absorber insertion fundamentally different from a rod.[2]

A concept developed for SNR-300 reactor [18] is to assemble the absorber material using a number of shorter bundles in an articulated joint configuration and allow a spring to thrust the absorber from its normal position below the core up into the core region upon demand. As illustrated in Fig. 14.1, the basic idea is to allocate one core assembly position for the purpose of pulling up this jointed

Table 14.3 Diversity in the reactivity shutdown system in the CRBRP design [17]

	Primary system	Secondary system
Control assembly (CA)		
Control rod	37-pin bundle	19-pin bundle
Guide geometry	Hexagonal	Cylindrical
Number of control rods	15	4
Control-rod-drive line (CRDL)		
Coupling to CA	Rigid coupling	Flexible collet latch
Connection to control-rod drive mechanism (CRDM)	CRDL lead screw to CRDM roller nuts	CRDL attached to CRDM carriage with pneumatic activation of CRDL latch through slender rod
Disconnect from CA for refueling	Manual	Automatic
Control-rod-drive mechanism (CRDM) Type	Collapsible rotor-roller nut	Twin ball screw with translating carriage
Overall stroke	0.94 m	1.75 m
Scram function		
Scram release	Magnetic, release CRDM roller nuts	Pneumatic, release CRDL, latch in CA
Scram assist	Spring in CRDM	Hydraulic in CA
Scram speed vs. flow rate	Increases with decreasing flow rate	Decreases with decreasing flow rate
Scram assist length	0.55 m	Full stroke
Scram deceleration	Hydraulic dashpot	Hydraulic spring
Scram motion through internals	Full stroke	6.4 m
Designer	Westinghouse	General electric

[2]For example, LWRs typically can add boron compounds to the coolant as an alternate way to shut the plant down in case the control rods fail to insert.

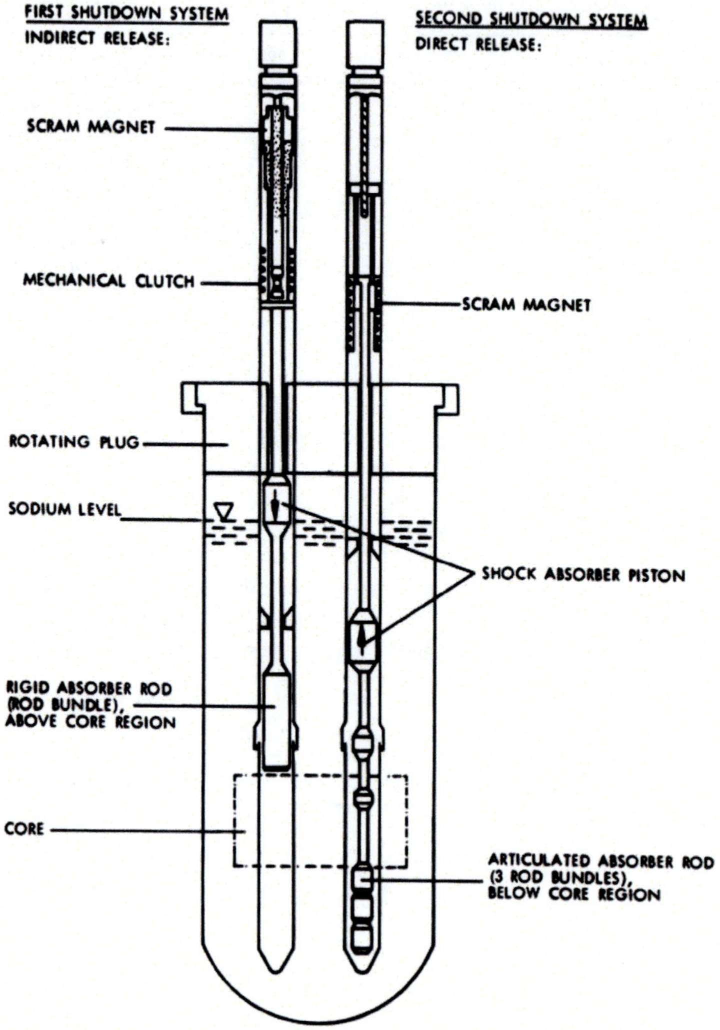

Fig. 14.1 Proposed SNR-300 backup shutdown system

absorber bundle, which is intended to facilitate insertion of the absorber bundles into the core and cause neutronic shut down even if the assembly guide tube was considerably distorted.

Another concept is that of "self-actuated shutdown systems"; namely, systems that do not require an external activation signal, but are automatically actuated based on reactor core conditions that would exist in off-normal events, such as high temperature (curie-point release of magnetically held shutdown rods) or upon loss of hydraulic pressure [19]. An example of the latter, illustrated in Fig. 14.2, is hydraulically suspended absorber balls. The coolant flow in this special assembly provides sufficient force to levitate the absorber balls hydraulically over the top of the core during normal operation. Should a reduction in primary loop flow occur the balls would fall into the core region and provide reactivity shutdown. While not a fast-acting shutdown mechanism as in a traditional scram system, such an approach would be very effective for loss-of-flow situations.

Fig. 14.2 Hydraulically suspended absorber ball concept for backup shutdown system

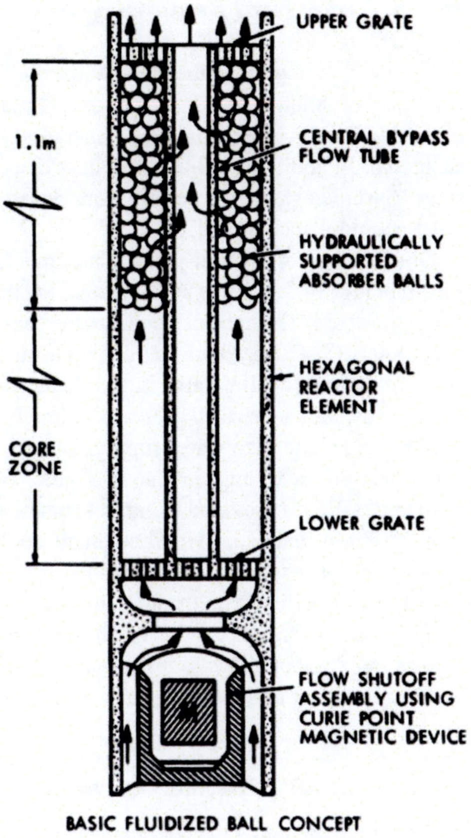

Such features can be combined with traditional scram devices, such as employing an articulated absorber section, similar to the SNR-300 design, but using a Curie point magnetic latch feature above the core. The absorber device is released downward into the core if either (1) holding coil current is lost, or (2) the coolant temperature rises beyond the Curie point, providing both activation in response to a scram signal and automatic activation in response to off-normal core conditions that would normally generate a scram signal.

Another interesting concept is the Gas Expansion Module, namely a specialized assembly located at the periphery of the core. In this concept, the assembly is closed at the top and filled with an inert gas. When the pumps are turned on, sodium is forced into the module, which remains so long as there is forced flow. When the pumps are turned off, the gas forces the sodium out of the assembly, increasing neutron leakage from the core and adding negative reactivity. Such devices were successfully tested in the FFTF reactor [20].

A very desirable concept, discussed more fully in the next chapter, is the incorporation of inherent characteristics of the core and coolant system to ensure reactor "shutdown" even were the PPS to fail. These characteristics depend upon large negative reactivity feedback from the core upon increase in coolant temperature and a smooth transition to natural convective cooling. Tests and analyses have been completed that demonstrate that the core can be protected for both loss of coolant flow and transient overpower events even if the PPS fails to function. This is discussed more fully in Chapter 15 [21].

14.3.3 Decay Heat Removal System

The other major part of the PPS is the Decay Heat Removal System, also sometimes called the Residual, or Shutdown, Heat Removal System [22]. Even if reactivity shutdown is successfully achieved, there is substantial heat generation by radioactive decay, especially immediately following shutdown. In addition to fission product decay, some decay heat is produced by beta decay of ^{239}U and ^{239}Np plus smaller amounts from decay of activation products (e.g. steel, sodium) and higher order actinides such as ^{242}Cm.

One example of reliability considerations for this system is afforded in the CRBRP design [17], as shown in Fig. 14.3. For this system, there are three backup systems to remove the decay heat should the normal heat sink (balance of plant steam/feed-water system) fail. The first is the Protected Air-Cooled Condensers (PACC) system that cools the steam drum. A second heat sink can be made available by opening the safety relief valve in the steam line between the steam drum and the turbine and venting steam to the atmosphere. A protected storage water tank is available for supplying make-up water, in addition to an alternate water supply. Two of the auxiliary feed-water pumps are electrically driven, but a steam-driven pump is also available. Finally, a completely separate Overflow Heat Removal System (OHRS) is provided to extract heat directly from the in-vessel primary loop position. The heat sink for this system is provided by an air-cooled heat exchanger, which not only represents diversity from the water system in the steam loop, but provides for decay heat removal in situations where the normal heat rejection path through the steam generator is not available.

The Super Phénix decay heat removal system included sodium-to-air heat exchangers on the secondary sodium loops plus four backup decay heat removal systems. Each backup system included an immersion cooler in the reactor tank, a closed sodium loop, and a sodium-to-air heat exchanger.

14.3.3.1 Natural Convection Operation

In addition to including the backup heat removal systems in the design, the components in the Reactor Heat Transport System are arranged in such a way that decay-heat can be removed by natural convection. Typically the pumps are tripped if the reactor is scrammed in a protected transient to reduce thermal shock. Natural circulation for decay heat removal is always required unless separated auxiliary pumps are used. For long-term cooling with protected transients, this requires the availability of sufficient heat sink capability to balance decay heat generation. The backup decay heat removal systems of SNR-300 and Super Phénix were designed in such a way that adequate heat removal is provided even with passive operation of the systems, i.e., with natural convection in the sodium-to-air heat exchangers as well as the sodium loops. It has been demonstrated under actual SFR operating conditions that natural convection can, in fact, be established rapidly enough to prevent fuel damage. Successful tests at EBR-II, FFTF, PFR and Phénix in which pumps were completely cut off at partial power demonstrated that natural convection can be readily established after loss of pumping capability at full power [21, 23].

14.3.4 Overall Reliability Goals

It is well recognized that the various aspects just discussed all contribute to the effectiveness of overall PPS reliability. A major question is the numerical level that the PPS must attain such that the consequences of PPS failure need not be considered for safety evaluation purposes [24, 25]. The NRC has reported goals (not requirements) that the core damage frequency should be less that 10^{-4} per reactor year and that the probability of a large radiation release should be less than 10^{-6} per reactor year. To

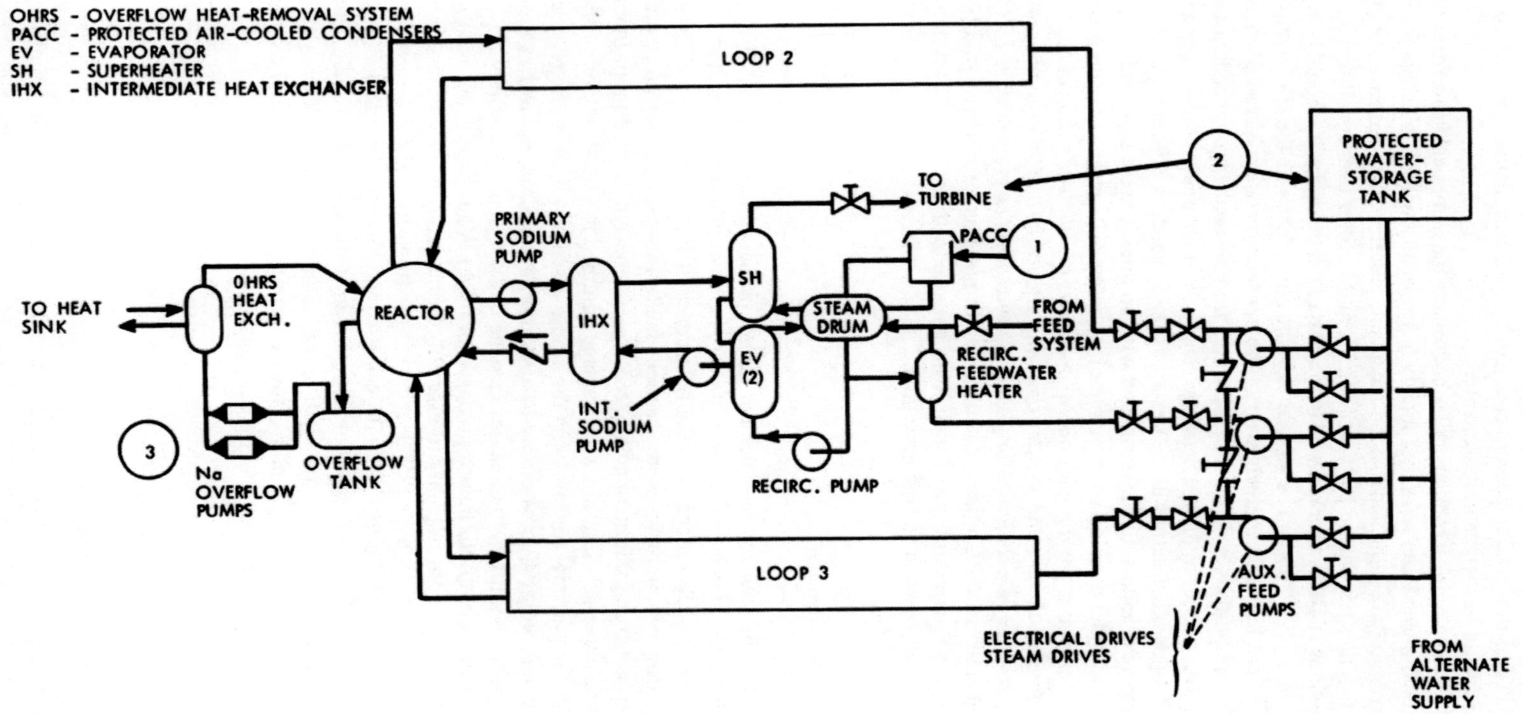

Fig. 14.3 Decay heat removal system for CRBRP design [17]

the extent that the PPS is designed to prevent core damage, a performance target of 10^{-4} per reactor year for PPS failure could be viewed as reasonable.

Another means for developing a framework for addressing the numerical value for sufficiency in the PPS acceptability is to observe the risks that society is apparently willing to accept for normal living patterns. Studies by Starr have indicated that the acceptable risk from the production of electricity is of the order of 10^{-6} fatalities per person per year [26]. From this, Starr later concluded that a target risk level for nuclear stations of 10^{-6}, which is well below that of natural hazards, should be well within the socially accepted range [27]. This can be viewed as being consistent with the NRC goals listed above. In practice, this goal can be easily met when all of the barriers to release of radiation are considered in a modern SFR plant. Further, it is not necessary that the PPS be the only mechanism contributing to achieving this goal. Indeed, higher PPS failure rates can be tolerated without compromising this safety goal, depending on other details of the design. As such, modern SFR designs can meet the goal of 10^{-6} without release of radioactive material and with no impact to the surrounding population.

The NRC, after much deliberation on the issue of Anticipated Transients Without Scram (ATWS) [24], concluded the following: *for a population of one thousand nuclear plants in the U.S., the safety objective will require that there be no greater than one chance in one million per year for an individual plant to experience an accident with potential consequences greater than 10 CFR Part 100 guidelines.* Subsequently, an NRC policy statement reaffirmed this quantitative goal [25]. Although no firm goals have yet been established for SFR acceptance, a figure of 10^{-6} system failures per reactor year certainly represents a conservative goal.

14.4 Local Fault Propagation

The evaluation of the potential for fuel element failure propagation (FEFP) in SFRs has been the subject of considerable research over a number of years in most countries involved in its development [28]. The study of FEFP is concerned with the occurrence of initially small local faults in the core that could possibly grow into more widespread serious faults. Pin-to-pin failure propagation is defined as a failure in one pin causing failure in an adjacent pin. In concept, if such propagation occurs, the propagation could be either self-limiting, with damage confined to a small region of the pin bundle, or progressive, with the potential for involving the whole assembly. If such progressive propagation occurs, the event could progress to assembly-to-assembly propagation, defined as the condition in which an assembly with an initial failure causes damage in a neighboring assembly.

The concern was that pin-to-pin failure propagation might begin slowly, but then proceed more rapidly and possibly involve even assembly-to-assembly failure propagation on a time scale that would be difficult to detect and arrest with the PPS. Hence, in assessing such possibilities, the two overriding questions were as follows: (1) if a fuel pin fails, can such failure propagation readily occur, and (2), if propagation can potentially occur, what instrumentation is necessary for timely activation of the PPS? There is overwhelming evidence that such propagation for current SFR designs is either very improbable or impossible, and that rapid propagation beginning with a single fuel pin failure does not occur. The basis for this determination is outlined below.

14.4.1 Pin-to-Pin Failure Propagation

14.4.1.1 Potential Initiating Mechanisms

There are three general categories of phenomena that could initiate a single pin failure sequence. First, random failures due to manufacturing defects could cause fuel pin failure, with the potential that any resulting fission gas release might provide a mechanism to induce failure propagation. Second, a local blockage formed by particulate matter (not credible unless a source of suitable particulate matter can be identified) could lead to higher local coolant temperatures leading to fuel pin failure (possibly including limited local boiling). Finally, a fuel pin loading enrichment error resulting in higher power generation in the fuel pin could cause local melting of fuel within the pin, followed by cladding failure, with the potential to release molten fuel from the pin. Other subcategories include insufficient fuel pin heat transfer and stochastic fuel pin failure with subsequent fuel-coolant interaction.

14.4.1.2 Pin Failure Detectability

Before investigating the potential of each of these mechanisms to cause pin-to-pin propagation, it is of interest to briefly discuss the issue of pin failure detectability. At least four methods currently exist for detecting initial fuel pin failure, depending on the cause and nature of the failure and any subsequent phenomena. First, an increased assembly outlet coolant temperature could be detected, but this method would only be useful where fuel pin failure can cause a sufficient assembly flow perturbation to measurably affect the outlet temperature, and each assembly would have to be monitored for outlet temperature. The second method is the use of delayed neutron detectors, in which delayed neutrons are detected from fission-product delayed-neutron precursors in the coolant as a result of fuel pin failure. This is a sensitive method but one that has a time delay associated with the transport of the fission-products or fuel from the failed pin to the vicinity of the detector. Further, it would not allow identification of the assembly with the failed fuel. A third related possibility is that of monitoring for any gaseous fission product release to the inert cover gas region(s) of the primary coolant circuit, which again is a sensitive method but has a time delay for transport of the fission products to the cover gas region and would not allow for identification of the assembly with the failed fuel. A fourth procedure, as a subset of this method, is designed to aid in identification of the assembly with failed fuel. Here a blend of inert gaseous isotopes is included as part of the inert gas loaded in the fuel pin at the time of manufacture, with each assembly having a unique mixture [29]. With this approach, the assembly in which the failed fuel pin resides can be uniquely identified. In general, the detection of a single pin failure via any of these techniques is either fairly slow, or is not highly sensitive. Hence, it is important to be able to prove convincingly that the propagating mechanisms identified above are not capable of leading to rapid propagation of fuel pin failure, or subsequent propagation from one assembly to another. More discussion of fuel pin failure detection is provided in Section 12.6.2.

Each of these detection techniques has been employed in fast reactor systems. In the Run-Beyond-Cladding-Breach program conducted at EBR-II, much was learned about the detection of fuel where the cladding has failed and the potential for local flow blockage. The result of that testing is that cladding failure, with either metal or oxide fuel, can be detected in a timely manner, more than sufficient to ensure the safety of the reactor. There was no potential for local flow blockage with metallic fuel, eliminating any possibility of pin-to-pin failure propagation. With oxide fuel, any reduction or blockage of the coolant flow channel at the failure location was shown to occur very slowly, and may be limited in size, providing ample time for corrective actions to be taken. Pin-to-pin propagation of cladding failure is not likely in this case either [30].

14.4.1.3 Fission Gas Release

The principal concern associated with fission gas release from irradiated fuel is the rapid local transient displacement of the liquid coolant with fission gas at the time of cladding failure, greatly reducing heat removal from the fuel pin and adjacent fuel pins and causing a cladding temperature increase (as well as causing a momentary localized reactivity increase similar to sodium boiling). The occurrence of such phenomena depends on the size of the cladding failure, where relatively larger failures would be required for a rapid fuel pin depressurization. Extensive in-core and out-of-core testing has demonstrated that there is little or no chance of pin-to-pin failure propagation due to such local transient fission gas displacement of the coolant, even with larger local failures. However, there may be the possibility that the initial cladding failure is small and does not lead to rapid expulsion of fission gas from the failed pin, but rather that the failed fuel pin depressurizes more slowly. If the fission gas release from such a small cladding failure resulted in a gas-jet impingement on the neighboring fuel pin for an extended period of time, and did lead to failure of this fuel pin as a result, the cladding failure would be directly across the coolant channel from the initially failed pin. Hence, if a subsequent fission gas jet from the second failed fuel pin occurs, it would tend to be directed back toward the initial failure. This results in a self-limiting pin-to-pin propagation process. Finally, it has been postulated that transitory mechanical loads associated with fission gas release might be high enough to mechanically fail neighboring fuel pins. However, fission gas release does not occur fast enough or with enough energy to cause the required high stresses for mechanical failure.

14.4.1.4 Local Coolant Blockage

As mentioned above, in order for local blockages to take place, foreign materials in the coolant stream must get to the site of the postulated blockage. However, it is now standard design practice to include strainers in the coolant inlet ports to prevent particles above a certain size from entering the assemblies. If fuel particles from a failed pin were to enter the coolant stream, they have been shown to be few in number prior to being detected by the fission gas release monitoring system. Blockage could conceivably occur due to swelling of failed pins, but this is a very slow process prior to "breakaway" swelling that can occur with high fast neutron fluence, a limit that is used to set exposure limits for core materials. For any planar blockage that is impervious to flow, nearly 50% blockage of the total flow area would need to occur in order to reduce the overall assembly flow by 5%. Hence, a local blockage would have to be very large to cause an appreciable disturbance to the coolant flow rate sufficient to cause damage to the fuel pins. Further, it has been shown that most blockages arising from the accumulation of particulate matter would be at least partially porous due to the blockage being formed from numerous small particles.

14.4.1.5 Molten Fuel Release

Fuel enrichment errors have the potential to create molten fuel if the result is either a sufficient increase in power or a reduction in solid fuel thermal conductivity. The primary protection against fuel enrichment errors is the strict quality assurance program that normally requires individual pin assay tests prior to assembly into a duct array. However, even if local enrichment errors were to occur, in-pile testing has indicated that the effects of the enrichment error would be readily detected and corrective actions can be taken before any damage threshold is reached.

A larger concern would be the potential thermal interaction between molten fuel and sodium coolant if it were possible for molten fuel to exit the fuel pin and contact the coolant, but this phenomenon is typically associated with lower-probability unprotected transients, and is discussed in Chapter 16.

In summary, pin-to-pin failure propagation as a result of a local fuel pin failure is viewed to be highly unlikely, and could only conceivably occur if the detected pin failures were ignored and conditions were conducive to local blockage formation. However, even if such pin-to-pin propagation were allowed to occur, it is sufficiently slow to allow subsequent detection before serious safety conditions arise [31].

14.4.2 Assembly-to-Assembly Propagation

14.4.2.1 Principal Propagation Concerns

There are four general categories of concerns regarding assembly-to-assembly failure propagation potential. These include (1) pin-to-pin failures that lead to full assembly involvement, (2) full assembly flow blockage, (3) pressure generation from a molten fuel–coolant interaction, and (4) the reactivity consequences of assembly meltdown. The first two are appropriate for a discussion on protected events. The latter two will be discussed in the chapters on unprotected events.

14.4.2.2 Flow Blockage

Fuel assembly failure propagation could conceivably result from a large flow blockage, one that is sufficient to cause overheating of the assembly resulting in fuel pin failure, cladding melting, and fuel melting. The potential for formation of such a large flow blockage and its detectability are related to design and operating practices, as was demonstrated in Fermi-1 [32]. Proper assembly designs (e.g. multiple coolant inlets, etc.), in addition to proper construction and operating practices prior to loading of the assembly into the reactor (e.g., gas flow testing), make a complete assembly blockage virtually impossible. An inlet blockage needs to become very large (~90% planar blockage for typical SFR designs) before the coolant flow is reduced sufficiently to result in the onset of coolant boiling at the assembly exit. Furthermore, recent progress in the understanding of the fluid mechanics and heat transfer in the wake behind blockages rules out the concern for developing significant internal blockages during operation prior to the occurrence of limited fuel pin failures that would be detected by fission product detectors (both solid and gaseous products) or delayed neutron detectors.

Fermi-1 Experience

A large coolant blockage did occur at the Fermi-1 reactor, when a plate below the core detached during startup testing and significantly blocked flow to at least two assemblies while the reactor was still not at full power [32]. Fuel melting and fuel pin failures occurred but did not propagate to assemblies that were not involved in the original blockage. The reactor was shut down and the damaged fuel assemblies along with the dislodged plate were removed prior to reloading with new fuel and restarting of the reactor. The event demonstrated that fuel melting on an assembly scale does not necessarily lead to propagation to other assemblies or an escalation in the severity of the event, but it should be cautioned that this conclusion is only valid for fast reactors with similar fuel and cladding designs, principally metal fuel. However, as a result of this event, design changes were made to the assembly coolant inlet such that multiple inlet coolant paths are available and a single flat object such as the dislodged plate in Fermi-1 cannot result in an essentially complete flow blockage.

14.5 Whole-Core Transients

Whole core transients are those events for which a reactivity or flow imbalance affects the entire reactor core. For any particular reactor design, extensive studies are required to be conducted to assure that none of these transients exceed the system design limits for events with successful PPS action.

Outlines are contained below for three sample calculations: one for a reactivity transient and the other two for transient under cooling events. Much of the detail will necessarily be omitted, but a brief discussion of the salient features is instructive to help visualize the types of analyses that must be done to assure effective PPS performance.

14.5.1 Reactivity Transients

The sample overpower transient chosen is a 3.4 ¢/s reactivity insertion rate into the FFTF reactor at full power. This reactivity insertion rate corresponds to the normal speed of control rod withdrawal (0.25 m/min), but assumes that the rod is continuously withdrawn and the transient is terminated by PPS action. The analytical tool used was the MELT-III Code [33]. Hot-channel factors were used to assure a 99.9% confidence level that the calculated fuel pin temperatures would not be exceeded.

Figure 14.4 shows the peak cladding temperatures for the case that assumes primary PPS action, as well the case where only secondary PPS action occurs, with the difference being in the time of the PPS action. The peak cladding temperatures (801°C for primary action and 831°C for secondary action) are well below the 870°C design limit determined for minor cladding damage. Hence, this calculation indicates there is considerable margin-to-failure (using the cladding temperature limit), even if the primary PPS fails to perform.

14.5.2 Transient Under Cooling Events

14.5.2.1 Loss of Primary System Flow

A transient under cooling event can occur with the loss of all off-site electric power and on-site standby electric power and is typically considered to be an Anticipated Operational Occurrence by the NRC, one that is expected to happen at least once during the lifetime of the plant. The sample case is taken for the FFTF where again the transient is assumed to be initiated from full power conditions. The IANUS Code [34] was used to perform the calculation and the same hot-channel factors were employed as in the case of the reactivity transient, reflecting the same high level of conservatism in the analyses. Complete loss of all forced convection in the primary and secondary sodium systems and in the sodium-to-air heat exchangers was assumed.

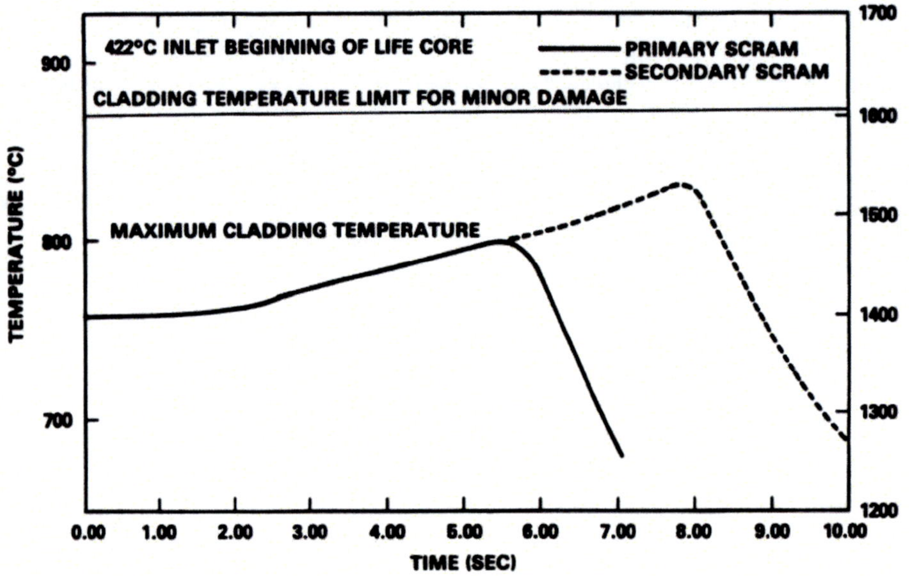

Fig. 14.4 Sample protected whole core reactivity transient (3.4 ¢/s insertion rate into full power FFTF)

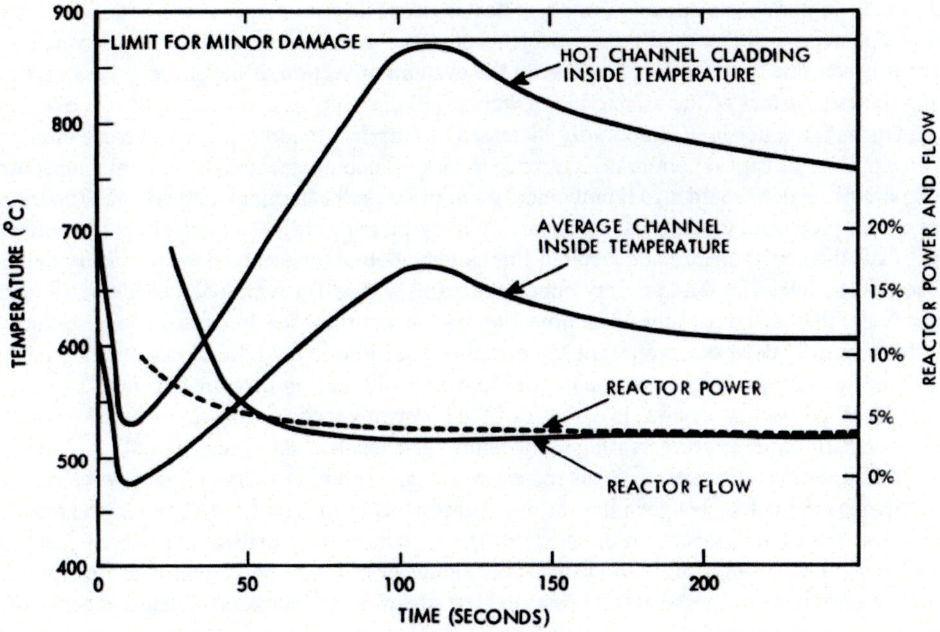

Fig. 14.5 Sample loss of all electric power accident calculated for FFTF. Note: Actual FFTF test results (obtained from a full power scram to complete natural convection with all electrical power off) yielded maximum cladding temperatures significantly below the full power steady-state values. It should be noted that the calculations plotted in this figure include all the safety hot channel factors with conservatisms and, therefore, are considerably higher than expected actual values for such an event. The actual test results, which are in close agreement with best-estimate pretest predictions, emphasize the safety margin imbedded in the conservative hot channel analyses required by the NRC

Figure 14.5 contains the key calculated results. Shown in this illustration is the inside wall temperature of the average cladding and the value for the hottest cladding in the core (including the hot-channel factors). PPS activation occurred at time = 0, shutting down the reactor. The cladding temperature drops immediately following the scram because the primary loop pump inertial coast down characteristics maintain the flow well above the power curve (which drops rapidly due to PPS action), overcooling the core. After about 50 s, however, the normalized power (all from decay heat) exceeds the normalized flow since the inertial coast down of the coolant pumps is over and transition to natural convection has not yet occurred. The normalized power/flow ratio becoming greater than 1.0 causes the cladding temperature to exceed its steady-state value. Core temperatures continue to rise until natural convection is fully established, after which cladding temperatures recede. As shown in Fig. 14.5, the maximum hot channel cladding temperature, including conservatisms, just approaches the 870°C limit. More realistic hot channel factors, noted in Ref. [35], reduce the peak cladding temperature by over 80°C from that shown. Hence, even for this event, cladding integrity is concluded to be maintained.

14.5.2.2 Flow Failure in the Steam System

As a second example of a transient under cooling event, we consider a transient initiated by a secondary-side flow failure in the CRBRP steam system. The specific failure is a breach in the saturated steam line connecting the evaporator to the super heater. For the CRBRP design, this event would cause an immediate cessation of super heater steam flow in that loop and initiation of a reactor trip. The super heater would then rapidly become isothermal at the sodium inlet temperature to the super heater due to the loss of cooling. In the evaporator section of the steam generator, sodium leaving the evaporators of the affected loop would initially drop in temperature due to overcooling that results as the water flow temporarily increased and flashed to atmospheric pressure through the pipe break. Then, as the evaporator dries out due to lack of incoming water, the sodium temperature in the evaporator section would also rapidly increase to the super heater inlet temperature. The transient temperatures propagate through the secondary cold leg piping, mitigated by the heat capacity of the intermediate loop components, and result in similar temperature transients in the secondary pump, the IHX secondary inlet, the IHX primary outlet, the primary cold leg piping and check valve, and the reactor vessel inlet nozzle. At the same time, the reactor scram causes the primary loop sodium temperatures to rapidly decrease, prompting temperature reductions in the IHX and the intermediate loop that propagate to the steam generator inlet, followed by additional temperature transients as the transition to natural convection occurs, as described in the previous section. Subsequently, all components would eventually experience decreasing temperatures as a result of the reactor scram.

As this discussion illustrates, the plant response to such an event is a very complex problem, involving numerous interactions between the various components of the reactor system and the balance of plant. It was realized very early in the development of nuclear reactors that analysis of such events would require the development of sophisticated computer-based simulation tools, and these have been in constant development for the past 50 years, taking advantage of increased computing capability for more accurate simulations.

14.6 Other Accident Considerations

In addition to the power/flow imbalance transients for which the PPS is designed to arrest, there are other transients for which protection must be provided. These include steam generator faults, sodium leaks and spills, fuel handling and storage faults, and external events (both natural catastrophic and man-made events). A brief summary of each class is contained below.

14.6.1 Steam Generator Faults

In the design of steam generators for a SFR system, the potential for water-to-sodium leaks is the major concern due to the exothermic chemical reaction that occurs between sodium and water. Small faults in steam generators can, if undetected, lead to wastage of surrounding tubes and ultimately to major damage in the steam generator. Reaction products could also contaminate the intermediate loop and possibly threaten the IHX, as water-side pressure is communicated to the intermediate loop. The possibility of large failures of tubes is a key consideration in the design of steam generator systems. Assuring the high integrity of the boundary between sodium and water/steam is therefore, an area of high priority in steam generator design. Designing steam generator systems to the ASME Code, Section III, Class 1, as well as requiring all tube welds to be inspectable after manufacture and crevice-free are among the measures taken towards minimizing the potential for tube leaks.

Nevertheless, provisions must be made to accommodate major leaks to avoid damaging the steam generator beyond repair or precipitating the evolution of a serious reactor accident. In systems using more than one steam generator in the same loop, the other steam generators in the loop could be severely damaged by the effects of a sodium-water reaction in one of the steam generator units if adequate relief is not provided. Also, the IHX could be damaged by pressurization, which could potentially open a path to the environment for radioactive primary loop sodium.

To design such a relief system, the current US practice is to assume a hypothetical failure of steam generator tubes. Replacement of the damaged steam generator is an acceptable design basis consequence, but in the case of multiple steam generators, loss of all the steam generators in the affected loop is likely to not be an acceptable consequence. ASME Code rules for Faulted Conditions are used in designing for this accident in the faulted unit, while the remaining steam generators in the same loop are designed to meet the ASME Code rules for Emergency Conditions.

Excessively high pressures are designed to be prevented by providing a pressure relief system for the intermediate loop that allows safe rapid expulsion of sodium and reaction products from the secondary sodium system. The pressure relief system is also designed to protect the IHX from the over-pressurization that may result from major leaks in the steam generator. As for the case of steam generators, the present approach in the US is to design the IHX structure for loads that could result from a major leak using ASME Code rules for Emergency Conditions.

14.6.2 Sodium Spills

Sodium is chemically reactive with air, oxidizing rapidly to create intense heat and fire. Large sodium spills and the associated fires are considered unlikely in a properly designed SFR since the hazard is recognized and essentially eliminated when designing the plant. The consequences of any radioactive sodium spills are mitigated by use of elevated piping (low static pressure) enclosed with metal-clad insulation to lessen leakage rate, enclosure of major equipment within secondary or "guard" vessels to provide an additional barrier between the sodium and air, and reduced oxygen (~1 volume percent) in primary sodium equipment cells to lessen the intensity of any fire that may occur. As a secondary precaution, the major primary sodium equipment cells are located in the sub-grade region of the reactor containment building away from the containment itself. Cells are enclosed with a gas-tight steel membrane or "liner" to contain the lower-oxygen atmosphere. The liner also protects structural concrete from reaction with sodium. Where the possibility and potential consequences of a sodium spill are considered significant, the cell liners are installed with provisions for thermal expansion, and the space between the cell liners and concrete is vented to limit buildup of steam pressure due to water release from the heated concrete behind the liners.

Secondary sodium systems are enclosed in air-filled cells since there is no radiological hazard associated with the sodium in the intermediate loop. Structural concrete for these cells is protected from sodium attack by lining the cell floor and lower side walls with a basin or "catch pan" of steel plate.

Even though no major sodium spills are expected in an SFR, safety analyses may consider limiting accidents such as large leaks and a partial loss of system inventory. Sodium fires result in heat generation accompanied by a pressure and temperature rise within the cells. Releases from appropriate postulated sodium pool and spray fires are evaluated as part of the license application for an SFR, and the resulting pressures and temperatures are compared to limiting structural values for the cells to ensure that design margins are sufficient. Additional detail regarding the nature of sodium fires is contained in Chapter 16.

There has been a lot of experience with sodium spills and their consequences in the course of fast reactor operation over the years. No personal injuries have occurred as a result of sodium spills, due to the design features discussed above and the fact that they are not high pressure systems. For example, a major spill involving ~100 gallons of sodium occurred at EBR-II upon failure of piping in a sodium sampling systems. The resulting fire was extinguished with no injuries. A more recent event was a major sodium leak at the MONJU reactor in Japan, involving failure of a thermocouple well in the secondary sodium piping. The resulting fire resulted in no injuries but did result in a reactor shutdown that extended over a decade, so politically it was quite serious. However, as more has been learned about the nature of sodium spills and fires, including their prevention and mitigation, there is a growing understanding that they are without significant consequence because of design and operating practices in the plants.

14.6.3 Fuel Handling and Storage Faults

The safety criteria and definition of faults for design of the fuel handling and storage systems consider five safety functions:

(1) remove residual (decay) heat from used fuel
(2) maintain subcriticality
(3) provide barriers to the release of radioactive material
(4) provide biological shielding
(5) prevent loss of safety functions of other fuel handling and storage systems.

Once the safety functions for the fuel handling and storage systems have been identified, generic types of faults are postulated (for example, loss of cooling of used fuel in each location where it is handled or stored). Typically, the NRC will require analysis of the worst accident of each generic class to predict the consequences. In some cases, several disparate accidents of a class may all have the possibility of exceeding the accident severity criteria, and all require analysis. The accidents are then analyzed in detail and the results compared to the allowable consequences. If the guideline exposure limits are approached, redesign to eliminate the fault or limit its consequences is implemented.

As a result of the above type of analysis, a number of specific requirements and design features have been developed for used fuel storage facilities and equipment that transports used fuel. These include in-vessel and ex-vessel transfer machines, grapples for fuel handling in hot cells, and used fuel shipping casks. Where needed, at least two means of cooling are provided, i.e. both a primary and a backup system, following the same defense-in-depth principles used for the reactor. All active and standby components required for maintenance of these two systems must be accessible for periodic inspection and/or checkout.

Guard vessels are provided around sodium vessels used for storage of used fuel, and the liquid sodium level in the vessel may not be less than the minimum safe level (that is, the level below which the fuel temperature would exceed the allowable criteria) following a vessel or cooling system leak, just as in the reactor. Siphon breakers are included in any vessel inlet or outlet lines where the possibility of siphoning exists. Either there are no permanently connected drain lines, or any such lines contain double valves locked closed. Unless the backup cooling system is inherent, the primary or backup means of cooling, or both, are connected to an emergency power supply. At least one system adequate for cooling used fuel is designed to Seismic Category I, which means that the cooling function will be maintained following a Safe Shutdown Earthquake.

One significant difference between SFR designs and light water reactors in handling of used fuel is that SFR individual used fuel assemblies are lifted out of the coolant during transfer from the reactor vessel. Considerable design and development effort has gone into the design of reliable and inherent cooling for the ex-vessel transfer machine that performs this operation, whether the machine contains sodium or an inert gas. Means of cooling for both the primary and backup systems can employ redundant blowers, thermal radiation, natural convection, or any other effective heat transfer mechanism capable of removing decay heat from the assembly.

Experience with SFRs has indicated that two of the most common means for release of radioactivity to containment were leakage through gas seals and spread of contamination through sodium leakage from the movable fuel transfer machine. To minimize the amount of radioactive gas leakage, double seals are used on fuel handling and storage equipment, with pressurized buffer gas in between if the contained level of radioactivity is high. Design features are also incorporated to minimize the amount of sodium leakage and to collect all such leakage in specified locations where it will not be released to the containment. Measures are also included to minimize the surface exposed to the contaminated interior of the transfer machine that could subsequently be exposed to the containment.

14.6.4 External Events

Natural phenomena and external, man-made conditions to be accommodated in the design of SFRs are the same as those for light water reactors; the guidelines and criteria for consideration of such events are established by regulatory authorities. A summary of events considered is given below.

14.6.4.1 Natural Phenomena

The plant must be designed so that it can be safely shut down and maintained in a safe condition following any natural phenomenon, including extremely unlikely natural phenomena, i.e., the most severe that have been reported or can be foreseen for the site and its surroundings. The natural phenomena that provide design bases for the plant include:

Probable Maximum Flood. This is determined by evaluating potential flooding conditions resulting from severe storms, ice conditions, potential failures of upstream dams, and tsunamis. In addition to addressing these external events individually, consideration should be given to combination of events, such as earthquake with tsunami, or tornado with loss of electric power.

Minimum Possible Low Water Level. Depending on the design features, this may affect the source of cooling water and/or the ability of the ultimate heat sink to perform its function. This low water level is determined from the most severe drought conditions or dam failure conditions considered possible at the site.

Tornado. A limiting tornado appropriate to the site must be factored into the design, including provisions to protect against associated tornado missiles.

Safe Shutdown Earthquake. The defined earthquake that could cause the maximum vibratory ground motion at the site must be considered.

14.6.4.2 Man-Made External Conditions

The plant must be designed so that it can be safely shut down and maintained in a safe condition for man-made external conditions that are considered to represent a significant risk to the plant. These conditions are highly site dependent and can be discussed here only in general terms. Missiles that could be generated within the plant must be considered if they could jeopardize the functioning of safety systems. For example, turbine missiles must either be included in the plant design bases or shown to be sufficiently improbable to be excluded. A specific orientation of the turbine shaft relative to the containment building is frequently used to minimize the probability of damage from turbine missiles. Examples of other potential conditions considered are an airplane crash, external explosions (from nearby industrial installations or transportation on nearby waterways, railways or highways), sabotage, terrorist activities, toxic chemical or radioactive releases from nearby industrial or nuclear plants, and fires.

References

1. U.S. Nuclear Regulatory Commission, *Safety Evaluation Report related to Operation of the Fast Flux Test Facility,* NUREG-0358, August 1978.
2. U.S. Nuclear Regulatory Commission, *Safety Evaluation Report Related to the Construction of the Clinch River Breeder Reactor Plant,* NUREG-0968, Vols. 1 and 2, March 1983.
3. U.S. Nuclear Regulatory Commission, *Preapplication Safety Evaluation Report for the Sodium Advanced Fast Reactor (SAFR) Liquid Metal Reactor,* NUREG-1369, December 1991.
4. U.S. Nuclear Regulatory Commission, *Preapplication Safety Evaluation Report for the Power Reactor Innovative Small Module (PRISM) Liquid Metal Reactor,* NUREG-1368, February 1994.
5. L. E. Strawbridge, "Safety-Related Criteria and Design Features in the Clinch River Breeder Reactor Plant," *Proceedings of the Fast Reactor Safety Meeting,* CONF-740401-P1, pp. 72–92, American Nuclear Society, Beverly Hills, CA, April 2–4, 1974.
6. R. J. Jackson, *Evaluation of FFTF Fuel Pin Design Procedures Vis-Vis Steady State Irradiation Performance in EBR-II,* Addendum to HEDL-TME 75-48, Hanford Engineering Development Laboratory, Richland, WA, October 1975.
7. R. E. Baars, *Evaluation of FFTF Fuel Pin Transient Design Procedure,* HEDL-TME 75-40, Hanford Engineering Development Laboratory, Richland, WA, September 1975.
8. R. B. Baker, F. E. Bard, and J. L. Ethridge, "Performance of Fast Flux Test Facility Driver and Prototype Driver Fuels, LMR: A Decade of LMR Progress and Promise," *ANS Winter Meeting*, Washington, D.C., pp. 184–195, November 11–15, 1990.
9. T. H. Baur, et al., "Behavior of Metallic Fuel in TREAT Transient Overpower Tests," *Proceedings of the International Topical Meeting on Safety of Next Generation Power Reactors,* Seattle, WA, p. 857, May 1988.
10. R. J. Forrester, et al., "EBR II High-Ramp Transients under Computer Control," *American Nuclear Society Annual Meeting*, Detroit, MI, June 12–17, 1983.
11. American Society of Mechanical Engineers, *Boiler and Pressure Vessel Code,* Section III, Division I, 1975 edition.
12. L. K. Chang and M. J. Lee, "Thermal and Structural Behavior of EBR II Plant During Unprotected Loss-of-Flow Transients," *8th International Conference, Structural Mechanics in Reactor Technology*, Brussels, Belgium, August 19–23, 1985.
13. J. J. Lorenz and P. A. Howard, *A Study of CRBR Outlet Plenum Thermal Oscillation during Steady State Conditions,* ANL-CT-76-36, Argonne National Laboratory, Argonne, IL, July 1976.
14. R. C. Berglund, et al., "Design of PRISM, an Inherently Safe, Economic, and Testable Liquid Metal Fast Breeder Reactor Plant," *Proceedings of the ANS/ENS International Conference on Fast Breeder Systems*, Vol. 2, Pasco, WA, September 13–17, 1987.
15. E. B. Baumeister, et al., "Inherent Safety Features and Licensing Plan of the SAFR Plant," *Proceedings of the International Conference on Fast Breeder Systems*, Vol. 1, p. 3.4-1, Pasco, WA, September 13–17, 1987.
16. D. J. Hill, "An Overview of the EBR II PRA," *Proceedings of the 1990 International Fast Reactor Safety Meeting*, Vol. IV, p. 33, American Nuclear Society, Snowbird, UT, August 12–16, 1990.
17. J. Graham, "Nuclear Safety Design of the Clinch River Breeder Reactor Plant," *Nuclear Safety*, 16, 5, September–October 1975.
18. F. H. Morgenstern, J. Bucholz, H. Kruger, and H. Rohrs, "Diverse Shutdown systems for the KNK-1, KNK-2, and SNR-300 Reactors," CONF-740401, *Proceedings of the Fast Reactor Safety Meeting*, Beverly Hills, CA, April 1974.
19. E. R. Specht, R. K. Paschall, M. Marquette, and A. Jackola, "Hydraulically Supported Absorber Balls Shutdown System for Inherently Safe LMFBRs," CONF-761001, *Proceedings of the International Meeting on Fast Reactor Safety and Related Physics*, Vol. III, p. 683, Chicago, IL, October 1976.
20. L. R. Campbell, et al., *FFTF Loss of Flow Without Scram Experiments with GEMS*, HEDL-TC-2947, Hanford Engineering Development Laboratory, Richland, WA, June 1987.
21. J. I. Sackett, et al., "EBR II Test Program," *Proceedings of the 1990 International Fast Reactor Safety Meeting*, Vol. III, p. 181, American Nuclear Society, Snowbird, UT, August 12–16, 1990.
22. A. K. Agrawal and J. G. Guppy, editors, *Decay Heat Removal and Natural Convection in Fast Breeder Reactors,* Hemisphere Publishing Corp., New York, NY, 1981.
23. T. R. Beaver, et al., "Transient Testing of the FFTF for Decay Heat Removal by Natural Convection," *Proceedings of the LMFBR Safety Topical Meeting*, Vol. II, pp. 525–534, European Nuclear Society, Lyon, France, July 19–23, 1982.
24. U.S. Nuclear Regulatory Commission, *Anticipated Transients Without SCRAM for Water-Cooled Power Reactors,* WASH-1270, September 1973.

25. 51 FR 30028, "Safety Goals for the Operation of Nuclear Power Plants; Policy Statement," Republication, 08/21/86.
26. C. Starr, "Social Benefit Versus Technological Risk," *Science*, 165, 1969, 1232–1238.
27. U.S. Congress, "Joint Committee on Atomic Energy," *Possible Modification or Extension of the Price-Anderson Insurance and Indemnity Act, Hearings Before the Joint Committee on Atomic Energy on Phase II: Legislative Proposals*, H.R. 14408, S. 3252, and S. 3254, 93rd Congress, 2nd Session, Pt. 2, Testimony of C. Starr, p. 617, May 16, 1974.
28. J. van Erp, T. C. Chawla, R. E. Wilson, and H. K. Fauske, "Pin-to-Pin Failure Propagation in Liquid-Metal-Cooled Fast Breeder Reactor Fuel Subassemblies," *Nuclear Safety*, 16, 3, May–June 1975, 391–407.
29. N. J. McCormick and R. E. Schenter, "Gas Tag Identification of Failed Fuel I. Synergistic Use of Inert Gases," *Nuclear Technology*, 24, 1974, 149–155. See also Part II. "Gas Tag Identification of Failed Fuel II. Resolution Between Single and Multiple Failures," *Nuclear Technology*, 24, 1974, 156–167.
30. J. D. B. Lambert, et al., "Performance of Breached LMFBR Fuel Pins During Continued Operation," *BNES Conference on Nuclear Fuel Performance*, March 25–29, 1985.
31. R. M. Crawford, et al., *The Safety Consequences of Local Initiating Events in an LMFBR*, ANL-75-73, Argonne National Laboratory, Argonne, IL, December 1975.
32. C. E. Branyan, "U.S. Experience with Fast Power Reactors, FERMI-I," *American Power Conference*, Chicago, IL, April 20–21, 1971.
33. A. E. Waltar, W. L. Partain, D. C Kolesar, L. D. O'Dell, A. Padilla, J. C. Sonnichsen, N. P. Wilburn, H. J. Willenberg (HEDL), and R. J. Shields (CSC), *MELT-III, A Neutronics Thermal-Hydraulics Computer Program for Fast Reactor Safety*, HEDL-TME 74-47, Hanford Engineering Development Laboratory, Richland, WA, December 1974.
34. S. L. Additon, T. B. McCall, and C. F. Wolfe, *Simulation of the Overall FFTF Plant Performance (IANUS—Westinghouse Proprietary Code)*, HEDL-TC 556, Hanford Engineering Development Laboratory, Richland, WA, December 1975.
35. *FFTF Final Safety Analysis Report*, HEDL-TI-75001, Hanford Engineering Development Laboratory, Richland, WA, December 1975.

Chapter 15
Unprotected Transients

John Sackett

15.1 Introduction

As indicated in the two previous chapters, there are several features that combine to make a SFR system safe and reliable. Only when a major off-normal condition is encountered, combined with a postulated failure of the Plant Protective System (PPS), can serious accident consequences be predicted. Even then, it has been demonstrated that a properly designed SFR can survive unprotected transients without damage to the fuel or other barriers to radiation release.

With this background, it is useful to address accidents in three categories as follows:

Protected transients. An event initiator occurs, such as a component failure, failure of a safety grade system (other than the reactor PPS), or an external event, followed by activation of the plant protection system to shut down the reactor.

Unprotected transients. An event initiator occurs as in the case for protected transients, but the reactor PPS fails to function. Such accidents may result in fuel damage, fuel melting, and even fuel pin failures. For some reactor designs, however, inherent features prevent damage to the core or reactor systems.

Severe accidents with core melting. These are typically hypothesized unprotected accidents where widespread melting and failure of the reactor fuel occurs.

This chapter provides an evaluation of the design choices important to achieve self-protecting response to unprotected transients. Our emphasis, therefore, will be on the principal physical processes and phenomenological interactions that tend to be most influential in the unprotected transients scenarios. Chapter 16 provides focus on hypothesized severe accidents with whole-core melting, major sodium fires, mechanical consequences of core disassembly, post-accident heat removal (PAHR), containment, and radiological consequences of leakage from containment.

Unprotected transients are generically categorized as transient overpower (UTOP) and transient under cooling (UTUC) events. The term UTUC refers to any under cooling transient, whereas ULOF refers specifically to losing flow and ULOHS refers to the loss of a heat sink. The leading symbol "U" in these descriptions indicates that the referenced transient is unprotected (i.e. the PPS fails). Absence of the "U" from these names would indicate a protected transient for which the PPS functions as designed.

J. Sackett (✉)
Argonne National Laboratory (Retired), Argonne, IL, USA
e-mail: johnflyer@3riversdbs.net

A.E. Waltar et al. (eds.), *Fast Spectrum Reactors*, DOI 10.1007/978-1-4419-9572-8_15,

Given a steady-state condition in which the rate of heat generation (nuclear fission) is just balanced by the rate of heat removal (coolant flow), a damaging condition could arise due to an increase in the heat generation term (with the heat removal term fixed), or due to a loss of heat removal capability (with the heat generation fixed). Hence, the UTOP transient refers to an event in which a reactivity insertion occurs (e.g., unlimited control rod withdrawal), leading to a power rise, but where normal coolant flow is maintained. Conversely, the UTUC transient refers to a class of events in which coolant flow is interrupted (e.g., loss of power to the pumps), but where fission heating is maintained.

Table 15.1 through Table 15.3 identify examples of the systems, subsystems, or components that may be involved in the transient and the phenomena that could potentially occur. All of these unprotected transients are Anticipated Transients without Scram (ATWS) scenarios.

Table 15.1 UTOP—reactivity addition ATWS

Reactor shutdown system failure resulting from:	Component or system failure	Relevant factors in transient response
Uncontrolled withdrawal of a single control rod Overcooling from pump speed increase	Reactor shutdown systems Control rod drive system Fuel and assemblies Primary pumps BOP heat rejection system	*Thermal-hydraulics* Heat removal path/capacity *Reactivity effects* Reactivity feedback at high power Fuel motion in intact fuel pins Coolant heating and margin to boiling Core reactivity feedback Core thermal and structural effects *Material behavior* Fuel cladding structural integrity at elevated temperatures Cooling systems structural integrity at elevated temperatures Containment structure integrity

Table 15.2 ULOF—loss of flow ATWS

Reactor shutdown system failure resulting from:	Component or system failure	Relevant factors in transient response
Electrical faults Mechanical faults Loss of site power Loss of piping integrity Internal flow blockage	Primary pump power supplies Pump mechanicals Off-site power Primary piping system Core and assembly coolant flow channels Core structure Fuel and assemblies Primary coolant system Inherent and passive safety systems Flow coast down extenders	*Thermal-hydraulics* Thermal inertia Pump-coast down profiles Sodium stratification Margin to boiling at peak temperature Core thermal and structural effects Heat removal path and capacity *Reactivity effects* Core reactivity feedback Fuel motion in intact fuel pins Core restraint system performance Reactor shutdown mechanism *Material behavior* Long-term performance of structures at elevated temperatures Fuel cladding integrity at elevated temperatures Containment structure integrity

Table 15.3 ULOHS—loss of heat sink ATWS

Reactor shutdown system failure resulting from:	Component or system failure	Relevant factors in transient response
Steam generator failure	Secondary sodium pumps	*Thermal-hydraulics*
Intermediate heat transport failure	Secondary system piping and IHX	Thermal inertia
Supercritical CO_2 system failure	Steam generators	Core thermal/structural effects
Decay heat removal system failure	Decay heat removal systems	*Reactivity effects*
	Sodium-CO_2 heat exchanger	Core reactivity feedback
		Fuel motion in intact fuel pins
		Core restraint system performance
		Reactor shutdown mechanism
		Material behavior
		Long-term performance of structures at elevated temperatures
		Fuel cladding structural integrity at elevated temperatures
		Containment structure integrity

15.2 Key Features Governing the Response to Unprotected Transients

As noted in Chapter 13, considerable gains have been achieved in the last two to three decades in assessing the consequences of unprotected SFR transients. The most dramatic is the recognition of the benefit of inherent passive safety features in well designed SFRs.[1]

15.2.1 Inherent Reactivity Feedback Mechanisms

As is widely known, the nuclear fission process in a fast reactor is affected by a number of physical phenomena that change core reactivity. The major components affecting core reactivity include both nuclear physics and materials phenomena as follows [1][2]:

Fuel Doppler. A net increase in the neutron absorption cross-section, caused by an increase in fuel temperature, introduces negative reactivity feedback, whereas a decrease in neutron absorption cross-section, caused by a decrease in fuel temperature, results in positive reactivity feedback. Different fuel types have different magnitudes of fuel Doppler feedback—or different *Doppler coefficients*, e.g., the Doppler coefficient for oxide fuel is more negative than for metallic fuel because of the presence of oxygen molecules and a softer neutron spectrum. However, Doppler reactivity feedback is affected by other inherent characteristics of the fuel, including thermo-physical properties such as thermal conductivity since it affects steady-state fuel temperature and the magnitude of temperature changes during transient conditions.

[1] It is conceivable that inherent passive safety features can be classified as a passive alternate to the PPS and can therefore reduce the number of unprotected accidents that must be considered. Such events can be classified as "accommodated" transients as the plant and fuel system design features are adequate to mitigate the consequences of the event initiator without actuation of the PPS. Such classification is very specific to the plant and fuel system design, and is not addressed here.

[2] As noted later in this section, these phenomena are the key inherent reactivity feedback mechanisms to compensate for changes in core power and temperature during transient conditions. They are the primary mechanisms for reactivity control at the designers' disposal when incorporating inherent passive safety features into SFR designs.

Coolant Density. An increase in coolant temperature reduces the coolant density, reducing the moderating and reflecting effects of the sodium coolant. The reactivity effect of a coolant density reduction varies across the core, being more positive in the core interior where the moderating effect is dominant and negative at the core boundaries where the leakage effect is more important. Depending on design choices affecting the relative importance of moderation and leakage, the reactivity feedback from increases in coolant temperature (reduction in coolant density) can be either positive or negative. Large reactor cores with similar core height and diameter are usually dominated overall by the moderating effect, so that increasing coolant temperatures causes positive reactivity feedback. Small cores or "pancaked" cores, where height is much smaller than the diameter, can be dominated by the leakage effect and increasing coolant temperatures results in negative reactivity feedback.

Fuel Axial Expansion. An increase in fuel temperature causes the fuel to expand axially based on the thermal expansion coefficient of either the fuel or cladding, or both, depending on the chemical and stress conditions at the fuel/cladding interface. Similarly, a decrease in fuel temperature will cause the fuel to contract axially, resulting in positive reactivity feedback. The reactivity feedback from fuel expansion or contraction depends on the thermal expansion coefficient(s) and the change in fuel and/or cladding temperature during transient conditions.

Core Radial Expansion. Fast reactor cores have a significant neutron leakage fraction due to the large neutron mean free path length, resulting in a large gradient of fuel reactivity worth at the edges of the core—making the reactivity of the core sensitive to changes in core geometry. If core assemblies move outward in the radial direction, increasing the effective diameter of the core and moving fuel from a region of higher worth to one of lower worth, negative reactivity feedback is generated. Conversely, if the core assemblies move inward, positive reactivity feedback results.

Control Rod Driveline Expansion. The relative motion between the core and the control rods is caused by a change in temperature of the control rod drivelines (CRDLs). The drivelines are normally located in the coolant outlet, or hot plenum, and respond to changes in core outlet temperature. An increase in CRDL temperature will cause the control rods to move further into the core, introducing negative reactivity feedback. A decrease in CRDL temperature will cause the control rods to move further out of the core, causing positive reactivity feedback.

Core Assembly Bowing. Changes in the duct wall temperature gradient across an assembly during a transient can cause assembly duct bowing to occur that can introduce either positive or negative reactivity. Bowing reactivity feedback is often difficult to accurately model. This is particularly true when assemblies are exposed to a flux gradient over a long time, since differential swelling can then result in bowing of assemblies at the perimeter of the core. Such changes are typically monitored through the life of the core and SFR cores are typically restrained by core clamping mechanisms near the top of the core and again near the top of the blanket/fission gas plenum region to minimize adverse effects of bowing.

Depending on the reactor design, there can be other components of reactivity feedback, such as in a pool plant where the control rods are suspended from the closure deck and control rod positions in the core are determined not only by the temperature of the CRDL, but by the vessel wall temperature as well (since the core is supported by the vessel wall). All such effects must be considered to determine the relevant inherent reactivity feedback mechanisms for evaluating their ability to either compensate or aggravate conditions in transients [2, 3].

The reactor core design choices can have a strong influence on the ability to use these inherent reactivity feedback mechanisms to favorably compensate for undesirable changes in core power and temperature during transient conditions. It is important to recognize that the reactivity feedback mechanisms can act in either a beneficial or detrimental manner, depending on the conditions. For example, some conditions require a response to an inadvertent introduction of positive reactivity in the core, as occurs in the UTOP accident. In this case, larger feedback from fuel Doppler can be beneficial.

However, in other conditions it is necessary to terminate the fission process and reduce core power to decay heat level, as in the ULOF and ULOHS, where lower feedback from fuel Doppler is preferred. It is also important to recognize that the change in core reactivity is the result of all of the reactivity feedback mechanisms, not any individual one. Proper incorporation of inherent reactivity feedback in reactor design requires a balanced approach to achieve the best overall performance in response to all potential accident conditions.

15.2.2 Natural Circulation Reactor Core Cooling

The primary coolant system in modern sodium-cooled fast reactors can easily be configured to provide natural circulation shutdown heat removal. The capability to remove shutdown decay heat with natural circulation provides a means to maintain reactor component temperatures at acceptable levels even in the event of loss of all off-site and emergency on-site power supplies. As a result, natural circulation heat removal can play an important role in the overall inherent safety approach.

Natural circulation flow arises due to the effect of gravity on a continuous fluid with a density difference along the elevation. Heavy fluid sinks to displace lighter fluid. Buoyancy-induced flow can be established when a fluid is heated, decreasing its density, at an axial position below the elevation at which the fluid is cooled (increasing its density). In a one-dimensional scenario, flow occurs when the buoyancy force is great enough to overcome form, friction, and shear losses. The natural circulation flow rate is regulated by the balance between the buoyancy force and the flow-related pressure losses. When the buoyancy force is provided by a thermally-driven density difference, the fluid flow rate will be determined by the fluid properties, the elevation difference between the heat sink and the heat source, and the density difference in the fluid between the heat source and the heat sink as caused by the heating and cooling. The flow rate and hot-to-cold coolant differential temperature relationship of single-phase natural circulation in steady-state conditions is very simple to derive as shown by Lewis [4].

$$\dot{m} = \left[\frac{2\bar{\rho}g\beta P}{\bar{K}c_p}\left(\bar{z}_{hx} - \bar{z}_c\right)\right]^{1/3} \tag{15.1}$$

$$\Delta T_c = \left(\frac{P}{\bar{\rho}c_p}\right)^{2/3}\left[\frac{\bar{K}}{2g\beta\left(\bar{z}_{hx} - \bar{z}_c\right)}\right]^{1/3} \tag{15.2}$$

where:

$\dot{m}$ = total natural circulation flow rate,
ΔT_c = core outlet minus core inlet temperature differential,
$\bar{\rho}$ = average coolant density in the core,
g = gravitational force vector,
β = volumetric coef. of thermal expansion of sodium,
P = power added/removed in the core/heat exchanger,
$\bar{K}$ = total pressure loss per area^2 around the loop,
c_p = specific heat for sodium,
$\bar{z}_{hx}$ = heat exchanger midplane elevation,
$\bar{z}_c$ = core midplane elevation.

Liquid sodium and its alloys are excellent fluids for natural circulation heat removal because of their thermo-physical properties. Due primarily to its high thermal conductivity, liquid sodium is capable of very high convective heat transfer rates, even at the modest fluid velocities characteristic of natural circulation. This tends to minimize the temperature differences between the heat source and the fluid, and between the fluid and the heat sink, and to reduce the overall source-to-sink temperature difference required for natural circulation cooling.

As illustrated in Eq. (15.1), the key design parameters to ensure natural circulation heat removal in SFR systems are (1) provision for a relatively free-flowing fluid natural circulation path, and (2) provision for sufficient elevation difference between the heat source and the heat sink.

In the primary coolant circuit, natural circulation flow may be established along the same flow path used for normal operation. Along this path, coolant is heated in the reactor, rises to the hot plenum, and flows through the intermediate heat exchangers (IHX) to the cold plenum and back to the reactor core. In accidents or emergency shutdown conditions in which no heat is removed in the IHX, heat can be removed by independent heat exchangers in series with the IHX for loop-type reactor designs. Alternatively, for pool-type reactor designs, direct reactor auxiliary cooling system (DRACS) heat exchangers may be located high in the cold pool. Primary coolant chilled in the DRACS heat exchangers falls to near the bottom of the cold pool, where it enters the primary coolant pump inlet and travels back to the reactor. For either loop- or pool-type reactor configurations, primary coolant natural circulation carries heat from the reactor to the IHX or auxiliary heat exchangers.

Heat transferred at the IHX may be removed by natural circulation in the intermediate coolant loop if sufficient elevation of the ultimate heat sink is provided. This mode of shutdown heat removal has been demonstrated in the loop-type FFTF reactor [5]. If the IHX path is not available, heat may be removed through the auxiliary heat exchangers to a second natural circulation loop. The working fluid in this second loop is often specified as sodium-potassium alloy (NaK) because of its low melting temperature compared to sodium. The NaK loop carries the heat through piping to a second heat exchanger located at a high elevation outside the containment building, where the heat is rejected to environmental air. Shutdown heat removal through such an auxiliary cooling system has been demonstrated in the pool-type EBR-II reactor [6].

15.3 Integrated Inherent Safety Approach

It is possible to design a reactor that has the ability to respond to initiators such as UTOP, ULOF, and ULOHS without damage, so long as the appropriate design choices are made. For such transients, it would be desirable to achieve a balance of reactivity feedback mechanisms and transition to natural convective cooling so that the response is favorable to all of the initiators. In practice, one may also have to rely on design features that limit the severity of the initiator to ensure a benign termination of the event, such as using control rod motion limiters to restrict the amount of reactivity that can be introduced into the core from an inadvertent control rod withdrawal, or extending the inertial coast down of the coolant pumps to provide sufficient time for the inherent reactivity feedback to reduce core power and limit the mismatch between the coolant flow and core power. If these and other similar considerations are included at the design phase of the reactor, it is possible for the reactor system to respond to a wide range of potential initiators without resulting in core damage or more serious consequences, even if the PPS fails to respond.

15.3.1 Modeling Tools for the Designer

There have been many tools developed and validated through testing to describe the dynamic behavior of fast reactor systems. The focus of the more recent work was to show that physically plausible sequences of events do not lead to *core disruption* (i.e. the large scale displacement of fuel pins or assemblies, including significant melting in severe scenarios), either because inherent reactivity feedback mechanisms maintain acceptable power and power-to-flow ratios (no coolant boiling or fuel melting) or because the system characteristics allow long times for corrective action to be taken (to remedy loss of heat sink). This resulted in considerable emphasis on modeling reactor behavior *prior* to core disruption, in contrast to the earlier, historical focus on core disruption discussed in Section 13.5.2.

The MELT series of codes [7] was developed primarily for use in analysis of the UTOP Hypothetical Core Disruptive Accident (HCDA) for FFTF. Later versions included a sophisticated internal fuel motion model, used for evaluation of molten fuel motion prior to pin failure in annular pins.

The SAS (for Safety Analysis System) series of codes were the most widely used for analysis of the initiating phase of HCDAs [8]. The SAS3D code [9] was used extensively for analysis of ULOF and UTOP accidents in reactors such as FFTF, CRBRP, SNR-300, and MONJU. The next version, SAS4A, included improvements in steady-state fuel pin characterization, modeling of fuel pin mechanics, molten cladding relocation, and motion of disrupted fuel [10].

The emphasis on inherent safety led to increased attention to system dynamics and reactivity feedback mechanisms that are active prior to core disruption. Consequently, the SASSYS-1 code was developed and refined [8, 11]. This code is capable of analyzing the full spectrum of transients from mild operational events through severe transients. It retains many of the features of SAS4A, including point kinetics and detailed thermal-hydraulic modeling of the core, inlet plenum, and outlet plenum. However, more detail is available in modeling the primary and intermediate heat transport systems, steam generators, and decay heat removals systems. Flexible arrangement of components and efficient numerical methods make SASSYS-1 very useful in the analysis of inherently safe reactor designs and in long-time transients involving decay heat removal systems.

In addition, much attention was given to the development of suitable fuel pin characterization and fuel pin mechanics under steady-state and transient conditions. Special emphasis was given to in-pin fuel motion prior to pin failure, since such motion is important in assessments of inherent safety.

With the advent of much more powerful computing capability, there is the opportunity to provide considerably more detailed models of reactor behavior for accident conditions. This area of research is quite promising, building upon the models that have been described here.

15.3.2 UTOP Transient

As discussed by Cahalan and Wigeland (Ref. [1]), the UTOP transient has historically been considered to be the limiting accident in a fast spectrum reactor due to the fact that the fuel in such a system is not arranged in its most reactive configuration during steady state operating conditions. Early studies assumed that fuel compaction would take place during a severe UTOP and that the ensuing positive reactivity insertion caused by such compaction would cause a very large power spike—leading to fuel vaporization and core disassembly.

However, experimental evidence has demonstrated just the opposite; namely, should an unprotected accident occur, fuel would move predominantly *away* from the middle of the core instead of compacting [12]. This is certainly true for the first phase of any overpower transient, where axial fuel expansion of solid fuel within the pin is a natural outcome of the increase in fuel and cladding temperatures. For cases when the transient proceeds to the point of fuel melting, axial movement has been experimentally observed under two conditions: (1) axial movement of molten fuel within the pin prior to cladding failure and, (2) axial fuel sweep out after fuel cladding failure [13, 14].

For the first case, experimental results show that fission gas can force molten fuel motion within the pin from the central region to the axial periphery of the core, introducing strong negative reactivity feedback prior to cladding failure. This may lead to termination of the accident if sufficient negative reactivity is inserted. This pre-clad failure fuel movement mechanism is sometimes called "fuel squirting" in Europe and "fuel extrusion" in the U.S.

Use of annular oxide pellets (i.e. fuel pellets incorporating a hole in the radial center of the pellet) provides an inherent shutdown mechanism during hypothetical SFR accidents by providing a pathway for molten fuel to be ejected from the active core region to the top and bottom blankets. The PINEX series of experiments (25% PuO_2) with FFTF type fuel pins (a central gap radius of 0.041 cm) verified the concept of axial internal motion within annular pins, representing an inherent shutdown mechanism for SFR accidents initiated by ramp rates of the order of 50 ¢/s [15–17]. The motion of melted fuel is driven by fission gas release, forcing the fuel in both axial directions away from the core midplane.

However, the question remained regarding whether this mechanism would still be present when slower ramp rates are encountered (such as ramp rates of the order of 5 ¢/s) and when solid fuel pellets are present. By observing the results of HUT-52A (25% PuO_2) experiments, it appears that this mechanism is expected to be effective even with solid fuel pellets and with lower ramp rates [16]. This is due to high fuel temperatures in the reactor fuel pins along the full length, combined with the central cavity having been formed (in mixed oxide fuel) via steady state irradiation. The fission gas driven fuel motion mechanism is expected to be more prominent in a fast spectrum, relative to that of a thermal spectrum, due to the more uniform fission gas release within the fuel pellets—as opposed to pellet surface fission gas releases under the self-shielded conditions of thermal spectrum reactors.

Experiments have been conducted for both fresh fuel pins and irradiated fuel pins (22% PuO_2 mixed oxide solid fuel pellets) in the Transient Reactor Test Facility (TREAT) for a reactivity insertion rate of 5 ¢/s. Pre failure axial molten fuel relocation was observed in both fresh fuel pins (TS-1 experiment) and irradiated fuel pins (TS-2 experiment) [18]. The molten fuel extrusion occurring in the tests lifted the upper pin components significantly prior to pin failure. From these experimental results, at least 15% of the total fuel inventory from the melt region appeared to undergo "fuel squirting"—providing a substantial negative reactivity feedback.

Should the accident progress to pin failure, there is experimental evidence that the molten fuel entering the coolant channel will be fragmented and swept upward and out of the active core by the coolant flow—again providing substantial negative reactivity feedback. Calculated results [19] performed for the regulatory review of the FFTF were substantiated by TREAT experiments [20].

15.3.3 ULOF Transient

For the ULOF transient, an example of the favorable performance using inherent features is shown in Figs. 15.1 through 15.3 (from Ref. [1]). Here the reactor response to a ULOF accident is shown for a 3,500 MWth pool-type reactor using metallic fuel. The reactor core is of a compact and efficient design, with a height of about 1 m and a core diameter of about 4.5 m, and a positive sodium void worth of $7.26. Figure 15.1 shows the flow coast down and the decrease in power that results from the inherent reactivity feedback, where the mismatch in core power and coolant flow reaches a maximum about 50 s after the start of the accident.

Figure 15.2 shows the change in peak core temperatures during the transient, where it is seen that the peak coolant temperature has a minimum margin to coolant boiling of about 125°C at 50–100 s after the start of the accident, which then increases as system temperatures drop back towards their steady-state values and natural circulation cooling is established. There is no fuel pin failure or fuel melting in this case. The peak fuel temperature is also relatively low due to the high thermal conductivity of metallic fuel, which is a significant advantage for this type of transient since core power must be reduced to decay heat levels in the absence of forced cooling from the coolant pumps.

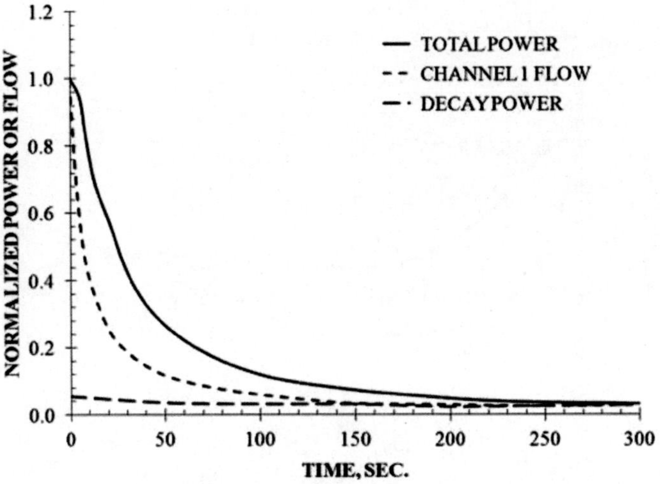

Fig. 15.1 Normalized power and flow in response to a ULOF transient for a large metallic-fueled fast reactor [1]

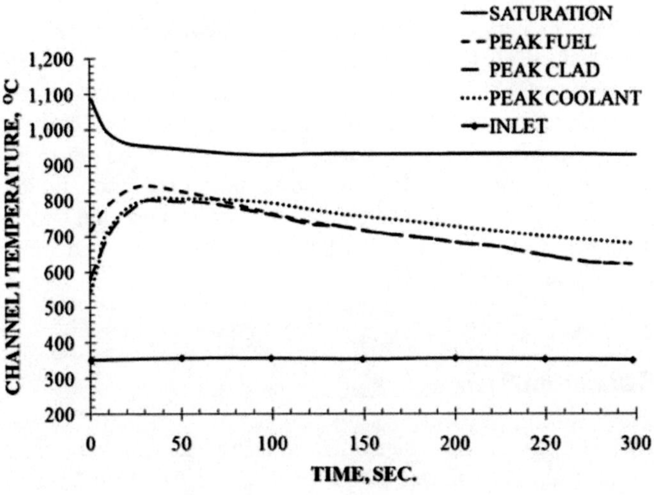

Fig. 15.2 Transient peak core temperatures in response to a ULOF transient for a large metallic-fueled fast reactor [1]

Figure 15.3 shows the reactivity feedback from each major mechanism as well as the total reactivity feedback. The dominant favorable feedbacks are from radial expansion of the reactor core and thermal expansion of the control rod drivelines (CRDLs). Early in the transient, unfavorable feedback is limited to the effect of decreasing coolant density. The net reactivity reaches a minimum of about –\$0.40 from about 50 to 150 s after the start of the accident, sufficient to reduce core power quickly enough to limit the mismatch in power and flow as shown in Fig. 15.1. Core power is reduced to decay heat levels, most reactivity feedback mechanisms result in positive reactivity feedback due to the lower fuel and coolant temperatures, but the total reactivity feedback is still negative. It is important to recognize that the reactor does not reach a neutronic shutdown for such events, but will essentially stabilize at zero fission power and zero reactivity. As the transient progresses, the final state will be in equilibrium between the core power and the heat removal capability, whether provided by the normal heat removal path or by an auxiliary decay heat removal system, until the reactor can be scrammed.

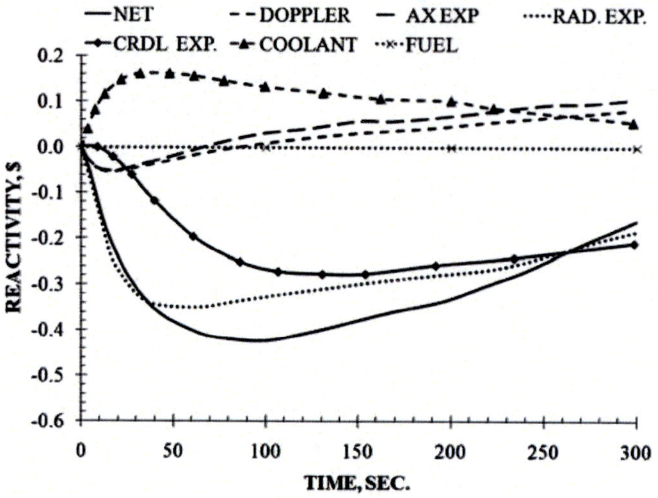

Fig. 15.3 Transient reactivity feedback in response to a ULOF transient for a large metallic-fueled fast reactor [1]

Similar analyses have been performed using oxide fuel, showing that it is more difficult to achieve the same inherent performance due to the different magnitude of the reactivity feedback mechanisms. The much higher steady-state fuel temperatures result in a less negative total reactivity as core power and temperatures are reduced, particularly resulting from the positive fuel Doppler effect upon cool-down. Achieving a benign termination to this transient requires design modifications such as extending the flow coast down to provide more time for the inherent power reduction, limiting the mismatch in power and flow. Other approaches are also beneficial, such as lowering the steady-state fuel temperature by lowering the power density by using smaller diameter fuel pins or adding additional devices such a gas expansion modules (GEMS) [21].

15.3.4 EBR-II Validation Tests

The models used in these analyses have been validated using data from tests conducted at the EBR-II reactor plant in 1986 [22]. EBR-II was subjected to all of the ATWS events without damage, demonstrating the potential for self-protecting response of a SFR (While anticipated TOP tests were conducted using EBR-II, more severe TOP events were simulated in the TREAT reactor).

The first of the EBR-II tests was loss of all pumping power with failure to scram—a ULOF, simulating a station blackout with failure to scram. The reactor was brought to 100% power and the pumps were turned off, allowing them to coast-down and the coolant flow to transition from forced to natural-convective flow. Testing and analysis over the previous 4 years had been conducted to accurately model the reactor for this event, and the system responded as expected. Special in-core temperature monitoring had been provided as a safety system to scram the reactor if temperatures rose to unexpectedly high levels, but they did not.

Temperatures initially rose rapidly as the cooling flow decayed, but the increase in temperature introduced sufficient negative reactivity feedback that the power was also reduced rapidly, resulting in peak core coolant temperatures that were higher than for normal operation (~704°C vs. 477°C at normal operation) but not high enough to damage the fuel. There was also significant margin to sodium boiling temperature, which would occur at ~893°C. The system power decreased rapidly, reducing peak core temperatures until they equilibrated at an average temperature very close to that at normal operation.

A point to be emphasized is that there was no fuel or core damage with this test, unlike what would occur in a conventional reactor system. In fact, this was the 45th test of ATWS events on this core and the reactor was restarted for a subsequent test that same afternoon.

The second test subjected the reactor to a ULOHS. The reactor was brought to 100% power and flow was stopped in the secondary heat transfer system, blocking the transfer of heat to the steam generators. As the reactor inlet temperature rose, negative reactivity feedback reduced power such that the temperature difference across the core was reduced; the peak coolant temperature never increased. The reactor temperatures equilibrated at an average temperature very close to that at normal operation.

The observed behavior in ULOHS EBR-II test resulted because of the very strong negative feedback associated with neutron leakage as the coolant temperature increased and the lack of a strong positive reactivity feedback from Doppler as fuel centerline temperatures fell. Metal fuel operates with a very low centerline temperature and, therefore, little Doppler feedback reactivity occurs as fuel temperature changes are relatively small. For ULOF or ULOHS events, coolant temperatures rise, power falls and if a high Doppler coefficient of reactivity exists in the system (such as for oxide fuel), the positive reactivity feedback will delay power reduction with the result that the sodium will quite possibly boil. Boiling sodium will insert significant positive reactivity, likely leading to a severe overpower transient, unless boiling begins near the top of the core where the void coefficient is negative. For this reason, a metal fuel core like used in EBR-II is generally self-protecting against under-cooling events while this may not be the case for oxide fueled cores. Such behavior is most easily achieved in smaller cores where neutron leakage is a significant contributor to reactivity feedback effects but becomes more difficult as the size of the core increases above ~1,000 MWe.

It was earlier thought that a high Doppler coefficient of reactivity was necessary to protect against severe UTOP events, so tests were done in TREAT with metal fuel to determine its performance under such conditions. Many metal fuel pins from EBR-II were subjected to severe UTOP events that took the pins to failure [23, 24]. It was found that the relatively low melting temperature of the fuel was important, since the fuel softened and then flowed like toothpaste in a tube before breaching the cladding. Fuel cladding failures, when they did occur, were at the top of the fuel column with the result that fuel escaped toward the top of the core. This flow of fuel occurred rapidly and would be effective in introducing large negative feedback during severe transients, acting as an effective self-protecting mechanism.

These results highlight the importance of design choices related to the fuel type in fast reactors. The influence of fuel on reactivity feedback is significant; oxide fuel provides higher Doppler feedback, which is beneficial for reducing a power rise but detrimental for avoiding sodium boiling upon ULOF. As noted in Chapter 14, an interesting option for overcoming the positive reactivity insertion associated with oxide fuel during an ULOF event is the introduction of GEMS—successfully demonstrated in tests conducted at the FFTF [21].

Further tests in EBR-II were conducted to determine the load-following characteristics of the reactor, which are very good. Metal fuel is not adversely affected by cyclic changes in power and temperature, which coupled with its strong tendency to maintain a constant average core temperature, greatly facilitates its ability to load follow. EBR-II could be easily controlled by fixing control-rod positions and controlling power demand at the steam turbine. A full range of safety and load following tests were conducted, including, for example, rapid run-up of the primary pumps to their maximum capacity, which would cool the core, and then inserting positive reactivity to raise the power level. No damage occurred to the fuel or core through all of these tests.

A level-1 probabilistic risk assessment (PRA) was completed to quantify these results [25]. It was shown that risks associated with EBR-II operation were substantially lower than typical LWR plants—an order of magnitude less. These attributes were incorporated into two notable plant designs, namely PRISM [26] (developed by GE) and SAFR [27] (developed by Rockwell International).

References

1. R. Wigeland, and J. Cahalan, "Mitigation of Severe Accident Consequences Using Inherent Safety Principals," *Fast Reactor Safety 2009*, Tokyo, Japan, December 2009.
2. Ph. Bergeonneau, et al., "Uncertainty Analysis on the Measurements and Calculation of Feedback Reactivity Effect in LMFBRs, Application of Super-Phenix-1 Startup Experiments," *30th NEACRP Meeting,* NEACRP-A-833, Helsinki, Finland, September 1987.
3. R. W. Schaefer, "Critical Experiment Tests of Bowing and Expansion Reactivity Calculations for Liquid-Metal-Cooled Fast Reactors," *Nuclear Science and Engineering*, 103, 196, 1989.
4. E. E. Lewis, *Nuclear Power Reactor Safety*, chapter 7, Wiley, New York, NY, 1977.
5. W. C. Horak, J. G. Guppy, and R. J. Kennett, *Validation of SSC Using the FFTF Natural-Circulation Tests*, BNL-NUREG-31437, Brookhaven National Laboratory Report, Upton, NY, December 1982.
6. L. K. Chang, et al., "Experimental and Analytical Study of Loss-of-Flow Transients in EBR-II Occurring at Decay Power Levels," *Conference on Alternative Energy Sources*, Miami Beach, FL, December 1985.
7. K. M. Tabb, et al., *MELT-III B – An Updated Version of the MELT Code*, HEDL-TME 78-108, Hanford Engineering Development Laboratory, Richland, WA, 1978.
8. J. E. Cahalan, and T. Y. Wei, "Modeling Developments for the SAS4A and SASSYS Computer Codes," *International Conference on Fast Reactor Safety*, American Nuclear Society, Snowbird, UT, August 1990.
9. M. G. Stevenson, et al., "Current Status and Experimental Basis of the SAS LMFBR Accident Analysis Code," *Proceedings of the Conference on Fast Reactor Safety*, CONF-740401-P3, p. 1303, Beverly Hills, CA, 1974.
10. A. M. Tentner, et al., "SAS4A Computer Model for the Analysis of Hypothetical Core Disruptive Accidents in Liquid Metal Reactors," *Eastern Computer Simulation Conference*, Orlando, FL, April 1987.
11. F. E. Dunn, and T. C. Wei, "The Role of SASSYS-1 in LMR Safety Analysis," *Proceedings of the International Topical Meeting on Safety of Next Generation Power Reactors*, Seattle, WA, May 1988.
12. L. E. Strawbridge, and G. H. Clare, "Exclusion of Core Disruptive Accidents from the Design Basis Accident Envelope in CRBRP," *Proceedings of the International Meeting on Fast Reactor Safety*, Vol. 1, pp. 317–327, American Nuclear Society, Knoxville, TN, April 21–25, 1985.
13. J. E. Cahalan, R. A. Wigeland, G. Friedel, G. Kussmaul, J. Moreau, M. Perks, and P. Royal, "Performance of Metal and Oxide Fuels During Accidents in a Large Liquid Metal Cooled Reactors," *Proceedings of the International Fast Reactor Safety Meeting*, Vol. IV, p. 73, Snowbird, UT, August 1990.
14. A. E. Wright, et al., "CAFÉ Experiments on the Flow and Freezing of Metal Fuel and Cladding Melts (2), Results, Analysis, and Applications," *International Conference on Fast Reactors and Related Fuel Cycles* (FR09), Kyoto, Japan, December 7–11, 2009.
15. D. E. Smith, F. J. Martin, and A. Padilla, *Internal Fuel-Motion Phenomenology: FUMO-E Code Analysis of PINEX Experiments*, HEDL-SA-2629-FP, June 1982; D. R. Porten, et al., "PINEX-2 Experiment, Concept Verification of an Inherent Shutdown Mechanism for HCDA's," *Proceedings of the International Meeting on Fast Reactor Safety Technology*, Seattle, WA, August 19–23, 1979.
16. P. C. Ferrell, D. R. Porten, and R. J. Martin, "Internal Fuel Motion as an Inherent Shutdown Mechanism for LMFBR Accidents: PINEX-3, PINEX-2, and HUT 5-2A Experiments," HEDL-SA-2264, *Fast Reactor Safety Meeting*, Sun Valley, Idaho, August 2, 1981.
17. E. T. Weber, et al., "Transient Survivability of LMFBR Oxide Fuel Pins," HEDL-SA-3349, *British Nuclear Energy Society Conference on Science and Technology of Fast Reactor Safety*, Guernsey, Channel Islands, May 12–16, 1986.
18. A. L. Pitner, et al., "TS-1 and TS-2 Transient Overpower Tests on FFTF Fuel," *Transactions of the American Nuclear Society*, 50, 351–352, 1985.
19. A. E. Waltar, N. P. Wilburn, D. C. Kolesar, L. D. O'Dell, A. Padilla, L. N. Stewart (HEDL), and W. L. Partain (NUS), *An Analysis of the Unprotected Transient Overpower Accident in the FTR*, HEDL-TME-75-50, Hanford Engineering Development Laboratory, Richland, WA, June 1975.
20. R. N. Koopman, et al., "TREAT Transient Overpower Experiment R12," *Transactions of the American Nuclear Society*, 28, 482, 1978.
21. T. M. Burke, "Summary of FY 1997 Work Related to JAPC-US DOE Contract 'Study on Improvement of Core Safety – Study on GEM (III)' ", HNF-2195-VA, *DOE Technical Exchange*, Tokyo Japan, February 10, 1998.
22. H. P. Planchon, et al., "The Experimental Breeder Reactor II Inherent Shutdown and Heat Removal Tests – Results and Analysis," *Proceedings of the International Meeting on Fast Reactor Safety*, Vol. 1, pp. 281–291, American Nuclear Society, Knoxville, TN, April 21–25, 1985.
23. T. H. Baur, *Behavior of Metallic Fuel in TREAT Transient Overpower Tests*, CONF-880506-14, TI88 010042, Argonne National Laboratory, Argonne, IL, May 17, 1988.

24. A. M. Tentner Kalimullah, and K. J. Miles, "Analysis of Metal Fuel Transient Overpower Experiments with the SAS4A Accident Analysis Code," *Proceedings of the International Conference on Fast Reactor Safety*, American Nuclear Society, Snowbird, UT, August 1990.
25. D. J. Hill, "An Overview of the EBR-II PRA," *Proceedings of the 1990 International Meeting on Fast Reactor Safety Meeting*, Vol. IV, p. 33, Snowbird, UT, August 12–16, 1990.
26. U.S. Nuclear Regulatory Commission, *Preapplication Safety Evaluation Report for the Power Reactor Innovative Small Module (PRISM) Liquid Metal Reactor*, NUREG-1368, Washington, DC, February 1994.
27. U.S. Nuclear Regulatory Commission, *Preapplication Safety Evaluation Report for the Sodium Advanced Fast Reactor (SAFR) Liquid Metal Reactor,* NUREG-1369, Washington, DC, December 1991.

Chapter 16
Severe Accidents and Containment Considerations

John Sackett

16.1 Introduction

The previous chapter considered transient sequences that were terminated without damage by self protecting features of the reactor system design. In this chapter, we provide a mostly qualitative approach to the sequential steps typically followed to evaluate hypothetical core-disruptive-accidents (HCDAs) with significant core damage. In addition, design consideration for containment and accommodation of large sodium fires are presented.

16.2 Unprotected Accidents with Severe Core Disruption

As noted in earlier chapters, because the fuel in fast reactors is not arranged in its most reactive configuration, early studies focused on the very conservative postulate that coherent fuel collapse during a major accident might lead to severe core disruption [1]. Recognition that this arbitrary assumption of coherent core collapse gave results that were much too conservative led to the development of a mechanistic approach to the analysis of core disruptive accidents [2]. Instead of postulating arbitrary conditions that lead to severe excursions, the mechanistic approach attempts to analyze accident sequences from a given initiating event up to the conclusion of the accident [3]. The previous chapter considered transient sequences that were terminated without damage to the core by self protecting features of the reactor system design. In this chapter, we provide a mostly qualitative approach to the sequential steps typically followed to evaluate hypothetical core disruptive accidents with significant core damage. A substantially more detailed review of the steps historically taken to understand the complex interactions involved for postulated HCDAs in oxide-fueled SFRs is contained in Appendix G.

16.2.1 Mechanistic Analysis Approach

As described by Marchaterre [3], the important feature of a comprehensive approach to mechanistic accident analysis is that it allows for an "early termination" path where accidents are terminated by early negative reactivity effects and with limited core damage. The core disruption phase is more commonly found to be gradual meltdown of the core and without an energetic excursion. As discussed

J. Sackett (✉)
Argonne National Laboratory (Retired), Argonne, IL, USA
e-mail: johnflyer@3riversdbs.net

A.E. Waltar et al. (eds.), *Fast Spectrum Reactors*, DOI 10.1007/978-1-4419-9572-8_16,

by Fauske [4] the consequences of such a meltdown may be described by a number of "general behavior principles" that can be verified experimentally.

There are three major focus areas in conducting a mechanistic analysis:

(1) Determine that the likely accident paths lead to early termination or non-energetic core disruption;
(2) Define the reactor design changes that must be made to insure that initiators lead to non-energetic core disruption; and
(3) Define the time scales and phenomena of interest and establish the validity of the application of the "general-behavior principles" in a given accident situation. In addition, parametric calculations using assumptions that deviate from the best estimate values can be carried out, including cases that lead directly to an energetic disassembly, and it is against these cases that reactor design margins are assessed.

The mechanistic assessment of the response of a reactor system leading to potentially core disruptive accidents generally proceeds through the following phases:

(1) Identifying potential initiators of importance;
(2) Conducting parametric analyses of the initiating phase;
(3) Evaluating transition phase behavior (should the accident proceed to this point);
(4) Conducting disassembly analyses for those parametric cases that lead to energetic disassembly;
(5) Performing damage evaluations;
(6) Evaluating post-accident heat removal; and
(7) Analyzing radiological consequences.

16.2.2 Initiating Phase Accident Analysis

The first step in accident analysis is to identify those initiators that will be the most important contributors to risk or will provide an envelope for possible consequences. In general, the accident sequences that dominate are unprotected loss of flow (ULOF) accidents. A large number of codes-have been developed for analyzing the initiating phase of these types of accidents. The SAS series of codes discussed in Chapter 15 are typical of codes developed for this purpose [5].

The initiating phase analyses involve a calculation of core neutronics and thermal behavior up to the point of loss-of-assembly geometry. Assemblies with similar power, flow, and irradiation conditions are lumped together into groups for the purpose of analysis. All of the assemblies within a given group or "channel" are assumed to behave identically. Phenomena such as transient thermal-hydraulics, sodium boiling, and fuel pin mechanics and failure, cladding and fuel motion, and fuel–coolant interactions are treated with one-dimensional models [6].

The reactivity effects of material motions are generally calculated using point reactor kinetics together with perturbation theory (cf. Chapter 6 for spatially-derived reactivity feedback coefficients) for computing reactivity changes associated with material motion. A point kinetics model is actually quite accurate for fast spectrum reactor analyses because the long mean free path of high energy neutrons prevents significant spatial power shifts. The models for the phenomena involved in various accident sequences have been verified and validated by both out-of-pile and in-pile tests.

Calculations with initiating phase accident codes are taken to the point where the reactor is either permanently shut down, with the core in a coolable configuration in its original location, or to the point where assembly geometry is lost.

A typical "best estimate" calculation [7] of a ULOF for the CRBRP would have the following sequence of events. The flow coastdown boiling begins in the assemblies with the highest power to flow ratios. This voiding of sodium increases reactivity and causes the power to rise but negative effects from fuel Doppler feedback and axial expansion prevent the reactor from reaching prompt critical. Cladding melts in the hottest assemblies, followed by fuel melting and fuel dispersal driven by fission gas release. The calculation is ended with the core neutronically shut down but uncoolable. Other parametric cases with more pessimistic assumptions give rise to mild hydrodynamic disassemblies.

In contrast, in typical unprotected transient overpower (UTOP) accidents, pin failures in the upper half of the fuel pins lead to molten fuel entering the coolant channel with subsequent sweep out of the fuel—resulting in a shut down reactor in a coolable configuration [8].

16.2.3 Transition Phase Analysis

Results of calculations for the initiating phase of accidents for FFTF and CRBRP led to the conclusion that the most likely path for postulated core-disruptive accidents would be either early termination with the core largely intact and coolable or a gradual meltdown with no large energy release but with the core uncoolable. For cases where the core is uncoolable at the end of the initiating phase accident analysis, the analysis must be continued until a stable coolable configuration is obtained. The analysis of this phase of the accident is termed the *transition phase*.

While the analysis of the transition phase is considerably less developed than the initiating phase accident realm, conclusions as to the behavior of the disrupted core can still be drawn. The course of the accident is expected to be as follows. Since the initial core disruption for this case is mild, the pressures generated are too low to cause a massive dispersal of molten fuel from the core region. The assembly duct walls are quickly melted and growing regions of molten fuel and steel begin to form in the hottest portions of the core. By this stage enough local axial fuel dispersal has taken place to bring the reactor subcritical. However, extended fuel dispersal may be inhibited by blockages that form because molten core materials refreeze in the colder regions near the axial ends of the core. Continued decay heating causes further core melting and previously melted core materials to boil up and fill the available volume. If paths are available through the upper or lower structure, material will begin to be ejected from the core region. If the passages are blocked, rapid melting attack on the structures will begin, allowing ejection of molten core materials from the core region. Because the high vapor velocities obtained by vaporization of core materials tend to lead to a dispersed system, a transition to a largely dispersed boiling pool will occur even at decay heat energy generation levels. Studies of the properties of a boiling pool of core materials indicate that the nature of the flow regime should be generally dispersive. In such a dispersed system, collapse and recriticality will be unlikely.

Detailed computer codes equipped to analyze the complex phenomena of the transition phase in an integrated fashion for FFTF and CRBRP reactor geometry included the TRANSIT-HYDRO code [9], developed at Argonne National Laboratory, and the SIMMER code [10], developed at Los Alamos National Laboratory.

The TRANSIT-HYDRO code provides detailed modeling of thermal-hydraulics, material behavior, and geometry on an assembly-by-assembly basis. Heat transfer and material motion between assemblies is modeled. It provides the capability to follow an accident sequence through heat up, fuel disruption and hex-can wall failure, after which fuel and cladding mass and energy transfer may occur. It is well suited to follow the details of the process of meltdown into a full molten core.

The SIMMER code is widely used for analysis of transition phase, disassembly and post-disassembly behavior. It provides a very general multi-phase, multi-component fluid dynamics capability, coupled with discrete time-dependent neutronics.

In modeling the dynamic behavior of a core under such conditions, continuity equations for mass, momentum, and energy must be solved. These basic equations, cast in both Eulerian and Lagrangian coordinates, are included in the original Fast Breeder Reactor book [11].

16.2.4 Disassembly Analysis

While the results of calculations for FFTF and CRBRP have led to the conclusion that the most likely path to core disruption is either early termination in a coolable configuration or a gradual non-energetic meltdown, analyses using more pessimistic assumptions have also been done to evaluate the reactor designs. In these cases, sequences of events that lead to an energetic core disassembly were postulated and the structural response of the system was assessed.

The disassembly phase of the calculation is entered if a prompt-critical excursion is induced. The rapid heating and vaporization of the fuel produce high pressures that disassemble the core and thereby end the power burst. The disassembly process was first described by a model developed by Bethe and Tait [1]. In their approach, the core was treated as a homogeneous fluid so that the material motion during disassembly could be calculated using a hydrodynamic approach. The calculations were done in spherical geometry. The reactor power was calculated using point kinetics with first-order perturbation theory to estimate the reactivity feedback associated with the material motion.

A number of improvements have subsequently been made to this basic approach. Specifically, Doppler feedback was included, improvements were made to the equation-of-state used to estimate the pressures, more accurate neutronics were implemented in some cases, and the capability for doing calculations in two-dimensional (r, z) geometry was added. A number of disassembly computer codes were developed in the United States as successive improvements were made [6]. The VENUS code [12, 13], developed at Argonne National Laboratory, constituted the first major step, followed by a substantially enhanced step in sophistication with the SIMMER code series [10] developed at the Los Alamos National Laboratory.

16.2.5 Damage Evaluation

During or following reactor disassembly, the rate at which the thermal energy released in the transient can be converted to work must be established. This is used to evaluate the potential damage that the excursion can cause to the system. Work can either come from the expansion of the core materials themselves or the interaction of these materials with the sodium coolant and its consequent vaporization and expansion.

Early studies of the damage potential of core-disruptive accidents assumed that the work on the surrounding reactor structures would be done by the expanding fuel materials. It was later recognized that the transfer of heat from the high temperature fuel to the sodium could considerably increase the potential work available to do damage.

However, it does not appear that the heat transfer to sodium can occur on a time scale short enough to significantly increase the work potential above that obtained by the expanding fuel vapor. Current practice, therefore, is to use the pressure volume curve determined by the expanding fuel vapor for damage evaluation—especially for situations where there is little sodium left in the disconfigured core. Should a major power excursion occur when the bulk of the sodium is still present, the sodium could provide an effective pressure source to terminate the excursion [14].

Once the pressure source has been established, the response of the system can be analyzed. This is usually done with a hydrodynamic calculation of the pressure propagation coupled with an analysis of the structural response of the important system components. The REXCO-HEP code [15] solves the basic conservation equations of mass, momentum and energy, together with the equations of state of the reactor materials, and calculates the response of the structures to the previously discussed pressure source term. The code treats not only the hydrodynamics but also the elastic and plastic deformations of the reactor structures and components.

16.2.6 Post-accident Heat Removal

Part of the analysis of core disruptive accidents is an evaluation of the post-accident heat removal (PAHR). The objective is to analyze the long-term decay heat removal from the fuel following a disruptive accident to show how the fuel comes to rest in various parts of the system where it can be permanently cooled. This has mainly taken the form of predicting the spectrum of post-disruptive dispositions that the fuel could have, and analyzing the subsequent coolability. The disposition of core materials and the required measures to assure cooling of the core debris are, of course, dependent on the reactor design under consideration.

In particular, whether one is dealing with cooling the disrupted core with an intact primary system or whether one must assume the primary system is no longer intact makes a significant difference in the relative importance of the key phenomena involved in post-accident cooling.

Considerable technology has been developed to analyze several of the heat transfer modes expected to be important in PAHR behavior. Predicting the heat removal from internally heated debris beds and molten pools are two of the most important technical problems. Considerable research in these areas has been done to understand the interactions of molten core materials with potential sacrificial materials for core retention systems and with reactor system structural materials such as concrete [16].

16.2.7 Radiological Consequences

The end objective of core-disruptive accident analysis is to estimate the radioactivity release and the corresponding potential exposure to members of the public. A typical analysis of radiological consequences involves four phases:

(1) Source term,
(2) Containment transient,
(3) Radioactivity transport and release, and
(4) Atmosphere dispersal.

The *source term* is defined as that quantity of radioactive materials released from the primary reactor system in a form capable of being spread through the atmosphere. Simulation experiments have shown qualitative evidence of processes that would tend to cause retention of radioactive materials, except noble gases, within the sodium system.

These processes include condensation, turbulent mixing, and particulate agglomeration and deposition within the reactor vessel. Comprehensive analytical models for these processes are under development. To date, only limited credit has been taken for calculated source term depletion by these processes, and conservative assumptions have been made based in part on the severity of the core disruption assumed. For severe excursions resulting in vaporization of a significant fraction of the core, typical assumed release fractions have been of the following-order:

Noble gases 100%
Halogens 1–25%
Volatile solids 0.1–10%
Nonvolatile solids 0.01–10%

Containment building temperature and pressure transients are determined primarily by the chemical reaction of sodium expelled from the reactor system together with decay heat of released radioactive materials. Individual design-dependent calculations are made to estimate sodium release from the reactor; for example, through leak paths in the reactor head, and the maximum temperature and pressure that could be produced by such a release.

The computer code SOFIRE [17] has been widely used to calculate long term temperature transients and the characteristics and effects of sodium pool fires. The CACECO code [18] was developed to calculate temperature and pressure transients and also the atmosphere chemical composition for interconnected spaces in the building. This code will also calculate leakage from the building for an assumed orifice leak path.

The release of noble gases from containment is calculated on the basis of complete mixing of the gases with the containment atmosphere and leakage at the same rate. The halogens and solids are subject to removal mechanisms within the containment, primarily dependent upon aerosol mechanics. Especially in the presence of a significant quantity of sodium oxide, aerosol agglomeration, settling, and plate-out can remove particulate radioactivity with a removal time constant of the order of hours. The computer code ABCOVE is used for analysis of aerosol concentrations and dispersal [19].

Radioactive aerosol materials are normally assumed to leak from the containment in proportion to their concentration in the atmosphere. No credit is taken for aerosol removal in leak paths, although current experimental programs are showing this to be a substantial effect.

The atmospheric dispersal of radioactivity after release from containment, and the corresponding potential exposure to individuals, is calculated by conventional methods used in LWR siting and safety analyses.

16.2.8 Containment

As noted above, early studies of hypothetical core disruptive accidents demonstrated that such events were self terminating with limited mechanical energy. However, such analysis is quite complicated and uncertainty in modeling is necessarily high. It is partly for this reason that attention has turned to aspects of design that result in fast reactor systems that are self-protecting in response to the ATWS events. But since a containment system is normally required for a nuclear power plant, the remaining question is that of determining appropriate design criteria for reactor containment.

Current design basis approaches emphasize two points of consideration: (1) external events (such as a tornado, fire, or plane crash), and (2) internal events associated with sodium spill and fire. Sodium spill and fire can occur from operator error, component failure, or as a result of a more severe reactor accident that includes partial core melt. It is generally only the later event, well beyond the design basis (referred to as the Hypothetical Core Disruptive Accident, HCDA), that can result in a significant radiological source term to containment. Hence, we start by discussing the impact of sodium fires on containment and then address the considerations generally taken into account in analyzing a HCDA.

16.3 Sodium Fires

If any boundary of the primary or secondary coolant system of a SFR has been ruptured, there is a possibility of a sodium *pool* or *spray* fire, or spray followed by pool fire (i.e., *combined fire*). A pool fire could occur if sodium is spilled in an equipment cell or on the floors of the Steam Generator Building or on the floor of the Reactor Containment Building. A spray fire could result from sodium being ejected from an equipment or pipe leak in primary or secondary coolant systems or from the vessel cover in an HCDA. It is important to understand the environmental conditions that allow sodium fires to exist so that design measures can be incorporated to reduce the likelihood of fires. A large amount of research has been done to determine the rate of sodium combustion, the pressure buildup from combustion gases, the physical properties of aerosols produced in sodium fires, and the development of computational models for analyzing the consequences of sodium fire and their validation.

Sodium fires occur according to the following main exothermic reactions:

Sodium Oxide: $4Na + O_2 \rightarrow 2Na_2O$
Sodium Peroxide: $2Na + O_2 \rightarrow Na_2O_2$

where the first reaction is the predominant mode. The main characteristics of sodium fire are low flames with dense white oxide smoke.

Based on the combustion process intensity, sodium is one of the least combustible among other conventional hydrocarbon fuels. Figure 16.1 represents temperature distributions [20] in the area above burning sodium and petrol pools of about 1 m^2 surface area, under the same conditions of combustion. At 1 m height above the sodium surface the temperature is lower than 100°C. In the case of petrol fire the time average temperature is more than 600°C at 2 m height above its surface. In sodium the reaction zone (flame zone) is situated close to the sodium surface (i.e., low flames) and in the case of petrol fires the flame is as high as 4 m. This is due to the difference in physical properties of sodium and petrol. The boiling temperature of sodium is 880°C at atmospheric pressure and the petrol boiling temperature is 80°C. The latent heat of evaporation of sodium is 4,340 kJ/kg and for petrol it is 12 times less compared to that of sodium. The heat of combustion of sodium is 10,900 kJ/kg, which is 4 times less than that of petrol. Because of these differences in properties, the petrol burning rate is nearly 4 times faster than that of sodium and consequently the energy release rate during sodium combustion is approximately 15 times less than that of petrol. Even though the energy release rate is considerably less compared to the petrol fire, the energy potential of sodium fire can jeopardize the safe operation of the reactor operation.

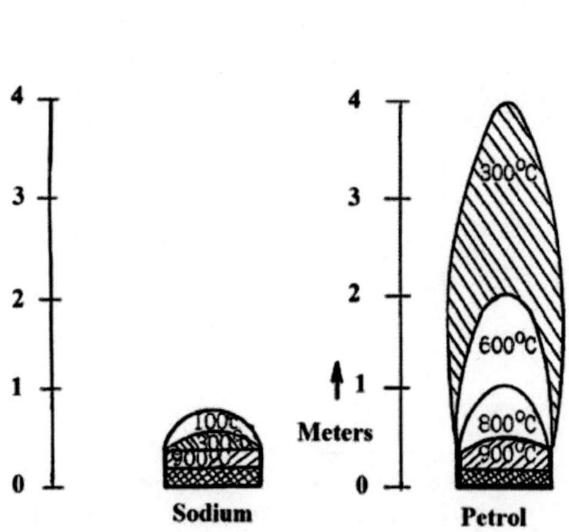

Fig. 16.1 Temperature distribution in the area above burning sodium and petrol [20]

The best design measures for reducing the possibility of sodium fires in the primary system is to provide double enveloped piping (i.e., guard piping) within inerted cells. The secondary piping system is also normally contained within inerted cells and constructed of high ductility materials—using the *Leak Before Break* (LBB) approach.[1] In inerted cells, a typical oxygen content of 2% or less will quench the sodium fire.

16.3.1 Pool Fires

A sodium pool fire will not normally ignite below about 250°C (although it could ignite as low as 200°C if the pool is disturbed). Since the fire occurs only at the pool surface, it is meaningful to characterize the combustion rate on an area basis. A fairly standard rate to use for a pool fire in air is 25 kg/m^2·h. Figure 16.2 illustrates a typical temperature profile for the sodium liquid below the surface and the atmosphere immediately above. Figure 16.3 typifies the pressure responses one would expect for a sodium pool fire in a given cell. The pressure would initially rise as the cell temperature increases, but would then diminish as the cell oxygen becomes depleted. In general, total sodium combustion will not take place in pool fires. Some 10–40% of un-burned sodium residue will remain below the sodium oxides crust of the pool, depending on the conditions of fire. For a given sodium inventory, the pressure would be lower for larger cell sizes. SOFIRE [21] is the principal U.S. code used for computing the pressure and temperature response of a sodium pool fire. ASSCOPS [22] and SPM [22] are other validated codes used for sodium pool fire analysis.

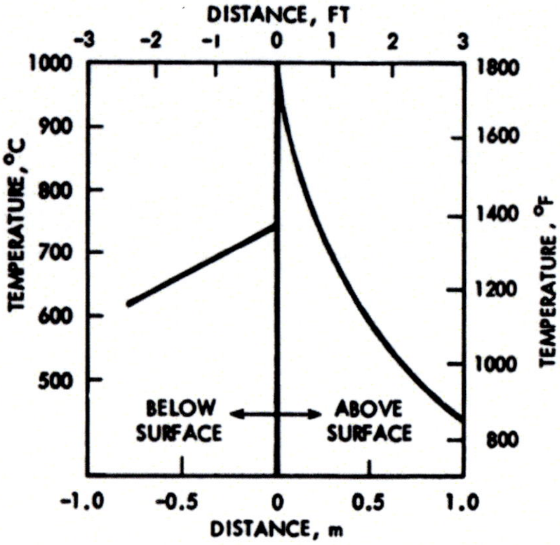

Fig. 16.2 Temperature in the pool and atmosphere near the pool surface in a sodium fire [23]

[1]Leak Before Break (LBB) defines a concept in which a leak will be detected and corrective actions taken prior to a large scale fracture or break occurs.

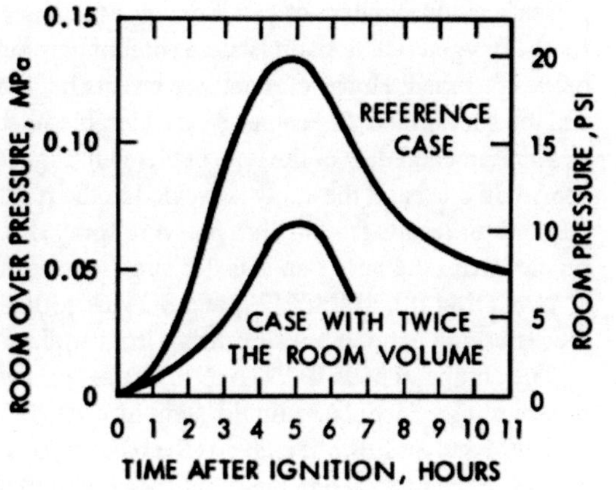

Fig. 16.3 Typical cell pressure for a sodium pool fire [24]

16.3.2 Spray Fires

The ignition temperature for a sodium spray fire is substantially lower than for a pool fire. If the sodium is in large droplet form, ignition may occur around 120°C, but ignition at lower temperatures could occur for fine mist conditions.

Peak cell pressures depend somewhat on the heat transfer conditions for the cell walls, but pressures in the 1 MPa range are theoretically possible. Experimentally measured pressures, however, are consistently lower than theoretical upper limit pressures. Figure 16.4 contains a plot of theoretical and measured pressures as a function of sodium-to-oxygen ratio for a sodium spray fire in air. (The

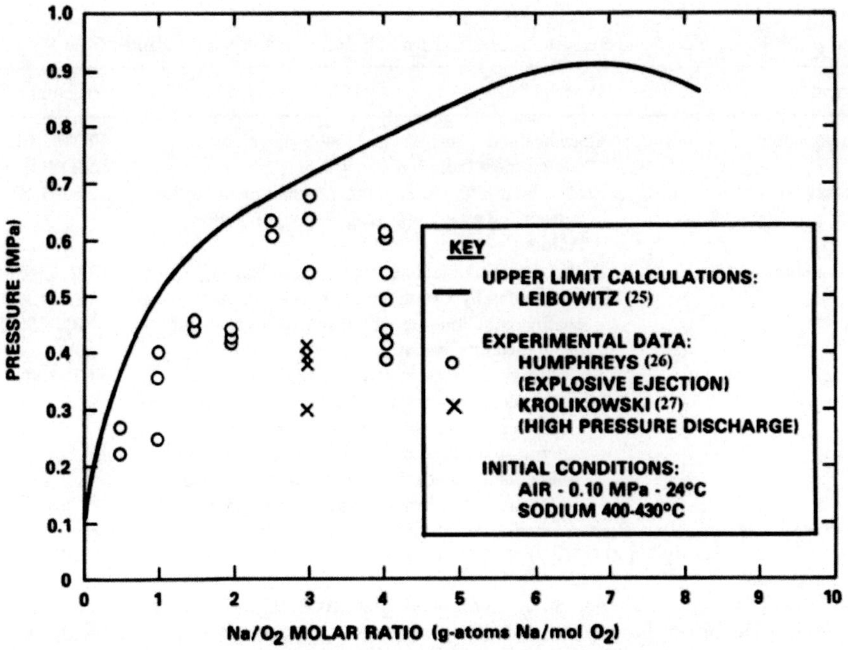

Fig. 16.4 Cell pressure for sodium spray fires in air as a function of Na/O_2 molar ratio [21]

pressure is independent of cell size since the basis is the ratio of Na to O_2.) The pressure rises rapidly for the oxygen rich domain (where sodium peroxide could be formed) and continues to rise as sodium oxide is formed. However, heat loss effects begin to reduce the pressure as oxygen becomes depleted and the sodium itself becomes a considerable heat sink [25–27].

Experimental data of the type depicted in Fig. 16.4 are generally gathered from two types of sodium sprays. In one case, the spray is created as the result of an explosive ejection, whereas in the other case the sodium results from a high pressure spray discharge. Since the former situation results in smaller droplet sizes, the burn rate is faster and the resulting pressure is higher (as shown in Fig. 16.4).

NACOM [28], SPRAY [29] and SOMIX [30] and are examples of U.S. codes used for determining pressures and temperatures resulting from sodium spray fire conditions.

The computer code CONTAIN-LMR [31] is a validated tool that provides an integral level analysis of potential accidents in liquid sodium cooled fast reactors. CONTAIN-LMR used the models of NACOM and SOFIRE-II codes as the base models for sodium spray and pool fire analysis respectively and this code was used in the design of containments of different reactors like ALMR [32], EFR [33], and KALIMER [34].

16.4 Containment Design and Transient Analysis

16.4.1 Containment Systems

A key decision that must be made early in the design of a fast spectrum reactor is the type of containment, confinement, or combination containment/confinement system to be used. Table 16.1 itemizes the types of containment systems selected for several fast reactor plants, together with a brief description of the salient features of each system. [35] Three of the systems are depicted schematically in Fig. 16.5.

Table 16.1 Types of containment systems used in current breeder reactor plants (adapted from Ref. [35])

Containment type	Description	Reactors
Single containment	Open head compartment and low-leakage outer containment building	FFTF, EBR-II, JOYO
Double containment	Sealed, inert high pressure inner containment barrier, surrounded by a low-leakage outer containment building[a]	FERMI, SEFOR
Containment/confinement	Sealed, low-leakage inner containment barrier, surrounded by a ventilated low pressure outer confinement building with discharge to stack via an air cleaning system	PFR, CRBRP,[b] SUPER PHÉNIX, BN-350, BN-600
Multiple containment with pump back	Sealed, high pressure inner containment barrier, surrounded by one or more outer barriers.[a] A negative pressure zone is maintained in the outermost space by pumping back leakage to the inner containment space. Eventual venting to a stack via the air cleaning system is provided	SNR-300

[a]The Reactor Containment Buildings (RCB's) in these cases are designed to withstand 0.17 MPa (10 psig) or greater differential pressures.

[b]The CRBRP design incorporated a concrete structure outside of the RCB. Its principal purposes are for shielding and missile protection. During certain postulated core disruptive accident situations, the RCB is designed to vent through a cleanup system into the gap between the building and the concrete structure. At other times, the gap is maintained at a negative pressure and any flow from the gap is filtered before release.

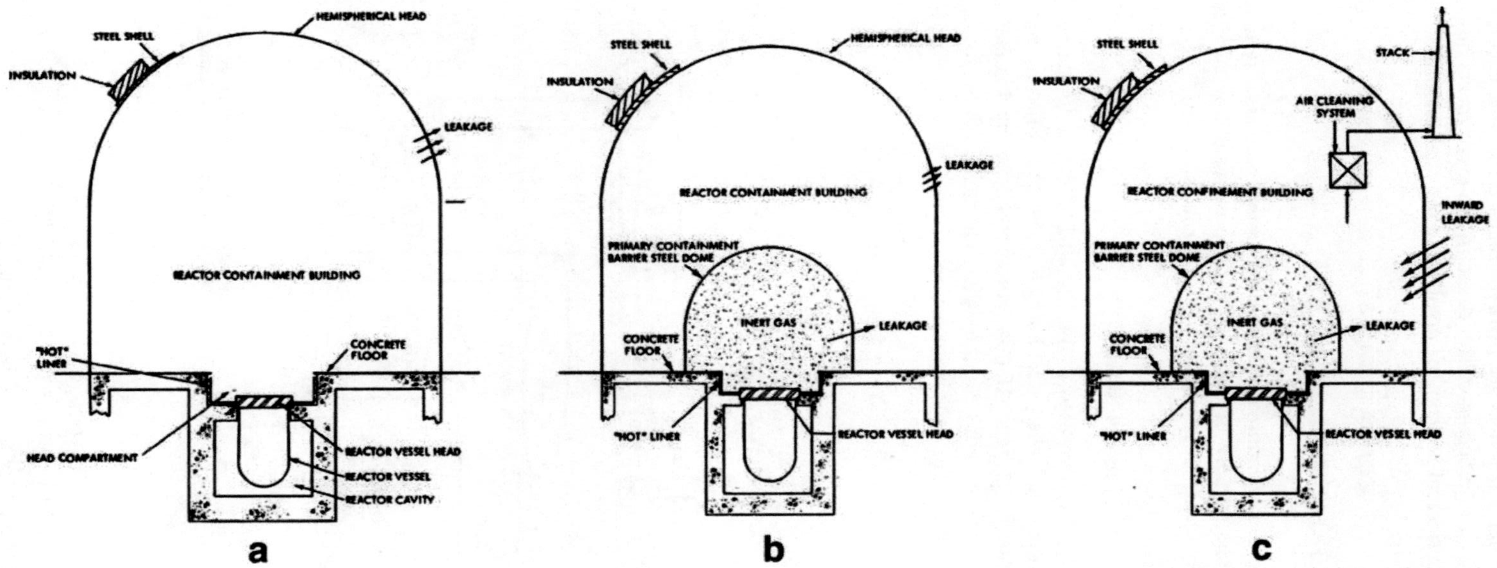

Fig. 16.5 SFR containment systems (multiple, with pump back, not shown) (**a**) Single containment, (**b**) Double containment, (**c**) Containment/Confinement

Selection of the particular scheme to use involves many trade-offs. Clearly, economics and the effectiveness of each system in reducing offsite doses in the event of various accidents are important factors. Some of the principal phenomena involved in the analysis of potential offsite doses are described below.

16.4.2 Containment Transient Analysis

Determining the consequences of failure of the primary system requires a systems treatment of many interacting phenomena. The purpose of this section is to indicate how such an analysis might be conducted, and to illustrate some key results for a typical analysis.

16.4.2.1 Cell Conditions and Leakage Rates

The first large computational system used in the United States to estimate cell temperatures, pressures, constituent composition, leak rates, etc., during a postulated SFR core melt-through and sodium spill was the CACECO code [36]. Figure 16.6 summarizes the principal interacting modes treated in CACECO. The chemical reactions considered by the code include those sodium reactions listed in Section 16.3, together with the exothermic hydrogen recombination reaction:

$$2H_2 + O_2 \rightarrow 2H_2O$$

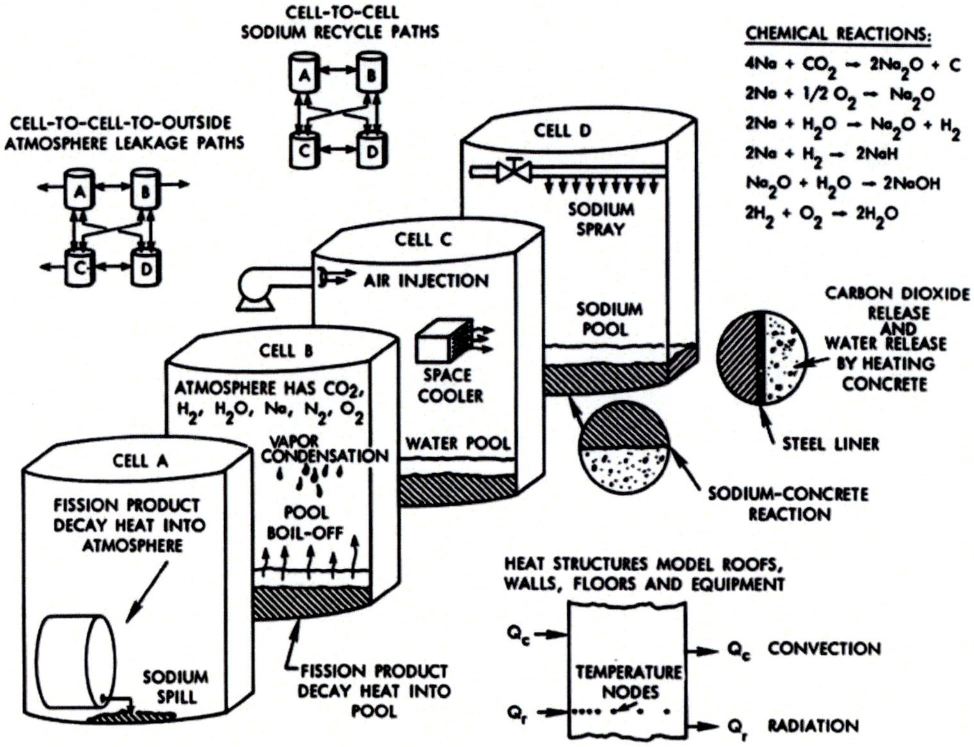

Fig. 16.6 Principal features of the CACECO cell transient computation system [36]

As indicated in Fig. 16.6, temperatures and pressures in interconnecting cells (e.g., reactor cavity, heat transport cells, and outer containment) are determined by the chemical reactions, heat transfer, and cell-to-cell mass transfer that might occur. The principal objectives of such a computational package are to determine sodium leakage rates from cell to cell and to estimate the time dependent pressure, temperature and material inventory in the outer containment. Updates of this computational system have been incorporated into the CONTAIN-LMR system [31] at Sandia.

16.4.2.2 Radiological Source Term

In order to estimate radiological leak rates to the environment and potential doses to the public, it is necessary to obtain a *radiological source term,* i.e., a concentration of radioactive materials in the reactor containment building (RCB). This source term includes principally fission products, plutonium, and sodium oxide.

A determination of how core debris might be transported to the RCB following a sodium fire or core melt is a difficult undertaking. One phase of the problem that is well resolved, however, is the fission product and actinide sources in the core itself. Computer codes [e.g., the U.S. codes ORIGEN [37], and RIBD [38]] are available to calculate detailed inventories of all fission product isotopes from the various fuel isotopes. Actinide inventories are calculated according to the methods addressed in Chapter 7.

The COMRADEX [39] code is an example of the type of model that can be used to estimate the source term. This code follows the radiological inventory through a series of four *chambers* within the containment. The initial chamber monitored is normally the reactor vessel or reactor cavity. This chamber then leaks to the second (e.g., the equipment cells) and/or to the outer containment. For each isotope in each chamber at each time step, the following differential equation is solved:

$$\begin{aligned}\frac{dN_i^k}{dt} &= \lambda_L^{(k-1)} N_i^{(k-1)} + \lambda_R^{(i-1)} N_{(i-1)}^k + \left[\lambda_R^{(i-1)} M_{(i-1)}^k\right] \\ &\quad -\lambda_R^i N_i^k - \lambda_C^k N_i^k - \lambda_L^k N_i^k,\end{aligned} \qquad (16.1)^2$$

where:

$i =$ isotope index (i is this isotope; $i-1$ is the preceding member of this decay chain)
$k =$ chamber index (k is this chamber; $k-1$ is the previous chamber)
$N =$ number of atoms (N_i^k is the number of atoms of this isotope in this chamber)
$\lambda_L =$ leak rate from the chamber indicated and is calculated by a CACECO-type code
$\lambda_R =$ radioactive decay constant for the isotope indicated
$\lambda_C =$ cleanup or fallout rate[3] for the chamber indicated and calculated by an aerosol code such as an updated version of HAA-3 (cf. below)
$M =$ number of atoms that have been captured (removed from the chamber by fallout or from the exhaust from the previous chamber by filtration) with decay of previously captured precursors adding to this value.

Although fission product actinide and radioactive sodium inventories at the start of a postulated accident can be determined accurately, an accurate calculation of the initial radiological atom

[2]This term is for the reintroduction of noble gases (e.g., Xe, Kr); thus, it is not considered if the isotope is not a noble gas.

[3]Fallout is not applied to noble gases. Sodium is normally sufficiently abundant that there are no free halogens; consequently, the sodium halides (e.g., NaI, NaBr) are usually treated as particulates with respect to fallout and filtration.

densities, *N*, in the COMRADEX code also requires a quantitative understanding of processes and mechanisms involved in the release and transport of radioactive material from the fuel and sodium into the chambers. The noble gases (Kr and Xe) are usually assumed to be released from both the fuel and the sodium.

Release of fuel aerosols and remaining fission products are more difficult to assess. There are several paths by which some of these materials can become airborne and, hence, contribute to the radioactive aerosol source term. First, in a core disruptive accident, they might move upward to the cover gas in a large bubble and then escape from the reactor vessel through openings in the head resulting from mechanical damage caused by the accident. The halogens (principally iodine) and some of the volatile fission products could be released by this path, and some of the fuel and solid fission products could be suspended as aerosols. Alternately, some fuel and fission products will escape from the resulting particulate bed or molten pool into the sodium. A tracing of the sodium through a computational model such as CACECO, combined with COMRADEX, could provide an estimate of the percentage of these elements that escape to the RCB. These source term release and transport processes (along with research programs directed toward their understanding) were reviewed in Ref. [40].

Condensation of vaporized fuel and fission products might produce small primary aerosol particles in a two-phase bubble, although the particles would rapidly agglomerate into larger clusters of particles. Some molten fuel may fragment into relatively small particles by hydrodynamic processes such as flashing or by two-phase flow through openings in the reactor head. Whether a large two-phase bubble can rise to the surface carrying aerosols with it or collapse and break up before reaching the cover gas is not clear. Several experimental programs are directed toward this phase of a potential accident. One example is the study at Oak Ridge National Laboratory [41] of partial vaporization of UO_2 pellets in a pool of sodium by capacitor discharge.

Since experimentally verified mechanistic models have not been developed to evaluate the release of solids from the reactor vessel and to follow them through the various barriers to the RCB, early SFR analyses were generally based on an arbitrary source term, in the RCB, such as 1% of the solids initially in the core.

16.4.2.3 Agglomeration and In-containment Settling

Substantial amounts of suspended particulate matter (e.g., sodium oxide or UO_2-PuO_2) can agglomerate and fall-out or plate-out on the floor and walls of the various cells through which such material must pass before escaping into the environment. Extensive experimental studies have provided good understanding of the behavior of nuclear aerosols.

The computer code HAA [42], together with its successors [e.g., HAARM [43]] was developed primarily for the purpose of calculating aerosol behavior during an HCDA. A similar model, PARDISEKO [44], was developed at Karlsruhe, and PARDISEKO-III compares well with HAARM-2. Other models include AEROSIM (U.K.) and ABC (Japan). These models, which are based on the assumption of a log-normal particle size distribution, will allow combinations of Brownian agglomeration, gravitational agglomeration, settling, wall plating, and leakage from the cell in question. The codes are used extensively to determine aerosol depletion rates in the reactor cavity, equipment cells, and the RCB. Useful reviews of aerosol modeling for fast spectrum reactor safety analysis appear in Refs. [45, 46].

16.4.2.4 Dose Calculations

Calculations of the dose that might be received by a person outside the RCB are essentially the same for fast reactors as for thermal reactors, with the possible exception that the effect of plutonium must be more carefully assessed for the fast reactor. As in thermal reactors, the dose can come initially from three sources: direct dose received by direct gamma radiation from sources within the RCB, external dose received from gamma radiation in clouds fed by leakage from the RCB, and internal dose received from inhalation of isotopes leaking from the RCB. An additional longer-term source is through various food and water chains.

Once the contaminated atmosphere leaks through or is exhausted from the outer containment (according to some determined leak or exhaust rate), local meteorological data are used to establish the dispersion characteristics of the airborne radioactivity in the same way as for thermal reactors. The COMRADEX code can also perform this calculation, and it integrates the three sources of dose listed above at any distance downwind from the RCB.

Whereas a code such as COMRADEX will determine doses, it is some-times of interest to determine integrated biological consequences to a population mass downwind from the reactor site. The CRAC code, originally developed for use in the WASH-1400 Reactor Safety [47], is often used for this purpose. Features of both the COMRADEX and CRAC codes have been incorporated into the CRACOME computational system [48].

16.4.2.5 Sample Cell Transient Calculation

An effective way to depict the principal parameters of interest during a vessel melt-through containment transient is to present results from a particular cell transient analysis. The sample case is taken from a study [35] of various containment systems with a single containment configuration (Fig. 16.5, concept a), i.e., an inert reactor cavity sealed off during normal operation from a 6.4×10^4 m^3 air-filled steel reactor containment building (RCB), which has a 0.1% leak rate per day [at an absolute pressure of 0.17 MPa (10 psig)]. The postulated containment transient is a core melt-through of the vessel (without disassembly), accompanied by a spill of 10^6 kg of primary loop sodium into the reactor cavity.

Figures 16.7 through 16.12 illustrate the principal features of the calculated transient. Figure 16.7 shows the heat source in the reactor cavity, which is principally the result of fission product decay heat in the core debris. Increasing sodium temperature in the reactor cavity leads to an increase in the sodium vapor pressure. This pressure buildup eventually causes seal failure and allows sodium to flow into the RCB where it can react with oxygen. For this case, the water released from the heated concrete is assumed to be vented to a condensing room and the steel liners do not fail. During the first 8 h sodium vapor begins to leak to the RCB, and increasing temperatures cause equipment cells to pressurize and nitrogen to leak to the RCB; thus the RCB begins to pressurize.

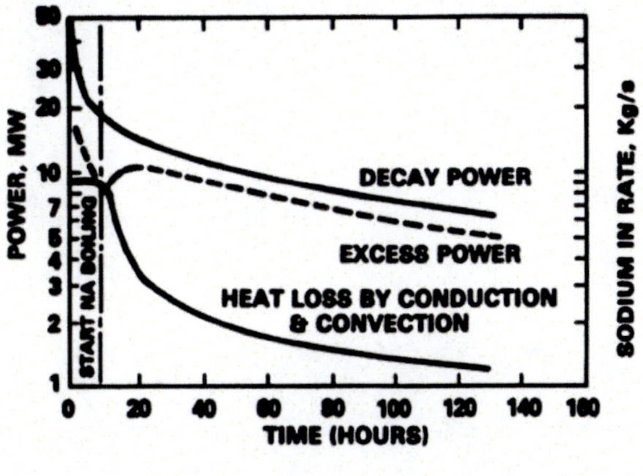

Fig. 16.7 Energy balance in cavity pool

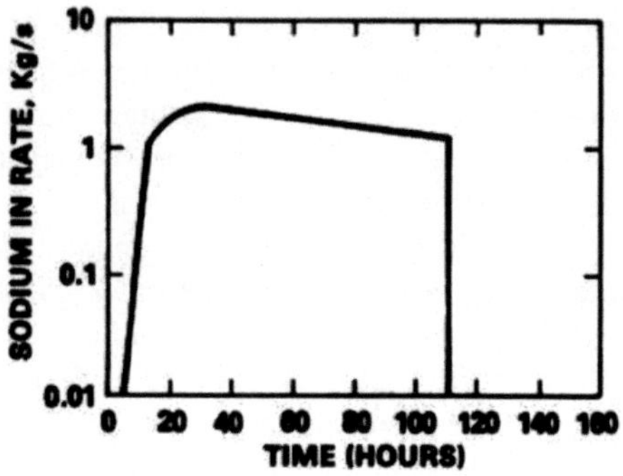

Fig. 16.8 Sodium entry rate to RCB

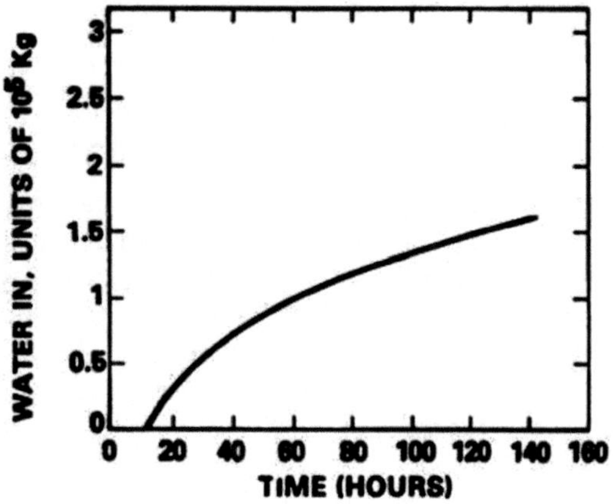

Fig. 16.9 Total amount of water entering RCB

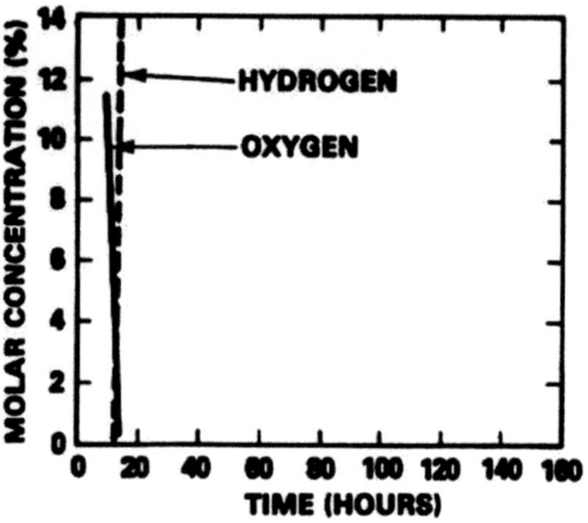

Fig. 16.10 RCB hydrogen and oxygen concentration

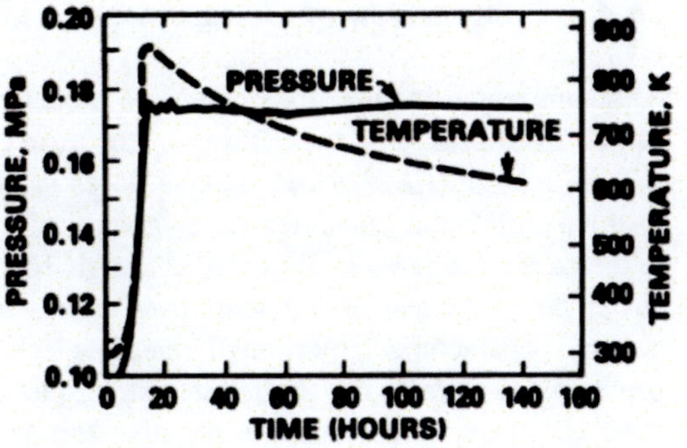

Fig. 16.11 RCB pressure and temperature

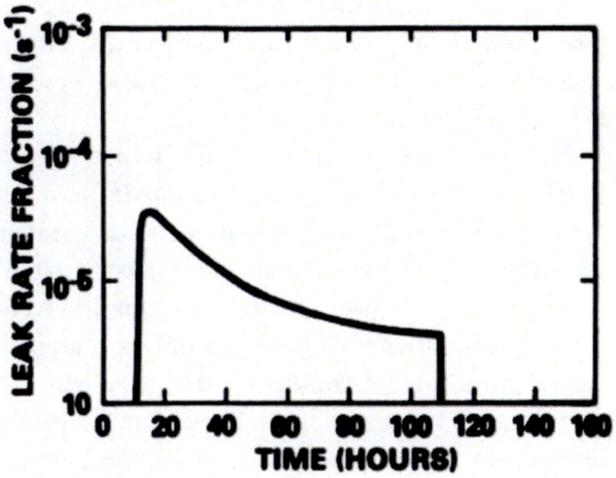

Fig. 16.12 Leak rate from the RCB

At about 8 h the sodium in the reactor cavity begins to boil, thus further pressurizing both the cavity and the RCB. This process continues for about 100 h, until all the sodium boils out. As noted from Fig. 16.8, the rate of sodium entry into the RCB becomes fairly high around 10 h, and this sodium reacts with the oxygen in the RCB, causing a rapid depletion of oxygen, as shown in Fig. 16.10. Heating of the concrete in the RCB releases water vapor (Fig. 16.9). This water vapor, together with hydrogen formed by the chemical reaction between water and sodium, further pressurizes the RCB. The rapid increase of hydrogen starts at about 14 h, as shown in Fig. 16.10, wherein hydrogen recombination was not taken into account.

At about 11 h the RCB pressure reaches 0.17 MPa (10 psig). For this example calculation, a pressure relief valve between the RCB and the outer atmosphere is activated at 0.17 MPa, thus holding the pressure constant at this value as shown in Fig. 16.11. The corresponding RCB leak rate is shown in Fig. 16.12.

Other scenarios are possible if different assumptions regarding system failures are made. For example, the reactor cavity could be linked to surrounding inert cells with rupture seals so that cavity pressures could be relieved by venting into these cells. Such a system has the advantage of retaining the bulk of the radioactive debris within inert areas with a minimized probability of venting to the RCB. Sample calculations for such a system [35] indicate substantially reduced dynamic pressures and temperatures in the RCB.

It should be noted that several mechanisms exist to cause a rapid pressure rise in the RCB (given the postulate of a massive sodium spill). Hence, it may be preferable to use a containment/confinement system—in preference to either a single or double containment system—to allow controlled and filtered release.

16.4.3 Comparison of Containment and Containment/Confinement Systems

Although many combinations of failure modes must be considered before any absolute statements can be made regarding the validity of the dose calculations attained for any particular containment system, a comparison of the *relative* doses calculated for the three systems shown in Fig. 16.5 is instructive. For the calculated doses given in Ref. [35], the single containment system corresponds to the case considered in Figs. 16.7 through 16.12. The double containment considered in the study has an inner low-leak containment over the reactor head area, whereas the containment/confinement system is characterized by a small inner low leakage containment surrounded by a high leakage outer confinement system. For the latter system, the atmosphere between the inner containment and the outer confinement is at a slight negative pressure (inward leakage from the environment), and this atmosphere is pumped through a high exhaust stack.

Regulatory guidelines in the United States provide bases for maximum allowable radiological doses from a design basis nuclear accident. These guidelines define dose limits for whole-body exposure and for thyroid exposure from iodine for a 2-h period in an exclusion area and for a longer period in a low population zone.

Relative doses as a function of distance from the reactor are shown for three containment systems in Fig. 16.13, taken from the analysis in Ref. [35]. All doses are normalized to the 2-h whole-body dose at 1,000 m for the containment/confinement system. Comparison of these doses indicates the relative effectiveness of each containment system in reducing offsite doses.

Figure 16.13a shows relative 2-h whole-body doses vs. distance for the three systems. Note that the double containment provides only a marginal improvement over the single containment system, whereas the containment/confinement system is slightly inferior—especially at large distances. The containment systems appear somewhat better for the 2-h dose because venting is not required in these cases for about 8 h, whereas some filtered exhaust is released to the outer atmosphere in the containment/confinement system.

For the 30-day whole-body and thyroid doses depicted in Fig. 16.13b, the relative improvement of the double vs. single containment configuration is still marginal. However, the containment/confinement system appears to represent a considerable improvement, particularly for

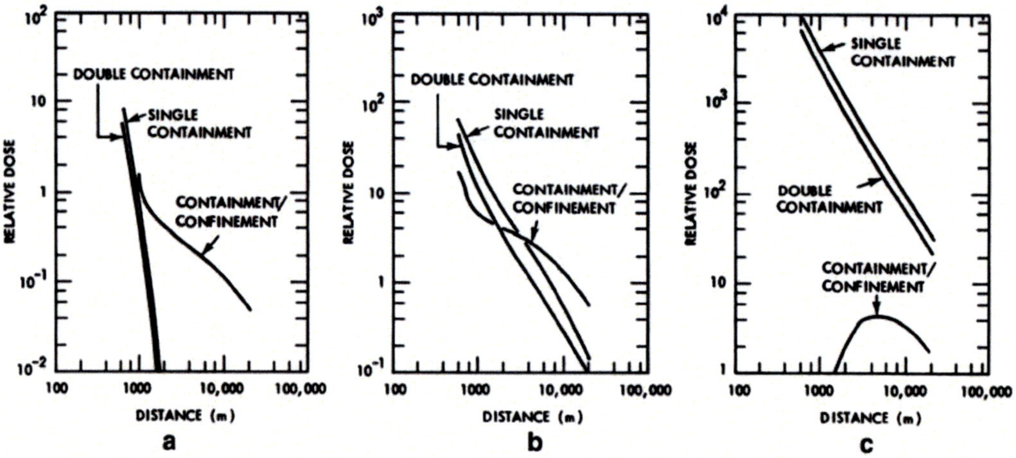

Fig. 16.13 Dose comparisons for three containment systems (relative to a 2-h whole body dose of 1 at 1,000 m for the containment/confinement system) **(a)** Two-hour whole body dose, **(b)** 30-day total whole body bose, **(c)** 30-day thyroid dose

the 30-day thyroid dose. The “S-shape” of the whole-body dose vs. distance curve for the containment/confinement system derives from the continual discharge of low-level radio nuclides, mostly noble gases. The much better thyroid characteristics of this system occur because of the filtration and cleanup of all discharged iodine particulates, plus high stack discharge. This marked improvement in 30-day thyroid dose is of considerable interest because this dose represents the limiting dose for low population distances in the majority of cases.

16.5 Severe Accident Perspectives

We emphasized in several sections of this book that sodium-cooled fast reactor systems contain inherent features that, properly incorporated, provide a very safe system—such that passive safety measures will prevent core disruption even for unprotected ATWS events. There are many scientists long associated with fast reactor safety development who consider it possible to demonstrate a negligible probability (and, indeed, impossibility for some phenomena) for many of the accident scenarios described in this chapter, as outlined in Chapter 15. Nonetheless, it is sometimes instructive to assess “end-of-spectrum” accidents, i.e. those well beyond the Design Basis, to determine the robustness of the containment system. Our objective, therefore, was to provide a qualitative description of the major problems in containment design toward which SFR safety research has been directed, together with an introduction to the selected analytical approaches for some of these problems.

In order to provide a perspective of where events such as interactions of core material and sodium with structures outside the primary system can occur, Fig. 16.14 shows the principal features of the reactor building for typical loop and pool designs. This figure illustrates the relative positions of the primary system, reactor cavity, equipment cells, refueling areas, and the large containment or confinement area in the reactor containment building.

Mechanical damage to the primary system might result from a two-phase expansion of either fuel or sodium. Fuel expansion could directly cause mechanical deformation of the primary system structures; or heat might first be transferred from fuel to sodium in a molten fuel/coolant interaction (MFCI) and the resulting sodium vaporization and expansion could lead to mechanical damage. The process of changing the form of energy generated during disassembly (i.e., thermal, or heat energy) to a form that can cause damage to primary system structures is called thermal/work energy conversion.

If primary system integrity is lost, several issues become important. Sodium fires, for example, become possible if a system rupture should occur in the vicinity of an atmosphere containing a sufficient oxygen concentration to support combustion. Hence, the conditions for ignition and sustainability of sodium fires require attention. Furthermore, sodium and/or core debris outside of the primary system could encounter concrete, causing water release due to concrete heating and producing direct chemical interactions. Such possibilities, resulting from primary system failure, lead us to consider overall ex-vessel containment transient analysis. This requires a description of the various systems for containment and/or confinement that are being considered for fast reactors in order to assure that radiological doses to the public are within acceptable levels.

Appendix G describes representative work done for cores containing mixed oxide fuel, given that most of the R&D in this area of safety analyses has been done for that fuel type. The transient behavior for major accidents based on different fuels, though different in detail, is generally expected to result in consequences bounded by the scenarios outlined in these two final safety chapters and that covered in Appendix G. In particular, the extensive analysis to date for metal fuel has demonstrated that there is no plausible entry into such events—at least for small to medium-sized reactors—even with the most conservative assumptions.

Perhaps this conclusion can best be summarized by the perspectives provided for the PRISM regulatory process: “These assessments have shown that essentially all severe ‘residual risk’ core accidents

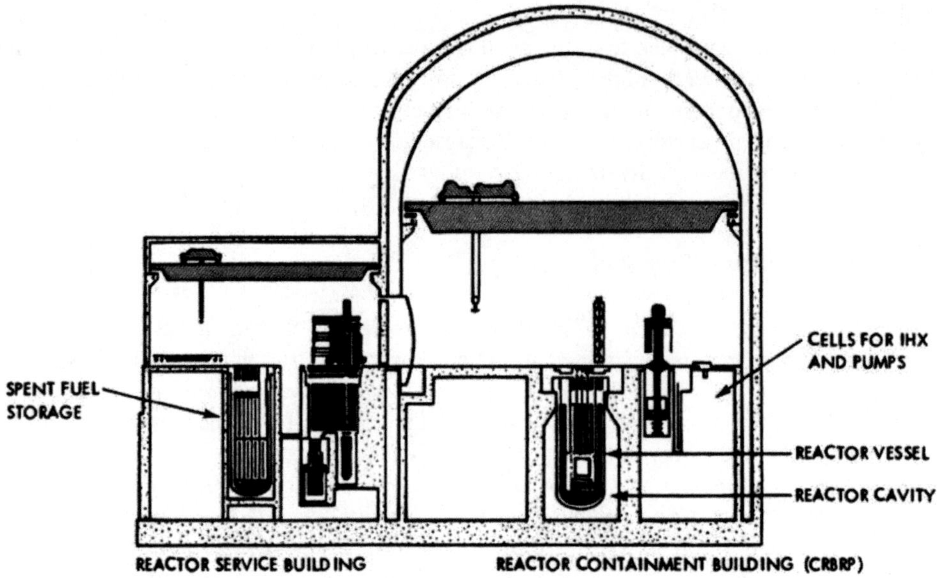

a

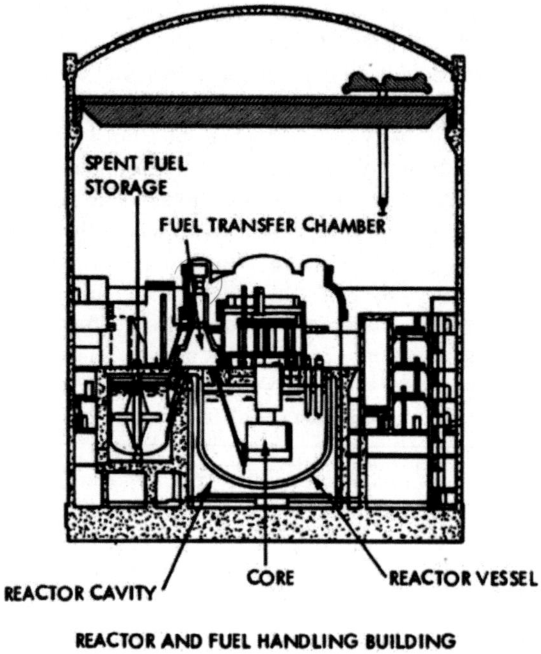

b

Fig. 16.14 Typical reactor building layout for SFR loop and pool designs (**a**) Loop design, (**b**) Pool design which could lead to core voiding result in zero or negligible energetics. None of the extremely severe voiding accident types appear to have the potential to bring a metal core to prompt critical (e.g., result in \$1 or more of net positive reactivity) as summarized below ... Unprotected transient overpower events, up to the maximum possible \$2.25 which could be inserted by the nine control rods at BOEC, will terminate with benign sweep-out of molten fuel and/or fuel-clad eutectic from the active core. There will be no energetics as the molten fuel does not interact with the sodium coolant. The relatively cold temperatures within the reactor vessel (other than in the core) result in the formation of subcritical, coolable, solid fuel 'stringers' within the reactor vessel" [49].

References

1. H. A. Bethe and J. H. Tait, *An Estimate of the Order of Magnitude of the Explosion When the Core of a Fast Reactor Collapses*, UKAEA-RHM(56)/113, 1956.
2. A. E. Waltar and A. Padilla, Jr., "Mathematical and Computational Techniques Employed in the Deterministic Approach to Liquid Metal Fast Breeder Reactor Safety," *Nuclear Science and Engineering, 64* (1977) 418–451.
3. J. F. Marchaterre, "Overview of Core Disruptive Accidents," *International Seminar on Containment of Fast Breeder Reactors*, San Francisco, CA, August 22–23, 1977.
4. H. Fauske, "The Role of Core-Disruptive Accidents in Design and Licensing of LMFBRs," *Nuclear Safety, 17* (5) (September–October 1976).
5. J. E. Cahalan and T. Y. Wei, "Modeling Developments for the SAS4A and SASSYS Computer Codes," *International Conference on Fast Reactor Safety*, American Nuclear Society, Snowbird, Utah, August 1990.
6. L. W. Deitrich and D. P. Weber, "A Review of U.S. Models and Codes for Analysis of Whole-Core Transients and Accidents in Fast Reactors," *Conference on the Science and Technology of Fast Reactor Safety*, Guernsey Islands, UK, May 12–16, 1986.
7. L. E. Strawbridge and G. H. Clare, "Exclusion of Core Disruptive Accidents from the Design Basis Accident Envelope in CRBRP," *Proceedings of the International Meeting on Fast Reactor Safety*, Vol. 1, pp. 317–327, American Nuclear Society, Knoxville, TN, April 21–25, 1985.
8. A. L. Pitner, et al., "Axial Relocation of Oxide Fuel During Overpower Transients," *Proceedings of the International Topical Meeting on Safety of Next Generation Power Reactors*, Seattle, WA, May 1988.
9. R. A. Wigeland and D. L. Graff, "Pool Boil-up Analysis Using the TRANSIT-HYDRO Code With Improved Vapor/Liquid Drag Models," *ANS/ENS International Conference*, Washington, D.C., November 11–16, 1984.
10. C. R. Bell, et al., "SIMMER-II Analysis of LMFBR Post-disassembly Expansion," *International Meeting on Nuclear power Reactor Safety, Topical Meeting*, Brussels, Belgium, October 16–19, 1978.
11. A. E. Waltar and A. B. Reynolds, *Fast Breeder Reactors*, chapter 15, Pergamon Press, New York, NY, 1981.
12. W. T. Sha and T. H. Hughes, "VENUS: A Two-Dimensional Coupled Neutronics-Hydrodynamics Computer Program for Fast Reactor Power Excursions," ANL-7701, Argonne National Laboratory, Argonne, IL, October 1970.
13. J. F. Jackson and R. B. Nicholson, "VENUS-II: An LMFBR Disassembly Program," AL-7951, Argonne, IL, Argonne National Laboratory, 1972.
14. L. D. O'Dell and A. E. Waltar, "Effect of Distributed Voids in LMFBR Core Disassembly Calculations," *Transactions of the American Nuclear Society, 15* (1972) 358.
15. Y. W. Chang and J. Gvildys, "Comparisons of REXCO Code Predictions with SRI SM-2 Experimental Results," *5th SMIRT Conference*, Berlin, Germany, August 13–17, 1979.
16. D. A. Powers, et al., *Exploratory Study of Molten Core Material/Concrete Interactions*, SAND 77-2042 Sandia National Laboratory, Albuquerque, New Mexico, February 1978.
17. J. Hopenfeld, M. Silberberg, and R. Johnson, "FFTF Sodium Fires and Blast Analysis Study," AI-68-MEMO-127, Rockwell International, February 1969.
18. W. T. Pratt, R. D. Gasser, and A. R. Marchese, *Containment Response to Postulated Ex-Vessel Core Meltdown Accidents in the Fast flux Test Facility*, BNL-NUREG-25522, Seattle, WA, 1978.
19. R. K. Hilliard, J. D. McCormack, and L. D. Muhlestein, *Results and code Predictions for ABCOVE Aerosol Code Validation with Low Concentration NAOH and NAI Aerosol – CSTF Test AB7*, HEDL-TME 85-1, Hanford Engineering Development Laboratory, Richland, WA.
20. *Fast Reactor Fuel Failures and Steam Generator Leaks: Transient and Accident Analysis Approaches*, pp. 200–201, IAEA-TECDOC-908, 1996.
21. L. Baurmash and R. L. Koontz, *SOFIRE-II User Report*, AI-AEC-13055, Rockwell International, 1973.

22. O. Miyake, et al., "Sodium Pool Combustion Codes for Evaluation of Fast Breeder Reactor Safety," *Journal of Nuclear Science and Technology, 28* (2) (1991) 107–121.
23. R. N. Lyon, ed., *Liquid Metals Handbook*, NAVEXOS P-733 (Rev), Atomic Energy Commission, DC, 1952.
24. J. Graham, *Fast Reactor Safety*, chapter 4, Academic Press, New York, NY, 1971.
25. L. Leibowitz, "Thermodynamic Equilibria in Sodium-Air Systems," *Journal of Nuclear Materials, 23* (1967) 233–235.
26. J. R. Humphries, Jr., "Sodium-Air Reactions as they Pertain to Reactor Safety and Containment," *Proceedings of Second International Conference on Peaceful Uses of Atomic Energy*, Vol. 22, p. 177, Geneva, 1958.
27. T. S. Krolikowski, L. Leibowitz, R. E. Wilson, J. C. Cassulo, and S. K Stynes, "The Reaction of a Molten Sodium Spray with Air in an Enclosed Volume, Part I. Experimental Investigations," *Nuclear Science and Engineering, 38* (1969) 156–160. See also T. S. Krolikowski, L. Leibowitz, R. O. Ivins, and S. K. Synes, same publication, "Part II. Theoretical Model," 161–166.
28. S. S. Tsai, *The NACOM Code for Analysis of Postulated Sodium Spray Fires in LMFBRs*, NUREG/CR-1405, BNL-NUREG-51180, Nuclear Regulatory Commission, Bethesda, MA, 1980.
29. P. R. Shire, *Spray Code Users Report*, HEDL-TME 76-94, Hanford Engineering Development Laboratory, Richland, WA, 1977.
30. M. P. Heisler and K. Mori, *SOMIX-1 Users Manual for the LBLCDC 7600 Computer*, N707TI130045, Rockwell International, 1976.
31. K. K. Murata, D. E. Carroll, K. D. Bergeron, and G. D. Vaidez, *CONTAIN-LMR/1B-Mod.1, A Computer Code for Containment Analysis of Accidents in Liquid-Metal-Cooled Nuclear Reactors*, SAND91-1490 UC-610, Sandia National Laboratory, Albuquerque, NM, 1993.
32. T. Chiao, S. A. Wood, P. K. Shen, and R. B. Baker, *Analyses of Postulated ALMR Containment and Steam Generator Building Accidents Using the CONTAIN-LMR Code*, WHC-SA-2121-FP, Richland, WA, 1994.
33. B. Carluec, S. Dechelette, and F. Balard, *Radiological Release Analysis for the EFR Project*, IAEA/IWGFR, Orai, Japan, 1996, 207–218.
34. S. Lee, D. Hahn, and S. Suk, "Preliminary Containment Performance Analysis for the Conceptual Design of KALIMER," *ICONE-7347*, Tokyo, Japan, 1999.
35. S. E. Seeman and G. R. Armstrong, *Comparison of Containment Systems for Large Sodium-Cooled Breeder Reactors*, HEDL-TME 78-35, Hanford Engineering Development Laboratory, Richland, WA, April 1978.
36. R. D. Peak, *User's Guide to CACECO Containment Analysis Code,* HEDL-TME 79-22, Hanford Engineering Development Laboratory, Richland, WA, June 1979.
37. M. J. Bell, *ORIGEN-The ORNL Isotope Generation and Depletion Code,* ORNL-4628, Oak Ridge National Laboratory, 1973.
38. R. O. Gumprecht, *Mathematical Basis of Computer Code RIBD,* DUN-4136, Douglas United Nuclear, Inc., 1968, and D. R. Man, *A User's Manual for Computer Code RIBD II, A Fission Product Inventory Code,* MEDL-TME 75-26, Hanford Engineering Development Laboratory, Richland, WA, January 1975.
39. G. W. Spangler, M. Boling, W. A. Rhoades, and C. A. Willis, *Description of the COMRADEX Code*, AI-67-TDR-108, Rockwell International, 1967.
40. A. B. Reynolds and T. S. Kress, "Aerosol Source Considerations for LMFBR Core Disruptive Accidents," *Proceedings of the CSNI Specialists Meeting on Nuclear Aerosols in Reactor Safety*, Gatlinburg, TN, April 1980.
41. A. L. Wright, T. S. Kress, and A. M. Smith, "ORNL Experiments to Characterize Fuel Release from the Reactor-Primary Containment in Severe LMFBR Accidents," *Proceedings of the CSNI Specialists Meeting on Nuclear Aerosols in Reactor Safety,* Gatlinburg, TN, April 1980.
42. R. S. Hubner, E. U. Vaughan, and L. Baurmash, *HAA-3 User Report,* AI-AEC-13038, Rockwell International, 1973.
43. L. D. Reed and J. A. Gieseke, *HAARM-2 User's Manual,* BMI-X-665, Columbus, OH, 1975.
44. H. Jordan and C. Sack, *PARDISEKO-III, A Computer Code for Determining the Behavior of Contained Nuclear Accidents,* KFK-2 15 1, Germany, 1975.
45. W. O. Schikarski, "On the State of the Art in Aerosol Modeling for LMFBR Safety Analysis," *Proceedings of the International Conference on Fast Reactor Safety and Related Physics,* CONF-761001, Vol. IV, pp. 1907–1914, Chicago, IL, 1976.
46. M. Silverberg, Chairman, *Nuclear Aerosols in Reactor Safety*, A State-of-the-Art Report by a Group of Experts of the OECD NEA Committee on the Safety of Nuclear Installations, Paris, June 1979.
47. N. Rasmussen, Director, *Reactor Safety Study, An Assessment of Accident Risk; in U.S. Commercial Nuclear Power Plants,* WASH-1400 (NUREG 75/014), Nuclear Regulatory Commission, Bethesda, MA, 1975.
48. M. G. Piepho, *CRACOME Description and Users Guide,* HEDL-TME 80-56, Hanford Engineering Development Laboratory, Richland, WA, 1981.
49. A. E. Dubberley, A. J. Lipps, C. E. Boardman, and T. Wu (GE Nuclear Energy), *Proceedings of ICONE 8, 8th International Conference on Nuclear Engineering*, Baltimore, MD, April 2–6, 2000.

Part V
Alternate Fast Reactor Systems

The gas-cooled fast reactor (GFR) and lead-cooled fast reactor (LFR) represent possible alternatives to the sodium-cooled fast reactor (SFR). Whereas many earlier parts of the book are equally relevant to all three fast spectrum systems (e.g. neutronic techniques and material properties), the use of a different coolant results in certain design and safety considerations that are fundamentally different from the SFR. This concluding section of the book is addressed to such differences. Chapter 17 is focused on GFR systems and Chapter 18 is focused on LFR systems (including both lead and lead-bismuth coolants).

Chapter 17
Gas-Cooled Fast Reactors

Kevan Weaver

17.1 Introduction

The gas-cooled fast reactor (GFR) represents an alternative to the sodium-cooled fast reactor (SFR) described throughout this book or the lead-cooled fast reactor (LFR) described in Chapter 18. Although many parts of the book are relevant to all fast reactors (e.g., neutronic techniques), the use of a gaseous coolant in the GFR results in certain design and safety considerations that are fundamentally different from other fast reactor systems. This chapter addresses such differences, is focused on historical and modern GFR design features, and provides a review of relevant GFR design considerations, especially as they pertain to safety.

Designs for a gas-cooled fast reactor, originally referred to as the GCFR, were developed in the United States and Europe as an alternative to liquid metal reactors during the 1960s through 1980s. The concept was revisited in 2002 through the Generation IV International Forum (GIF) assessment,[1] and the acronym was changed to GFR. However, some of the design goals and subsequent design choices within the GIF were much different than those pursued for the GCFR.

As noted in Section 4.8, gas cooling leads to a harder neutron spectrum than sodium or heavy liquid metal cooling, thus allowing a higher breeding ratio and a shorter doubling time relative to the liquid metal cooled reactors, if high power densities can be maintained. It was also thought that the GCFR had potential for lower capital costs. Since helium is a chemically and neutronically inert single-phase gas, with low radioactivity from activation (mainly from impurities in the gas), the GCFR would not require an intermediate heat transport loop, special cell liners or fire protection, or special cleanup and decontamination systems.

Initial conceptual design work on the GCFR was originated in the United States at General Atomics circa 1962. For over a decade, the studies included preliminary designs of a 300-MWe demonstration plant and a 1,000-MWe commercial power plant. These were based as a variation of liquid metal cooled core designs with metal clad pins using oxide or carbide fuel. Essentially all of the studies prior to 1978 were based on downward coolant flow through the core. In 1979 the reference design was modified to provide upward coolant flow through the core in order to incorporate natural circulation cooling capability in the event of loss of all forced coolant circulation.

In the late 1960s the program proceeded with considerable international cooperation from the German national laboratories at Karlsruhe (GM) and Jülich (KfA), the German industrial firm Kraftwerk Union (KWU), and the Swiss National Laboratory at Wuerenlingen (EIR). An independent

[1] See Section 1.2 for a discussion of the GIF. The "Generation IV" reactors associated with GIF are often referred to as "Gen IV" designs.

K. Weaver (✉)
TerraPower, LLC, Bellevue, WA, USA
e-mail: kweaver@terrapower.com

A.E. Waltar et al. (eds.), *Fast Spectrum Reactors*, DOI 10.1007/978-1-4419-9572-8_17,

GCFR study program of a commercial size GCFR plant was begun in the late 1960s under the auspices of the European Association for Gas Cooled Breeder Reactors (GBRA).

Several potential gas coolants, including steam, carbon dioxide, and helium, were examined prior to the selection of helium. Advantages and disadvantages were noted for each coolant. Steam was rejected due to cladding compatibility problems, positive coolant reactivity effects, and reduced breeding performance. Several factors led to the rejection of carbon dioxide, including the higher pressure drop and associated forces across components, increased acoustic loadings, and the economic penalty assessed to providing the increased primary coolant pumping power required. An intercomparison of coolants, including helium and steam, is provided in Table 11.11 of Chapter 11.

A novel concept based on N_2O_4 coolant was pursued in the former Soviet Union. This cycle involved partial dissociation of N_2O_4 into NO_2 during heating, and then recombination during the turbine expansion phase. It is possible that N_2O_4 could be liquefied in the cold-leg return, thus reducing pumping power requirements.

Research and design studies performed under the GIF, starting in 2002 to the present, took a different approach to the gas cooled fast reactor. First, as previously mentioned, the acronym was shortened from GCFR to GFR. Second, safety was given a high priority in the GIF, and in particular, the use of inherent or passive safety systems for all Gen IV designs was one of the main goals. The use of inherent or passive safety systems in the GFR is a formidable task given the heat transfer characteristics of the gases involved, and especially using the historical designs. Loss-of-coolant/depressurization accidents are most likely not survivable using the historical GCFR designs because they would result in core damage, if passive/inherent safety systems are relied upon for post transient decay heat removal. The exception would be if they were operated at low power densities, which would incur economic penalties, as well as a lowering of the doubling time. The coolants considered under the GIF were helium and supercritical CO_2 (S-CO_2).[2]

One potential advantage for a helium-cooled GFR is the use of the same balance-of-plant (BOP) equipment, components, and systems as the Very High Temperature Reactor (VHTR).[3] Although the VHTR is a thermal spectrum reactor, its proposed outlet temperatures are close to or higher than those of the GFR (~800°C), and most of the systems outside of the nuclear island can be drawn from those developed for the VHTR (e.g., pumps/circulators, turbine-generators, piping, etc.).

One of the advantages for an S-CO_2 cooled GFR are lower outlet temperatures (500–550°C) than a helium-cooled GFR, while maintaining high thermal efficiencies. These, and other differences, and their advantages/disadvantages will be discussed in Section 17.2 below.

17.2 Gas Coolants

Of all the basic design choices to be made in developing a fast reactor system, selection of the coolant probably has the largest influence on the resulting plant configuration and performance. Even between the different gas coolants there are several options to consider.

This is particularly evident when comparing the implications of helium versus liquid metal cooled systems. Table 17.1 contains a condensed set of advantages and disadvantages of employing helium as the cooling medium. The table is constructed to emphasize design related factors first, followed by safety related considerations. Note that for other gas coolants, e.g., S-CO_2, these advantages and disadvantages are somewhat different.

[2] CO_2 operated above its critical point, either as a direct or indirect Brayton power conversion cycle.

[3] The Very High Temperature Reactor (VHTR) was one of the six systems chosen in the GIF for further study, and is a thermal neutron spectrum, graphite moderated reactor similar to the DRAGON (UK), AVR (Germany), Peach Bottom (US), Fort St. Vrain (US), and THTR (Germany) reactors that were operated between 1963 and 1989.

Table 17.1 Relative advantages of helium vs. liquid metal coolants

	Advantages	Disadvantages
	• High breeding ratio, low doubling time: Good fuel conservation; small burnup reactivity swing, low control requirements	• High neutron leakage: Large/commercial core size required for good economics
Design Related	• Nonradioactive coolant: No intermediate loop (lower capital cost); maintenance and inspection access	• Relatively poor heat transfer properties: Cladding surface roughening required (increased pumping power required); or high temperature ceramic material (long development time)
	• Low coolant neutronic interaction: More room to accommodate metal swelling; small coolant coefficient of reactivity	• High pressure required for economic power generation (8–10 MPa): Low coolant heat capacity: high pumping power required for gas
Safety Related	• Inert and transparent coolant: No corrosion; nonflammable coolant; remote visual inspection	• Reduced heat removal capability upon depressurization: Natural circulation inadequate for depressurized decay heat removal (reliable forced circulation systems required); or back-up pressure for natural circulation (requires guard containment or injection of "heavy" gas, i.e., gas heavier than helium)
	• Single phase coolant: No abrupt loss of cooling potential; no potential for mechanical damage from flashing coolant: no interruption of natural circulation possible	• Lower negative Doppler constant: Slower response to mitigate reactivity transients

17.2.1 Advantages

One of the neutronic advantages of gas cooling, and particularly helium, is a harder neutron spectrum. A harder spectrum will increase the fission-to-capture ratio, and increase the number of fission neutrons released for each fission event (ν). The result is a higher conversion ratio, and from an actinide management standpoint, a harder spectrum will result in more of the actinides fissioning. Helium has a very small neutron absorption cross-section, and at the atom densities present, the macroscopic neutron scattering cross-section of helium is small. This feature of helium neutronic insensitivity allows the core lattice to be more open. Such a configuration can be used to accommodate a lower pressure drop across the core. An additional consequence of the small neutronic sensitivity is that a He-cooled reactor does not exhibit a strong positive void reactivity coefficient. As helium is inert, in-core materials are typically lacking in any incompatibility challenges, which can allow for higher outlet temperatures. In addition, other chemical reactions are absent with the primary coolant. Finally, since helium is always employed in the gas phase (i.e., single phase coolant), there is no possibility of an abrupt change in cooling characteristics. Other advantages of gas cooling include an optically transparent coolant, allowing for visual inspections during maintenance, and a coolant that does not become radioactive as previously mentioned.

Two other advantages of gas cooling, not indicated in the table, deserve brief mention. Because the coolants are always in the gas phase, there is the design potential of employing a direct cycle (Brayton cycle), i.e., the primary coolant is ducted directly to the turbine without using an intermediate loop. Again, this is what has been proposed as one of the power conversion cycles for the VHTR using helium, which would apply directly to the GFR. Finally, without an intermediate loop, capital costs should be lower.

17.2.2 Disadvantages

We now touch on the principal disadvantages of using a gas coolant. Whereas the lack of neutronic interaction of helium was previously noted to have advantageous features, this property also has a negative design feature for small systems. A low power GFR (of the prototype range of 300 MWe or less), with its relatively small core, has substantially higher core neutron leakage than liquid metal reactors of the same power level, and results in a higher core fissile mass loading. For the historical GCFR designs (i.e., metal clad pin fuel), it was necessary to induce turbulence by roughening the surface, or adding fins, on the fueled portion of the fuel pin cladding exterior walls because of the inherently poorer heat transfer properties of helium relative to other coolants. The result is an increase of the coolant axial pressure drop over the core and a more complicated thermal-hydraulic design, especially for natural convective cooling. Another direct result of the poor heat transfer properties of helium is the high pumping power required for adequate circulation. One of the bigger challenges of a GFR system is that helium must be pressurized up to the 7–10 MPa (1,050–1,500 psi) range in order to provide sufficient cooling for the high power density cores of interest for economic power generation. For the S-CO_2 system, the pressures are approximately 20 MPa (2,900 psi). High system pressures for helium systems are required, in part, to compensate for the low heat capacity of the coolant. For the S-CO_2 cycle, the high pressures are required for the high thermal efficiency. The direct consequence of the need for high-pressure gas is to require assurance that depressurization cannot occur rapidly, or that adequate safety systems are engaged during transient conditions that will remove the decay heat during transient conditions. Indeed, it is the concern about the consequences of depressurization—despite the remote probability of the occurrence—which has provided the major safety concern for GFR concepts. This requirement of reliable system pressurization was the principal driving force for employing a prestressed concrete reactor vessel (PCRV) for the historical metal pin fuel designs, and a guard containment/vessel (GC) for the more recent Gen IV designs.

Due to the potential for accidental depressurization of the GFR, provisions must be made to cool the core at the resulting reduced pressures. In addition, reliable core cooling must also be provided for planned depressurizations such as equipment maintenance and refueling. At the low pressures expected for both operational and accidental depressurizations, natural circulation cannot provide adequate heat removal for the historical metal pin designs. For the Gen IV designs, this is partly mitigated by using high temperature in-core materials (e.g., ceramics), and by operating at lower power densities during steady state. Nonetheless, reliable decay heat removal systems must be provided to accommodate depressurized operation.

Finally, the fuel Doppler coefficient, which is clearly a parameter of considerable safety importance, has a smaller negative value for the GFR relative to other fast reactors. There are two factors leading to this result. First, the neutron spectrum for a GFR is harder, thus leaving fewer neutrons in the Doppler resonance region. Second, the GFR has a larger fissile mass fraction, which is required in order to achieve criticality in the more loosely coupled core.

Perhaps the strongest barrier to widespread adoption of the GFR concepts is that there is very little practical experience with such systems; no prototypes or commercial units have ever been built. Nonetheless, there are enough attractive features inherent to GFRs to warrant careful study of such systems.

17.3 Core Design

17.3.1 Core Neutronics

The GFR has several unique neutronic characteristics, which result from the neutronic transparency of the coolant and the high void fraction in the lattice. The low coolant density results in a hard neutron spectrum and a high leakage of neutrons to the axial and radial blanket. The hard spectrum reduces the importance of absorption in the resonance energy range, therefore reducing the impact of resonance broadening with temperature. As a result, the fuel Doppler coefficient is typically half as large as compared to other fast reactors. When using S-CO_2 as a coolant in the primary system, the difference is not so dramatic.

Calculational methods for neutronic analysis of the GFR are very similar to other fast reactors. In fact, many of the same computer code systems can be employed, both deterministic and stochastic. A notable difference is the need to account for neutron streaming in the GFR when diffusion theory methods are employed. This is particularly important for reactivity coefficient determinations. The use of transport theory or Monte Carlo techniques will more properly account for neutron streaming in the low-density regions (e.g., coolant channels) as compared to diffusion theory, and are becoming more viable as design tools due to faster computers and code parallelization.

Fuel type will also influence the neutron spectrum, and thus design parameters. Metal and metal alloy fuel (e.g., U-Zr, U-Mo, etc.) provide the hardest spectrum and highest heavy metal density, while oxides produce the lowest heavy metal density and a softer spectrum. Other fuel types lie between the metal and oxide fuels in terms of density and spectral effects. These include the nitride (UN) and carbide (UC) fuels, and the fertile free/inert matrix fuels proposed under the GIF for actinide burning. Note that nitride fuels also influence core neutronics by the isotope of nitrogen that is used; ^{14}N (the predominant isotope) results in (n, p) reactions that serve as a neutron sink and production of ^{14}C, while ^{15}N appears to alleviate this effect. This would require the use of enriched ^{15}N.

In addition to fuel, cladding and core structural materials will also influence the neutronic performance of GFR's. Historical GCFR cores were designed to simply replace sodium coolant with helium, while opening the lattice to reduce core pressure drop. These designs employed metal cladding, grid spacers, and ducts, resulting in temperature limits similar to their liquid metal counterparts. The neutronic effects are mostly similar to those found in liquid metal cooled designs. The Gen IV designs considered in-core materials other than metal due to the higher outlet temperatures. These mostly included ceramics such as SiC (both monolithic and fiber-fiber composites) in different geometric forms. The resulting neutronic effect includes a softer spectrum (mainly due to the presence of carbon), resulting in a larger Doppler coefficient than other fast reactors.

The neutronic design of a GFR core is thus largely dependent on the fuel type and in-core materials that are used, and on the purpose of the reactor. If high breeding ratios are desired, then metal fuel and metal in-core materials would be the best choice. If high temperatures are desired, then ceramic fuel and ceramic in-core materials would be the best choice. In either case, however, the designer always needs to keep in mind that these choices will affect more than just the neutronics of the reactor system. Material choices will affect the reactivity coefficients (void, expansion, etc.), which will feed back into the thermal hydraulics, etc., which will affect the operating and transient limits of the reactor.

17.3.2 Historical GCFR Fuel and Assembly Design

The historic GCFR designs were based on fuel loading schemes similar to those of the sodium-cooled reactors; namely, a three-batch loading with a refueling interval of approximately 1 year. Up to four radial enrichment zones were proposed to flatten the radial peak-to-average power.

The historic GCFRs were designed to employ variable orificing capability in each fuel assembly. This allowed for the potential to maintain a constant coolant temperature rise over the core as the power shifted during each burn cycle.

As discussed earlier, the fuel pins of the historic GCFR were similar in design to the fuel pins of the sodium-cooled reactors. The one exception would be the cladding surface roughening and the pressure equalization system. Figure 17.1 illustrates a typical GCFR fuel pin and the associated axial temperature distribution.

The cladding surface of the historic GCFR fuel pins were roughened over the entire core length to enhance heat transfer. Whereas it was recognized early in the development of the GCFR that the thermal performance of the GCFR could be improved by surface roughening, such a design measure directly increases the cooling system pressure drop, and in turn affects the coolability during transient conditions. This last point will be discussed further in the section on system and plant design (Section 17.4).

It is useful to define a performance index, I, which relates the improved thermal performance and increase in pressure drop as:

$$I = \frac{\mathrm{St}^3}{f},$$

where

St = Stanton number = $\mathrm{Nu}/\mathrm{Re} \cdot \mathrm{Pr}$,
f = friction factor.

The performance index is proportional to the ratio of the thermal power to the pumping power. Hence, it is desirable to maximize I.

For a typical two-dimensional roughness [1], the Stanton number is about twice that for a smooth surface and the friction factor is about four times that for a smooth surface. Hence, the performance index for a rough surface is about twice that of a smooth surface. A three-dimensional roughening with a performance index of about 4 times that of a smooth surface was also proposed for the GCFR [2].

Several designs for cladding surface roughening were developed, with the basic goal of keeping the cladding hot spot below an acceptable limit while minimizing the core pressure drop. The design most seriously considered was a ribbed design as illustrated in the inset of Fig. 17.1. The design is characterized by the rib height, the rib width, and the rib pitch (i.e., distance between ribs). Also considered as alternative to roughening the clad were the placement of fins along the surface of the cladding to enhance the heat transfer of the pins.

Pressure inside the fuel pins was equalized to that of the reactor coolant by collective venting of fission gas through the fission product trap and through a vent manifold to the helium purification system. This pressure equalization relieved the cladding from mechanical stresses due to external gas coolant pressure and internal fission product gas pressure. The pressure equalization feature also eliminated the need for a large fission gas plenum in the fuel pin. The use of pressure equalized and vented fuel pins in the GCFR eliminated one of the primary failure mechanisms in other fast reactor fuel pins, namely, creep rupture of the cladding due to internal pressure.

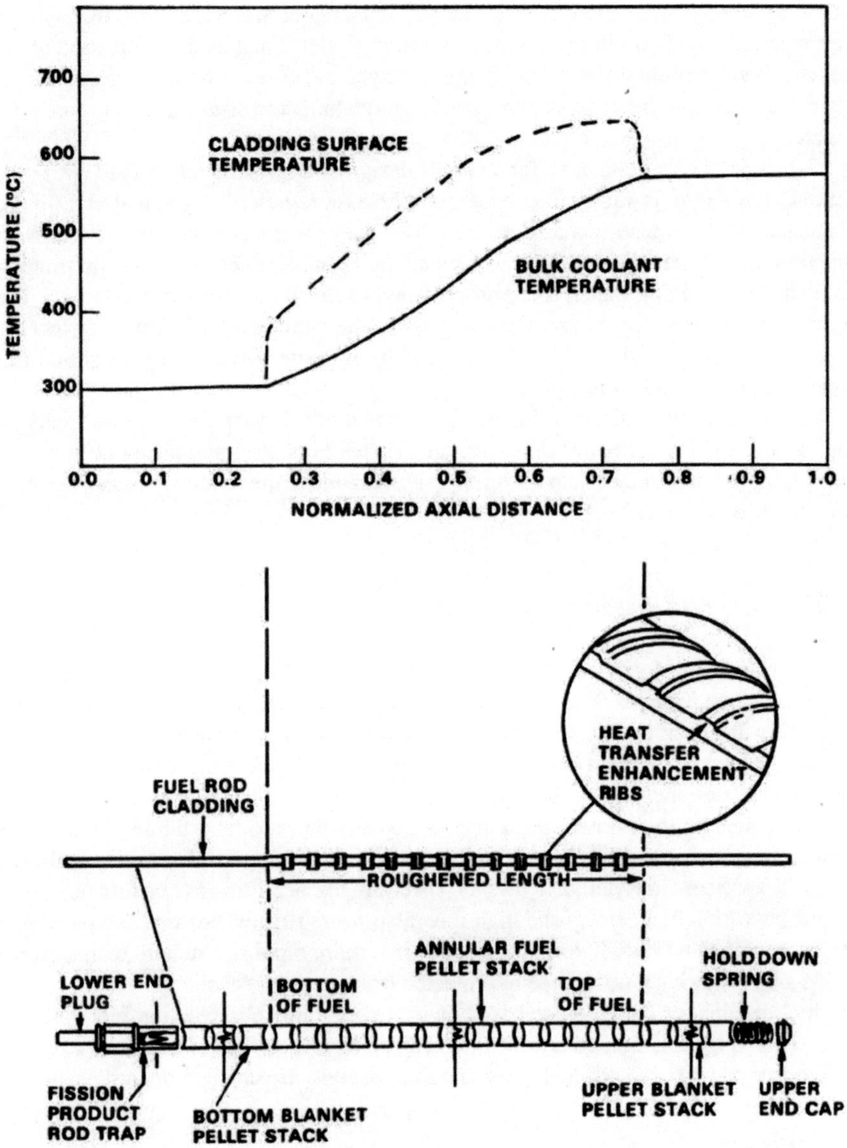

Fig. 17.1 GCFR fuel pin and axial temperature distribution

The reactor core design of an upflow GCFR was similar in many ways to that of sodium cooled reactors. For a comparable power output, the GCFR core tended to be slightly larger since a more open lattice was required for cooling. The similarity of the GCFR core design to that of liquid metal cooled reactors was not by coincidence, but rather by intent. The GCFR design was based upon utilizing fuel and in-core material technologies that were developed for liquid metal cooled reactors to minimize development costs. As a consequence, the fuel design utilized the same fuel and structural materials operating under similar temperature and irradiation conditions typical of other fast reactor fuel designs. Mixed oxide fuel was employed for the active core, although either carbide or possibly metal fuel could be used. Depleted UO_2 was used in the axial and radial blankets.

The GCFR required a much more open lattice (i.e., spacing between fuel pins) than what is typical of other fast reactors. The fuel pin spacing in a GCFR was determined by a combination of factors, the most significant being a trade-off between pumping power and core neutronic performance. However, the open lattice appeared to be an advantage in accommodating void swelling for the core pin cladding and duct materials.

Figure 17.2 shows a core schematic for a GCFR design of approximately 360 MWe.

The central core region is surrounded by a radial blanket region of fertile material (blanket pins). Fertile blanket material is also included in the driver fuel assemblies, above and below the active core. Reflector/shield assemblies are located beyond the radial blanket assemblies to protect the core restraint structures. Shielding material is also provided in the core assemblies below the lower axial blanket to protect the core support grid plate and above the upper axial blanket to protect the control rod guide structures. Due to the harder neutron spectrum, more shielding was required in a GCFR than in typical metal cooled fast reactors.

A fixed area replaceable orifice was included in the exit nozzle to maintain uniform outlet temperatures from each assembly. The lateral spacing of the fuel pins was maintained by means of spacer grids. Spacer grids are used rather than a wire wrap system due to the relatively large pitch-to-diameter ratio of the fuel pins.

17.3.3 GFR Fuel and Assembly Designs

The GIF goals of improved sustainability, economics, safety and reliability, proliferation resistance and physical protection over previous design generations caused scientists and engineers to rethink the fuel and assembly design of GFR's. Increased outlet temperatures result in higher thermal efficiencies, but also challenge the use of metal in-core materials due to the small margin to melting.

The neutronic requirements of a fast spectrum converter (low parasitic absorption), the need for high fissile atom density to maintain a reasonable conversion ratio, and the requirement for fission product containment at high temperature, limit the number of fuel types that can be realistically considered for GFR application. Relative to thermal spectrum temperature gas reactors (e.g., the VHTR), GFR concepts have high fuel density and high specific power (10× or greater). The resultant low core heat capacity combined with high decay heat result in a more rapid rise in core temperature to levels in excess of 1,600°C during a postulated unprotected loss of coolant accident.

Current fuel designs for the GFR are based on more standard solid solution fuel (pellet) clad in a refractory ceramic (e.g., SiC/SiC composites), and dispersion fuels (either as fibers or particles) in an inert plate/block type matrix. Both fuel types and their designs are detailed below.

17.3.3.1 GFR Pin-Type Fuel

Pin-type fuel is attractive due to the large operating database accumulated for similar geometry fast reactor fuels and cladding materials. The high outlet temperature for some of the GFR designs, and the desire to exclude core restructuring as a result of clad melting for safety reasons, however, will eliminate most, if not all, steel-based alloys from consideration. This would include the dispersion strengthened steels, e.g., oxide-dispersion-strengthened (ODS) alloys. The remaining choices are limited to ceramics or refractory metals for high temperature survivability, which are described in more detail below.

Fuel response to severe transients, e.g., loss-of-coolant, is the overriding factor in the design of pin-type fuels. Either a gas plenum must be built into the fuel to accommodate fission gas release, or fuel venting will need to be employed as had been suggested for the GCFR. During normal reactor operation at a 7 MPa coolant pressure, fuel can be designed such that the net stress on the cladding is compressive; i.e., the coolant pressure is larger than the pin internal pressure. For fuel designs using

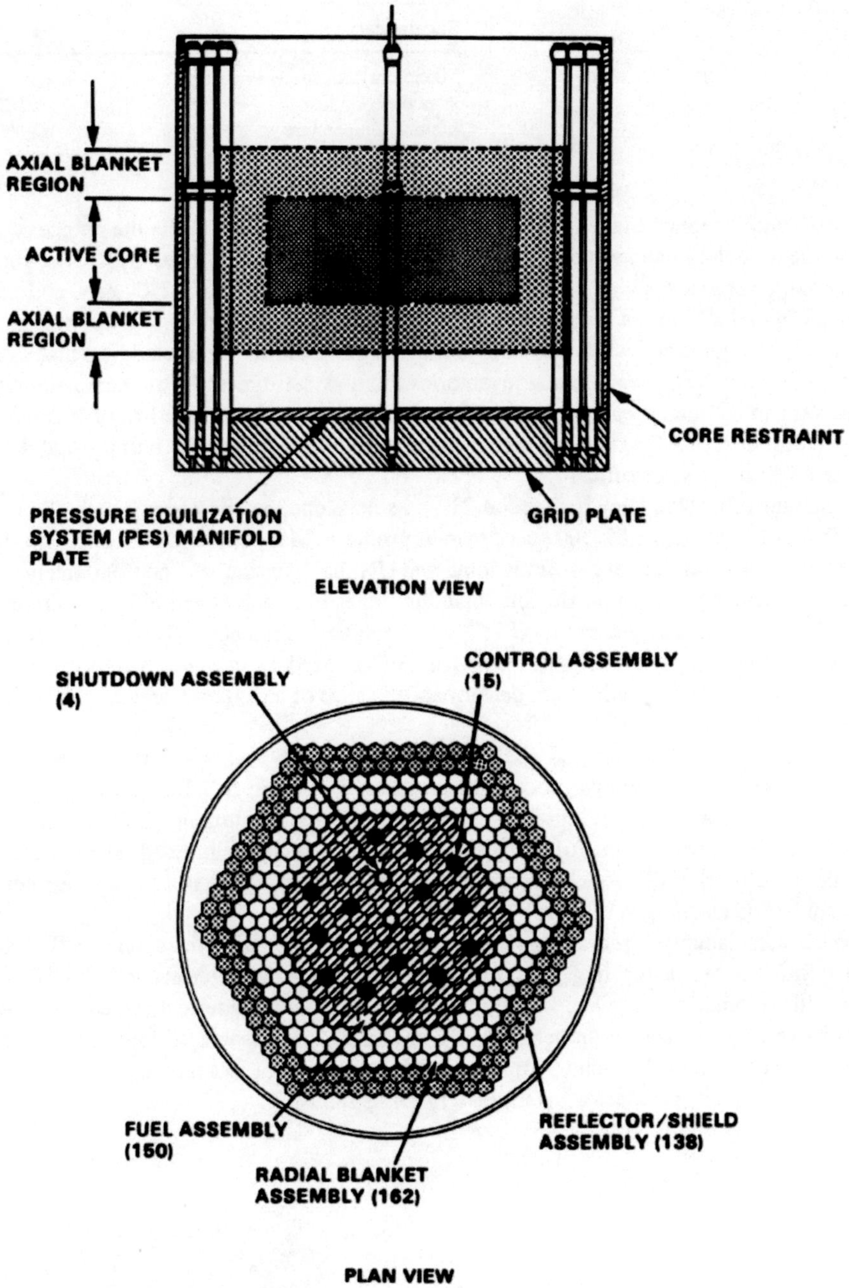

Fig. 17.2 GCFR schematic for 360 MWe design (Note: Numbers in parenthesis indicate the number of assemblies.)

a plenum, the length is then a compromise between internal gas pressure that can be tolerated during core depressurization and shutdown, and the coolant pressure drop through the core during operation.

Specific requirements for pin-type fuel are, in general, similar to those defined for liquid metal-cooled fast reactors, and are listed in Table 17.2. The exception is the required cladding melt

Table 17.2 Pin-type fuel requirements

Requirement	Reference value
Maximum diametrical swelling	<2%
Peak dose	>80 dpa
Melting temperature	>2,000°C

temperature, which is much higher. The high temperature requirement limits the choice of cladding material to the refractory elements and metalloids (B, C, Nb, Mo, Ru, Hf, Ta, W, Re, Os, and Ir), and high temperature ceramics, e.g., the binary carbides and nitrides (SiC, ZrC, TiC, ZrN, and TiN).

Obviously B and Hf are not practical materials due to their effect on core neutron economy, and Ru, Os, and Ir are not practical due to cost and availability. In addition, cladding made of Ta, W, and Mo requires very large heavy metal inventories to allow self breeding, or conversion ratios near unity. Carbon and carbon/carbon composites are excluded due to a service life of less than 15 dpa [3], compared to the ~80 dpa required during the service life of fuel at 5% burnup, and due to their moderating effect on core neurotics.

The remaining allowable alloy base metal, Nb, was developed in alloy form as Nb-1Zr and PWC-11 (Nb-1Zr-0.06C) during the SP-100 space reactor program. Many fuel irradiation experiments have been conducted using variants of Nb-Zr cladding and UN fuel to burnups of approximately 6% in fast spectrum environments [4]. A thin rhenium cladding inner liner was found to be important for limiting fuel-cladding-chemical-interaction (FCCI). Neutronic performance of Nb-1Zr clad fuel encased in Nb-1Zr ducts is marginal, but improves when SiC is used as the duct material [5]. Concerns related to the use of Nb-based alloys are performance in case of air ingress, and sensitivity to coolant impurities.

Silicon carbide is another potential cladding choice, although at this writing it is in a much earlier stage of development. Various processes have been developed to fabricate SiC, e.g., the nano infiltration transient eutectic (NITE) process, and chemical vapor infiltration (CVI) of a fiber-preform woven over a monolithic SiC tube [6], as shown in Fig. 17.3. A SiC reinforced ferritic steel cladding is also under development [7]. Joining of SiC to SiC, or other materials, is a major issue confronting development of SiC cladding.

Silicon carbide cladding is most compatible with carbide (UC) and mixed carbide (U-Pu-C) fuel. Reaction of SiC with nitride fuel (e.g., UN) is thermodynamically favorable above 550°C, although the reaction could presumably be prevented through the use of a cladding liner. Silicon carbide composites have demonstrated good irradiation behavior [8], and should be capable of maintaining mechanical properties at dose levels well beyond 50 dpa, and at temperatures of less than 1,000°C.

Specific requirements for matrix material are listed in Table 17.3.

Fig. 17.3 Test specimens for fabricated SiC tube; a potential clad material for GFR's

Table 17.3 GFR fuel matrix material reference requirements

Requirement	Reference value
Melting/decomposition temperature	>2,000°C
Radiation induced swelling	<2% over service life
Fracture toughness	>12 MPa $m^{1/2}$
Thermal conductivity	>10 W/m·K
Neutronic properties	Materials that allow low core heavy metal inventory and maintain goods safety parameters

17.3.3.2 GFR Particle and Matrix Fuels

For GFR matrix and dispersion fuel designs, UC and UN are desirable for their high heavy metal density, high conductivity, and minimal impact on neutron spectrum, although limited irradiation data exists for these types of fuel.[4] The matrix materials are dependent on the coolant and operating temperatures, and can be classified into three categories: ceramic (for high temperatures), refractory metal (for modest to high temperatures), and metal (for modest temperatures). One possible variant of these concepts is shown in Fig. 17.4, where Fig. 17.5 shows a micrograph of the ceramic coatings.

As in more conventional dispersion fuel, fuel particles are embedded in an inert matrix. In this case, a buffered particle fuel is used to isolate the matrix from mechanical stress caused by swelling fuel particles. Because the matrix is not exposed to fission fragments or mechanical stress, it retains its integrity throughout irradiation, and acts as a barrier to fission products. Finally, an outer coating acts as a second barrier to fission gas release. Hexagonal plate-type elements containing cooling channels are stacked to form fuel elements, as shown in Fig. 17.6.

A more complete list of dispersion fuel geometries is shown in Fig. 17.7. Particle fuel considered for the GFR is shown in Fig. 17.8.

The process for choosing the matrix material is similar to those of the high temperature cladding described above, where the monocarbide and mononitride refractory materials appear to give the best performance.

A sample core face map for a large, 2,400 MWth GFR using pins/blocks/plates in hexagonal assemblies is shown in Fig. 17.9. Note the use of Zr_3Si_2 as the reflector material rather than a standard blanket pin with depleted uranium. This is one possible option that can be used to minimize

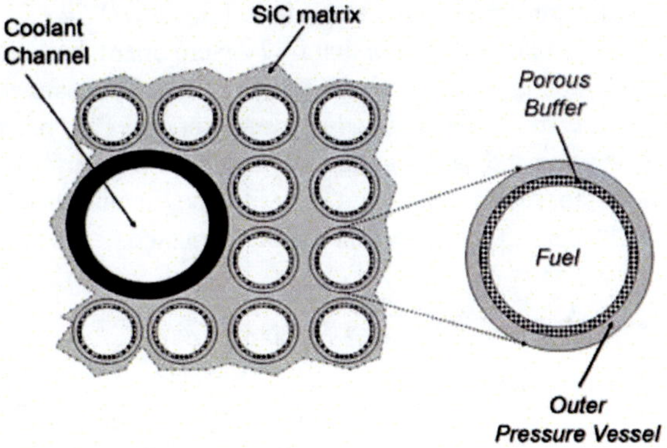

Fig. 17.4 GFR dispersion fuel concept

[4] An exception to the carbide fuel irradiation experience exists in India, where the Fast Breeder Test Reactor (FBTR) has used UC fuel for several decades.

Fig. 17.5 Micrograph of SiC coatings that could be used for GFR particle fuel

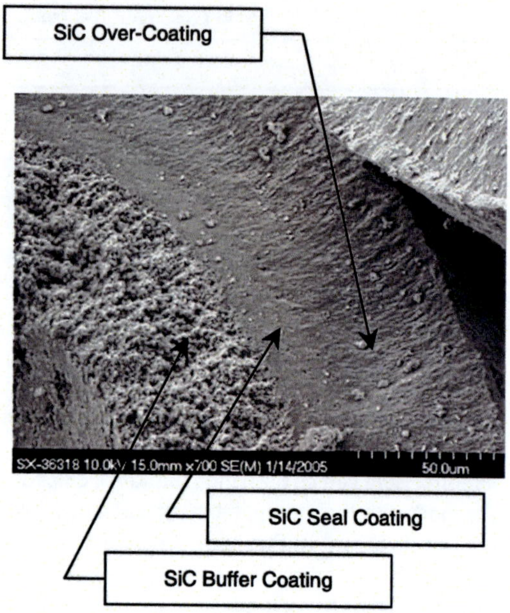

Fig. 17.6 GFR plate-type fuel element concept. (Plates are stacked to form an element or assembly.)

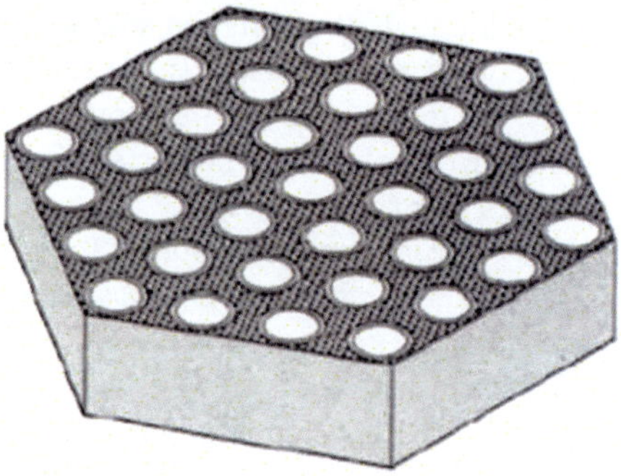

plutonium production by not using a uranium-based blanket. The corresponding parameters for a sample pin-based core can be found in Table 17.4.

It is important to note that fuel development, including fabrication and irradiation performance, is a key viability issue for the GFR, and cannot be separated from the safety design and performance of the GFR. Fuel mechanical and thermal properties are needed from beginning to end-of-life of the reactor to support the safety case, and will have a significant impact on the safety system design work. In addition, fuel development may include the viability of using minor actinide bearing material, which will have further affects on the performance of the GFR.

17.3.4 Core Thermal Hydraulics

Although correlations will not be fully developed in this section, considerable attention must be given to the thermal hydraulic design of a GFR. This aspect of the design is important for at least two reasons:

- The temperature drop from the cladding outer surface or coolant channel inner surface to the coolant is relatively large, thus there is a strong incentive to minimize this temperature drop.

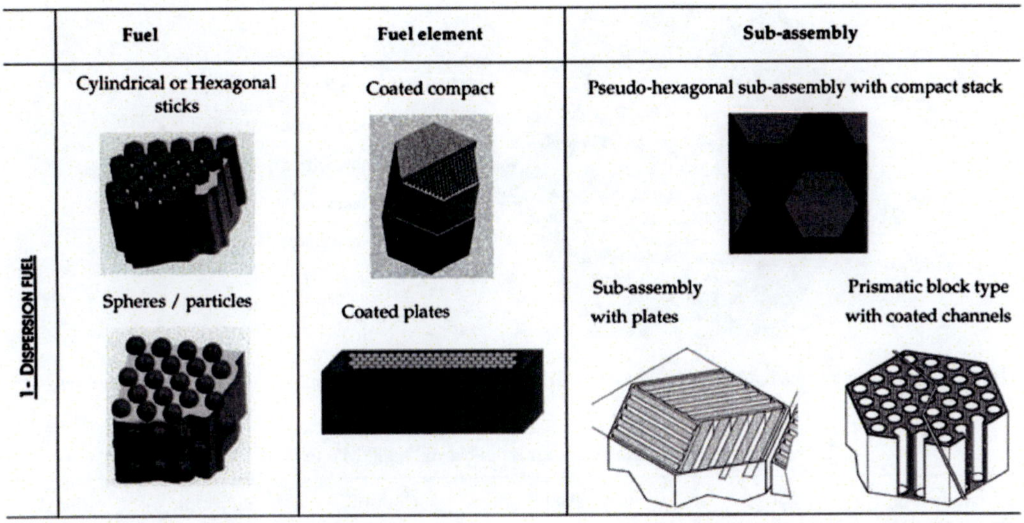

Fig. 17.7 GFR dispersion fuel in plates and blocks

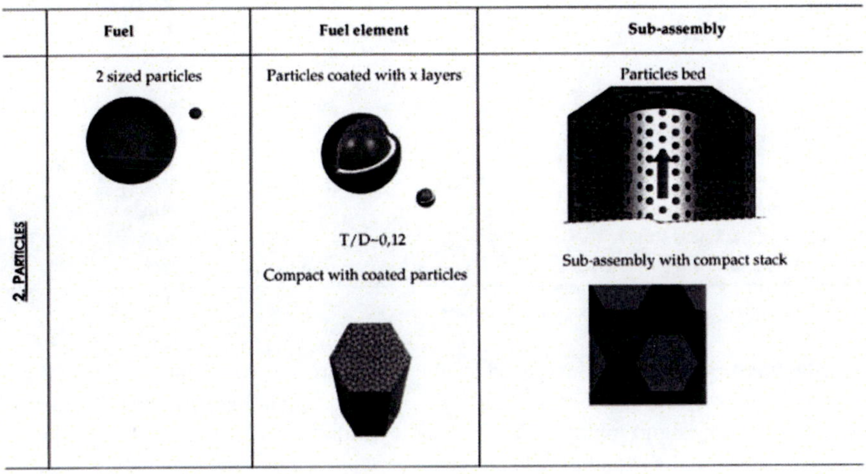

Fig. 17.8 GFR particle fuel alternatives

- Surface roughening on the cladding or coolant channels leads to turbulent flow, but laminar flow prevails for low flow conditions or smooth channels, and must likewise be understood. Furthermore, the thermal hydraulic conditions may be considerably different for the smooth versus roughened portions of the clad or coolant channel.

For these reasons, and their safety implications, determining accurate heat transfer to the coolant is important. However, the reader should be cautioned that correlations tend to change as new data become available, particularly for turbulent flow conditions. In addition to the complications mentioned above, hot channel factors must also be included. These calculations, however, are very similar to those described in Section 10.4, and will not be developed here.

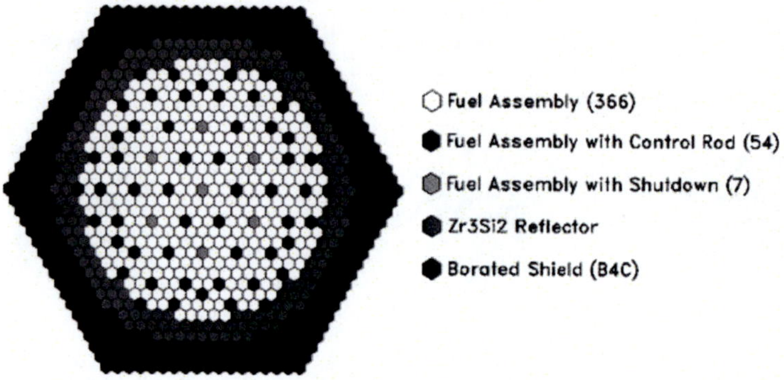

Fig. 17.9 Sample core face map for a 2,400 MWth GFR reactor (number of assemblies is shown in parentheses)

Table 17.4 Sample parameters for a pin based, 2,400 MWth GFR reactor

Assembly type	Fuel	Control	Reflector	Shield
Assembly pitch (mm)	222	222	222	222
Assembly flat-to-flat (mm)	215	215	215	215
Duct thickness (mm)	3.7	3.7	3.7	3.7
Duct material	SiC	SiC	Zr_3Si_2	W
Number of pins	271	234	19	19
Pin diameter (mm)	9.57	9.57	40.1	40.1
Clad thickness (mm)	1.00	1.00	N/A	0.019
Clad material	SiC	SiC	N/A	W
Pellet outer diameter (mm)	7.37	7.37	N/A	38.9
Pellet inner diameter (mm)	3.02	3.02	N/A	N/A
Pellet material	(U, Pu)C	(U, Pu)C	Zr_3Si_2	B_4C
Control rod outer diameter (mm)	N/A	80.5	N/A	N/A
Control rod cladding thickness (mm)	N/A	1	N/A	N/A
B_4C diameter (mm)	N/A	78.5	N/A	N/A

Thermal-hydraulic correlations developed for the GCFR surface-roughened cladding for both laminar and turbulent flow are given in Chapter 17 of the original Fast Breeder Reactor book [9]. Fortunately, the GFR design employs smooth pins, cylindrical channels, or plates. This somewhat simplifies the core thermal hydraulics calculations.

17.4 System and Plant Design

The GFR can be designed for different operating conditions and power conversion cycles. For lower temperature operation (500–600°C), such as that used in the historical GCFR design, at least two different power conversion cycles can be used—an indirect Rankine steam cycle, or a gas Brayton cycle using S-CO_2 as either a direct or indirect cycle. The indirect steam cycle was used in the GCFR designs where the helium primary coolant was pumped through the steam generator. For the GFR, design options exist to use S-CO_2 both for direct and indirect Brayton cycles, i.e., in a direct cycle where the CO_2 is pumped through the core directly to a turbine, and indirect where helium is pumped through an intermediate heat exchanger (IHX) with S-CO_2 as the working fluid on the other side. As thermal decomposition of CO_2 accelerates at around 700°C, and the oxidation/corrosion rate increases dramatically, there is a practical limit to the maximum operation temperature operation for S-CO_2 of approximately 600–650°C.

A further option for the GFR is high temperature operation (800–850°C) using direct or indirect power conversion cycles, and/or production of process heat for non-electrical use. The cycles can be gas-to-turbine, gas-to-gas (e.g., helium–helium, helium–air, etc.), or gas-to-steam. Additionally, process heat can be utilized through an intermediate heat exchanger (IHX), and an appropriate working fluid for the process heat plant. This is similar to what has been proposed by the VHTR for hydrogen production, or other process heat products. The following sections describe the generic design considerations for both the historical GCFR, and the Gen IV GFR, since it is clear that many plant combinations can be used from both generations of gas-cooled reactors that are dependent on the end use of the reactor.

17.4.1 Historical GCFR Design Considerations

The overall features of the historical GCFR plant are presented below for a demonstration plant design delivering 360 MWe. A plant layout is presented in Fig. 17.10. The plan view and side view are presented in parts (a) and (b), respectively.

Figure 17.11 illustrates the nuclear steam supply system (NSSS) of a proposed demonstration GCFR. It is apparent from the illustration that the primary coolant system (core, main steam generator and auxiliary heat exchanger, main and auxiliary circulators, etc.) is located entirely within the massive prestressed concrete reactor vessel (PCRV) structure.

In the GCFR, high primary system helium pressures were necessary to provide the operating conditions required for economical electric power generation. Typical GCFR primary system pressures were in the range of 7–10 MPa (1,050–1,500 psi), which is comparable to the pressure of steam supplied to light water reactor (LWR) steam turbines. In order to keep the possibility of sudden system depressurization to an acceptably low value, the PCRV concept was adopted. This system, originally developed for the high temperature gas reactor, consists of a massive block of concrete that is prestressed by longitudinal tensions and by circumferential cable wrapping so as to maintain the PCRV in compression at all times. The PCRV for the demonstration GCFR is approximately 32 m high by 33 m in diameter.

The reactor core and its associated structural support and shielding components are located in the central PCRV cavity. The three main cooling loops, each of which includes a steam generator and a helium circulator, are located in three peripheral cavities in the vessel walls. Three auxiliary loops are contained in smaller cavities located in the PCRV walls between the steam generator cavities. Each of these cavities has a concrete closure plug at the top, which can be opened for initial installation of components and for their repair and replacement if required. The inlet to the auxiliary heat exchanger is via a circumferential duct from the adjacent steam generator inlet plenum. The purpose for this design, as opposed to directly ducting hot gas from the core outlet through the PCRV, is to preserve the temperature profile when transferring from forced circulation in the main loop to natural circulation in the auxiliary loop.

All cavities and ducts within the PCRV are lined with a 13 mm thick steel liner to provide leak-tightness. This liner is protected from the high helium temperatures by a specially engineered insulation system on the coolant side, and by cooling water circulating in embedded piping on the concrete side. Liner stresses are compressive and they are relatively low because the effect of helium pressure is more than offset by the effect of the prestressing system.

The schematic shown in Fig. 17.12 illustrates the GCFR heat transport systems. The normal operating or power generation system consists of the primary coolant system using high pressure helium, the steam and feed water system and the plant heat rejection system. The primary coolant system uses motor-driven centrifugal compressors to circulate helium upward through the core where it receives

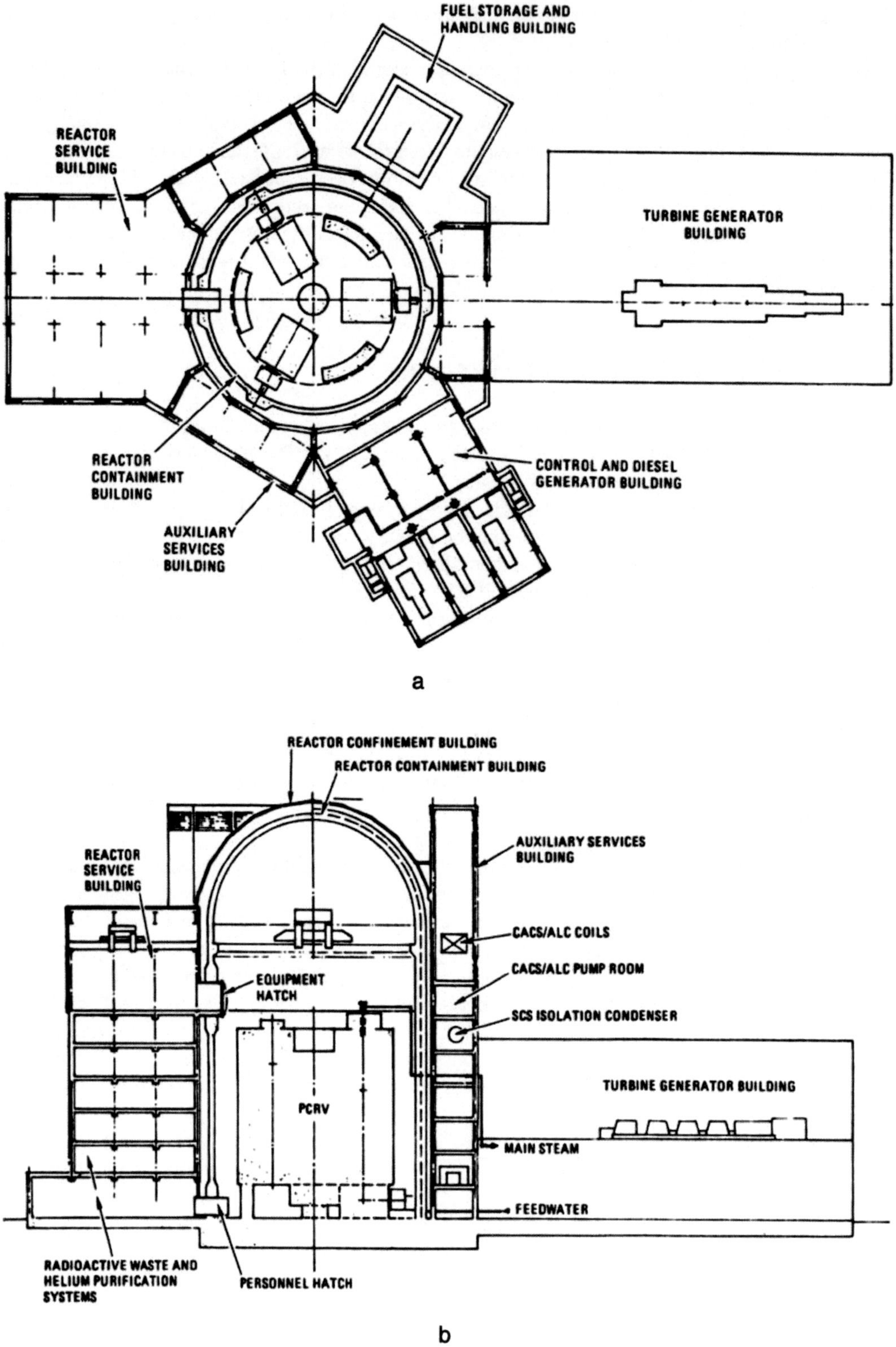

Fig. 17.10 Plant layout for 360 MWe: (**a**) plan view, (**b**) side view

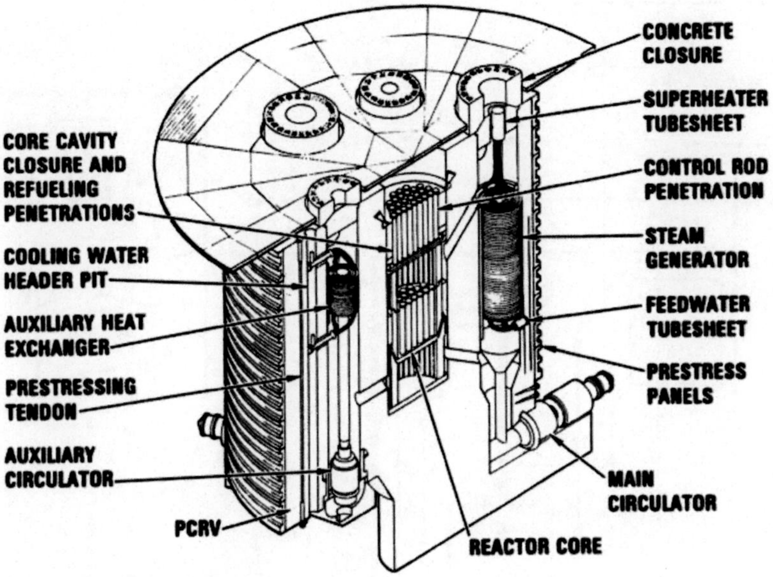

Fig. 17.11 NSSS for demonstration GCFR

nuclear heat and downward through the steam generators where it transfers heat to the entering feed water to produce superheated steam for use in the main turbine generator. The primary coolant system and all of its associated components are housed within the PCRV.

The plant heat rejection system includes a circulating water system (not shown in Fig. 17.12), an evaporative cooling tower, a water makeup system and a cooling tower blow down system. An intermediate loop is not necessary for the GCFR.

There are three separate systems for residual (decay) heat removal in the GCFR:

(1) Steam bypass to the condenser using the normal power conversion heat transfer system components,
(2) Operation of the three Shutdown Cooling System (SCS) loops, which make use of the steam generators and main helium circulators (driven by pony motors) with safety-class feed and heat dump components, and
(3) Operation of the three Core Auxiliary Cooling System (CACS) loops, which employ separate safety-class circulators, heat exchangers and heat dump circuits. The CACS system is also designed to transfer residual heat from the core by natural convection to the ultimate heat sink, ambient air.

Steam bypass is used for residual heat removal when the normal heat transfer system components are operable. If the normal heat transfer system components are unavailable, plant cool down is automatically initiated in the safety-class SCS, as shown in Fig. 17.12. Should both the above systems be unavailable, plant cool down is accomplished using the CACS as shown in Fig. 17.12.

17.4.2 GFR Design Considerations

The current designs of the GFR operate at a relatively high power density (50–100 kW/l), but have very little thermal inertia to aid in decay heat removal during accident conditions. The low thermal

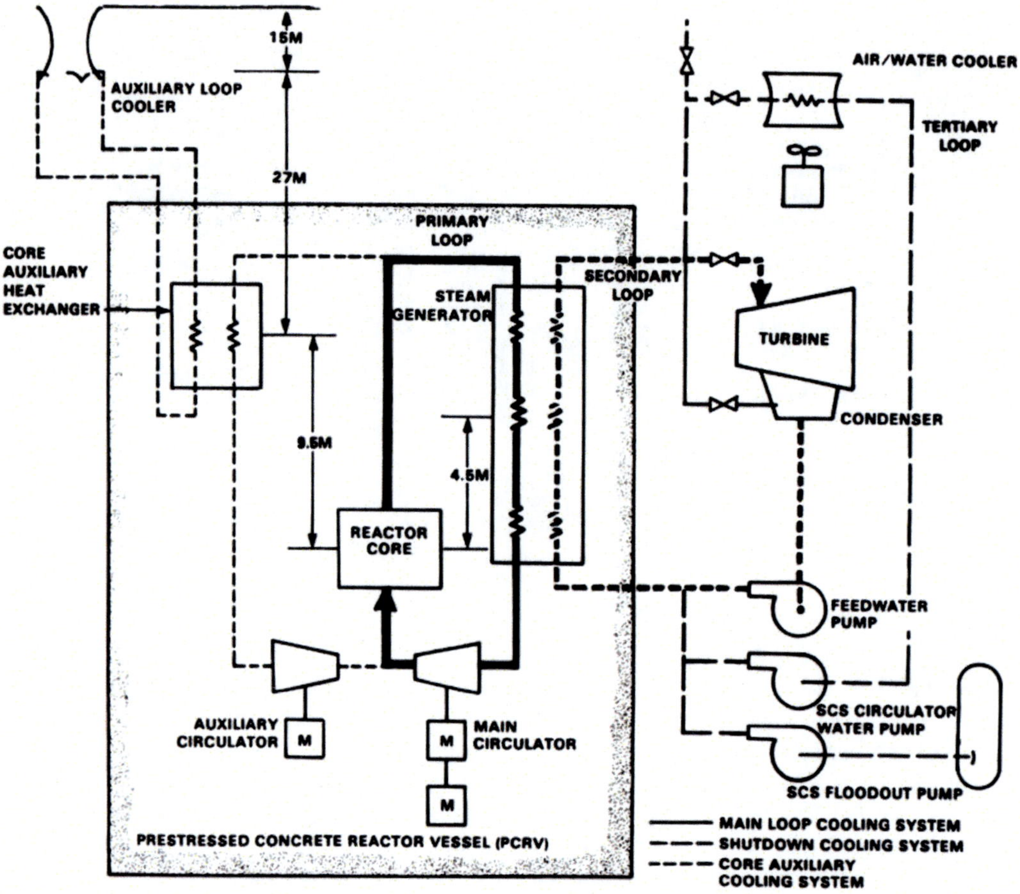

Fig. 17.12 GCFR heat transport systems

inertia, coupled with a working fluid that has poor heat transfer characteristics, creates a major challenge for the GFR with respect to exclusive use of passive/inherent decay heat removal during accident conditions; specifically for loss-of-coolant and loss-of-flow accidents. Conductive and radiative heat transfer alone cannot effectively remove the decay heat; convective cooling is also needed if passive/inherent systems are to remain as a viable option.

That is not to say that it is impossible to design a GFR with complete passive/inherent safety, e.g., reliance solely on conductive and radiative heat transfer for decay heat removal. However, such a design would require power densities that are more typical of thermal spectrum gas-cooled reactors (<10 kW/l). Unfortunately, GIF studies have shown that the low power densities result in unacceptable fuel cycle costs—on the order of the capital cost of the reactor itself [10–12]. However, increasing power density results in higher decay heat rates, and an attendant temperature increase in the fuel and core during off normal events. A balance is needed between the economics and sustainability of high power densities, and low power densities that will more easily meet inherent safety goals. For example, use of blowers/fans is possible during accident conditions, which may only require 3% of nominal flow to remove the decay heat based on assumed pressure conditions. This, however, requires reliance on active systems. In order to incorporate passive/inherent systems, innovative designs have been studied, and a mix of passive and active systems may meet the requirements for decay heat removal during accident conditions while satisfying economic objectives.

As was previously mentioned, gas coolants provide a relatively poor heat transfer medium with low thermal inertia at low pressure (0.1 MPa, or ~1 atmosphere). But using good design practices, one can show that at even high power densities (>100 kW/l), natural convection can successfully remove core decay heat at pressurized conditions (e.g., 7 MPa) using acceptable primary system elevation differences. Even the short-term decay transient, from 6 to 3% power, can be accommodated. Previous studies in the GIF have shown that natural convection is well suited for the block/plate and pin fuel forms. The pebble-bed/particle fuel form would be the most challenging form for this option. Pebble beds have inherently high resistance to core flow, which is difficult to overcome through design. It is important to note that for this option to work (i.e., use of natural convection), a guard confinement (GC) or double containment/vessel would most likely be required.

The approach is to only permit depressurization of the primary system through design means to a secondary backup pressure, which we will refer to as the backpressure. The option of using backpressure is a major design choice; the lower the pressure, the worse the natural convection heat removal capability, while higher backpressures require robust structures and assured leak tightness that will affect the economics of the plant. Several design options that utilize a backpressure are discussed below.

17.4.2.1 Emergency Heat Exchangers (EHX) for Decay Heat Removal

Consider a pin-type GFR using helium as the primary coolant at 7 MPa, with a direct power conversion cycle where the turbine is included in the primary loop. Let us assume that during normal shutdown conditions, there is a powered shutdown heat removal system to extract the decay heat and maintain safe temperatures within the reactor core.

Here we describe an abnormal (emergency) shutdown condition where it is assumed that there are no available power sources to drive the compressor. As one option to mitigate this challenge, let us assume the plant is to have emergency heat exchangers (EHXs) that are connected via piping in series with the reactor core and elevated with respect to the core (see Section 15.2.2, where the same concept is described for SFR natural circulation reactor core cooling). During normal reactor operation, check valves isolate the EHXs from the primary loop. When the EHXs are needed, the check valves open and the core and emergency heat exchangers are part of a closed flow loop. An ultimate heat sink on the secondary side (e.g., cold water) is used to directly or indirectly cool the secondary side of the EHXs. This heat extraction cools the helium as it passes through the EHXs. Natural circulation drives the helium flow, which is heated as it passes up through the reactor core and cooled as it passes down through the EHXs. An important parameter is the helium pressure in the EHX loop, which could be much less than 7 MPa due to a depressurization event (e.g., loss-of-coolant). As the pressure decreases, the amount of coolant in the loop decreases, along with the ability of the coolant to circulate and extract heat. It is essential that the behavior of this loop and its thermal-hydraulic limitations be well understood. Note that a fixed decay power level and system pressure in some instances can result in two possible steady states—one at a relatively high flow rate and low reactor coolant outlet temperature, and another at a very low flow rate with very high reactor coolant outlet temperature. Also note that when the system pressure is too low, there may be no steady state condition for the system. Hence, a thorough understanding of this thermal-hydraulic behavior is essential in order to develop an effective design. A schematic of a sample EHX is shown in Fig. 17.13, for simplicity, check valves are not shown.

A good understanding of the thermal-hydraulic behavior of the EHX loop can be obtained from relatively simple steady-state models that are easy to solve on a computer spreadsheet, such as shown in Eq. (15.1). In the modeling, one can assume that the primary coolant temperature exiting the EHX is a known value (e.g., 50 or 100°C). This assumption helps to keep the model simple by avoiding the need to include the secondary-side coolant of the EHX. Since all of the emergency heat exchangers

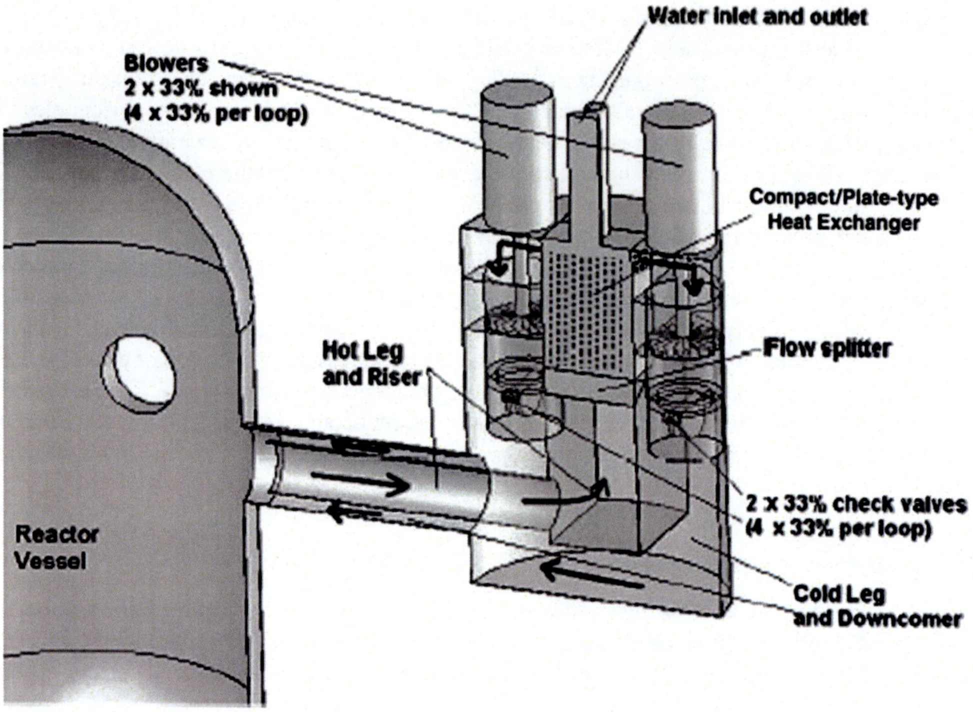

Fig. 17.13 Detailed schematic of an EHX

are in parallel, the model can assume one large heat exchanger. In addition, the core includes an upper and lower region, but the only heat producing region is in the core itself.

There are two basic equations that must be solved to analyze the primary-side EHX loop. First, the energy balance equation mathematically requires that the power generated in the core (which is the same as that extracted by the EHX) is the product of the core temperature rise, the helium flow rate, and the specific heat capacitance of the helium coolant. Second, the momentum balance equation requires that the friction pressure drop and the buoyancy pressure rise around the loop must be equal. The friction pressure drop is due to hydraulic resistance in the core, the parallel EHX's, and connection piping, plena, and valves. For simplicity, one can ignore the hydraulic resistances of the latter three. These can be made sufficiently small by making their flow areas sufficiently large. For more detailed analyses, these losses can be approximated by increasing the entrance and exit losses of the core and the EHX's. The buoyancy pressure rise is due to the helium in the downward flow through the EHX, and the coolant below it being denser than the upward flow through the core. The greater the vertical separation between the core and the EHX, the greater the buoyancy pressure rise.

17.4.2.2 Injection of Heavy Gas to Enhance Decay Heat Removal

The GCFR of the 1970s employed three auxiliary cooling loops, as was discussed in Section 17.4.1 above. Based upon a PRA assessment for the GFR, it has been confirmed that the use of three 50% capable loops is to be preferred due to the limitation imposed by common mode failures.

To further reduce system failure probability, passive gas accumulators, which can power the gas turbines used to drive auxiliary loop gas blowers, have been added, as shown in Fig. 17.14. This approach was employed to increase the diversity of emergency power sources, since failure to provide service

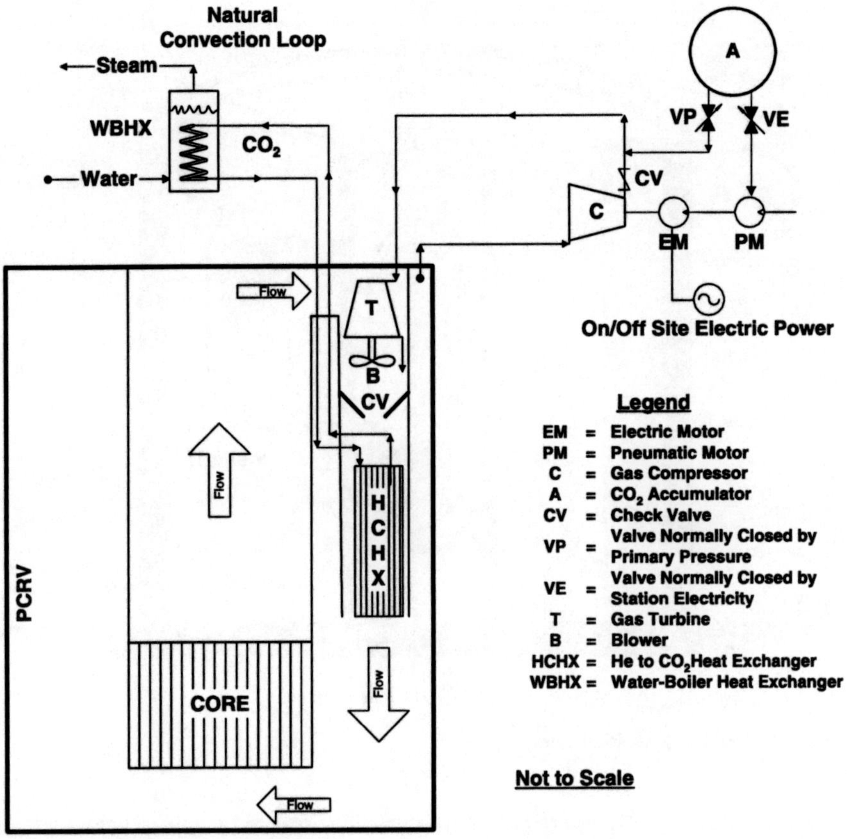

Fig. 17.14 A proposed pneumatically-powered decay heat removal system for GFR's

by diesel generators is typically the dominant event in LWR severe accident scenarios. Using this particular system design during a loss of coolant accident, the CO_2 accumulators discharge through a valve normally held shut by primary loop pressure. The turbine exhausts into the primary system, thereby providing CO_2 makeup replacing He; the former is much more effective under natural convection. In the event of station blackout without primary system depressurization, the CO_2 accumulators discharge through a valve normally held shut by station electricity, to power a pneumatic motor on the compressor driveshaft; in this case the CO_2 is vented to the atmosphere.

17.4.2.3 Design of Hybrid Active/Passive Decay Heat Removal Systems

From a technological risk viewpoint, the option that was identified in the GIF as the most effective was a reactor system with a guard confinement (GC) around the primary vessel to maintain the primary system pressure at a high enough level for good natural circulation, coupled to a hybrid passive/active approach using heat exchangers and blowers. The GC would be sized for an LWR containment range backpressure of 0.5–0.7 MPa, with an initial pressure of 0.1 MPa. Again, significantly higher backpressure is required for use of total natural convection for decay heat removal at the GFR target power densities (50–100 kW/l).

The lower backpressure, plus whatever natural convection is available at this pressure, would be utilized to significantly reduce the blower power of the active decay heat removal, where the system

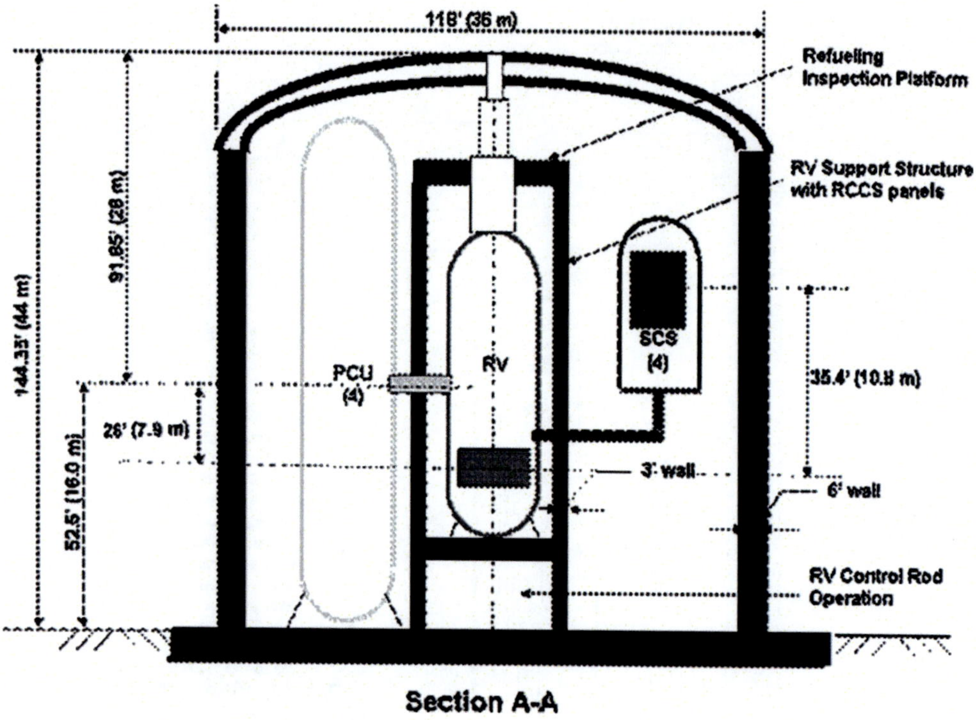

Fig. 17.15 Guard containment elevation view

would be sized to remove 2–3% decay power. The objective was to be able to have such low power requirements that power supplies such as batteries, without the need for startup, could be utilized. This lower backpressure should be sufficient to support natural convection for removal of 0.5% decay heat, which would occur at ~24 h after a depressurization event. Thus, there should be no need for active systems/power supplies after 1 day. Furthermore, since the decay heat will drop from 2–3 to 0.5% in this time period, credit should be taken in probability space for loss of active systems during the 24 h.

For a large, 2,400 MWth GFR core, the GC option with a 0.5 MPa backpressure was selected. The volume and cost analysis described in Ref. [13] showed that, at these lower pressures, the GC option has a lower cost than the corresponding PCRV planned for the GCFR design. This GC option is shown in elevation view in Fig. 17.15. The pre-stressed concrete containment is 36 m in diameter and 44 m high. The power conversion unit (PCU) is located outside the GC, and the GC penetrations for connecting to the PCU and shutdown cooling system (SCS; equivalent to the EHX discussed earlier) ducts must be separated 4.5 m vertically and 1 m horizontally to meet ASME code. Use of smaller spacing would lead to a thicker GC wall, and may introduce fabrication problems.

The overall plant dimensions are driven by the reactor vessel (RV) and PCU requirements. In turn the RV radial dimensions are based upon the core diameter, and the reflector and shielding thicknesses needed. The RV height is based on the SCS height and ducting, the IHX height and location, the core height, the PCU cross vessel location, and refueling reach concerns.

17.4.2.4 GFR System and Decay Heat Removal Design Summary

Based on the GIF studies, the following conclusions can be made for GFR system design and decay heat removal:

(1) A fully passively safe GFR design is possible with low power density, but the economics would be prohibitive.
(2) To reduce the economic burden of a fully passively/inherently safe system, a backpressure is needed to enhance natural convective cooling during off-normal events. The higher the backpressure, the more effective the cooling based on the main driving force for natural convection—mass flow. Thus, increased mass (or higher pressure) enhances natural convection. However, the containment must withstand these pressures for extended periods of time, which would require a higher cost for the containment. Conversely, a guard (or secondary) vessel could be designed to withstand the needed backpressure at a lower cost.
(3) Heavy gas injection enhances natural convection, and reduces the backpressure needed to remove the decay heat due to the increased mass of the working fluid. This could include gases such as CO_2.
(4) Fully active systems require very little power ($\sim$100 kW $\times$ 3 blowers), where their reliability may be better than their passive/inherent counterparts.
(5) Active/passive combined systems require a minimal backpressure (0.5–0.7 MPa), and only $\sim$16 kW ($\times$ 3) for the active side. This scenario would require that blowers are run for the first 24 h of an accident, at which point they can be shut down to allow only natural convective cooling to occur for the remainder of the accident period.

Items 2 and 5 above seem to best maintain the ideals of using passive/inherent safety as much as is reasonably achievable given the other GIF goals. However, other issues will need further study, e.g., cooling of the GC reduces the pressure of the working fluid, which decreases the effective cooling of the core.

References

1. M. Dalle Donne and L. Meyer, "Turbulent Convective Heat Transfer from Rough Surfaces with Two-Dimensional Rectangular Ribs," *Int. J. Heat Mass Transfer*, 20 (1977) 581–620.
2. C. B. Baxi and M. Dalle Donne, *Fluid Flow and Heat Transfer in the Gas Cooled Fast Breeder Reactor*, GA-A 15941, General Atomics, San Diego, CA, July 1980.
3. L. L. Snead and J. W. Klett, "Ceramic Composites for Structural Applications," *Proc. GLOBAL 2003*, American Nuclear Society (2003) 1077–1078.
4. See for example B. J. Makenas, J. W. Hales, and A. L. Ward, "Fuels Irradiation Testing for the SP-100 Program," *Proc. 8th Symposium on Space Nuclear Power Systems*, American Institute of Physics (1991) 886–891.
5. E. A. Hofman and T. A. Taiwo, "Physics Studies of Preliminary Gas-Cooled Fast Reactor Designs," *Proc. GLOBAL 2003*, American Nuclear Society (2003) 82–91.
6. Gamma Engineering Proposal to DOE for FY-2003 SBIR, Phase 1 Solicitation, "Development of a Hybrid SiC/SiC Ceramic Composite for Gas Cooled Fast Reactor Fuel Cladding and Core Structurals," Department of Energy, SBIR Program, January 2003.
7. D. Smith, P. McIntyre, B. Basaran, and M. Yavuz, "SiC Composite: A New Fuel Cladding for High-Temperature Cores," *Proc. GLOBAL 2003*, American Nuclear Society (2003) 1821–1823.
8. S. J. Zinkle, "Nonfissile Ceramics for Future Nuclear Systems," *Proc. GLOBAL 2003*, American Nuclear Society (2003) 1066–1067.
9. A. E. Waltar and A. B. Reynolds, *Fast Breeder Reactors*, Pergamon Press New York, NY (1981).
10. K. D. Weaver et al., *Gas-Cooled Fast Reactor: FY03 Annual Report*, INEEL/EXT-03-01298, Idaho National Laboratory, September 2003.
11. K. D. Weaver et al., *Gas-Cooled Fast Reactor: FY04 Annual Report*, INEEL/EXT-04-02361, Idaho National Laboratory, September 2004.
12. K. D. Weaver et al., *Gas-Cooled Fast Reactor: FY05 Annual Report*, INL/EXT-05-00799, Idaho National Laboratory, September 2005.
13. T. W. C. Wei et al., "System Design Report," I-NERI Project #2001-002-F, Report GFR 023, Argonne National Laboratory, February 2005.

Chapter 18
Lead-Cooled Fast Reactors

James Sienicki

18.1 Introduction

This chapter describes fast neutron reactors utilizing either of two so-called Heavy Liquid Metal Coolants; namely, lead (Pb) and lead-bismuth eutectic (LBE), both having significantly higher densities and boiling temperatures than sodium. At the current time, the main driver for interest in such reactors is the potential for reductions in the nuclear power plant capital cost per unit electrical power, realized by taking advantage of the particular properties of the Heavy Liquid Metal Coolant. A lead-cooled fast reactor (LFR) is not a sodium-cooled fast reactor (SFR) with a different coolant. Effective LFR designs differ in significant ways from SFR designs and the coolant technologies for the Heavy Liquid Metal Coolants are significantly different from the coolant technology for sodium. The present chapter discusses the distinctive features of the Heavy Liquid Metal Coolants and their consideration in LFR design. Three examples of LFR concepts are described.

18.2 Design Considerations for Heavy Liquid Metal Coolants

18.2.1 Distinctive Features

The term lead-cooled fast reactor (LFR) usually applies to a fast reactor utilizing either of two Heavy Liquid Metal Coolant (HLMC) materials. The first is Pb itself, which has a melting temperature of 327.45°C and an atmospheric boiling temperature of 1,743°C. The second is LBE, which is composed of 45.0 at.% Pb and 55.0 at.% Bi with a melting temperature of 124.5°C and an atmospheric boiling temperature of 1,670°C. The Pb and LBE densities at 480°C are 10,470 and 10,100 kg/m^3, respectively. Heavy Liquid Metal Coolant properties and HLMC technology are discussed in detail in the *Handbook on Lead-bismuth Eutectic Alloy and Lead Properties, Materials Compatibility, Thermal-hydraulics and Technologies 2007 Edition* [1].

Lead-bismuth eutectic-cooled reactors were developed and built in the former Soviet Union as naval submarine propulsion reactors and supporting land prototypes. However, those reactors were strictly for military purposes. To date, there has not been a LFR demonstration or prototype built for commercial purposes. The LFR is thus lagging behind the SFR in terms of a technology base and an experience base.

J. Sienicki (✉)
Innovative Systems Development Section, Argonne National Laboratory, Argonne, IL, USA
e-mail: sienicki@anl.gov

A.E. Waltar et al. (eds.), *Fast Spectrum Reactors*, DOI 10.1007/978-1-4419-9572-8_18,

Heavy Liquid Metal Coolants have several key inherent properties, which if taken advantage of by the designer together with specific design options, can potentially reduce the plant cost and improve safety. LFR designs are not simply SFR designs with a different fluid substituted for sodium; LFR designs can differ in fundamental ways from SFR designs.

Heavy Liquid Metal Coolants do not have the combustibility of sodium (or of the sodium-potassium eutectic coolant, NaK). In contrast to these alkali-based coolants, the HLMCs do not react vigorously with water, steam, or air. Thus, in designing a reactor cooled by HLMC, it is not necessary to incorporate the specific design features (as is the case for SFRs) that address the threat of coolant combustibility such as measures for accommodating or avoiding coolant fires following leaks. In addition, the designer does not need to incorporate features similar to those needed to accommodate sodium-water reactions and sodium-water reaction products including combustible hydrogen gas. One power conversion option considered for LFRs is the supercritical carbon dioxide (S-CO_2) Brayton cycle. Lead is calculated not to react chemically with a CO_2 working fluid of the S-CO_2 Brayton cycle above about 250°C, which is well below the Pb melting temperature of 327.45°C. Lack of vigorous interactions of Pb with the working fluid—whether water/steam or CO_2—makes it possible to consider eliminating the need for an intermediate cooling circuit, thereby substantially reducing the plant complexity and capital cost while enhancing plant reliability.

18.2.2 High Temperature Operation

The high boiling temperatures of the HLMCs allow for potential operations at system temperatures that are not limited by coolant boiling but by the integrity of the steel cladding and structures; that is, the HLMC boiling temperatures exceed the temperatures at which steel cladding and structural steels lose their strength and melt. The operating pressure for Pb or LBE LFRs is low, and so flashing does not occur should a leak develop in the primary coolant system boundary, thus enabling the traditional significant advantages of a low pressure compact liquid metal system. One LFR primary coolant system configuration in which the intermediate coolant system is eliminated is a pool-type reactor configuration. In this configuration, all major primary system components including the core and HLMC-to-working fluid heat exchangers (HXs) are installed inside of the reactor vessel, which is surrounded by a guard vessel.

The coolant level inside of the reactor vessel is such that in the event of a reactor vessel leak, the faulted level of coolant contained by the guard vessel always exceeds the minimal HLMC flow entrance level to the HXs. The lack of coolant flashing or boiling due to the high HLMC boiling temperature, combined with the pool system configuration and guard vessel, preclude the loss of primary coolant. It also assures that heat removal from the core and heat transfer to the normal HXs, in-vessel emergency decay heat removal heat exchangers, or heat removal by natural circulation of air over the outside of the guard vessel continues by means of natural circulation of single-phase primary HLMC coolant. The same could be said for a SFR for which the sodium temperatures remain below the sodium boiling point. However, the high boiling temperature of HLMC, which far exceeds that of sodium, in principle provides a pathway to operation at higher temperatures not limited by coolant boiling. This would offer the opportunity to better take advantage of the efficiency benefits of advanced energy conversion systems operating at higher temperatures such as the S-CO_2 Brayton cycle that can be exploited more effectively.

Operation of a LFR at higher temperatures faces the same challenges for materials as operation of SFRs at higher temperatures, including the need to handle creep and creep-fatigue of structural materials as well as the current lack of high temperature design methodologies in design codes such as the ASME Boiler and Pressure Vessel Code. In addition, operation at higher temperatures with existing materials poses greater material corrosion challenges for HLMCs, which are not present or are significantly less severe with those same materials at lower temperatures. A solution may be the development of new materials for high temperature service with Pb or LBE.

18.2.3 Neutronics and Fluid Flow

Lead has a slightly lower absorption of neutrons than does sodium, although the neutron absorption of both coolants is low. Lead also has a slightly greater elastic scattering of neutrons than sodium but the energy loss per collision is lower with Pb. The lower neutron absorption permits the fuel pin lattice with Pb coolant to be opened up by increasing the coolant volume fraction without a significant reactivity penalty. Increasing the coolant volume fraction increases the hydraulic diameter for coolant flow through the core with a corresponding reduction in the core frictional pressure drop. As a result, natural circulation is more effective and can transport a greater fraction of the core power. One way of opening up the core is to increase the spacing between adjacent fuel pins. Another approach, which may be carried out in conjunction with greater fuel pin spacing, is to increase the fuel pin diameter. However, opening up the core reduces the power density such that a LFR core may be larger in diameter than a SFR core. A designer seeking to reduce the overall coolant volume of a LFR relative to a SFR of the same thermal power level must overcome the consequences of having to accommodate a larger core diameter.

A core having a large pin spacing and large pin diameter leads to a different approach for supporting the fuel pins than is typically employed for SFRs; that is, hexagonal assembly ducts with wire-wrapped fuel pins on a assembly hexagonal lattice with the pins attached at the bottom of the assembly. For the larger pin spacing and pin size typical of a LFR, the fuel pins can be attached at the bottom of an assembly and held so as to maintain the proper interpin spacing using one or more levels of grid spacers similar to the approach followed in LWR fuel assemblies. The larger pin diameter and grid spacers provide resistance against deformation of the pins.

Fuel assembly flow ducts are not required for structural support in the LFR. Thus, the steel associated with the ducts is eliminated from the core region. By eliminating fuel assembly ducts, an open lattice core configuration can be achieved similar to Western-design Pressurized Water Reactors (PWRs). An open lattice core configuration with wide openings for cross flow between the fuel pins eliminates flow blockage accident initiators in which the coolant flow entering at the bottom of the core is postulated to be locally blocked. However, the open lattice core with unvarying lattice dimensions abandons the design feature to locally control the power-to-flow ratio by means of orificing of individual fuel assemblies available in the SFR. To some extent, the power-to-flow ratio can be locally changed by modifying the local fissile enrichment or by changing the local hydraulic diameter in the core by varying the local lattice dimensions.

It is common to find statements, particularly in some older publications, that the high HLMC density ($\rho_{Pb} = 10{,}400\ \mathrm{kg/m^3}$) raises the pumping power requirements (due to the dependency of frictional pressure drop upon density for an assumed fixed coolant velocity) implying a negative feature of LFRs due to the HLMC density. The pumping power requirement is generally equal to a summation over the local pressure drop multiplied by the volumetric flow rate. If the major pressure drop contribution is concentrated in the core region, then for a fixed core thermal power and temperature rise, the core contribution to the pumping power is equal to the pressure drop multiplied by the core thermal power divided by the product of the coolant density, specific heat, and core temperature rise. However, in an appropriately designed LFR, the pumping power requirements may be lower than the SFR in spite of the high density. This is because the fuel pins can be spaced more widely apart, reducing the coolant velocity in the core and the core pressure drop, and because the intermediate circuit loops and the associated intermediate circuit coolant pumps can be eliminated.

18.2.4 High Coolant Density Considerations

The high HLMC density compared with the density of sodium results in a need for thicker structures to accommodate structural loads due to the HLMC mass as well as forces exerted by the HLMC fluid. In this regard, the HLMC density is a negative in that the greater structural thicknesses required could result in greater costs for structural steel. This motivates the LFR designer to seek to take advantage of other HLMC features to reduce the number and size of structures required versus the SFR. In practice, the high HLMC density is a significant factor in limiting the size of a LFR due to the need to accommodate seismic events. For example, the European Lead-Cooled System (ELSY) concept [2–4] shown in Fig. 18.1 achieves 600 MWe (1,500 MWth) using two-dimensional seismic isolation and could likely be designed for even higher power. More is provided on this concept in Section 18.4.2.

Because the HLMC density exceeds that of steel structures, the structures inside of the reactor vessel or any other vessel in which they are immersed in HLMC are buoyant and would float in the coolant if unrestrained. The density of oxide fuels is less than that of the coolant such that fuel assemblies containing oxide fuel would float in the coolant if unrestrained. The density of metallic and nitride fuels exceeds that of the HLMCs. Thus, it may be possible to design removable fuel assemblies containing metallic or nitride fuel having a net effective density greater than that of the coolant such that the fuel assemblies would sink in the coolant.

The high HLMC density is a benefit in some respects, including improved safety. Significantly, the high density limits void growth and downward penetration following a postulated in-vessel HX tube rupture such that, for the analysis of specific design concepts, a significant void may not be transported to the core but instead rises benignly to the HLMC free surface [5] to be removed from the vessel by a passive pressure relief system. In addition, the designer may incorporate specific design features

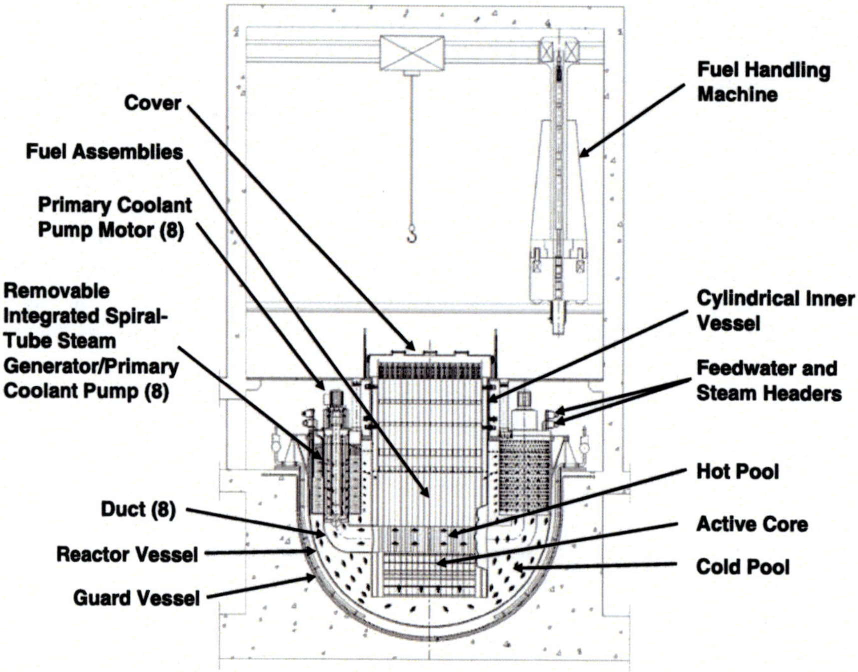

Fig. 18.1 Illustration of the ELSY primary coolant system showing the pool configuration with reactor and guard vessels, reduced Pb coolant volume, removable modular integral steam generator-main coolant pump units, and cylindrical inner vessel

to further reduce the likelihood of void transport to the core and assure that the void rises to the free surface instead. This further contributes to the feasibility of eliminating the intermediate coolant circuit.

In ELSY, only ~1.5 m of elevation head is needed to drive flow through the steam generators and the core due to the high Pb density. This reduces the required coolant level enabling the primary coolant volume/mass to be significantly reduced. Reducing the primary coolant mass reduces the amount of reactor vessel steel required. Due to the high density, a lower column height of coolant is needed to provide the required net positive suction head (NPSH) for the pumps, making it possible to place the mechanical pumps inside the hot leg portion of the circuit above the elevation of the core—significantly decreasing the required vertical shaft height as well as further reducing the required system volume and requirements for internal structures. The pumps feeding the steam generators in the hot leg raise the nominal elevation of the cold pool above that of the hot pool such that in the event of a station blackout and pump coast down, the head corresponding to the height difference between the cold and hot pools maintains flow through the low pressure drop core as the cold and hot pool levels tend to equilibrate, minimizing the initial thermal transient. In contrast, when the primary coolant pumps are installed in the cold leg of the circuit of a pool-type reactor (i.e., the traditional configuration for SFRs), the nominal hot pool level exceeds that of the cold pool and the head corresponding to the hot pool-cold pool height difference tends to oppose the continued flow of coolant through the core as the pool levels tend to equilibrate.

18.2.5 Corrosion Resistance

Heavy liquid metal coolant dissolves Fe, Cr, and Ni from unprotected steels at rates that increase with temperature [6]. Dissolution rates in Bi typically exceed those in Pb such that LBE usually attacks unprotected steels more aggressively than Pb coolant [7]. Active maintenance of the dissolved oxygen potential in the coolant within a proper concentration window and limiting HLMC velocities to values of ~1 m/s or less has emerged as a well established means for forming protective oxide layers (Fe_3O_4 at temperatures below ~570°C), significantly retarding the dissolution rate and avoiding the formation of solid PbO particulate. The formation of a protective oxide layer does not eliminate attack upon the structure; it reduces the rate of attack. Thus, for structures that may be exposed to HLMC for a long time, it is necessary to design for wastage of a portion of the structural thickness over time.

The current understanding of attack upon steel structures is that at temperatures below 425°C, there is minimal attack upon stainless steel even in the absence of measures to control the oxygen concentration in the coolant. Thus, those structures at temperatures of about 425°C or less would not be subject to any significant attack by HLMC. Above about 425°C, active oxygen control is required. Stainless steel can be utilized with active oxygen control up to a limiting temperature of about 500°C. For higher temperatures, stainless steel is subject to significantly more rapid attack. It has been found that ferritic/martensitic steels such as 9Cr and 12Cr steels have a greater resistance to attack with active oxygen control at temperatures above about 500°C. For example, T91 (a 9Cr alloy) may be usable somewhat above 500°C to about 550°C. This upper temperature is close to the temperature of ~570°C at which Fe_3O_4 is no longer stable and decomposes. At the time of this writing, there have been and are approaches under investigation or development for materials or material preparations that may enable service in HLMC at temperatures of about 550°C or greater. One approach is the development of ferritic alloys containing a greater proportion of Si, which may provide improved resistance to a corrosive attack by HLMC. One particular promising approach for the cladding may be weldment of a surface layer of Si-enhanced steel upon ferritic/martensitic steel to form a layered billet that is co-extruded [8]. The Si-enhanced layer provides improved resistance to corrosive attack but has poor irradiation stability. The substrate provides structural strength and irradiation stability.

It has recently been demonstrated that T91 ferritic/martensitic (F/M) steel cladding is protected from Pb corrosive attack at velocities at least as high as 3 m/s and temperatures as high as 550°C by "aluminizing" using the treatment currently applied in the GESA IV apparatus at the Forschungszentrum Karlsruhe [9]. The ELSY concept incorporates aluminized T91 cladding. Aluminizing might offer protection at higher temperatures and velocities as well. The ELSY concept is also being developed to limit the peak temperature of cladding exposed to Pb to ~550°C with inlet/outlet temperatures of 400/480°C, and to limit the coolant velocities to ~2 m/s.

The ELSY configuration (Fig. 18.1) has mechanical axial flow pumps located in the hot leg in a configuration in which the hot pool is isolated from the reactor vessel. This enables the entire inner surface of the austenitic stainless steel reactor vessel to face the cold pool, which is nominally at 400°C, minimizing the corrosion risk to the reactor vessel. Pump impellers at the nominal core outlet temperature of 480°C would be subjected to significantly higher velocities than other structures. A promising material under investigation by the ELSY project for pump impellers is Ti_3SiC_2 manufactured by 3-ONE-2, LLC [10, 11]. It can be pressed, slip cast, injection molded, machined or thermally sprayed onto metals to form coatings. It is described as stiff, thermal shock resistant, damage tolerant, tough, and fatigue resistant. Its corrosion resistance has been tested by exposure to Pb with low oxygen potential at 800 and 650°C for 1,000 h in a quartz Harp loop [12]. No evidence of Ti_3SiC_2 attack by Pb was observed.

To decrease the oxygen concentration in the reactor, bubbling hydrogen gas in a hydrogen-helium mixture (for inerting against hydrogen burning) is a means of efficiently removing oxygen through the reaction of hydrogen and oxygen forming water vapor that is carried away with the gas stream. A means of increasing the oxygen concentration is dissolution of lead oxide (PbO) spheres/particles by the coolant flowing through a packed bed of PbO spheres. In order to maintain the oxygen potential at the desired level, it is necessary to measure the dissolved oxygen potential with sufficient accuracy. Oxygen sensors can be used for this purpose. An oxygen sensor incorporates a solid electrolyte that conducts oxygen ions. If the electrolyte is installed in an electrical circuit such that current can flow into and from the HLMC coolant, a voltage difference is created that depends upon the oxygen potential.

Dissolved elements enter the coolant where they can give rise to the formation of contaminants. For example, dissolved Fe can oxidize in a HLMC resulting in oxide particles. In a LFR, these contaminants need to be removed from the coolant to avoid the potential for plugging of narrow passages (e.g., coolant flow channels at grid spacer locations in the core). Approaches for filtering of contaminants from LBE and Pb coolants have been developed in Russia [13–15].

The oxide layers that form on stainless steel walls provide an electrical resistance. Together with the lower electrical conductivity of HLMC relative to sodium, they make electromagnetic (EM) pumping of HLMC much less efficient than it is with sodium coolant. Due to the resulting low pump efficiency, EM pumps are not seriously considered for LFR primary or intermediate circuit pumps.

18.2.6 Coolant Freezing

As noted earlier, lead (Pb) coolant has a high melting temperature of 327.45°C, while LBE melts at 124.5°C. The LBE-cooled reactors developed in the former Soviet Union and some LFR concepts utilize LBE mainly because of its lower melting temperature. For HLMC, a means of heating the reactor system must be provided to melt the coolant and maintain it in a molten state prior to startup. The same is the case for sodium coolant but the melting temperature is higher with LBE and significantly higher with Pb. Refueling, inspection, and maintenance operations are typically carried out at a temperature providing sufficient margin above the coolant freezing temperature to assure that localized coolant freezing does not occur anywhere in the system. For sodium-cooled reactors, a refueling temperature of 200°C is typical. This temperature may also suffice for providing sufficient margin (i.e., 75°C) above the freezing temperature of LBE. For Pb coolant, a refueling temperature of 400°C is likely needed. The higher Pb melting temperature thus means that refueling, inspection, and maintenance operations must be carried out at about 400°C.

Freezing and thawing of the coolant that could result in damaging stresses in fuel pins and structures must be avoided. This has also been a consideration for sodium-cooled reactors. For example, if coolant is frozen to fill the cross-sectional volume of a channel and is heated such that a mass of liquid is trapped, then the thermal expansion of the melting material and liquid can stress the surrounding structure, potentially resulting in failure. For a fresh core, the non-nuclear heating system must avoid freezing until a significant decay heat source builds up in the fuel. The use of a pool-type configuration with its low surface-to-volume reduces the risk of freezing. LBE has a low shrinkage during solidification and rather high plasticity and low strength in the solid state that are favorable properties in limiting or eliminating damage as the alloy solidifies and is cooled down to ambient temperature. Thus, the stresses developing when LBE surrounds a structure and freezes might be controlled during the freezing-thawing procedure by assuring a very low freezing rate (i.e., a low rate of temperature decrease) and melting the solidified coolant downwards from the free surface. Lead coolant has a greater shrinkage than LBE upon solidification. The safety evaluation of operational transients and postulated accidents must include events involving overcooling of HLMC and assessment of the potential for local freezing of the coolant (e.g., on the outside of heat exchanger tubes) and its effects.

18.2.7 Heat Transfer Correlations

Correlations for liquid metal heat transfer similar to those described in Section 9.3 apply to HLMC with appropriate consideration for the properties of Pb. In general, such correlations have been developed on the basis of data from liquid metal heat transfer experiments. However, there are data indicating that there are other considerations in the application of liquid metal heat transfer correlations to HLMCs. One is the oxide layer formed upon the structure, which contributes a small thermal resistance. Heat transfer experiments carried out in the DELTA loop at Los Alamos National Laboratory revealed that the overall heat transfer coefficient depends upon the dissolved oxygen concentration, increasing with oxygen content [16]. This behavior was attributed to the wetting behavior of the LBE; the heat transfer data suggests that the degree of non-wetting decreases with increasing oxygen amount. The HLMCs do not wet structure or wet to a lower extent than alkali metal coolants such as sodium, which will wet steel given time. Thus, a detailed analysis of heat transfer will include accounting for the effects of the dissolved oxygen potential upon the overall thermal resistance between the bulk coolant and the channel surface.

18.2.8 Health Hazards

Neutron capture in ^{209}Bi leads directly to production of the isotope, ^{210}Po ($^{209}\mathrm{Bi}+n \rightarrow {}^{210}\mathrm{Po} + e^-$). The ^{210}Po is an alpha emitter with a half-life of 138 days. Elemental Po is a solid at room temperature and has a melting temperature of 254°C and an atmospheric boiling temperature of 962°C. Despite the high boiling temperature, polonium has a high volatility. In HLMC, the ^{210}Po is retained in the form of the compound, lead polonide, PbPo. However, contact with moisture can result in the formation of polonium hydride, PoH_2, with a melting temperature of -36.1°C and an atmospheric boiling temperature of 35.3°C, which is volatile [17]. Thus, the health hazards are gaseous compounds and aerosols containing ^{210}Po including polonium hydride, which represent a hazard in terms of transport and inhalation.

The generation of ^{210}Po with LBE is the main reason for selecting Pb coolant for some LFR design concepts for which the amount of ^{210}Po is calculated to be reduced by two to four orders of magnitude relative to LBE. However, ^{210}Po is not completely eliminated due to Bi impurities as well as creation of ^{209}Bi from transmutation reactions $\left(^{208}\mathrm{Pb} + n \rightarrow ^{209}\mathrm{Pb} \rightarrow ^{209}\mathrm{Bi} + \mathrm{e}^-\right)$.

Operations such as refueling or repair, as well as responses to events involving coolant leakages, must protect workers from the effects of ^{210}Po and minimize its release to the environment. The Russians have developed measures [13] including radiation control zones at work sites, drawing air out of areas with spills or high alpha activity combined with the use of fine fiber aerosol filters, protective clothing and respirators, use of caissons with pressurized air, removal of spilled or frozen LBE, applying easy-to-remove polymer films to polonium-contaminated surfaces, and prohibiting cutting, welding, etc. in areas of high polonium contamination. Initial studies have also been carried out in Russia and elsewhere for potential means of removing Po from LBE during operation [18, 19].

Lead is a toxic substance and a health hazard, which adversely affects numerous body systems and causes forms of health impairment and disease that arise after periods of exposure as short as days or as long as several years. Regulations and procedures for protecting industrial and construction workers from the health hazards of lead are well established and routinely implemented in the U.S. Regulations for working with lead have not impeded the charging, startup, operation, or modification of any LBE or lead experiment loops or facilities.

18.2.9 Fuel Materials

Utilization of an advanced fuel such as nitride fuel together with lead coolant can potentially further contribute to the inherent safety features of a LFR, provided that the nitride fuel can be demonstrated to perform well in steady state and transient irradiation tests and can be reliably manufactured to meet the performance requirements. Relative to oxide fuel, nitride fuel has a higher atom density, making it possible to reduce the fuel volume relative to that of oxide fuel. This further enables an increase in the coolant volume fraction without loss of the ability to achieve a core internal conversion ratio of unity and a low burnup reactivity swing, which in turn reduces the need for excess reactivity—thereby reducing the effects of rod withdrawal accident initiators. Relative to metallic fuel, which undergoes eutectic liquefaction reactions with steel cladding at temperatures of 650°C and higher, nitride fuel is expected to be compatible with both the Pb bond and Si-enhanced steel and ferritic/martensitic steel cladding materials to significantly higher temperatures at which the cladding melts or the nitride decomposes. Transuranic nitride fuel has a high decomposition temperature estimated to exceed 1,350°C, such that the fuel maintains its integrity at temperatures above which the cladding and steel structural materials lose their strength or melt.

Nitride has a high thermal conductivity, which when combined with bonding of the fuel pellets to the cladding by means of liquid Pb between the pellets and cladding, reduces the peak fuel temperatures during normal operation and accidents. This reduces the stored energy in the fuel and decreases the positive reactivity contribution resulting from cool down of the fuel as the fuel and coolant temperatures equilibrate during accidents as the core power decreases. Nitride fuel has the potential for low fission gas release per unit volume, reducing the thermal creep of the cladding resulting from the hoop stress loading due to internal pressurization of the fuel pin by released fission gas. The fast neutron spectrum cores of the Small Secure Transportable Autonomous Reactor (SSTAR) and Secure Transportable Autonomous Reactor with Liquid Metal (STAR-LM) concepts [20–24] with Pb coolant and transuranic nitride fuel have strong reactivity feedbacks, which provide significant negative reactivity upon heatup or equilibration of system temperatures. The strong reactivity feedback reduces the core power to match the heat removal from the reactor system, inherently shutting down the reactor fission power in the event that the two active shutdown systems fail to scram the reactor. More details about the SSTAR and STAR-LM concepts are provided in Section 18.4.3.

18.3 Russian Experience

The current experience base for HLMCs includes 80 reactor years operating experience in the Soviet Union and subsequently the Russian Federation with LBE-cooled reactors for strictly military purposes. Two 73 MWth and eight 155 MWth submarine reactors were operated with 40 reactor years operation underway together with 70 and 155 MWth land prototypes [13, 25–27]. The Russians chose to not first construct and operate a smaller-scale demonstration of a LBE-cooled reactor to demonstrate and prove the technology (e.g., a technology pilot plant) due to the military urgency for deployment. The submarine reactors were put into service and filled the role of test reactors as well as operational systems. Technical difficulties and accidents occurred during reactor operation and lessons were learned as experience was gained in coolant technology, corrosion performance, mass transfer in LBE systems, and means to deal with ^{210}Po produced during operation. Limited descriptions of the reactors and incidents have been presented in the open literature. Detailed design information about the LBE-cooled submarine reactors has never been provided. For example, drawings of the designs have never appeared in the open literature.

Following the breakup of the Soviet Union, Russian efforts were devoted to developing land-based LFR power plants for electricity generation. Two main paths have been followed by two different organizations. One is the development of the 300 MWe BREST-OD-300 and 1,200 MWe BREST-OD-1200 Pb-cooled reactors led by the Research and Design Institute of Power Engineering (NIKIET) in Moscow. The other is the LBE-cooled SVBR-75/100 led by the Institute of Physics and Power Engineering in Obninsk. The BREST concepts [28] utilize a pool primary system configuration involving a single bottom-supported large tank surrounded by bricks. It is argued that rupture of the tank would be followed by freezing of the Pb in the gaps adjacent to the bricks thereby self-plugging the leak. Seismic resistance is not a feature of the BREST concepts. To provide sufficient margin above the Pb freezing point, the BREST inlet temperature is 420°C; the outlet temperature is 540°C. The maximum temperature of Pb in contact with cladding is about 630°C. The energy conversion system is a Rankine supercritical steam cycle operating at 27 MPa with steam generator inlet and outlet temperatures of 355 and 525°C, respectively.

Significant information about the Russian LBE-cooled reactors, LFR concepts, and HLMC technology can be found in the proceedings from three HLMC conferences held in Obninsk in 1998 [13], 2003 [14], and 2008 [15]. The latter meeting commemorated the 50th anniversary of the first LBE-cooled reactor, the 27/VT land prototype, in Obninsk, Russia.

18.4 Example LFR Concepts

At the time of this writing, only LFR concepts (i.e., studies) have been developed in investigations of LFRs for commercial purposes. To illustrate features that designers have incorporated into LFR concepts, three examples are given below.

18.4.1 SVBR-75/100

The SVBR-75/100 (Lead-Bismuth Fast Reactor) [29–31] shown in Figs. 18.2 and 18.3 is a 75–100 MWe (210–280 MWth) concept utilizing LBE coolant in which the power level depends upon the specific design parameters chosen. Conditions and dimensions are also provided in Table 18.1 (included at the end of this chapter). The SVBR-75/100 concept was originally developed at the Institute of Physics and Power Engineering (IPPE) in Obninsk, Russia, and is stated to be based on the existing Project 705/705 K 155 MWth submarine reactor technology. Inlet and outlet temperatures for the

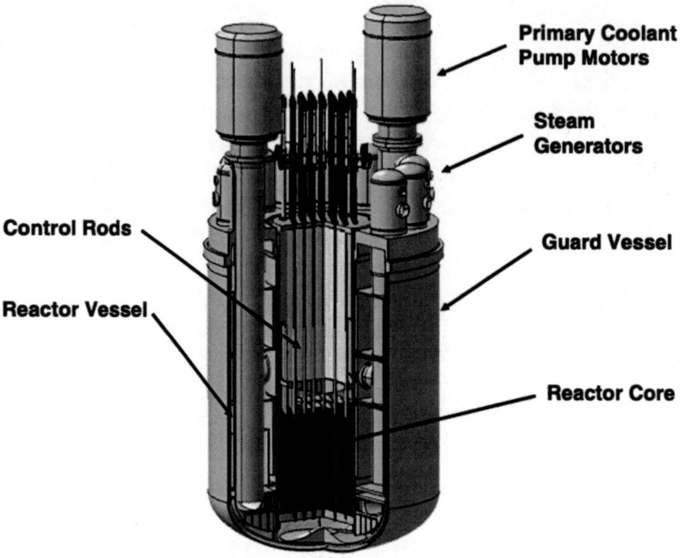

Fig. 18.2 SVBR-75/100 reactor [29]

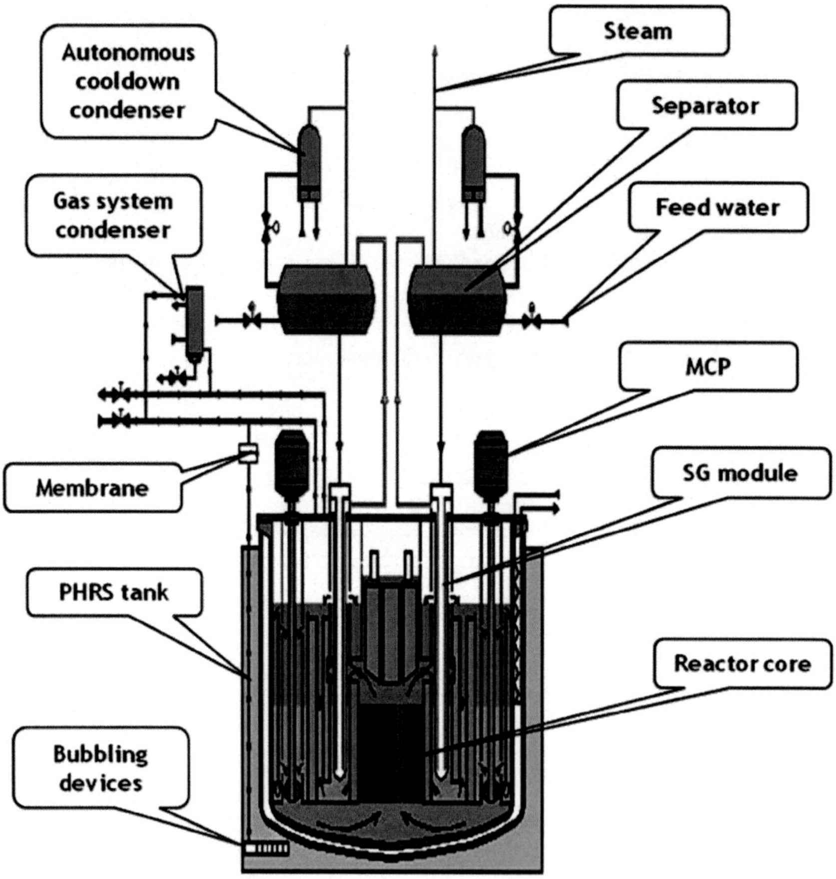

Fig. 18.3 Schematic illustration of SVBR-75/100 reactor PHRS = Primary Heat Removal System [29] (MCP = Main (Primary) Coolant Pump; SG = Steam Generator)

SVBR-75/100 are 320 and 482°C, respectively, relative to the LBE freezing point of 125°C. The peak cladding temperature is below 550°C. The reactor is a pool-type system with a reactor vessel incorporating steam generators and mechanical pumps having their motors located above the reactor vessel upper head. The steam generators are immersed inside of the reactor vessel; there is no intermediate coolant circuit. The steam generators and pumps are located in separate compartments inside of the vessel. The flow path is such that following a steam generator tube rupture, the steam bubbles/voids can rise to a LBE free surface and escape to the cover gas region. Pressure relief occurs at a pressure of 1.0 MPa following a steam generator tube rupture. The safety rods incorporate "fusible locks" that provide passive shutdown if the temperature exceeds 700°C. Although initial SVBR-75/100s would be fueled with uranium dioxide having a uranium enrichment of 16.1%, it is envisioned that in the future, transitions could be made to the usage of plutonium and minor actinides from LWR spent fuel and later from SVBR-75/100s.

The reactor would be refueled by removing one assembly at a time because of the decay heat level. Each assembly would be immersed in lead, which subsequently freezes and the assembly goes to dry storage. In contrast, the new core can be installed as a single cartridge. For emergency decay heat removal, the SVBR-75/100 incorporates a water tank surrounding the guard vessel that surrounds the reactor vessel. Water cooling circuits remove heat from the water tank to air heat exchangers that transfer the energy to the atmosphere heat sink. For early times following neutronic shutdown, boiling of water may occur off of the guard vessel surface. Although the water saturation temperature is less than the LBE freezing temperature, the LBE is calculated not to freeze for a very long time due to the temperature drops across the reactor and guard vessels and the gap between them. Potential applications have been investigated in which a number of SVBR-75/100 reactors would be installed in a single reactor building to deliver steam to one or two turbo generators.

18.4.2 European Lead-Cooled System (ELSY)

The ELSY concept is a 600 MWe (1,500 MWth) central station LFR for electricity generation [2–4]. The ELSY project involves a consortium of nineteen organizations with funding from the Italian government, other European governments, and the European Union 6th Framework Program. Preconceptual system design for ELSY has the goal of taking advantage of specific attributes of Pb coolant to reduce the plant costs per kilowatt of electric to less than those of SFRs and LWRs. The main objectives of ELSY include demonstration of the technical feasibility of a LFR, economics, and innovation for simplification, cost reduction, and safety. In developing a preconceptual design for ELSY, approaches and configurations based upon traditional SFR designs have not been presumed. Instead, the ELSY concept has been developed anew to take advantage of the specific features of Pb coolant, existing knowledge of materials compatibility with HLMCs, existing knowledge of HLMC coolant technology, and experience with SFR design features.

Figure 18.1 shows an illustration of the ELSY concept while Table 18.1 provides conditions and dimensions. Lead (Pb) has been selected for the ELSY coolant to minimize the formation of ^{210}Po. The fuel is selected to be mixed oxide; mixed U-Pu nitride fuel is also considered as a more advanced option. Several innovations contribute to a reduction of the HLMC volume and mass resulting in a short-height (~9 m) reactor vessel having a diameter of ~12.5 m. The reactor vessel is fabricated of austenitic stainless steel and is surrounded by a guard vessel, which is anchored to the reactor cavity/pit. ELSY utilizes a superheated steam power converter operating at a steam temperature of 450°C and pressure of 18 MPa. The core inlet and outlet temperatures are selected to be 400°C and 480°C, respectively, enabling the use of existing codified materials such as Type 316 or 321 austenitic stainless steel or T91 ferritic steel to the largest extent plus aluminized T91 for fuel pin cladding and steam generator tubes. The maximum fuel pin cladding temperature is about 550°C, with the aim of increasing this temperature up to 600°C if the aluminized cladding can be shown to perform reliably in Pb coolant.

The intermediate heat transport circuit is eliminated, which greatly reduces the size and cost of the nuclear steam supply system and the reactor building. Eight main coolant Pb pumps are installed in the hot leg of the primary circuit. Each pump also incorporates a pony motor for removal of decay heat. No pump bearings are present beneath the Pb free surface; all pump bearings supporting the shaft are installed in the gas space.

Each pump is part of a removable modular integral steam generator and main coolant pump, which further reduces the required coolant volume. The steam generators incorporate horizontal spiral tubes; the Pb discharged from the pump impeller generally flows outward in radial directions between and over adjacent horizontal spiral tubes containing water/steam. Since the Pb is free to exit a steam generator at any height, the steam generators remain effective for partial core powers at partial submergence—reducing the required additional free surface height to accommodate postulated vessel failure events. Therefore, it is not necessary to assure a faulted Pb coolant level that exceeds the elevation of a fixed inlet to the steam generator to assure adequate coolant circulation through the core for decay heat removal following a postulated vessel failure event. This further contributes to reduction of the HLMC volume and mass. The steam generator preconceptual design incorporates an innovative passive design feature to prevent steam from being transported radially into the Pb coolant cold pool following a postulated tube rupture. Pressure relief is provided to vent the steam from a steam generator to prevent over pressurization of the primary coolant system. Ducts containing two rupture discs are installed on the top of each steam generator to vent steam from the reactor cover gas to the above-reactor enclosure.

To further minimize the potential flow rate of water and steam following a postulated rupture, the headers for feed water and steam are installed outside of the primary coolant boundary such that only individual steam generator tubes are located inside of the primary boundary. Thus, the discharge of water and steam into Pb following a postulated header rupture is eliminated.

The reactor vessel (Fig. 18.1) has a cover that is an annular steel plate with a central opening accommodating an extended Cylindrical Inner Vessel (CIV). The cover incorporates penetrations including large openings for the insertion or removal of modular integral steam generator-main coolant pump units. The cold collector portion of the cold pool is located in the annular region between the reactor vessel and the CIV. The CIV extends upward above the cover and contains the fuel and shielding assemblies. The hot collector includes the region immediately above the core and the eight ducts that deliver heated coolant from the core to the main coolant pumps. This arrangement minimizes the volume of the hot pool and isolates it inside the central portion of the reactor vessel away from the reactor vessel wall.

Because a fuel assembly can be designed to be buoyant in Pb, the ELSY preconceptual design eliminates the need for a core support plate. Each fuel assembly is a tall structure having a part immersed in the coolant and an overlying part above the Pb free surface. The upper part of an assembly above the cover of the reactor vessel is fixed and supported from above. This configuration simplifies refueling. Fuel assemblies are installed and removed from the core using a simple handling machine operating in the cover gas at ambient temperature conditions. The fuel handling operations are thus carried out with full visibility. There is no in-vessel refueling machine, which further contributes to a reduction of the Pb coolant volume. Because there is no need for an underlying core support plate, there is no below-core structure for structural support.

Figure 18.4 illustrates the layout of a 21 by 21 square fuel assembly containing 428 mixed oxide fuel pins, a central square support tube, and four supporting structural rods located at the four corners. The central square support tube and four corner structural rods provide the required structural support and stability for the fuel assembly. The fuel pin outer diameter is 1.05 cm and the fuel pin square pitch

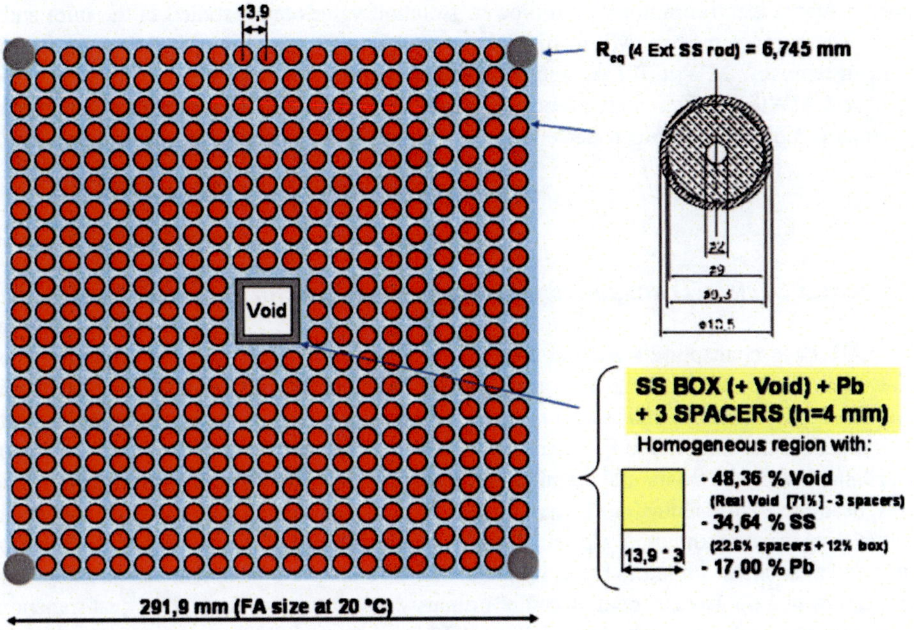

Fig. 18.4 Illustration of ELSY 21 by 21 square fuel assembly with 428 fuel pins, a central square support tube, and four corner support rods

is 1.39 cm, corresponding to a square-lattice pitch-to-diameter ratio of 1.32. Velocities through the core are limited to ~2 m/s.

A possible open-lattice core configuration incorporates 162 fuel assemblies in three enrichment zones and eight large safety and shutdown rod locations. The fuel assembly square pitch is 29.4 cm. Three different enrichment zones are employed to flatten the radial power distribution across the core. The PuO_2 enrichments of the three zones in atom percent are 14.61, 16.65, and 19.61 at.%; the core average enrichment is 16.75 at.%. To further flatten the power distribution across the core, forty fuel assemblies incorporate slim control rods called Finger Absorber Rods (FARs) that are inserted into the void of the central square column tube. Thirty-two different fuel assemblies incorporate FARs as safety rods. The control rods incorporate B_4C absorber enriched to 90% ^{10}B. The active core has an effective diameter of about 5 m and a height of 0.9 m. The core is surrounded by shielding assemblies, which act as a reflector and provide shielding for the reactor vessel.

The ELSY approach to in-service inspection is to make all in-vessel components removable so that they can be inspected outside of the reactor vessel. There is no in-vessel inspection under Pb coolant.

For emergency decay heat removal, ELSY provides independent, redundant, and diverse pathways for heat removal. The outside of the guard vessel is cooled with a Reactor Vessel Auxiliary Cooling System (RVACS) in which air flows through a number of pipes. The RVACS removes 2 MWth (0.13% nominal power) at 430°C, which is sufficient only in the long term following neutronic shutdown. ELSY is also provided with a Reactor Pit Cooling System (RPCS) for use during in-service inspection of the reactor vessel. ELSY also incorporates a Direct Reactor Cooling (DRC) system with heat exchangers directly immersed inside of the Pb coolant cold pool. There are four cooling circuits consisting of two W-DHR loops and two WA-DHR loops. Each W-DHR loop provides for drainage of water into a bayonet tube dip cooler. Vaporization of water and venting of the steam provide for heat removal. Each WA-DHR loop can also operate with water is this fashion, or it can utilize forced convection air cooling or natural circulation air cooling. Forced air circulation is provided by a battery-powered electric motor-driven blower. Isolation valves are installed in the inlet and outlet air ducts. Water can remove 5 MWth (0.33% nominal power) per loop. Air driven by forced convection can remove 2 MWth (0.13% nominal power) per loop, and air driven by natural circulation can remove 1 MWth (0.067% nominal power) per loop. A WA-DHR loop can also be operated with connections to both the air and water cooling circuits to enable cooling by air enhanced by water cooling.

18.4.3 Small Secure Transportable Autonomous Reactor (SSTAR)

The SSTAR LFR concept [20–24] has been developed jointly by Argonne National Laboratory and Lawrence Livermore National Laboratory under the U.S. Department of Energy Generation IV Nuclear Energy Systems Initiative as a small reactor for international deployment (this initiative is the U.S. aspect of the Generation IV International Forum (GIF) described in Section 1.2). SSTAR is a small (19.8MWe/45 MWth) natural circulation fast reactor plant incorporating proliferation resistance features, fissile self-sufficiency for efficient utilization of uranium resources, autonomous load following suitable for small or immature electric grids, and a high degree of passive safety. Loss-of-flow and loss-of-coolant accident initiators are eliminated. The SSTAR pre-conceptual design integrates three major features in a pool vessel configuration: primary coolant natural circulation heat transport; lead (Pb) coolant; and transuranic nitride fuel. The SSTAR concept with primary coolant natural circulation heat transport can be scaled up to 181 MWe (400 MWth) as a variant called the STAR-LM with optional potable water production using a portion of the reject heat.

SSTAR has been developed for the future global energy economy and not for near-term development. The SSTAR preconceptual design assumes that a number of advanced technologies will be successfully developed, including as yet to be developed and code qualified advanced cladding and structural materials enabling service in Pb for 15–30 years at peak cladding temperatures up to about 650°C at a core outlet temperature of 570°C, qualified transuranic nitride fuel meeting fuel performance requirements, S-CO_2 Brayton cycle energy conversion, whole-core cassette refueling, and in-service inspection approaches for components immersed in Pb coolant. If SSTAR were to be developed for near-term deployment, the operating system temperatures would need to be reduced (e.g., 480°C core outlet temperature like ELSY) to enable the use of existing codified materials and an existing fuel type such as metallic fuel would have to be qualified and utilized.

For the manufacture of H_2, STAR-H2 (Hydrogen) is a 400 MWth STAR-LM version for hydrogen generation using a Ca-Br thermo chemical ("water cracking") cycle. Deployment would follow SSTAR and STAR-LM when cladding and structural materials for operation in Pb up to 800°C are developed. One material worthy of investigation is Ti_3SiC_2. STAR-H2 operates with core inlet and outlet temperatures of 664°C and 793°C, respectively. The balance of plant involves a cascade of H_2 production (and O_2 production) via the Ca-Br cycle, in-house electricity generation using the S-CO_2 Brayton cycle, and water desalination.

Figure 18.5 is a cross section illustration of SSTAR. Conditions and dimensions are included in Table 18.1. The Pb coolant is contained inside a reactor vessel surrounded by a guard vessel. The Pb coolant flows through a perforated flow distributor head located beneath the core; this structure provides an essentially uniform pressure boundary condition at the inlet to the core. The Pb flows upward through the core and a chimney above the core formed by a cylindrical shroud. The vessel has a height-to-diameter ratio large enough to facilitate natural circulation heat removal at all power levels up to and exceeding 100% nominal. The coolant flows through openings near the top of the shroud and enters four modular Pb-to-CO_2 heat exchangers located in the annulus between the reactor vessel and the cylindrical shroud. Inside each heat exchanger, the Pb flows downward over the exterior of

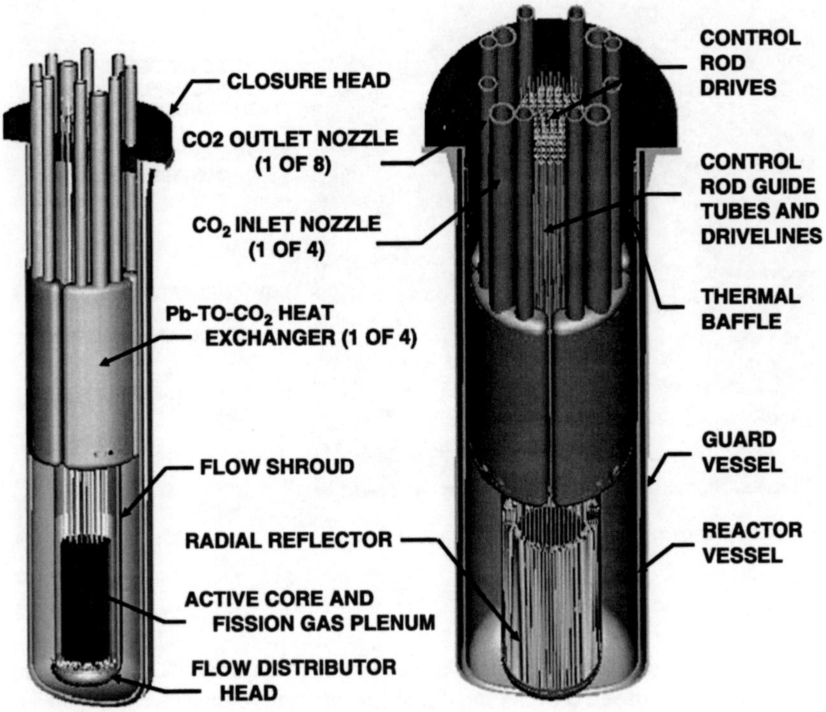

Fig. 18.5 Small secure transportable autonomous reactor (SSTAR)

tubes through which the CO_2 flows upwards. The CO_2 enters each heat exchanger through a top entry nozzle, which delivers the CO_2 to a lower plenum region in which the CO_2 enters each of the vertical tubes. The CO_2 is collected in an upper plenum and exits the heat exchanger through two smaller diameter top entry nozzles. The Pb exits the heat exchangers and flows downward through the annular downcomer to enter the flow openings in the flow distributor head beneath the core. A thermal baffle is provided in the vicinity of the Pb free surface. The thermal baffle is necessary to protect the reactor vessel from thermal stresses that would otherwise result from exposure to the Pb during startup and shutdown transients when the Pb temperature varies between the cold shutdown temperature of 420°C and the nominal core outlet temperature of 567°C.

The SSTAR core (Fig. 18.6) is an open lattice of large diameter (2.5 cm OD) fuel pins on a triangular pitch ($P/D = 1.185$) that does not consist of removable fuel assemblies. The fuel pins are permanently attached to an underlying grid plate. This configuration restricts access to fuel and eliminates fuel assembly blockage accident initiators. The compact active core, which is 1.22 m in diameter by 0.976 m in height, is removed as a single cassette during refueling and replaced by a fresh core. The active core diameter is selected to minimize the burnup reactivity swing over the 30-year core lifetime and the power level of 45 MWth is conservatively chosen to limit the peak fluence on the cladding to 4×10^{23} n/cm^2, which is the maximum exposure for which HT9 cladding has been irradiated. The core has a low power density, which facilitates the 30-year core lifetime during which the burnup and peak fluence are attained. The core incorporates two low enrichment zones in the core central region, which assists in reducing the burnup reactivity swing. Three driver enrichment zones reduce the peak-to-average power. Primary and secondary sets of control rods are uniformly dispersed throughout the active core. The control rods incorporate boron carbide absorber with enhancement of the isotope, ^{10}B. Small adjustments of the control rods are carried out to compensate for small changes

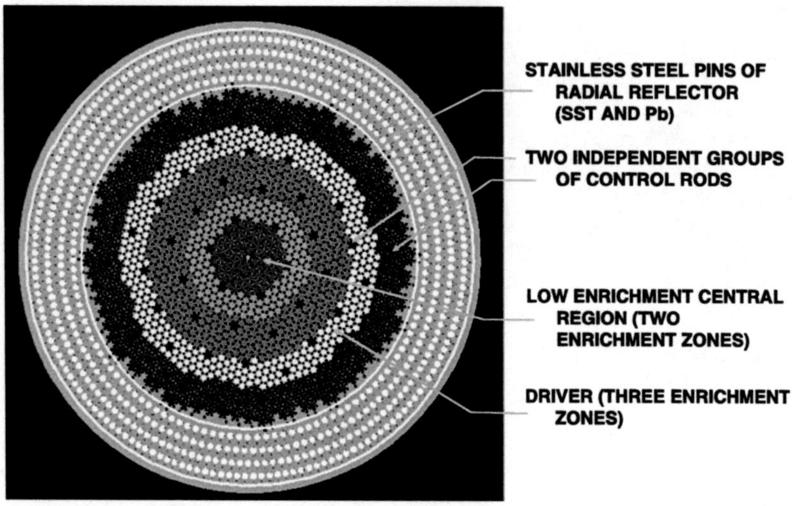

Fig. 18.6 SSTAR 30-year open-lattice core with all fuel pins shown

in the burnup reactivity swing. The radial reflector consists of an annular "box" containing 50 vol.% stainless steel rods and 50 vol.% Pb with a small flow rate of Pb to remove the small power deposition in the reflector. Stainless steel is necessary to act as a shield to reduce the neutron fluence in the reactor vessel.

The cladding incorporates an as yet to be developed and code qualified co-extruded functionally graded material incorporating a surface layer of Si-enhanced steel to retard the oxidation rate of the cladding resulting from exposure to dissolved oxygen in the coolant and a substrate providing structural strength and irradiation stability. The transuranic nitride fuel pellets are bonded to the cladding by molten Pb to reduce the temperature difference between the pellet outer surface and cladding inner surface. The fuel pins are held by two levels of grid spacers.

Conversion of the core thermal energy to electricity is accomplished using a S-CO_2 Brayton cycle power converter providing higher plant efficiencies and expected lower balance of plant costs than the traditional Rankine steam cycle operating at the same reactor core outlet temperature. The interest in higher plant efficiencies has driven interest in operation of SSTAR at higher Pb temperatures to take advantage of the increase in plant efficiency with temperature of the S-CO_2 Brayton cycle. The coupled reactor-S-CO_2 Brayton cycle plant system thermal hydraulic pre-conceptual design has been optimized to maximize the cycle efficiency. For a 650°C peak cladding temperature, a reactor core outlet temperature of 567°C is achieved resulting in a Brayton cycle efficiency of 44.2% and a net plant efficiency of 44.0%, subtracting off the electrical power requirements for the circulating water pumps.

A control strategy has been developed for the S-CO_2 Brayton cycle energy converter whereby the cycle is automatically controlled such that the heat removal from the Pb-to-CO_2 heat exchangers is adjusted to ensure that the generator power matches the load demand from the electric grid. This autonomous load following feature is made possible since the core power automatically adjusts itself due to the strong reactivity feedbacks of the fast spectrum core with Pb coolant and transuranic nitride fuel such that the core power matches the heat removal from the Pb-to-CO_2 heat exchangers. In particular, it is not necessary for the reactor core power to be changed through the motion of control rods except for startup and shutdown operations.

Table 18.1 Conditions and dimensions for the SVBR-75/100, ELSY, and SSTAR concepts

Characteristic \ LFR concept	SVBR-75/100 (Lead-Bismuth fast reactor-75/100)	ELSY (European Lead-Cooled system)	SSTAR (small secure transportable autonomous reactor)
Power, MWe (MWth)	101.5(280)	600(1,500)	19.8(45)
Coolant	LBE	Pb	Pb
Reactor vessel dimensions height/Diameter, m	6.92/4.53	9.5/12.5	12.0/3.23
Fuel	Uranium dioxide	Mixed oxide	Transuranic nitride (TRUN) enriched to ^{15}N
Enrichment, wt.%	16.1	14.54/17.63/20.61 PuO_2, 3 radial zones	1.7/3.5/17.2/19.0/20.7 TRU/HM, 5 radial zones
Refueling interval or core lifetime, years	~8	5 (open 4-batch cycle)	30
Core inlet/outlet temperature, °C	320/482	400/480	420/567
Active core dimensions height/diameter, m	0.9/1.645	0.9/4.649	0.976/1.22
Fuel/coolant volume fractions	~0.615/0.28	0.304/0.537	0.45/0.35
Average power density in active core, W/cm^3	140	117	42
Average (peak) discharge burnup, MWd/kg heavy metal	67(100)	78(94)	81(131)
Burnup reactivity swing, $	–	~2.6	<1
Peak fuel temperature, °C	–	2,100	841
Cladding	Si-enhanced ferritic/martensitic steel	Aluminized T91	Si-enhanced ferritic/martensitic steel bonded to fuel pellets by Pb
Peak cladding temperature facing LBE or Pb, °C	600	550	650
Fuel assemblies	Hexagonal with 22.545 cm hexagonal distance –central support tube and 6 corner tie rods	Square with 29.19 cm side – 428 fuel pins in 21 by 21 square array with central square support tube and 4 corner support rods	Whole-core cassette
Fuel pin lattice	Triangular	Square	Triangular
Fuel pin diameter, cm	1.20 with four helical spacer ribs	1.05	2.50

Table 18.1 (continued)

LFR concept / Characteristic	SVBR-75/100 (Lead-Bismuth fast reactor-75/100)	ELSY (European Lead-Cooled system)	SSTAR (small secure transportable autonomous reactor)
Fuel pin pitch-to-diameter ratio	1.133	1.32	1.185
Core hydraulic diameter, cm	~0.50	~1.28	1.371
Control and safety rods	37 = 29 control rods + 6 safety rods + 2 regulating control rods moving inside of central tubes in 37 fuel assemblies	Finger absorber rods – 40 primary (control rods) + 32 secondary (safety) rods moving inside of central square tubes in fuel assemblies – B_4C enhanced in ^{10}B	24 primary + 30 secondary rods uniformly distributed on core triangular lattice – B_4C enhanced in ^{10}B
Primary coolant flow rate, kg/s	11,760	126,200	2,107
Power conversion cycle	Saturated steam	Superheated steam	Supercritical CO_2 Brayton
Steam generator type (number)	Regenerative shell-and-tube with bayonet tubes having central and annular tube channels in a triangular lattice (6 in each of two steam circuits)	Spiral-tube integrated with primary coolant pump (8)	
Steam pressure, MPa	9.5	18	–
Feed water/steam temperature, °C	241/307	335/450	–
Steam flow rate, kg/s	161	960	–
Pb-to-CO_2 heat exchanger type (number)	–	–	Shell-and-tube (4)
CO_2 turbine inlet and cycle minimum temperatures, °C	–	–	552/31.25
Max/Min CO_2 pressures in cycle, MPa	–	–	20/7.4
CO_2 flow rate, kg/s	–	–	245
Cycle efficiency, %	–	42	44.2
Net plant efficiency, %	36.25	40	44.0

References

1. *Handbook on Lead-bismuth Eutectic Alloy and Lead Properties, Materials Compatibility*, Thermal-hydraulics and Technologies 2007 Edition, OECD NEA No. 6195, ISBN 978-92-64-99002-9. Available online at http://www.nea.fr/html/science/reports/2007/nea6195-handbook.html
2. L. Cinotti, C. F. Smith, J. J. Sienicki, H. Aït Abderrahim, G. Benamati, G. Locatelli, S. Monti, H. Wider, D. Struwe, A. Orden, and I. S. Hwang, "The Potential of the LFR and the ELSY Project," *International Congress on Advances in Nuclear Power Plants* (ICAPP 2007), Nice, France, May 13–18, 2007, Paper 7585.
3. L. Cinotti, G. Locatelli, E. Malambu, H. Aït Abderrahim, S. Monti, G. Benamati, C. Artioli, and H. Wider, "The ELSY Project for a Sustainable Deployment of Nuclear Energy," *10th ISTC SAC Seminar, Nizhny Novgorod*, Russia, September 24–27, 2007.
4. L. Cinotti, G. Locatelli, H. Aït Abderrahim, S. Monti, G. Benamati, K. Tucek, D. Struwe, A. Orden, G. Corsini, and D. Le Carpentier, "The ELSY Project," *International Conference on the Physics of Reactors "Nuclear Power: A Sustainable Resource"* (PHYSOR'08), Interlaken, Switzerland, September 14–19, 2008.
5. M. T. Farmer and J. J. Sienicki, "Analysis of Transient Coolant Void Formation During a Guillotine-Type Tube Rupture Event in the STAR-LM System Employing a Supercritical CO_2 Brayton Cycle," *12th International Conference on Nuclear Engineering* (ICONE12), Arlington, VA, April 25–29, 2004, Paper ICONE12-49227.
6. C. Fazio, A. Almazouzi, L. Soler Crespo, J. Henry, F. Roelofs, and P. Turroni, "Lead, Lead-bismuth Eutectic Technologies and Materials Studies for Transmutation Systems," *Proceedings of Global 2009*, Paris, France, September 6–11, 2009, Paper 9019.
7. G. Ilinčev, "Research Results on the Corrosion Effects of Heavy Liquid Metals Pb, Bi and Pb-Bi on Structural Materials with and without Corrosion Inhibitors," *Nuclear Engineering and Design, 217* (2002) 167–177.
8. J. Y. Lim and R. G. Ballinger, "Alloy Development for Lead-Cooled Reactor Service," *MIT-Tokyo Tech Symposium on Innovative Nuclear Energy Systems*, Massachusetts Institute of Technology, Cambridge, MA, November 2–4, 2005.
9. A. Weisenburger, A. Heinzel, G. Mueller, H. Muscher, and A. Rusanov, "T91 Cladding Tubes with and without Modified FeCrAl Coatings Exposed in LBE at Different Flow, Stress and Temperature Conditions," *IV International Workshop on Materials for HLM Cooled Reactors and Related Technologies*, Rome, Italy, May 21–23, 2007.
10. M. W. Barsoum and T. El-Raghy, "The MAX Phases: Unique Carbide and Nitride Materials," *American Scientist, 89* (July–August 2001) 334.
11. See http://www.3one2.com
12. L. A. Barnes, N. L. Dietz Rago, and L. Leibowitz, "Corrosion of Ternary Carbides by Molten Lead," *Journal of Nuclear Materials, 373* (2008) 424–428.
13. "Heavy Liquid Metal Coolants in Nuclear Technology, Proceedings Consist of Two Volumes," *Proceedings of the Conference Heavy Liquid Metal Coolants in Nuclear Technology*, State Scientific Center of the Russian Federation Institute for Physics and Power Engineering named after Academician A. I. Leipunsky, Obninsk, Editor-in-Chief B. F. Gromov, Editors A. V. Zhukov, G. I. Toshinsky, O. I. Komkova, and S. V. Boodarina, October 5–9, 1998.
14. "Russian Forum for Science and Technology Fast Neutron Reactors," Obninsk, December 8–12, 2003.
15. Heavy Liquid-Metal Coolants in Nuclear Technologies (HLMC-2008), Obninsk, September 15–19, 2008.
16. F. Niu, R. Candalino, and N. Li, "Effect of Oxygen on Fouling Behavior in Lead-Bismuth Coolant Systems," *Journal of Nuclear Materials, 366* (2007) 216–222.
17. J. Buongiorno, C. Larson, and K. R. Czerwinski, "Speciation of Polonium Released from Molten Lead Bismuth," *Radiochimica Acta, 91* (2003) 153–158.
18. E. I. Yefimov, D. V. Pankratov, and S. V. Ignatiev, "Removal and Containment of High-Level Radioactive Polonium from Liquid Lead-Bismuth Coolant," *Materials Research Society Symposium Proceedings Volume 506: Scientific Basis for Nuclear Waste Management XXI*, Davos, Editors I. G. McKinley and C. McCombie, Switzerland, September 28–October 3, 1997.
19. J. Buongiorno, E. P. Loewen, K. Czerwinski, and C. Larson, "Studies of Polonium Removal from Molten Lead-Bismuth for Lead-Alloy-Cooled Reactor Applications," *Nuclear Technology, 147* (2004) 406–417.
20. J. J. Sienicki and A. V. Moisseytsev, "SSTAR Lead-Cooled, Small Modular Fast Reactor for Deployment at Remote Sites – System Thermal Hydraulic Development," *Congress on Advances in Nuclear Power Plants* (ICAPP 2005), Seoul, May 15–19, 2005, Paper 5426.
21. W. S. Yang, M. A. Smith, S. J. Kim, A. V. Moisseytsev, J. J. Sienicki, and D. C. Wade, "Lead-Cooled, Long-Life Fast Reactor Concepts for Remote Deployment," *Congress on Advances in Nuclear Power Plants* (ICAPP 2005), Seoul, May 15–19, 2005, Paper 5102.
22. J. Sienicki, D. Wade, A. Moisseytsev, W. S. Yang, S.-J. Kim, M. Smith, G. Aliberti, R. Doctor, and D. Matonis, "STAR Performer," *Nuclear Engineering International* (July 2005) 24.

23. J. J. Sienicki, A. Moisseytsev, D. C. Wade, and A. Nikiforova, "Status of Development of the Small Secure Transportable Autonomous Reactor (SSTAR) for Worldwide Sustainable Nuclear Energy Supply," *International Congress on Advances in Nuclear Power Plants* (ICAPP 2007), Nice, France, May 13–18, 2007, Paper 7218.
24. J. J. Sienicki, D. C. Wade, and A. Moisseytsev, "Role of Small Lead-Cooled Fast Reactors for International Deployment in Worldwide Sustainable Nuclear Energy Supply," 2007 *International Congress on Advances in Nuclear Power Plants* (ICAPP 2007), Nice, France, May 13–18, 2007, Paper 7228.
25. G. Toshinsky, "A. Leipunsky and Nuclear Power Installations with Lead-Bismuth Liquid Metal Coolant for Submarines," *Yaderna Energetika*, Centenary of A. I. Leipunsky, Supplement to No. 4, 2003, in English, ISSN 0204-3327.
26. M. Rawool-Sullivan, P. Moskowitz, and L. Shelenkova, "Technical and Proliferation-Related Aspects of the Dismantlement of Russian Alfa-Class Nuclear Submarines," *The Nonproliferation Review* (Spring 2002).
27. A. S. Pavlov, *Warships of the USSR and Russia 1945–1995*, Translated by G. Tokar, Editor (English-language Edition) N. Friedman, Naval Institute Press, Annapolis, MD, 1997.
28. *Power Reactors and Sub-Critical Blanket Systems with Lead and Lead-Bismuth as Coolant and/or Target Material, Utilization and Transmutation of Actinides and Long Lived Fission Products*, IAEA-TECDOC-1348, International Atomic Energy Agency, May 2003.
29. G. I. Toshinsky, A. V. Zrodnikov, V. I. Chitaykin, O. G. Grigoriev, U. G. Dragunov, V. S. Stepanov, N. N. Klimov, I. I. Kopytov, V. N. Krushelnitsky, and A. A. Grudakov, "Small Modular Lead-Bismuth Cooled Fast Reactor for Multi-Purpose Use: SVBR 75/100," *Annex 10 of Innovative Small and Medium Sized Reactors: Design Features, Safety Approaches and R&D Trends*, Final Report of a Technical Meeting Held in Vienna, June 7–11, 2004, IAEA-TECDOC-1451, International Atomic Energy Agency, May 2005.
30. A. V. Zrodnikov, G. I. Toshinsky, O. G. Komlev, Yu. G. Dragunov, V. S. Stepanov, N. N. Klimov, I. I. Kopytov, and V. N. Krushelnitsky, "Nuclear Power Development in Market Conditions with Use of Multi-Purpose Modular Fast Reactors SVBR 75/100," *Nuclear Engineering and Design, 236* (2006) 1490–1502.
31. "Lead-Bismuth Cooled Fast Reactor SVBR 75/100," *Annex XIX of Status of Small Reactor Designs Without On-Site Refueling*, IAEA-TECDOC-1536, pp. 511–550, International Atomic Energy Agency, January 2007.

Appendix A
Fast Reactor Data

Pavel Tsvetkov and Alan Waltar

This appendix summarizes principal design data for fast spectrum reactors being developed, designed and constructed worldwide as well as for historical fast reactor designs. The included fast reactor data are derived from several data sources:

- Appendix A of "Fast Breeder Reactors" by Alan E. Waltar and Albert B. Reynolds (Pergamon Press, 1981). The international team of fast reactor experts assisted in compiling and checking the original data:
  - France—J. Petit and A. P. Schmitt,
  - Germany—J. Gilles and R. Fröhlich,
  - Italy—F. Granito,
  - Japan—K. Aizawa, Y. Kani, and A. Watanabe,
  - Russia—M. F. Troyanov and Y. Bagdasarov,
  - United Kingdom—A. M. Judd, A. G. Edwards, M. R. Hayns, and H. Teague,
  - United States—E. L. Fuller, J. I. Sackett, J. M. Keeton, A. E. Klickman, and R. B. Rothrock.
- *Fast Reactor Database 2006*, Update, IAEA-TECDOC-1531, Vienna, Austria (2006);
- *Liquid Metal Cooled Reactors: Experience in Design and Operation,* IAEA-TECDOC-1569 (2007);
- Journal papers and national government technical reports some of which are listed in Bibliography section.

It must be emphasized that conceptual designs are being continuously revised. Included design data provide details of specific design configurations as they were published prior to this book release date. Current design configurations and construction details may differ from the provided design data compilations.

Abbreviations

There are several abbreviations commonly used in reactor data tables:

- Layout
  - L—Loop
  - NSG—no steam generator (dump heat exchanger at turbine)
  - P—Pool
  - SG—Steam Generator,

A.E. Waltar et al. (eds.), *Fast Spectrum Reactors*, DOI 10.1007/978-1-4419-9572-8,

- Core geometry
  - AA—Axial blanket above core
  - AB—Axial blanket below core
  - C—Approximately cylindrical prism
  - H—Hexagonal prism
  - Het—heterogeneous
  - R—Radial blanket
  - S—Square prism
- Core characteristics
  - N—Configuration not for breeding
- Design and performance
  - F—Fins on pin cladding
  - G—Grids
  - W—Wire-wrapped design
- Control system
  - Group 1—fine rods
  - Group 2—coarse rods
- Pump type
  - E—Electromagnetic
  - C—Centrifugal
- Safety and containment
  - C—Cylindrical with dome
  - R—Rectangular building
  - SP—Sphere
  - SQ—Square with dome
  - V—Vented
  - NV—Not vented

Bibliography

1. *Fast Reactor Database 2006*, Update, IAEA-TECDOC-1531, Vienna, Austria (2006).
2. *Liquid Metal Cooled Reactors: Experience in Design and Operation,* IAEA-TECDOC-1569, Vienna, Austria (2007).

Experimental Fast Reactors

Rapsodie, *France:*

3. G. Vendries, "RAPSODIE", *Proceedings of the 3rd United Nations International Conference on the Peaceful Uses of Atomic Energy*, Geneva, 1964, Vol. 6, United Nations, New York, NY (1965).

KNK-II *(Kompakte Natriumgekuhlte Kernreaktoranlage), Germany:*

4. "KNK-II—Operating Experience and Fuel Cycle Activities", *Proceedings of the Conference on Nuclear Power Experience*, Vol. 5, IAEA, Vienna (1983).

FBTR *(Fast Breeder Test Reactor), India:*

5. G. Srinivasan, K. V. Suresh Kumar, B. Rajendran, and P. V. Ramalingam, "The Fast Breeder Test Reactor—Design and Operating Experiences", *Nuclear Engineering and Design*, **236**, pp. 796–811 (2006).

JOYO *(Eternal Light), Japan:*

6. T. Aoyama, S. Maeda, Y. Maeda, and S. Suzuki, "Transmutation of Technetium in the Experimental Fast Reactor JOYO", *Journal of Nuclear and Radiochemical Sciences*, **6**(3), pp. 279–282 (2005).

DFR *(Dounreay Fast Reactor), UK:*

7. H. Gartwright, et al., "The Dounreay Fast Reactor—Basic Problems in Design", *Proceedings of the Second United Nations International Conference on the Peaceful Uses of Atomic Energy*, Geneva, 1958, Vol. 9, United Nations, New York, NY (1959).
8. J. L. Philips, "Operating Experience with the Dounreay Fast Reactor", *Nuclear Power*, **7** (1962).
9. *Proceedings of the Symposium on the Dounreay Fast Reactor*, December 1960, BNEC, London (1961).

BOR60 *(Bystrij Opytnyj Reactor = Fast Experimental Reactor), Russia:*

10. "List of Research Reactors, Critical and Subcritical Assemblies Supervised by Gosatomnadzor", 13 July 1992, Gosatomnadzor, Russia (1992).
11. A. P. Naumov, O. K. Nickolaenko, and N. V. Markina, et al., "Analytical Possibilities of the BOR-60 Reactor", *Journal of Radioanalytical and Nuclear Chemistry*, **167**(1), pp. 23–30 (1993).

EBR-II *(Experimental Breeder Reactor II), USA:*

12. "The Physics Design of EBR-II", *Physics of Fast and Intermediate Reactors, Proceedings Series*, Vol. III, Vienna, IAEA (1962).

FERMI, *USA:*

13. E. P. Alexanderson, "Fermi-I: New Age for Nuclear Power", LaGrange Park, IL, American Nuclear Society (1979).
14. A. A. Amorosi and J. G. Yevick, "An Appraisal of the Enrico Fermi Reactor", *Proceedings of the 2nd United Nations International Conference on the Peaceful Uses of Atomic Energy, Geneva*, 1958, Vol. 9, United Nations, New York, NY (1959).

FFTF *(Fast Flux Test Facility), USA:*

15. E. R. Astley, "Progress Report, Fast Flux Test Facility Reference Concept", BNWL-470, AEC, USA (1967).
16. W. A. Dautel, "Fast Flux Test Facility Final Safety Analysis Report", WHC-TI-75002, DOE, USA (1999).

BR10 *(Bystrij Reactor = Fast Reactor), Russia:*

17. A. L. Leipunski, et al., "Experimental Fast Reactors in the Soviet Union", *Proceedings of the Second United Nations International Conference on the Peaceful Uses of Atomic Energy*, Vol. 9, Geneva, 1958, United Nations, New York, NY (1959).

18. "Experimental Fast Reactors in the Soviet Union", *Physics of Fast and Intermediate Reactors, Proceedings Series*, Vol. III, IAEA, Vienna (1962).

CEFR *(China Experimental Fast Reactor), China:*

19. Mi Xu, "The Status of Fast Reactor Development in China", IAEA (2000). Link: http://www.iaea.org/inisnkm/nkm/aws/fnss/fulltext/29033624.pdf

Demonstration or Prototype Fast Reactors

Phénix, *France:*

20. X. Elie and J. M. Chaumont, "Operation Experience with the Phénix Prototype Fast Reactor", *Proceedings of the International Conference*, Kyoto, Vol. 1, pp. 5.1-1–5.1-10 (1991).
21. J. Guidez, P. Le Coz, L. Martin, P. Mariteau, and R. Dupraz, "Lifetime Extension of the Phénix Plant", *Nuclear Technology*, **150**(1), pp. 37–43 (2005).

PFBR *(Prototype Fast Breeder Reactor), India:*

22. S. C. Chetal, et al., "The Design of the Prototype Fast Breeder Reactor", *Nuclear Engineering and Design*, **236**, pp. 852–860 (2006).

BN350 *(Bystrie Neytrony = Fast Neutrons), Russia:*

23. S. Golan, et al., "Comparative Analysis of the Arrangement and Design Features of the BN-350 and BN-600 reactors", *International Symposium on Design, Construction and Operating Experience of Demonstration Liquid Metal Fast Breeder Reactors*, IAEA SM-225/64, Vienna (1978).

BN600 *(Bystrie Neytrony = Fast Neutrons), Russia:*

24. "Operating Experience with Beloyarsk Fast Reactor BN-600", IAEA-TECDOC-1180, IAEA, Vienna (2000).

KALIMER150 *(Korean Advanced Liquid MEtal Reactor), Korea*:

25. J. E. Cahalan and D. Hahn, "Passive Safety Optimization in Liquid Sodium-Cooled Reactors, Final Report", ANL-GenIV-095, Argon National Laboratory, USA (2005).

SVBR75/100 *(Svinetc-Vismuth Bystriy Reactor = Lead-Bismuth Fast Reactor), Russia:*

26. A. V. Zrodnikov, et al., "Multipurposed Reactor Module SVBR-75/100", *Proceedings of ICONE 8*, April 2–6, 2000, Baltimore, MD, USA, ASME (2000).

BREST300 *(Bystriy Reactor ESTestvennoy Bezopasnosti = Fast Reactor Natural Safety), Russia:*

27. V. V. Orlov, et al., "The Closed On-Site Fuel Cycle of the BREST Reactors", *Progress in Nuclear Energy*, **47**(1–4), pp. 171–177 (2005).

Commercial Size Fast Reactors

BN800 *(Bystrie Neytrony = Fast Neutrons), Russia*:

28. A. I. Kiryushin, et al., "BN-800 – Next Generation of Russian Federation Sodium Reactors", *International Conference on Innovative Technologies for Nuclear Fuel Cycle and Nuclear Power*, June 23–26, 2003, Vienna, Austria (2003).

EFR *(European Fast Reactor), EU*:

29. W. Marth, "The Story of the European Fast Reactor Cooperation", KfK 5255, Kernforschugszentrum Kalsruhe GmbH, Karlsruhe, Germany (1993).

ALMR *(Advanced Liquid Metal Reactor), US*:

30. E. L. Gluekler, "U.S. Advanced Liquid Metal Reactor (ALMR)", *Progress in Nuclear Energy*, **31** (1–2), pp. 43–61 (1997).
31. "Preapplication Safety Evaluation Report for the Power Reactor Innovative Small Module (PRISM) Liquid-Metal Reactor", NUREG-1368, Nuclear Regulatory Commission, USA (1994).

Table A.1 Experimental fast reactors

General \ Reactor	Rapsodie	KNK-II	FBTR	PEC	JOYO®	DFR	BOR60	EBR-II	FERMI	FFTF	BR10	CEFR
	France	Germany	India	Italy	Japan	UK	Russia	USA	USA	USA	Russia	China
Date critical	1967	1972	1985	Never	1977	1959	1968	1961	1963	1980	1958	2010
Date full power	1967	1978	–	Never	1977	1963	1970	1965	1970	1980	1959	–
Thermal rating (MW_{th})	40	58	40	120	140	60	55	62.5	200	400	8	65
Elect. rating (MW_e)	0	20	13	0	0	15	12	20	61	0	0	23.4
Driver fuel	MOX	MOX	PuC-UC	MOX	MOX	U-Mo	MOX	U-Zr	U-10Mo	MOX	UN	UO_2
Core geometry												
Core shape	H	H	C	C	H	H	H	H	S	H	H	C
Number of assemblies												
Inner core	64–73	7	76	78	19–25	153	80–114	127	105	28–34	86–90	81
Outer core	–	22	0	0	58–60	189	0	0	0	45–48	0	–
Radial blanket	276	5	342	0	0	300	138	366	531	0	–	0
Outside	211	49	294	199–262	223	1572	–	144	222	108	30–34	622
Inner core diameter (mm)	–	358	–	–	–	–	–	–	–	767	–	–
Outer core diameter (mm)	446	824	492	833	800	530	460	697	831	1202	206	600
Fissile zone height (mm)	320	600	320	650	500	530	450	343	775	914	400	450
Radial blanket OD (mm)	1,270	–	1,260	1,551[a]	0	1,980	770	1,562[a]	2,030	1,778[a]	–	–
Radial blanket height (mm)	1,077	980	1,000	2,419[a]	0	2,490	900	1,397[a]	1,650	1,198[a]	–	–
Up Ax. blanket Thi.(mm)	0	200	0	180	0	142	100	0	356	144[a]	0	100
Low. Ax. blanket Thi.(mm)	0	200	0	225	0	0	150	0	356	144[a]	0	250
Core characteristics												
No. enrichment zones	1	2	1	1	2	1	1	1	1	2	1	1
Inner core enrich. (%)	30 (Pu)	88–95	55	28.5	30	75	56–90	67	25.6	20.3	90	–
Outer core enrich. (%)	–	37	–	–	34	–	–	–	–	24.6	–	64.4
Fissile content, ^{235}U (kg)	79.5	312	0.7	79	110	247	95	229	484	14	113	235.4
Fissile content, ^{239}Pu (kg)	31.5	28	85.6	175	–	3	53	4.5	0	536	–	0.41
Fissile total, All Pu (kg)	–	39	124.4	310	160	3.5	58	5	0	616	–	0.414

®Mark-III core MOX = PuO-UO_2 [a]Reflector

Table A.1 (continued)

Reactor	Rapsodie	KNK-II	FBTR	PEC	JOYO®	DFR	BOR60	EBR-II	FERMI	FFTF	BR10	CEFR
	France	Germany	India	Italy	Japan	UK	Russia	USA	USA	USA	Russia	China
Core characteristics (continued)												
Core volume fractions												
Fuel	0.425	0.32	0.374	0.346	0.37	0.4	0.48	0.318	0.279	0.31	0.445	0.374
Coolant	0.396	0.43	0.354	0.376	0.37	0.4	0.29	0.487	0.472	0.39	0.287	0.376
Steel	0.136	0.21	0.238	0.248	0.23	0.2	0.23	0.195	0.249	0.26	0.218	0.19
Void or FG space	0.023	0.04	0.034	0.03	0.03	0	0	0	0	0.04	0.05	0.06
Peak flux (10^{15} n/cm^2 -s)	3.2	1.9	3.4	4	5.7	2.5	3.7	2.7	4.5	7.2	0.86	3.1
Ave. flux (10^{15} n/cm^2 -s)	2.3	1.3	2.5	2.6	3.5	1.9	3	1.6	2.6	4.2	0.63	2.1
Peak lin. power (kW/m)	43	45	35	36.5	42	37	54	34.8	28	40.1	44	40
Ave. lin. power (kW/m)	31	24	27	24.5	–	25	40	23	17	27.6	32	26.1
Max blanket power (kW/m)	–	5	–	2.1	–	–	–	4.9	14	–	–	–
Max. power density (kW/l)[d]	3,060	1,280[a]	2,344	1,384	2,500	1,250	2,300	2,704	2,774	656	2,182	1,867
Av. power density (kW/l)[d]	2,210	985[a]	1,806	930	1,600	900	1,900	1,610	1,642	387	1,588	1,132
Breeding gain (core)	N	N	N	N	N	N	N	N	–	N	N	N
Breeding gain (total)	N	N	N	N	N		N	N	0.16	N	N	N
Reactivity coefficients												
Temp. (pcm/°C)[b]	–4.5	–5/–4.7	–4.8/–4.5	–3.5/–3.3	–3.1	–5.4	–4	–3.6	–0.39	–1.08	–2.2	–4.57
Power (pcm/MW_{th})[b]	–6	–8/–7.9	–19/–35	–2.5/–4.3	–4.2	–6.7	–6.5	–4.2	–0.2	–0.4	–8.2	–6.54
Max coolant void ($)	–	–2.4/–3.2	–20.57	+0.022	–41	0	–8	–		+~1.5	–6.1	–4.99
Doppler (pcm/°C)	0	–0.003	0	–0.003	–0.0017	0.0002	0.0015	–	–0.00026	–0.005	0	–0.0025
Doppler (void) (pcm/°C)	0	–	0	–0.002	–0.00095	–	–	–0.003	–	–0.003	0	–0.0021
Λ(μs)	0.1/0.24[c]	0.38	–	–	0.23/0.3	0.14	–	0.1	0.14	0.5	–	–
β_{eff}	0.005	0.006	–	0.0036	.005/.004	0.007	–	0.0068	0.007	0.0032	–	–

® Mark-III core [a] Test Loop [b] Free fuel/Bonded fuel [c] Rapsodie/Rapsodie fortissimo [d] kW/l = kW/litre of fuel volume within cladding

Table A.1 (continued)

Reactor / Fuel system	Rapsodie	KNK-II	FBTR	PEC	JOYO®	DFR	BOR60	EBR-II	FERMI	FFTF	BR10	CEFR
	France	Germany	India	Italy	Japan	UK	Russia	USA	USA	USA	Russia	China
No. pins/assembly, core	61	169[b]/211[c]	61	91	127	1	37	91	140	217	7	61
Fuel pin OD (mm)	5.1	6[b]/8.2[c]	5.1	6.7	5.5	20	6	4.42	4.01	5.84	8.4	6
Fuel pin length (mm)	320	1540	531.5	1,935	1,533	1,228	1,100	343	833	2380	615	1,622
Fuel pellet form	Solid	Annul.	–	Annul.	Solid	Annul.	Annul.	Fissium	Fissium	Solid	–	–
Fuel pellet density (%TD)	92.0	86.5	86	95.0	94.0	100	75/84	93	100	90.4	90	96.5
Fuel smear density (%TD)	88	80	78	87.6	87.0	95	–	75	100	85.5	82	77.6
Cladding thickness (mm)	0.37	0.38	0.37	0.45	0.35	2.3	0.3	0.305	0.127	0.38	0.4	0.3
Cladding material	316	1.4970	316[d]	316[d]	316[d]	Nb	Cr15Ni15	316	Zr	316[d]	[g]	[h]
Duct material	SS	SS	316L(CW)	316	316[d]	–	SS	SS	SS	SS	[g]	[i]
Duct thickness (mm)	1.0	2.6	–	2.4	1.9	–	11.0	1.0	2.4	3	–	–
Duct flat-to-flat (mm)	49.8	108	49.8	85.5	78.5	–	44	58.17	67.2	116	26.1	59
Fuel pin spacers	W	G	W	W	W	F	W	–	G	W	W	W
Assembly pitch (mm)	50.8	129	50.8	81.5	81.5	23.4	45	58.93	68.4	120	27	61
Fuel pin pitch (mm)	7.1/5.9[a]	10.1[b]/7.9[c]	–	7.9	7.6/6.5[e]	–	6.7	5.7	5.1	7.3	–	–
Fuel pin pitch/diameter	1.06/1.16[a]	1.23[b]/1.32[c]	–	0.18	1.21/1.18[e]	–	1.12	1.29	1.26	1.24	–	–
Fission gas location	Top/Top& bottom	–	–	Bottom	Top	Top	Bottom	Top	none	Top	–	–
FG volume/pin (cm^3)	2.5	16	1.9	15.6	10	0	7.3	2.4	0	19.0	4.8	10.3
Max FG pressure (MPa)	12.8	2.6	6.0	5.0	7.3	–	10.0	12.4	0	4.28	5.0	2.8
Reflector/Blankets												
No. blanket pins/assembly	7	121	7	–	0	1	–	19	25	–	–	–
Blanket pin OD (mm)	16.5	9.15	16.5	–	0	34[f]	–	12.5	11.3	–	–	–
Blanket pin length (mm)	1,079	1,363	1,079	–	0	2,490[f]	–	1,397	1,650	–	–	–
Blanket pellet den. (%TD)	96	94	95	–	–	–	–	93	100	–	–	–
Blanket smear den. (%TD)	91	89	90	–	–	–	–	90	98	–	–	–
Blanket clad thick. (mm)	0.5	0.5	0.5	–	0	0.9[f]	–	0.457	0.25	–	–	–
Blanket clad material	016	1.49814[a]	316	–	0	18/8/1[i]	Cr16Ni15	304 L	304	316[c]	–	–

®Mark-III core [a]Rapsodie/Rapsodie fortissimo [b]Driver [c]Test zone [d]20% Cold worked SS [e]Mark I and Mark II cores [f]Reflector [g]Crl6Nil5Mo3Nb [h]06Cr16Ni15Mo2Mn2TiVB [i]08Cr16Ni11Mo3Til

Table A.1 (continued)

Reactor	Rapsodie France	KNK-II Germany	FBTR India	PEC Italy	JOYO® Japan	DFR UK	BOR60 Russia	EBR-II USA	FERMI USA	FFTF USA	BR10 Russia	CEFR China
Reflector/Blankets (continued)												
Blanket pin spacers	W	W	W	–	–	F	–	–	W	–		W
FG volume/pin (cm^3)	2	16	–	–	0	0	–	12.8	–	–		7
Refueling												
Mean run length (days)	80	–	45–60	60	60	55	100	49	14	107	100	73
Mean refueling (days)	10	–	7	15	16	–	45	7	–	25/46	12	14
Max. BU att. (MWd/tHM)	102,000	172,000	–	–	143,900	3,000	176,000	80,000	4,000	155,000[a]	62,300	–
Ave. BU att. (MWd/tHM)	–	75,000	–	–	68,500	2,500	73,000	66,000	3,000	60,000	45,500	–
Goal BU (MWd/tHM)	–	–	50,000	65,000	200,000	–	260,000	–	10,000	45,000	–	100,000
Storage positions	40	–	–	76	20	–	none	75	35	57	–	–
Control system												
No. safety rods	6	8	6	11	6	15	3	2	8	9	2 +2	11
OD safety rods (mm)	45.0	10.3	–	17.7	–	23.0	12.0	–	15.9	12.0	–	14.9
Safety rod Material	BC90	–	BC90	–	–	B80	BC80	Fuel	BC	B20	–	BC91
No. pins/safety rod	1	55	1	7	–	1	7	–	6	–	–	7
No. group 1 control rods	6	–	6	11	–	0	2	–	2	3	2 (Ni)	2
No. group 2 control rods	5	–	0	–	0	6	2	–	–	6	1 MRR	3
No. pins/group 1 control	–	–	–	7	7	–	4	–	19	61	–	7
No. pins/group 2 control	–	55	–	7	–	10	7	–	–	61	–	7
OD group 1 rods (mm)	–	10.3	–	17.7	18.5	–	12.0	–	7.9	12.0	–	14.9
OD group 2 rods (mm)	–	–	–	17.7	None	20.0	12.0	–	–	–	–	14.9
Group 1, Group 2 materials	BC90	BC90	BC90	BC90	BC90	Fuel	BC80	Fuel	BC	B20	–	BC20, 91
Rector vessel												
Inside diameter (mm)	2,350	1,870	2,350	3,080	3,600	3,200	1,400	7,920	4,800	6,170	338	7,960
Thickness (mm)	15	16	15	30	25	12	20	19	50	65	7	25/50
Height (mm)	–	10,150	–	10,300	10,000	6,300	6,200	3,960	11,000	13,130	4,500	12,195
Material	316	1.6770	316	316	304	18/8/1	Cr18Ni9	304	304	304	Cr18 Ni9	316

®Mark-III core [a]230,000 MWd/tHM was achieved with an advanced MOX annular fuel design, HT-9 clad, 169 pin/assembly

Table A.1 (continued)

Reactor / Containment	Rapsodie	KNK-II	FBTR	PEC	JOYO®	DFR	BOR60	EBR-II	FERMI	FFTF	BR10	CEFR
	France	Germany	India	Italy	Japan	UK	Russia	USA	USA	USA	Russia	China
Rad. shield (inside vessel)	SS	Gray Fe	SS	Ni+B_4C	SS	SS+BG	SS	Graphite	SS	SS	–	SS+B_4C
Rad. shield (out of vessel)	Con.	HD Con.	Con.	HD Con.	G+Con.	Con.	Fe+Con.	BG+Con.	Con.	Con+B_4C	Fe+Con.	Con.
Ax. shield (inside vessel)							SS	–	–	–	SS+B_4C	SS
Ax. shield (above vessel)							SS+CS+G	–	–	–	P+B_4C+SS	SS+Con.
Containment geometry	C	C	C	C	C	SP	R	C	C	C	R	SQ
Containment material	SS	SS	Con.	CS	CS	SS	Con.	CS	CS	CS	Con.	Con.+SS
Containment volume (m^3)	15,000	5,000	15,000	18,000	18,600	11,500	–	14,000	7,900	64,100	–	17,000
Design pressure (MPa)	0.235	0.25	0.025	0.15	0.15	0.125	–	0.166	0.32	0.069	–	0.1
Design leak rate (%/day)	10[a]	1.0		0.5	3.0[b]	0.075		0.2	0.1	0.1		
Heat transport												
Primary												
Coolant, cover gas	Na, Ar/He	Na, Ar	Na, Ar	Na, Ar	Na, Ar	NaK,Ar	Na,Ar	Na.Ar	Na,Ar	Na,Ar	Na,Ar	Na,Ar
Type/No, loops	L/2	L/2	L/2	L/2	L/2	L/24	L/2	P/2	L/3	L/3	L/2	P/2
Pump type/position	C/ Cold	C/Hot	–	C/Cold	C/Cold	E/Cold	C/Cold	C/Cold	C/Cold	C/Hot	E	
Coolant inventory (t)	36.8	27	26.7	118	126	51	22	286	160	406	1.7	260
Flow rate/loop (kg/s)	115	140	115	315	380	19	135	250	395	727	24	200
Total flow rate (Kg/s)	230	280	230	630	750	450	270	500	1,185	2,180	48	400
Max/ave. coolant vel.(m/s)	5.5	–	6.2/5.4	6.1/5.0	6.1/5.3	6.0/6.0	11/8	8/~0.5	–4.8	7.4/6.8	4.0/–	4.7/3.7
Core Pressure drop (MPa)	–	–	0.3	–	0.33	–	0.35	–	–	–	0.1	0.28
Inlet reactor temp. (°C)	400	360	380	400–450	350	230	330	371	288	360	350	360
Outlet reactor temp (°C)	510	525	544	550	500	350	545	473	427	565	470	516
Hot/cold leg piping mat.	316	–	316	316	304	18/8/1	Cr18 Ni9	304	304	316/304	Cr18 Ni9	–/304
OD hot leg piping (mm)	302	200	300	609	510	101	325	356	760	710	127	–
Thick. hot let pipe (mm)	4	–	4	9.5	9.5	3.5	12	6.35	9.5	10	8	–
OD cold leg piping (mm)	300	200	300	355.6	450/300	101	325/219	324	760	405	127	127
Thick. cold leg pipe (mm)	–	–	4	8	7.9/6.5	–	12/10	10.3	9.5	10	8	8
Max. fuel/clad temp. (°C)	2,180/635	2,055/600	–/600	2,340/700	2,500/620–675	650/400	–/710	688/580	602/566	2,250/680	–/565	–/670

®Mark-III core [a]At 0.025 MPa [b]At 0.13 MPa, 360 C BG = Borated graphite; Con. = Concrete; HD Con. = High density concrete; CS = Carbon Steel; SS = Stainless steel; P = Parafin

Table A.1 (continued)

Secondary \ Reactor	Rapsodie	KNK-II	FBTR	PEC	JOYO®	DFR	BOR60	EBR-II	FERMI	FFTF	BR10	CEFR
	France	Germany	India	Italy	Japan	UK	Russia	USA	USA	USA	Russia	China
Coolant/no. loops	Na/2	Na/2	Na/2	Na/2	Na/2	NaK/12	Na/2	Na/1	Na/3	Na/3	Na/2	Na/2
Pump type	C/Hot	C/Cold	–	C/Cold	C/Cold	E/Cold	C/Cold	E/Cold	C/Cold	C/Cold	–	–
Coolant inventory (t)	20	50	44	67	73	63	20	41	102	199	5	48.2
Flow rate loop (kg/s)	102	130	69	312	330	38	110	297	400	727	25	137
Total flow rate (kg/s)	204	260	138	624	670	450	220	297	1,200	2,180	50	274
Hot leg temp. (°C)	485	504	510	495	470	335	480	467	408	459	380	495
Cold leg temp. (°C)	360	322	284	350	300	195	210	270	269	316	270	310
Hot leg piping	316	1.6770	316 LN	316	$2\frac{1}{4}$ Cf-1Mo	18/8/1	Cr18 Ni9	304[a]	$2\frac{1}{4}$ Cr-1Mo	316	Cr18 Ni9	304 H
Cold leg piping	316	1.6770	316 LN	316	$2\frac{1}{4}$ Cr- 1Mo	18/8/1	Cr18 Ni9	304[a]	$2\frac{1}{4}$ Cr-1Mo	304	Cr18 Ni9	304 L
OD hot leg pipe (mm)	208	200	200	355.6	320	152	325/219	305	305	405	127	219
Thick. hot leg pipe (mm)	4	–	8	8	10.3	3.5	12/10	6.35	9.5	10	8	10
OD cold leg pipe (mm)	200	200	200	355.6	300/250/200	152	325/219/108	324	460/305	405	127	325
Thick, cold leg pipe (mm)	–	–	8	8	10.3/9.3/8.2	–	8/6	6.35	9.5	10	8	12
Num. inter heat exch.	2	2		2	2	24	4	1	3	3		
Steam generator												
Number	–	2	–	–	–	12		8	3	–	–	–
Outlet steam temp. (°C)	NSG	200	200	NSG	NSG	200	200	301	171	NSG	NSG	190
Inlet steam temp. (°C)	NSG	485	490	NSG	NSG	270	430	433	407	NSG	NSG	470
Steam pressure (MPa)	NSG	7.85	16.7	NSG	NSG	–	8	8.79	4.1	NSG	NSG	13
Turbine												
Number	0	1	1	0	0	–	1	1	1	0	0	1
Inlet pressure (MPa)	–	7.85	–	–	–	1.0	8.82	8.62	3.97		–	–
Inlet temp. (°C)	–	485	–	–	–	270	460	435	404		–	–
Num. dump heat exch.	2		–	–	4	12	–	1	0	12	–	–

®Mark-III core [a]21/4 Cr-1Mo used for connection to steam generator components

Table A.2 Demonstration or prototype fast reactors

Reactor / General	Phénix	SNR300	PFBR	MONJU	PFR	CRBRP	BN350	BN600	KALIMER150	SVBR75/100	BREST300
	France	Germany	India	Japan	UK	US	Kazakhstan	Russia	Korea	Russia	Russia
Date critical	1973	–	–	1994	1974	–	1972	1980	–	–	–
Date full power	1974	–	–	1996	1977	–	1973	1981	–	–	–
Thermal rating (MW_{th})	563[a]	762	1,250	714	650	975	750	1,470	392.2	265	700
Elect, rating (MW_e)	255	327	500	280	250	380	130[c]	600	162.2	80	300
Driver fuel			PuO_2–UO_2				UO_2	UO_2[d]	U-TRU-Zr	UO_2	PuN-UN-MA
Core geometry											
Core Shape	H	C	H	H	H	H, Het	C	C	Het	C	C
Number of assemblies											
Inner core	55	109	85	108	28	156/82[b]	61/48[1]	136/94[1]	54	55	45
Outer core	48	90	96	90	44	0	113	139	–	none	64/36[2]
Radial blanket	90	96	120	172	41	126	350	362	72	none	64/36[3]
Outside	1,317	186	419	324	94	312	107	190	241	none	148
Inner core Diameter (mm)	960	1,353	1,353	1,368	933	–	880/1,100[1]	1,270/1,650[1]	1,559	1,645	1,280
Outer core dia. (mm)	1,390	1,780	1,970	1,800	1,470	2,020	1,580	2,050	–	–	1,990/2,296[2]
Fissile zone height (mm)	850	950	1,000	930	910	914	1,000	1,030	1,000	900	1,100
Radial blanket OD (mm)	1,880	2,130	2,508	2,400	1,840	2,850	2,490	3,000	1,931	2,090	–
Rad. blanket height (mm)	1,668	1,750	1,600	1,600	1,460	1625	1,580	1,580	1,000	–	–
Up Ax. blanket thi (mm)	0	400	300	300	102	356	300	300	–	–	–
Low. Ax. blanket thi (mm)	300	400	300	350	450	356	400, 350	–	–	–	–
Core characteristics											
No. enrichment zones	2	2	2	2	2	1	3	3	1	4	1
Inner core enrich. (%)	18	25 (Pu)	20.7	16	22.0	–	17 (UO_2)	17 (UO_2)	21.1	16.1	14.6 Pu+MA
Outer core enrich (%)	23	36 (Pu)	27.7	21	28.5	32.8	26 (UO_2)	26 (UO_2)	–	–	–
Intermed core enrich. (%)	–	–	–	–	–		21 (UO_2)	21 (UO_2)	–	–	–
Fissile content, ^{235}U (kg)	35	57	17.3	13.5	50	7.6	1,220 UO_2	2,020 UO_2	20.48	1,470	–
Fissile content, ^{235}Pu (kg)	717	1,058	1,361	870	760	1,468	75	110	1,090	–	–
Fissile total. All Pu (kg)	931	1,536	1,978	1,400	950	1,705	77	112	1,519.78	–	2,260

[a]Limited to 350 MWth since 1993 [b]Internal blanket [c]150 MWth desalination [d]first, partly PuO_2-UO_2 later [e]first, PuO_2-UN later (PuO_2-UO_2 backup)
[1]Inner zone/intermediate zone of the inner core [2]Inner zone/outer zone of the outer zone [3]Inner zone/outer zone of the outer zone MA = Minor actinides

Table A.2 (continued)

	Phénix	SNR300	PFBR	MONJU	PFR	CRBRP	BN350	BN600	KALIMER150	SVBR75/100	BREST300
Reactor	France	Germany	India	Japan	UK	US	Kazakhstan	Russia	Korea	Russia	Russia
Core characteristics (continued)											
Core volume fractions											
Fuel	0.37	0.295	0.297	0.335	0.35	0.325	0.380	0.375	0.376	0.55	0.30
Coolant	0 35	0.50	0.410	0.400	0.41	0.419	0.33	0.34	0.3747	0.285	0.60
Steel	0.25	0.19	0.239	0.245	0.21	0.234	0.22	0.215	0.249	0.14	0.10
Void or FG space	0.03	0.015	0.054	0.020	0.03	0.022	0.07	0.07	0	0.025	0
Peak flux (10^{15} n/cm^2 -s)	6.8	6.7	8.1	6.0	7.6	5.5	5.4	6.5	3.01	1.7	3.8
Ave. flux (10^{15} n/cm -s)	–	4.9	4.5	3.6	5.0	3.6	3.5	4.3	2.2	1.15	2.35
Peak lin. power (kW/m)	45	36	45	36	48	40.3	40	47	28.7	36	42/39/33
Ave. lin. power (kW/m)	27	23	28.7	21	27.0	26.7	24	28	20.12	24.3	–
Max blanket power (kW/m)	41	23	35	27	50	54.1	48	48	28.49	–	–
Max. power density (kW/l)	1,950	1,613	1,763	–	1,720	1,983	1,995	1,587	342.9	382	835
Av. power density (kW/l)	1,200	1,016	1,247	–	1,160	1,023	1,155	940	240.4	140	510
Breeding gain (core)	–	–	Neg.	–	–	–	–	–	–	0.04 (MOX)	–
Breeding gain (total)	0.16[a]	0.10	0.05	0.2	–0.05	0.24	0	–0.15	0.05	–0.13 (UO_2)	0.05
Reactivity coefficients											
Temp. (pcm/°C)	–2.7	–2.3	–1.8/–1.2[b]	–2.0	–3.3	–0.63	–1.9	–1.7	–	–2.2	–1.9
Power (pcm/MW_{th})	–0.5	–0.3	–0.64/–0.57[b]	–0.94	–1.7	–0.2	–0.7	–0.6	–	–3.1	–0.3
Max. coolant Void ($)	–	+2.9	+4.3	–	+2.6	+2.29	–0.6	–0.3	2.6	–2.9	–1.6
Doppler (pcm/°C)	–0.006	–0.004	–7E–3	–8E–3	–7E–3	–3F–3	–0.007	–0.007	–0.0042	–	–0.0066
Doppler (void) (pcm/°C)	–0.004	–0.003	–5E–3	–4E–3	–	–2E–2	–0.0049	–0.0044	–	–	–
$\wedge(\mu s)$	0.33	0.4	–	0.44	0.49	0.41	–	–	–	–	–
β_{eff}	0.0032	0.0035	–	0.0036	0.0034	0.0034	0.0035	–	–	–	–

[a]Total breeding gain of 1.16 was experimentally defined a time of reprocessing [b]Fresh/equilibrium

Table A.2 (continued)

Fuel System \ Reactor	Phénix	SNR300	PFBR	MONJU	PFR	CRBRP	BN350	BN600	KALIMER150	SVBR75/100	BREST300
	France	Germany	India	Japan	UK	US	Kazakhstan	Russia	Korea	Russia	Russia
No. pins/assembly, core	217	127	217	169	325/265/169	217	127	127	271	220	156/160
Fuel pin OD (mm)	6.6	7.6	6.6	6.5	5.8/6.6/8.5	5.84	6.9	6.9	7.4	12	9 4/9 8/10.5
Fuel pin length (mm)	850	2,475	2,580	2,800	2,250	2,906	2,445	2,445	3,708.1	1,638	2,250
Fuel-pellet form	Solid	Solid	–	Solid	–	Solid	Annul.	Annul.	–	–	–
Fuel pellet density (%TD)	95.0	86.5	94.6	85.0	100	91.3	100	100	15.8	100	95.0
Fuel smear density (%TD)	85.0	80.0	90.0	–	100	83.2	100	100	75.0	100	80.0
Cladding thickness (mm)	0.45	0.38	0.45	0.47	0.38	0.38	0.4	0.4	0.55	0.4	0.5
Cladding material	[a]	[b]	[c]	316	316	316 (20%CW)	Cr16Ni15Mo2+ MnTiSi (CW)		HT9	EP-823 (12% Cr)	
Duct material	316	Aust. SS	[c]	316	PE16/FV4-18	–	Cr13Mn Nb		HT9	–	[d]
Duct thickness (mm)	–	2.6	–	3.0	2.9	3.0	2.0		–	–	–
Duct flat-to-flat (mm)	124	110	131.3	105	142	116	96	96	157	225.45	166.5
Fuel pin spacers	W	G	W	W	G	W	W	W	W	G	G
Assembly pitch (mm)	127	115	135	116	145.3	121	98	98.4	161	223.88	167.7
Fuel pin pitch (mm)	7.7	7.9	–	7.9	7.4	7.3	7.0	8.0	–	–	–
Fuel pin pitch/diameter	1.18	1.32	–	1.22	1.26	1.26	1.15	1.17	–	–	–
Fission gas location	Top & Bottom	Bottom	–	Top	Bottom	Top	Top	Bottom	–	–	–
FG volume/pin (cm^3)	13	25	25.7	–	14	21.1	20.6	20.6	–	44.3	47/52/60
Max FG pressure (MPa)	–	3.1	5.8	6.9	5.6	4.93	4.4	5.0	7.6	3.0	3.0
Reflector blankets											
No. blanket pins/assembly	61	61	61	61	85	61	37	37	127	none	none
Blanket pin OD (mm)	13.4	11.6	14.33	12.0	13.5	12.85	14.0	14.0	12.0	none	none
Blanket pin length (mm)	1,668	2,475	2,370	2,800	1,900	2,959	1,980	1,980	3,708	none	none
Blanket pellet den. (%TD)	–	95.0	94	93.0	100	95.6	93.0	93.0	100	none	none
Blanket smear den. (%TD)	–	91.0	90.7	90.0	100	93.2	90.0	90.0	85.0	none	none
Blanketed clad thick. (mm)	0.45	0.55	0.6	0.5	1.0	0.38	0.4	0.4	0.55	none	none
Blanket clad material	–	1.4970	[c]	316	316	316 (20%CW)	Cr16Ni15Mo2+ MnTiSi (CW)		HT9	–	–

[a]Cr17Ni13Mo2.5Mn1.5TiSi [b]X10CrNiMoTiB15 [c]15Cr 15Ni MoTi (CW) [d]Cr12 Ni06Mo0.9

Table A.2 (continued)

	Phénix	SNR300	PFBR	MONJU	PFR	CRBRP	BN350	BN600	KALIMER150	SVBR75/100	BREST300
Reactor	France	Germany	India	Japan	UK	US	Kazakhstan	Russia	Korea	Russia	Russia
Reflector/blankets (continued)											
Blanket pin spacers	W	W	W	W	G	w	W	W	W	–	none
FG volume/pin (cm^3)	12	89	93.4	–	34	133	46	46	–	–	–
Refueling											
Mean run length (days)	90	588[b]	240	148	90	275	105	160	547	2,200	300
Mean refueling (days)	7	–	22	30	21	90	10	15	–	60	25
Max. BU Att. (MWd/tHM)	150,000[a]	–	–	–	200,000	–	97,000	97,000	–	–	–
Ave. BU Att. (MWd/tHM)	100,000	–	–	–	150,000	–	58,000	60,000	–	–	–
Goal BU (MWd/tHM)	170,000	86,000	113,000	94,000	250,000	74,200	120,000	120,000	120,670	106,700	91,700
Storage positions	41	–	–	–	20	–	41	124	–	–	–
Control system											
No. safety rods	6	12	3	6	5	15	3	6	1USS	6	8
OD safety rods (mm)	28.0	15.5	21.4	17.0	22.0	–	23.0	23.0	–	40.0	20.5
Safety rod material	BC48	BC47	BC65	BC90	BC40	BC92	BC80	BC80	BC	BC50	BC20
No. pins/safety rod	7	19	19	19	19	–	7	7	–	1	30
No. group 1 control rods	–	1	9	3	0	9	2	2	–	2	12
No. group 2 control rods	–	8	–	10	5	6	7	19	–	29	8
No. rapid shutdown rods	–	–	–	–	10	–	5	8	–	13	–
Additional shutdown rods	–	–	–	–	–	–	–	–	6 GEM	–	45HSR+ 12GEM
No. pins/group 1 control	–	19	19	19	–	37	7	31	61	7	30
No. pins/group 2 control	–	19	–	19	19	31	85	8	–	7	–
OD group 1 rods (mm)	–	15.5	22.4	17.0	–	15.3	9.5	9.5	–	12.0	20.5
OD group 2 rods (mm)	–	15.5	–	17.0	22.0	14.0	6.9	23.0	–	12.0	–
Group 1, Group 2 materials	BC48	BC47	BC65	BC39	BC20	BC92	BC60, UO_2 enr./ UO_2 depl.	BC20	BC	BC50	Er_2O_3

[a]These levels were reached with 8 cores of fuel which was 166,000 fuel pins GEM = Gas expansion module; HSR = Hydraulically-suspended rod [b]588 days or 441 equivalent full power days

Table A.2 (continued)

Reactor	Phénix	SNR300	PFBR	MONJU	PFR	CRBRP	BN350	BN600	KALIMER150	SVBR75/100	BREST300
	France	Germany	India	Japan	UK	US	Kazakhstan	Russia	Korea	Russia	Russia
Reactor vessel											
Inside-diameter (mm)	11,820	6,700	12,850	7,100	12,200	6,170	6,000	12,860	6,920	4,130	6,800
Thickness (mm)	15	–	25/40	50	25/50	60	50	30	50	35	40
Height (mm)	12,000	15,000	12,920	17,800	15,200	17,920	11,900	12,600	18,425	7,000	14,140
Material	316	304	316	304	321	304	Cr18N19		316	Cr18Ni9	Cr16Ni10
Containment											
Rad. shield (inside vessel)	G+SS	SS	SS+B_4C	SS	G+SS	316	SS	G+SS	304+B_4C+SS	SS+B_4C+PbBi	SS
Rad. shield (out of vessel)	Con.	Con.	Con.	Con.+SS	Con.+SS	Con.	Con.+SS	Con.	Con.	H_2O+SS+Con.	Con.
Ax. shield (inside vessel)	SS+B_4C	–	SS+B_4C+G	SS	–	–	SS	SS	SS	SS+B_4C+PbBi	SS+Pb
Ax. shield (above vessel)	Con.+SS	–	Con.	Con.+SS	–	–	SS+Con.	SS+Con.+G.	–	SS+B_4C	Con.+SS
Containment geometry	R	R	R	C	R	C	R	R	C	C	R
Containment material	Con.	–	Con.	Carb. S	Con.+S	Carb. S	Con.	Con.	2.25Cr1Mo	Con	Con.+SS
Containment volume (m^3)	31,000	323,000	87,000	130,000	74,000	170,000	–	–	1,036	80,000	–
Design pressure (MPa)	0.040	0.024	0.25	0.03	0.005	0.170	–	–	0.254	0.03	–
Design leak rate (%/day)	–	3.2	–	1.0	–	–	–	–	–	–	–
Heat transport											
Primary											
Coolant, cover gas	Na, Ar	Na, Ar	Na, Ar	Na, Ar	Na, Ar	Na, Ar	Na, Ar	Na, Ar	Na, Ar	Pb-Bi	Pb
Type/No. loops	P/3	L/3	P/2	L/3	P/3	L/3	L/6	P/3	P/4	P/2	P/4
Pump type/position	C/Cold	C/Hot	–	C/Cold	C/Cold	C/Hot	C/Cold	C/Cold	–	–	–
Coolant inventory (t)	800	550	1,100	760	850	630	470	770	–	193	8,600
Flow rate/loop (kg/s)	1,000	1,180	3,540	1,420	1,030	1,747	790	2,200	536	5,880	10,400
Total flow rate (kg/s)	3,0000	3,550	7080	4,250	3,090	5,240	3,950	6,600	2,143	11,760	14,600
Max/ave coolant vel.(m/s)	12/9	–/5	8.0/7.7	6.9/5.8	9/7.3	7.3/6.7	7.4/6.5	8.0/7.5	5.1/4.2	–/2.0	1.67/–
Core pressure drop (MPa)	0.45	–	0.54	0.25	–	–	0.69	0.70	<0.6	0.4	0.155
Inlet reactor temp. (°C)	395	377	397	397	399	388	280	365	386	286	420
Outlet reactor temp (°C)	560	546	544	529	550	535	430	550	530	435	540
Hot/cold leg piping mat.	316	–	316	304	321	316	Cr18Ni9	Cr18Ni9	316	EP302/Cr18Ni9	–

Table A.2 (continued)

Reactor / Primary	Phénix	SNR300	PFBR	MONJU	PFR	CRBRP	BN350	BN600	KALIMER150	SVBR75/100	BREST300
	France	Germany	India	Japan	UK	US	Kazakhstan	Russia	Korea	Russia	Russia
OD hot leg piping (mm)	–	610	–	810	–	914	630	–	–	–	–
Thick. hot let pipe (mm)	–	–	–	11	–	13	13	–	–	–	–
OD cold leg piping (mm)	–	560	620	610	–	610	630/529	636	–	none	none
Thick. cold leg pipe (mm)	–	–	10	9.5	–	13	13/12	16	–	none	none
Max. fuel clad temp. (°C)	2,300/650	1,850/600	–/697	2,200/675	–/670	2,350/732	1,800/600	2,500/695	–	–/600	–/644
Secondary											
Coolant/No. loops	Na/3	Na/3	Na/2	Na/3	Na/3	Na/3	Na/6	Na/3	Na/2	None/–	None/–
Pump type/position											
Coolant inventory (t)	381	402	410	760	240	580	450	830	–	–	–
Flow rate/loop (kg/s)	773	1,090	2,900	1,030	975	1,612	880	2,030	902	–	–
Total flow rate (kg/s)	2,319	3,270	5,800	3,090	2,925	4,836	4,400	6,090	1,804	–	–
Hot leg temp. (°C)	550	520	525	505	540	502	415	510	511	none	none
Cold leg temp. (°C)	343	335	355	325	370	344	260	315	340	none	none
Hot/cold leg piping mate.	321/304	–	316	304	321	316/304	Cr18Ni9	Cr18Ni9	316	none	none
OD hot leg pipe (mm)	510	610	558.8	560	360	610	529	630	356	none	none
Thick. hot leg pipe (mm)	6	–	8	9.5	10	13	12	13	7.9	none	none
OD cold leg pipe (mm)	510	560	813	560	610	457	529	820	356	none	none
Thick, cold leg pipe (mm)	7	–	10	9.5	12	13	12	13	7.9	none	none
Num. inter heat exch.	6	6	–	3	6	3	6	6	–	–	–
Steam generator											
Outlet steam temp. (°C)	512	495	493	487	515	482	410	505	483.2	260	525
Inlet steam temp. (°C)	246	230	235	240	342	242	158	240	230	225	355
Steam pressure (MPa)	16.8	16.7	16.7	12.5	12.8	9.81	4.5	13.2	15.5	4.7	27
Turbine											
Number	1	1	1	1	1	1	4	3	1	1	1
Inlet pressure (MPa)	16.3	16.0	–	12.5	12.8	10.0	4.9	13.7	–	–	–
Inlet temp. (°C)	510	495	–	483	513	482	435	505	–	–	–

Table A.3 Commercial size fast reactors[a]

General \ Reactor	Super–Phénix1	Super–Phénix2	SNR2	DFBR	CDFR	BN800	BN1600	BN1800	EFR	ALMR	JSFR1500	BREST1200
	France	France	Germany	Japan	UK	Russia	Russia	Russia	EU	US	Japan	Russia
Date critical	1985	–	–	–	–	2015	–	–	–	–	–	–
Date full power	1986	–		–	–	–	–	–	–	–	–	–
Therm. rating (MW_{th})	2,990	3,600	3,420	1,600	3,800	2,100	4,200	4,000	3,600	840	3,530	2,800
Elect, rating (MW_e)	1,242	1,440	1,497	660	1,500	870	1,600	1,800	1,580	303	1,500	1,200
Coolant												
Primary/secondary	Na	Na	Na	Na	Na	Na	Na	Na	Na	Na	Na	Pb
Temp., primary (°C)	542	544	540	550	540	544	550	575	545	499	550	540[e]
Temp., second. (°C)	525	525	510	520	510	505	515	540	525	477	520	–
Steam												
Temp. (°C)	487	495	495	495	490	490	495	525	490	429	495	525
Pressure (MPa)	17.7	17.7	17.2	16.6	17.4	13.7	13.7	26.0	18.5	15.2	18.0	27.0
Primary circuit	P	P	P	L	P	P	P	P	P	P	L	P
Driver fuel			PuO_2–UO_2					PuN–UN	PuO_2–UO_2	UPuZr[d]	PuO_2–UO_2	PuN–UN–MA
Core geometry												
Core shape	H	H	C	H	C	C	C	C	C	H, Het	–	C
Radial blanket	R	R	R	R	R	R	R	–	R	R	–	–
Top axial blanket	AA	–	AA	AA	AA	–	AA	–	AA	–	–	–
Bottom axial blanket	AB	AB	AB	AB	AB	AB	AB	–	AB	–	–	–
Number of assemblies												
Inner core	193	208	252	199	193	211/156[c]	258	642	207/108[c]	84	–	148
Outer core	171	180	162	96	156	198	216	–	72	8	–	108/76[f]
Radial blanket	234	78	120	138	234	90	84		78	0	–	
Outside	1,288	270[b]	450	1,237	–	546	1,087	1,001	873	180	–	208
Equivalent diameter												
Inner core (mm)	2,600	2,900	–	2450	2,250	1,630/2092[c]	3,160	–	2,948/3,688[c]	–	–	3,350
Outer core (mm)	3,700	3,970	4,130	2,990	3000	2,561	4,450	5,167	4,051	2,164	–	4,150/4,750[f]

[a]Differences between Tables A.3 and A.1/A.2 are due to availability of data for conceptual designs vs. operating and historical systems [b]reflector
[c]inner/intermed. zones–of the inner core [d]PuO_2–UO_2 backup [e]mix. primary coolant temp. at SG inlet [f]inner/outer zones MA = Minor Actinides

Table A.3 (continued)

	Super-Phénixl	Super-Phénix2	SNR2	DFBR	CDFR	BN800	BN1600	BN1800	EFR	ALMR	JSFR1500	BREST1200
Reactor	France	France	Germany	Japan	UK	Russia	Russia	Russia	EU	US	Japan	Russia
Core geometry (continued)												
Fiss. Zone Height (mm)	1,000	1,200	1,000	1,000	1,150	880	780	800	1,000	1,070	–	1,100
Radial blanket												
OD (mm)	4,700	4,325	5,080	3,570	3,800	2,750	4,800	–	4,383	2,427	–	–
Height (mm)	1,600	1,510	1,600	1,700	1,800	1,580	1,150	–	1,000	1,473	–	–
Thickness of axial blanket												
Upper (mm)	300	0	500	350	300	–	0	–	–	203	–	–
Lower (mm)	300	300	500	350	300	350	350		250	0	–	–
Core characteristics												
Assembly pitch (mm)	179	–	185	158	147	100	188	188	188	161.4	–	231.2
Duct–flat–to–flat (mm)	173	–	180	145	141	94.5	184	189.3	183	157.1	–	230
Assem length (mm)	5,400	4,850	–	4,600	4,000	3,500	4,500	4,500	4,800	4775	–	3,850
No. enrichm. zones	2	2	–	2	2	3	2	1	3	1	–	1
Enrichment (%)												
Inner core	16	–	18 Pu	11	15.0	19.5	18.2	–	18.3	23.2	–	–
Outer core	19.7	–	23 Pu	16	20.5	24.7	21.1	14.8	26.9	–	–	13.8 (Pu + MA)
Intermed. core	–	–	–	–	–	22.1	–	–	22.4		–	–
Fissile content, ^{235}U (kg)	142	–	210	40	60	30	80	–	81	30	–	–
Fissile content ^{239}Pu (kg)	4,054	–	4,800	2,430	3,000	1,870	5,400	–	–	–	–	6,060
Fissile total, all Pu (kg)	5,780	–	8,000	4,130	3,400	2,710	7,900	12,070	8,808	2,800	–	8,560
Core volume fraction												
Fuel	0.37	0.37	0.364	0.39	0.25	0.340	0.415	0.446	0.361	0.378	–	0.26
Coolant	0.34	0.37	0.39	0.33	0.51	0.390	0.306	0.294	0.329	0.366	–	0.635
Steel	0.24	0.24	0.22	0.23	0.18	0.220	0.229	0.228	0.235	0.257	–	0.105
Void	0.05	0.02	0.026	0.05	0.06	0.05	0.05	0.032	0.075	0	–	0

MA = Minor actinides

Table A.3 (continued)

Reactor	Super-Phénix1	Super-Phénix2	SNR2	DFBR	CDFR	BN800	BN1600	BN1800	EFR	ALMR	JSFR1500	BREST1200
	France	France	Germany	Japan	UK	Russia	Russia	Russia	EU	US	Japan	Russia
Core characteristics (continued)												
Power density												
Max. (kW/l)	1,250	1,200	800	–	2,400	1,796	1,130	925	1,100	950	–	690
Average (kW/l)	785	755	500	–	1,750	1,152	670	536	670	610	–	550
Run length (days)	640	270	365	456	270	140	330	500	425	595	–	300
Refueling (days)	120	15 or 45	30	60	28	13.7–17	35	–	20	105	–	–
Burnup												
Max. (MWd/tHM)	90,000	–	150,000	110,000	–	98,000	170,000	118,000	190,000	150,000	–	–
Aver (MWd/tHM)	60,000	–	120,000	90,000	–	66,000	115,000	66,000	134,000	100,000	–	–
Peak flux (10^{15} n/cm^2 s)	6.1	5.0	5.4	–	10	8.8	5.5	–	5.3	3.3	–	3.8
Breeding gain. Total	0.18	–	0.12	0.2	0.15	–0.02	0.1	–	0.02	0.23	–	0.05
Reactivity coefficients												
Temp. (pcm/°C)	–2.75	–	–	–	–0.20	–1.7	–1.6	–	–1.1	–	–	–1.9
Power (pcm/MW$_{th}$)	–0.1	–	–	–	–0.16	–0.36	–0.1	–	–0.12	–	–	–0.3
Max. coolant void ($)	+5.9	–	–	+4.0	+5.7	~0[c]	~0[c]	~0	+6.4	+6.5	–	–1.6
Doppler, 10^{-3} (pcm/°C)												
Voided core	–7.0	–	–	–	–5.6	–4.0	–	–	–5.0	–2.6	–	–
Unvoided core	–9.0	–	–	–8.0	–8.0	–7.0	–7.0	–	–6.5	–4.4	–	–6.6
Λ (μS)	0.42	–	–	–	–	–	–	–	–	–	–	–
β_{eff}	0.004	–	0.004	–	0.003	–	–	–	–	–	–	–
Fuel system												
No. pins/assembly	271	271	271	217	325	127	331	331	331	271	–	272
Fuel pin OD (mm)	8.5	8.5	8.5	8.5	6.6	6.6	8.5	8.6	8.2	7.44	–	9.1/9.6/10
Clad. thickness (mm)	0.56	0.56	0.565	0.5	0.52	0.4	0.55	0.55	0.52	0.56	–	0.5
Clad. material	[a]	–		[b]	PE16	[d]	[d]		AIMI[f]	HT-9	–	EP823 (12Cr)
Pin length (mm)	2,700	2,690	2,900	3,100	2,500	2,000	2,410	2,300	2,645	3,842	–	–
Duct material	–	–	–	[b]	–	[a]	[e]		EM10[g]	HT-9	–	Cr13Ni06Mo0.9
Fuel pin spacers	W	W	G	W	G	W	W	W	W	W	–	–

[a]Cr 17 Ni 13 Mo 2.5 Mn 1.5 Ti Si [b]Advanced austenitic [c]Core and upper part of assemblies [d]Cr16Ni15Mo2MnTiSi(CW) [e]Cr13MnNb [f]or PE16 [g]or euralloy

Table A.3 (continued)

Reactor	Super-Phénix1 France	Super-Phénix2 France	SNR2 Germany	DFBR Japan	CDFR UK	BN800 Russia	BN1600 Russia	BN1800 Russia	EFR EU	ALMR US	JSFR1500 Japan	BREST1200 Russia
Fuel system (continued)												
Peak lin. power (kW/m)	48	48	45	41	43	48	48.7	41	52	31	–	42/40/33
Ave. lin power (kW/m)	30	30	–	25	28	31	30	24	26	19	–	–
Max. clad. temp.(°C)	620[a]	627[a]	570[a]	700[a]	670	700	675	–	636[a]	609	–	650
FG volume/pin (cm^3)	43	–	52	–	–	18	50	–	47	31.6	–	–
Fuel pellet dens. (%TD)	95.5	95.5	93.0	95.0	100	100	100	–	96.0	100	–	92.0
Fuel smear dens.(%TD)	82.6	–	87.0	83.7	100	100	100	100	82.7	75.0	–	75 0
Reflector/blankets												
No blanket pins/assem.	91	127	127	127	85	37	91	–	169	127	–	–
Fuel pin OD (mm)	15.8	13.6	15.8	11.3	13.5	14.0	17.5	–	11.5	12.0	–	–
Clad. thickness (mm)	0.57	0.57	0.6	0.4	0.5	0.4	0.5	–	0.6	0.54	–	–
Clad. material	–	–	–	[b]	PE10	[d]	[d]	–	AIM1[f]	HT-10	–	–
Pin length (mm)	1,944	2,480	2,900	3,100	2,000	1,980	2,000	–	2,645	3,842	–	–
Fuel pin spacers	W	W	W	W	W	W	W	–	W	W	–	–
Peak lin. power (kW/m)	48	48	–	–	63	48	39.6	–	41	34	–	–
FG volume/Pin (cm^3)	40	–	150	–	–	46	–	–	100	–	–	–
Pellet den.(%TD)	95.5	–	96.0	95.0	10.8[c]	10.6[c]	10.6[c]	–	96	15.7[c]	–	–
Smear den. (%TD)	91.6	–	90.0	–	9.7[c]	9.7[c]	10.0[e]	–	89	85	–	–
Max FG pressure (MPa)	4.0	–	5.0	–	–	5.0	–	–	6.2	6.7	–	–
Control system												
No. safety rods	24	27	37	30	12	12	12	18	33	–	–	–
No. group 1 rods	21	–	–	–	0	2	2	2	5+12	9	–	–
No. group 2 rods	–	–	–	–	18	16	23	17	4+12	–	–	–
No. rapid shutdown rods	21	–	–	–	–	12	37	18	33	–	–	–
Additional shutdown	3	–	–	–	–	3HSRs[e]		5HSRs[e]	–	6GEM[g]	–	–
No. pins/safety rod	31	31	55	31	19	7	–	19	37	–	–	–
No. pins/group 1 rod	31	–	61	–	–	7	–	19	55	61	–	–
No. pins/group 2 rod	–	–	–	–	19	7	–	19	55	–	–	–

[a]Best est.; no hot-spot factors [b]Advanced austenitic [c]Absolute density values in g/cm^3 [d]Cr16Ni15Mo2MnTiSi(CW) [e]HSR: Hydraulically–suspend. rods [f]or PE17 [g]GEM: Gas expansion module

Table A.3 (continued)

Reactor	Super-Phénix1	Super-Phénix2	SNR2	DFBR	CDFR	BN800	BN1600	BN1800	EFR	ALMR	JSFR1500	BREST1200
	France	France	Germany	Japan	UK	Russia	Russia	Russia	EU	US	Japan	Russia
Control system (continued)												
OD safety rods (mm)	26.7	–	–	20.0	22.0	23.0	–	31.0	22.78	–	–	–
OD group 1 rods (mm)	21.0	–	17.6	–	–	23.0	–	31.0	22.78	16.7	–	–
OD group 2 rods (mm)	–	–	–	–	22.0	23.0	–	31.0	22.78	–	–	–
Safety rod material	BC90	BC90	–	BC92	BC30	BC92	BC80	BC92	BC30,45,90	–	–	–
Gr. 1, 2 rod materials	BC90	BC90	B90	–	–	BC20	BC80	–	BC30,45,90	BC20	–	–
Reactor vessel												
Inside diam. (mm)	21,000	20,000	15,000	10,400	19,220	12,900	17,000	17,000	17,200	9,118	10,700	9,000
Thickness (mm)	25/60	20/35	–	50	25	30	25	25	35	51	30	50
Height (mm)	17,300	16,200	–	16,000	18,100	14,000	14,000	19,950	15,900	19,355	21,200	~18,600
Material	316	316	304	316FR	316	Cr18Ni9	Cr18Ni9	316	316	316 FR	Cr 16 Ni 10	
Containment												
Radial shield												
Inside vessel	SS	SS + B	SS	SS/B_4C	SS	SS,C, B	SS	SS, C[c]	SS+B_4C	304+B	SS+Zr–H	SS
Outside vessel	Concrete	Concrete		[b]	[b]	Concrete		Concrete	Concrete		[b]	Concrete
Axial shield												
Inside vessel	SS+B_4C pins	–	–	B_4C	–	SS	SS	SS	SS+B	–	SS+B_4C	SS,Pb
In/above reactor	CS	–	–	[b]	–	SS,C	SS,C	SS,CS,gr.	SS	–	[b]	Concrete,CS
Containment building												
Material	Concrete	Concrete	Concrete	[b]	[b]			Concrete		SS	[b]	Concrete,CS
Ventilation	NV	–	–	NV	–	NV	–	NV	V	NV	V	V
Containment (m^3)	170,000	–	180,000	27,000	40,200	–	–	–	136,000	–	20,000	–
Pressure (MPa)	0.004	–	[a]	0.05	0.1	–	–	–	0.05	–	0.18	–

[a]Structures are determined by large airplane crash considerations [b]Concrete and steel [c]SS+graphite and borated graphite SS = Stainless steel

Table A.3 (continued)

Reactor	Super-Phénix1	Super-Phénix2	SNR2	DFBR	CDFR	BN800	BN1600	BN1800	EFR	ALMR	JSFR1500	BREST1200
	France	France	Germany	Japan	UK	Russia	Russia	Russia	EU	US	Japan	Russia
Heat removal												
Primary loops	4	4	4	3	4	3	3	3	3	1	2	4
Secondary loops	4	4	8	3	4	3	6	6	6	1	2	none
Primary inventory (t)	3,200	3,300	3,300	1,700	3,000	820	2,600	2,620	2,200	700	1,333	–
Second. inventory (t)	1,500	800	1,250	570	1,600	1,100	2,700	–	1,300	30	862	–
Coolant flow rate (kg/s)												
Primary/loop	–	4,925	4,500	2,720	3,860	2,900	6,500	–	6,433	4,762	9,002	–
Secondary/loop	3,270	3,920	4,000	2,260	3,747	2,780	2,970	–	2,550	4,409	7,511	–
Max. core vel. (m/s)	7.7	–	–	–	7.0	7.3	5.7	–	7.8	5.3	4.1	<2.0
Av. core vel. (m/s)	6.1	–	–	–	6.5	6.7	5.3	–	6.7	4.7	2.9	–
Primary coolant system												
Press. drop (MPa)	0.47	–	–	0.5	–	0.68	0.45	–	0.5	0.5	0.3	–
Outlet temp. (°C)	542	544	540	550	540	544	550	575	545	498	550	540
Inlet temp. (°C)	395	397	390	395	370	354	395	410	395	358	395	420
Secondary coolant system												
Outlet temp. (°C)	525	525	510	520	510	505	515	540	525	477	520	–
Inlet temp. (°C)	345	345	340	335	335	309	345	370	340	324	335	–
Steam generator												
Outlet temp. (°C)	490	490	490	495	490	490	495	525	490	454	497	525
Inlet temp. (°C)	237	237	240	240	196	190	240	270	240	215	240	355

Appendix B
GE Pool-Type Prism Fast Reactor

Russell Stachowski

B.1 Introduction

The US DOE sponsored Advanced Liquid Metal Reactor (ALMR) program in the 1990s developed a reactor concept with design features including passive reactor shutdown, passive shutdown heat removal, and passive reactor cavity cooling [1–3]. That reactor concept was called Power Reactor Innovative Small Module, or PRISM. The PRISM reactor design is an advanced fast reactor plant consisting of three identical power blocks of 465 MWe, for a total plant electrical rating of 1395 MWe. Each power block is comprised of three reactor modules with individual thermal ratings of 471 MWth. The individual reactor modules are small, modular, pool-type, sodium-cooled fast reactors. Following the completion of the ALMR program in March of 1995, General Electric (GE) proceeded with the development of a more advanced modular fast reactor design called Super PRISM (S-PRISM) [4]. S-PRISM is a design that leverages the same compact modular pool-type reactor concept, sized to enable factory manufacturing and modular construction.

B.2 Overall Plant Description

The reference S-PRISM plant described here (other variants are possible) utilizes six reactor modules arranged in three identical 760 MWe (net) power blocks consisting of two reactor modules each for an overall plant electrical rating of 2,280 MWe. Plants with only one or two power blocks having ratings of 760 and 1,520 MWe, respectively, could also be deployed. A plot plan showing the overall layout of a 2,280 MWe (net) S-PRISM plant is shown in Fig. B.1. The plant is physically divided into three areas: (1) the nuclear island, which consists of one, two or three power blocks, each consisting of two 1000 MWth reactor modules, (2) the optional Fuel Cycle Facility (FCF) and (3) balance of plant areas. This division allows a separation of safety grade construction to minimize the area that requires the highest level of security. The reactor power blocks are separated by 40 m for sequential construction of the first, second and third power blocks.

Each of the 1,000 MWth reactor modules is connected to a single steam generator, which is heated by twin secondary sodium loops that are connected to the intermediate heat exchangers (IHX) in the reactor module. Figure B.2 provides a flow diagram of the power train and Table B.1 summarizes the key performance characteristics of a S-PRISM plant.

Figure B.3 shows how both Nuclear Steam Supply Systems (NSSS) in each power block have been placed on a single horizontally isolated platform to simplify and reduce the size of the reactor building. Horizontal seismic isolation is provided to enhance seismic safety margins and to allow the application of a standard nuclear island design to sites with different soil and seismic characteristics. Both complete NSSS, including the reactor systems, the intermediate heat transport systems (IHTS),

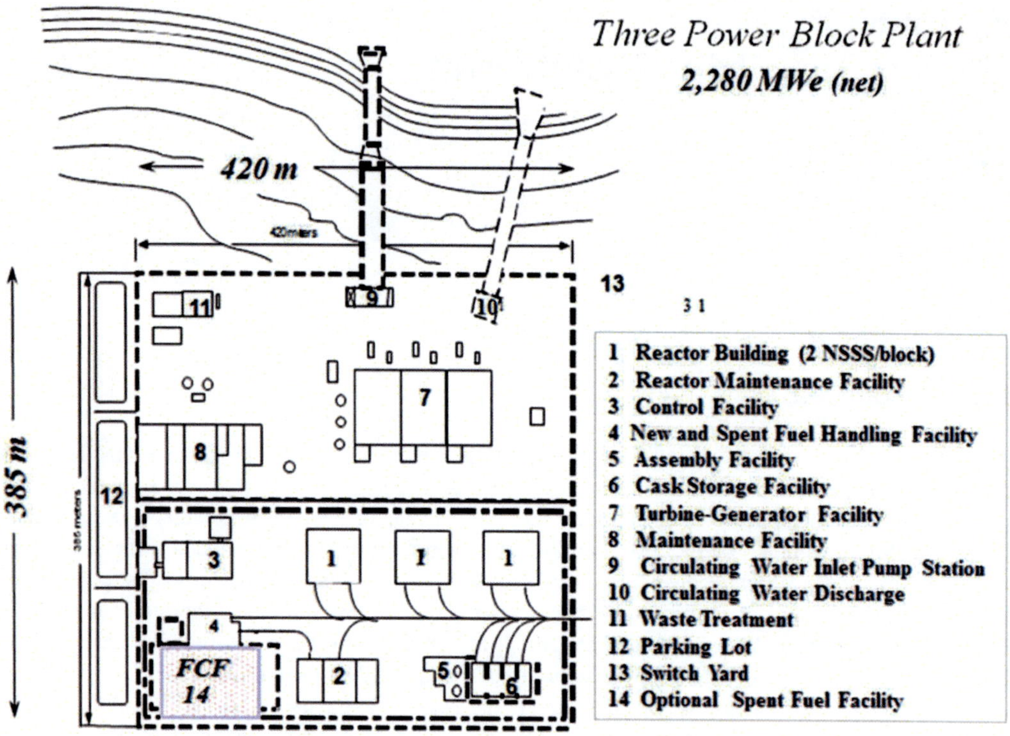

Fig. B.1 Plot plan for 2,280 MWe (net) S-PRISM plant. Ref. [5] ©ASME 2000

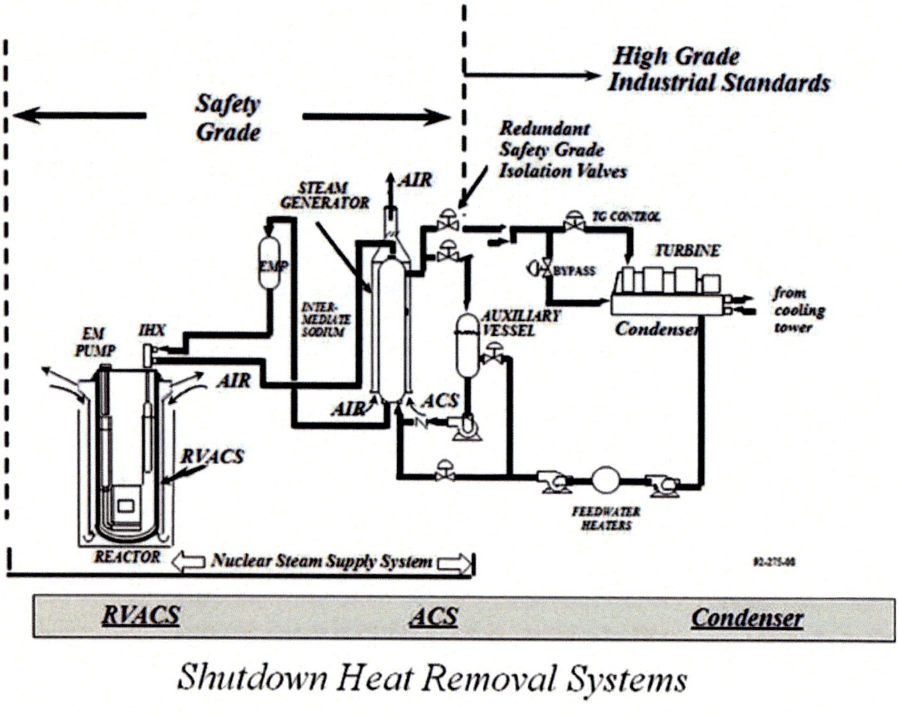

Fig. B.2 S-PRISM main power train and heat removal. Ref. [5] ©ASME 2000

Table B.1 S-PRISM plant design parameters

Overall plant	
Net electrical output, MWe	2,280
Net steam cycle efficiency, %	38.0
Number of power blocks	3
Number of reactor/plant	6
Plant availability, %	93
Power block	
Number of reactor modules	2
Gross/net electrical output, MWe	825/760
Number of steam generators	2
Steam generator type	Helical coil
Steam cycle	Superheated
Turbine type	TC-4F 3,600 rpm
Turbine throttle conditions	171 atg/468°C
Feed water temperature	215°C
Reactor module	
Core thermal power, MWth	1,000
Primary Na inlet/outlet temperature, °C	363/510
Secondary Na inlet/outlet temperature, °C	321/496

steam generator systems, as well as all safety systems, including the reactor vessel auxiliary cooling system (RVACS), reactor protection system, EM pump coastdown system, and the sodium service facility are located on a common horizontally isolated platform, which is supported by 112 seismic isolation bearings in one power block. Figure B.4 provides a three dimensional view of the NSSS, which shows how the two IHX units in each reactor are connected to a single 1,000 MWth steam generator in a close coupled arrangement that minimizes the length of IHTS piping. Figure B.5 provides a schematic illustrating how natural circulation of air is used to remove the shutdown heat load

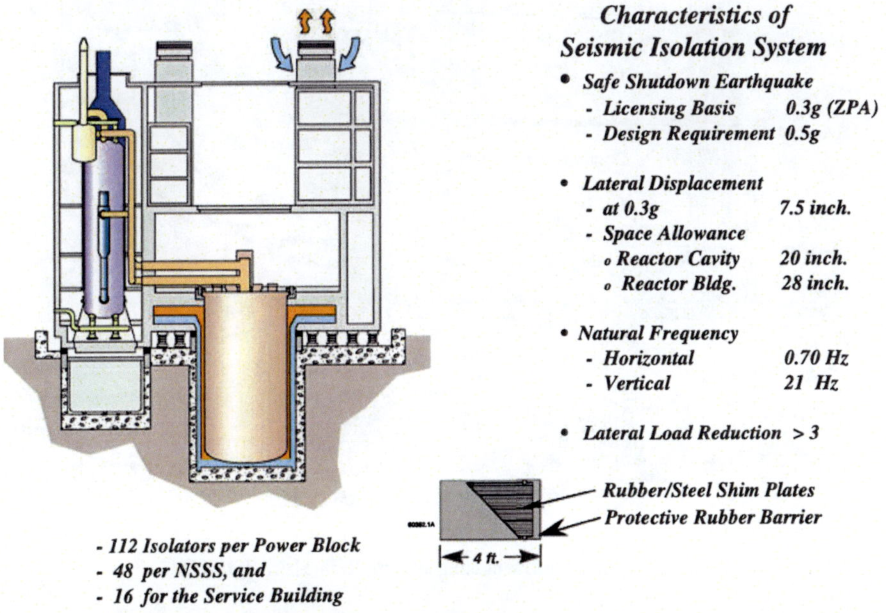

Fig. B.3 S-PRISM seismic isolation system. Ref. [5] ©ASME 2000

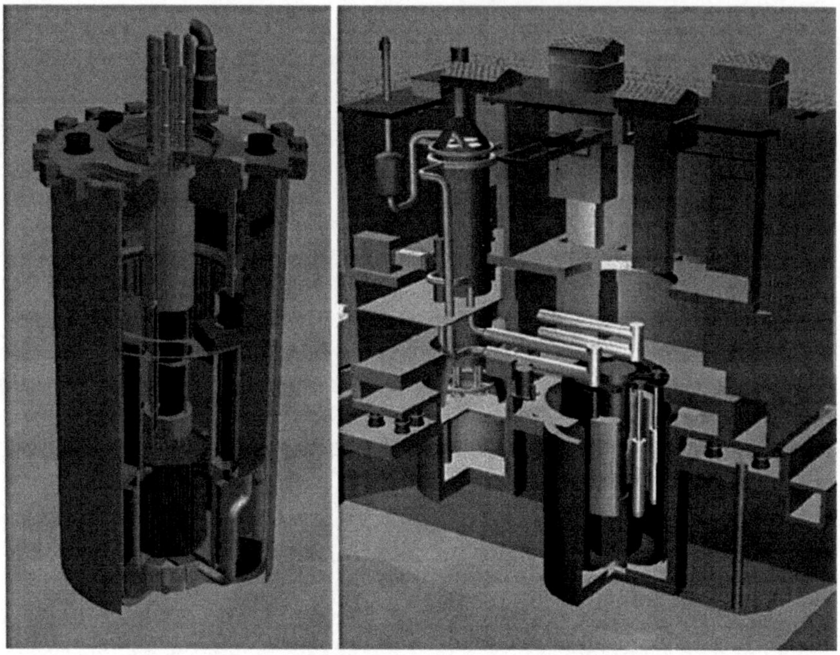

Fig. B.4 S-PRISM reactor and nuclear steam supply systems. Ref. [5] ©ASME 2000

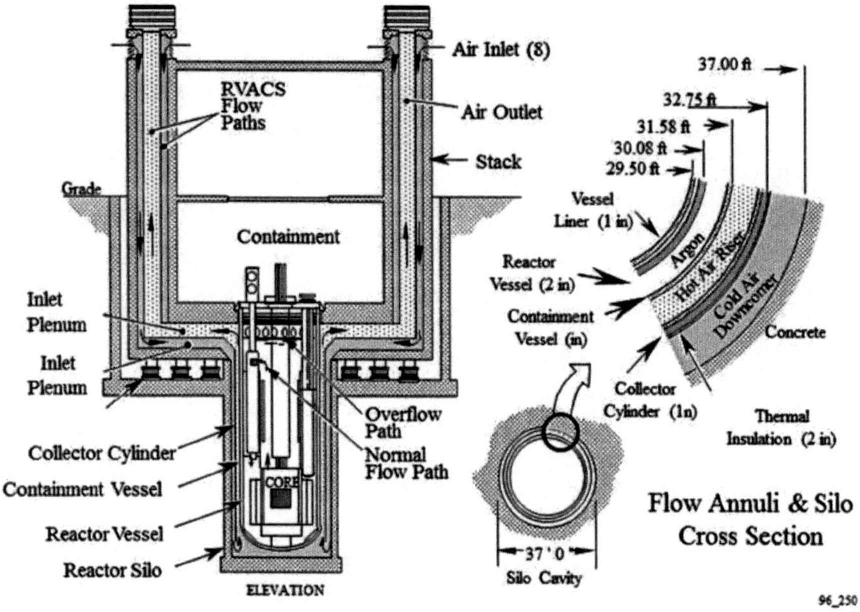

Fig. B.5 S-PRISM natural circulation containment cooling system. Ref. [5] ©ASME 2000

from the reactor system. Because passive reactor shutdown and shutdown heat removal systems are utilized, S-PRISM does not require active safety grade support systems such as diesel generators and hardened circulating water cooling systems. By eliminating the need for active safety related support systems the size of the safety-related envelope was minimized.

B.2.1 Nuclear Island

The nuclear island contains all of the facilities that require high security and/or contain radioactive materials or provide support to those facilities. The nuclear island is inside a protected area enclosed by a double security fence and a vehicle barrier. All personnel, vehicle, and railroad access to the protected area are controlled from hardened security facilities. The power blocks are located within the nuclear island. Each power block contains two completely independent NSSSs.

A wide gauge railroad system serves to transport the fuel transfer casks on cask transporters between the reactor facilities and the fuel service facility during refueling operations, to the reactor maintenance facility for servicing, or to the cask transporter garage for storage. The nuclear island warehouse, reactor maintenance facility, and radwaste facility are grouped together, and the structures, systems and access provisions are integrated. The control building includes the main control room from which all three-power blocks are operated.

B.2.2 Fuel Cycle Facility

The FCF is a design option that integrates spent light water reactor and S-PRISM fuel receiving, processing, waste conditioning and storage, and new fuel assembly fabrication operations into a single co-located facility. It is located in a dedicated area in the southwest corner of the Nuclear Island within its own security area and is provided with its own support facilities to permit independent ownership and operation. All S-PRISM fuel assemblies are fabricated and recycled in the FCF and are not shipped off-site. Core assemblies are transferred between the reactors and the facility inside fuel transfer casks via rail through the railroad gate.

B.2.3 Balance of Plant Area

The balance of plant area includes the main turbine generator facility that houses the turbine generators and their supporting auxiliary systems, including the condensers, condensate polishers, feed heaters, and cycle pumps. This facility is located in the center of the yard next to the power blocks that contain the NSSSs. The turbine building is surrounded by external condensate storage tanks, main step up transformers and auxiliary transformers, chemical storage tanks, and the turbine lube oil storage tanks. Two separated railroad spurs and roadways provide maintenance access to the turbine generator buildings.

B.3 Reactor

B.3.1 Reactor Module

Figure B.6 shows the S-PRISM reactor module. It is composed of the reactor vessel, reactor closure, containment vessel, internal structures, internal components, reactor module supports, and reactor.

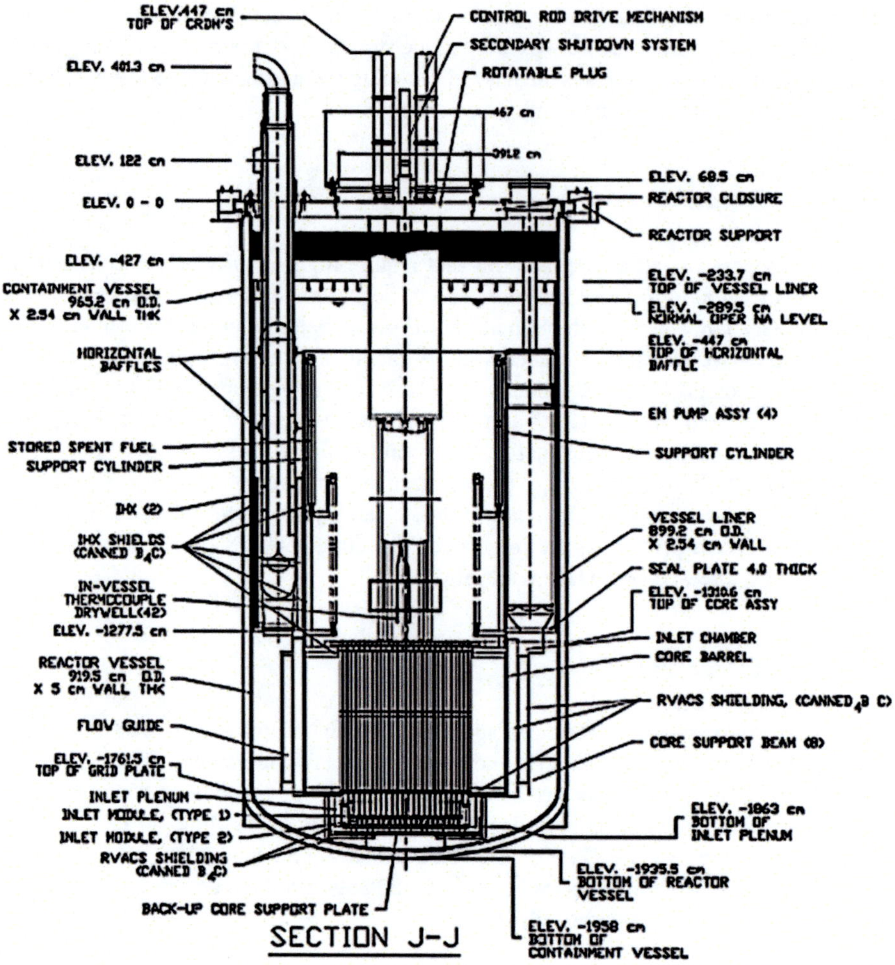

Fig. B.6 S-PRISM reactor module. Ref. [5] ©ASME 2000

The outermost structure is the guard vessel, which surrounds the reactor vessel. The guard vessel also forms the lower portion of the primary containment boundary. It is made of 2.5 cm thick 2-1/4Cr-1Mo steel. The reactor vessel is made of 5.0 cm thick 316 SS. The 20-cm gap between the reactor vessel and the containment vessel is filled with argon gas, as illustrated in the cross section in Fig. B.5. To facilitate the use of a tag gas type failed fuel leak detection and location system, helium is used as the reactor cover gas. During power operation, the reactor is hermetically sealed.

The reactor closure is a 45 cm thick 304 SS plate with a single rotatable plug and penetrations for the reactor equipment and primary sodium and cover gas service lines. There are no penetrations in the reactor vessel or the containment vessel. The reactor vessel is butt-welded to a skirt that is integral with the underside of the closure. The containment vessel is bolted to the closure and seal welded. The reactor module is supported entirely from the closure, which is held in place by eighteen hold down brackets.

The reactor core is supported by a redundant beam structure attached at the bottom and the sides of the reactor vessel. A core barrel and support cylinder, extending from the core inlet plenum to an elevation above the core, has storage racks attached to its inner surface for storage of spent fuel assemblies. A redundant structure designed to retain molten fuel is located immediately below the core inlet plenum. Two intermediate heat exchangers (IHX) and four electromagnetic (EM) primary pumps are suspended from the reactor closure. Control rod drives (CRD), in-vessel instrumentation, and an in-vessel fuel transfer machine (IVTM) are also suspended from the rotatable plug in the closure. The reactor module is approximately 20-m tall and 10 m in diameter.

B.3.2 Reactor Core and Fuel

Layouts of the reference oxide and metal-fueled cores are shown on Fig. B.7. The reactor design is fully transparent to the type of fuel selected in that either core can be used without changes to the reactor structure or refueling system and either core will meet all performance goals. Oxide cores may be utilized until metal fuel with its simpler, lower cost, and more proliferation resistant dry (pyro-metallurgical processing system) is developed. Both cores are designed to produce 1,000 MWth with an average primary coolant temperature rise of 147°C. The ferritic alloy HT9 will be used for cladding and assembly ducts in order to minimize swelling associated with high neutron fluence. At the end

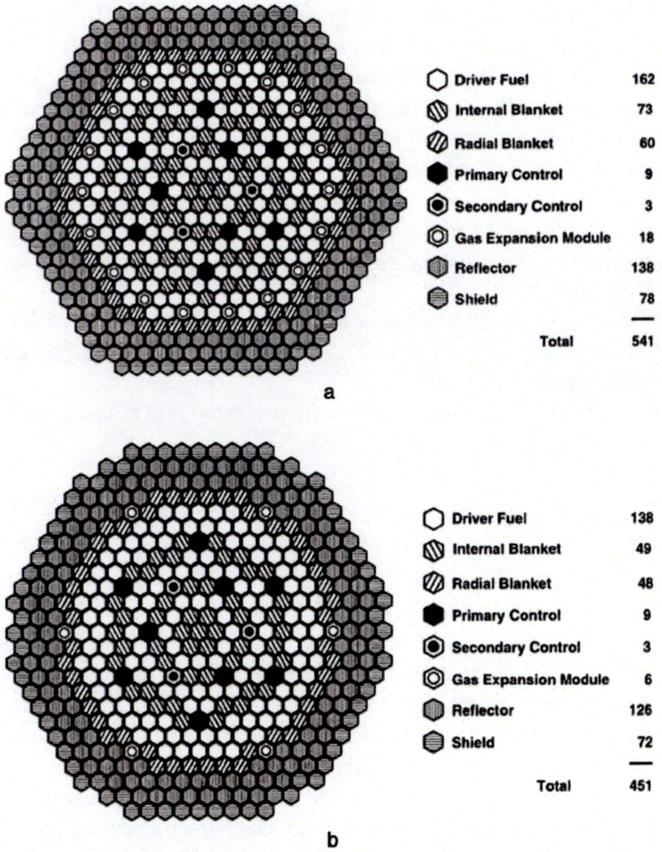

Fig. B.7. (a) Oxide core layout for S-PRISM, (b) Metal core layout for S-PRISM. Ref. [5] ©ASME 2000

of each operating cycle, one third of the driver fuel is replaced and the blankets are shuffled from the center of the core toward the perimeter of the core to reduce radial peaking.

Shield assemblies are provided to prevent excessive irradiation damage to reactor structures and components surrounding the core and to limit the activation of intermediate sodium and particles carried in the reactor vessel auxiliary cooling system (RVACS) air circuit.

Gas Expansion Modules (GEMs) are located at the periphery of the active cores. The GEMs enhance the S-PRISM's ability to safely withstand severe loss of primary flow events by providing supplementary negative reactivity insertion.

B.3.3 Reactivity Control and Safety Performance

Reactivity control for normal operations of startup, load following, and shutdown is accomplished by a system of nine identical control rods. Each control unit consists of a drive mechanism, a driveline, and a control assembly (absorber bundle and outer duct). A stepping motor, controlled by the plant control system (PCS), actuates a lead screw to insert and withdraw the absorber for normal operation.

The nine control rods have scram diversity and shutdown redundancy. Each control rod unit provides two diverse means of scramming the absorber bundle (release and drive-in). A Class 1E, electronically positioned mechanical rod stop system (RSS) prevents unprotected control rod withdrawal and excessive reactivity insertion.

A redundant and diverse secondary shutdown system is also provided. It is composed of three rods. In the event that the primary control rods fail to operate to terminate an event, the secondary system will be scrammed by a separate independent Reactor Protection System (RPS). A magnetic curie point latch will automatically release the secondary rods during under cooling or over power event in the unlikely event that both scram systems fail to act as intended.

In the event that both shutdown systems fail, the inherent negative reactivity feedback response of the core to temperature increases will bring the core to a safe, stable, zero fission power state at an elevated temperature. S-PRISM has a unique capability for accommodating severe, but extremely unlikely ATWS accidents with very benign consequences. These events include:

(1) Inadvertent withdrawal of all control rods without scram (unprotected transient overpower).
(2) Loss of primary pump power and loss of all cooling by the IHTS without scram (unprotected loss of flow/loss of cooling).
(3) Loss of all cooling by the IHTS without scram (unprotected loss of cooling).

B.3.4 Reactor Refueling System

Reactor refueling occurs every 23 months. Within the reactor, used fuel assemblies are moved to storage racks located at the periphery of the core where they are allowed to decay for one cycle prior to removal. This reduces the decay heat level in the used fuel assemblies to allow them to be passively cooled by natural circulation of air, thereby avoiding the need for under Na handling. A used assembly is moved from the core to an in-vessel storage position by the IVTM. The fuel from the fuel transfer cask is lowered to the in-vessel transfer position, then moved by the IVTM to an empty position within the core. The IVTM then moves a spent assembly from the core to an in-vessel storage position. The IVTM then moves a decayed used fuel assembly from the in-vessel storage to the transfer position located directly below the port in the reactor closure to which the transfer cask is attached and sealed.

An ex-vessel transfer mechanism (EVTM) located within the used fuel transfer cask is used to exchange used fuel and other core assemblies in the reactor with new assemblies from the transfer cask, which is designed to handle six assemblies. The transfer cask is then returned to the fuel service facility by a cask transporter where the used fuel will be cleaned, inspected, and stored in air cooled storage rack and the cask re-loaded with fresh fuel. Finally, the decayed, used fuel assembly is raised into the fuel transfer cask. This cycle is repeated for all fuel assemblies. The movement of the non-fuel assemblies (blanket, control, and radial shield assemblies) is similar, but they are not stored in the in-vessel storage for a cycle before being removed from the core.

B.4 Reactor Coolant System and Connected Systems

The primary system boundary includes the reactor vessel, reactor closure, closure penetrations, below-head ducting of the two IHX units, and the primary sodium and cover gas clean-up system piping up to and including the first isolation valve located immediately outboard of the reactor closure. During power operation, all sodium and cover gas service lines are closed with double isolation valves and all other penetrations in the reactor closure are seal-welded. Thus, the primary system operates in a totally sealed manner during power operation.

B.4.1 Intermediate Heat Transport System (IHTS)

The IHTS is a closed loop system that transports the reactor generated heat to the steam generator system by circulating non-radioactive sodium between the IHX units and the steam generator. The hot leg sodium at 485°C, is transported in separate 72-cm OD 316 stainless steel pipes from the two IHX units to a single 1,000 MWth SG. Two high temperature EM pumps located in the cold legs return the sodium to the IHX units at 325°C. The high temperature secondary EM pumps are similar to the ones used in primary system.

B.4.2 Steam Generator System

The steam generator system is comprised of the steam generator, startup recirculation tank/pump, leak detection subsystem, and steam generator isolation valves. There is one steam generator system for each reactor module. Two steam generators feed a single turbine-generator in each power block through a header arrangement.

Each steam generator is a vertically oriented, helical coil, sodium-to-water counter flow shell-and-tube exchanger. The unit is rated for 1,000 MWth and generates superheated steam at 165 atg and 462°C. The feed water temperature is 215°C and the sodium inlet temperature is 485°C. Steam/water flows upwards on the tube side and sodium flows downwards on the shell-side. The steam generator material is Mod 9 Cr—1 Mo steel.

B.4.3 Reactor Shutdown Heat Removal System

The Shutdown Heat Removal System (SHRS) provides post-shutdown decay heat removal. The turbine condenser using the turbine bypass normally removes reactor shutdown heat. Two safety grade auxiliary cooling systems are provided for cases when an alternative method of shutdown heat removal is required. The RVACS and the Auxiliary Cooling System (ACS) may be used during maintenance or repair operations when the SHRS is unavailable. Together, the RVACS and ACS systems have the capability of maintaining the reactor temperatures well below design limits following a loss of the normal heat sink. Both systems are designed to perform their decay heat removal function without electric power. The main condenser cooling system is constructed to high quality industrial standards; whereas the RVACS, the IHTS, and associated ACS are designed, constructed, and maintained in accordance with the requirements for safety-related equipment.

B.4.4 Reactor Vessel Auxiliary Cooling System (RVACS)

The operation of the RVACS is illustrated in Fig. B.5. This system can dissipate all of the reactor's decay heat through the reactor vessel and containment vessel walls by radiation and convection to naturally circulating air outside the containment vessel without exceeding structural temperature limits. A description of the system and its performance is presented in this section. The RVACS operates continuously, but functions at its intended high heat removal rate only when all other reactor heat removal systems are inoperative and the temperature has increased. With the reactor shut down and the EM

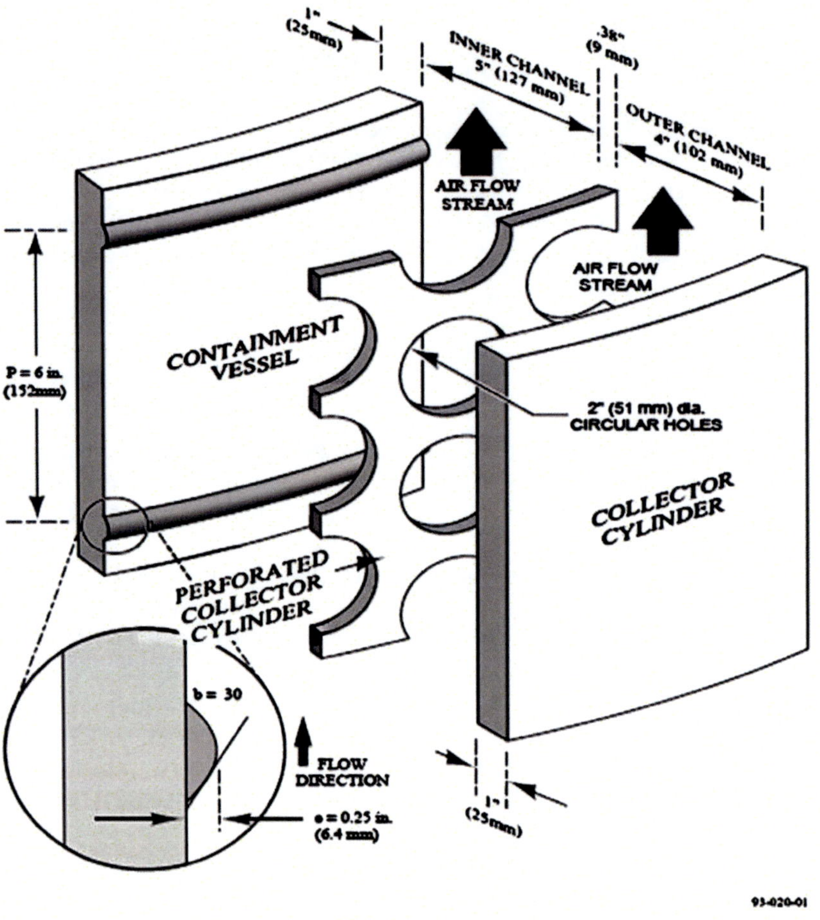

Fig. B.8 RVACS Hot air riser with boundary layer trips. Ref. [5] ©ASME 2000

pumps stopped, primary sodium flow through the reactor core is maintained by natural circulation. The decay heat generated in the core is removed by the primary sodium and transferred to the reactor vessel. The heat is radiated (97%) and convected (3%) from the reactor vessel to the containment vessel. The heat is then deposited into air between the containment vessel and the collector cylinder by convection. The heated air column drives natural circulation to exhaust the decay heat into the atmosphere.

Figure B.8 illustrates how boundary layer trips and a perforated collector cylinder have been utilized to increase the heat removal capability. When RVACS is required for decay heat removal, natural circulation of primary sodium moves heat from the core to the reactor vessel. As the temperature of the reactor sodium and reactor vessel rise, the radiant heat transfer across the argon gap to the containment vessel (97% by thermal radiation) increases to accommodate the heat load. With the increase in containment vessel temperature, the heat transfer from the containment vessel to the atmospheric air surrounding the containment vessel increases. Because RVACS uses natural circulation of sodium and air and is always in operation, a small amount of heat (<0.5 MWth) is lost from the reactor modules during normal operation. This loss is small because radiant heat transfer is employed to transfer heat from the reactor vessel and the containment vessel and radiant heat transfer varies in direct proportion the absolute temperature to the fourth power. This allows the RVACS heat removal rate to increase during heat up transients associated with protected loss of the normal heat sink events.

In the highly unlikely event that the IHTS becomes unusable due to a loss of secondary sodium, the safety-related RVACS will passively remove the residual heat without the ACS. Figure B.9 provides a temperature versus time plot that shows how the primary system temperatures will vary following a protected loss of the normal heat sink. Note that when the ACS is available the loss of normal heat sink transient is extremely mild—peaking at less than 565°C (1,050°F).

B.4.5 Auxiliary Cooling System (ACS)

The ACS uses natural circulation of the primary coolant to move decay heat from the core to the IHX and into the IHTS coolant. Natural circulation of the IHTS coolant transports the heat to the steam generator, where natural circulation of atmospheric air past the shell side of the steam generator

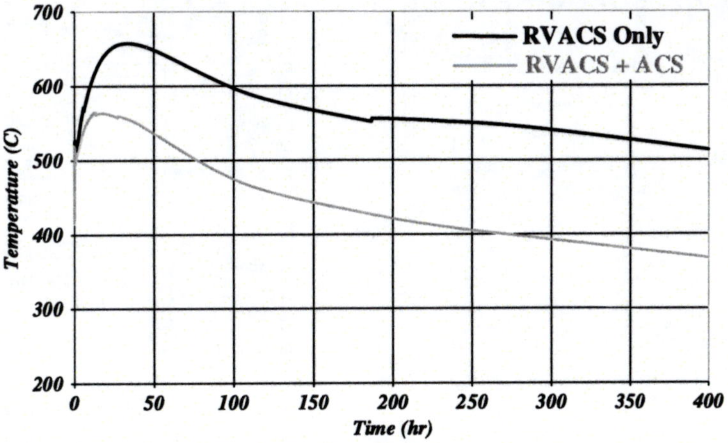

Fig. B.9 Nominal peak coolant temperature, RVACS only and RVACS + ACS

removes the decay heat. The safety-related ACS consists of an insulated shroud around the steam generator shell with an air intake at the bottom and an isolation damper above the steam generator to prevent heat loss during normal operation. Actuation of the ACS is initiated by opening the exhaust damper. ACS heat removal is supplemented by the continuously operating RVACS.

To increase the ACS heat removal rate and reduce the time to cool down the system for maintenance outages, an auxiliary fan, located in the exhaust stack, may be activated. In the natural circulation mode, the exhaust stack damper is open and the auxiliary fan does not operate.

B.5 Containment

The S-PRISM containment utilizes three successive barriers (fuel cladding, primary coolant boundary and a containment boundary) to protect the public from postulated release from the reactor vessel. The containment consists of a lower containment vessel that surrounds the reactor vessel and a low leakage pressure-retaining boundary composed of a steel lined concrete upper containment structure that encloses the reactor closure.

The upper portion of the S-PRISM containment shown in Fig. B.10 is a large (20 by 22-m) room with a height of 10-m. The steel lined upper containment structure is designed to limit leakage to less than 1 vol.% per day at 0.35 kg/cm^2 (5 psig) to mitigate postulated design basis accidents. The auxiliary service room that is located between the upper containment of two modules is 8 m tall, 9 m wide, and 34 m in length. It contains the primary Na service and cover gas systems and the primary sodium storage tanks and is also steel lined.

The lower portion of the containment also serves as a guard vessel. It consists of a 25-mm (one-inch) thick, 9.6-m diameter steel vessel made of 2-1/4 Cr-1Mo. The lower containment vessel has no

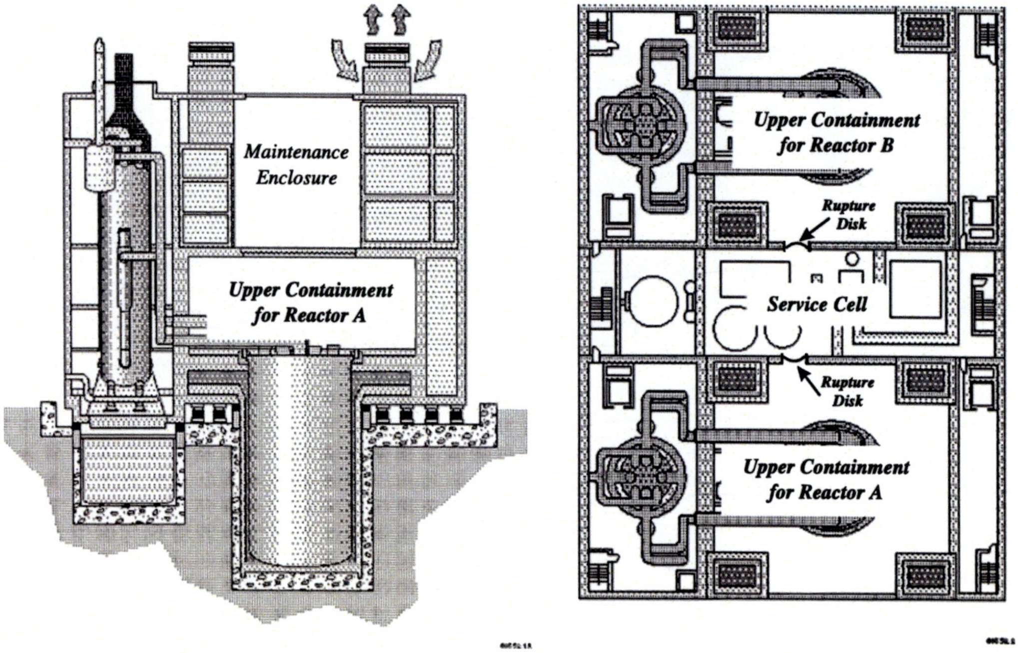

Fig. B.10 S-PRISM containment. Ref. [5] ©ASME 2000

penetrations and is designed to remain essentially leak tight. The 20 cm annulus between the reactor vessel and the containment vessel is sized to retain the primary sodium in the unlikely event of a reactor vessel leak such that the reactor core, the stored spent fuel, and the inlets to the intermediate heat exchangers remain covered with sodium.

A maintenance enclosure located above the upper confinement serves as a secondary containment. A standby gas treatment system with emergency gas treatment filters is used to maintain a negative pressure during maintenance and refueling activities in the maintenance enclosure.

Controlled venting from the containment region above reactor "A" to the service cell and, if necessary, to the containment region above reactor "B" would occur if needed to limit the peak pressures generated by severe accidents. The use of multiple containment volumes and controlled venting reduces the peak containment pressure produced by large Na pool and spray fires by a factor of two.

To provide defense in-depth, the upper containment structure has been designed to contain Na spray and pool fires that would occur if a reactor closure breach occurred as a result of a hypothetical core disruptive accident (HCDA). Analysis has shown that the peak pressures and structural temperatures that would be produced as a consequence of postulated Na spray and pool fires are within the containment design basis 0.4 bar (6 psig). The innovative S-PRISM containment systems use of controlled venting to the service vault and, if necessary, to the second containment reduces the peak pressures that occur under conservative design basis events by a factor of two. This makes it feasible to provide a relatively open and easily accessible upper containment space without an excessively thick and costly upper containment structure.

The use of rupture disks to limit the peak pressures in the upper containment structure represents an acceptable risk to the plant investment because the probability of a large pool or an HCDA induced spray fire is extremely low (less than one event in 10 million years). And CONTAIN-LMR analysis has shown that the deposition from either pool or sodium spray fires in the second containment would be extremely small; on the order of 0.07% for the pool fire case and 0.8% for the 136 kg (300 pound) spray fire case.

Preliminary assessments indicate that the S-PRISM containment system will meet all US and Japanese requirements, including large Na pool (2,000 kg) and spray fire (100 kg) events without exceeding the design basis containment pressure or the structural temperature limits associated with the steel lined upper containment structure.

B.5.1 Sodium Water Reaction Pressure Relief System

The Sodium Water Reaction Pressure Relief System (SWRPRS) rupture disk and relief line is located in the cover gas space at the top of the SG to:

(1) allow the use of a lower rupture disk set point (~ one versus 20 bar) for faster detection and isolation of a leak in the event that the chemical and acoustic leak detection systems fail to act.
(2) allow the plant operator to retain the secondary Na within the IHTS/SG until the plant can be cooled, and
(3) prevent the generation of damaging rupture disk initiated pressure waves in the IHTS and the generation of a high velocity slug of sodium and reaction products in the relief piping.

Continued circulation of the reaction products within the IHTS also reduces the concentration of the contaminants and allows the plant operator to clean up the secondary sodium before the loop is drained. This approach allows the IHTS and ACS to function after minor Na/water leaks thereby reducing the number of higher temperature RVACS-only cooling events.

B.6 Power Generation System

A S-PRISM power block consists of two reactor modules, each with one steam generator that collectively supply one turbine generator set. Steam from the two steam generators in a power block is combined and supplied at near-saturated conditions to the high-pressure inlet of the turbine generator. The exhaust steam enters the low-pressure turbine sections after it passes through moisture separators and reheaters. The steam is exhausted from the low-pressure sections to the condenser, the steam-jet air-ejector (SJAE) condensers, the steam packing exhauster (SPE) condensers, and the steam generator system (SGS) blow down coolers, before it enters the feed water and condensate system. The turbine-generator for each power block is an 804 MW, tandem compound, four-flow, 3,600-rpm superheat machine with rated inlet steam conditions of 165 atg 462°C. The turbine exhausts to a longitudinal shell surface condenser operating at an average back pressure of 2.5 inches Hga, while extracting steam for six stages of feed water heating. There are five low-pressure heating stages and one high pressure heating stage for feed water heating.

B.7 Instrumentation and Control

Normal reactor operations are conducted using the plant control system (PCS). The PCS contains a high level of automation for plant control, protection of the plant investment, and data handling transmission. The PCS functions utilize highly reliable redundant digital equipment and reliable power supplies. The nine nuclear steam supply systems, three turbine generators, and associated balance-of-plant (BOP) equipment in the standard S-PRISM plant can be controlled by means of the PCS from a single control center.

The S-PRISM design includes a reactor protection system (RPS) that is independent of the PCS. The RPS, in response to changes in monitored parameters, initiates reactor module safety-related trips to shut down the reactor. There are nine local and independent RPSs, one per reactor. Each local RPS consists of four identical sensor and electronic logic divisions, each located immediately adjacent to the reactor in equipment vaults. The RPS performs independent Class 1E conditioning and monitoring of sensors to determine plant status during and after an accident. All safety-related data handling and information transmission are provided locally by the RPS for the individual module.

References

1. Magee, P.M., Dubberley, A.E., Lipps, A.J., and Wu, T., 1994, "Safety Performance of the Advanced Liquid Metal Reactor," ARS'94 Topical Meeting-Advanced Reactor Safety Conference, Pittsburgh, PA.
2. Magee, P.M., 1994, "Status of NRC Licensing Review of the U.S. Advanced Liquid Metal Reactor," The International Topical Meeting on Sodium Cooled Fast Reactor Safety, Obninsk, Russian Federation.
3. Quinn, J.E., 1994, "Realizing the World Economic, Environmental and Non-Proliferation Benefits of the ALMR Actinide Recycle System," The International Symposium on Global Environmental and Nuclear Energy Systems, Shizuoka, Japan.
4. Boardman, C.E., Carroll, D., and Hui, M., 1999, "A Fast Track Approach to Commercializing the Sodium Cooled Fast Reactor," Proceedings of the 7th International Conference on Nuclear Engineering (ICONE-7), Tokyo, Japan.
5. Boardman, C.E., et al., 2000, "Description of the S-PRISM Plant," Proceedings of the 8th International Conference on Nuclear Engineering (ICONE-8), Baltimore, MD.

Appendix C
Japan Loop-Type Sodium-Cooled Fast Reactor

Toshikazu Takeda and Yoshitaka Chikazawa

C.1 Introduction

Japan has a long track record of maintaining a commitment to fast reactor technology—recognizing the need for a reliable, long-term energy source compatible with environmental stewardship. Accordingly, considerable efforts have been focused on innovative design approaches to achieve optimal safety, waste minimization, enhanced proliferation resistance, and economical competiveness. Studies carried out over the past decade [1, 2] have concluded that these objectives could be achieved, with the possible exception of acceptable cost. Hence, a great deal of the recent emphasis has been on innovate methods to reduce capital cost and minimize operational expenses. All types of fast reactor designs were considered, including advanced loop and pool configurations, differing fuel types and coolants, and complete costs over the entire life cycle. Such studies have led to a conclusion in Japan that loop-type systems [3] may be the best route to achieving these goals.

The system chosen for detailed design is the Japan Sodium-Cooled Fast Reactor (JSFR). It is a two-loop, oxide fueled system rated at 3,530 MWth and 1,500 MWe. Mitsubishi Heavy Industries, Ltd. has been selected as the core company for development of this project [4, 5].

C.2 Design Principles

C.2.1 Innovative Technology

The principal innovative design features that are included in the JSFR to achieve the goals stated above include the following:

(1) A two-loop primary heat transfer system with large diameter piping
(2) Simplified L-shaped piping, using 9Cr low-thermal expansion steel, to substantially reduce piping lengths
(3) Double-walled piping throughout to eliminate the possibility of sodium leaks and subsequent sodium-air fires
(4) An integral IHX/primary pump component to eliminate a second vessel housing and associated piping
(5) Oxide dispersion strengthened (ODS) steel for fuel cladding to achieve high burn-up
(6) Double-walled tube in the steam generator to minimize any sodium-to-water interactions
(7) Simplified fuel handling system
(8) Advanced seismic isolation

Table C.1 provides a summary of these main innovative technologies.

Table C.1 JSFR innovative technologies

Items	Innovative technologies
Core and fuel	– High burnup fuel with ODS cladding material – Safety enhancement technologies; SASS, re-criticality free core
Reactor system	– Compact reactor system
Cooling system	– Two-loop cooling system of large diameter piping made of Mod. 9Cr-1Mo steel – Integrated pump-IHX component – More highly reliable SG with double-walled straight tube
DHRS	– Capability of natural circulation of decay heat removal
BOP	– Simplified fuel handling system
Reactor building	– CV made of steel plate reinforced concrete (SCCV) – Advanced seismic isolation system for SFR

C.2.2 Safety Principles

Enhancement of safety, based on defense-in-depth principles, has been determined to be a key factor in the core and fuel design. The reactor shutdown system consists of two independent subsystems—each of which is designed to prevent fuel failure for design basis accidents (DBA). Further, a Self-Actuated Shutdown System (SASS) is provided for JSFR as a passive safety feature for reactor shutdown to enhance the prevention capability against core disruptive accidents (CDA) from Anticipated Transients without Scram (ATWS). Since JSFR adopts a restrained core with a core former to meet the severe seismic design requirements in Japan, absorber rods are capable to be inserted by gravity only with a prompt and passive cutoff mechanism.

Given the large size of the JSFR, the positive sodium void worth represents a classic safety issue. Accordingly, the magnitude of the positive sodium void worth is limited to six dollars and the reactor includes a fuel assembly with an inner duct structure designed to allow molten fuel to be rapidly expelled from the core. This feature is included to assert a re-criticality free core concept.

Since Core Disruptive Accidents (CDA) have been historically analyzed for beyond design basis events (BDBE), JSFR is acquiescing to the traditional conservative hypothesis. However, JSFR is designed to mitigate CDA consequences to eliminate severe energetics. In-pile tests using the Impulse Graphite Reactor (IGR) in the Republic of Kazakhstan, together with an out of pile test series, have shown the effectiveness of the inner duct structure noted above. As a result, severe energetics due to positive void worth and molten fuel compaction appear to be eliminated [6].

The assurance of decay heat removal has been improved by enhancing natural circulation capability both for design basis events and also beyond DBE such as ATWS and protected loss of heat sink (PLOHS). One Direct Reactor Auxiliary Cooling System (DRACS), a sodium to sodium heat exchanger, is installed in the reactor vessel and two Primary Reactor Auxiliary Cooling Systems (PRACS) are sodium to sodium heat exchangers installed in the primary side of the intermediate heat exchanger vessel. They are designed for fully passive operation as well as redundant capacity to achieve very high reliability.

C.2.3 In-service Inspection and Repair

A key part of reducing operational costs in the Japanese program is to apply rigorous In-Service Inspection and Repair (ISI&R) strategy throughout the design [7]. Sodium-cooled reactors have the well known maintenance challenges consisting of chemical activity, opacity, and high temperature operation. On the other hand, low pressure operation provides the capability of continuous monitoring for sodium leakage. Further, sodium has excellent compatibility with the structural materials. With double-walled piping throughout, sodium leaks are of little consequence as long as the coolant level in the reactor vessel is adequately maintained. However, continuous monitoring is exercised to ensure that a Leak Before Break strategy can be successfully implemented.

A comprehensive ISI program for JSFR has been developed. Inspection methods and frequency have been established based on the ISI program of "Monju," the ISI code for the light water reactor (Japan Society of Mechanical Engineers, JSME S NA1), and for the liquid metal cooled reactor (American Society of Mechanical Engineers, ASME section XI division 3).

A compact reactor system is indispensable for the purpose of achieving the economic competitiveness for JSFR. This includes compact in-vessel fuel handling and advanced shielding. This compactness challenges the quest for maintaining an adequate in-service inspection and repair (ISI&R) capability, but JSFR accommodates in-service inspection programs on the reactor vessel and in-vessel structures by allowing access for inspection devices.

For the repair program, anticipated failures have been listed and categorized, based on frequency and required repair levels. For such events, each component is required to be capable of the repair work on an expedited basis. Access and maintenance space are carefully allocated in the component structural design. For some small components, a whole component removal and overhaul in the reactor building is being planned.

C.3 Overall Plant Description

The major parameters [8, 9] for the JSFR are contained in Table C.2. Figure C.1 is a bird's eye view of the reactor coolant loops and Figs. C.2 and C.3 provide an overview of the reactor building and plant layout, respectively.

C.4 Reactor Design

C.4.1 Fuel System

The core design conditions [10] are tabulated in Table C.3. The initial fuel system for the JSFR is MOX with recycle fuel; some minor actinides are assumed to be included. This is being done in recognition for the need to smoothly transfer from a LWR to a fast reactor system [11].

Table C.2 JSFR major parameters

Items	Specifications
Reactor type	Sodium-cooled loop type
Electricity output	1,500 MWe
Thermal output	3,530 MWth
Number of loops	2
Primary sodium temperature	550/395°C
Primary sodium flow rate	3.24×10^7 kg/h/loop
Secondary sodium temperature	520/335°C
Secondary sodium flow rate	2.70×10^7 kg/h/loop
Main steam temperature and pressure	495°C 18.7 MPa
Feed water temperature and flow rate	240°C, 5.77×10^6 kg/h
Plant efficiency	Approx. 42%

Fig. C.1 Bird's-eye view of JSFR

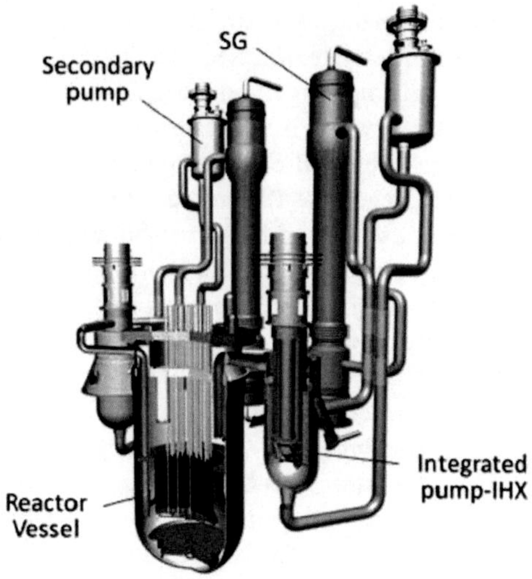

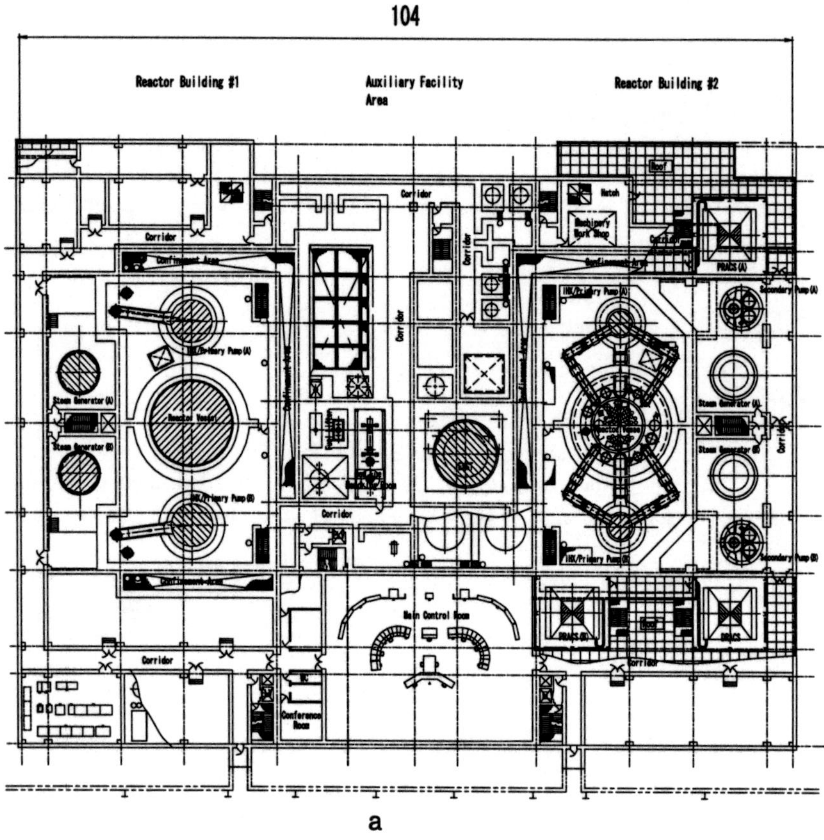

Fig. C.2 Sketch of JSFR reactor building

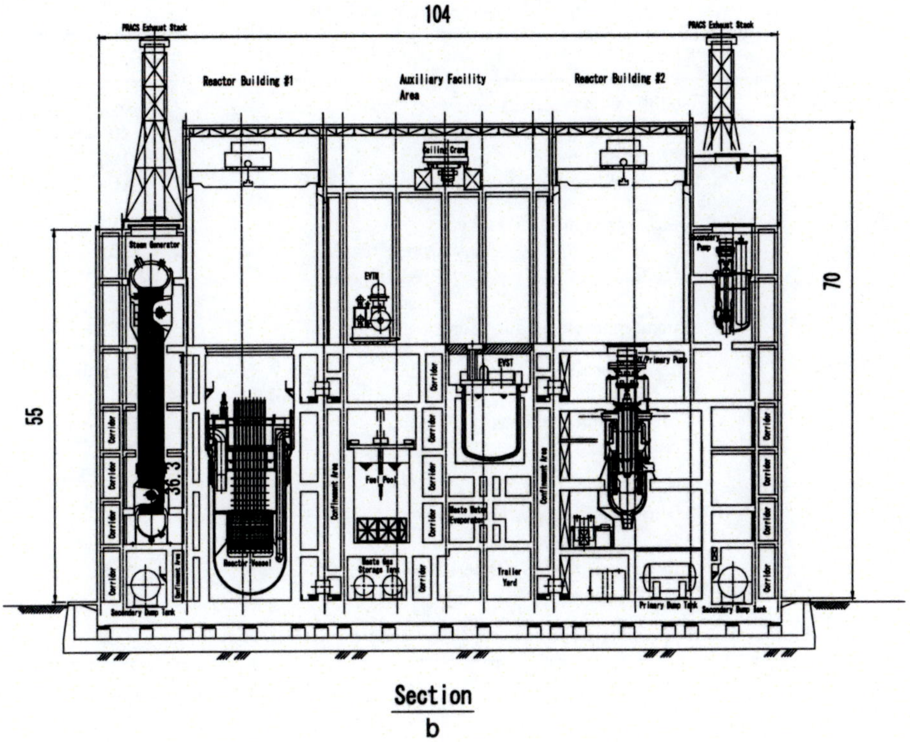

Fig. C.2 (continued)

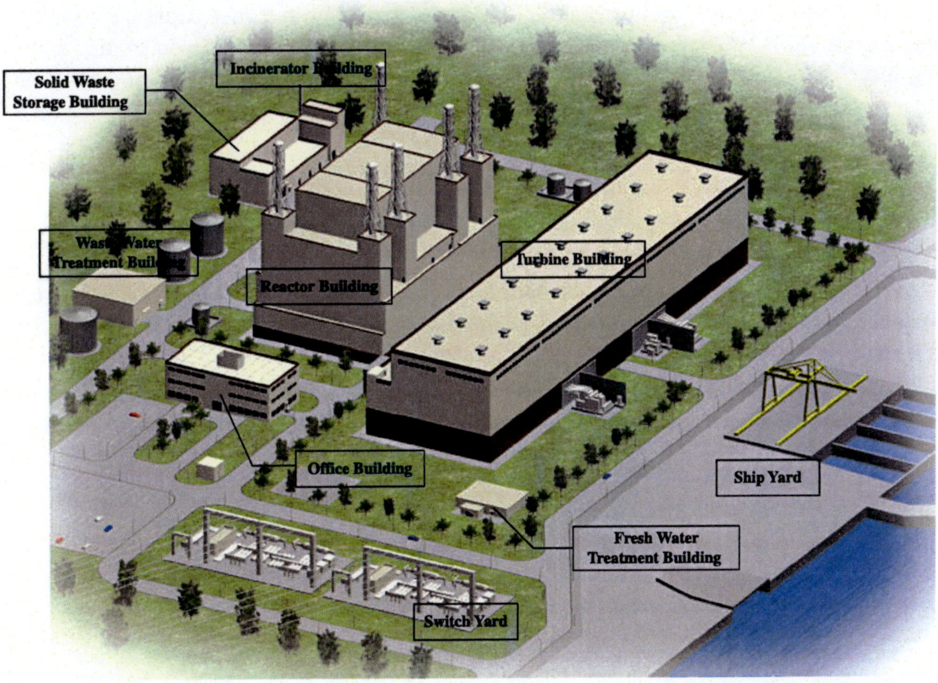

Fig. C.3 JSFR plant layout

Table C.3 Core design conditions

	Items		Conditions
Safety requirement	Sodium void reactivity, $		≤6
	Core specific power, kW/kg-MOX		≥40
	Core height, cm		≤100
	Recriticality-free		FAIDUS type assembly
Design target	Discharge burnup, GWd/t core		150
		Core + blanket	≥60
	Breeding ratio	Breeding core	1.1
		Break even core	1.03
	Operation cycle length, months		≥18
Core and fuel spec.	Fuel composition	TRU	FR multi-recycle
		FP containment, vol.%	0.2
	Core fuel smeared density, %TD		82
	Core material	Cladding	ODS
		Wrapper tube	PNC-FMS
Design limit	Maximum linear power, W/cm		≤430
	Maximum neutron dose[a], n/cm^2		$\leq 5 \times 10^{23}$
	Cladding maximum temperature[b], °C		≤700
	CDF [steady state]		≤0.5
Others	Pin bundle pressure drop, MPa		≤0.2

[a]E > 0.1 MeV.
[b]Mid-wall.

The target core average burnup at discharge is 150 GWd/t and the total average discharge burnup (including blankets) is about 60 GWd/t. The high burnup is intended to reduce fuel cycle cost and also reduce the fuel mass capacity requirement for the fuel cycle facilities. Consistent with overall economics, the JSFR is being designed to accept fuel with relatively little decontamination required during recycle to allow for economies in the fuel cycle facilities. This is likewise consistent with enhanced proliferation resistance.

C.4.2 Core Configuration

The core is being designed to have flexible breeding capability to be consistent with uranium resource utilization requirements. The maximum breeding ratio, targeted from 1.1 to 1.2, is sized not only for fuel breeding but also to enhance economic advantages and to reduce the environmental burden. The core with high breeding capability is characterized by a high internal conversion ratio, which allows flexibility for fewer blankets, longer cycle times, and less environmental burden.

Table C.4 contains the core and fuel specifications for the JSFR. The only substantive difference between the breeding core and the core with a BR equal to unity is absence of the radial blanket. Figure C.4 contains cross sections of both reference cores.

The advanced oxide-fueled design incorporates a large diameter fuel pin. This leads to a considerably higher internal conversion ratio and allows a break even breeding core, in which the breeding ratio is just above 1.0 without a radial blanket. The total discharge average burnup (including blanket) can be as high as about 100 GWd/t for a refueling batch average. Much higher average burnup levels

Table C.4 Core and fuel specifications for the JSFR

Items	Breeding core	Break even core
Nominal full power, MWe/MWth	1.500/3.570	Same
Coolant temperature [outlet/inlet], °C	550/395	Same
Primary coolant flow, kg/s	18,200	Same
Core height, cm	100	Same
Axial blanket thickness [upper/lower], cm	20/20	15/20
Number of fuel assemblies [core/radial blanket]	562/96	562/–
Envelope diameter of radial shielding, m	6.8	Same
Fuel pin diameter [core], mm	10.4	Same
Fuel pin cladding thickness [core], mm	0.71	Same
Number of fuel pins per assembly [core]	255	Same
Wrapper tube outer flat–flat width, mm	201.6	Same
Wrapper tube thickness, mm	5.0	Same

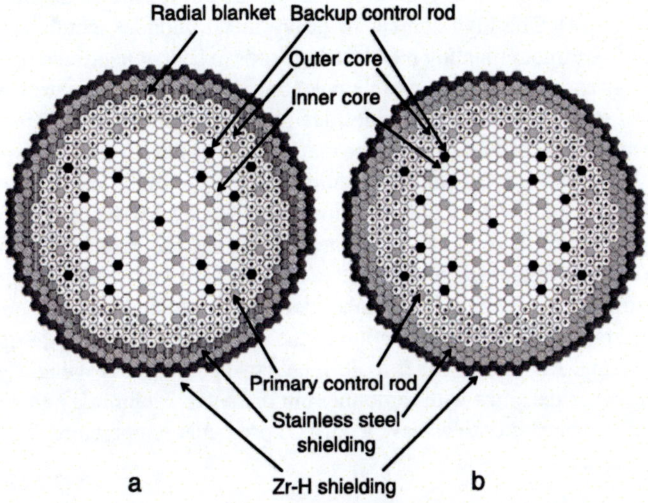

Fig. C.4 Core configurations of the JSFR reference cores: (**a**) Breeding core, (**b**) Break even core

are possible compared to a conventional design with small diameter fuel pins. The resulting core has the flexibility to be modified to achieve a breeding ratio with 1.1 by simply adding a radial blanket. This advanced concept is expected to offer a considerable economical advantage by consistently achieving high burnup and breeding with a small blanket.

As noted earlier, a special fuel assembly has been designed for incorporation into the JSFR to accommodate severe, highly unlikely accident conditions. Figure C.5 shows a conceptual view of a FAIDUS (Fuel Assembly with Inner Duct Structure) type assembly [12]. An inner duct is installed at the corner of the assembly and a part of the upper shielding element is removed. Molten fuel generated in a major accident would enter the inner duct channel and be expelled out of the core without interference from the upper shielding. The FAIDUS component is expected to offer superior performance for molten fuel release out of the core during a CDA.

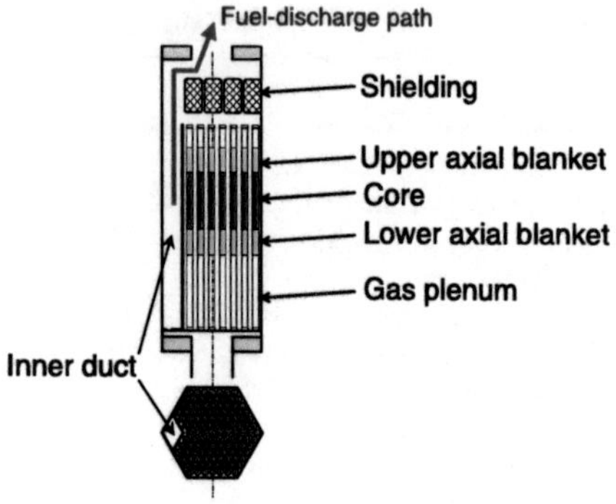

Fig. C.5 FAIDUS type fuel assembly

C.4.3 Metal Fuel

A metal fuel core concept is also being investigated for the JSFR core as an alternative to the reference oxide core [13, 14]. The high content of heavy metal nuclides, combined with appropriate irradiation experience of zirconium alloy metal fuel, affords excellent neutronic performance, including breeding capability and long operation cycle potential. A significant drawback of metal fuel is the limiting constraint of steel cladding temperature due to fuel cladding compatibility concerns (about 50°C lower than desired). The primary goal of the metal fuel core design study is to obtain a core outlet temperature of 550°C, which is identical to the outlet temperature of the reference oxide fuel core.

The JSFR metal fuel core has been able to achieve the desired 550°C core outlet temperature by applying a new core concept of single Pu enrichment to reduce the radial power peaking factor. By specifying a single Pu enrichment of around 12%, a breeding ratio of about 1.0 can be obtained with two or three radial zones with different fuel number densities to obtain low radial power peaking. This optimized design offers stable power distribution in the core through the reactor operation cycle and hence low power peaking. Another option is to designate two core regions with different Zr content to give different fuel number densities with same fuel pin diameter. Preliminary analyses indicated that this concept also has the capability to achieve a 550°C core outlet temperature.

C.5 Plant System Descriptions

C.5.1 Reactor structure

A compact reactor system is indispensable for the purpose of achieving the economic competitiveness desired for JSFR. This compactness must be achieved while maintaining in-service inspection and repair (ISI&R) capability and robustness even against a safety shutdown level earthquake. Figure C.6 shows a schematic illustration of the JSFR reactor structure [15].

One key to achieving a compact reactor vessel for the JSFR is the adoption of high performance Zr-H radial shielding, taking advantage of high neutron moderation features, which reduces the diameter of the core barrel and thus the reactor vessel diameter.

In a liquid metal reactor, the reactor structure strongly depends on in-vessel fuel handling, since in-vessel handling is generally conducted under the reactor vessel plug due to sodium chemical activity with air. The JSFR in-vessel fuel handling approach consists of a single rotating plug, an upper inner structure with a slit (slit UIS), and a pantograph fuel handling machine (FHM) with flexible positioning. The UIS slit provides FHM arm access to core components while minimizing the diameter of the rotating plug.

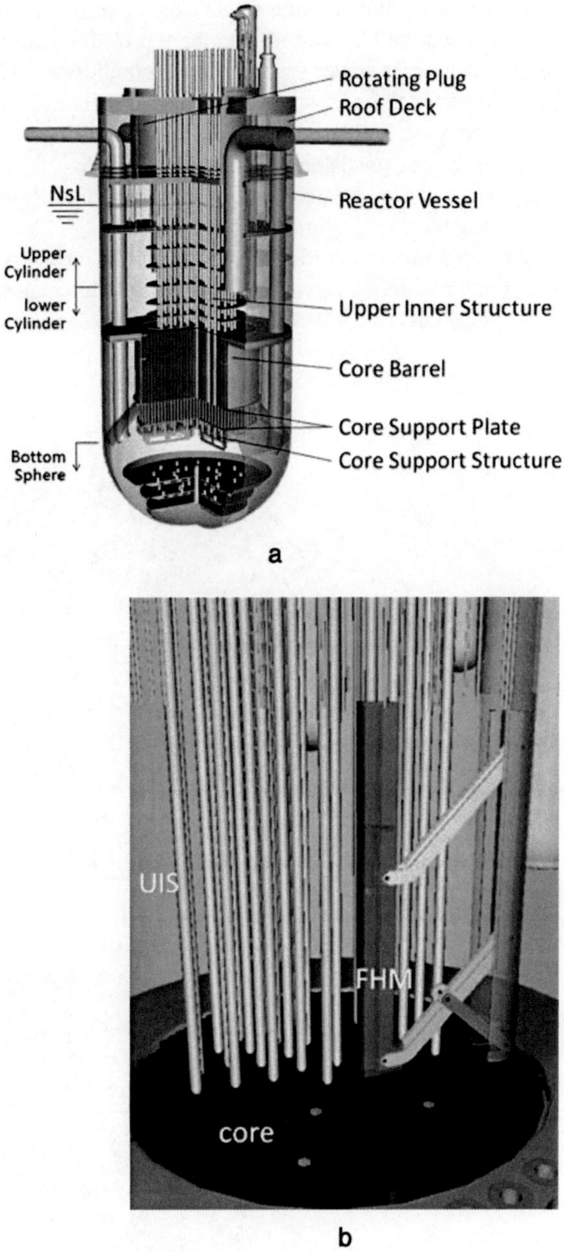

Fig. C.6 JSFR reactor structure

Another feature of the JSFR is the adoption of a "Hot vessel" concept, i.e. no vessel cooling system or coolant over-flow system—both reducing the reactor vessel diameter. Structural integrity against the stress caused by the change of axial temperature profile is ensured by establishment of the advanced elevated temperature structural design standard, which makes it possible to use an inelastic analysis, and by a direct assumption method of the thermal load [16].

A major advantage of achieving a compact design of the reactor vessel is enabling the whole reactor structure to be assembled in a factory: the reactor vessel, roof deck, UIS, and core support structure—all achieving fine dimensional precision and quality of welding. The tolerance of center axes between the UIS bottom and the top of the core barrel is extremely important to ensure control rod insertion even under severe seismic conditions. The allowable tolerance from manufacturing has been estimated to be less than a few tens of millimeters based on the mockup control rod experiment.

The compact JSFR reactor vessel can apply ring-shaped forging manufacturing techniques, which provides the fine precision required. The ring-shaped forging of the reactor vessel eliminates a welding line around the high stress parts near the sodium free surface in normal operation condition, thus enhancing reliability against thermal stress.

As noted earlier, the JSFR design team has focused considerable attention on in-service inspection and repair capability. Accordingly, several maintenance holes exist at the reactor vessel plug to install Under Sodium Area Monitors (USAM) as shown in Fig. C.7, providing access to the upper, middle

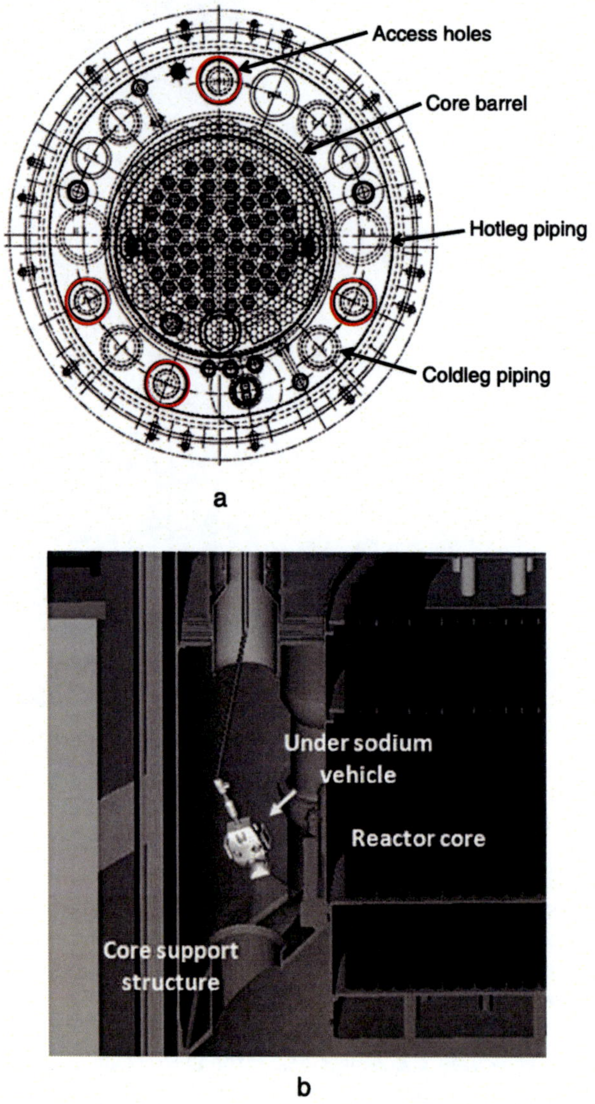

Fig. C.7 Access holes for under sodium inspections

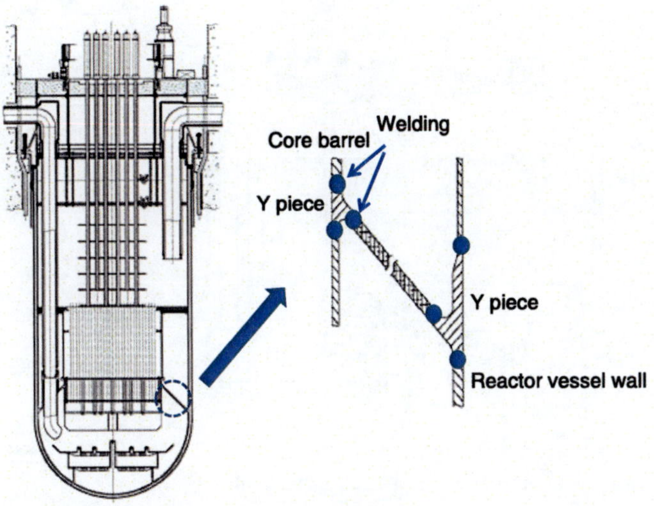

Fig. C.8 Single-piece-"Y" forging ring at core support structure

and lower plenums. The USAM contains an under sodium viewer/volumetric examination sensor to conduct examinations in 200°C sodium. The in-vessel structure is also designed to accommodate an ISI program. As shown in Fig. C.8, the core support structure consists of a pair of single piece "Y junction" forgings to reduce welding lines and provide ISI device access.

C.5.2 Cooling System

Reducing the cooling system to two loops is a key design feature of advanced Japanese fast reactors to enhance economic efficiency [17, 18]. Since a two-loop configuration is adopted even for the large-size JSFR plant design at 1,500 MWe power output, the coolant flow rate per loop increases and the diameter of piping is necessarily enlarged.

Figure C.9 is a schematic illustration of the JSFR loop-type cooling system. The JSFR primary and secondary piping systems are drastically simpler than conventional loop type reactors, taking advantage of Modified 9Cr-1Mo steel, which has high strength at elevated temperature and a low thermal expansion coefficient. The simple piping system reduces the amount of piping material and provides a compact component arrangement.

All primary pipes are covered with a guard pipe to maintain the coolant level within the RV with no coolant leakage into the containment vessel (CV), even if a guillotine break of a primary boundary is not totally eliminated and assumed as a beyond design basis event (BDBE). The volume between the coolant boundary piping and the guard pipe is minimized to mitigate the loss of primary coolant flow rate in the event of a pipe break.

The two loop cooling system does, however, raise certain hydraulic and safety issues [17, 18]. The increased flow rate per loop challenges hydraulic and structural integrity on primary piping. One of the major issues is flow induced vibration due to flow separation at the L-shaped elbow, since the curvature radius of the elbow is equivalent to the piping diameter in order to provide the compact component arrangement. Another concern is swirl flow near the primary hot-leg intake. The swirl flow might enhance vortex cavitation caused by local depressurization at the vortex center and the vortex cavitation might affect structural integrity. To address these concerns, flow dynamics in the primary piping and hot-leg intake have been experimentally investigated.

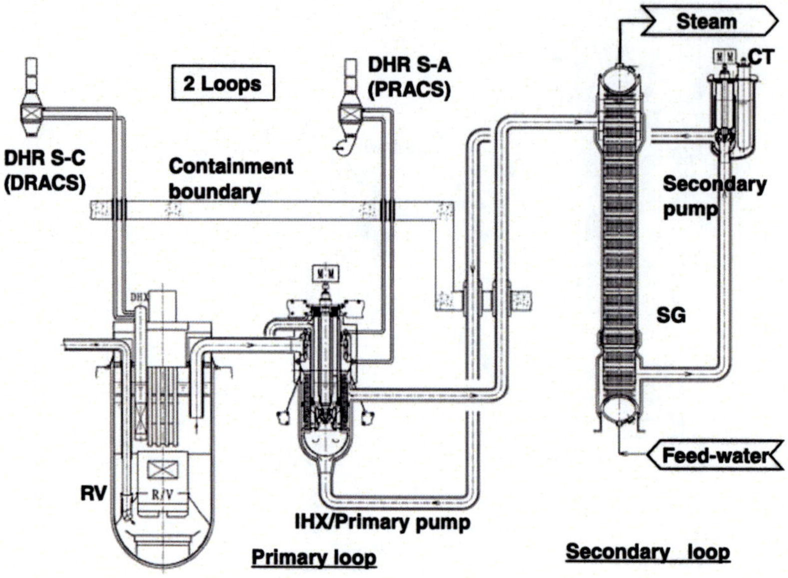

Fig. C.9 Schematic illustration of the JSFR cooling system

From the safety point of view, a primary pump seizure accident in the two-loop system is regarded as one of the critical safety issues of a DBE, since this results in more severe consequences than a pipe break accident (given the double coolant boundary concept). In the case of a primary pump seizure accident, the core flow rate suddenly reduces. To accommodate such an accident, redundant and diverse safety protection systems are installed in JSFR to provide secure, rapid reactor shutdown and reduce the fuel pin bundle pressure drop to under 0.2 MPa to maintain coolant inertia and allow natural convection flow. Transient calculations with such robust core designs have provided assurance of core safety for such an accident.

As for the modified 9Cr-1Mo piping, pipes without longitudinal welding lines are being investigated to enhance reliability. 9Cr-1Mo creep strength at welded joints generally decreases relative to the base metal at high temperature and long time conditions. This is known as Type-IV damage. This damage is taken into account in the hot-leg piping design of primary cooling system, although such reduction has not been observed at 550°C [19]. Electromagnetic flow meters cannot be utilized to monitor flow rate since Mod. 9Cr-1Mo steel is ferromagnetic material. Hence, the JSFR design has adopted an ultrasonic flow meter system and its performance is basically confirmed [20].

C.5.3 Decay Heat Removal System (DHRS)

The decay heat removal system (DHRS) consists of a combination of one loop of DRACS and two loops of PRACS as shown Fig. C.10 [17, 21]. The DRACS heat exchanger is located in the reactor vessel upper plenum. Each heat exchanger of PRACS is located in the IHX upper plenum. The DHRS operates fully by natural convection and is activated only by action of DC-power-operated dampers of air coolers. The air cooler dampers have redundancy in order to perform their function even assuming a single-failure criterion.

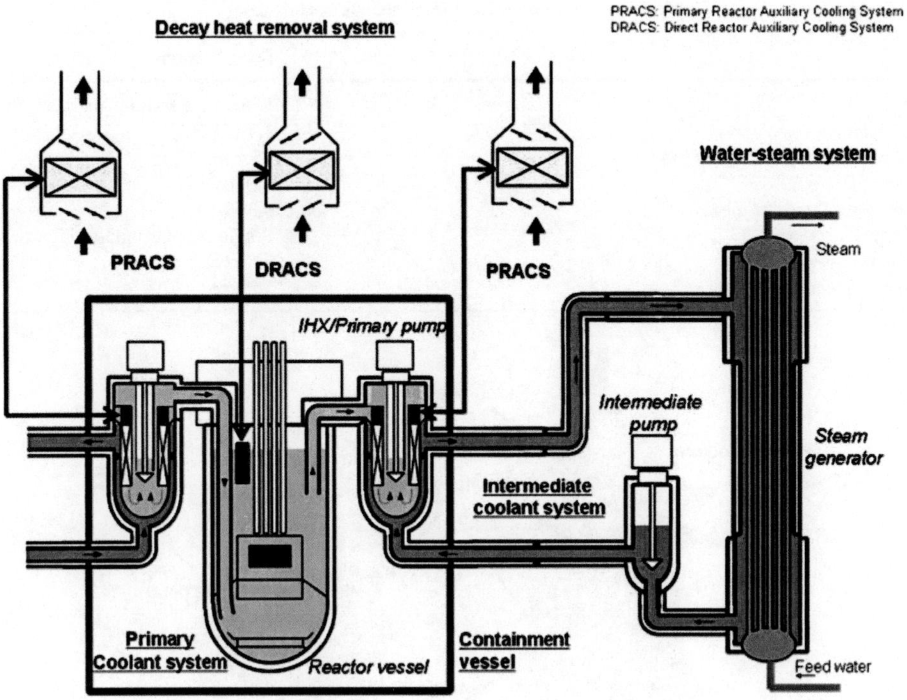

Fig. C.10 Decay heat removal system for JSFR

After conducting safety analyses for the DHRS, it was determined that the PLOHS accident would be the limiting event (i.e. failure to maintain decay heat removal after reactor shutdown). Even so, PSA analyses indicates that the probability of such an accident is extremely low (less than 10^{-7}/reactor-year) [22].

C.5.4 Intermediate Heat Exchanger

The IHX with a built-in primary pump component consists of a homocentric pump and IHX in one vessel [23, 24]. Table C.5 and Fig. C.11 provide IHX specifications and a schematic illustration. One inlet and two outlets for primary sodium are located on the casing. A flash-board structure in the upper plenum provides uniform flow distribution and prevents temperature differences at the heat exchanger tubes. Primary sodium flows through the inside of the heat exchanger tubes and exits into the lower plenum at the pump suction. Gas insulation exists between primary hot and cold sodium to prevent thermal loss. A secondary sodium inlet is located coaxially at the integrated IHX/Pump bottom. Secondary coolant flows along the outside of the heat exchanger tube bundle. Baffle plates are placed in the IHX, inducing zigzag flow to improve heat transfer performance.

The pump shaft is installed in the center, surrounded by the heat exchanger tubes. Supports for the heat exchanger tubes and pump are arranged at different floor levels: lower floor for the tubes and higher floor for the pump to isolate the heat exchanger tubes from pump vibration. A convoluted expansion joint absorbs any thermal expansion difference between the shell and pump separating wall. An annulus exists between the IHX and pump, with two seal rings to diminish flow leakage through

Table C.5 Specifications of intermediate heat exchanger

Items	Specifications
IHX type	Cross flow sodium—straight tube type
Heat transfer capacity	1,765 MWth
Outer diameter of heat transfer tube	25.4 mm
Heat transfer length	6.0 m
Number of heat transfer tubes	9,360
Pump type	Single-stage, single-suction type
Pump head	79 mNa
Pump rated speed	550 rpm

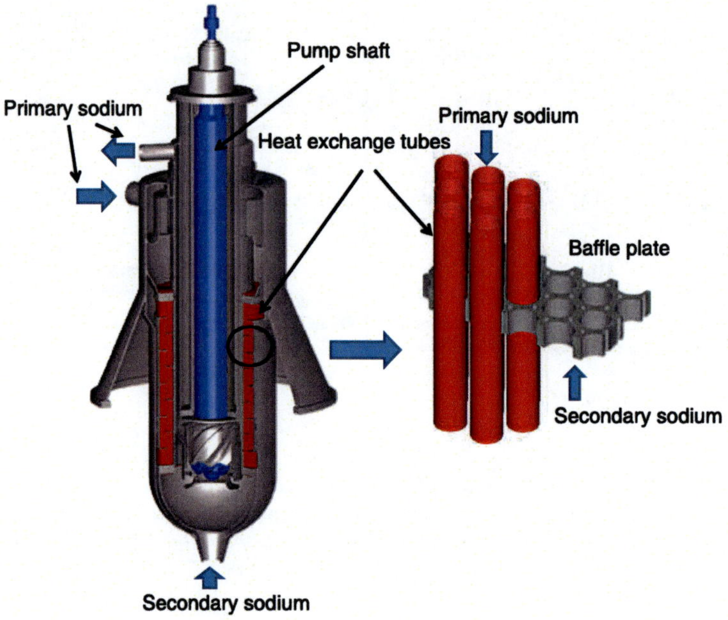

Fig. C.11 JSFR intermediate heat exchanger

the gaps. Any flow leakage through the seal ring and the hydrostatic bearing is discharged into the suction area to control the sodium surface level in the pump.

The integral IHX/primary pump component has certain advantages from the ISI&R viewpoint. The pump shaft can be removed for maintenance and the pump shaft hole provides access to the IHX tubes for volumetric inspection and repair.

C.5.5 Steam Generator

In the design of the steam generator (SG) unit, prevention and mitigation of water/steam leakage accidents is given high priority. Since JSFR is a two-loop system, one SG unit has to handle a very large heat exchange capacity of 1,765 MWth (half the total plant output of 3,530 MWth). The large SG requires a high tube reliability, since the larger number of tubes results in a higher possibility of failure per unit. Further, the heavier sodium inventory causes longer leak detection time, and the repair of the large SG causes longer reactor outage relative to more standard designs.

Table C.6 Steam generator specifications

Items	Specifications
SG type	Double-walled straight tube type
Thermal capacity	1,765 MWth
Tube diameter	19.0 mm
Tube pitch	40.0 mm
Effective tube length	29.0 m
Tube material	Mod.9Cr-1Mo steel
Sodium flow rate	2.70×10^7 kg/h
Steam/water flow rate	2.884×10^6 kg/h
Sodium temperature	520°C/335°C
SG outlet steam	497.2°C (19.2 MPa)

Therefore, JSFR employs a double-walled straight tube design, which consists of mechanically contacted inner and outer tubes to essentially prevent sodium/water reactions [25]. Table C.6 and Fig. C.12 provide SG specifications and a schematic illustration. A previous double-walled tube SG concept [26] successfully provided detection of inner or outer wall failure via monitoring annuli of double tubes, since outer tube integrity cannot be inspected under reactor operation. The mechanically contacted tubes allow inspections on both inner and outer tubes using eddy current tests on inner tubes and ultrasonic tests of outer tubes. The mechanically contacted double-walled tubes also mitigate

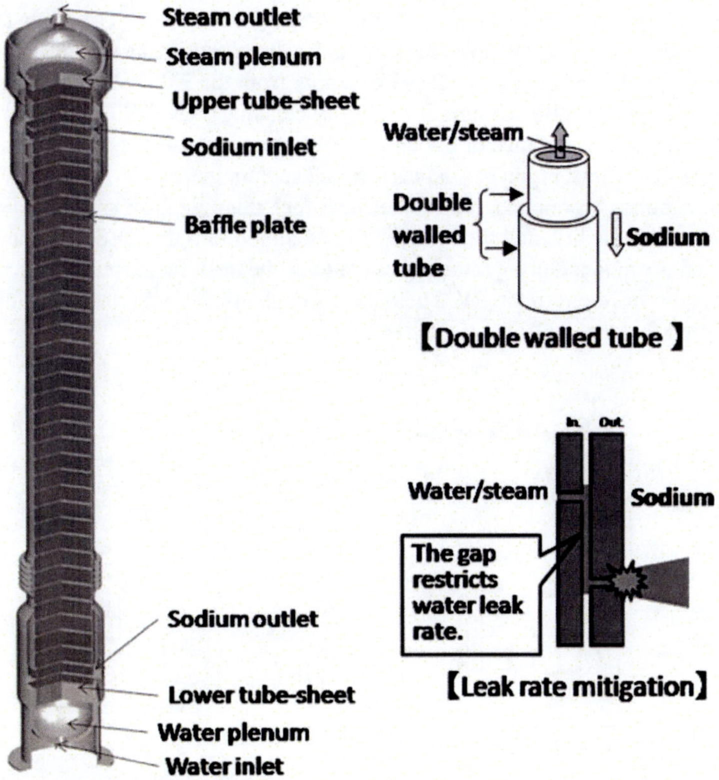

Fig. C.12 JSFR steam generator

sodium water reactions in case of double boundary failures, since the annular gap restricts the water leak flow rate below the tube failure propagation limit.

C.5.6 Fuel Handling System

An advanced fuel handling system (FHS) that enhances a compact in-vessel fuel handling system, allows short plant outage and low construction cost, has been proposed [27]. The schematic diagram of FHS is shown in Fig. C.13. The FHS consists mainly of a fuel handling machine (FHM), an ex-vessel fuel transfer machine (EVTM), ex-vessel fuel storage facilities (the ex-vessel fuel storage tank (EVST) and the used fuel storage pool), used fuel cleaning facilities and fresh fuel handling facilities.

In the reactor vessel during refueling, the FHM discharges used fuel from the core to a sodium pot. The sodium pot has two positions to accommodate core components and is transferred by the EVTM to the EVST. When the sodium pot arrives at the RV, it contains two fresh fuel assemblies. Then the FHM exchanges the two fresh fuel assemblies for two used fuel assemblies. The EVTM carries the sodium pot containing the two used assemblies from the RV to the EVST. No active cooling device is required during transportation from the RV to EVST thanks to the heat capacity of the sodium pot in the normal refueling operation. The EVTM is, however, equipped with a sodium pot cooling system for abnormal conditions. The sodium pot cooling system is only activated in case of trouble with the transportation mechanism and has two independent cooling lines, which are connected by an emergency power supply system.

The used fuel assemblies are stored in the EVST to reduce decay heat. Before the next refueling, used fuel assemblies are transported to the used fuel storage pool by the EVTM. From the EVST to the used fuel storage pool, the EVTM handles the assemblies directly utilizing forced circulation of argon gas. When the EVTM discharges the used fuel assembly from the EVST, the assembly is held at the EVST guide tube to blow out the residual sodium (dry cleaning). The EVTM transfers the assembly to the assembly lifter at the entrance of the used fuel storage pool. At the assembly lifter, argon gas cooling is provided and moist argon gas makes residual sodium inert.

Fresh fuel assemblies accommodated by the fresh fuel shipping cask are unloaded and checked at a new fuel handling facility, and then moved to and stored into the EVST by the EVTM during the reactor normal operation. During refueling operations, the fresh fuel assemblies accommodated in the sodium pot are transferred to the RV. The FHM loads the fresh fuel assemblies into the core and discharges the used fuel assemblies to the sodium pot.

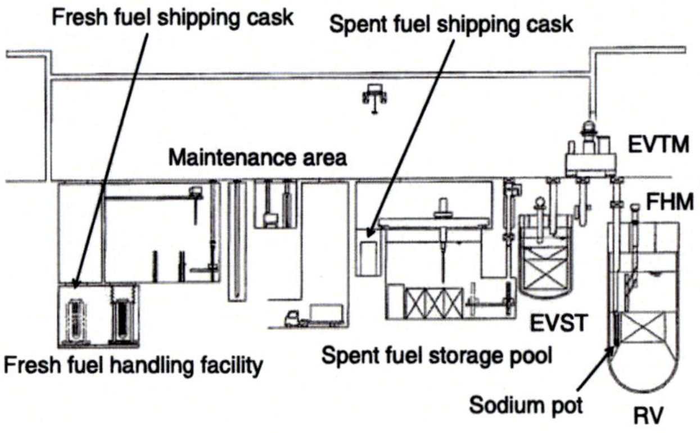

Fig. C.13 JSFR fuel handling system

C.5.7 Reactor Building

The layout plan of the reactor building has been investigated with attention to maintaining adequate seismic reliability and allowing enough space to optimally arrange the components and equipment in the reactor building for plant operability and maintainability [28]. Sketches of the reactor building and plant layout were shown in Figs. C.2 and C.3. Because of the double-walled piping system, sodium fires have a low probability, thus enabling the design pressure of the containment vessel (CV) to be low. This allows the adoption of a rectangular CV with a relatively small volume. However, to adopt the rectangular CV, Steel plate reinforced Concrete structure (SC structure) is essential. The SC structure is composed of a steel plates partly assembled at a factory, which contributes to improving the quality of the structure. Such a structure has the potential to shorten the construction period, contributing to high economic performance.

To bear earthquake forces, heavy components such as tanks or gas turbine generators are located in lower floor. The DHRS units are located on the upper floor to ensure natural circulation capability. Maintenance work spaces are also carefully considered in the layout design.

A seismic isolation system is carefully adopted in the reactor building. For sodium-cooled fast reactors, a seismic isolation system is attractive since the component walls are generally thinner than those of light water reactors. The JSFR design includes a horizontal seismic isolation system consisting of laminated rubbers and oil dampers.

References

1. K. Aizawa, "R&D for Fast Reactor Fuel Cycle Technologies in JNC", *Proceedings of Global 2001*, No. 050, Paris, France, September (2001).
2. K. Ito and T. Yanagisawa, "Last Twenty Years Experiences with Fast Reactor in Japan", *Proceedings of the International Conference on Fast Reactors and Related Fuel Cycles*, IAEA-CN-176-INV-07, Kyoto, Japan, December (2009).
3. Y. Chikazawa, S. Kotake, and S. Sawada, "Comparison of Pool/Loop Configurations in the JAEA Feasibility Study 1999–2006", *Proceedings of the International Conference on Fast Reactors and Related Fuel Cycles*, IAEA-CN-176-08-08, Kyoto, Japan, December (2009).
4. Y. Sagayama, "Launch of Fast Reactor Cycle Technology Development Project in Japan", *Proceedings of Global 2007*, Boise, ID, September (2007).
5. H. Niwa, "Current Status and Perspective of Advanced Loop Type Fast Reactor in Fast Reactor Cycle Technology Development Project", *Proceedings of Global 2007*, Boise, ID, September (2007).
6. S. Kotake, N. Uto, K. Aoto, and S. Kubo, "Safety Design Features for JSFR – Passive Safety and CDA Mitigation", *Proceedings of Annual Meeting on Nuclear Technology*, Dresden, Germany, May (2009).
7. N. Nishiyama, "Japan JSFR Design Study and R&D Progress in the Fact Project", *Proceedings of the International Conference on Fast Reactors and Related Fuel Cycles: Challenges and Opportunities*, IAEA-CN-176-02-16P, Kyoto, Japan, December (2009).
8. S. Kotake, et al., Feasibility Study on Commercialized Fast Reactor Cycle Systems, Current Status of the FR System Design, *Proceedings of Global 2005*, No. 435, Tsukuba, Japan (2005).
9. K. Aoto, et al., "Japan JSFR Design Study and R&D Progress in the Fact Project", *Proceedings of the International Conference on Fast Reactors and Related Fuel Cycles*, IAEA-CN-176-01-07, Kyoto, Japan, December (2009).
10. T. Mizuno, et al., "Advanced Oxide Fuel Core Design Study for SFR in the 'Feasibility Study' in Japan", *Proceedings of Global 2005*, Paper No. 434, Tsukuba, Japan, October 9–13 (2005).
11. S. Maruyama, K. Kawashima, S. Ohki, T. Mizuno, and T. Okubo, "Study on FBR Core Concepts for the LWR-to-FBR Transition Period", *Proceedings of Global 2009*, Paper 9316, Paris, France, September (2009).
12. T. Mizuno and H. Niwa, "Advanced MOX Core Design Study of Sodium-Cooled Reactor in Current Feasibility Study on Commercialized Fast Reactor Cycle System in Japan", *Nuclear Technology*, Vol. 146, No. 2, 143–145 (2004).
13. K. Sugino and T. Mizuno: "A New Concept of Sodium Cooled Metal Fuel Core for High Core Outlet Temperature", *Proceedings of 2004 International Congress on Advances in Nuclear Power Plants*, Pittsburgh, PA (2004).

14. T. Mizuno, T. Ogawa, K. Sugino, and M. Naganuma, "Advanced Core Design Studies with Oxide and Metal Fuels for Next Generation Sodium Cooled Fast Reactors", *Proceedings of 2005 International Congress on Advances in Nuclear Power Plants*, Paper 5195, Seoul, Korea, May (2005).
15. Y. Sakamoto, S. Kubo, S. Kotake, and Y. Kamishima, "Development of Advanced Loop-Type Fast Reactor in Japan (3): Easy Inspection and High Reliable Reactor Structure in JSFR", *Proceedings of 2008 International Congress on Advances in Nuclear Power Plants*, Paper 8227, Anaheim, CA, June (2008).
16. N. Kasahara, K. Nakamura, and M. Morishita, "Recent Developments for Fast Reactor Structural Design Standard (FDS)", SMiRT18, Beijing, China, August 7–12 (2005).
17. H. Yamano, S. Kubo, K. Kurisaka, Y. Shimakawa, and H. Sago, "Development of Advanced Loop-Type Fast Reactor in Japan (2): Technological Feasibility of Two-Loop Cooling System in JSFR", *Proceedings of 2008 International Congress on Advances in Nuclear Power Plants*, Paper 8231, Anaheim, CA, USA, June (2008).
18. H. Yamano, et al., "Unsteady Elbow Pipe Flow to Develop a Flow-Induced Vibration Evaluation Methodology for JSFR", *Proceedings of the International Conference on Fast Reactors and Related Fuel Cycles*, IAEA-CN-176-08-09, Kyoto, Japan, December (2009).
19. M. Tabuchi and Y. Takahashi, "Evaluation of Creep Strength Reduction Factors for Welded Joints of Modified 9Cr-1Mo Steel (P91)", *Proceedings of ASME Pressure Vessels and Piping Division Conference*, PVP2006-ICPVT-11-93350 Vancouver, Canada (2006).
20. T. Hiramatsu, et al., "Ultrasonic Flowmeter for JSFR", *Proceedings of the International Conference on Fast Reactors and Related Fuel Cycles*, IAEA-CN-176-02-11P, Kyoto, Japan (2009).
21. S. Kubo, Y. Shimakawa, H. Yamano, and S. Kotake, "Safety Design Requirements for Safety Systems and Components of JSFR", *Proceedings of the International Conference on Fast Reactors and Related Fuel Cycles*, IAEA-CN-176-03-10, Kyoto, Japan (2009).
22. K. Kurisaka, "Probabilistic Safety Assessment of Japanese Sodium Cooled Fast Reactor in Conceptual Design Stage", *Proceedings of the 15th Pacific Basin Nuclear Conference*, Sydney, Australia, October 15–20 (2006).
23. H. Hayafune, et al., "Development of the Integrated IHX/Pump Component 1/4-scale Vibration Testing", ICONE14, Paper 89745, Miami, FL, USA, July (2006).
24. T. Handa, et al., "Japan Research and Development for the Integrated IHX/Pump", *Proceedings of the International Conference on Fast Reactors and Related Fuel Cycles*, IAEA-CN-176-08-07, Kyoto, Japan (2009).
25. K. Kurome, et al., "Japan Steam Generator with Straight Double-Walled Tube – Development of Fabrication Technologies of Main Structures Made of High Chrome Steel-Made ?", *Proceedings of the International Conference on Fast Reactors and Related Fuel Cycles*, IAEA-CN-176-08-22P, Kyoto, Japan (2009).
26. N. Kisohara, et al., "Feasibility Studies on Double-Wall-Tube Type Primary Steam Generator", *Proceedings of FR'91*, Kyoto, Japan (1991).
27. S. Usui, T. Mihara, H. Obata, and S. Kotake, "Development of Advanced Loop-Type Fast Reactor in Japan (4): An Advanced Design of the Fuel Handling System for the Enhanced Economic Competitiveness", *Proceedings of 2008 International Congress on Advances in Nuclear Power Plants*, No. 8223, Anaheim, CA, USA, June (2008).
28. H. Hara, et al., "Japan Conceptual Design Study of JSFR (4) – Reactor Building Layout", *Proceedings of the International Conference on Fast Reactors and Related Fuel Cycles*, IAEA-CN-176-08-13P, Kyoto, Japan (2009).

Appendix D
Economics Calculational Approach

Pat Owen and Ronald Omberg

... the ideas of economists and political philosophers, both when they are right and when they are wrong, are more powerful than is commonly understood. Indeed, the world is ruled by little else. Practical men, who believe themselves quite exempt from any intellectual influences, are usually the slaves of some defunct economist. I am sure the power of vested interests is vastly exaggerated compared to the gradual encroachment of ideas ... soon or late, it is ideas, not vested interests, which are dangerous for good or evil ...
John Maynard Keynes, The General Theory of Employment, Interest, and Money

D.1 Introduction

Any enterprise, if it is to be viable for an extended period of time, must be able to sell a product for more than its cost. Or, insofar as the enterprise is concerned, revenues must be sufficient to cover expenditures. Since the very existence of the enterprise depends on this principle, an economic assessment of any project is an essential part of the decision to pursue it. This is particularly true for a project such as a power plant which entails a large investment.

The decision to construct a power plant involves both the disciplines of engineering analysis and economic analysis. Engineering analysis is used to determine the design of the plant, the efficiency of the plant, the life of components within the plant, and the time required to construct the plant. Economic analysis, on the other hand, is used to determine the cost of constructing the plant as well as the cost of operating the plant over an extended period of time.

Both engineering analysis and economic analysis should be employed in a balanced fashion in order to ensure the success of a large engineering project. For example, if engineering is emphasized at the expense of economics, the result will probably be a well-designed but economically uncompetitive plant. In contrast, if economics is emphasized at the expense of engineering, the result will probably be a plant which is initially inexpensive but which will subsequently incur large operating costs. Thus, a comprehensive understanding of both disciplines is desirable. Such an understanding of the engineering aspects is developed in other chapters; this appendix will develop a fundamental understanding of the economic aspects.

In the case of a nuclear power plant, the product that is ultimately sold in the marketplace is electricity. Consequently, it is important to know the average amount to charge for electricity since this determines revenues, and revenues must be sufficient to cover costs. Estimating the charge for electricity is particularly difficult in the case of a power plant because costs associated with the plant are incurred over several decades. The principal costs which are incurred are

---

This appendix was Chapter 3 of the original *Fast Breeder Reactor* book (Waltar and Reynolds, Pergamon Press, 1981).

(1) the *capital cost,* which is incurred while constructing the plant,
(2) the *fuel cost,* which is incurred while operating the plant,
(3) the *operation and maintenance cost,* which is incurred over the life of the plant,
(4) *income taxes,* which are incurred as a consequence of engaging in a private enterprise, and
(5) *other costs,* which are incurred as a consequence of engaging in any commercial enterprise.

Capital and fuel costs are often the major costs and are also the costs that can be strongly influenced by engineering design. For example, there is a large difference in capital costs between a current reactor such as a light-water reactor (LWR) and an advanced reactor such as a fast reactor. The standard LWR on the once-through fuel cycle has a relatively low capital cost; however, the fuel cost will increase as the price of uranium increases. In contrast, the fast spectrum reactor has a relatively high capital cost but a fuel cost which is insensitive to uranium price. Thus, the LWR tends to be fuel intensive while the fast spectrum reactor tends to be capital intensive.

In order to calculate a single average charge for electricity for each of these plants, it is necessary to develop a method that handles capital and fuel expenditures in a consistent manner. Since capital expenditures occur during the construction period and fuel expenditures occur over the life of the plant, it is necessary to compare expenditures that occur at different times. A method for accomplishing this will be developed in the next section.

D.2 Basic Concept of the Time Value of Money

The stock which is lent at interest is always considered as a capital by the lender. He expects that in due time it is to be restored to him, and that in the meantime the borrower is to pay him a certain annual rent for the use of it. The borrower may use it either as a capital, or as a stock reserved for immediate consumption. If he uses it as a capital, he employs it in the maintenance of productive labourers, who reproduce the value with a profit. He can, in this case, both restore the capital and pay the interest without alienating or encroaching upon any other source of revenue. If he uses it as a stock reserved for immediate consumption, he acts the part of a prodigal, and dissipates in the maintenance of the idle, what was destined for the support of the industrious. He can, in this case, neither restore the capital nor pay the interest without either alienating or encroaching upon some other source of revenue, such as the property or the rent of land.
Adam Smith, An Inquiry into the Nature and Causes of the Wealth of Nations

Money is a valuable asset—so valuable, in fact, that individuals and organizations are willing to pay additional money in order to have it available for their use. This is indicated by the continual willingness of banks and savings institutions to pay for the use of money. These institutions provide a market that determines the interest rate. We will use this interest rate to develop the time value of money. For example, consider \$100 borrowed from a bank for 1 year at an interest rate of 5% per year. Clearly, at the end of 1 year the bank will expect to receive \$105. This sum includes both a *return of* the investment (\$100) plus a *return on* the investment (\$5).

More generally, consider an amount C which is borrowed for 1 year at an annual interest rate i, as shown in Fig. D.1.

At the end of 1 year, the investment C must be repaid. In addition, an amount iC representing a return on the investment must also be paid. Therefore, the total amount that must be paid back at the end of the year is

$$\text{Pay back} = C + iC = C(1 + i)\,. \tag{D.1}$$

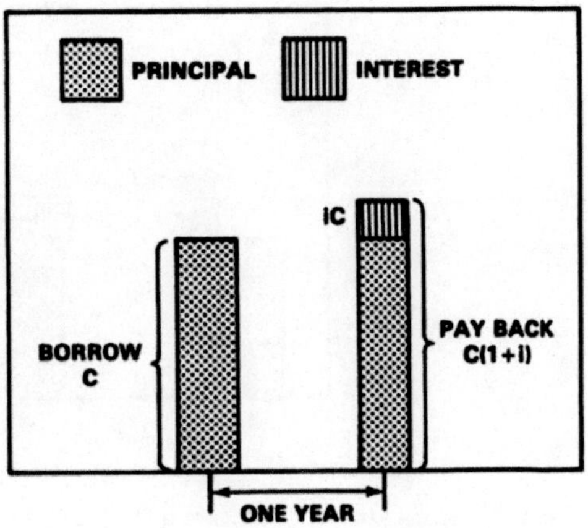

Fig. D.1 Borrowing money for 1 year

Let us now consider a more complex example. Suppose it is known that an expense of $100 will be incurred 1 year in the future. In order to pay the expense in 1 year, one may wish to set aside the appropriate amount of money now. This amount could, of course, be invested such that the future expense of $100 is covered exactly. If the interest rate on the investment is 5% per year, then $95.24 must be invested now. That is, $95.24 (1 + 0.05) = $100.00.

This concept, of course, can be generalized for any amount of money, as shown in Fig. D.2.

If an expense C will be incurred 1 year in the future, an amount C' may be set aside such that the future expense C will be covered. If C' is invested at an interest rate i, the amount available to pay out at the end of 1 year is

$$\text{Pay out} = C' + iC' = C'(1 + i). \tag{D.2}$$

Since we intend to use the amount $(1 + i)C'$ to cover the expense C, we would like

$$C'(1 + i) = C. \tag{D.3}$$

Therefore, the amount C' that must be invested at the beginning of the year in order to cover an expenditure C at the end of the year is

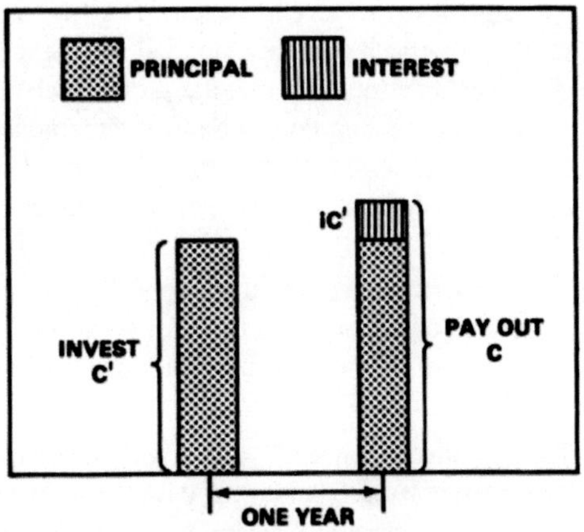

Fig. D.2 Meeting a known expense at the end of 1 year

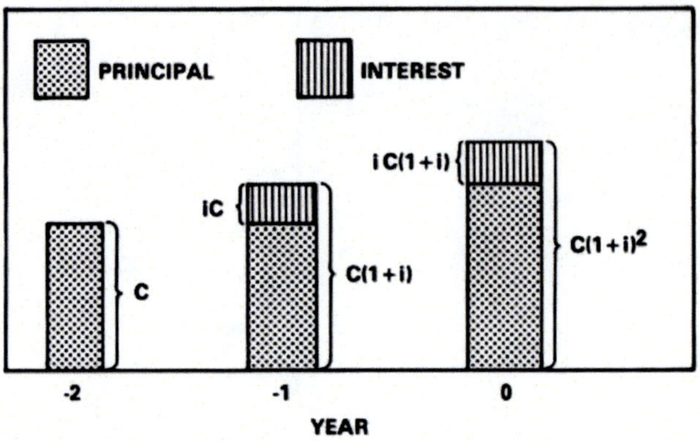

Fig. D.3 Present value of an expense in the past

$$C' = \frac{C}{1+i}. \tag{D.4}$$

These two examples will now be extended to develop the concept of the *current value* of an expense or revenue occurring at any point in time. With this concept, one can find an amount of money in any year that is equivalent to an amount in any other year. This concept, known as the *present value concept*, is very useful when analyzing the economics of a nuclear power plant.

It is useful to be able to determine the current value of either a cost that has occurred in the past or one that will occur in the future. Consider first the case of finding the present value of a cost that has occurred in the past. Assume for example, that a cost C was incurred 2 years ago, as shown in Fig. D.3. If this cost had not been incurred, the amount C would have been available for investment 2 years ago. If it had been invested, the amount C would increase in value as follows:

Available for investment 2 years ago: C
Value of investment 1 year ago: $C + iC = C(1 + i)$
Value of investment today: $C(1 + i) + i\,[C(1 + i)] = C(1 + i)^2$.

Thus, an *investment* of C 2 years ago has increased in value to the amount $C(1 + i)^2$ today. Alternatively, we may say that the current value of an *expenditure* C that occurred 2 years ago is $C(1 + i)^2$. More specifically, if we did not have that expenditure 2 years ago and invested it instead, we could have met an expenditure $C(1+i)^2$ today. This is known as the present value of a past expense C. This approach may be extended to an expense C that has occurred n years in the past. The reader should be able to verify that the present value of such a past expense is $C(1 + i)^n$.

A similar technique can be used for finding the present value of an expense that will occur in the future. For example, assume that the amount C must be paid out in 2 years, as shown in Fig. D.4. This expense could be met by investing an amount C' as follows:

Amount of investment today: C'
Value of investment in 1 year: $C' + iC' = C'(1 + i)$
Value of investment in 2 years: $C'(1 + i) + i\left[C'(1 + i)\right] = C'(1 + i)^2$.

Thus, an investment of C' today will increase in value to the amount $C'(1 + i)^2$ in 2 years. If we want an investment of C' today to cover an expenditure C in 2 years, then we would like $C'(1 + i)^2$ to equal C. The amount C' is known as the present value of the future expense C, and

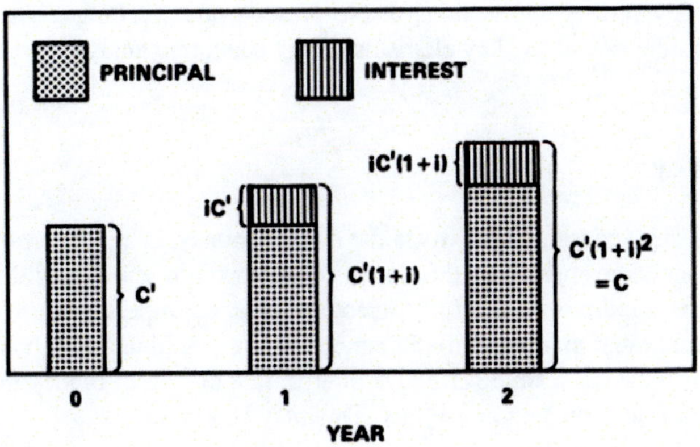

Fig. D.4 Present value of an expense in the future

$$C' = \frac{C}{(1+i)^2}. \tag{D.5}$$

This approach may be extended to an expense C that will occur n years in the future. The reader should be able to verify that the present value of such a future expense $C/(1+i)^n$.

To improve our understanding of the present value concept, let us consider a simple example involving nuclear fuel. The costs associated with this fuel involve (1) a purchase which occurs before the fuel is loaded into the reactor, and (2) an expenditure which occurs after the fuel is discharged from the reactor. The timing of a typical set of expenditures is shown in Fig. D.5. Since the expenditures occur at different times, the complete cost of the fuel is determined by both the magnitude and the timing of the expenditures. In order to calculate the present value of the expenditures, it is necessary to choose a reference point in time. For convenience, we choose to present-value all expenditures to the time at which the fuel is loaded into the reactor. With this as a reference point, the front-end expenditure F has occurred 1 year in the past while the back-end expenditure B will occur 4 years in the future. The complete fuel cost is the sum of the present value of the expenditures, or $F(1+i)^1 + \mathrm{B}/(1+i)^4$.

The present value concept can be used to find a single cost for fuel, as shown above. In addition, the present value concept may be applied to other components of the cost of a nuclear power plant. For example, with some additional analysis, it can be used to determine the amount that must be charged for electricity in order to cover the capital cost of the plant.

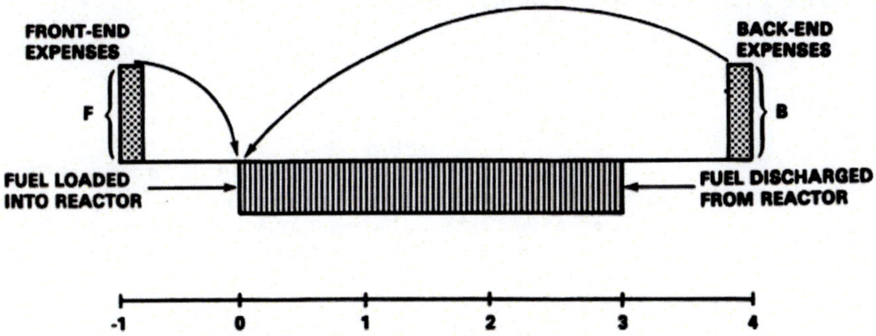

Fig. D.5 Present value of nuclear fuel expenditures

However, before any further applications of the present value technique are developed, we will digress briefly to consider one of the key elements of any economic analysis.

D.3 Cost of Money

A key component of any economic analysis is the cost of money, or the interest rate i. This value is a function of the financial arrangements developed for the project, and this will vary from project to project. Therefore it is important to carefully select the most appropriate value for the project under consideration. Nuclear power plants are often financed with a combination of bonds (debt) and stocks (equity). When this is the case, a single effective interest rate or cost of money may be defined as the weighted mean rate of return on debt and equity. This may be expressed as

$$i = (b \times i_b) + (e \times i_e)\,, \tag{D.6}$$

where

$i =$ effective interest rate
$b =$ fraction of funds obtained by debt (or bonds)
$i_b =$ debt (or bond) interest rate
$e =$ fraction of funds obtained by equity (or stocks)
$i_e =$ equity (or stock) interest rate.

Let us calculate an effective interest rate using typical values for bond and equity interest rates. For example, a *deflated* bond interest rate of 2.5% per year is often used in economic calculations. Deflated, in this case, implies a real rate of return of 2.5% per year and does not include the effect of inflation. Inflation, when included, simply raises the rate of return by the inflationary rate itself. A similarly deflated value for the equity interest rate would be 7% per year. It is common for utilities to finance nuclear power plants with a structure of 55% bonds and 45% stocks. In this case, the effective deflated interest rate would be

$$i = (0.55 \times 0.025) + (0.45 \times 0.07) = 0.045, \text{ or } 4.5\%\,. \tag{D.7}$$

An effective interest rate calculated in this fashion will be used in the subsequent development of the major cost components, including capital cost and the taxes associated with capital expenditures, and fuel costs and the taxes associated with fuel expenditures.

D.4 Capital Cost

All production is for the purpose of ultimately satisfying a consumer. Time usually elapses, however- and sometimes much time-between the incurring of costs by the producer and the purchase of the output by the ultimate consumer. Meanwhile the entrepreneur has to form the best expectations he can as to what the consumers will be prepared to pay when he is ready to supply them after the elapse of what may be a lengthy period; and he has no choice but to be guided by these expectations, if he is to produce at all by processes which occupy time.
John Maynard Keynes, The General Theory of Employment, Interest, and Money

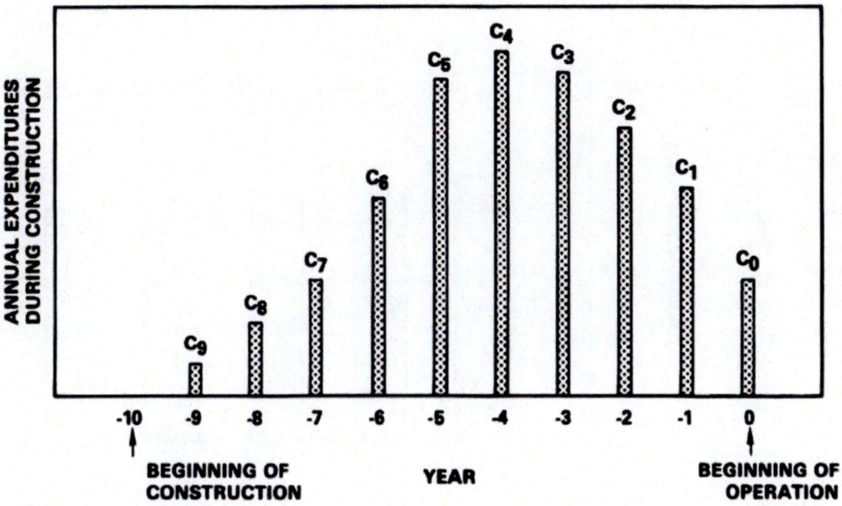

Fig. D.6 Annual construction costs

D.4.1 Construction Cost

The construction of a power plant occurs over an extended period of time and involves a sequence of nonuniform expenditures, as shown in Fig. D.6. The complete capital cost of the plant is determined by both the magnitude and timing of these expenditures. In order to apply the present value concept, it is necessary to choose a point in time at which to value the expenditures. For convenience, we will choose the time at which the reactor begins to generate electricity as a reference point.

The total capital cost C is the sum of the present value of all expenditures. If the construction period extends for N years, the total capital cost including interest during construction may be expressed as

$$C = C_0 + C_1(1+i) + C_2(1+i)^2 + \cdots + C_{N-1}(1+i)^{N-1}, \tag{D.8}$$

or

$$C = \sum_{k=0}^{N-1} C_k(1+i)^k, \tag{D.9}$$

where we have assumed that the payment is made at the end of the year.

D.4.2 Paying Back a Capital Investment

After the plant is constructed, it is necessary to recover revenues sufficient to repay the original investment. Let us develop the technique for calculating this revenue requirement with a simple example. For the sake of illustration, let us consider an initial investment of \$5,000. Suppose we intend to pay back this investment over a 5 year period with an annual interest rate of 5%. There are several methods that could be employed to pay back the investment. One such method would involve five equal payments of \$1,000 in order to return the principal. In addition to returning the principal, the interest on the outstanding balance must also be repaid each year. The total payment at the end of any year k is the sum of the two, and is given by

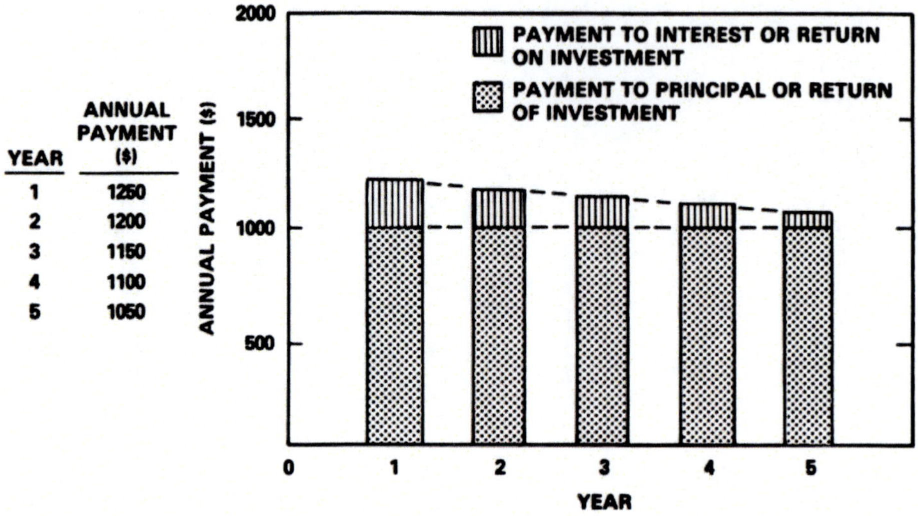

Fig. D.7 Repayment of a $5,000 capital investment with equal payments to principal

$$\text{Payment at end of year } k = \begin{pmatrix}\text{payment to}\\ \text{principal}\end{pmatrix} + \begin{pmatrix}\text{interest on out}-\\ \text{standing balance}\end{pmatrix}$$
$$= \frac{5000}{5} + 0.05\left[5000 - (k-1)\,\frac{5000}{5}\right]. \quad \text{(D.10)}$$

The annual payments for this method of paying back a capital investment are shown in Fig. D.7. The reader will note that the total annual payment decreases as the outstanding balance decreases.

In some cases, a uniform-rather than a decreasing-set of payments is more convenient. If $5,000 is borrowed for 5 years at an interest rate of 5%, the original investment could be paid back with uniform payments, as shown in Fig. D.8. The reader will note that, while the total annual payment is constant, the payment to principal increases in each succeeding year while the payment to interest simultaneously decreases.

Since uniform annual payments are often preferred, let us examine this method in more detail. Again, consider the previous example in which the total annual payment is $1,155. This can be regarded as a uniform annual cost to the project, and let us call this quantity C_u. Since both principal and interest must be paid, each payment C_u, is divided between return *of* the investment and return *on* the investment. The division between payments to principal and interest is shown in Table D.1. This table was constructed by first considering the outstanding debt in any year k, denoted here as D_k. In the same year k, the payment to interest I_k is $0.05 \times D_k$. The payment to principal P_k is then given by

$$P_k = C_u - I_k = C_u - 0.05 \times D_k. \quad \text{(D.11)}$$

Consequently, the outstanding debt at the beginning of year $k + 1$ is

$$D_{k+1} = D_k - P_k = D_k - (C_u - 0.05 \times D_k)\,, \quad \text{(D.12)}$$

and one must continue to make the annual payment C_u until the outstanding debt D_{k+1} is reduced to zero. In this case, the original capital investment will be completely paid back at the end of 5 years.

This example has verified that a uniform annual payment of $1,155 will completely return the original capital investment over 5 years. However, it is not evident how one initially determines this payment. This can be accomplished by considering the problem from the viewpoint of the lender.

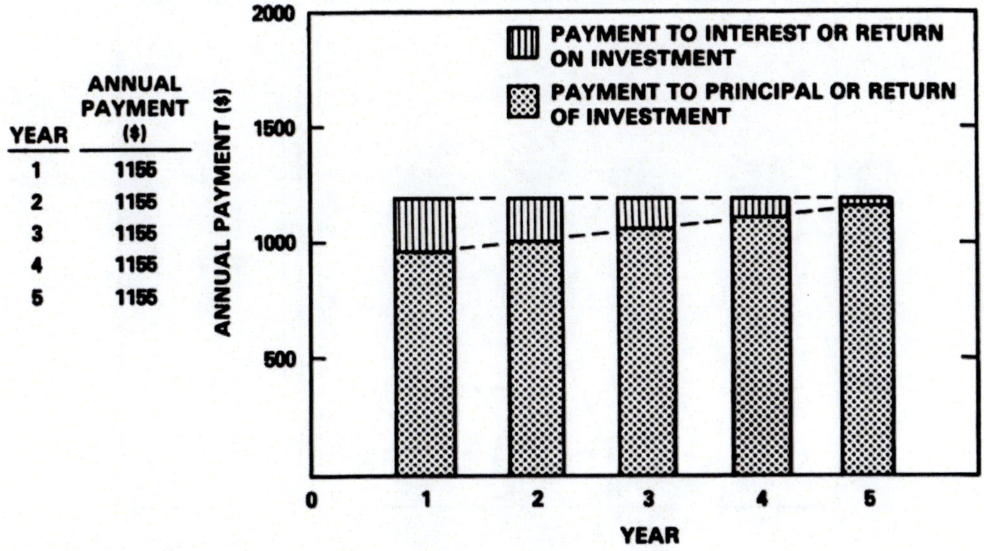

Fig. D.8 Repayment of a \$5,000 capital investment with equal annual payments

Table D.1 Payments to principal and interest with the uniform annual payment method

Year (k)	Outstanding debt at beginning of year (D_k)	Payment to interest ($I_k = 0.05 \times D_k$)	Payment to principal ($P_k = C_u - 0.05 \times D_k$)	Outstanding debt at end of year (D_{k+1})
1	5,000	250	905	4,095
2	4,095	205	950	3,145
3	3,145	157	998	2,147
4	2,147	108	1,047	1,100
5	1,100	55	1,100	0

Recall that the lender originally loaned \$5,000, which was to be returned in five equal annual payments. The lender, considering the time value of money, will require \$5,000 $(1.05)^5$ at the end of 5 years. This is because, by lending the money, he foregoes the opportunity to invest it. Thus, the sum of the present value of the payments C_u must be \$5,000 $(1.05)^5$ at the end of 5 years—or, as shown in Fig. D.9,

$$5000(1.05)^5 = C_u + C_u\,(1.05) + C_u\,(1.05)^2 + C_u\,(1.05)^3 + C_u\,(1.05)^4\,. \tag{D.13}$$

Solving for C_u,

$$5000 = \left(\frac{1}{(1+0.05)^5} + \frac{1}{(1+0.05)^4} + \frac{1}{(1+0.05)^3} + \frac{1}{(1+0.05)^2} + \frac{1}{(1+0.05)}\right) C_u \tag{D.14}$$

and the annual payment C_u, is found to be \$1,155.

In more general terms, if an initial investment C is to be paid back uniformly over K years at an annual interest rate of i, then

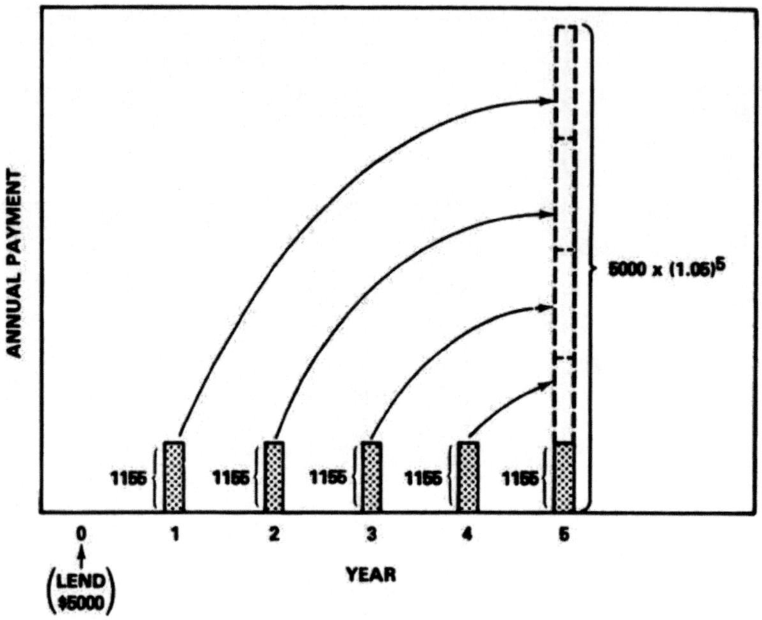

Fig. D.9 Uniform annual payment for a capital investment

$$C = \left[\frac{1}{(1+i)^K} + \frac{1}{(1+i)^{K-1}} + \cdots + \frac{1}{(1+i)^3} + \frac{1}{(1+i)^2} + \frac{1}{(1+i)}\right] \times C_u. \tag{D.15}$$

Solving for the uniform annual payment C, gives

$$C_u = \frac{C}{\sum_{k=1}^{K} \frac{1}{(1+i)^k}}. \tag{D.16}$$

Regarding $1/(1+i)$ as a term in a geometric progression, one can show with elementary algebra that C_u can be expressed as

$$C_u = C \times \left[\frac{i \times (1+i)^K}{(1+i)^K - 1}\right]. \tag{D.17}$$

This expression is commonly known as the *sinking-fund* repayment equation or the *amortization equation.*

The annual payment C_u, which represents a cost to the project, must be covered by revenues obtained from the sale of electricity. Revenues are determined by the amount charged for a unit of electricity multiplied by the number of units produced. If we define L_{cap} as the amount charged per unit of electrical energy in order to cover capital expenditures and E as the amount of electrical energy produced in any year, then

$$\text{Revenue} = L_{cap} \times E. \tag{D.18}$$

The revenue in each year will cover the expenditure in the year if $L_{cap} \times E = C_u$. Thus, the constant, or *levelized,* charge for electricity sufficient to cover capital expenditures is

$$L_{cap} = \frac{C_u}{E}, \tag{D.19}$$

and recalling the sinking fund repayment equation

$$L_{cap} = \frac{C}{E} \times \left[\frac{i \times (1+i)^K}{(1+i)^K - 1}\right]. \tag{D.20}$$

D.4.3 Fixed Charges Associated with a Capital Investment

In the previous section, a method was developed for calculating the return of and the return on an investment. There are, however, other annual expenses associated with capital investments such as property insurance, property taxes, and replacement costs. These are collectively known as *fixed* charges and are usually expressed as a fixed percent of the original capital investment. In fact, since the value of the property decreases as the power plant ages, these annual costs may actually decrease with time. However, for simplicity we will consider that the fixed charges are a constant proportion, f, of the initial capital investment throughout the life of the plant.

For an original investment of C, the constant annual fixed charges may be expressed as

$$\text{Annual fixed charges } = f \times C.$$

While the value of f will vary with location, tax structure, and type of plant, a value of 5% is typically used for a nuclear power plant. Since annual fixed charges represent a cost to the project, they must be covered by revenues obtained from the sale of electricity. If we define L_{fc} as the levelized charge for a unit of electricity sufficient to cover fixed charges, then $L_{fc} \times E$ is the revenue collected in any single year. If this revenue is to cover expenditures, then

$$L_{fc} \times E = f \times C, \tag{D.21}$$

or

$$L_{fc} = f \times C/E. \tag{D.22}$$

D.4.4 Taxes Associated with Capital Expenses

If a power plant is owned by a private utility, then taxes must be paid on revenues collected to cover capital expenses. The calculation of taxes is complicated by provisions in the tax law, which allows certain costs to be deducted from revenues before taxes are assessed. Although deductible costs are ultimately determined by the law, typical deductions for a nuclear power plant include items such as the interest paid on bonds and the amount of depreciation allowed on capital equipment. Since taxes are paid on revenues after deductions, the tax payment in any year may be represented as revenues, in this case, include money collected to pay back the original capital investment. In addition, revenues must also be sufficient to pay the taxes associated with this investment.

Table D.2 Interest payment to debt and equity

Year (k)	Payment to interest $0.05 \times D_k$	Payment to bond interest $\frac{i_b b}{i_b b + i_e e} \times 0.05 \times D_k$	Payment to equity interest $\frac{i_e e}{i_b b + i_e e} \times 0.05 \times D_k$
1	250	76	174
2	205	62	143
3	157	48	109
4	108	33	75
5	55	17	38
	$i_b b/(i_b b + i_e e) = 0.304$	$i_e e/(i_b b + i_e e) = 0.696$	

$$\begin{pmatrix}\text{taxes on}\\ \text{capital}\end{pmatrix} = (\text{tax rate}) \times \left[(\text{revenues}) - \begin{pmatrix}\text{interest}\\ \text{paid on}\\ \text{bonds}\end{pmatrix} - (\text{depreciation})\right]. \tag{D.23}$$

As noted previously, the annual revenue required to cover a capital investment is given by $L_{\text{cap}} \times E$ where L_{cap} is the levelized charge for a unit of electricity necessary to cover the return of and the return on the investment. If we define L_{ctax} as the levelized charge for a unit of electricity sufficient to cover taxes, then the annual revenue which will be obtained to cover both capital investment and taxes is

$$\text{Revenue} = \left(L_{\text{cap}} + L_{\text{ctax}}\right) \times E. \tag{D.24}$$

Note that, while we have defined L_{ctax} we have yet to calculate it.

As indicated in Eq. (D.23), interest paid on bonds is a deductible cost. In general, it is somewhat difficult to calculate exactly the interest paid on bonds. It can, however, be calculated exactly with the sinking-fund repayment equation [Eq. (D.17)], which was developed in the previous section. This expression allowed us to calculate the annual cost to the project necessary to cover the return of and the return on a capital investment, where the return on a capital investment includes payment to both debt and equity.

In order to calculate the interest payment to debt alone, let us return to the example shown in Table D.1. Since the total interest payment in that example is given by $0.05D_k$, then the bond interest payment will be

$$\text{bond interest payment} = \frac{i_b b}{i_b b + i_e e} \times 0.05D_k. \tag{D.25}$$

Similarly, the equity interest payment would be

$$\text{Equity interest payment} = \frac{i_e e}{i_b b + i_e e} \times 0.05D_k. \tag{D.26}$$

Therefore, if the debt and equity interest rates and fractions are known, a series of payments to bond interest can be constructed. This is shown in Table D.2 for the example from the previous table. Note that the payment to bond interest varies year by year.

In addition to bond interest, depreciation can also be deducted from revenues prior to the payment of taxes. Two commonly used methods, the straight-line method and the sum-of-years digits method, will be discussed here although any reasonable and consistent depreciation method can be used. The straight-line method assumes deductions due to depreciation will be taken in a uniform

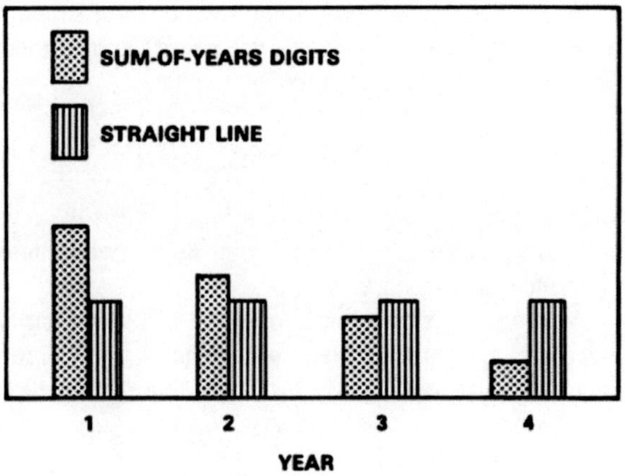

Fig. D.10 Comparison of the depreciation allowance using the straight line or sum-of-years digits method

manner throughout the lifetime of the plant. If C represents the initial capital investment, K the plant lifetime, and $c\mathrm{dep}_k$ the depreciation allowance in year k, then

$$c\mathrm{dep}_k = \frac{C}{K}. \tag{D.27}$$

Alternatively, in the sum-of-years digits method, it is assumed that the largest deductions will be taken early in the plant life and that the depreciation deduction decreases with time. In this case, the depreciation allowance in any year k is given by

$$c\mathrm{dep}_k = \frac{C\,(K+1-k)}{\sum_{j=1}^{K} j}. \tag{D.28}$$

The depreciation allowance which results from these two methods of depreciating capital costs is shown in Fig. D.10. The sum-of-years digits method is usually preferred since it reduces taxes in the early years of the project. Both methods, however, produce the same total depreciation over the life of the project. The advantage of the sum-of-years digits method is not that it foregoes tax payments, but rather that it defers them.

Since both the bond interest payment and the depreciation allowance vary throughout the lifetime of the plant, it should be clear that the tax payment will also vary. Thus, the revenue required to cover taxes on capital will vary from year to year. It is still desirable, however, to compute a single levelized value that can be charged for electricity over an extended period of time in order to cover taxes on capital. This can be done using a fundamental equation in economics that equates expenditures and revenues accounting for the time value of money in both cases. This may be expressed as

$$\sum_{k=1}^{K} \frac{\mathrm{Exp}_k}{(1+i)^k} = \sum_{k=1}^{K} \frac{\mathrm{Rev}_k}{(1+i)^k} \tag{D.29}$$

where Exp_k is the expenditure in year k and Rev_k is the revenue allocated in year k to cover the expenditure. Note that it is not necessary for the revenue in year k to precisely equal the expenditure in that year. Rather, since money has a time value, it is only necessary that the sum of the present value of the revenues equal the sum of the present value of the expenditures.

Given this fundamental economic equality, suppose that the expenditures are taxes and that we wish to determine the revenues necessary to cover them. Recall from Eq. (D.23) that the expenditure for taxes in any year k is given by

$$\mathrm{Exp}_k = t \times \left[\left(L_{\mathrm{cap}} + L_{\mathrm{ctax}}\right) \times E - \mathrm{cbin}_k - \mathrm{cdep}_k\right] \tag{D.30}$$

where t is the tax rate and $\left(L_{\mathrm{cap}} + L_{\mathrm{ctax}}\right) \times E$ is the revenue required to cover both capital investment and the taxes. We have defined cbin_k as the bond interest in any year k and cdep_k as the depreciation allowance in any year k.

Although the expenditure for taxes will vary from year to year, the levelized charge for a unit of electricity L_{ctax} will produce the revenue $L_{\mathrm{ctax}} \times E$ in any year. This revenue, when present-valued, should be sufficient to cover tax expenditures over the life of the plant. Or, in the form of an equation,

$$\sum_{k=1}^{K} \frac{t \times \left[\left(L_{\mathrm{cap}} + L_{\mathrm{ctax}}\right) \times E - \mathrm{cbin}_k - \mathrm{cdep}_k\right]}{(1+i)^k} = \sum_{k=1}^{K} \frac{L_{\mathrm{ctax}} \times E}{(1+i)^k}. \tag{D.31}$$

It is now possible to solve for the levelized charge for electricity sufficient to cover taxes L_{ctax}. Doing so gives

$$L_{\mathrm{ctax}} = \frac{t}{1-t} \frac{\displaystyle\sum_{k=1}^{K} \frac{L_{\mathrm{cap}} \times E - \mathrm{cbin}_k - \mathrm{cdep}_k}{(1+i)^k}}{\displaystyle E \times \sum_{k=1}^{K} \frac{1}{(1+i)^k}}. \tag{D.32}$$

This complicated expression for taxes is often very difficult to evaluate. It can be simplified somewhat by rewriting the expression to separate the bond interest term as follows:

$$L_{\mathrm{ctax}} = \frac{t}{1-t} \left[\frac{\displaystyle\sum_{k=1}^{K} \frac{L_{\mathrm{cap}} \times E - \mathrm{cdep}_k}{(1+i)^k}}{\displaystyle E \times \sum_{k=1}^{K} \frac{1}{(1+i)^k}}\right] - \frac{t}{1-t} \left[\frac{\displaystyle\sum_{k=1}^{K} \frac{\mathrm{cbin}_k}{(1+i)^k}}{\displaystyle E \times \sum_{k=1}^{K} \frac{1}{(1+i)^k}}\right]. \tag{D.33}$$

If we make the simplifying assumption that the depreciation allowance is constant each year, then the quantity $L_{cap} \times E - \mathrm{cdep}_k$ is also constant. In this case, the numerator of the first term may be factored through the summation to give

$$L_{\mathrm{ctax}} = \frac{t}{1-t} \frac{\left(L_{\mathrm{cap}} \times E - \mathrm{cdep}\right)}{E} - \frac{t}{1-t} \left[\frac{\displaystyle\sum_{k=1}^{K} \frac{\mathrm{cbin}_k}{(1+i)^k}}{\displaystyle E \times \sum_{k=1}^{K} \frac{1}{(1+i)^k}}\right]. \tag{D.34}$$

We will now develop a method for evaluating the bond interest term. Recall that the payment to bond interest varies from year to year, and so it cannot be factored through the summation. It is possible, however, to develop a relatively simple expression for the sum of the present value of the payments to bond interest. The development is based on the fundamental principles involved in repaying a capital investment.

Recall that a capital investment may be repaid in a uniform fashion with an annual payment C_u, and this quantity is calculated using the sinking-fund repayment equation. The annual payment is divided between principal and interest, and in any year k

$$C_u = P_k + I_k, \text{ or}$$

$$P_k = C_u - I_k. \tag{D.35}$$

Recall also that the recursion relation for the outstanding debt, D_k, is

$$D_{k+1} = D_k - P_k. \tag{D.36}$$

We can now calculate the quantities D_k, I_k, and P_k for each year. Note that the outstanding debt at the beginning of the first year is the initial investment C, so

$$P_1 = C_u - I_1 = C_u - iC, \tag{D.37}$$

and

$$D_2 = D_1 - P_1 = C - (C_u - iC) = C(1+i) - C_u. \tag{D.38}$$

This sequence of calculations can be repeated for years two and three, as shown in Table D.3, and for each additional year of the plant lifetime. It can be seen that payment to principal in any year k can be expressed in terms of the annual payment C_u and the initial investment C as follows:

$$P_k = C_u(1+i)^{k-1} - iC(1+i)^{k-1}.$$

We can now express the total payment to interest in a single year k as

$$\begin{aligned} I_k &= C_u - P_k \\ &= C_u - C_u(1+i)^{k-1} + iC(1+i)^{k-1}. \end{aligned} \tag{D.39}$$

The fraction of this payment that is made to bond interest is given by

$$\begin{aligned} \text{cbin}_k &= \frac{bi_b}{i} I_k \\ &= \frac{bi_b}{i}\left[C_u - C_u(1+i)^{k-1} + iC(1+i)^{k-1}\right]. \end{aligned} \tag{D.40}$$

We can now use this expression for cbin_k to develop a simple formula for the sum of the present value of the payments to bond interest.

$$\begin{aligned} \sum_{k=1}^{K} \frac{\text{cbin}_k}{(1+i)^k} &= \sum_{k=1}^{K} \frac{bi_b}{i} \frac{\left[C_u - C_u(1+i)^{k-1} + iC(1+i)^{k-1}\right]}{(1+i)^k} \\ &= \sum_{k=1}^{K} \frac{bi_b}{i}\left[\frac{C_u}{(1+i)^k} - \frac{C_u}{(1+i)} + \frac{iC}{(1+i)}\right] \\ &= \frac{bi_b}{i}\left[C - \frac{KC_u}{(1+i)} + \frac{KiC}{(1+i)}\right]. \end{aligned} \tag{D.41}$$

Table D.3 Calculation of annual payments to principal and interest

		Year			
		1	2	3	k
Outstanding debt at beginning of year	D_k	C	$C(1+i)-C_u$	$C(1+i)^2$ $-C_u(2+i)$	–
Annual payment to interest	$I_k = iD_k$	iC	$iC(1+i)-iC_u$	$iC(1+i)^2$ $-iC_u(2+i)$	–
Annual payment to principal	$P_k = C_u - I_k$	$C_u - iC$	$C_u(1+i)$ $-iC(1+i)$	$C_u(1+i)^2$ $-iC(1+i)^2$	$C_u(1+i)^{k-1}$ $-iC(1+i)^{k-1}$
Outstanding debt at end of year	$D_{k+1} = D_k - P_k$	$C(1+i)-C_u$	$C(1+i)^2$ $-C_u(2+i)$	$C(1+i)^3$ $-C_u[(1+i)^2$ $+(2+i)]$	–

This expression may now be substituted into Eq. (D.34) to obtain the following equation for L_{ctax}

$$L_{\text{ctax}} = \frac{t}{1-t}\frac{(L_{\text{cap}} \times E - \text{cdep})}{E} - \frac{\left(\frac{t}{1-t}\right)\left(\frac{bi_b}{i}\right)\left[C - \frac{KC_u}{(1+i)} + \frac{KiC}{(1+i)}\right]}{E\left[\frac{(1+i)^K - 1}{i \times (1+i)^K}\right]}. \tag{D.42}$$

In summary, we have developed the three levelized charges necessary to cover all expenditures associated with capital. They are

L_{cap}: The levelized charge for electricity necessary to pay for the return on and return of a capital investment.

L_{fc}: The levelized charge for electricity necessary to pay for fixed charges, such as property taxes and insurance.

L_{ctax}: The levelized charge for electricity necessary to pay for taxes associated with a capital investment.

D.5 Fuel Cycle Expenses

> ... the immense strength [of science] lies in its power of accurate prediction. There is a minimum of fumbling. Trial-and-error methods are reduced to a level which was quite unknown in former phases of human activity. An aero-engineer who, applying scientific principles, designs a new aircraft not only knows, even in the drawing board stage, that his aircraft will be capable of flying, but he also can forecast, with a fair degree of accuracy, its performance. ... The value of [scientific] prediction lies not so much in allowing us to see into the future, but in permitting us to exclude many avenues of progress which have no future. The scientific method directs human effort, not by helping us to see farther afield, but by limiting our progress along lines which have a reasonable chance of leading us where we hope to go. ... The ability to limit progress to profitable avenues carries in itself that economy of effort which has made science so successful.
> *Kurt Mendelssohn, The Secret of Western Domination*

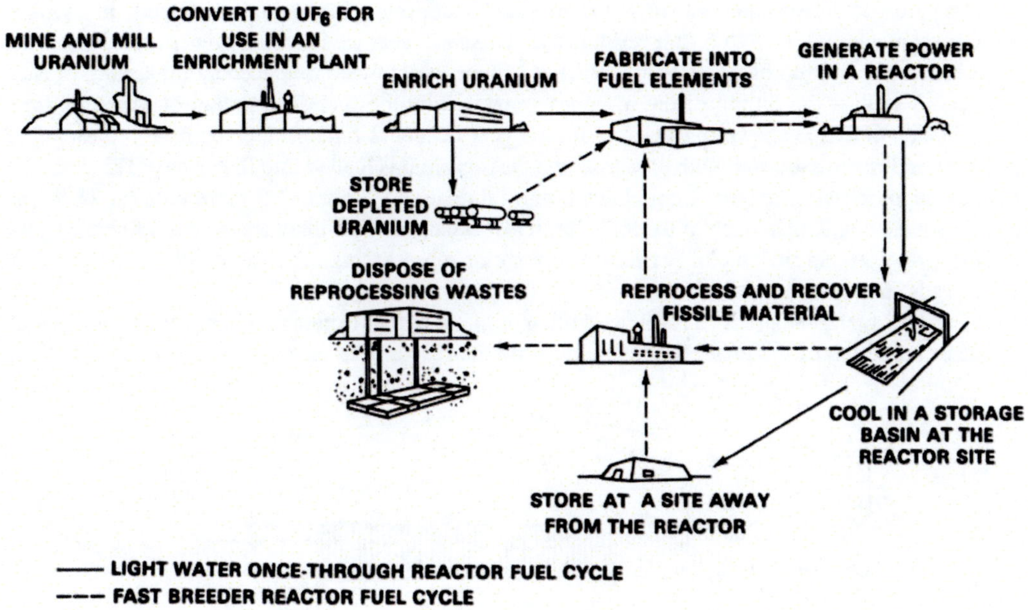

Fig. D.11 Nuclear fuel cycle

The ability to accurately assess the fuel cost of a power plant is often crucial to the success of the project. A plant that is destined to encounter ever-increasing fuel costs might be considered an avenue of progress with little future. In fact, the very health of the economy of the industrial nations is contingent upon identifying and avoiding such avenues. Nuclear power, with its relatively stable fuel costs, then becomes an attractive avenue of progress.

The evaluation of a nuclear fuel cost, however, is complex since the nuclear *fuel cycle* is composed of many processes that occur at different points in time. The processes that comprise the fuel cycle for both an LWR and a fast spectrum reactor are shown in Fig. D.11. Consider, for a moment, a single fuel element or assembly that produces power within a nuclear reactor. This fuel element must first be prepared for use in the reactor-that is, fissile material must be purchased and must then be fabricated into a fuel pin. The time required for fuel preparation is known as the *lead time,* and is generally about 1 year. Many such fuel elements, which comprise a *batch* of fuel, are loaded into the reactor and generate power for a number of years. In light water reactors and fast reactors, a fuel element typically resides in the reactor for a period of 2–6 years. Refueling generally occurs once a year through once every two years when a new batch is loaded and an irradiated batch is discharged. The irradiated element is then stored in a cooling basin at the reactor site and remains there for at least 1 year. The fuel element may then be shipped to a storage site or to a reprocessing plant. If it is reprocessed, the residual fissile material is extracted and recycled for use in another or in the same reactor. Waste products from the reprocessing stream are then permanently stored. The time between discharge from the reactor and either reprocessing or ultimate disposal is known as the *lag time*.

D.5.1 Direct Expenses for a Single Batch of Fuel

In calculating the fuel expense incurred during the operation of a nuclear power plant, it is common to consider all the expenses associated with a single batch of fuel. Recall that the expenses associated with preparing and disposing of a single batch of fuel occur throughout a period of years. In a typical light water reactor with a 3 to 6 year residence time and a 1-year lead and lag time, the fuel expenses will occur over a period of 5 to 8 years. Thus, it is necessary to take into account the timing of each expense as well as the timing of the revenues generated by the fuel. The timing of expenses for a typical batch of fuel is shown in Fig. D.12 for a 3 year residence time. In this figure, F represents all the front-end costs associated with preparing the fuel before it is loaded into the reactor. This includes the cost of purchasing uranium, enrichment, and fabrication. Similarly, B represents the back-end costs associated with disposing of the fuel after it is discharged. This includes the cost of shipping the used (spent) fuel and the cost of permanent storage or reprocessing. The timing of these individual expenses is shown in Fig. D.13.

A batch of fuel must produce revenues sufficient to cover fuel expenses. Each year, a single batch produces a fraction of the total energy produced by the reactor. For example, in a reactor with three

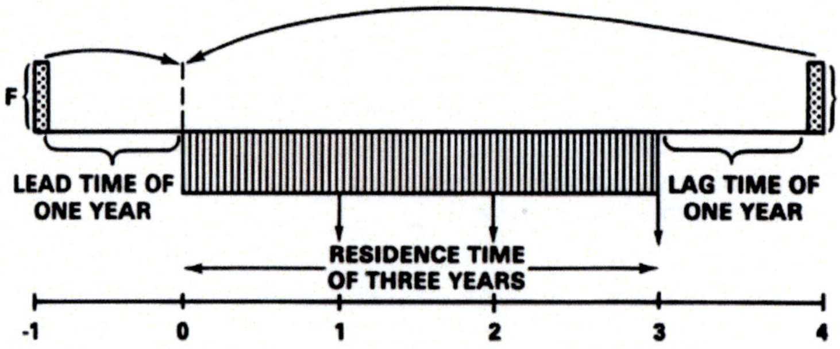

Fig. D.12 Timing of expenses of a typical batch of fuel

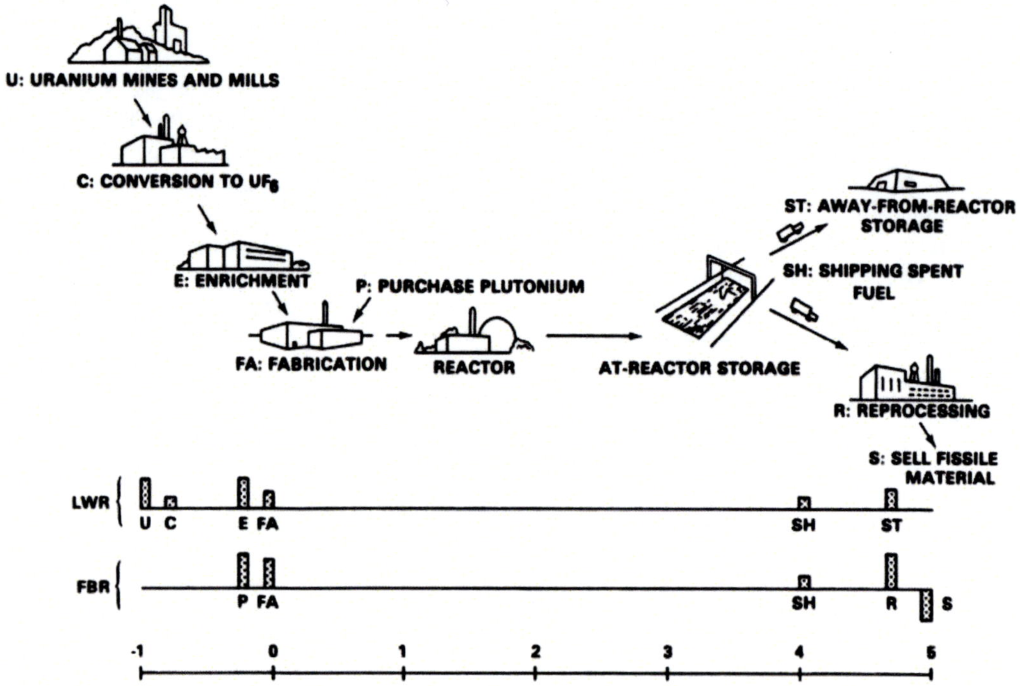

Fig. D.13 Timing of expenses for a batch of fuel

batches, a single batch produces approximately one-third of the total energy production in each year. It is possible to calculate a levelized charge for electricity sufficient to cover fuel expenses from this fraction of the total energy production. The levelized charge may be found by performing the following calculations:

(1) First, identify all revenues and expenses associated with the batch of fuel in each year. Revenues will, of course, depend upon the amount charged for electricity produced by the batch.
(2) Second, using the time at which the batch is loaded into the reactor as a reference point, calculate the present value of all expenses. Similarly, calculate the present value of all revenues. The choice of a reference point is arbitrary, but the start of operation is generally chosen for convenience.
(3) Next, equate the sum of the present value of the revenues to the sum of the present value of the expenses.
(4) Finally, solve this expression for the levelized charge for electricity sufficient to cover the expenses associated with the batch of fuel.

Let us consider a reactor that contains N batches of fuel. Assume that each batch resides in the reactor for a period of N years with a lead time ld and a lag time lg. Also assume that each batch generates the same fraction of the total energy produced by the reactor in each year. If we define L_b as the levelized charge for electricity necessary to cover the expenses associated with a batch of fuel, then the revenues obtained from the sale of electricity each year will be $L_b\, E/N$. We can now equate the present value of the revenues to the present value of the expenses, as shown in Fig. D.14.

$$\frac{L_b\, E/N}{(1+i)} + \frac{L_b\, E/N}{(1+i)^2} + \cdots + \frac{L_b\, E/N}{(1+i)^N} = F(1+i)^{ld} + \frac{B}{(1+i)^{lg+N}}. \tag{D.43}$$

Solving for the levelized charge for electricity necessary to cover the expenses associated with a batch of fuel L_b gives

$$L_b = \frac{F(1+i)^{ld} + \dfrac{B}{(1+i)^{lg+N}}}{\dfrac{E}{N}\left[\dfrac{1}{(1+i)} + \dfrac{1}{(1+i)^2} + \cdots + \dfrac{1}{(1+i)^N}\right]}. \tag{D.44}$$

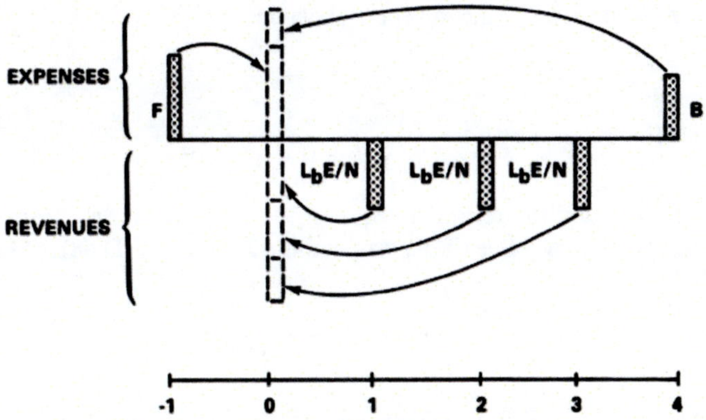

Fig. D.14 Present value of expenses and revenues for a batch of fuel

D.5.2 Taxes Associated with a Single Batch of Fuel

For some nuclear plants, tax law requires fuel to be considered a capital investment rather than an operating expense. In this case, additional revenues must be collected in order to pay the taxes associated with fuel. Let us denote the levelized charge for electricity to cover taxes associated with a batch of fuel as L_{btax}. The revenues obtained from the sale of electricity required to cover both direct expenses and taxes on a batch each year will then be $(L_b + L_{btax})\,E/N$. In this case, we must again equate the present value of revenues and expenditures. If T represents the total tax payment present-valued to the beginning of energy production by the batch, then

$$\frac{(L_b + L_{btax})\,E/N}{(1+i)} + \frac{(L_b + L_{btax})\,E/N}{(1+i)^2} + \cdots + \frac{(L_b + L_{btax})\,E/N}{(1+i)^N}$$

$$= F(1+i)^{ld} + \frac{B}{(1+i)^{lg+N}} + T. \tag{D.45}$$

Subtracting Eq. (D.43) from Eq. (D.45) gives

$$L_{btax}\,E/N\left[\frac{1}{(1+i)} + \frac{1}{(1+i)^2} + \cdots + \frac{1}{(1+i)^N}\right] = T. \tag{D.46}$$

Our problem now becomes one of calculating the present value of the tax payment T. The present value of the tax payment T can be calculated by recalling that (1) taxes are paid on revenues, and (2) revenues can be reduced by depreciation prior to the payment of taxes. Thus, if t represents the tax rate and D the depreciation allowance in a single year, the present value of the tax payment is

$$T = t\,(\text{Revenues} - \text{Depreciation}) \tag{D.47}$$

$$= t\left[\frac{\{(L_b + L_{btax})\,E/N\} - D}{(1+i)} + \frac{\{(L_b + L_{btax})\,E/N\} - D}{(1+i)^2} + \cdots + \frac{\{(L_b + L_{btax})\,E/N\} - D}{(1+i)^N}\right]$$

where $(L_b + L_{btax})\,E/N$ represents the revenue generated in any year. Collecting terms in this equation gives

$$T = t\left[\{(L_b + L_{btax})\,E/N\} - D\right]\left[\frac{1}{(1+i)} + \frac{1}{(1+i)^2} + \cdots + \frac{1}{(1+i)^N}\right]. \tag{D.48}$$

Recall that Eq. (D.46) equated the revenues required to pay taxes to the present value of the tax payment. Substituting the present value of the tax payment T from Eq. (D.48) into Eq. (D.46) gives

$$L_{btax}\,E/N\left[\frac{1}{(1+i)} + \frac{1}{(1+i)^2} + \cdots + \frac{1}{(1+i)^N}\right] = T.$$

$$= t\left[\{(L_b + L_{btax})\,E/N\} - D\right]\left[\frac{1}{(1+i)} + \frac{1}{(1+i)^2} + \cdots + \frac{1}{(1+i)^N}\right]. \tag{D.49}$$

Solving this equation for L_{btax} gives

$$L_{btax}\, E/N = t\left[\{(L_b + L_{btax})\, E/N\} - D\right], \tag{D.50}$$

or

$$L_{btax} = \frac{t}{1-t}\left(L_b - DN/E\right). \tag{D.51}$$

We now have an explicit expression for the levelized charge for electricity necessary to cover the taxes associated with a single batch of fuel in terms of L_b and D. We have previously calculated the levelized charge L_b required to cover the direct expenses associated with a single batch, as shown in Eq. (D.44). We have yet to calculate the depreciation allowance D.

The depreciation allowance for a batch of fuel is often assumed to be directly proportional to the amount of energy produced by the batch in any year. A batch produces a cumulative amount of energy E throughout its residence time of N years. We will assume it will produce the same amount of energy E/N in any year, and the amount of depreciation that may be deducted from revenues in any year is

$$D = \text{Dep}\left(E/N\right)/E \tag{D.52}$$

$$= \text{Dep}/N \tag{D.53}$$

where Dep is the total depreciation allowance. This approach is often called the *unit* of *production method.*

The value of Dep depends upon current tax law, and hence may change with time. Currently all direct expenses associated with a batch can be depreciated. One must be very careful, however, when incorporating the time value of money into a calculation of depreciation. At present, the time value of money can only be considered when determining the value of expenses incurred either before or after operation. It cannot, as indicated in Fig. D.15, be used to present-value all expenditures to a common reference time.

With this approach,

$$\text{Dep} = F(1+i)^{ld} + \frac{B}{(1+i)^{lg}}. \tag{D.54}$$

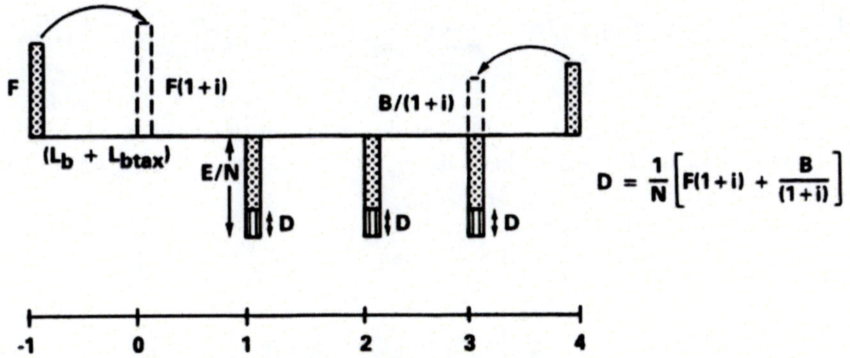

Fig. D.15 Calculation of depreciation for a batch of fuel

We may now substitute Eq. (D.44) for L_b and Eqs. (D.53) and (D.54) for D into Eq. (D.51) to obtain L_{btax}, which is

$$L_{btax} = \frac{t}{1-t}\left[\frac{F(1+i)^{ld} + \dfrac{B}{(1+i)^{lg+N}}}{\dfrac{E}{N}\left[\dfrac{1}{(1+i)} + \dfrac{1}{(1+i)^2} + \cdots + \dfrac{1}{(1+i)^N}\right]} - \frac{\left(F(1+i)^{ld} + \dfrac{B}{(1+i)^{lg}}\right)}{E}\right]. \tag{D.55}$$

Recall now we have developed two components associated with the charge for electricity necessary to cover fuel expenses. They are

L_b: The levelized charge for electricity necessary to cover the direct expenses associated with a batch of fuel.

L_{btax}: The levelized charge for electricity necessary to cover the taxes associated with a batch of fuel.

The total levelized charge to cover all expenses associated with a batch of fuel is then $L_b + L_{btax}$.

D.5.3 Levelized Charge to Cover Fuel Expenses

Often, the ability to calculate the levelized charge for electricity from a single but typical batch is adequate for design analysis. The levelized charge, however, can vary from batch to batch. In particular, the batches that comprise the initial core and the final core differ from equilibrium batches. If it is desirable to calculate an average charge for electricity over the life of the reactor, then all batches must be considered. Let us define L_{fuel} the levelized charge for electricity necessary to cover all fuel expenses for all batches. Then equating the revenue for all batches to the sum of the revenues for each batch, as shown in Fig. D.16, gives

$$\sum_{m=1}^{M}\left[\left(\sum_{n=1}^{N} L_{fuel}\frac{E_{mn}}{(1+i)^n}\right)\frac{1}{(1+i)^{b_P}}\right] = \sum_{m=1}^{M}\left[\left(\sum_{n=1}^{N}(L_b + L_{btax})_m E_{mn}\frac{1}{(1+i)^n}\right)\frac{1}{(1+i)^{b_P}}\right], \tag{D.56}$$

where M is the number of batches, E_{mn} is the amount of energy generated by batch m in year of residence n, and b_p is the time at which a batch begins to generate power. Solving for L_{fuel} gives

$$L_{fuel} = \frac{\sum_{m=1}^{M}\left[\left(\sum_{n=1}^{N}(L_b + L_{btax})_m E_{mn}\frac{1}{(1+i)^n}\right)\frac{1}{(1+i)^{b_P}}\right]}{\sum_{m=1}^{M}\left[\left(\sum_{n=1}^{N}\frac{E_{mn}}{(1+i)^n}\right)\frac{1}{(1+i)^{b_P}}\right]}. \tag{D.57}$$

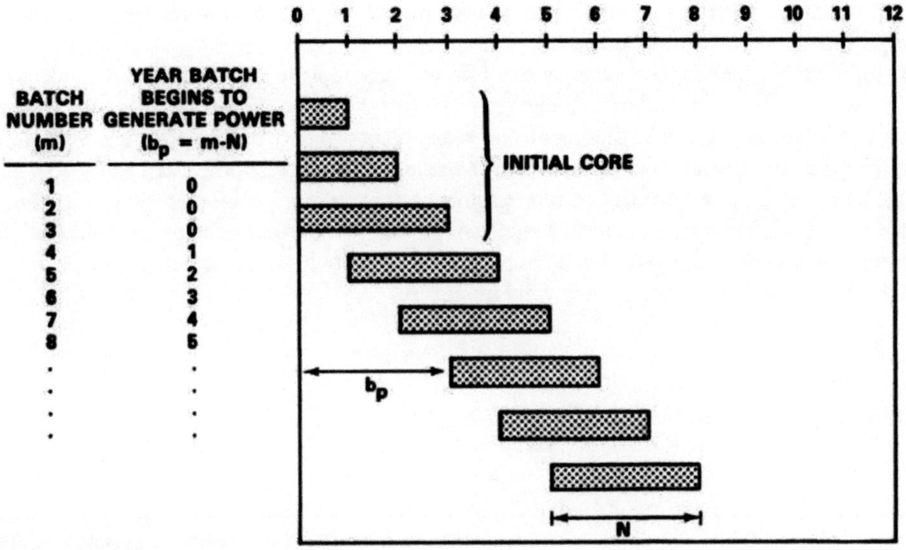

Fig. D.16 Calculation of a fuel cost considering all batches of fuel

D.6 Total Power Cost

The total levelized charge for electricity necessary to cover capital, fuel, and operating and maintenance expenditures may be expressed as

$$L_{\text{total}} = L_{\text{ctot}} + L_{b\text{tot}} + L_{om}$$

where

$$L_{\text{ctot}} = L_{\text{cap}} + L_{\text{ctax}} + L_{fc}$$

$$L_{b\text{tot}} = L_b + L_{b\text{tax}}$$

and L_{om} is the levelized charge necessary to cover operating and maintenance expenses.

D.7 An Illustrative Example

Figures have two important properties; they say a lot and they appear to be accurate. In both respects there is a real disadvantage connected with the apparent advantage. Figures say so much that without a detailed discussion it is impossible to understand them. When the discussion is finished, the substance is in the words rather than in the figures and the latter are only a device by which to remember the discussion. Figures are also much too accurate. The trouble is that with regard to the future this accuracy can never be attained.
Edward Teller, Energy from Heaven and Earth

An example, unfortunately, has all the properties of a figure. An example has the ability to convey information in a very efficient manner while simultaneously appearing to be more accurate than the situation warrants. With this in mind, we will attempt to exploit the former while relying on the maturity of the reader to avoid the latter.

The input values used for the example that follows were representative of conditions occurring in the 1980 time frame. Such values will certainly change with time. For example, at the time of this writing, one might assume LWR lifetimes of 60 years, plant capacity factors of around 90%, refueling intervals of 1.5 to 2.0 years, etc. In addition, the example is targeted towards fast breeder reactors (FBRs). The fuel cycle economics of fast reactors in the "burner" configuration differ from FBRs because of their different role in the fuel cycle. Accordingly, the reader should be cautioned that the following numerical example is offered simply to demonstrate how the equations are used. One can reliably employ these equations for any valid input.

D.7.1 Reactor Characteristics

		Light-water reactor (once-through cycle)	Homogeneous oxide fast breeder reactor		
Power level (MWe)		1,300	1,000		
Capacity factor (%)		70	70		
Refueling interval (years)		1	1		
Plant life time (years)		30	30		
Lead and lag time (years)		1	1		
			Core	Axial blanket	Radial blanket
Number of batches		3	2	2	5
Uranium requirement[a] (short tons U_3O_8/yr)		240	0	0	0
Enrichment requirement[a] (10^3 SWU/yr)		145	0	0	0
Plutonium (kg/yr)	Charge	0	1,480	0	0
	Discharge	215	1,410	160	135
Heavy metal (kg/yr)	Charge	33,300	12,600	9,360	8,480
	Discharge	31,500	11,600	9,150	8,290

[a]Tails composition of 0.2%

D.7.2 Economic Parameters

Bond Fraction	0.55
Equity Fraction	0.45
Uninflated Bond Interest Rate	0.025
Uninflated Equity Interest Rate	0.07
Fixed Charge Fraction of Capital Cost	0.03

	Light-water reactor (once-through cycle)	Homogeneous oxide fast breeder reactor		
Tax rate	0.50	0.50		
U_3O_8 ($/lb)	40			
Enrichment ($/SWU)	100			
Plutonium ($/g)		30		
Capita cost of plant[a] ($/kWe)	800	1,000		
Operation and Maintenance costs ($/kWe·yr) Fixed	13	17		
Variable	1	1		
		Core	Axial blanket	Radial blanket
Fabrication ($/kg)	150	1,150	50	150
Spent fuel shipping and disposal ($/kg)	150	–	–	–
Reprocessing, including shipping and waste disposal ($/kg)	–	600	600	600

[a]Includes interest during construction

D.7.3 Effective Interest Rate

$$\begin{aligned} i &= (b \times i_b) + (e \times i_e) \\ &= (0.55 \times 0.025) + (0.45 \times 0.07) \\ &= 0.045 \text{ \$/\$} \cdot \text{yr}. \end{aligned}$$

D.7.4 Return of and Return on a Capital Investment

$$\begin{aligned} C_u &= C \times \left[\frac{i \times (1+i)^K}{(1+i)^K - 1}\right], \\ L_{\text{cap}} &= \frac{C_u}{E}. \end{aligned}$$

LWR

$$\begin{aligned} C(\text{LWR}) &= 800 \text{ \$/kWe} \times 1300 \times 10^3 \text{kWe} \\ &= 1.04 \times 10^9 \text{\$} \\ C_u(\text{LWR}) &= 1.04 \times 10^9 \text{ \$} \times \left[\frac{0.045 \times (1.045)^{30}}{(1.045)^{30} - 1}\right] \text{\$/\$} \cdot \text{yr} \\ &= 1.04 \times 10^9 \text{\$} \times 0.061 \text{ \$/\$} \cdot \text{yr} = 6.4 \times 10^7 \text{ \$/yr} \\ E(\text{LWR}) &= 1300 \times 10^3 \text{kWe} \times 0.70 \times 8760 \text{ hr/yr} \\ &= 8.0 \times 10^9 \text{kWh/yr} \\ L_{\text{cap}}(\text{LWR}) &= \frac{6.4 \times 10^7 \text{ \$/yr} \times 10^3 \text{mills/\$}}{8.0 \times 10^9 \text{kWh/yr}} \\ &= 8.0 \text{ mills/kWh}. \end{aligned}$$

FBR

$$\begin{aligned}
C(\text{FBR}) &= 1000\ \$/\text{kWe} \times 1000 \times 10^3\ \text{kWe} \\
&= 1.0 \times 10^9\ \$ \\
C_u(\text{FBR}) &= 1.0 \times 10^9\ \$ \times 0.061\ \$/\$\ \cdot \text{yr} \\
&= 6.1 \times 10^7\ \$/\text{yr} \\
E(\text{FBR}) &= 1000 \times 10^3\ \text{kWe} \times 0.70 \times 8760\ \text{hr/yr} \\
&= 6.1 \times 10^9\ \text{kWh/yr} \\
L_{\text{cap}}(\text{FBR}) &= \frac{6.1 \times 10^7\ \$/\text{yr} \times 10^3 \text{mills}/\$}{6.1 \times 10^9\ \text{kW/yr}} \\
&= 10.0\ \text{mills/kWh}.
\end{aligned}$$

D.7.5 Fixed Charges on a Capital Investment

$$L_{fc} = f \times C/E.$$

LWR

$$\begin{aligned}
L_{fc}(\text{LWR}) &= \frac{0.03\ \$/\$ \cdot \text{yr} \times 1.04 \times 10^9\ \$ \times 10^3\ \text{mills}/\$}{8.0 \times 10^9\ \text{kWh/yr}} \\
&= 3.9\ \text{mills/kWh}.
\end{aligned}$$

FBR

$$\begin{aligned}
L_{fc}(\text{FBR}) &= \frac{0.03\ \$/\$ \cdot \text{yr} \times 1.00 \times 10^9\ \$ \times 10^3\ \text{mills}/\$}{6.1 \times 10^9\ \text{kWh/yr}} \\
&= 4.9\ \text{mills/kWh}.
\end{aligned}$$

D.7.6 Taxes on a Capital Investment

$$\begin{aligned}
L_{\text{ctax}} &= \left(\frac{t}{1-t}\right)\frac{(L_{\text{cap}} \times E - \text{cdep})}{E} - \left(\frac{t}{1-t}\right)\left(\frac{bi_b}{i}\right)\left(\frac{1}{E}\right) \\
&\quad \times \left[\frac{i \times (1+i)^K}{(1+i)^K - 1}\right]\left[C - \frac{KC_u}{(1+i)} + \frac{KiC}{(1+i)}\right] \\
\text{cdep}_k &= \frac{C}{K}.
\end{aligned}$$

LWR

$$c\text{dep(LWR)} = \frac{1.04 \times 10^9\ \$}{30\,\text{yr}}$$
$$= 3.5 \times 10^7\ \$/\text{yr}$$

$$L_{\text{ctax}}(\text{LWR}) = \left(\frac{0.50}{1-0.50}\right)$$
$$\times \left(\frac{8.0\ \text{mills/kWh} \times 8.0 \times 10^9\ \text{kWh/yr} - 3.5 \times 10^7\ \$/\text{yr} \times 10^3\ \text{mills/}\$}{8.0 \times 10^9\ \text{kWh/yr}}\right)$$
$$-\left(\frac{0.50}{1-0.50}\right)\left(\frac{0.55 \times 0.025}{0.045}\right)\left(\frac{0.061\ \$/\$ \cdot \text{yr}}{8.0 \times 10^9\ \text{kWh/yr}}\right)\Bigg[1.04 \times 10^9\ \$$$
$$-\frac{30\ \text{yr} \times 6.4 \times 10^7\ \$/\text{yr}}{1.045} + \frac{30\ \text{yr} \times 0.045\ \$/\$ \cdot \text{yr} \times 1.04 \times 10^9\ \$}{1.045}\Bigg](10^3\ \text{mills/}\$)$$

$$L_{\text{ctax}}(\text{LWR}) = 3.6\ \text{mills/kWh} - 1.3\ \text{mills/kWh}$$
$$= 2.3\ \text{mills/kWh}.$$

FBR

$$c\text{dep(FBR)} = \frac{1.0 \times 10^9\ \$}{30\,\text{yr}} = 3.3 \times 10^7\ \$/\text{yr}$$

$$L_{\text{ctax}}(\text{FBR}) = \left(\frac{0.50}{1-0.50}\right)$$
$$\times \left(\frac{10.0\ \text{mills/kWh} \times 6.1 \times 10^9\ \text{kWh/yr} - 3.3 \times 10^7\ \$/\text{yr} \times 10^3\ \text{mills/}\$}{6.1 \times 10^9\ \text{kWh/yr}}\right)$$
$$-\left(\frac{0.50}{1-0.50}\right)\left(\frac{0.55 \times 0.025}{0.045}\right)\left(\frac{0.061\ \$/\$ \cdot \text{yr}}{6.1 \times 10^9\ \text{kWh/yr}}\right)\Bigg[1.0 \times 10^9\ \$$$
$$-\frac{30\ \text{yr} \times 6.1 \times 10^7\ \$/\text{yr}}{1.045} + \frac{30\ \text{yr} \times 0.045\ \$/\$ \cdot \text{yr} \times 1.0 \times 10^9\ \$}{1.045}\Bigg](10^3\ \text{mills/}\$)$$

$$L_{\text{ctax}}(\text{FBR}) = 4.6\ \text{mills/kWh} - 1.7\ \text{mills/kWh}$$
$$= 2.9\ \text{mills/kWh}.$$

D.7.7 Operation and Maintenance (fixed O&M)

$$L_{om} = \frac{(\text{fixed O\&M}) + \left[(\text{variable O\&M}) \times (\text{capacity factor})\right]}{E}.$$

LWR

$$L_{om}(\text{LWR}) = \frac{\begin{bmatrix}(13\ \$/\text{kWe} \cdot \text{yr} \times 1300 \times 10^3\ \text{kWe} \\ +(1\ \$/\text{kWe} \cdot \text{yr} \times 0.70 \times 1300 \times 10^3\ \text{kWe})\end{bmatrix} \times 10^3\ \text{mills/}\$}{8.0 \times 10^9\ \text{kWh/yr}}$$
$$= 2.2\ \text{mills/kWh}.$$

FBR

$$L_{om}(\mathrm{FBR}) = \frac{\begin{bmatrix}(17\,\$/\mathrm{kWe}\cdot\mathrm{yr}\times 1000\times 10^3\,\mathrm{kWe}\\ +(1\,\$/\mathrm{kWe}\cdot\mathrm{yr}\times 0.70\times 1000\times 10^3\,\mathrm{kWe})\end{bmatrix}\times 10^3\,\mathrm{mills}/\$}{6.1\times 10^9\,\mathrm{kWh/yr}}$$

$$= 2.9\ \mathrm{mills/kWh}.$$

D.7.8 Fuel Cycle Expenses (for a Single Equilibrium Batch)

Direct Expenses

$$L_b = \frac{F(1+i)^{ld} + \dfrac{B}{(1+i)^{lg+N}}}{\dfrac{E}{N}\left[\dfrac{1}{(1+i)} + \dfrac{1}{(1+i)^2} + \cdots + \dfrac{1}{(1+i)^N}\right]}.$$

LWR

$$\begin{aligned}
F(\mathrm{LWR}) &= (\mathrm{uranium}) + (\mathrm{enrichment}) + (\mathrm{fabrication})\\
&= (240\ \mathrm{tons}\ U_3O_8/\mathrm{yr}\times 2000\ \mathrm{lb/ton}\times 40\,\$/\mathrm{lb}\ U_3O_8)\\
&\quad + \left(145\times 10^3\ \mathrm{SWU/yr}\times 100\,\$/\mathrm{SWU}\right) + (33\,300\ \mathrm{kg/yr}\times 150\,\$/\mathrm{kg})\\
&= 3.9\times 10^7\ \$/\mathrm{yr}.\\
B(\mathrm{LWR}) &= (\text{spent fuel shipping and disposal})\\
&= 31\,500\,\mathrm{kg/yr}\times 150\,\$/\mathrm{kg}\\
&= 4.7\times 10^6\ \$/\mathrm{yr}
\end{aligned}$$

$$\mathrm{L}_b(\mathrm{LWR}) = \frac{\left[3.9\times 10^7\ \$/\mathrm{yr}\times 1.045 + \dfrac{4.7\times 10^6\ \$/\mathrm{yr}}{(1.045)^4}\right]\times 10^3\ \mathrm{mills}/\$}{\dfrac{8.0\times 10^9\ \mathrm{kWh/yr}}{3}\left[\dfrac{1}{1.045} + \dfrac{1}{(1.045)^2} + \dfrac{1}{(1.045)^3}\right]}$$

$$= 6.1\ \mathrm{mills/kWh}.$$

FBR

Fuel costs will be calculated individually for each of the three zones in the FBR. In addition, costs due to fabrication and reprocessing will be separated from those due to the purchase or sale of plutonium.

D.7.8.1 Fabrication and Reprocessing

– Core

$$\begin{aligned}
L_b\,(C - F\&R) &= \left[12600\ \mathrm{kg/yr}\times 1150\,\$/\mathrm{kg}\times 1.045 + 11\,600\,\mathrm{kg/yr}\times 600\,\$/\mathrm{kg}/(1.045)^3\right]\\
&\quad \times 10^3\ \mathrm{mills}/\$ \div \left[\frac{6.1\times 10^9\ \mathrm{kWh/yr}}{2}\times\left(\frac{1}{1.045} + \frac{1}{(1.045)^2}\right)\right]\\
&= 3.7\ \mathrm{mills/kWh}.
\end{aligned}$$

– Axial Blanket

$$L_b(A - F\&R) = \left[9360\ \text{kg/yr} \times 50\,\$/\text{kg} \times 1.045 + 9150\,\text{kg/yr} \times 600\,\$/\text{kg}(1.045)^3\right] \times 10^3\ \text{mills/\$}$$

$$\div \left[\frac{6.1 \times 10^9\ \text{kWh/yr}}{2}\left(\frac{1}{1.045} + \frac{1}{(1.045)^2}\right)\right]$$

$$= 0.93\ \text{mills/kWh.}$$

– Radial Blanket

$$L_b(R - F\&R) = \left[8480\ \text{kg/yr} \times 150\,\$/\text{kg} \times 1.045 + 8290\,\text{kg/yr} \times 600\,\$/\text{kg}/(1.045)^6\right] \times 10^3\ \text{mills/\$}$$

$$\div \left[\frac{6.1 \times 10^9\ \text{kWh/yr}}{5}\left(\frac{1}{1.045} + \frac{1}{(1.045)^2} + \cdots + \frac{1}{(1.045)^5}\right)\right]$$

$$= 0.96\ \text{mills/kWh.}$$

– Subtotal

$$\begin{aligned} L_b\,(\text{FBR} - F\&R) &= L_b\,(C - F\&R) + L_b\,(A - F\&R) + L_b\,(R - F\&R) \\ &= 3.7 + 0.93 + 0.96\ \text{mills/kWh} \\ &= 5.6\ \text{mills/kWh.} \end{aligned}$$

D.7.8.2 Purchase and Sale of Plutonium

– Core

$$L_b(C - P) = \frac{\left[1480\ \text{kg/yr} \times 1.045 - 1410\ \text{kg/yr}/(1.045)^3\right] \times 10^3\ \text{g/kg} \times 30\,\$/\text{g} \times 10^3\ \text{mills/\$}}{5.7 \times 10^9\ \text{kWh/yr}}$$

$$= 1.6\ \text{mills/ kWh(cost).}$$

– Axial Blanket

$$L_b(A - P) = \frac{\left[0 - \dfrac{160\ \text{kg/yr} \times 10^3\ \text{g/kg} \times 30\,\$/\text{g}}{(1.045)^3}\right] \times 10^3\ \text{mills/\$}}{5.7 \times 10^9\ \text{kWh/yr}}$$

$$= -0.74\ \text{mills/kWh (credit).}$$

– Radial Blanket

$$L_b(R - P) = \frac{\left[0 - \dfrac{135\ \text{kg/yr} \times 10^3\ \text{g/kg} \times 30\,\$/\text{g}}{(1.045)^6}\right] \times 10^3\ \text{mills/\$}}{5.4 \times 10^9\text{kWh/yr}}$$

$$= -0.58\ \text{mills/kWh (credit).}$$

– Subtotal

$$\begin{aligned} L_b(\text{FBR}{-}P) &= L_b\,(C - P) + L_b\,(A - P) + L_b\,(R - P) \\ &= 1.6 - 0.74 - 0.58\ \text{mills/kWh} \\ &= 0.3\ \text{mills/kWh (cost).} \end{aligned}$$

Total cost for a Batch of FBR Fuel

$$
\begin{aligned}
L_b(\text{FBR}) &= L_b(\text{FBR}-F\&R) + L_b(\text{FBR}-P) \\
&= 5.6\,\text{mills/kWh} + 0.3\,\text{mills/kWh} \\
&= 5.9\ \text{mills/kWh}.
\end{aligned}
$$

Summary of FBR Fuel Costs (all values given in mills/kWh)

	Fabrication and reprocessing	Plutonium purchase or sale	Subtotal
Core	3.7	1.6	5.3
Axial blanket	0.9	−0.7	0.2
Radial blanket	1.0	−0.6	0.4
Subtotal	5.6	0.3	5.9
Total fuel cost: 5.9 mills/kWh			

Fuel Taxes

$$
L_{b\text{tax}} = \left(\frac{t}{1-t}\right)\left[L_b - \frac{F(1+i)^{ld} + \dfrac{B}{(1+i)^{lg}}}{E}\right].
$$

LWR

$$
\begin{aligned}
L_{b\text{tax}}(\text{LWR}) &= \left(\frac{0.50}{1-0.50}\right) \\
&\times \left[6.1\,\text{mills/kWh} - \frac{\left(3.9\times10^{7}\,\$/\text{yr}\times1.045 + \dfrac{4.7\times10^{6}\,\$/\text{yr}}{1.045}\right)\times10^{3}\,\text{mills/\$}}{8.0\times10^{9}\,\text{kWh/yr}}\right] \\
&= 0.4\,\text{mills/Wh}.
\end{aligned}
$$

FBR

$$
\begin{aligned}
L_{b\text{tax}}(\text{FBR}) &= \left(\frac{0.50}{1-0.50}\right) \\
&\times \left[5.9\,\text{mills/kWh} - \frac{\left(6.0\times10^{7}\,\$/\text{yr}\times1.045 - \dfrac{3.4\times10^{7}\,\$/\text{yr}}{1.045}\right)\times10^{3}\,\text{mills/\$}}{6.1\times10^{9}\,\text{kWh/yr}}\right] \\
&= 1.0\,\text{mills/kWh}.
\end{aligned}
$$

Total Levelized Power Cost

$$
L_{\text{total}} = L_{\text{cap}} + L_{\text{ctax}} + L_{fc} + L_b + L_{b\text{tax}} + L_{om}.
$$

LWR

$$L_{\text{total}}(\text{LWR}) = (8.0 + 3.9 + 2.3 + 2.2 + 6.1 + 0.4)\ \text{mills/kWh}$$
$$= 22.9\,\text{mills/kWh}.$$

FBR

$$L_{\text{total}}(\text{FBR}) = (10.0 + 4.9 + 2.9 + 2.9 + 5.9 + 1.0)\ \text{mills/kWh}$$
$$= 27.6\,\text{mills/Wh}.$$

D.8 Economic Analysis and the Transition to the Fast Spectrum Reactor

> *Commercialize: to cause something, having only a potential income-producing value, to be sold, manufactured, displayed, or utilized so as to produce income. Webster's Dictionary.*

The principal goal of a civilian reactor development program is to develop a power plant that can supply electricity on a commercial basis. It is, therefore, important to understand thoroughly the factors that influence the time at which the fast spectrum reactor can be considered a commercial enterprise. Several such factors were qualitatively discussed in Chapter 3.

One factor that definitely affects the commercialization date is the size of the uranium resource base. In particular, the amount of electrical energy that can be produced by the LWR is ultimately limited by the amount of ^{235}U contained in that base. In contrast, the fast spectrum reactor is a technological advancement that will allow energy to be obtained from the ^{238}U in the uranium resource base. Since the amount of ^{238}U is considerably larger than the amount of ^{235}U, the amount of energy obtainable with the fast spectrum reactor is also considerably larger. Or, in other words, the energy recovered from the uranium resource base will be incomplete without the fast spectrum reactor.

It is important to note, however, that incomplete energy recovery is by no means unique to the LWR. In fact, it is not unusual to extract approximately one-third of the oil in any given oil reservoir, with the other two-thirds remaining in the ground because the cost of extraction is prohibitive. As oil prices increase, however, the employment of an enhanced recovery technique becomes increasingly attractive. In comparison, one can say that the fast spectrum reactor is an enhanced recovery technique that becomes increasingly attractive as U_3O_8 prices increase. Thus, the fast spectrum reactor is ultimately subject to the same requirements imposed upon any enhanced recovery technique—viz, the additional resource will be extracted when it is economically attractive.

The time at which the fast spectrum reactor, or any enhanced recovery technique for that matter, becomes economically attractive is determined by the price of the resource undergoing depletion as well as the additional capital investment required to obtain an additional amount of the resource. Thus, for the fast spectrum reactor, the time for commercialization is determined primarily by the price of the U_3O_8 and the capital cost differential between the fast spectrum reactor and the LWR. The price of U_3O_8 is in turn determined by the extent of the proven reserve as well as the cost of extracting U_3O_8 from this reserve. In addition, the price of U_3O_8 is affected by the costs associated with the additional exploration required to convert probable uranium resources into proven uranium reserves.

The price of U_3O_8 has a large influence on the cost of energy from the LWR, as illustrated in Fig. D.17. In contrast, the price of U_3O_8 has little influence on the cost of energy from the fast spectrum reactor. While LWR energy costs will be lower than fast spectrum reactor energy costs when U_3O_8 prices are low, the reverse will be true when U_3O_8 prices are high. Thus, for some U_3O_8 price, a *point of economic indifference* exists. This U_3O_8 price, at which the total cost of power from either an LWR or an fast spectrum reactor is equal, is often called the *transition point*. The components of

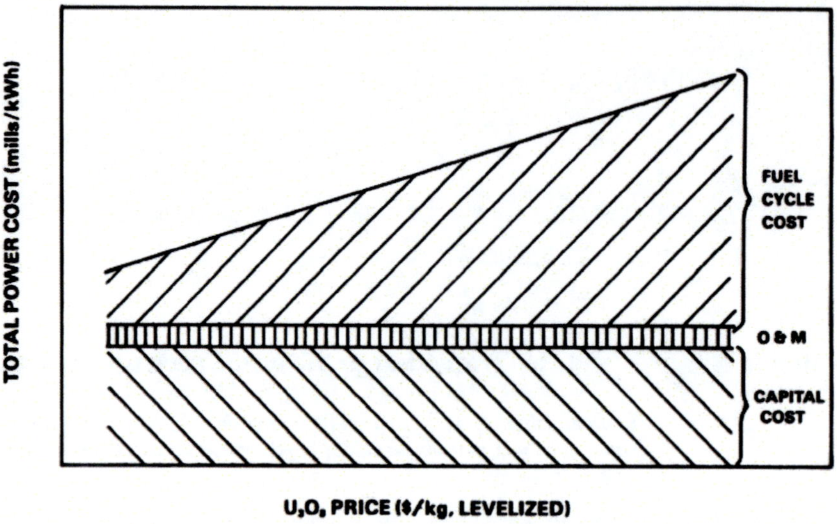

Fig. D.17 Total power cost components for an LWR as a function of U_3O_8 price

the total cost of power both prior to, and at, the transition point are illustrated in Fig. D.18. While the transition point is conceptually clear, it is often difficult to calculate uniquely. Thus, the timing of the breeder, and hence planning of an fast spectrum reactor program, is subject to uncertainty.

The reasons one has difficulty in calculating a unique economic transition point are several. First, the total cost of power from the fast spectrum reactor is a very strong function of its capital cost-and

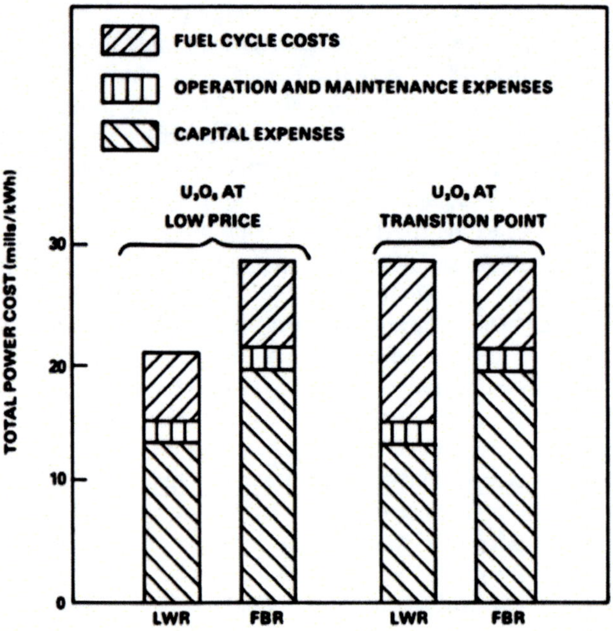

Fig. D.18 Comparison of power cost components for the LWR and the fast spectrum reactor at low U_3O_8 price and at the transition point

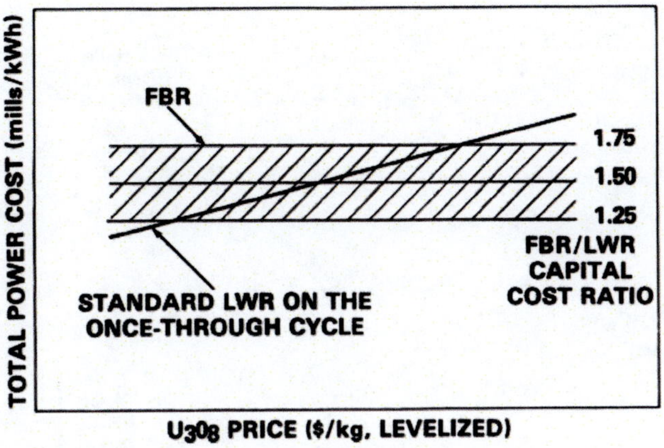

Fig. D.19 Transition from the standard LWR on the once-through cycle to the fast spectrum reactor. The *horizontal lines* represent fast spectrum reactor total power cost for the assumed ratio of fast spectrum reactor/LWR capital costs

the capital cost is somewhat uncertain. It is possible that the capital cost of the fast spectrum reactor could be as low as 1.25 times that of an LWR, or it is possible that it could be as high as 1.75 times that of an LWR.[2] This implies that the total cost of power from the fast spectrum reactor is commensurately uncertain. The effect of this uncertainty on the transition point is illustrated in Fig. D.19.

The price of U_3O_8 in the future is a second uncertainty in an analysis of the timing of the breeder. Since uranium is a depletable resource, the price of U_3O_8 can be expected to increase as the supply is depleted. As with most minerals, the initial supply of U_3O_8 was probably obtained from rich deposits with a low marginal cost of production. As this supply is depleted, it becomes necessary to work poorer deposits with a higher marginal cost of production. Purchasers must then be willing to pay higher prices before a continuous supply of uranium will be available from these deposits. Thus, in order to accurately assess the rate at which U_3O_8 prices will increase, it is necessary to accurately assess the rate at which the uranium supply is consumed.

Since the price of U_3O_8 is an increasing function of the cumulative amount of uranium consumed, the ability to estimate future U_3O_8 prices is directly dependent upon the ability to estimate future uranium consumption. Future uranium consumption, in turn, depends directly upon the number of light water reactors. The number of light water reactors expected over the next two or three decades, often called the nuclear growth projection, is itself subject to uncertainty.

Although uncertainty exists, it is still possible to estimate a commercialization date for the fast spectrum reactor. Given a reasonable estimate for the nuclear growth projection, one can calculate the amount of U_3O_8 that would be committed without the fast spectrum reactor, as shown in Fig. D.20. If the amount of both high and low grade uranium resources can be estimated, it is possible to estimate a U_3O_8 price as a function of the cumulative amount committed. One such estimate is shown in Fig. D.21. When combined with Fig. D.20, which indicates the cumulative amount committed as a function of time, one obtains an estimate of the price of U_3O_8 as a function of time, as shown in Fig. D.22. As previously noted, the price of U_3O_8 strongly affects the total cost of power from the LWR, while the converse is true for the fast spectrum reactor. In contrast, the total cost of power from the fast spectrum reactor is strongly affected by the capital cost uncertainty. Both effects are illustrated in Fig. D.23. If Fig. D.23 is combined with Fig. D.22, it is possible to estimate the total

[2]From the discussion in Chapter 3, is conceivable that the capital cost differential between the fast spectrum reactor and the LWR could eventually come closer to parity with future fast spectrum reactor enhancements.

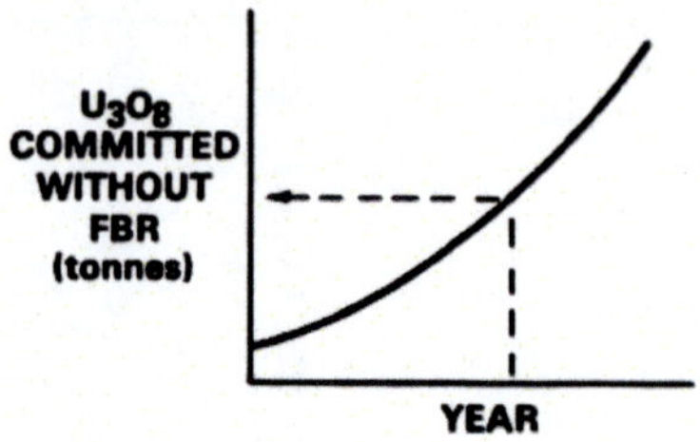

Fig. D.20 The amount of U_3O_8 committed as a function of time without the fast spectrum reactor

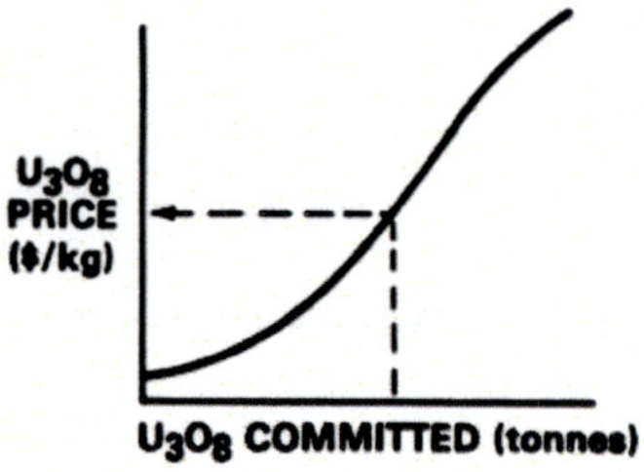

Fig. D.21 The price of U_3O_8 as a function of the amount committed

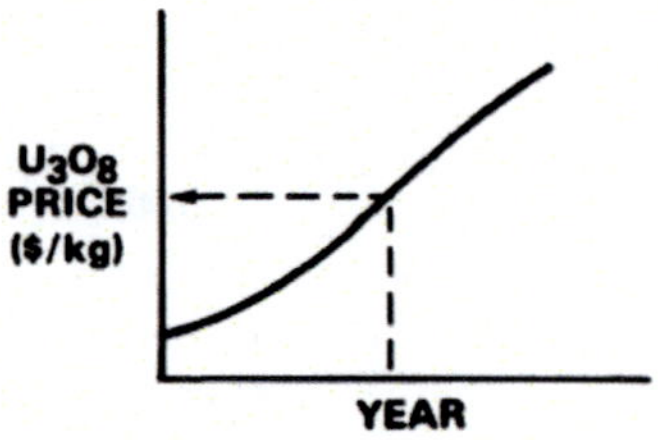

Fig. D.22 The price of U_3O_8 as a function of time obtained by combining Figs. D.20 and D.21

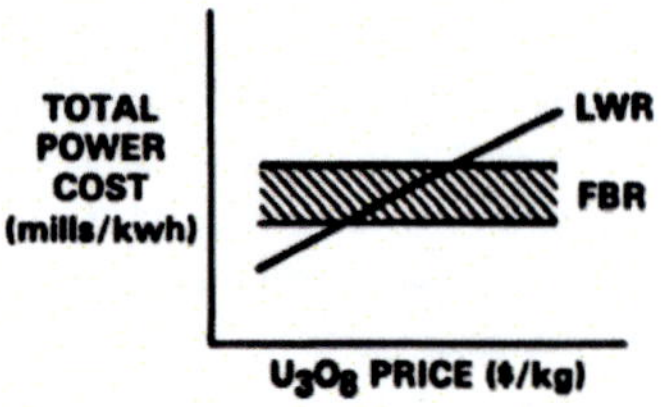

Fig. D.23 The total cost of power from the LWR and the fast spectrum reactor as a function of U_3O_8

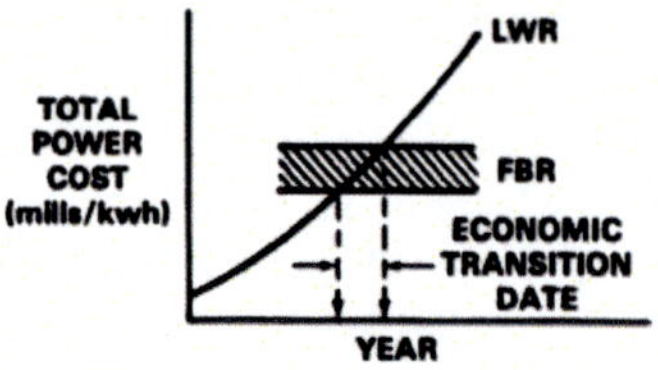

Fig. D.24 The time at which a transition would be economic

cost of power from both the LWR and the fast spectrum reactor as a function of time as shown in Fig. D.24. The economic transition date is then the time at which the total cost of power from the fast spectrum reactor becomes less than that from the LWR.

Since the development of a commercial fast spectrum reactor requires decades—at least two, perhaps three—it is not surprising to find that different nations regard the fast spectrum reactor differently. Some nations, usually resource poor, are willing to run the risk of commercializing the fast spectrum reactor prior to the time of economic attractiveness. Other nations, usually resource rich, attempt to time the fast spectrum reactor program such that commercialization and economic attractiveness are coincident.

D.9 Symbols Used in This Appendix

Levelized Charges for Electricity

L_b = fuel costs for a single batch of fuel
L_{btax} = taxes associated with a single batch of fuel
L_{btot} = total fuel expenses
L_{cap} = return of and return on a capital investment
L_{ctax} = taxes associated with a capital investment
L_{ctot} = total capital expenses
L_{fc} = fixed charges associated with a capital investment
L_{fuel} = fuel costs based on all batches of fuel
L_{om} = operation and maintenance expenses
L_{total} = total power cost

Money Rates

b = fraction of funds obtained by bonds
e = fraction of funds obtained by equity
f = fixed charge fraction
i = effective interest rate or weighted mean rate of return
i_b = bond interest rate
i_e = equity interest rate
t = tax rate

Capital Investment

C = initial capital investment
$c$bin = payment to bond interest
$c$dep = depreciation allowance on a capital investment
C_u = uniform annual payment for the return of and return on a capital investment
D = outstanding debt on a capital investment
E = electrical output level
Exp = expenditures
I = payment to interest
K = number of years for payback of capital investment
P = payment to principal
Rev = revenues

Fuel

B = all back-end costs associated with a single batch of fuel
D = depreciation allowance in a single year for a batch of fuel
Dep = total depreciation allowance for a batch of fuel
F = all front-end costs associated with a single batch of fuel
ld = lead time for front-end costs
lg = lag time for back-end costs
N = residence time for a batch of fuel (also number of fuel batches)
T = total tax payment for a single batch of fuel

Subscript

k = year index.

Appendix E
Fast Reactor Simulations

Pavel Tsvetkov and Won Sik Yang

E.1 Introduction

Substantial progress has been made since the early days of fast reactor analyses, particularly in the area of neutronics. Highlights in this area are summarized in Section E.2. Similar efforts are needed for the remainder of fast reactor design issues, such as robust validated sets of computational tools and databases to simulate neutronics, thermal hydraulics and mechanical/structural behavior in steady-state and transient conditions. The design features of Generation IV fast reactor designs that are associated with the Generation IV International Forum (GIF),[1] which often incorporate innovation well beyond the available experimental bases, will have to rely on advanced simulation capabilities to confirm their performance characteristics.

Viability of innovative system design features will require confirmation by credible analyses verified with experimental data. Validated performance and safety models will be required as the basis for regulatory reviews and licensing. The design of future nuclear systems will require simulation capabilities that provide considerably more accurate predictions of system performance than available in current systems [1]. More will be said on this matter in Section E.3.

Computer based analyses have already been used to predict in detail quantities that could not be readily measured. Examples include the aging of structures, power distributions in cores, transient safety behavior, etc. Complementing or replacing testing with high-fidelity computer simulation will make it possible to collect simulated data that can, in conjunction with a sound experimental validation program, be used to understand fundamental processes and uncertainties that affect facility efficiency, safety, and cost. One can, for example, imagine virtual prototyping of reactor cores yielding data that leads to more accurate identification of design margins and the primary contributors to their uncertainties, allows early experimentation with novel design concepts, and ultimately significantly reduces plant licensing timelines. In other areas, such as advanced fuel fabrication, atomistic fuel simulations could ultimately make it possible to target a small subset of promising candidate fuel types for further experimentation, greatly reducing the number of experiments to be performed [2]. Efforts in this direction are summarized in Section E.4.

One important concept that must be acknowledged when attempting to supplement or replace experimental validation programs with computational validation is the need to fundamentally examine and understand the underlying derivations, assumptions, and inputs to the codes as they relate to the problem at-hand. This is because the development of computer codes (particularly those used in thermal-hydraulics and materials simulations) is generally analogous to deriving the neutron transport equation based on engineering judgment and intuition without the benefit of analytical derivations

Pavel Tsvetkov with principal contributions from Won Sik Yang

[1] See Section 1.2 for a discussion of the GIF. The "Generation IV" reactors associated with GIF are often referred to as "Gen IV" designs

from first principles for confirming the result. One may neglect terms thinking they are not important (or they were not important in application to earlier problems) or completely missing terms because they are subtle and do not readily manifest. In either case, the result can produce misleading or erroneous validation conclusions due to neglected or missed physical modeling in the codes. Several instances are noted within this appendix where this "fundamental examination" has been proposed in order to support advances in computational methods. Such endeavors are important to the advancement of computational tools in preparation of applying them to the issues described earlier in this introduction.

E.2 Advances in Neutronics Simulation

Many computational tools exist for performing stand alone analyses of core reactivity, power production, fuel burn-up, shielding design, and ex-core criticality safety. The challenge is to address each of these analyses in an integrated, self-consistent framework.

The analysis of neutronics for complex systems requires a very high degree of sophistication for the description of the geometry and for physically realistic representation of the energy dependence of neutron cross section data used for modeling. The availability of accurate cross sections and their temperature dependence is necessary for modeling the various aspects of neutron transport and interaction within highly heterogeneous reactor cores. An assessment of pertinent cross section data available through ENDF (Evaluated Nuclear Data Files) and other sources should be performed to determine additional needs for experimental and theoretical nuclear physics work to support the establishment of an adequate nuclear data base. A comprehensive assessment would also include photonuclear reaction cross sections and cross sections for relevant charged-particle interactions.

E.2.1 Deterministic vs. Monte Carlo Approaches

Neutronics modeling has traditionally relied on both stochastic (Monte Carlo) simulations and deterministic transport and diffusion theory approaches:

- Monte Carlo techniques incorporate the basic physics at the level of stochastic particle tracking with the general system geometry and material cross sections governing the particle track histories. Monte Carlo offers the strong conceptual advantage of keeping a close correspondence between computational and physical geometric and cross section energy dependence models. At the same time, as numerical methods, Monte Carlo techniques also rely on various numerical approximations such as interpolation of cross sections, representation of scattering angles, Doppler broadening of cross sections, etc. Various physical approximations are also used. For example, thermal motion of nuclides is taken into account only in the thermal energy region, although its impacts on low-energy resonances (in tens of electron volts range) are not negligible. Except for a few materials, the thermal motion is also approximated with the free gas model. Although Monte Carlo methods are often used as "numerical experiments," their own numerical approximations, as those mentioned above, need to be noted.
- Monte Carlo can become computationally impractical for several different classes of problems. These include calculations of small reactivity coefficients, some types of sensitivity/uncertainty propagation studies, time-dependent solutions, and some types of burn up calculations. For these applications, as well as for several other aspects of neutronics analysis, computational advantages can be achieved with deterministic transport and diffusion theory approaches that complement the Monte Carlo approach.

The two basic computational methods, taken together, can provide a much more comprehensive picture of the neutronics aspects of nuclear reactor system behavior than either method alone.

The deterministic approach may also be favored by the need for coupling the neutronics models with other deterministic models (e.g., thermal hydraulics system models) with which data is dynamically shared. In fact, neutronics models are ideally organized from the outset to take advantage of inherent couplings of neutronics behavior to the thermal hydraulic, structural, and possibly the radiochemical behavior of the reactor system. Existing and improved stand alone models should be adapted to ultimately serve as sub-models within a multi-physics, multi-scale, and probably multi-processor approach.

Both deterministic and Monte Carlo tools are being employed for analyzing fast reactor systems. The deterministic tools are more suitable for design optimization studies. However, because they require preparation of multigroup cross sections, more efforts might be required for exploratory conceptual studies. The Monte Carlo tools are more robust in their exploratory capabilities but they are more demanding with respect to computation resource. Systematic approaches for applying both types of tools in complementary exploratory and design optimization studies are of paramount importance to the next generation of reactor designs. A particular focus is on the development of Gen IV designs taking into account safety and performance of reactor units as well as fuel cycle characteristics.

E.2.2 Summary of Codes for Fast Reactor Neutronics Calculations

To support the development of fast reactors in the U.S., an extensive suite of computer codes has been developed and validated at Argonne National Laboratory (ANL). The following is a list of codes that are applicable to fast spectrum reactors (The latest information can be found on the web at http://www.ne.anl.gov/codes/):

- **ETOE/MC^2-2/SDX** [3, 4] is a code system for generating broad-group, cell-average microscopic cross sections on the basis of ENDF/B basic nuclear data. The condensation process takes into account resonance self-shielding effects and the heterogeneity of the unit cell geometry and core arrangement. The resulting composition-, temperature- and region-dependent microscopic cross sections are suitable for use in diffusion or transport theory calculations for a fast reactor core. Recently, MC^2-2 and SDX were integrated into a new code MC^2-3 with enhanced capabilities for resonance self-shielding, inelastic scattering and fission spectrum representations, and slowing-down and transport calculations [5]. The unit cell transport calculation, which was previously done in SDX in a fine group level (230 groups), has been extended to ultrafine (~2,000) and hyperfine (~300,000) group levels.
- **DIF3D** [6, 7] is a versatile code system for carrying out multigroup, multi-dimensional whole core neutronics calculations. It is used to compute the reactor multiplication factor, flux and power distributions, and other functionals of the neutron flux. The system models Cartesian, curvilinear, and hexagonal core geometries either by means of the finite difference approach or by a highly efficient nodal option.
- **VARIANT** [8] is a multigroup nodal transport theory code derived by using a variational nodal approach. VARIANT is implemented as a solution option in the DIF3D code system; it provides highly accurate transport theory results at a fraction of the computing cost of alternative transport methods (e.g., discrete ordinates or Monte Carlo methods).
- **DIF3D-K** [9] and **VARIANT-K** [10] are codes for executing three-dimensional reactor kinetics calculations using the nodal diffusion and variational nodal transport solution options of DIF3D for Cartesian and hexagonal core geometries. These space-dependent kinetics codes

have been coupled to the SAS4A/SASSYS-1 fast reactor safety and accident analysis code system [11].

- **VIM** [12] is a Monte Carlo code for simulating neutral particle transport. It employs continuous-energy representations of cross sections and enables precise modeling of problem geometry and collision physics. Because long computing times are typically needed to obtain acceptably low statistical uncertainties in predicted quantities, VIM is used primarily for benchmarking less rigorous multigroup diffusion or transport models. To reduce its running time, a parallel computing option has been implemented for both networks of workstations and scalable multiprocessor computers.
- **REBUS-3** [13] is a reactor burnup and fuel cycle analysis code that makes use of DIF3D as a flux calculation module. This code models nuclide transmutations on a three-dimensional, region-dependent basis and provides considerable flexibility for specifying operational constraints and fuel management strategies for both in- and out-core portions of the fuel cycle. A unique and powerful technique for simulating equilibrium core characteristics can be invoked as an alternative to discrete, cycle-by-cycle core follow calculations.
- **RCT** [14] is a code that post-processes the depletion-dependent results of REBUS-3/DIF3D nodal calculations to reconstruct intra-assembly distributions of multigroup fluxes, power densities, burnup, and nuclide number densities.
- **ORIGEN-RA** is a modified version of the ORIGEN [15] code developed by Oak Ridge National Laboratory. This code is used to perform isotopically detailed nuclide transmutation calculations based on the flux history computed with REBUS-3 and RCT. In addition to nuclide inventories, this code is used to estimate radiation emission characteristics and decay power for irradiated reactor constituents.
- **VARI3D** [16] is a generalized perturbation theory code that allows calculation of the effects on reactivity and reaction rate ratios of alterations in microscopic cross sections and/or material number densities. VARI3D is most frequently used to compute the reactivity coefficient distributions and kinetics parameters employed in reactor dynamics and safety analyses. The relevant neutron and adjoint flux distributions required to compute these quantities are provided by DIF3D.
- **DPT** [17] is a depletion perturbation theory code that calculates the sensitivity coefficients of depletion-dependent reaction rate ratios with respect to cross sections and initial nuclide densities using the depletion-dependent nuclide densities of REBUS-3 and the relevant neutron and adjoint flux distributions provided by DIF3D.
- **GMADJ** [18] is a code that computes the uncertainties of integral parameters by propagating nuclear data uncertainties by the use of the sensitivity coefficients from VARI3D and DPT calculations and evaluated nuclear data covariance. It provides a formal procedure based on the Bayesian statistics to quantify the uncertainties associated with predicted reactor performance parameters with full applications of integral experimental values. The GMADJ code and ANL integral experiment database proved very useful in the past for reducing integral parameter uncertainties and providing best predictions of integral parameters for proposed designs [19].

Many of these codes are also applicable to other reactor types. Figure E.1 provides a simplified overview of the existing ANL code system for fast reactor analysis.

The simulation of spectral transitions within the fast reactor core is identified as an issue requiring special attention when deterministic methods are used. The resolution of this issue might lead to more advanced homogenization techniques in combination of broader use of stochastic methods. The spectral transition effects at the core-reflector interfaces are particularly important in the fast reactor configurations without the blanket zones.

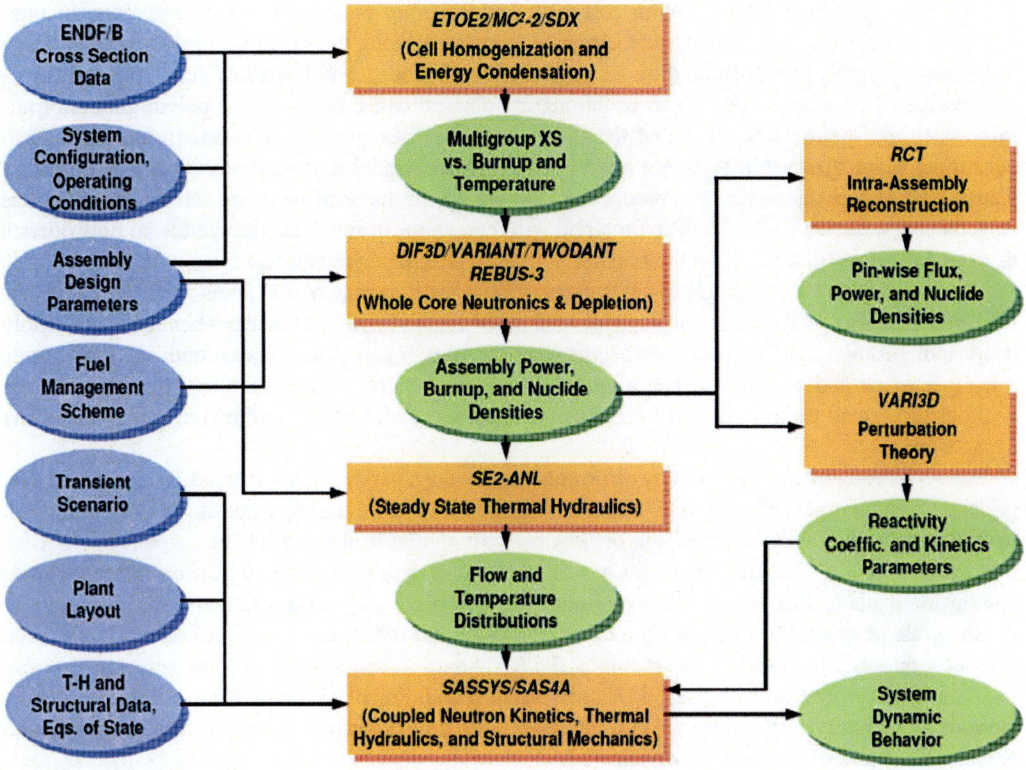

Fig. E.1 ANL code system

E.2.3 DIF3D: A Code to Solve Multi-Dimensional Diffusion and Transport Theory Problems

The DIF3D code system is a multi-group steady-state neutron diffusion and transport theory solver. It provides three flux solution options: finite-difference diffusion theory [6], nodal diffusion theory [7], and variational nodal transport [8] theory methods. Eigenvalue, adjoint, fixed source and criticality (concentration) search problems are permitted. Flux and power density maps by mesh cell and region-wise balance integrals are provided. Although primarily designed for fast reactor problems, up-scattering and (for finite difference option only) internal black boundary conditions are also treated. The variational nodal transport solver VARIANT is included from the release of DIF3D8.0/VARIANT8.0.

The finite difference option solves one-, two- and three-dimensional orthogonal (rectangular and cylindrical) and triangular geometry diffusion theory problems. Anisotropic diffusion coefficients are permitted. Mesh-centered finite-difference equations are solved by optimized iteration methods. A variant of the Chebyshev semi-iterative acceleration technique is applied to outer (fission source) iterations, and an optimized block-successive-over-relaxation method is applied to the within-group iterations. Optimum over-relaxation factors are pre-computed for each energy group prior to the initiation of the outer iterations. The forward sweep of the LU decomposition algorithm for the resulting tri-diagonal matrices is computed prior to outer iteration initiation in orthogonal non-periodic geometry cases.

The nodal diffusion theory option solves the multi-group neutron diffusion equations in two- and three-dimensional hexagonal and Cartesian geometries using the nodal expansion method. Each hexagonal assembly is represented by a single mesh cell (node), and Cartesian geometry node sizes are specified by the user. The nodal equations are derived using higher order polynomial approximations to the spatial dependence of the flux within the (hexagonal or Cartesian) node. The final equations, which are cast in response matrix form, involve spatial moments of the node-interior flux distribution plus surface-averaged partial currents across the faces of the node. These equations are solved using a fission source iteration method with coarse-mesh rebalance acceleration. Equivalence theory parameters (discontinuity factors) are permitted with hexagonal nodal models [20].

VARIANT solves the multigroup neutron transport equations in two- and three-dimensional hexagonal and Cartesian geometries using a variational nodal method with one node per hexagonal assembly (Cartesian geometry node sizes are specified by the user). Variational nodal methods incorporate a number of attractive features. These include a standard hierarchy of space-angle approximation, well behaved small mesh limits, and the absence of both ray effects and artificial diagonal streaming depressions.

The nodal equations are derived by reformulating the even parity transport equation as a variational principle and incorporating the odd parity flux at interface as a Lagrange multiplier [21]. The even and odd parity fluxes are expanded in orthogonal polynomial trial functions of space and spherical harmonics of angle. The final equations are cast in a set of response matrix equations relating partial current moments to flux and source moments. Dimensionless parts of the response matrices involving integrals in space and angle are pre-computed once and for all using MATHEMATICA for each geometry option. The results are stored in FORTRAN data statements and used to generate response matrix sets for unique nodes (defined by cross section and dimension data). Flux and source expansions of up to sixth order and partial current expansions up to second order are allowed. Angular and scattering expansions of up to P_5 are allowed.

The equations are solved by fission source iteration in conjunction with a coarse mesh rebalance acceleration scheme. The inner iterations are accelerated by a partitioned matrix scheme equivalent to a synthetic diffusion acceleration method. The computational resources required for evaluation and storage of response matrices for problems involving large numbers of unique node types impose a practical limit on problem complexity. For highly heterogeneous problems involving thousands of different node types, calculation and storage of response matrices represents the primary computational cost.

Problem dimensions are all variable. Sufficient memory must be available to contain all data for at least one energy group. In three dimensional finite difference option problems, a concurrent inner iteration strategy permits the specification of an unlimited number of planes.

DIF3D reads and writes the standard interface files specified by the Committee on Computer Code Coordination (CCCC) [22]. DIF3D is included in the REBUS-3 code package and can thus be used to provide the neutronics solutions required in REBUS-3 depletion calculations.

E.2.4 REBUS-3 Fuel Cycle Analysis Capability

REBUS-3 is a system of codes designed for the analysis of reactor fuel cycles. Two basic types of problems are solved [23].

1. The equilibrium conditions of a reactor operating under a fixed fuel management scheme,
2. The explicit cycle-by-cycle, or non-equilibrium, operation of a reactor under a specified periodic or non-periodic fuel management program.

For the equilibrium type problems, the code uses specified external fuel supplies to load the reactor. Optionally, reprocessing may be included in the specification of the external fuel cycle and discharged fuel may be recycled back into the reactor. For non-equilibrium cases, the initial composition of the reactor core may be explicitly specified or the core may be loaded from external feeds and discharged fuel may be recycled back into the reactor as in equilibrium problems. Four types of search procedures may be carried out in order to satisfy user-supplied constraints:

1. Adjustment of the reactor burn cycle time to achieve a specified discharge burnup,
2. Adjustment of the fresh fuel enrichment to achieve a specified multiplication constant at a specified point during the burn cycle,
3. Adjustment of the control poison density to maintain a specified value of the multiplication constant throughout the reactor burn cycle,
4. Adjustment of the reactor burn cycle time to achieve a specified value of the multiplication constant at the end of the burn step.

The total reactor burn cycle time is divided into one or more subintervals, the number of which is specified by the user. An explicit nuclide depletion computation is performed in each region of the reactor over each of these subintervals using the average reaction rates over the subinterval. These average reaction rates are based on fluxes obtained from explicit multi-dimensional diffusion or transport theory neutronics solutions computed at both the beginning and end of the subinterval using DIF3D or TWODANT. (In a version of REBUS-3 named REBUS-PC, the Monte Carlo code MCNP can also be invoked for flux calculation.) The nuclide transmutation equations are solved by the matrix-exponential technique. The isotopes to be considered in the burnup equations, as well as their transmutation reactions, are specified by the user.

Microscopic cross sections are permitted to vary as a function of the atom density of designated reference isotopes in the problem. The user may specify control rod positions at each time node in the problem. A number of relational database datasets containing various types of summary results are available for use in tailoring reports. REBUS-3 provides a general external cycle modeling capability: flexible reprocessing scheme with user-specified allocation of discharged fuels to multiple reprocessing plants and isotope-dependent recovery factors, flexible re-fabrication scheme with multiple reprocessing plant outputs and external feeds and user-specified multi-level priority scheme for distributing available atoms to different fuel types, and modeling of time delays between various processes and radioactive decays. A fully automatic restart capability is also available, but it is rarely used in modern computing environments.

These analysis capabilities for critical reactors can also be applied to subcritical accelerator-driven systems [24]. In this case, the distribution of spallation neutron sources is specified by the user as an external fixed source [25]. The fuel depletion calculation is performed at a specified power level by scaling the spallation source intensity to compensate the reactivity and source multiplication variations during an irradiation cycle. The enrichment search capability automates the adjustment of the transuranic element loading in the fresh fuel such that a specified multiplication factor is achieved for a fixed source problem at a specified point during the burn cycle.

The fuel cycle model used in REBUS-3 may be subdivided into an in-reactor cycle and an ex-reactor or external cycle. Figure E.2 indicates the main logical flow for REBUS-3.

The in-reactor cycle is concerned with the location of each discrete "fuel bundle" (or "composition") in space over the period during which it resides in the reactor. A bundle may be loaded at a particular position, and then irradiated for some specified time in a flux that is determined by the reactor composition as well as the control and power requirements. After this initial burn, the bundle may or may not be repositioned in space. In any case, if any of the fuel bundles in the reactor are altered, the irradiation rate will undergo a discontinuous change at this

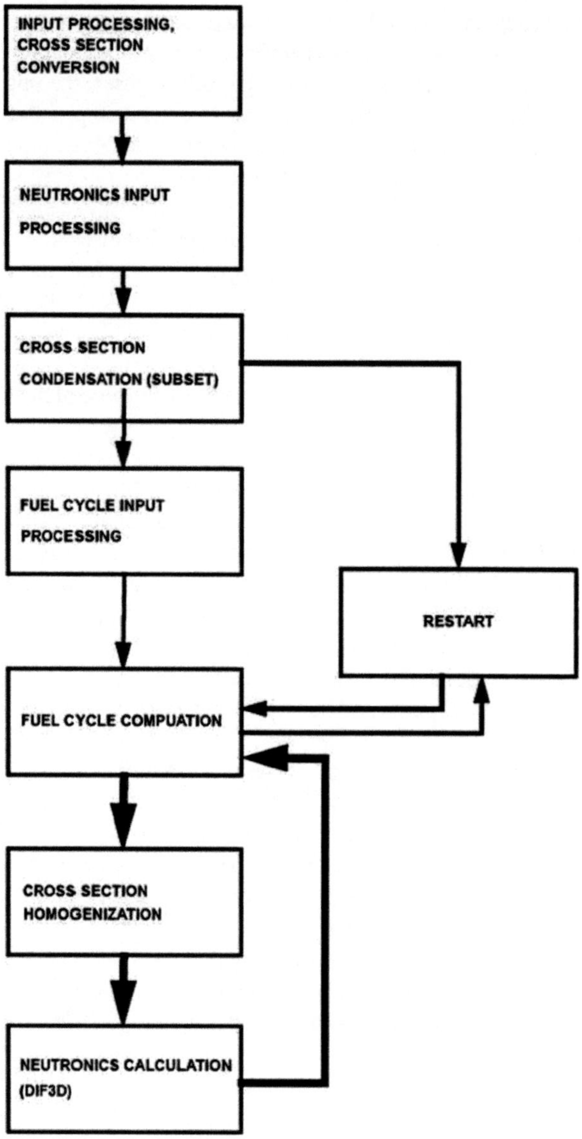

Fig. E.2 Simplified REBUS-3 logical flow

so-called "fuel management time." After several burn/shuffle sequences, the bundle is discharged from the reactor. Spatial movement of any given bundle is, of course, constrained by the requirement of volume preservation. In-reactor movement of a single bundle requires movement of other bundles:

- One bundle must be moved from its position and inserted in another location, or discharged from the reactor to make room for the original transfer.
- The space vacated by the original bundle must be filled by fresh fuel, or by a bundle moved from another location.

This linked sequence of fuel bundle motions is referred to as a *fuel management path* since it describes the path a specific bundle takes through the reactor as a function of time. Each fuel management path is defined by an index triplet (material type, burn stage, and region), which specifies the in-core location (region) for every burn stage (i.e., fresh, once-burned, twice-burned, etc.) of each fuel bundle (material type).

If the as-charged compositions of all material types are held constant and the burn/refuel process is repeated, the composition of the discharged fuel bundles will approach constant values as time increases, assuming that control and criticality constraints are not violated. This reactor condition is actually calculated (or approximated) as one step in obtaining the solution to the equilibrium recycle problem.

Two possibilities arise in the case that fuel management procedures change after a burn cycle:

- The first is when a particular move sequence is to be carried out only once. Each fuel bundle in the reactor must be accounted for separately at the end of each burn cycle.
- The second possibility is the case in which a number of burn cycles elapse without any fuel movement within a particular material type, then one discharge-charge move is made, and so on with a fixed repetition factor.

Clearly no equilibrium condition can be defined for the first of these possibilities. The second one however, is repetitive and thus may admit an equilibrium condition.

The amount of numerical work required for each pass toward the solution is equal to that for a single burn cycle multiplied by the least common multiple of the repetition factors of all the material types in the reactor. It is for this reason that the equilibrium search procedure is limited to problems with a repetition factor of one, i.e. those in which the same fuel management procedure is carried out at the end of every burn cycle.

It should be noted that the above indexing system is most convenient for describing repetitive partial refueling schemes such as the scatter-reloading scheme often considered for fast reactors. Individual fuel bundles, however, can be accounted for separately at the end of each burn cycle through direct reference to the index triplet.

The burn stage index is redundant in a simple non-equilibrium problem in which a different fuel management procedure is carried out after each burn cycle. In an equilibrium problem, however, this stage index has a definite value. Since by definition the fuel management procedure is held fixed, the sequence of stages of a material type present in the reactor at any time represent different phases in the burnup history of a fuel bundle. Hence, the equilibrium solution can be obtained by calculating only one burn cycle explicitly.

The external cycle model used in REBUS-3 is intended to represent the actual course of events following the discharge of fuel from the reactor. Hence, this ex-reactor model consists of the following successive steps:

- Cooling,
- Delivery to a reprocessing plant,
- Re-fabrication using both reprocessed and external feed supplies,
- Preloading storage,
- Reactor charge.

The normal procedure following the discharge of a fuel bundle from the reactor is assumed to be storage in a cooling pond followed by transfer to one or more reprocessing plants. Each discharged label has a cooling time associated with it. This cooling time may logically also include the time required for delivery to the reprocessing plants. For each discharge label, any number of reprocessing plants may be specified as destinations. One also specifies here the fraction of the discharge to be delivered to each plant. All of the reactor discharge material not delivered to a reprocessing plant will be considered to be sold. Provision is made for sale of discharged fuel both directly after cooling and following reprocessing. The discharged fuel material of each type may be cooled for a different period, and then divided before dispatch to several different reprocessing plants. The cooled fuel may be divided and/or combined for input to one or more reprocessing plants. Each of these plants has an associated reprocessing time. Each reprocessing plant is assigned a set of "recovery fractions" for each active isotope being processed. The re-fabrication phase is difficult to arrange into a fully determined yet flexible scheme that is easy for the user to specify in terms of input data requirements. The correct enrichment ratio of each charge material is unknown at problem preparation time, so that specific high- and low-reactivity fuel requirements are unknown. Since the model is of a recycle system, the discharged fuel, and therefore part of the re-fabrication fuel inventory, is undefined until the reactor performance is known. The method adopted in REBUS-3 is a priority ordering scheme [23], which is relatively easy to specify and is not unduly restrictive in scope. The basis of the model is as follows. At some minimum time before the refueling date, a choice of fuel for re-fabrication is made. Fuel is available in the form of high-reactivity (CLASS 1) and low-reactivity (CLASS 2) batches, each with fixed isotopic ratios. There may be a number of each of these, some of which represent reprocessed fuel and some of which come from external feeds.

A complex integrated scheme of the fast reactor design development is shown in Fig. E.3. The analysis procedure addresses the need for proper selection of new materials that should assure material

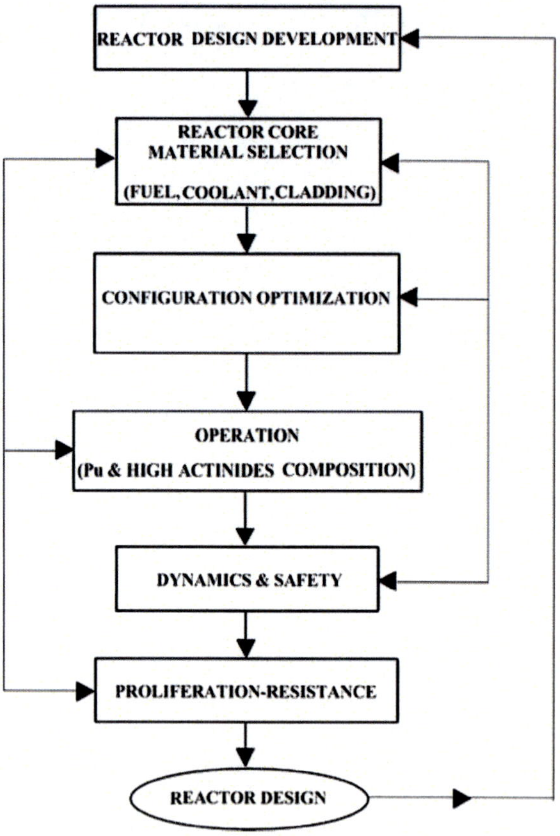

Fig. E.3 Fast reactor design development

compatibility and provide attainability of the design-targeted system performance characteristics. The final material selection is based on physical properties of the candidate materials, their anticipated inter-compatibility in the system, and on the detailed reactor physics analysis that takes into account a wide variety of performance characteristics including overall reactivity swing, achievable burnup level, attainable fast fluence and material limits, power peaking and its dynamics during reactor operation, temperature reactivity coefficients, and kinetics parameters. The applied detailed 3D models consider spatial locations of each fuel element bundle during the entire reactor operation. To increase discharge burnup levels and decrease fluence-to-burnup ratios, material selection is followed by optimization of the core configuration. As illustrated in Fig. E.3, the design development is an extensive iterative process because of complex interactions and interdependence between various characteristics of the fast spectrum system.

The overall fast reactor modeling procedure is shown in Fig. E.4. Criticality calculations with DIF3D and whole-core burnup calculations with REBUS-3 provide the performance characteristics. The obtained burnup-dependent data can be used to calculate reactivity coefficients and kinetics

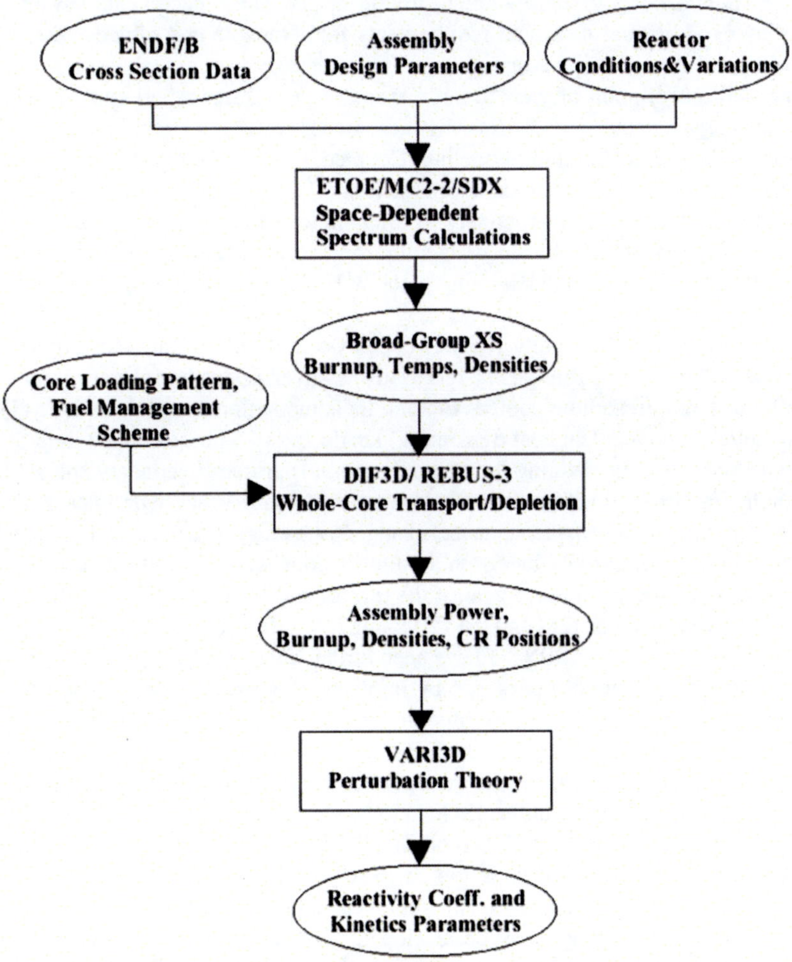

Fig. E.4 Reactor physics analysis of a fast reactor using the ANL code system (DIF3D, REBUS-3, and VARI3D)

parameters with VARI3D. The following set of the power unit performance characteristics can be obtained using REBUS-3:

- Material (nuclide) mass flow characteristics;
- Peak and average discharge burnup levels;
- Peak fast fluences and peak burnup levels;
- Initial (charge) enrichment (fissile content) and the final (discharge) fissile content;
- Core multiplication;
- Breeding or conversion ratios.

E.3 Systems and Safety Considerations

The passive safety of Gen IV designs must be confirmed in simulations and eventually in prototypic experiments. The requirement of passive safety leads to transients dominated by natural reactivity feedbacks and coolant flows driven by small driving heads. These necessitate advanced tools and confirmation by experimental data, and requirements for a reliable and effective thermo-structural reactivity feedback for the fast-spectrum systems. It is necessary to approach fast reactor simulations from an integrated model point of view to make sure design features relied upon to achieve passive safety are adequately characterized. Addressing these issues requires a simulation approach that integrates neutronics, T-H, and structural modeling. Coupled T-H and 3-D spatial kinetics capability is required to provide an accurate (reference or benchmark) simulation of physical phenomena. The ability to perform such calculations routinely would help to reduce margins required for calculation uncertainties and help to increase the accuracy of current core monitoring and surveillance capabilities. With the detailed solutions, models using point or 1D kinetics can be normalized and used with confidence.

The SAS4A/SASSYS-1 [26] safety analysis code system developed at ANL provides a coupled T-H and 3-D spatial kinetics capability [27]. It is used to analyze design basis accidents and anticipated transients without scram in sodium cooled reactors. Its computational models include (1) single and multiple-pin fuel/cladding/coolant heat transfer, (2) single and two-phase coolant fluid dynamics, (3) fuel/cladding mechanics with cladding failure prediction, (4) primary/secondary coolant loops liquid metal thermal hydraulics, and (5) balance-of-plant water/steam thermal hydraulics. It also has component models for pumps, valves, plena, heat exchangers, steam generators, turbines, and condensers. The neutronics model was initially based on a point kinetics with space-dependent reactivity feedbacks. This code system was validated against the reactor and plant data from the EBR-II shutdown heat removal test (SHRT) series [28] and the FFTF inherent safety test series [29].

The capabilities of SAS4A/SASSYS-1 have been further enhanced by coupling with the DIF3D-K and VARIANT-K spatial kinetics codes and by implementing external neutron source kinetics for accelerator-driven system analysis [11]. A three-dimensional sub-channel coolant T-H model has also been implemented and validated against the EBR-II steady state and SHRT transient temperature measurements [30]. Efforts are ongoing to incorporate computational fluid dynamics (CFD) models for enhanced prediction of T-H quantities [31].

New capabilities based on detailed CFD models will be developed in the coming decades as development progresses. Development of these tools will require attention to the coupling approach and interface, coupled code convergence, use of improved algorithms and multiprocessing. A formal uncertainty evaluation methodology must be developed. There is also a need to couple the containment response to the systems calculation (due to passive safety features involving system wide behavior). Validation of these coupled models will be required. There is a clear advantage in developing multi-scale simulation capabilities for fast reactors that would facilitate development of new materials and also guide design solutions.

Another code system developed at ANL to couple neutronics and thermal-hydraulics performance is the SE2-ANL system [32]. It is a modified version of the SUPERENERGY-2 thermal-hydraulic code, which is a multi-assembly, steady-state sub-channel analysis code developed at MIT for application to fast reactor (wire-wrapped and ducted) rod bundles. At ANL, the code was coupled to heating calculation methods based on the DIF3D code system, and models were added for hot spot analysis, fuel element temperature calculations, and allocation of coolant flow subject to thermal performance criteria.

E.4 The Move Toward High-Fidelity Simulation Tools

High fidelity simulation-based methods are now within the reach of modern supercomputers, which currently are approaching a peak theoretical performance of one peta-FLOP/s, about five orders of magnitude faster than a standard desktop computer. In fact, other related industries, such as automotive, airline, chip manufacturers, etc. have demonstrated the effective use of high-fidelity modeling as an integral part of the conceptual design and optimization process. Related industries are making remarkable progress in reducing experiments in favor of high-end simulation—e.g. pharmaceuticals, engine design, turbines, etc. Viability of new technologies and design features will require confirmation by credible analyses verified with experimental data. Credible analyses will also be required as the basis for regulatory reviews and licensing of new designs. The required simulation capabilities include computer codes and databases for simulating neutronics, thermal-hydraulics and structural behavior in steady-state and transient conditions. For each system and type of analysis, the adequacy of existing analysis tools will need to be assessed and the required enhancements to their capabilities implemented and qualified.

A formal validation process needs to be established to bring the full benefits of modern simulation capabilities to the design and safety of the facilities and the research and development process. The purpose of the validation process is to provide for each calculated quantity an estimated uncertainty that takes into account all possible sources of uncertainties in the analysis, and integrates them for each particular situation.

Several approaches are currently available and have successfully been used in the past in the nuclear area:

- The rudimentary approach has historically consisted of taking a code—developed to whatever level of sophistication—and comparing its output to data from a set of representative experiments. The uncertainties are then estimated on the basis of the comparison. This process does not distinguish the origins of uncertainties and biases, and gives only a vague definition of the domain of validity.
- More advanced approaches have been developed, for example in neutronics and structural mechanics, where reference codes (that provide exact solutions to the basic equations) exist and can be used to estimate numerical biases; uncertainty propagation techniques have been developed, and have been used to estimate the final uncertainties due to nuclear data or mechanical properties. Finally, formal statistical processes have been developed to formally compare integral experimental data and predictive results.

Many of the required modeling capabilities are crosscutting in that they are applicable to multiple nuclear systems [33]. Examples are Monte Carlo and deterministic transport methods for neutronics modeling, modern CFD methods for heat transfer and fluid flow simulation, and modular code systems for fuel cycle evaluations and simulation of transients and accidents. Consistency of cross-model assumptions, data, and approaches will be essential for the success of the long-term, large-scale modeling efforts. This is especially true when data and models address phenomena across vastly different scales, with the results of one calculation becoming the input of another. A set of sub-model requirements based on physical constraints, prediction accuracy, and technical and computational consistency within an integrated approach must be defined. Advances in these capabilities will help reduce uncertainties in predicted system behavior, which can be exploited in system development by targeting the best performance achievable without exceeding the capabilities or limits of the technologies employed by the system [2].

Current reactor physics analysis capabilities and nuclear data are generally adequate for the initial phases of the development of new nuclear systems. Due to distinct features of some of these systems, however, the available tools must be tested to establish their accuracy and characterize uncertainties in their predictions. Existing state-of-the-art tools provide a basis for future development and improvement.

The long-term goal for the simulation program is the development of an architectural model that will facilitate modeling the entire fuel cycle from mining through final disposition of waste material, taking into account interacting factors (e.g., market forces, socio-political effects, technology risk). This will facilitate the implementation of a comprehensive suite of simulation tools for the design, analysis and engineering of next-generation nuclear energy systems with enhanced safety, reduced environmental impact, optimal deployment of facilities, and reduced construction cost. Goals include [34]:

- Integrated 3D reactor core simulations with rigorous propagation of uncertainty;
- Coupled thermal hydraulic and primary loop simulation;
- Advanced fuel design and performance;
- Fuel behavior engineering;
- Advanced secondary loop and balance of plant engineering and analysis;
- Advanced fuel cycle design;
- Separations facility engineering optimization;
- Repository design including seismic, geological, chemical, and thermal studies;
- Overall nuclear energy systems model development suitable for economic analysis.

E.5 Sources of Code Systems

The Radiation Safety Information Computational Center (RSICC) at Oak Ridge National Laboratory maintains a comprehensive collection of literature, computer codes, and data libraries contributed by specialists from around the world. RSICC may be reached at http://www-rsicc.ornl.gov/.

References

1. "A Science-Based Case for Large-Scale Simulation," Vol. 1, Office of Science, U.S. Department of Energy (July 30, 2003).
2. "A Science-Based Case for Large-Scale Simulation," Vol. 2, Office of Science, U.S. Department of Energy (September 19, 2004).
3. B. J. Toppel, H. Henryson II, and C. G. Stenberg, "ETOE-2/MC^2-2/SDX Multi-group Cross-Section Processing," *Proceedings of RSIC Seminar-Workshop on Multi-group Cross Sections*, Conf-780334-5, Oak Ridge, TN, March 1978.

4. H. Henryson II, B. J. Toppel, and C. G. Stenberg, *MC^2-2: A Code to Calculate Fast Neutron Spectra and Multi-group Cross Sections*, ANL-8144, Argonne National Laboratory, Argonne, IL (1976).
5. W. S. Yang, M. A. Smith, C. H. Lee, A. Wollaber, D. Kaushik, and A. S. Mohamed, "Neutronics Modeling and Simulation of SHARP for Fast Reactor Analysis," *Nuclear Engineering and Technology*, **42**, 475 (2010).
6. K. L. Derstine, *DIF3D: A Code to Solve One-, Two-, and Three-Dimensional Finite-Difference Diffusion Theory Problems*, ANL-82-64, Argonne National Laboratory, Argonne, IL (1984).
7. R. D. Lawrence, *The DIF3D Nodal Neutronics Option for Two- and Three-Dimensional Diffusion Theory Calculations in Hexagonal Geometry*, ANL-83-1, Argonne National Laboratory, Argonne, IL (1983).
8. G. Palmiotti, E. E. Lewis, and C. B. Carrico, *VARIANT: VARIational Anisotropic Nodal Transport for Multidimensional Cartesian and Hexagonal Geometry Calculation*, ANL-95/40, Argonne National Laboratory, Argonne, IL (1995).
9. T. A. Taiwo and H. S. Khalil, "DIF3D-K: A Nodal Kinetics Code for Solving the Time-Dependent Diffusion Equation," *Proceedings of the International Conference on Mathematics and Computations, Reactor Physics, and Environmental Analyses*, Portland, OR (1995).
10. T. Taiwo, R. Ragland, G. Palmiotti, and P. J. Finck, "Development of a Three-Dimensional Transport Kinetics Capability for LWR-MOX Analyses," *Transactions of the American Nuclear Society*, **79**, 298 (1999).
11. J. E. Cahalan, T. Ama, G. Palmiotti, T. A. Taiwo, and W. S. Yang, "Development of a Coupled Dynamics Code with Transport Theory Capability and Application to Accelerator-Driven Systems Transients," *Proceedings of ANS International Topical Meeting on Advances in Reactor Physics and Mathematics and Computation into the Next Millennium, PHYSOR 2000*, Pittsburgh, PA (2000).
12. R. N. Blomquist, "VIM – A Continuous Energy Neutronics and Photon Transport Code," *Proceedings of the International Topical Meeting on Advances in Mathematics, Computations and Reactor Physics*, Pittsburgh, PA (1992).
13. B. J. Toppel, *The Fuel Cycle Analysis Capability REBUS-3*, ANL-83-2, Argonne National Laboratory, Argonne, IL (March 1983 revised October 26, 1990).
14. W. S. Yang, P. J. Finck, and H. Khalil, "Reconstruction of Pin Power and Burnup Characteristics from Nodal Calculations in Hexagonal Geometry," *Nuclear Science and Engineering*, **111**, 21 (1992).
15. M. J. Bell, *ORIGEN – The ORNL Isotope Generation and Depletion Code*, ORNL-4628, Oak Ridge National Laboratory, Oak Ridge, TN (1973).
16. C. H. Adams, Private Communication, Argonne National Laboratory (1975).
17. W. S. Yang and T. J. Downar, "Generalized Perturbation Theory for Constant Power Core Depletion," *Nuclear Science and Engineering*, **99**, 353 (1988).
18. W. P. Poenitz and P. J. Collins, "Utilization of Experimental Integral Data for Adjustment and Uncertainty Evaluation of Reactor Design Quantities," NEACRP-L-307, *Proceedings of the Nuclear Energy Agency Committee on Reactor Physics (NEACRP) Specialists Meeting*, Jackson Hole, WY (1988).
19. P. J. Collins, S. E. Aumeier, and H. F. McFarlane, "Evaluation of Integral Measurements for the SP-100 Space Reactor," *Proceedings of the 1992 Topical Meeting on Advances in Reactor Physics*, Charleston, SC (1992).
20. P. J. Finck and K. L. Derstine, "The Application of Nodal Equivalence Theory to Hexagonal Geometry Lattice," *Proceedings of the International Topical Meeting on Advances in Mathematics, Computations, and Reactor Physics*, Pittsburgh, PA, Vol. **4**, 16.14-1 (April 28–May 2, 1991).
21. C. B. Carrico, E. E. Lewis, and G. Palmiotti, "Three-Dimensional Variational Nodal Transport Methods for Cartesian, Triangular, and Hexagonal Criticality Calculations," *Nuclear Science and Engineering*, **111**, 168 (1992).
22. R. D. O'Dell, *Standard Interface Files and Procedures for Reactor Physics Codes, Version IV*, UC-32, Los Alamos National Laboratory, Los Alamos, NM (1977).
23. R. P. Hosteny, *The ARC System Fuel Cycle Analysis Capability, REBUS-2*, ANL-7721, Argonne National Laboratory, Argonne, IL (1978).
24. W. S. Yang and H. S. Khalil, "Analysis of the ATW Fuel Cycle Using the REBUS-3 Code System," *Transactions of the American Nuclear Society*, **81**, 277 (1999).
25. W. S. Yang, J. C. Beitel, E. Hoffman, and J. A. Stillman, "Source Coupling Interface between MCNP-X and Deterministic Codes for ADS Analyses," *Transactions of the American Nuclear Society*, **88**, 592 (2003).
26. J. E. Cahalan, et al., "Advanced LMR Safety Analysis Capabilities in the SASSYS-1 and SAS4A Computer Codes," *Proceedings of the International Topical Meeting on Advanced Reactors Safety*, Pittsburgh, PA (1994).
27. H. S. Khalil, et al., "Coupled Reactor Physics and Thermal-Hydraulics Computations with the SAS-DIF3DK Code," *Proceedings of the Joint International Conference on Mathematical Methods and Super-Computing for Nuclear Applications*, Saratoga Springs, NY (1997).
28. S. H. Fistedis, ed., "The Experimental Breeder Reactor-II Inherent Safety Demonstration," *Nuclear Engineering and Design* (special issue), **101**, 1–90 (1987).
29. T. M. Burke, et al., "Results of the 1986 Inherent Safety Tests," *Transactions of the American Nuclear Society*, **54**, 249 (1987).

30. F. E. Dunn, J. E. Cahalan, D. Hahn, and H. Jeong, "Whole Core Sub-Channel Analysis Verification with the EBR-II SHRT-17 Test," *Proceedings of 2006 International Congress on Advances in Nuclear Power Plants,* Reno, NV (2006).
31. T. H. Fanning and T. Sofu, "Modeling of Thermal Stratification in Sodium Fast Reactor Outlet Plenums During Loss of Flow Transients," *Proceedings of the International Conference of Fast Reactors and Related Fuel Cycles (FR 2010)*, Kyoto, Japan (2009).
32. W. S. Yang and A. M. Yacout, "Assessment of the SE2-ANL Code Using EBR-II Temperature Measurements," *Proceedings of the 7th International Meeting on Nuclear Reactor Thermal Hydraulics*, NUREG/CP-0142, Saratoga Springs, NY (1995).
33. "The Path to Sustainable Nuclear Energy: Basic and Applied Research Opportunities for Advanced Fuel Cycles," Office of Science, U.S. Department of Energy (September 2005).
34. "Nuclear Physics and Related Computational Science R&D for Advanced Fuel Cycles," Office of Science, U.S. Department of Energy, Bethesda, MD (August 10–12, 2006).

Appendix F
4-Group and 8-Group Cross Sections

Alan Waltar and Pavel Tsvetkov

The cross sections in this appendix are characteristic of a typical neutron spectrum and material composition for the core of an oxide-fueled sodium-cooled fast reactor (SFR). They were obtained from 4- and 8-group cross sections supplied by D. R. Haffner, R. W. Hardie, and R. P. Omberg of the Hanford Engineering Development Laboratory (1978). These cross section sets were collapsed from a 42-group cross section set developed from ENDF/B using the shielding-factor method. Doppler cross sections for ^{238}U appearing at the end of the appendix were estimated from earlier calculations. The cross sections in this appendix are useful for educational purposes and for rough estimates and comparisons; they are not intended, however, for any actual design applications.

All cross sections are in units of barns. "Fission products" refers to fission product pairs, such that one fission product pair is produced per fission. The cross sections are presented as follows:

- 4-group cross sections,
- 8-group cross sections,
- ^{238}U Doppler cross sections.

Table F.1 4-group structure

Group	Δu	Lower energy of group	χ
1	2.5	820 keV	0.76
2	2.0	110 keV	0.22
3	2.0	15 keV	0.02
4	–	0	0

Table F.2 4-group cross sections

Material	Group	σ_{tr}	σ_c	σ_f	$\sigma_{er} + \sigma_{ir}$	ν_f
Boron (natural)	1	1.6	0.06	–	0.45	–
	2	3.3	0.2	–	0.33	–
	3	2.8	0.6	–	0.15	–
	4	3.9	2.0	–	–	–
^{10}B	1	1.8	0.3	–	0.45	–
	2	4.2	1.1	–	0.33	–
	3	5.2	3.0	–	0.15	–
	4	12.2	10.3	–	–	–
Carbon	1	1.8	0.001	–	0.39	–
	2	3.4	0	–	0.38	–

Table F.2 (continued)

Material	Group	σ_{tr}	σ_c	σ_f	$\sigma_{er} + \sigma_{ir}$	ν_f
	3	4.3	0	–	0.27	–
	4	4.4	0	–	–	–
Oxygen	1	2.2	0.007	–	0.40	–
	2	3.8	0	–	0.26	–
	3	3.6	0	–	0.16	–
	4	3.6	0	–	–	–
Sodium	1	2.0	0.002	–	0.51	–
	2	3.6	0.001	–	0.17	–
	3	4.0	0.001	–	0.13	–
	4	7.0	0.009	–	–	–
Chromium	1	2.4	0.006	–	0.34	–
	2	3.6	0.005	–	0.13	–
	3	4.6	0.02	–	0.05	–
	4	11.1	0.07	–	–	–
Iron	1	2.2	0.007	–	0.40	–
	2	2.8	0.005	–	0.08	–
	3	5.1	0.010	–	0.03	–
	4	7.2	0.03	–	–	–
Nickel	1	2.3	0.073	–	0.31	–
	2	4.4	0.010	–	0.11	–
	3	12.7	0.03	–	0.06	–
	4	21.1	0.05	–	–	–
Molybdenum	1	3.6	0.02	–	0.75	–
	2	7.0	0.06	–	0.11	–
	3	8.2	0.13	–	0.06	–
	4	9.5	0.7	–	–	–
^{232}Th	1	4.5	0.08	0.07	1.20	2.34
	2	7.4	0.19	0	0.20	–
	3	11.8	0.41	0	0.05	–
	4	15.3	1.48	0	–	–
^{233}U	1	4.4	0.03	1.81	0.93	2.69
	2	6.4	0.17	2.05	0.12	2.52
	3	11.5	0.30	2.74	0.04	2.50
	4	17.9	0.74	6.49	–	2.50
^{235}U	1	4.6	0.06	1.2	0.79	2.69
	2	7.4	0.3	1.3	0.20	2.46
	3	12.5	0.6	1.9	0.04	2.43
	4	18.7	2.0	5.0	–	2.42
^{238}U	1	4.6	0.06	0.32	1.39	2.77
	2	7.8	0.13	0	0.22	–
	3	12.1	0.35	0	0.05	–
	4	12.9	0.9	0	–	–
^{239}Pu	1	4.9	0.02	1.83	0.83	3.17
	2	7.5	0.16	1.55	0.13	2.92
	3	12.0	0.45	1.63	0.07	2.88
	4	17.4	2.4	3.25	–	2.87

Table F.2 (continued)

Material	Group	σ_{tr}	σ_c	σ_f	$\sigma_{er} + \sigma_{ir}$	ν_f
^{240}Pu	1	4.8	0.07	1.59	0.74	3.18
	2	7.4	0.18	0.27	0.22	2.95
	3	12.0	0.5	0.07	0.05	2.88
	4	17.2	2.1	0.13	–	2.87
^{241}Pu	1	4.8	0.08	1.65	0.82	3.23
	2	8.0	0.20	1.72	0.31	2.98
	3	12.6	0.48	2.48	0.05	2.94
	4	19.8	1.74	6.32	–	2.93
^{242}Pu	1	4.5	0.04	1.46	0.65	3.12
	2	7.1	0.12	0.17	0.18	2.89
	3	12.6	0.33	0.04	0.05	2.81
	4	22.0	1.54	0.02	–	2.81
Fission products (pairs)	1	7.8	0.05	–	1.83	–
	2	11.4	0.17	–	0.20	–
	3	14.7	0.50	–	0.09	–
	4	19.1	1.88	–	–	–

Table F.3 4-group scattering matrices (elastic & inelastic), $\sigma_{h \to g}$

Material	$\sigma_{1\to2}$	$\sigma_{1\to3}$	$\sigma_{2\to3}$	$\sigma_{3\to4}$
Boron (and ^{10}B)	0.45	0	0.33	0.15
Carbon	0.39	0	0.38	0.27
Oxygen	0.40	0	0.26	0.16
Sodium	0.51	0	0.17	0.13
Chromium	0.32	0.02	0.13	0.05
Iron	0.37	0.03	0.08	0.03
Nickel	0.29	0.02	0.11	0.06
Molybdenum	0.71	0.04	0.11	0.06
^{232}Th	1.15	0.05	0.21	0.05
^{233}U	0.87	0.06	0.12	0.04
^{235}U	0.77	0.02	0.20	0.04
^{238}U	1.32	0.07	0.22	0.05
^{239}Pu	0.79	0.04	0.13	0.07
^{240}Pu	0.72	0.02	0.22	0.05
^{241}Pu	0.78	0.04	0.31	0.05
^{242}Pu	0.59	0.06	0.18	0.05
Fission Products (Pairs)	1.75	0.08	0.20	0.09

Table F.4 8-group structure

Group	Δu	Lower Energy of Group	χ
1	1.5	2.2 MeV	0.365
2	1.0	820 keV	0.396
3	1.0	300 keV	0.173
4	1.0	110 keV	0.050
5	1.0	40 keV	0.012
6	1.0	15 keV	0.003
7	3.0	750 eV	0.001
8	–	0	0

Table F.5 8-group cross sections

Material	Group	σ_{tr}	σ_c	σ_f	$\sigma_{er} + \sigma_{ir}{}^a$	ν_f
^{10}B	1	1.4	0.3	–	0.4401	–
	2	2.0	0.3	–	0.6502	–
	3	3.7	0.7	–	0.79	–
	4	4.6	1.5	–	0.57	–
	5	4.7	2.3	–	0.39	–
	6	5.8	3.8	–	0.34	–
	7	10.9	9.1	–	0.06	–
	8	31.5	29.8	–	–	–
Carbon	1	1.6	0.003	–	0.4760	–
	2	1.9	0	–	0.57	–
	3	3.0	0	–	0.65	–
	4	3.8	0	–	0.67	–
	5	4.2	0	–	0.39	–
	6	4.4	0	–	0.61	–
	7	4.4	0	–	0.10	–
	8	4.5	0	–	–	–
Oxygen	1	1.2	0.02	–	0.3023	–
	2	2.8	0	–	0.58	–
	3	3.7	0	–	0.69	–
	4	3.8	0	–	0.46	–
	5	3.6	0	–	0.39	–
	6	3.6	0	–	0.37	–
	7	3.6	0	–	0.06	–
	8	3.6	0	–	–	–
Sodium	1	1.5	0.005	–	0.623	–
	2	2.2	0.0002	–	0.6908	–
	3	3.6	0.0004	–	0.4458	–
	4	3.5	0.001	–	0.29	–
	5	4.0	0.001	–	0.35	–
	6	3.9	0.001	–	0.30	–
	7	7.3	0.009	–	0.04	–
	8	3.2	0.008	–	–	–
Chromium	1	2.3	0.006	–	0.9998	–
	2	2.5	0.006	–	0.40	–
	3	2.6	0.005	–	0.1201	–
	4	4.6	0.005	–	0.22	–
	5	5.5	0.012	–	0.28	–
	6	3.1	0.033	–	0.12	–
	7	11.5	0.069	–	0.02	–
	8	4.5	0.027	–	–	–
Iron	1	2.2	0.02	–	1.0108	–
	2	2.1	0.003	–	0.46	–
	3	2.4	0.005	–	0.12	–
	4	3.1	0.006	–	0.14	–
	5	4.5	0.008	–	0.28	–
	6	6.1	0.012	–	0.07	–
	7	6.9	0.032	–	0.04	–
	8	10.4	0.020	–	–	–
Nickel	1	2.2	0.02	–	0.994	–
	2	2.4	0.016	–	0.304	–

Table F.5 (continued)

Material	Group	σ_{tr}	σ_c	σ_f	$\sigma_{er} + \sigma_{ir}$[a]	ν_f
	3	3.2	0.008	–	0.19	–
	4	5.5	0.012	–	0.20	–
	5	6.9	0.019	–	0.26	–
	6	21.3	0.049	–	0.13	–
	7	21.4	0.053	–	0.07	–
	8	17.0	0.037	–	–	–
Molybdenum	1	2.9	0.005	–	1.5708	–
	2	4.1	0.024	–	0.878	–
	3	6.1	0.046	–	0.273	–
	4	7.9	0.063	–	0.17	–
	5	8.1	0.088	–	0.14	–
	6	8.2	0.20	–	0.13	–
	7	9.3	0.57	–	0.03	–
	8	–	–	–	–	–
^{232}Th	1	4.2	0.13	0.14	2.627	2.47
	2	4.6	0.11	0.04	1.105	2.13
	3	5.7	0.18	0	0.3704	–
	4	9.0	0.20	0	0.3634	–
	5	11.6	0.33	0	0.25	–
	6	12.1	0.51	0	0.08	–
	7	14.5	1.15	0	0.02	–
	8	27.7	6.59	0	–	–
^{233}U	1	4.6	0.02	1.72	1.667	2.91
	2	4.4	0.05	1.85	0.88	2.59
	3	5.2	0.13	1.91	0.2883	2.53
	4	7.6	0.21	2.16	0.202	2.51
	5	10.4	0.25	2.39	0.15	2.50
	6	13.3	0.36	3.19	0.07	2.50
	7	17.1	0.65	5.89	0.01	2.50
	8	28.5	2.04	15.82	–	2.50
^{235}U	1	4.2	0.04	1.23	0.1394	2.90
	2	4.8	0.09	1.24	0.853	2.59
	3	6.2	0.18	1.18	0.4746	2.48
	4	8.7	0.32	1.40	0.312	2.44
	5	11.7	0.53	1.74	0.15	2.43
	6	13.9	0.79	2.16	0.08	2.42
	7	17.7	1.71	4.36	0.01	2.42
	8	33.0	5.76	15.06	–	2.42
^{238}U	1	4.3	0.01	0.58	2.293	2.91
	2	4.8	0.09	0.20	1.49	2.58
	3	6.3	0.11	0	0.3759	–
	4	9.4	0.15	0	0.2935	–
	5	11.7	0.26	0	0.20	–
	6	12.7	0.47	0	0.09	–
	7	13.1	0.84	0	0.01	–
	8	11.0	1.47	0	–	–
^{239}Pu	1	4.5	0.01	1.85	1.495	3.40
	2	5.1	0.03	1.82	0.826	3.07
	3	6.3	0.11	1.60	0.3709	2.95
	4	8.6	0.20	1.51	0.1905	2.90

Table F.5 (continued)

Material	Group	σ_{tr}	σ_c	σ_f	$\sigma_{er} + \sigma_{ir}{}^{a}$	ν_f
	5	11.3	0.35	1.60	0.15	2.88
	6	13.1	0.59	1.67	0.09	2.88
	7	16.5	1.98	2.78	0.01	2.87
	8	31.8	8.54	10.63	–	2.87
^{240}Pu	1	4.3	0.02	1.61	1.534	3.40
	2	5.1	0.09	1.58	0.723	3.07
	3	6.1	0.15	0.51	0.3713	2.96
	4	8.5	0.20	0.09	0.2929	2.90
	5	11.0	0.34	0.06	0.22	2.88
	6	13.6	0.77	0.08	0.09	2.87
	7	17.1	1.85	0.13	0.02	2.87
	8	19.7	5.92	0.16	–	2.87
^{241}Pu	1	4.5	0.05	1.61	1.838	3.46
	2	5.0	0.11	1.67	0.642	3.13
	3	6.6	0.15	1.53	0.8246	3.01
	4	9.2	0.25	1.87	0.504	2.96
	5	11.9	0.40	2.31	0.22	2.94
	6	13.7	0.58	2.70	0.07	2.94
	7	18.6	1.47	5.5	0.01	2.93
	8	37.7	5.88	19.23	–	2.93
^{242}Pu	1	4.3	0.01	1.67	1.033	3.34
	2	4.5	0.05	1.36	0.68	3.00
	3	5.8	0.11	0.34	0.2846	2.89
	4	8.2	0.13	0.04	0.3002	2.84
	5	11.2	0.24	0.03	0.23	2.82
	6	14.6	0.45	0.05	0.10	2.81
	7	21.0	1.25	0.02	0.03	2.81
	8	36.9	6.10	0	–	–
Fission Products	1	6.5	0.02	–	3.767	–
(pairs)	2	8.6	0.07	–	1.908	–
	3	10.4	0.11	–	0.499	–
	4	12.3	0.21	–	0.343	–
	5	13.9	0.38	–	0.20	–
	6	15.8	0.65	–	0.21	–
	7	18.5	1.58	–	0.04	–
	8	28.3	6.49	–	–	–

[a]Values for removal cross sections are accurate to two digits only. Digits beyond two are provided in order to maintain consistency with the scattering matrices so that a neutron balance can be achieved.

Table F.6 8-group scattering matrices (elastic & inelastic), $\sigma_{h \to g}$

	h \ g	2	3	4	5	6	7	8
^{10}B	1	0.43	0.008	0.002	0.0001	0	0	0
	2	–	0.65	0.0002	0	0	0	0
	3	–	–	0.79	0	0	0	0
	4	–	–	–	0.57	0	0	0
	5	–	–	–	–	0.39	0	0
	6	–	–	–	–	–	0.34	0
	7	–	–	–	–	–	–	0.06
Carbon	1	0.47	0.005	0.0009	0.0001	0	0	0
	2	–	0.57	0	0	0	0	0
	3	–	–	0.65	0	0	0	0
	4	–	–	–	0.67	0	0	0
	5	–	–	–	–	0.39	0	0
	6	–	–	–	–	–	0.61	0
	7	–	–	–	–	–	–	0.10
Oxygen	1	0.30	0.002	0.0003	0	0	0	0
	2	–	0.58	0	0	0	0	0
	3	–	–	0.69	0	0	0	0
	4	–	–	–	0.46	0	0	0
	5	–	–	–	–	0.39	0	0
	6	–	–	–	–	–	0.37	0
	7	–	–	–	–	–	–	0.06
Sodium	1	0.52	0.09	0.003	0.009	0.001	0	0
	2	–	0.69	0	0.0004	0.0004	0	0
	3	–	–	0.44	0.005	0.0008	0	0
	4	–	–	–	0.29	0	0	0
	5	–	–	–	–	0.35	0	0
	6	–	–	–	–	–	0.30	0
	7	–	–	–	–	–	–	0.04
Chromium	1	0.79	0.16	0.04	0.009	0.0008	0	0
	2	–	0.31	0.06	0.02	0.01	0	0
	3	–	–	0.12	0	0	0.0001	0
	4	–	–	–	0.22	0	0	0
	5	–	–	–	–	0.28	0	0
	6	–	–	–	–	–	0.12	0
	7	–	–	–	–	–	–	0.02
Iron	1	0.75	0.20	0.05	0.01	0.0008	0	0
	2	–	0.33	0.10	0.02	0.01	0	0
	3	–	–	0.12	0	0	0	0
	4	–	–	–	0.14	0	0	0
	5	–	–	–	–	0.28	0	0
	6	–	–	–	–	–	0.07	0
	7	–	–	–	–	–	–	0.04
Nickel	1	0.67	0.22	0.08	0.02	0.004	0	0
	2	–	0.25	0.04	0.01	0.004	0	0
	3	–	–	0.19	0	0	0	0
	4	–	–	–	0.20	0	0	0
	5	–	–	–	–	0.26	0	0

Table F.6 (continued)

	g → / h ↓	2	3	4	5	6	7	8
	6	–	–	–	–	–	0.13	0
	7	–	–	–	–	–	–	0.07
Molybdenum	1	1.09	0.39	0.08	0.01	0.0008	0	0
	2	–	0.62	0.20	0.05	0.008	0	0
	3	–	–	0.23	0.04	0.003	0	0
	4	–	–	–	0.17	0	0	0
	5	–	–	–	–	0.14	0	0
	6	–	–	–	–	–	0.13	0
	7	–	–	–	–	–	–	0.03
^{232}Th	1	1.20	1.01	0.34	0.07	0.007	0	0
	2	–	0.86	0.20	0.04	0.005	0	0
	3	–	–	0.36	0.008	0.002	0.0004	0
	4	–	–	–	0.36	0.003	0.0004	0
	5	–	–	–	–	0.22	0.03	0
	6	–	–	–	–	–	0.08	0
	7	–	–	–	–	–	–	0.02
^{233}U	1	0.61	0.73	0.26	0.06	0.007	0	0
	2	–	0.58	0.24	0.05	0.01	0	0
	3	–	–	0.28	0.008	0.0003	0	0
	4	–	–	–	0.20	0.002	0	0
	5	–	–	–	–	0.14	0.01	0
	6	–	–	–	–	–	0.07	0
	7	–	–	–	–	–	–	0.01
^{235}U	1	0.72	0.48	0.16	0.03	0.004	0	0
	2	–	0.72	0.12	0.01	0.003	0	0
	3	–	–	0.43	0.04	0.004	0.0006	0
	4	–	–	–	0.29	0.02	0.002	0
	5	–	–	–	–	0.14	0.01	0
	6	–	–	–	–	–	0.08	0
	7	–	–	–	–	–	–	0.01
^{238}U	1	1.28	0.78	0.20	0.03	0.003	0	0
	2	–	1.05	0.42	0.01	0.01	0	0
	3	–	–	0.33	0.04	0.005	0.0009	0
	4	–	–	–	0.29	0.003	0.0005	0
	5	–	–	–	–	0.18	0.02	0
	6	–	–	–	–	–	0.09	0
	7	–	–	–	–	–	–	0.01
^{239}Pu	1	0.66	0.60	0.19	0.04	0.005	0	0
	2	–	0.64	0.15	0.03	0.006	0	0
	3	–	–	0.31	0.05	0.01	0.0009	0
	4	–	–	–	0.18	0.01	0.0005	0
	5	–	–	–	–	0.13	0.02	0
	6	–	–	–	–	–	0.09	0
	7	–	–	–	–	–	–	0.01
^{240}Pu	1	0.75	0.58	0.17	0.03	0.004	0	0
	2	–	0.60	0.11	0.01	0.003	0	0

Table F.6 (continued)

	h ↓ / g →	2	3	4	5	6	7	8
	3	–	–	0.33	0.04	0.001	0.0003	0
	4	–	–	–	0.29	0.002	0.0009	0
	5	–	–	–	–	0.21	0.01	0
	6	–	–	–	–	–	0.09	0
	7	–	–	–	–	–	–	0.02
^{241}Pu	1	0.62	0.77	0.34	0.10	0.008	0	0
	2	–	0.57	0.06	0.01	0.002	0	0
	3	–	–	0.76	0.06	0.004	0.0006	0
	4	–	–	–	0.45	0.05	0.004	0
	5	–	–	–	–	0.19	0.03	0
	6	–	–	–	–	–	0.07	0
	7	–	–	–	–	–	–	0.01
^{242}Pu	1	0.47	0.39	0.14	0.03	0.003	0	0
	2	–	0.40	0.21	0.05	0.02	0	0
	3	–	–	0.26	0.02	0.004	0.0006	0
	4	–	–	–	0.30	0.0002	0	0
	5	–	–	–	–	0.21	0.02	0
	6	–	–	–	–	–	0.10	0
	7	–	–	–	–	–	–	0.03
Fission products	1	2.12	1.26	0.32	0.06	0.007	0	0
(pairs)	2	–	1.40	0.42	0.08	0.008	0	0
	3	–	–	0.48	0.01	0.007	0.002	0
	4	–	–	–	0.32	0.02	0.003	0
	5	–	–	–	–	0.20	0	0
	6	–	–	–	–	–	0.21	0
	7	–	–	–	–	–	–	0.04

Tabulated below are approximate increases in ^{238}U effective capture cross sections due to the Doppler effect resulting from an increase in average fuel temperature in an SFR core from 700 to 1,400 K.

Table F.7 ^{238}U Doppler cross sections

4-group		8-group	
Group	$\Delta\sigma_c$ (barns)	Group	$\Delta\sigma_c$ (barns)
1	0	1	0
2	0	2	0
3	0.004	3	0
4	0.05	4	0.0001
–	–	5	0.0008
–	–	6	0.004
–	–	7	0.045
–	–	8	0.20

Appendix G
End of Spectrum HCDA Perspectives for SFRs

Alan Waltar

G.1 Introduction

Substantial work has been done over the past several decades to assess "end-of-spectrum" accidents in oxide-fueled sodium-cooled fast reactors (SFRs). Such accidents have often been referred to as Hypothetical Core Disruptive Accidents (HCDAs) due to the extremely improbable/hypothetical nature of accidents proceeding to this degree of severity. Whereas it is highly doubtful, as expressed in Chapters 13 through 16, that realistic accident scenarios in well designed SFRs could ever reach the point of energetic disassembly, many safety analysts have been motivated to search for "cliffs", i.e. points beyond which severe containment challenges could occur. Accordingly, in this appendix, we begin with the assumption that an energetic core disassembly process has taken place and we proceed to review how the energy generated in such a disruption could cause physical damage to the barriers established to prevent harm to plant personnel and the public.

G.2 Fuel Expansion

Two approaches have been used to determine the mechanical work that could be done by a nuclear excursion that results in a melted and partially vaporized core. The first is a bounding approach in which the energy generated from the expanding fuel vapor that is theoretically available as work energy is used to do work on the environs. The second is a mechanistic approach in which the expansion is analyzed mechanistically in order to determine the extent to which this work potential can be dissipated within the structural surroundings.

G.2.1 Bounding Approach

One way to estimate the mechanical work that can be done is to compute the *PdV* work due to expansion of the two-phase fuel bubble down to some post-accident system pressure. A final pressure of one atmosphere has often been used to provide an upper bound to this potential, although more realistic assessments are based on the recognition that pressures inside the reactor vessel would likely be considerably higher at the time the bubble expansion process is complete.

A useful way to determine the maximum theoretical work energy possible for an expanding mixture of two-phase (liquid–vapor) fuel is to follow the accident sequence on a temperature-entropy (*T-s*) diagram for the fuel, as illustrated in Fig. G.1. If the core disassembly calculation is initiated for a core operating at normal steady-state conditions, the prompt burst begins at State 1 with the fuel in solid form and in equilibrium with fuel vapor at that low temperature.

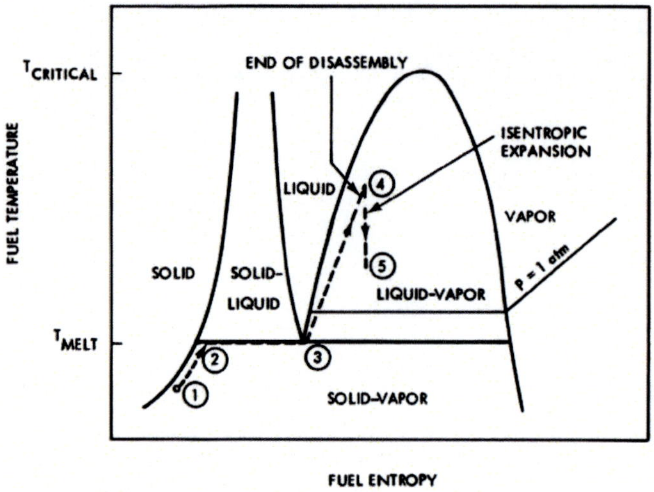

Fig. G.1 Sequence of core disruptive accident on a fuel *T-s* diagram

The melting temperature shown on this diagram is about 3,040 K for UO_2-PuO_2: the saturation temperature at 1 atmosphere is about 3,500 K, and the critical point is of the order of 8,000 K.

The reactivity input causing the power excursion leads to an increase in fuel temperature. The fuel melts between States 2 and 3. Subsequent heating raises the fuel temperature while partially vaporizing the fuel, causing the pressure rise that ultimately disassembles the core. At the end of the nuclear excursion, the fuel would either remain under the dome on the *T-s* diagram (though near the saturation line, since expansion volumes and qualities are small), or would be in the liquid region if the liquid expands to fill all the space and produces a *hard system.*[1] The former path is the case most often encountered in calculated transients. Hence, in our example in Fig. G.1, we illustrate the fuel state at the end of disassembly by State 4.

It is then assumed that the two-phase fuel expands isentropically (i.e., adiabatically and reversibly) to a lower pressure State 5. This expansion process constitutes the work energy phase. This expansion of the "HCDA bubble" could damage the vessel internals, strain the vessel and pipes, and accelerate the sodium above the core upward—where it might impact the closure head, thus damaging the head and further straining the vessel. These processes are discussed further in Section G.4. The expansion would cease after the sodium impacted the head and the bubble filled all the space left by the rising sodium and strained vessel, a state represented by State 5 in Fig. G.1.

To determine the work done by the fuel in the expansion process, we can write the first law of thermodynamics for the control mass of two-phase fuel:

$$dQ = dU + dW = 0, \tag{G.1}$$

where dQ represents the heat transferred from the fuel during the expansion, dU is the change in internal energy, and dW is the work done on the surroundings (e.g., on the sodium, and ultimately, the

[1] A two-phase system is commonly referred to as a "soft system" (due to the "softness" of the gas pressure) while an all-liquid system is called a "hard system." If the core does not disassemble while it is still a soft system, the high liquid pressures of the hard system will disassemble it very quickly.

vessel). If we assume the expansion process to be isentropic, then $dQ = 0$, and the work calculated will be the maximum possible from the expansion of the fuel alone. Making this assumption,

$$W_{max} = U_4 - U_5. \tag{G.2}$$

In an accident calculation, some thermodynamics property at State 5 is generally given-either P_5 (pressure), or v_5 (specific volume) from the available expansion volume. The fuel properties (hence, U_4 and the specific entropy s_4) at the end of disassembly are calculated. From s_4 (which equals s_5) and either P_5 or v_5 together with the fuel equation of state, the final internal energy U_5 can be calculated. Hence, this work can be obtained.

A simple expression for the work can be derived from Eq. (G.2) in terms of quantities more easily obtained than U_4 and U_5 by making the following approximations: the vapor acts as a perfect gas; h_{fg} is constant; v_l is negligible relative to v_g; and c_p for liquid fuel is constant. Thus,

$$W_{max} = M_f \left[c_p \left(T_4 - T_5 \right) - h_{fg} \left(x_5 - x_4 \right) + R(x_5 T_5 - x_4 T_4) \right], \tag{G.3}$$

where

M_f = fuel mass expanded (kg)
c_p = specific heat for liquid fuel (J/kg · K)
h_{fg} = fuel latent heat of vaporization (J/kg)
x = quality
R = gas constant for fuel, 8,317/270 J/kg · K for UO_2-PuO_2

The quality at State 4 is determined from the expression

$$x_4 = \frac{\dfrac{V_4}{M_f} - v_{l4}}{v_{g4} - v_{l4}}, \tag{G.4}$$

where V_4 is the appropriate core volume that contains M_f at the end of disassembly, and the v's are saturated vapor and liquid specific volumes at temperature T_4.

The quality at State 5 can be determined from the relation

$$x_5 = x_4 \frac{T_5}{T_4} + \frac{c_p T_5}{h_{fg}} \ln \frac{T_4}{T_5}. \tag{G.5}$$

If P_5 is known, T_5 can be found from the vapor pressure equation of state, which for oxide fuel can be determined from

$$p = \exp\left(-\frac{78847}{T} + 53.152 - 4.208 \ln T \right), \tag{G.6}$$

Then x_5 and, finally, W_{max} can be calculated.

If the available volume, or V_5 is the final known condition instead of P_5, the problem is more complicated. Neglecting x_5 and assuming $v_{g5} = RT_5/P_5$ (perfect gas),

$$v_5 = \frac{V_5}{M_f} = x_5 v_{g5} = x_5 \frac{RT_5}{P_5}.$$

Substituting this expression for x_5 into Eq. (G.5) gives

$$\frac{v_5 P_5\,(T_5)}{RT_5} = x_4 \frac{T_5}{T_4} + \frac{c_p T_5}{h_{fg}} \ln \frac{T_4}{T_5}, \tag{G.7}$$

where $P_5(T_5)$ is obtained from the vapor pressure relationship noted above. This equation must be solved for T_5 in order to find x_5 and, finally, W_{max}.

The actual expansion problem has an additional degree of complexity not yet addressed. We have presented Eqs. (G.2) through (G.7) as if the entire core could be represented by a single temperature at State 4 and another at State 5. The actual condition at State 4, however, is a distribution of temperatures, with the hottest temperature at the center. If one were to expand each fuel volume (e.g., each mesh cell in an actual disassembly calculation) independently to the final pressure, the expansion work would be significantly greater than if the core were first thermally mixed to obtain a uniform (average) temperature before expansion occurred. Some examples of the magnitude of this effect are provided in Ref. [1]. The reason for the large difference is that thermal mixing of the fuel is a highly irreversible process. The effect of this irreversibility is to decrease the potential of the fuel to do expansion work. This mixing prior to expansion (or during the expansion process) is called self-mixing. If significant self-mixing does not take place before (or during) expansion, and if average fuel temperatures are used in Eq. (G.3), the work calculated from this procedure would be too low.

G.2.2 Mechanistic Approach

Several phenomena would be present in an actual expansion process in a SFR HCDA that would significantly reduce the deliverable work below the theoretically available work. Heat and frictional losses to the surrounding structures would lower the potential for mechanical damage. Self-mixing in the core (referred to above) and pressure gradients in the expanding fluid would lower the efficiency of the conversion of thermal energy to work energy. Abrupt area changes resulting from structural geometry variations in the expansion path would lead to a further reduction. These reductions are particularly large if the internal structure above the core (including axial blanket and fission gas plenum in the fuel pins) remains in place. If above-core structures are forced out of position by high core pressures, the reduction would be less. Analysis of work done with these phenomena taken into account is an example of the mechanistic approach (cf. Sections 13.6.1 and 16.2.1). Early calculations of fuel expansion and two-dimensional coolant hydrodynamics were made with the REXCO computer code (introduced in Section 16.2.5), and some examples of REXCO results are described in Section G.4.2 below. In addition, experimental investigations of bubble expansion with simulant materials have been carried out at a number of facilities both in the United States and Europe, and with actual fuel at the TREAT (ANL-Idaho Falls), ACPR (Sandia-Albuquerque), and CABRI (CEA-Cadarache) reactors.

One sophisticated computational tool for analyses of mechanistic effects during fuel expansion is the SIMMER code, which was described briefly in Chapter 16, Sections 16.2.3 and 16.2.4. Early calculations with the SIMMER code indicated that the phenomena described above might reduce the actual work potential of an expanding fuel bubble by an order of magnitude. Figure G.2 shows the result of a SIMMER calculation for a SFR HCDA in which mechanistic results of damage potential (work) are compared with isentropic fuel expansion results [2]. Damage potential is plotted as a function of average fuel temperature at the end of the disassembly phase. The lower curves (model uncertainties) show the range of values calculated by SIMMER to account for the phenomena described above for reducing damage potential. On top of these uncertainties are added the uncertainties resulting from fuel-sodium mixing and the subsequent vaporization of sodium (molten fuel/coolant interaction, as discussed in the next section). The calculations indicate that the work

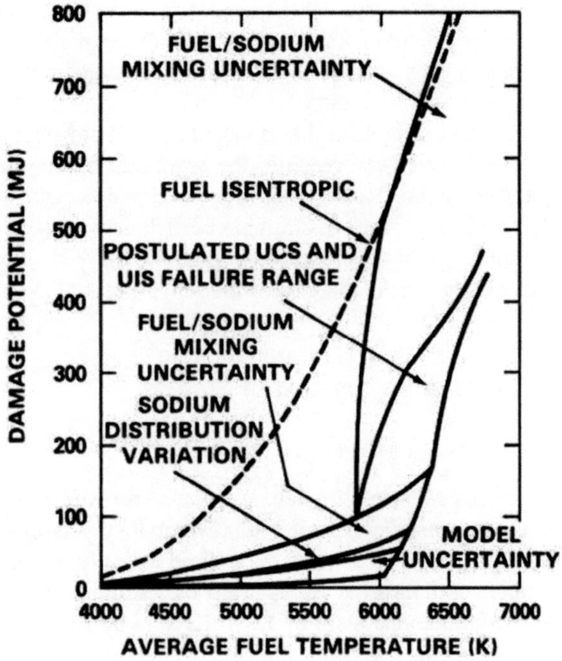

Fig. G.2 Overall influence of the mechanistic treatment and uncertainties on damage potential [2] (UCS and UIS signify upper core structure and upper internal structure)

potential remains low relative to the results for isentropic fuel expansion as long as the average fuel temperature does not approach 6,000 K, a value that can probably be ruled out by core design.

G.3 Molten Fuel/Coolant Interaction

The possibility of molten fuel interacting with sodium during a SFR accident sequence has long been a concern because of the potential for sufficiently rapid heat transfer to trigger a sodium vapor explosion. As readily evident from the preceding discussions, there are several phases of an unprotected accident sequence where a molten fuel/coolant interaction (MFCI) could significantly influence the course of events. However, it is in the mechanical consequences phase that attention was first focused on the role of sodium during a power excursion.

A MFCI is said to occur if fuel mixes with sodium in such a manner that the rate of heat transfer from the fuel to the sodium is much faster than the rate which occurs during normal boiling. The consequence of this interaction is a rapid rise in pressure due to sodium vaporization, followed by expansion of high pressure two-phase sodium. Since the boiling temperature of sodium is well below the melting temperature of mixed oxide fuel, such a process would enable the sodium to be a more efficient working fluid than expanding fuel. Hence, for a given energy deposition at the end of the nuclear excursion, more damage might be imparted to the core surroundings if sodium, rather than fuel, were the expanding medium.

Frequently MFCI's are subdivided into two categories—energetic MFCI's (or vapor explosions) and low energy MFCI's (or vigorous boiling). In a vapor explosion, the time scale for heat transfer between the fuel and the sodium is small compared with that for the expansion of the fuel/coolant mixture. In this case, a significant fraction of the thermal energy in the fuel can be converted into mechanical work on the surroundings. In low energy MFCI's, heat transfer rates are lower and mechanical work from sodium expansion may be quite low. If the MFCI zone is constrained, however, by the large pool of sodium above it so that expansion cannot occur rapidly, enough sodium might be vaporized even in a low energy MFCI to generate more mechanical work than would have resulted from fuel expansion alone.

Efforts to determine the damage potential of MFCI's in an HCDA have taken two directions. First, estimates have been made of the work that can be done by sodium expansion in SFR geometries based on various assumptions or approximations with regard to specific phenomena in the interaction. Second, a large research effort has been directed toward a more complete understanding of the physical phenomena that occur in a vapor explosion. We shall describe several sodium expansion models and then discuss briefly the more important aspects of the physics of vapor explosions.

G.3.1 Sodium Expansion Models

G.3.1.1 Hicks-Menzies

Hicks and Menzies [3] were the first to provide a quantitative analysis of MFCI's in a SFR. They offered a bounding calculation in which they determined the thermodynamically maximum possible work that could be done by the sodium for a given amount of energy generated in the fuel. Figure G.3 contains a plot of the Hicks-Menzies results for an initial fuel temperature of 3,450 K and an initial sodium temperature of 1,150 K. This plot shows the expansion work as a function of both the sodium-to-fuel mass ratio and the final pressure.[2] As expected, the work potential increases as the final

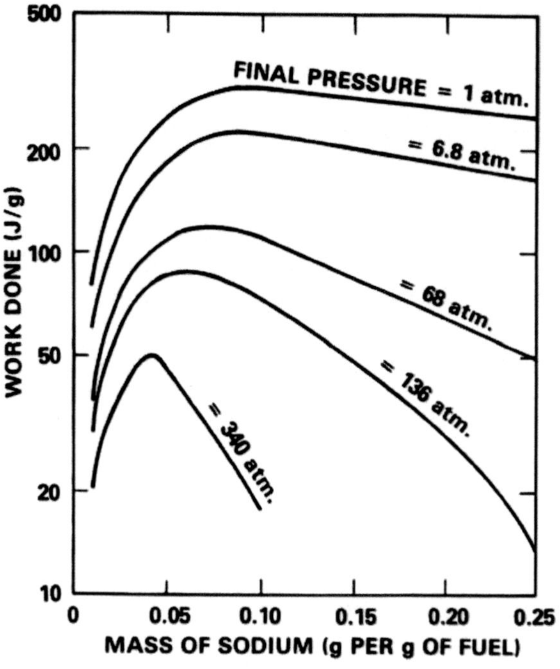

Fig. G.3 Hicks-Menzies curve for the work done by sodium expansion for an initial fuel temperature of 3,450 K and an initial sodium temperature of 1,150 K [3]

[2]An updated calculation was reported by Judd [4] which used an improved (Himpan) equation of state for sodium instead of perfect gas behavior assumed by Hicks and Menzies. Maximum work values were increased by about 30%.

pressure is lowered. The nonlinear dependence upon the sodium-to-fuel mass ratio can be explained as follows. For a small sodium mass there is not enough working fluid to cause much *PdV* work to be done, whereas for a high sodium mass the heat transferred from the fuel is absorbed by so much sodium that the peak sodium temperature remains low. The maximum work occurs for a sodium-to-fuel mass ratio slightly less than 0.1, which also corresponds roughly to the mass ratio in a typical undisturbed SFR core.

The principal approximations made by Hicks and Menzies were (1) thermal equilibrium, (2) temperature-independent material properties, (3) neglect of the heat of fusion for fuel, and (4) perfect gas behavior for the sodium. The first approximation is by far the most important. It implies that perfect heat transfer between fuel and sodium exists throughout the sodium expansion process. Hence, the calculation is purely thermodynamic; no potentially limiting rate processes are considered.

The Hicks-Menzies model is a two-step process. First, fuel and sodium are mixed and energy is instantaneously transferred as heat from molten fuel to liquid sodium until thermal equilibrium is reached, i.e., until the temperatures of the two fluids are equalized. In the second step, the sodium vaporizes and expands doing *PdV* work on the surroundings. Throughout the expansion process, as the sodium temperature is reduced by the expansion, heat transfer from the liquid fuel to the sodium is allowed to continue so that the two fluids remain in thermal equilibrium. (Physically this would require an infinitely rapid heat transfer rate across an infinitesimal temperature difference—which is the reason that this model represents a bounding calculation.)

As was done for the fuel expansion, the sodium heating and expansion processes can be illustrated on a *T-s* diagram for sodium, shown in Fig. G.4. Saturated liquid sodium initially at State 1 is heated by the molten fuel to State 2, or possibly up through the critical point ($T_c = 2,509$ K) to State 3.[3] An expansion process accompanied by heat transfer from the fuel to the sodium (as in the Hicks-Menzies model) is indicated by paths 2→2B and 3→3B. On the other hand, after the sodium is heated, the expansion might occur so rapidly that it could follow an isentropic path, as indicated by 2→2A and 3→3A, in which case the work done would be significantly reduced below the Hicks-Menzies result.

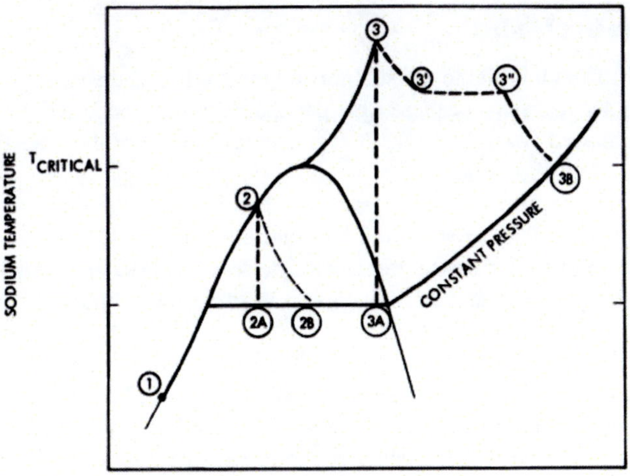

Fig. G.4 Schematic HCDA sequences on a sodium *T-s* diagram

[3]The saturated vapor pressure equation for sodium was given in Section 11.4.6.

The path 3→3′ represents the expansion while heat is transferred from molten fuel; 3′→3″ represents a constant temperature expansion as the fuel solidifies; and 3″→3B represents heat transfer from solidified fuel.

Assuming no phase changes during the initial mixing process and constant values for specific heats, the equilibrium sodium temperature prior to the expansion is simply

$$T_{eq} = \frac{c_f T_f + m c_{\mathrm{Na}} T_{\mathrm{Na}}}{c_f + m c_{\mathrm{Na}}}, \tag{G.8}$$

where the subscripts f and Na refer to the initial fuel and sodium conditions. The quantity m represents the sodium-to-fuel mass ratio. During the expansion, the sodium is the fluid that does work on the surroundings. The fuel transfers its heat to the sodium and is always at the same temperature as the sodium. Based on a control mass consisting of fuel plus sodium, the expansion is adiabatic. The first law for the expansion of the combined fuel and sodium control mass is

$$dW + dU = 0, \tag{G.9}$$

where $dW = PdV$ of the expanding sodium, and $dU = dU_{\mathrm{Na}} + dU_{\mathrm{fuel}}$. This is the calculation that was performed by Hicks and Menzies. The calculation of W depends strongly on the equilibrium temperature, T_{eq} at the beginning of the expansion.

It is unlikely that the sodium temperature will ever rise to the sodium critical temperature, T_c. If a vapor explosion occurs, the sodium temperature will likely be limited to the spontaneous nucleation temperature, which is about 0.9 T_c. For low energy MFCI's there are several barriers to heat transfer from the fuel to the sodium that tend to limit the sodium temperature and prevent the instantaneous heat transfer assumed in the Hicks-Menzies model and, hence, to limit the work potential of the sodium. These barriers include low thermal conductivity of the fuel, vapor blanketing by sodium vaporization, and blanketing by fission gas.

G.3.1.2 Time-Dependent Models

Both the heat transfer from the fuel to the sodium and the motion of the expanding sodium are time-dependent processes. Accounting for these time-dependent effects reduces the work potential of the sodium below that obtained from the Hicks-Menzies model. Some of the essential features of time-dependent models are described below.

The rate of heat-transfer from fuel to sodium is reduced first by the fuel conductivity of the fuel droplets. A temperature gradient exists in the fuel that prohibits the fuel and sodium from being at identical temperatures. Resistance to heat transfer may also exist at the fuel-sodium interface, particularly after sodium vaporization begins or fission gas is released. In addition, the time required for fuel to fragment and mix with the sodium would influence the time-dependence of the expansion process.

The motion of the expanding sodium in the MFCI zone is constrained by the material surrounding the zone, and particularly by the colder sodium pool above the core. The constraint of the MFCI zone is modeled in two stages, an early *acoustic constraint* followed later by an *inertial constraint.*

During the first part of the expansion, which is controlled by the acoustic constraint, the constraining sodium above the MFCI zone is compressed so that pressure from the MFCI zone is transmitted through the sodium pool by an acoustic wave. For this acoustic constraint time domain, the MFCI pressure, P, is related to the velocity, V, of the upper surface of the expanding zone by the relation

$$P - P_0 = \rho_0 c_0 V, \tag{G.10}$$

where P_o, ρ_o, and c_o are initial values of pressure, density, and sonic velocity in the constraining sodium.

Later, the sodium above the MFCI zone moves upward as an incompressible inertial mass; the constraint is then said to be inertial. The relation between the MFCI pressure and the expansion velocity of the zone is then obtained directly from $F = ma$, or

$$P - P_0 = \rho H \left(\frac{dV}{dt} + g \right), \tag{G.11}$$

where H is the height of the sodium being accelerated above the MFCI zone, P_0 is the pressure in the cover gas over the sodium, and g is the gravitational acceleration.

G.3.1.3 SOCOOL Model

Padilla [5] pioneered the first effort to calculate the reduction in work energy of expanding sodium due to the time-dependent effects of heat transfer and constraint. The SOCOOL model assumed an acoustic constraint on the MFCI zone until the pressure wave traversed the distance to the top of the sodium pool and back to the MFCI zone—a time called the *unloading time.* Uninhibited heat flow was allowed from the droplet surface to the liquid sodium prior to the unloading time, but the low thermal conductivity of the oxide fuel was taken into account by considering the time dependence of the heat transfer process within the spherical droplet. After unloading, heat transfer from the fuel to the sodium was stopped entirely based on the assumption that the droplet would then be vapor blanketed, and a thermodynamic, non-time-dependent expansion was then calculated to obtain the work done. In this kind of model, the first law would be written separately for the sodium in terms of time derivatives as

$$\dot{Q} = \dot{U} + \dot{W}, \tag{G.12}$$

where $\dot{Q}$ represents the rate of heat transfer from the fuel to the sodium.

The rate of heat transfer and the expansion work calculated by SOCOOL were strongly affected by the fuel droplet size. The thermal equilibrium case (as in Hicks-Menzies) corresponds to a droplet size of zero. Experimental data on fragmentation size of UO_2 fuel in sodium suggests mean particle diameters of the order of 100–1,000 μm (cf. Fig. G.8, at the end of this section) for which calculated work energies are significantly lower than for the thermal equilibrium case. Smaller sizes cannot be ruled out, however, if a detonation wave were possible as discussed later for the Board-Hall model.

G.3.1.4 ANL Parametric Model

Cho and Wright [6] expanded and generalized Padilla's model. First, they allowed a more general constraint by using both an acoustic constraint followed by an inertial constraint on the MFCI zone. They also added further heat transfer models and included a time constant to account for the time delay associated with fuel fragmentation and mixing with sodium. Inclusion of these effects led to further reductions in the calculated work energy potential of expanding sodium. Additional modifications were made to Padilla's model in parametric models elsewhere.

For some time, a problem with the parametric transient models was the unavailability of appropriate experiments with which to compare the models. Some measure of success was achieved in the late 1970s by Jacobs with the Karlsruhe (Germany) version of a parametric-type code, MURTI [7]. Jacobs, together with French and American colleagues, was reasonably successful in calculating MFCI's between molten UO_2 and sodium in experiments at both the Centre d'Études Nucléaires de Grenoble [8] and the Sandia Laboratories [9].

G.3.2 Vapor Explosions

The primary concern with regard to an MFCI in SFR safety analysis is whether a large energetic vapor explosion could ever occur. Considerable progress has been made toward an understanding of vapor explosions, and the major results are reviewed in Refs. [10–12].

Many experiments have been conducted in which molten UO_2 fuel and sodium have been brought into contact in order to observe under what conditions, if any, a vapor explosion could occur between these materials. In nearly every case, no vapor explosion occurred. In a few cases involving small quantities of sodium, energetic small-scale explosions have taken place [13]. In several experiments [8, 9], significant pressure pulses occurred that may be more accurately described as low energy or small-scale MFCI's rather than energetic or large-scale vapor explosions. This extensive experience indicates that a large-scale vapor explosion between oxide fuel and sodium may be either impossible or extremely unlikely in a SFR.

Even more experiments on vapor explosions have been conducted with materials other than UO_2 and sodium. These tests have contributed to an understanding of specific mechanisms involved in vapor explosions. More-over, this experience has led to substantial agreement about many, though not all, aspects of vapor explosions.

G.3.2.1 Stages in a Vapor Explosion

It is generally agreed that a large-scale vapor explosion involves several necessary stages, usually described as (1) coarse premixing, (2) triggering, (3) escalation, and (4) propagation. Each of these stages is described below. This discussion is followed by an outline of the two theories around which most of the interest in vapor explosions is centered, namely: (1) spontaneous nucleation theory, proposed by Fauske [14], and (2) detonation theory, pro- posed by Board and Hall [15].

Coarse Premixing

The first stage in a vapor explosion is the coarse mixing of the hot liquid throughout the cold liquid. Coarse mixing refers to a mixture of relatively large particles of molten fuel as opposed to the tiny droplets later needed for the very rapid heat transfer required for a vapor explosion. Film boiling appears to be necessary to allow coarse mixing, and it prevents large-scale contact of molten fuel and sodium for a sufficient time to allow significant premixing to occur.

The severity of a vapor explosion is often associated with whether the mixing and the resulting interaction are *coherent.* In a coherent interaction, rapid premixing of hot and cold liquids occurs throughout a large volume so that the ensuing explosion can involve a large fraction of the two liquids almost instantaneously. In a SFR, coherent mixing of large quantities of fuel and sodium would be required for there to be any chance of a sufficiently energetic interaction to breach the vessel.

Triggering

Another requirement for a vapor explosion is liquid-liquid contact between the hot and cold liquids. A triggering mechanism must exist to cause this contact. Liquid-liquid contact would occur upon collapse of the vapor blankets present during film boiling. Mechanisms that could cause this collapse are a pressure pulse (shock wave) and cooling of the hot liquid.

Escalation

The coarse fuel droplets must break up into smaller droplets in order to provide the large heat transfer surface required for the rapid heat transfer rates in a vapor explosion. Some questions related to droplet fragmentation mechanisms that might lead to escalation of a MFCI are still unresolved. Local pressurization on contact between the hot and cold liquids may provide mechanisms for droplet breakup. Fauske [14] has presented the theory that the initial contact temperature between the fuel and the sodium must be at least as high as the spontaneous nucleation temperature of the sodium in order for sufficient local pressurization to take place to escalate the interaction into a vapor explosion. Further discussions of both spontaneous nucleation theory and fragmentation are provided below.

Propagation

The final stage in a vapor explosion is the propagation of the interaction through the coarse mixture of fuel and coolant. Early theories assumed that when a vapor explosion took place, collapse of the vapor blankets and subsequent fragmentation would occur essentially simultaneously throughout the mixture. Colgate [16] suggested the first time-dependent propagation mechanism. Board and Hall [15] then proposed that the mechanism responsible for propagation is a detonation wave. Their theory is described briefly below, following the discussion of spontaneous nucleation theory.

G.3.2.2 Spontaneous Nucleation Theory

Fauske [14] proposed that the following conditions are required for a large-scale vapor explosion:

(1) coarse premixing, allowed by film boiling;
(2) liquid–liquid contact;
(3) initial contact temperature equal to or greater than the spontaneous nucleation temperature of the coolant; and
(4) adequate constraints.

Above a temperature called the *spontaneous nucleation temperature*[4] embryonic vapor bubbles with a critical radius sufficient to maintain stability nucleate at an extremely rapid rate, and very rapid phase transformation follows. An impressive number of experiments have been performed, particularly with water and organic liquids, to show that a vapor explosion occurs only when the contact temperature is at least as high as the spontaneous nucleation temperature of the cold liquid.

When two fluids are brought into contact, the interface, or contact, temperature, T_i, is given by

$$\frac{T_{h0} - T_i}{T_i - T_{c0}} = \sqrt{\frac{(\rho c_p k)_c}{(\rho c_p k)_h}}, \qquad \text{(G.13)}$$

where the subscripts h and c refer to the hot and cold fluids and subscripts 0 refer to initial values. The quantities ρ, c_p, and k are density, specific heat, and thermal conductivity, respectively.

Figure G.5 compares the spontaneous nucleation temperature, T_{SN}, for sodium with the contact temperature, T_i, between oxide (UO_2) fuel and sodium for a particular set of initial fuel and sodium temperatures. The differences between T_{SN} or T_i and the sodium saturation temperature, T_{sat}, at the system pressure (i.e., the *degrees* of *superheat*) are plotted against system pressure. Note that the interface temperature is well below the spontaneous nucleation temperature. Hence, according to Fauske's

[4]In a pure uniform liquid, the spontaneous nucleation temperature is called the homogeneous nucleation temperature. The value of the homogeneous nucleation temperature is roughly 0.9 of the absolute critical temperature. (For sodium, $T_c = 2{,}509$ K). If nucleation sites (e.g., impurities or surfaces) are present in the liquid, bubbles may form spontaneously on these sites at temperatures somewhat lower than the homogeneous nucleation temperature.

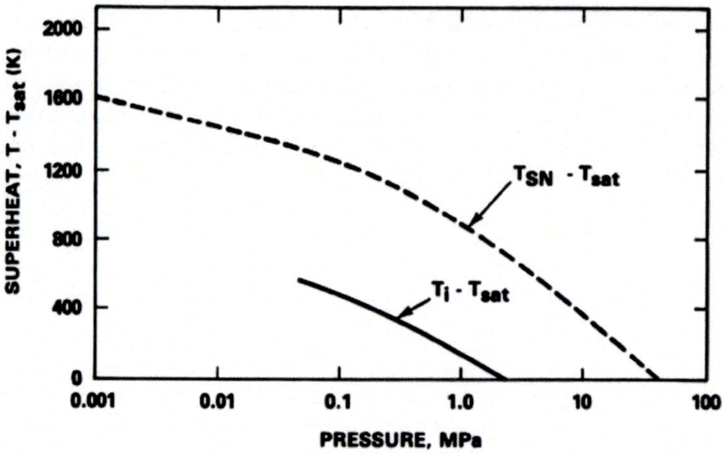

Fig. G.5 Comparison between spontaneous nucleation temperature for sodium and sudden contact temperature between UO_2-PuO_2 fuel initially at 3,470 K and sodium initially at 1,070 K

hypothesis, no large-scale vapor explosion would occur when oxide fuel is mixed with sodium at these temperatures.[5] A second figure, Fig. G.6, shows the domain of initial temperatures for which $T_i < T_{SN}$, for mixtures of UO_2 and sodium, which includes virtually all temperatures that might be encountered in a HCDA. The contact angle α refers to the angle for wetting between sodium and impurities or surfaces that might serve as nucleation sites; an angle of zero corresponds to homogeneous nucleation.

Even if spontaneous nucleation theory rules out a large-scale vapor explosion between oxide fuel and sodium, this is not necessarily the case for carbide fuel. Equation (G.13) yields a higher interface

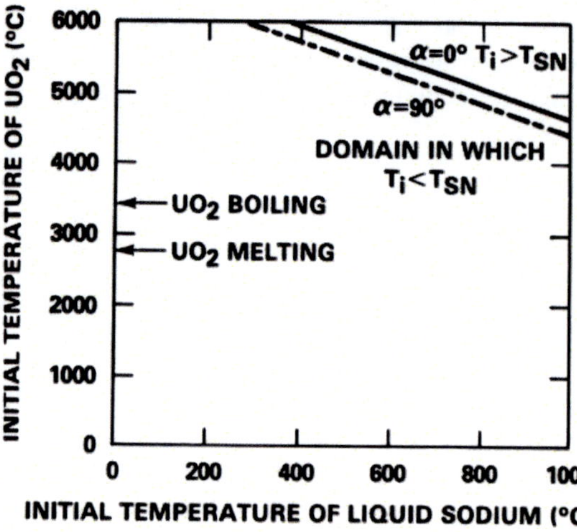

Fig. G.6 Illustration that the spontaneous nucleation temperature is unlikely to be reached on contact between molten oxide fuel and liquid sodium for temperatures of interest [17]. $T_i < T_{SN}$ for all initial fuel and sodium temperatures below the *diagonal lines*. α is theangle between a sodium embryonic bubble and a nucleation site, $\alpha = 0$ corresponds to homogeneous nucleation

[5]A small-scale, delayed explosion might occur if sodium were entrapped in fuel and heated to the spontaneous nucleation temperature. Fauske suggests that this is what happened in the small vapor explosions reported in Ref. [13].

temperature between carbide fuel and sodium due to the higher thermal conductivity of carbide fuel.[6] Even in the oxide fuel case, contact temperatures between high-temperature molten steel cladding and sodium can be above the spontaneous nucleation temperature of sodium. In these systems, however, the other conditions listed above must also be satisfied in order for a large-scale vapor explosion to occur.

G.3.2.3 Detonation Theory

Coherence for efficient large-scale MFCI's implies the existence of a propagation mechanism that couples the regions of explosive energy release to the adjacent unexploded regions. Colgate [16] proposed that explosive expansion of the cold fluid would cause further mixing between hot and cold fluids and, hence, propagation of the interaction. Board and Hall [15] expanded on Colgate's ideas by developing a treatment of the explosion dynamics for the case of a steadily propagating one-dimensional interaction. Three stages are postulated in the model. For the first stage, it is assumed that fuel and coolant become coarsely intermixed. In the second stage, an unidentified trigger mechanism is assumed to result in a shock wave. In the third stage, the shock wave traveling through the coarse mixture causes fine fragmentation of the fuel and further mixing, which in turn produces the rapid heat transfer necessary for sustaining the wave.

These conditions are illustrated in Fig. G.7, in which the coarse premixture is shown, followed by the fragmentation due to passage of the shock wave. As in the case of spontaneous nucleation theory, experimental verification has been obtained for various aspects of this model. Details of the fuel fragmentation mechanisms behind the shock front are not resolved, however.

G.3.3 Fuel Fragmentation

For any of the MFCI models, a prerequisite to the rapid heat transfer rate is a large heat transfer surface and, hence, extensive fragmentation of the molten fuel. Numerous fragmentation models have been suggested, and different ones may apply to different types of MFCI's.

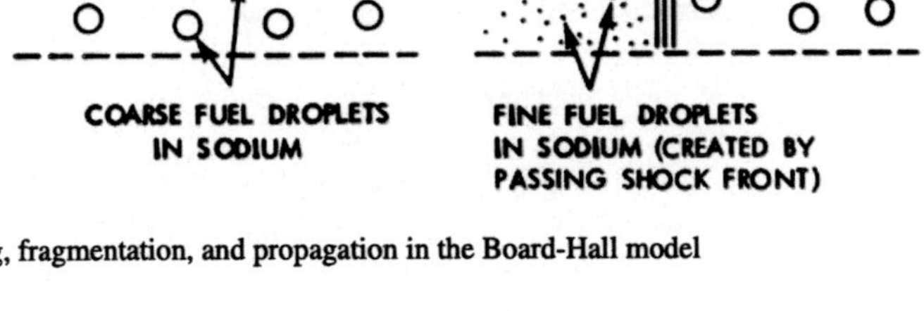

Fig. G.7 Premixing, fragmentation, and propagation in the Board-Hall model

[6]This could be a concern for metal and nitride fuel as well, given their high thermal conductivity.

One important category of breakup process is hydrodynamic fragmentation caused by high relative velocities between the fuel and the coolant. This mechanism may account for the high-velocity breakup of the coarse premixture by a shock wave as postulated in the Board-Hall model. Hydrodynamic fragmentation mechanisms are categorized according to a dimensionless parameter called the Weber number:

$$\mathrm{We} = \frac{\rho D V^2}{\sigma}, \tag{G.14}$$

where

ρ = density of liquid being fragmented,
D = droplet diameter,
V = relative velocity,
σ = surface tension of the liquid being fragmented.

Although breakup can occur with We as low as ~12, the processes that can produce the small particles needed for the Board-Mall model in a short enough time period are boundary layer stripping and catastrophic fragmentation from Taylor instabilities, both of which occur at higher We and, hence, very high relative velocities.

Weber number correlations for these fragmentation processes, together with breakup time correlations as functions of Bond number (Bo) and drag coefficients(C_D) (where $\mathrm{Bo} = 3/8 C_D \mathrm{We}$), have been developed for gas–liquid systems. Experimental data [18] indicate that the correlations hold also for liquid–liquid systems, though the entire area of liquid–liquid hydrodynamic fragmentation is one for which further experimental and theoretical development is needed.

Another class of fragmentation mechanisms is broadly identified by Bankoff [11] under the category of boiling mechanisms. These include many suggested mechanisms variously called violent boiling, compression waves, bubble collapse, jet penetration, coolant entrapment, and vapor blanket collapse. Some of these could be cyclic mechanisms, e.g., coolant could be trapped with fuel, followed by explosive vaporization, collapse of the resulting two-phase bubble, jet penetration of fuel by the coolant, and repeat of the entrapment cycle. Much research has been addressed to the relation between collapse of vapor blankets and the strength of the subsequent interaction. If the liquid–liquid contact temperature on collapse of the vapor blanket is greater than the spontaneous nucleation temperature, experiments indicate that the interaction is strong, an observation that lends support to the spontaneous nucleation theory of Fauske.

Other fragmentation theories have also been proposed that generally lead to slower and less violent breakup. Thermal stress fragmentation is one such mechanism [19]. This model is based on rapid freezing of the outer shell of the fuel droplet which causes high thermal stresses and causes the fuel surface to shatter. This mechanism could explain much of the fragmentation data obtained from the relatively quiescent injection of fuel into sodium. However, the time required to solidify fuel (~50 ms) is too long to allow this mechanism to contribute to a vapor explosion. Another mechanism suggested is the rapid release of gas (such as fission product gas) from within a molten fuel droplet. This mechanism is likely not very important for oxide fuel due to the low solubility of inert gases in UO_2.

Many experiments have been performed, both in-pile and out-of-pile, in which oxide fuel has been quenched by sodium. Fragmentation always occurs, and the resulting particle sizes vary over a wide range. Example particle size distributions are shown in Fig. G.8. The in-pile tests were run in the TREAT reactor in which a power transient causes the ejection of molten fuel from fuel pin test samples into sodium. These particles are generally larger than the sizes needed for propagation according to the Board-Hall model. Further work on particle breakup in shock tubes is needed to study the potential for smaller size particles.

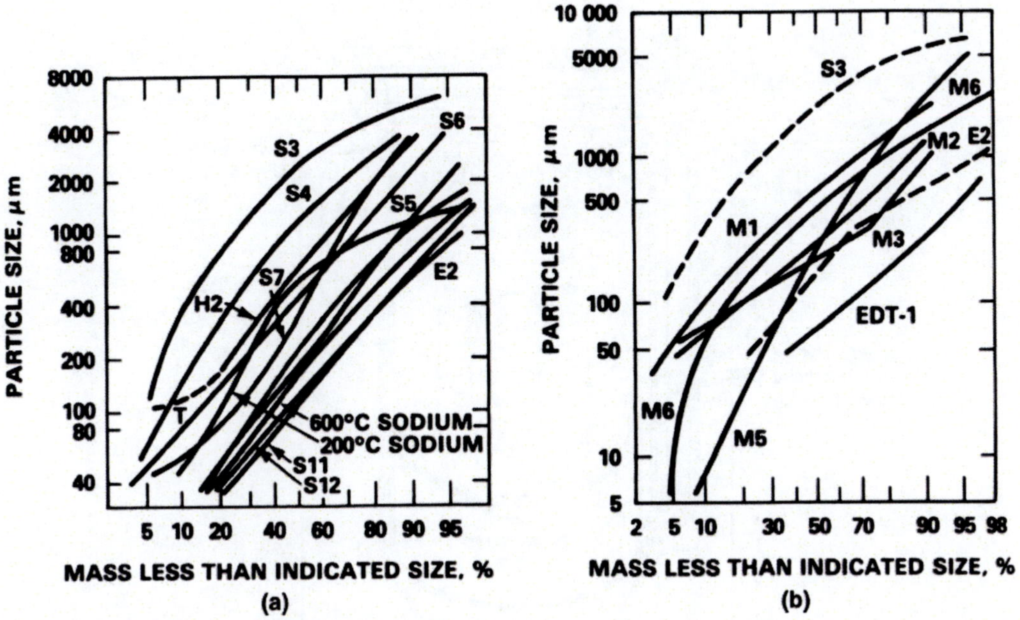

Fig. G.8 Oxide fuel particle size distributions after fragmentation in sodium [20]: (**a**) Treat and small-scale tests, (**b**) Large scale out-of-pile experiments

G.4 Energy Partitioning and Mechanical Consequences

Once the energy deposition source term of the HCDA has been defined, the problem becomes that of determining the consequences to the in-vessel components and to the vessel and its associated piping. Figure G.9 is a sketch showing the key factors that must be considered for a loop-type system.

An expanding high temperature-pressure bubble would be felt on the reactor vessel internals and would be transmitted through the fluid to the boundaries of the system. The downward force against the core support structure might be transmitted through the fluid to the lower head of the reactor vessel and, in any case, would be transmitted via the vessel walls to the reactor vessel support system. The radially propagating pressure wave would be felt at the wall of the reactor vessel. The pressure, if sufficiently high, could cause deformation of the reactor vessel wall radially outward at approximately the same elevation as the maximum pressure.

In the upward direction, the sodium slug (i.e., the pool above the high temperature-pressure bubble) would be accelerated upward although the upper core and upper internal structures would provide significant resistance to the bubble expansion. The acceleration of the sodium slug would impart a kinetic energy to the slug. Upon impact of this slug with the reactor vessel head, the momentum of the slug would be transferred to the head, imparting some upward velocity. This impact would be inelastic, and the excess energy would be dissipated radially by deformation of the upper vessel wall. The upward motion of the vessel head would generally be resisted by its supporting structure and holddown system.

Effects of an energetic accident on the remainder of the heat transport system would result from displacement of the reactor vessel, with corresponding nozzle displacement that could result in a deflection of piping, and from pressure waves that could be propagated throughout the primary heat transport system through the outlet and inlet sodium nozzles.

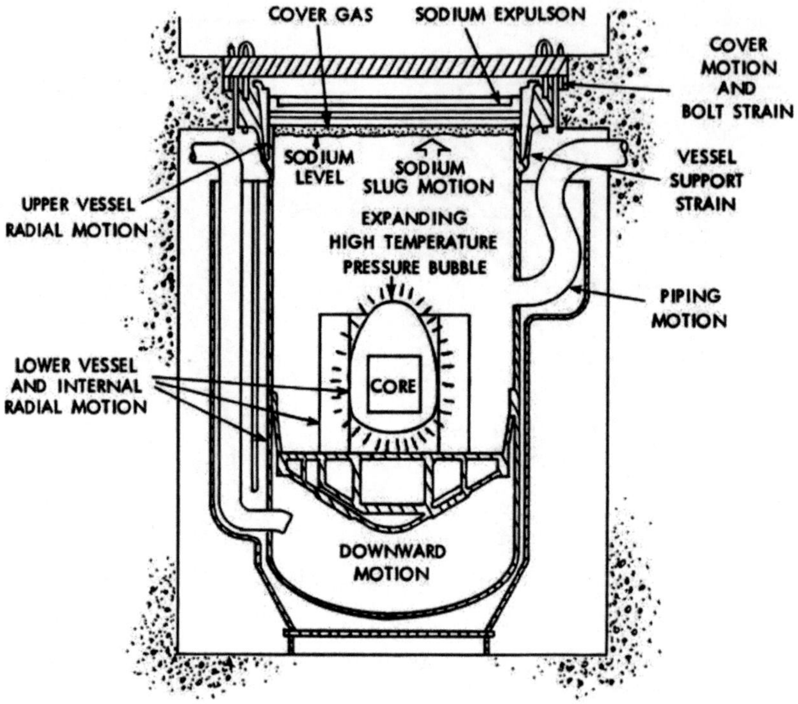

Fig. G.9 Reactor geometry and mechanics of potential damage for a loop-type system

The section below begins with a historical perspective in early scoping methods for correlating mechanical damage. This is followed by a summary of experimental support, including both the SL-1 accident and subsequent simulant experiments. Computational methods are then reviewed, together with a few results selected from a vast amount of literature on the subject to illustrate the nature of the problem involved.

G.4.1 Experimental Data

G.4.1.1 Early Studies

For the early fast reactors the fraction of energy generated during disassembly that was assumed to be converted to mechanical work was based on extremely conservative assumptions. Later research succeeded in reducing this conservatism to some extent, and there is research still underway to establish the degree to which additional physical phenomena may reduce this fraction still further.

Early assessments of damage potential due to such energy release were based on equating this potentially available work energy to an equivalent charge of TNT, i.e.,

$$2 \text{ MJ} \simeq 1 \text{ lb TNT}.$$

Hence, every MJ of work potential was assumed to represent the explosive potential of about 1/2 lb TNT. The principal reason for using this equivalence was that substantial experience existed, particularly within the U.S. Naval Ordnance [21], regarding the damage potential for detonating TNT.

Any attempt to predict the actual mechanical response of a SFR system to an HCDA excursion using the TNT energy equivalence model is ambiguous because the pressure-time characteristics of a TNT detonation are considerably different from those of a nuclear excursion. Mechanical damage from an explosion or pressure transient can be caused by both a shock wave, which is transmitted rapidly to a structure, and the more slowly expanding bubble of reaction products or vaporized material. Pressures in a TNT detonation build up on a microsecond time scale and reach the order of 5,000 MPa. Hypothesized energy releases from an HCDA excursion, on the other hand, build up over a millisecond time scale, and peak pressures are orders of magnitude lower. As a consequence, much of the damage potential of a TNT detonation to surrounding structures comes from shock wave effects, whereas longer-term bubble expansion would be the predominant damage mode for the slower time-scale pressure buildup of a SFR excursion.

G.4.1.2 SL-1 Accident

The SL-1 accident [22], involving a small military thermal reactor at the Idaho Falls test site in 1961, provides an interesting data base upon which to estimate the potential mechanical damage from a major power burst. The energy source term for SL-1 is substantially different (aluminum-clad metal fuel in water coolant) from that of a SFR; consequently, very little insight is possible with regard to the magnitude or time scale regarding a postulated SFR HCDA event. However, given the total work energy potential, some of the resulting mechanical deformation effects are quite instructive.

The total energy released from the nuclear excursion in SL-1 was estimated [22] to be 130 MJ. Of this total, some 50–60 MJ was believed to be generated in the central fuel elements and transferred to water in less than 30 ms in an MFCI between the aluminum and water. It was this prompt release that caused the damage to the core, thermal shields, and reactor vessel.

Proctor [23] attempted to reconstruct the sequence of major events, given the 50–60 MJ prompt energy generated, by observing the permanent vessel strain and the vessel jump. Figure G.10 is a sketch of the post-accident vessel (exaggerated for emphasis), along with the actual vessel strains measured. Proctor deduced that about 2.5 MJ of energy was deposited in the lower vessel, core structure and thermal shield as a result of the initial pressure burst in the core. Another 3.6 MJ of energy was attributed to the water slug, which caused the vessel to lift about 3.3 m, thereby damaging the vessel head and causing permanent strain near the top of the vessel. This total work energy, 6.1 MJ, represented about 12% of the prompt nuclear release (50 MJ) or about 5% of the total nuclear release (130 MJ).

The major goal of Proctor in conducting this analysis was to see whether the earlier naval studies on containment of explosives could be employed for nuclear reactor HCDA mechanical consequences assessment. Much of this earlier work relating to the damaging effects of underwater explosions had been documented in a classic book by Cole, entitled Underwater Explosions [24]. This work was augmented in the late 1950s and early 1960s by U.S. Naval Ordnance Laboratory experiments conducted with models of the Fermi reactor to determine the effects of a hypothetical explosion in that system. Additional experiments, using a variety of test vessel sizes and configurations, resulted in a set of empirical explosion "containment laws" [21].

A key observation in these tests was that the radial vessel strain at the point of vessel failure was always greater than one-third the ultimate elongation of failure in an ordinary tensile test for the reactor vessel material. In these explosion containment laws, the observed vessel wall strain was correlated to the high energy release for characteristic explosions similar to chemical high explosives (i.e., TNT).

In his SL-1 evaluations [23], Proctor showed that the vessel strain correlated reasonably well with the reactor explosion containment laws. The reactor vessel wall strain was considerably less than the predicted failure strain, and no wall failure was observed. In the SL-1, the liquid surface was well below the reactor vessel head at the time of the accident. When the water slug struck the reactor

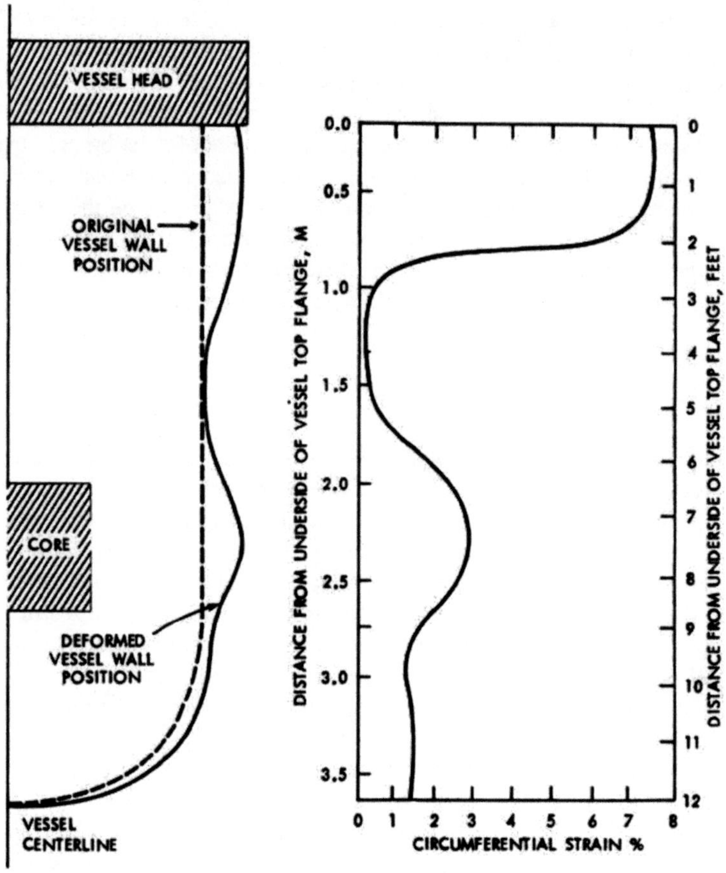

Fig. G.10 Vessel strains following the SL-1 accident (adapted from Ref. [23])

vessel head, the impact on the lower face of the head lifted the entire reactor vessel from its original position and also radially deformed the upper wall of the vessel.

G.4.1.3 Simulant Experiments

Experimental programs to assess mechanical damage in a SFR HCDA have been conducted with simulant materials both in the United States [25–27] (e.g., at SRI International and Purdue University) and in Europe [28, 29]. (e.g., at Foulness, Winfrith, and Ispra). Since expansion times would be much slower (e.g., in the millisecond range) in a SFR HCDA than in a chemical explosion, and since bubble expansion would cause more damage than shock waves in an HCDA, a considerable effort was expended to develop simulant explosives with a burning rate more compatible with the millisecond time frame [30]. In addition to the vessel damage experiments referenced above, tests with simulant explosives under sodium have been conducted at Cadarache to study breach of the cover and sodium transport through the cover in an HCDA [31].

Small scale in-pile tests involving fuel and/or sodium expansion and energy partitioning have been carried out at several transient test reactors, e.g., TREAT, ACPR, and CABRI.

G.4.2 Computational Procedures

Several computational systems have been developed to assess the structural consequences of an HCDA. Of these systems, the REXCO series [32] has been the most widely used, presumably because it was developed specifically for reactor accident analysis.[7] In order to more fully analyze systems that encounter large geometric distortions, Eulerian codes, such as ICECO [33], have also been developed.

Figure G.11 represents a typical model set up for a REXCO calculation of a loop-type reactor system, and Fig. G.12 illustrates the time sequence of events. An expanding high temperature, high pressure HCDA gas bubble in the core region is assumed to become manifest at time zero. The expansion of this high pressure region is clearly seen in the first 15.9 ms, and by 38.3 ms the sodium slug has impacted the vessel head. In this REXCO calculation the sodium slug impacts the vessel head uniformly. Some analyses with the SIMMER code, e.g., Ref. [2], indicate that the gas bubble tends to jet upward so that the slug strikes the head in a nonuniform manner, impacting first at the center of the cover. This results in a decreased upward force on the head. The vessel plenum length/diameter ratio is an important factor in determining the extent of nonuniform slug impact.

Following impact of the sodium slug on the reactor vessel head, REXCO can then be used to calculate the further partition of energy. The results of an early REXCO calculation [34], shown in Fig. G.13, illustrate the concept. Time zero in this figure corresponds to the time when the sodium slug strikes the head. An initial triangular shaped force on the head is followed by a longer and smaller residual force. The axial kinetic energy of the sodium is first transferred into internal energy and radial

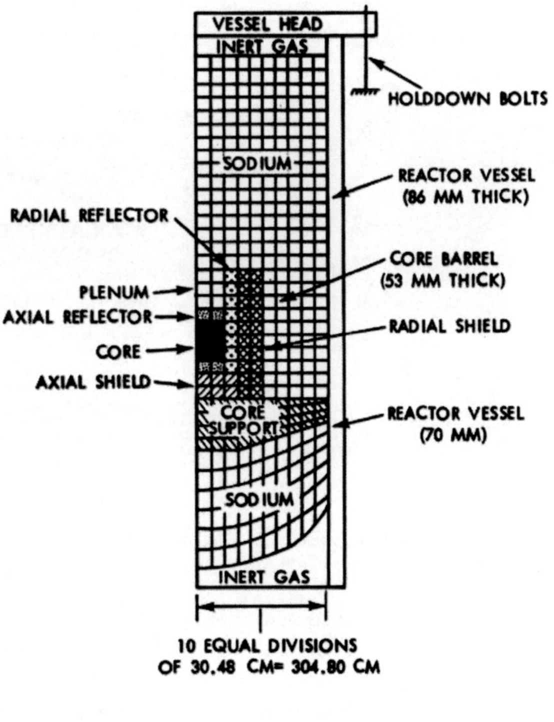

Fig. G.11 REXCO model of a SFR loop system

[7]REXCO, as well as other Lagrangian codes, grew out of common ancestry, i.e., the HEMP code at the Lawrence Livermore Laboratory or F-MAGEE at Los Alamos National Laboratory.

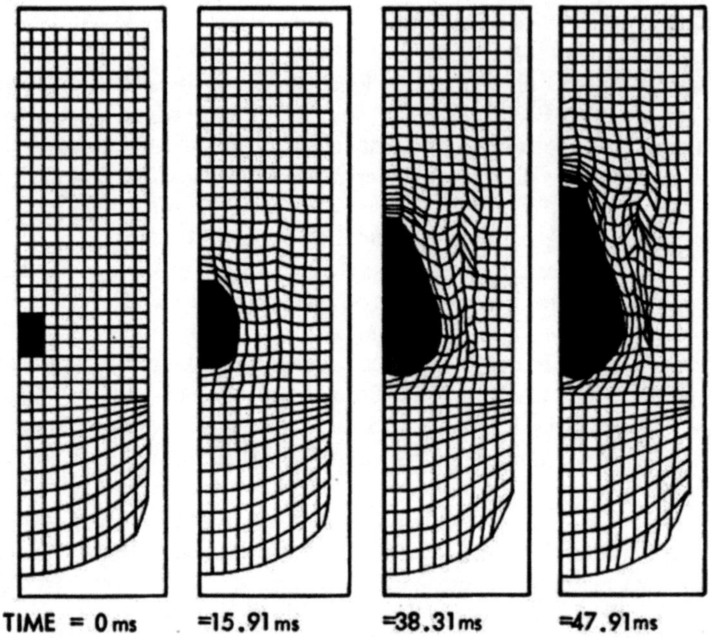

Fig. G.12 Example of reactor configurations predicted by the REXCO code at various times during the mechanical consequence phase of an HCDA

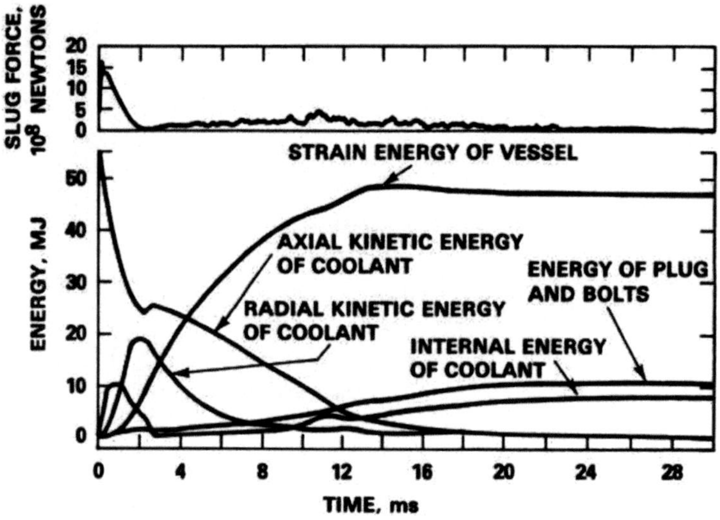

Fig. G.13 Example calculation of slug force and energy partition after sodium slug impacts the vessel head in an HCDA [34]

kinetic energy of the sodium. Some of the original energy is converted to strain energy in the head and hold-down bolts, but the bulk eventually becomes manifest as strain energy in the vessel. Some energy remains as internal energy of the sodium. The initial conditions for the calculations (i.e., at head impact) were 30 m/s velocity, 54 MJ slug kinetic energy (1.2×10^5 kg sodium slug), 86 mm vessel-wall thickness, 7×10^5 kg head, 0.32 m^2 total area of holddown bolts, and 1.4 MPa core vapor pressure.

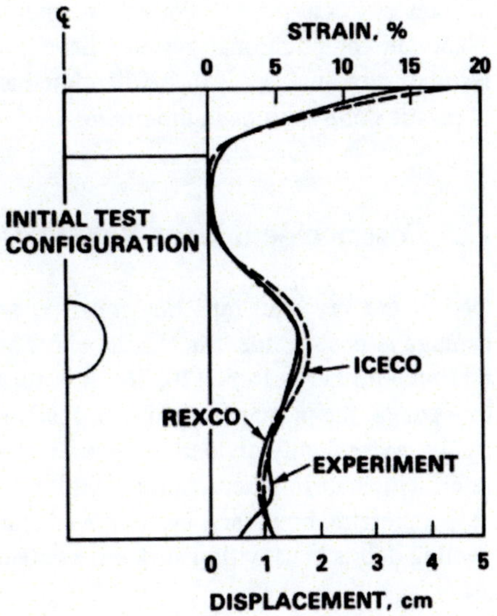

Fig. G.14 Comparison of REXCO and ICECO radial strain calculations [35] with SNR-300 simulant experiments [29]

Both REXCO and ICECO calculations are compared [35] to radial strains measured in the SNR-300 simulation experiments at Ispra [29] in Fig. G.14. The similarity with the SL-1 results in Fig. G.10 is evident.

Figure G.15 gives examples of typical permanent strains calculated as compared to the strain levels anticipated for failure (in parenthesis). In addition to the vessel and vessel support features, it is also

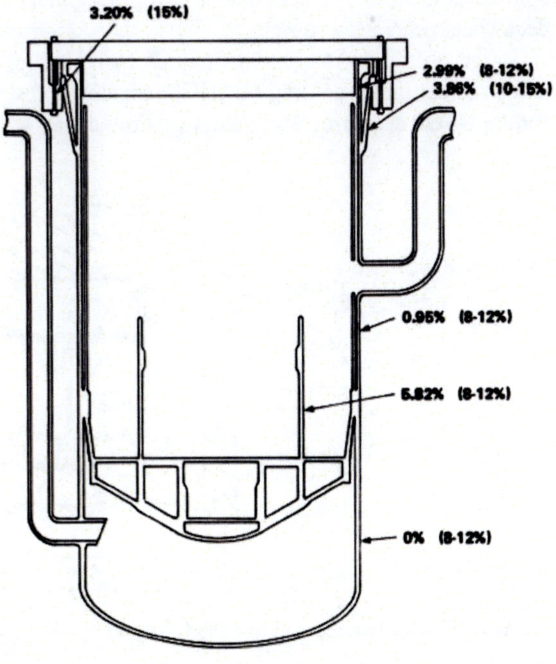

Fig. G.15 Prototype vessel strains predicted for upper-limit HCDA at the end of the mechanical consequences phase in FFTF (values in parenthesis are failure strains)

of interest to consider the effects of pressure wave propagation around the primary system piping. Elaborate computational systems have also been developed to treat these problems. An important result of such analyses is the realization that substantial pressure peak attenuation may occur by means of plastic deformations in the piping.

G.5 Post-accident Heat Removal

One of the key questions that must be addressed in studying accidents involving significant core damage is post-accident heat removal. The principal issues are outlined in this section; the topic is reviewed further in Refs. [20, 36]. It is appropriate first to define the heat generation sources and then to focus on the principal geometric locations where heat transfer and cooling occur. These locations can be conveniently divided into *in-vessel* and *ex-vessel.* As illustrated in Fig. G.16, in-vessel cooling refers either to in-place cooling (within the core) or to cooling debris scattered within the reactor vessel pressure boundary, but removed from the original core location. Ex-vessel cooling focuses on cooling debris beds within structures outside the reactor vessel.

G.5.1 Post-accident Heat Sources

The first step in assessing post-accident coolability is defining the distribution of heat sources following reactor shutdown. This involves determining the location of the fuel inventory and the distribution of heat sources.

G.5.1.1 Fuel Relocation

Knowledge of the fuel location following a whole-core accident is necessary since the fuel matrix comprises most of the nonvolatile fission fragments. These radionuclides make up the bulk of the decay heat generation source.

Based on the discussion in Chapter 15 regarding the unprotected transient overpower (UTOP) and transient under cooling (UTUC) accidents, most of the fuel would remain essentially in-place for the UTOP scenario. Some axial fuel relocation would occur in scattered assemblies and some of

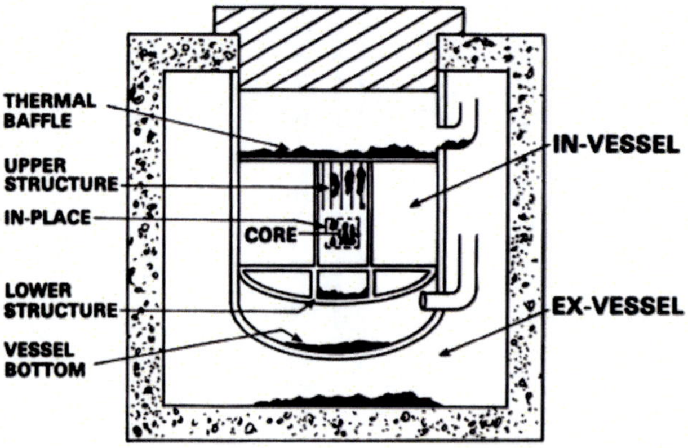

Fig. G.16 Potential locations of fuel debris for post-accident heat removal analysis (loop-type system)

these assemblies may contain partial blockages; most of the fuel would remain at or near its original location.

It is in the unprotected UTUC event that substantial fuel relocation has been hypothesized to take place. Historically, a considerable inventory of fuel was hypothesized to be ejected out of the top of the core into the upper core structure region in the UTUC event; and some fuel could penetrate into the upper sodium pool region and come to rest in the thermal baffle area or even reach the outlet piping (cf. Fig. G.16).

With regard to downward penetration, large heat sinks below the reactor core would likely cause fairly thick steel blockages to form as a result of molten cladding relocation. In-vessel retention devices might be used in some SFR designs in order to prevent penetration of the vessel by molten fuel. Such a device, consisting of a series of trays that could be cooled by the sodium pool, was illustrated for Super Phénix in Fig. 12.7. Without in-vessel retention capability, molten fuel, if present in large amounts, could melt through the bottom of the vessel and fall, together with the sodium still in the vessel, into the reactor cavity.

Recriticality Potential

One of the concerns associated with loss of original core geometry in a fast reactor is the potential for recriticality of core material. Long-term post-accident heat removal can be successful only if assurance is provided that recriticality cannot occur. Hence, the geometry of potential locations for particulate beds or molten pools must be considered. Although a discussion of critical fuel geometries is beyond the scope of this book, the topic is reviewed in Refs. [20, 36].

Designs for structures below the core have been developed for the specific purpose of distributing molten fuel or core debris formed in an HCDA into non-critical configurations. An example of an in-vessel design is the Super Phénix core retainer (Fig. 12.7), in which each tray in the retainer is designed to catch a quantity of core material too small to achieve criticality.

An ex-vessel design is illustrated in Fig. G.20 in Section G.5.3 on ex-vessel phenomena. This figure shows a distribution device that spreads out core debris into a subcritical geometry on the Core Retention System.

G.5.1.2 Heat Sources

The principal heat source in a damaged subcritical core is decay heat from fission products. Beta decay of ^{239}U (23.5 min half life) is important for several hours and beta decay of ^{239}Np (2.35 day half life) provides a significant heat source for several days. Smaller sources include decay heat from activated steel and sodium and higher order actinides such as ^{242}Cm. Heat from the fission process itself decays rapidly after neutronic shutdown, following the longest (~80 s) decay period of the delayed neutrons.

Decay heat from the fission products (including the noble gases) follows the curve contained in Fig. G.17. Computer codes such as the U.S. codes ORIGEN [37], CINDER [38], and RIBD [39] can be used to obtain the contributions of individual radionuclides to the decay heat.

The low volatility fission products are sometimes subdivided further according to whether oxides are formed in oxide fuel. This distinction is made since the oxides tend to remain with the fuel, whereas the metals tend to concentrate with the steel. The noble gases and halogens and some of the volatile fission products are normally assumed to separate from the fuel upon loss of fuel pin integrity, but the Group II and III elements do not necessarily escape from the primary sodium because the halogens may react with sodium and the Group III elements are soluble in sodium.

Fission products have been categorized in various ways for convenience. The four groups listed in Ref. [20] are given in Table G.1, together with the elements in each group and the principal form in which they are present in the oxide fuel.

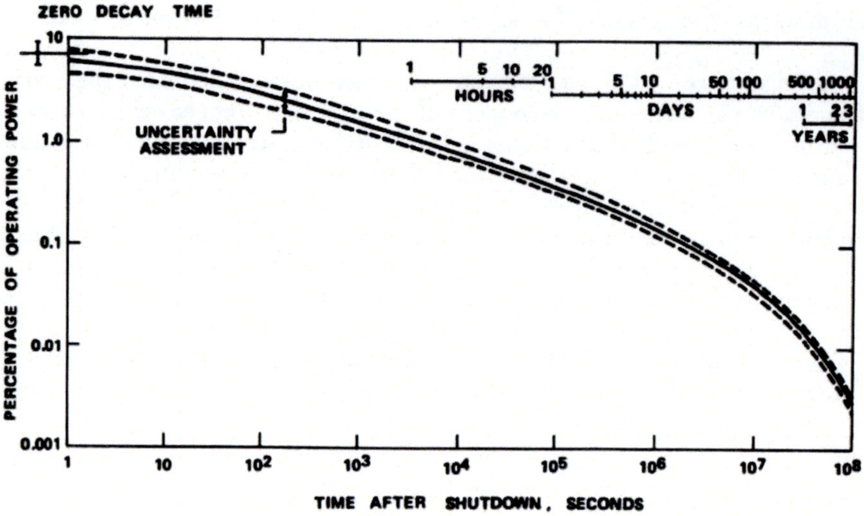

Fig. G.17 Fission product decay heat for FFTF [4]

Table G.1 Fission product categories

Group	Category	Family	Principal form	Element
I	Noble gases	Noble gas	Elemental	Xe, Kr
II	Halogens	Halogen	Elemental	I, Br
III	Volatile solids	Alkali metal	Metal	Cs, Rb
		Transition	Metal	Ag, Cd
		–	Metal	As, Se, In, Sn, Sb, Te
IV	Low volatility (or "Nonvolatile") solids	Transition	Metal	Tc
		Noble metal	Metal	Ru, Rh, Pd
		Alkali earth	Oxide	Sr, Ba
		Transition	Oxide	Mo, Y, Zr, Nb
		Rare earth (Lanthanides)	Oxide	La, Ce, Pr, Nd, Pm, Sm, Eu, Gd

Fission yields for ^{239}Pu are shown in Fig. G.18. Differences are noted around A = 90 (e.g., Sr) and between 105 and 110, but the effects of the differences are not large. The fission product decay heat level as a fraction of operating power in SFR fuel is similar to that of LWR fuel. However, the SFR decay heat density is substantially higher due to the higher operating power density.

G.5.2 In-vessel Cooling

Sodium is a very effective coolant—and this applies also for cooling a damaged core or core debris. It maintains a single phase liquid condition up to 880°C at atmospheric pressure and has an adequate heat capacity. Furthermore, as noted in Section G.3, there is ample evidence that molten ceramic fuel (and also molten steel) will fragment upon contact with sodium, thereby exposing large heat transfer areas for removal of decay heat.

Such properties can be used to advantage in a SFR system to provide long-term cooling of a damaged core or of core debris within the reactor vessel. Topics of importance include in-place cooling

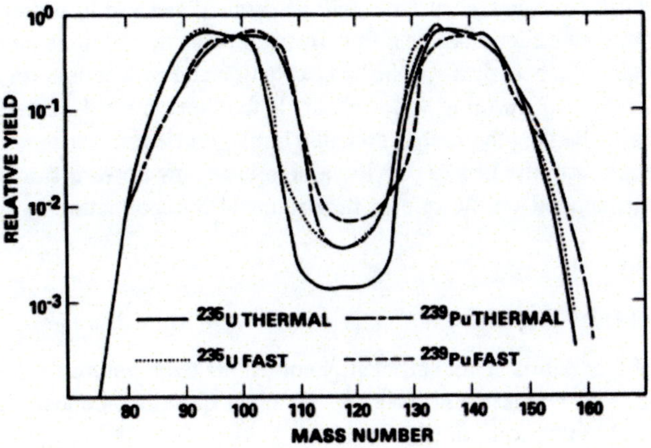

Fig. G.18 Fission product yields for ^{235}U and ^{239}Pu

(i.e., where the damaged core is basically intact) and particulate bed cooling (where debris has relocated to other locations within the vessel). The ability of the sodium to cool the debris, however, is predicated on open flow paths to the heat generating sources. If large flow blockages occur, or if sodium boils dry from the debris bed, then a molten core-debris pool could form. Hence, cooling of each of these configurations must be considered.

G.5.2.1 In-place Cooling

Four conditions have been proposed for cooling fuel material remaining in a substantially damaged core region, each entailing a different cooling configuration and associated heat transfer mechanism: (1) cooling the outer rows of pins within each assembly, even though the central region may be blocked, (2) cooling voided assemblies (possibly to the point of flow recovery) by radial heat transfer to neighboring intact assemblies, (3) cooling particulate debris in blocked assemblies via sodium leakage through the debris, and (4) cooling molten debris in the lower shielding section via radial heat transfer to neighboring assemblies.

Studies indicate that Condition 1 can be satisfied with natural convection flow for heat loads below 10–20% of full power, and that Condition 2 can be satisfied if power in the local voided assembly is below 3% of operating power and flow in neighboring assemblies is maintained by pony motor flow. Natural convection cooling of debris Conditions 3 and 4 is possible only at low power levels (perhaps 0.5% full power). Hence, as would intuitively be expected, in-place cooling is readily achieved for essentially original geometry, even under natural convection conditions alone, but cooling capability is increasingly jeopardized as the degree of fuel concentration increases.

G.5.2.2 Particulate Bed Cooling

Since molten fuel and steel will fragment upon contact with liquid sodium, there is considerable interest in determining the cooling characteristics of packed beds of particulate core debris. Potential locations of particulate beds are shown in Fig. G.16. The first step in assessing the cooling of such a bed is to determine its depth and composition. Given a horizontal surface upon which such a bed can form, the depth is determined by knowing the total debris mass available, the particular surface area, and the debris porosity (typically about 50%).

Debris beds that form on in-vessel surfaces will likely be immersed in sodium. Experiments have shown that the predominant direction of heat flow from coolable beds will be upward to the overlying sodium. For shallow beds the heat flow will be by conduction and convection; for deeper beds, sodium boiling occurs within the bed, forming vapor channels that vent from the interior of the bed to the overlying sodium. As the bed depth exceeds coolable limits, *particulate burnout* occurs, meaning that sodium cannot penetrate into the bed to provide a high heat flow upward from the bed. Coolability beyond this point will depend on the rate of heat removal by conduction in both the upward and downward directions.

G.5.2.3 Molten Pool Heat Transfer

If a particulate bed of fuel debris is not sufficiently cooled, its temperature rises until a bed of molten fuel and steel is obtained. For this situation, the principal questions concern the direction of heat flow—whether up-ward, downward, or sideward.

Natural convection occurs in internally heated molten pools, with hot fluid rising and colder fluid moving downward. The onset of motion in a fluid layer occurs when the buoyancy forces caused by temperature gradients exceed the viscous forces. Inertial forces also affect the transport process after flow develops. The Grashof number (Gr) combines inertial, viscous, and buoyancy forces. Since natural convection is a combined momentum and energy exchange process, the Prandtl number (Pr) also influences the heat transfer. Hence, natural convection is correlated with the product of Gr and Pr, a product known as the Rayleigh number (Ra). For an *internally heated fluid* between two horizontal plates at equal temperatures and separated by a distance L, the Rayleigh Number, labeled $\mathrm{Ra_I}$, is[8]

$$\mathrm{Ra_I} = \frac{g\beta Q L^5}{\nu\alpha k}, \tag{G.15}$$

where

β = coefficient of thermal expansion
g = gravitational acceleration
Q = volumetric heat generation rate
ν = kinematic viscosity
α = thermal diffusivity
k = thermal conductivity
L = distance between plates.

Figure G.19 contains data and correlations from several sources showing the fraction of heat from a molten pool flowing downward (for equal boundary temperatures) [36]. The abscissa is $\mathrm{Ra_I}/64$ where $\mathrm{Ra_I}$ is defined by Eq. (G.15). Values of $\mathrm{Ra_I}$ of interest for post-accident heat removal are generally in the higher range of values on Fig. G.19 for which only 10–30% of the heat generated would be transferred downward.

Sideward heat transfer correlations are less well defined. Work on sideward as well as other conditions for upward and downward heat transfer from molten pools is an area occupying increasing attention for both LWR and SFR safety analysis.

Heat transfer correlations for upward, downward, and sideward heat transfer by natural convection from molten pools with internal heat sources are generally of the form:

---

[8] The Rayleigh number for natural convection driven by a temperature difference ΔT instead of an internal heat source is given by $g\beta\Delta T L^3/\nu\alpha$. To obtain $\mathrm{Ra_I}$, the ΔT is replaced by QL^2/k.

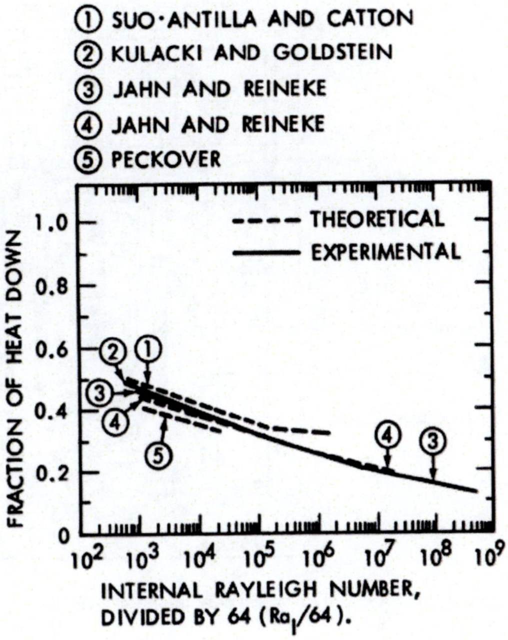

Fig. G.19 Comparison of heat transfer correlations (fraction of heat flowing downward) for internally heated horizontal pools with equal boundary temperatures [36]. Numbers *encircled* indicate experimenters as found in Ref. [36]

$$\mathrm{Nu} = C \cdot \mathrm{Ra}_{\mathrm{I}}^{m}. \tag{G.16}$$

The use of a correlation of this form is described in work by G. Fieg, who developed correlations to determine the directional heat transfer in fuel and steel pools [41].

G.5.3 Ex-vessel Phenomena

If the core debris melts through the vessel, heat transfer and interaction of the debris in the reactor cavity must be examined. Two topics are of principal interest: first, techniques that might be employed to reduce the consequences of melt-through such as engineered core retention concepts (i.e., devices specifically designed to stop the flow of molten fuel debris), and, second, the interaction of sodium and core debris with structural material below the reactor vessel.

G.5.3.1 Engineered Ex-vessel Core and Sodium Retention Concepts

The engineered concepts so far devised to mitigate the consequences of vessel melt-through include *steel liners, crucibles,* and various types of *sacrificial beds.*

An important objective of reactor cavity design is to prevent contact of sodium with concrete since water released from concrete could interact with sodium to form hydrogen. The concrete reactor cavity walls are lined with a stainless *steel liner* to separate sodium from the concrete. This liner is sufficient to contain sodium but not core debris after particulate bed dry out.

Conceptually, a *crucible* retention system is designed to contain core debris indefinitely. It consists of a container or structure in which a sacrificial material is located. Cooling is generally provided by a system completely independent of the reactor coolant system. Figure G.20 shows such a system, as was proposed for the SNR-300 reactor. The distribution device is designed to distribute the core debris

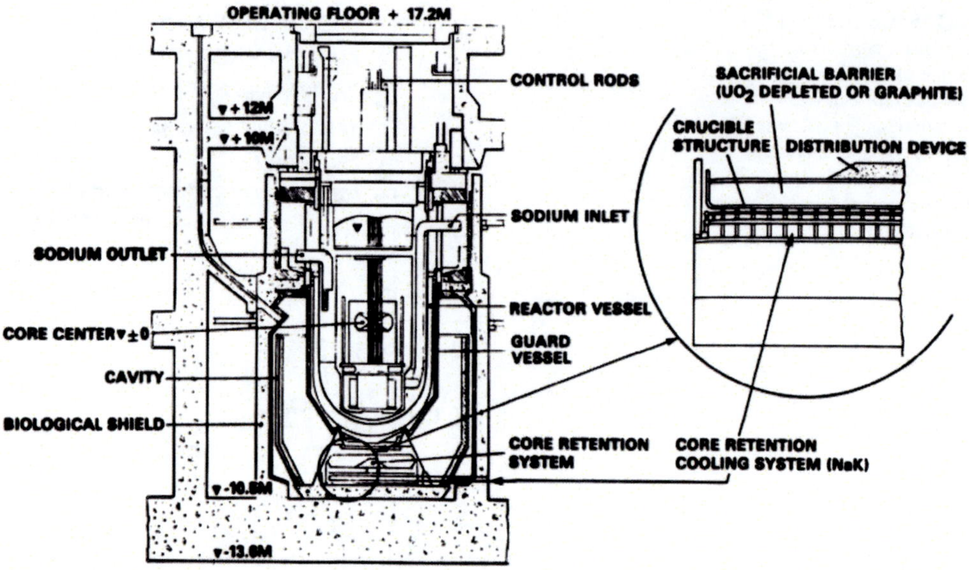

Fig. G.20 SNR-300 ex-vessel core retention system

over the crucible surface both to prevent recriticality and to spread out the heat source. The SNR-300 crucible structure was designed to be cooled by NaK.

Three types of sacrificial beds have been proposed as ex-vessel core retention systems according to the extent of cooling designed. The *completely passive* sacrificial bed represents a device with no cooling system, constructed of a high thermal capacity/low thermal conductivity material and arranged such that hot core debris can penetrate only at a slow rate. Materials considered for such a bed include ThO_2, depleted UO_2, magnesium oxide (MgO), and graphite. Penetration is especially slow in MgO due to its large specific heat and heat of fusion, and almost no gas or aerosols evolve from MgO beds during penetration.

Passively cooled sacrificial beds have been proposed where the outer surface of the bed is cooled by natural convection. This cooling would provide improved potential for retaining core debris and also enhanced thermal protection of the concrete cavity walls. An obvious deviation from the above concept is an *actively cooled* sacrificial bed.

Table G.2 summarizes the debris accommodation (core retention) provisions employed in selected SFR systems.

G.5.3.2 Interaction of Sodium and Core Debris with Concrete

Concrete Thermal Response

Although it is common practice to provide steel liners on the inner surface of all concrete cells where sodium could spill (in order to prevent direct contact between sodium and concrete), hot sodium on a steel-lined slab of concrete could still cause appreciable heating of the concrete.[9]

[9]Some designs include insulating material, such as firebrick, between the concrete and the steel liners in order to minimize concrete heating in the event of a sodium spill. Such liners are often referred to as "hot" liners.

Table G.2 Debris accommodation (core retention) provisions in SFR's (adapted from Ref. [36])

Reactor	Provisions
United States	
EBR-I	None
EBR-II	None
FERMI-1	In-vessel: Zirconium meltdown pan Ex-vessel: Graphite crucible below vessel with primary shield tank
SEFOR	Ex-vessel: Sodium catch tank 45 ft below vessel. Fuel dispersion cones below vessel and in catch tank
FFTF	In-vessel: Debris cooling in core support structure dome and above thermal baffle, but no special provision Ex-vessel: Steel lined cavity
CRBRP	In-vessel: Some small amount of debris cooling on reactor internal structures, but not special design provisions Ex-vessel: Lined and insulated reactor cavity, vented pipe chase and ex-containment cooling-venting purge system
United Kingdom	
DFR	Fuel dispersion cone and melt tubes to bedrock
PFR	Single layer of trays within tank; capable of retaining seven assemblies
CDFR (proposed)	Three layers of trays within tank; retention capacity for entire core
France	
Rapsodie	None
Phénix	Externally cooled outer vessel
Super-Phénix	In-vessel catch trays, external cooling of safety vessel
Germany/Netherlands/Belgium	
SNR-300	In-vessel: Catch trays in lower plenum Ex-vessel: High-temperature crucible with NaK cooling

Substantial amounts of water can be liberated from structural concrete if it is heated, as evidenced from the water release data in Fig. G.21, for any kind of structural concrete [42]. Note that the free water (or capillary water) is readily released at fairly low temperatures. At about 450°C, chemically bound water is released. There are some data to suggest that at even higher temperatures, additional release may occur due to complete decomposition and dehydration of the aggregate material. The important point, however, is that a large quantity of water can be released from concrete at an elevated temperature, and some accommodation (such as vents) must be provided to relieve the resulting pressure.

In addition to water release, substantial amounts of CO_2 gas could be released from heated concrete via the dissociation of $CaCO_3$ if limestone aggregates are present:

$$CaCO_3 + \text{heat} \rightarrow CaO + CO_2.$$

Another concern for unlined concrete is the breaking away of large pieces of concrete by cracking or through a process known as *spalling*. The thermal shock caused by the direct contact of molten core debris would almost certainly cause some surface spalling. There has been some concern that large concrete sections could be penetrated via fissures without the need to expend the energy required to melt through the section.

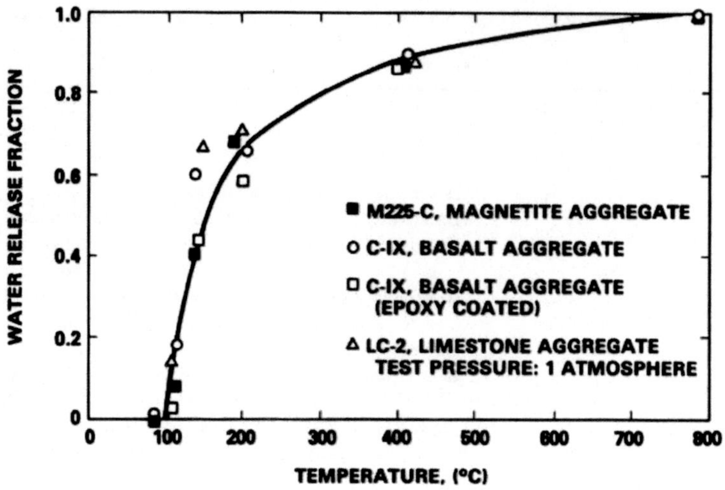

Fig. G.21 Water release from concrete [42]

Sodium Concrete Reactions

Direct interaction of hot sodium on bare concrete produces chemical reactions that liberate hydrogen gas. Most of this hydrogen is generated by exothermic sodium-water reactions:

$$2Na + H_2O \rightarrow Na_2O + H_2,$$

$$Na + H_2O \rightarrow NaOH + \frac{1}{2}H_2.$$

In a series of sodium-concrete reaction tests [43], hydrogen was initially produced at a rate of approximately 5 $kg/m^2 \cdot h$. Unless the hydrogen is vented or oxygen is provided to remove the hydrogen by recombination,[10] a sustained sodium-water reaction could lead to severe over pressurization of the containment.

Data from such tests indicate sodium initially heats the concrete, driving off water, and then the chemical reaction of sodium with water and aggregate material releases both hydrogen gas and energy. The released energy can accelerate sodium-concrete reactions since it causes increased sodium and concrete heating, further liberation of water, and further chemical reaction with sodium.

On the other hand, the initial sodium attack rates on concrete can subside in a few hours due to buildup of heavy, viscous reaction products that settle on any horizontal surfaces and impede the reaction. A similar saturation effect is noted for vertical surfaces, but the attack rate is greater (presumably since the reaction products tend to gravitate away, leaving a better interaction geometry). This saturation effect could be negated if sufficient concrete cracking could occur to expose fresh concrete surfaces. Quantitative rates of sodium reaction with concrete are not fully resolved.

Two preliminary expressions have been derived [43] to fit the sodium penetration depth for horizontal surfaces of magnetite and limestone concrete:

$$\text{Magnetite: } d = 17.5\left(1 - \exp^{-0.2t}\right)$$
$$\text{Limestone: } d = 10.4\left(1 - \exp^{-0.4t}\right)$$

[10] Devices provided to perform such a function are called hydrogen recombiners.

where

d = penetration depth (mm)
t = time (hours).

Figure G.22 compares these curves with actual penetration data for horizontal concrete surfaces. Similar equations were also fit to a more limited data base for sodium penetration into vertical surfaces, and this fit is also included in the figure. A penetration rate for a horizontal surface of 13 mm/h for 4 h is often used for containment transient calculations.

Other important exothermic sodium reactions can occur during sodium interaction with concrete (in addition to sodium burning in oxygen). These are listed in Fig. 16.6.

Core Debris–Concrete Interactions

In addition to potential chemical attack by sodium, the possibility of direct concrete contact with molten steel and molten fuel must be considered for an ex-vessel accident. The H_2O and CO_2 releases from concrete due to heating are capable of oxidizing the steel content of the penetrating core debris. Such oxidation could lead to the generation of H_2 and CO to add to the hydrogen liberated from a sodium/concrete interaction.

Prediction of the rate of concrete penetration by a core melt is a difficult problem that has not yet been fully resolved, though preliminary analytical methods are available (e.g., the GROWS [44] and the USINT [45] codes in the U.S.). The problem is complicated by the addition of molten products from the concrete to the pool as the melt penetrates. This addition of low-density material reduces

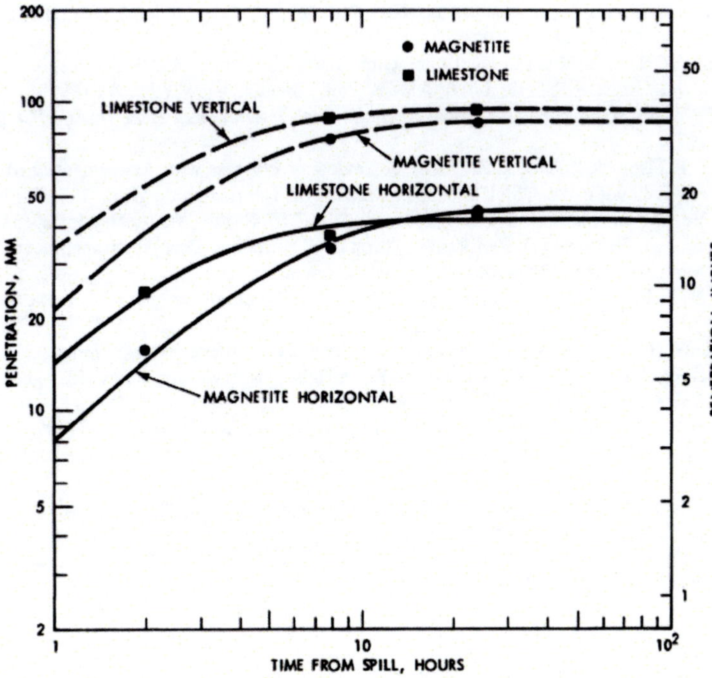

Fig. G.22 Concrete penetration by sodium [43]

the density of the molten pool to the extent that it might eventually float above molten steel, thus reversing the layering of material in the pool during the penetration process. In addition, gas rising from the concrete will remove heat and may spurge some fission products from the pool.

References

1. M. Kirbiyik, P. L. Gamer, J. G. Refling, and A. B. Reynolds, "Hydrodynamics of Post-Disassembly Fuel Expansion," *Nuclear Engineering and Design, 35* (1975) 441–460.
2. C. R. Bell, J. E. Boudreau, J. H. Scott, and L. L. Smith, "Advances in the Mechanistic Assessment of Postdisassembly Energetics," *Proceedings of the International Meeting on Fast Reactor Safety Technology, Vol. I,* Seattle, WA (1979), 207–218.
3. E. P. Hicks and D. C. Menzies, "Theoretical Studies on the Fast Reactor Maximum Accident," *Proceedings of the Conference on Safety, Fuels, and Core Design in Large Fast Power Reactors,* ANL-7120, Argonne National Laboratory, Argonne, IL (1965) 654–670.
4. A. M. Judd, "Calculation of the Thermodynamic Efficiency of Molten-Fuel-Coolant Interactions," *Transactions of the American Nuclear Society, 13* (1970) 369.
5. A. Padilla, Jr., "Analysis of Mechanical Work Energy for LMFBR Maximum Accidents," *Nuclear Technology, 12* (1971) 348–355.
6. D. H. Cho, R. O. Ivins, and R. W. Wright, "Pressure Generation by Molten Fuel-Coolant Interactions Under LMFBR Accident Conditions," *Proceedings of the Conference on New Developments in Reactor Mathematics and Applications,* CONF-710302, *Vol. I,* Idaho Falls, ID (1971) 25–49.
7. H. Jacobs, "Computational Analysis of Fuel-Sodium Interactions with an Improved Method," *Proceedings of the International Conference on Fast Reactor Safety and Related Physics,* CONF-761001, *Vol. III,* Chicago, IL (1976) 926–935.
8. M. Amblard and H. Jacobs, "Fuel Coolant Interaction; The CORRECT II UO_2-Na Experiment," *Proceedings of the International Meeting on Fast Reactor Safety Technology, Vol. III,* Seattle, WA (1979) 1512–1519.
9. H. Jacobs, M. F. Young, and K. O. Reil, "Fuel-Coolant Interaction Phenomena Under Prompt Burst Conditions," *Proceedings of the International Meeting on Fast Reactor Safety Technology, Vol. III,* Seattle, WA (1979) 1520–1528.
10. S. J. Board and L. Caldarola, "Fuel Coolant Interactions in Fast Reactors," *ASME Symposium on the Thermal and Hydraulic Aspects of Nuclear Reactor Safety,* ASME Winter Meeting, New York, NY (1979).
11. S. G. Bankoff, "Vapor Explosions: A Critical Review," *Sixth International Heat Transfer Conference,* Toronto, ON, Canada (1979).
12. A. J. Briggs, T. F. Fishlock, and G. J. Vaughan, "A Review of Progress with Assessment of MFCI Phenomenon in Fast Reactors Following the CSNI Specialist Meeting in Bournemouth, April 1979," *Proceedings of the International Meeting on Fast Reactor Safety Technology, Vol. III,* Seattle, WA (1979) 1502–1511.
13. D. R. Armstrong, F. J. Testa, and D. Raridon, Jr., "Molten UO_2-Sodium Dropping Experiments," *Transactions of the American Nuclear Society, 13* (1970) 660.
14. H. K. Fauske, "The Role of Nucleation in Vapor Explosions," *Transactions of the American Nuclear Society, 15* (1972) 813.
15. S. J. Board, R. W. Hall, and R. S. Hall, "Detonation of Fuel Coolant Explosions," *Nature, 254* (1975) 319–321. See also M. Baines, S. J. Board, N. E. Buttery, and R. W. Hall, "The Hydrodynamics of Large Scale Fuel Coolant Interactions," *Nuclear Technology, 49* (June 1980) 27–39.
16. S. A. Colgate and T. Sigurgeirsson, "Dynamic Mixing of Water and Lava," *Nature, 244* (1973) 552–555.
17. H. K. Fauske, "The Role of Core-Disruptive Accidents in Design and Licensing of LMFBR's," *Nuclear Safety, 17* (1976) 550–567.
18. M. Baines and N. E. Buttery, "Differential Velocity Fragmentation in Liquid-Liquid Systems," RD/B/N4643, Berkeley Nuclear Laboratories (September 1979).
19. A. W. Cronenberg and M. A. Grolmes, "Fragmentation Modeling Relative to the Breakup of Molten UO_2 in Sodium," *Nuclear Safety, 16* (6) 1975.
20. E. L. Gluekler and L. Baker Jr., "Post Accident Heat Removal in LMFBR's," O. C. Jones and S. G. Bankoff, eds., *Liquid Metal Fast Breeder Reactors, Vol. 2,* The American Society of Mechanical Engineers, New York, NY (1977) 287–325.
21. W. R. Wise and J. F. Proctor, "Explosion Containment Laws for Nuclear Reactor Vessels," NOLTR-63-140, U.S. Naval Ordnance Laboratory (1965).

22. J. F. Kunze (ed.), "Additional Analysis of the SL-1 Excursion," USAEC Report IDO-19313 (TM-62-11-707), General Electric Company, 1962, and "Final Report of SL-1 Recovery Operation," USAEC Report IDO-19311, General Electric Company, Idaho Test Station (July 27, 1962).
23. J. F. Proctor, "Adequacy of Explosion-Response Data in Estimating Reactor-Vessel Damage," *Nuclear Safety, 8* (6) (1967) 565–572.
24. R. H. Cole, *Underwater Explosions*, Princeton University Press, Princeton, NJ (1948).
25. A. L. Florence, G. R. Abrahamson, and D. J. Cagliostro, "Hypothetical Core Disruptive Accident Experiments on Single Fast Test Reactor Model," *Nuclear Engineering and Design, 38* (1976) 95.
26. D. D. Stepnewski, G. L. Fox, Jr., D. E. Simpson, and R. D. Peak, "FFTF Scale Model Structural Dynamics Tests," *Proceedings of the Fast Reactor Safety Meeting,* CONF-740401-P2, Beverly Hills, CA (1974).
27. M. Saito and T. G. Theofanous, "The Termination Phase of Core Disruptive Accidents in LMFBR's," *Proceedings of the International Meeting on Fast Reactor Safety Technology, Vol. III,* Seattle, WA (1979) 1425–1434.
28. H. Holtbecker, N. E. Hoskin, N. J. M. Rees, R. B. Tattersall, and G. Verzeletti, "An Experimental Programme to Validate Wave Propagation and Fluid Flow Codes for Explosion Containment Analysis," *Proceedings of the International Meeting on Fast Reactor Safety and Related Physics,* CONF-761001, *Vol. III,* Chicago, IL (1976) 1304–1313.
29. M. Egleme, N. Brahy, J. B. Fabry, H. Lamotte, M. Stievenart, H. Holtbecker, and P. Actis-Dato, "Simulation of Hypothetical Core Disruptive Accidents in Vessel Models," *Proceedings of the International Meeting on Fast Reactor Safety and Related Physics,* CONF-761001, *Vol. I,* Chicago, IL (1976) 1314–1323.
30. D. J. Cagliostio, A. L. Florence, G. R. Abrahamson, and G. Nagumo, "Characterization of an Energy Source for Modeling HCDA's in Nuclear Reactors," *Nuclear Engineering and Design, 27* (1974) 94.
31. J. P. Breton, A. Lapicofe, A. Porrachia, M. Natta, M. Amblard, and G. Berthoud, "Expansion of a Vapor Bubble and Aerosols Transfer," *Proceedings of the International Meeting on Fast Reactor Safety Technology, Vol. III,* Seattle, WA (1979) 1445–1454.
32. Y. W. Chang and J. Gvildys, "REXCO-HEP: A Two-Dimensional Computer Code for Calculating the Primary System Response in Fast Reactors," ANL-75-19, Argonne National Laboratory, Argonne, IL (1975).
33. Chung-yi Wang, "ICECO-An Implicit Eulerian Method for Calculating Fluid Transients in Fast-Reactor Containment," ANL-75-81, Argonne National Laboratory, Argonne, IL (1975).
34. Y. W. Chang, J. Gvildys, and S. H. Fistedis, "A New Approach for Determining Coolant Slug Impact in a Fast Reactor Accident," *Transactions of the American Nuclear Society, 14* (1971) 291.
35. A. H. Marchertas, C. Y. Wang, and S. H. Fistedis, "A Comparison of ANL Containment Codes with SNR-300 Simulation Experiments," *Proceedings of the International Meeting on Fast* Reactor *Safety and Related Physics,* CONF-761001, *Vol. III,* Chicago, IL (1976) 1324–1333.
36. M. S. Kazimi and J. C. Chen, "A Condensed Review of the Technology of Post-Accident Heat Removal for the Liquid-Metal Fast Rreeder Reactor," *Nuclear Technology, 38* (1978) 339–366.
37. M. J. Bell, *ORIGIN- The ORNL Isotope* Generation *and Depletion Code*, ORNL-4628, Oak Ridge National Laboratory, Oak Ridge, TN (1973).
38. T. R. England, R. Wilaynski, and N. L. Whittemore, *CINDER-7: An Interim Report for Users*, LA-5885-MS, Los Alamos Scientific Laboratory, Los Alamos, NM (1975).
39. R. O. Gumprecht, *Mathematical Basis of Computer Code RIBD*, DUN-4136, Douglas United Nuclear, Inc. (1968), and D. R. Marr, *A User's Manual* for *Computer Code RIBD-II, A Fission Product Inventory Code*, HEDL-TME 75-26, Hanford Engineering Development Laboratory, Richland, WA (January 1975).
40. D. R. Marr and W. L. Bunch, *FTR Fission Product Decay Heat,* HEDL-TME 71-72, Hanford Engineering Development Laboratory, Richland, WA (February 1971).
41. G. Fieg, "Heat Transfer Measurements of Internally Heated Liquids in Cylindrical Convection Cells," *Proceedings of the 4th PAHR Information Exchange Meeting*, Varese, Italy (October 10–12, 1978) 144.
42. J. D. McCormack, A. K. Postma, and J. A. Schur, *Water Evolution from Heated Concrete*, HEDL-TME 78-8, Hanford Engineering Development Laboratory, Richland, WA (February 1979).
43. J. A. Hassberger, "Intermediate Scale Sodium-Concrete Reaction Tests," HEDL TME 77-99, Hanford Engineering Development Laboratory, Richland, WA (March 1978). See also J. A. Hassberger, "Intermediate Scale Sodium-Concrete Reaction Tests with Basalt and Limestone Concrete," HEDL-TME 79-55, Hanford Engineering Development Laboratory (September 1980); R. P. Colburn, et al., "Sodium Concrete Reactions," *Proceedings of the International Meeting on Fast Reactor Safety Technology, Vol. IV,* Seattle, WA (August 1979) 2093.
44. L. Baker, F. B. Chenug, R. Farhadieh, R. P. Stein, J. D. Gabor, and J. D. Bingle, "Thermal Interaction of a Molten Core Debris Pool with Surrounding Structural Materials," *Proceedings of the International Conference on Fast Reactor Safety Technology, Vol. I,* Seattle, WA (1979) 389–399.
45. R. L. Knight and J. V. Beck, "Model and Computer Code for Energy and Mass Transport in Decomposing Concrete and Related Materials," *Proceedings of the International Conference on Fast Reactor Safety Technology, Vol. IV,* Seattle, WA (1979) 2113–2121.

Appendix H
Internet Resources

Pavel Tsvetkov

- Software resources
  - OECD Nuclear Energy Agency, Data Bank Computer Program Services,
    http://www.nea.fr/dbprog/
  - U.S. Department of Energy, Energy Science and Technology Software Center,
    http://www.osti.gov/estsc/
  - U.S. Nuclear Regulatory Commission, NRC Developed Computer Codes,
    http://www.nrc.gov/about-nrc/regulatory/research/comp-codes
  - U.S. ORNL, Radiation Safety Information Computational Center,
    http://rsicc.ornl.gov/
- Nuclear data
  - International Atomic Energy Agency, Nuclear Data Services,
    http://www-nds.iaea.org/
  - OECD Nuclear Energy Agency, Nuclear Data Services,
    http://www.nea.fr/dbdata/
  - U.S. National Nuclear Data Center,
    http://www.nndc.bnl.gov/
- Materials
  - International Atomic Energy Agency, Thermophysical Properties of Materials for Nuclear Engineering,
    http://www-pub.iaea.org/MTCD/publications
  - International Nuclear Safety Center, Material Properties Database,
    http://www.insc.anl.gov/matprop/
- Safety
  - European Commission, Nuclear Energy Safety,
    http://ec.europa.eu/energy/nuclear/safety/safety_en.htm
  - International Nuclear Safety Center,
    http://www.insc.anl.gov/
  - OECD Nuclear Energy Agency, Safety Joint Research Project Databases,
    http://www.nea.fr/dbprog/safety-joint-research-databases.html
  - U.S. Nuclear Regulatory Commission,
    http://www.nrc.gov/

- Fast spectrum reactors
  - International Atomic Energy Agency, Fast Reactors,
    http://www.iaea.org/NuclearPower/FR/
  - International Atomic Energy Agency, Fast Reactors Database,
    http://www.iaea.org/inisnkm/nkm/aws/frdb/index.html
  - International Atomic Energy Agency, Fast Reactors Database, Small and Medium Sized Reactors,
    http://www.iaea.org/NuclearPower/SMR/
- Generation IV reactors
  - Generation IV nuclear energy systems (Gen-IV),
    http://nuclear.gov/genIV/neGenIV1.html
  - International Project on Innovative Nuclear Reactors and Fuel Cycles (INPRO),
    http://www.iaea.org/INPRO/
  - The Generation IV International Forum (GIF),
    http://www.gen-4.org/
- Fuel cycle
  - Fuel cycle research and development,
    http://nuclear.gov/fuelcycle/neFuelCycle.html
  - International Atomic Energy Agency, Integrated Nuclear Fuel Cycle Information Systems,
    http://www-nfcis.iaea.org/
  - International Atomic Energy Agency, Nuclear Fuel Cycle and Materials,
    http://www.iaea.org/OurWork/ST/NE/NEFW/nfcms_home
- Sustainable development of nuclear energy
  - American Nuclear Society,
    http://www.ans.org/
  - European Commission, Energy,
    http://ec.europa.eu/energy
  - European Union,
    http://europa.eu/
  - International Atomic Energy Agency,
    http://www.iaea.org/
  - International Atomic Energy Agency, International Nuclear Information System (INIS),
    http://www.iaea.org/inisnkm/
  - International Energy Agency,
    http://www.iea.org/
  - International Energy Agency, Nuclear Energy Handbook,
    http://www.iaea.org/inisnkm/nkm/ws/index.html
  - International Energy Agency, World Energy Outlook,
    http://www.worldenergyoutlook.org/
  - U.S. National Science Foundation,
    http://www.nsf.gov
  - Organization for Economic Cooperation and Development (OECD), Nuclear Energy Agency (NEA),
    http://www.nea.fr/

- **U.S. Department of Energy, the Office of Nuclear Energy,**
  http://nuclear.gov
- **U.S. Department of Energy,**
  http://www.energy.gov/
- **U.S. Nuclear Regulatory Commission,**
  http://www.nrc.gov/
- **World Factbook,**
  https://www.cia.gov/library/index.html
- **World Nuclear Association,**
  http://www.world-nuclear.org

Index

D

M